FACE PROCESSING

Advanced Modeling and Methods

FACE PROCESSING

Advanced Modeling and Methods

Edited by

Wenyi Zhao

Vision Technologies Lab
Sarnoff Corportaion
Princeton, NJ 08540
Email: wzhao@sarnoff.com

and

Rama Chellappa

Center for Automation Research
University of Maryland
College Park, MD 20740
Email: rama@umiacs.umd.edu

ELSEVIER

AMSTERDAM • BOSTON • HEIDELBERG • LONDON
NEW YORK • OXFORD • PARIS • SAN DIEGO
SAN FRANCISCO • SINGAPORE • SYDNEY • TOKYO

Academic Press is an imprint of Elsevier

Academic Press is an imprint of Elsevier
30 Corporate Drive, Suite 400, Burlington, MA 01803, USA
525 B Street, Suite 1900, San Diego, California 92101-4495, USA
84 Theobald's Road, London WC1X 8RR, UK

This book is printed on acid-free paper. ♾

Library of Congress Cataloging-in-Publication Data
Application submitted

British Library Cataloguing-in-Publication Data
A catalogue record for this book is available from the British Library.

ISBN 13: 978-0-12-088452-0
ISBN 10: 0-12-088452-6

For information on all Academic Press publications
visit our Web site at www.books.elsevier.com

Printed in the United States of America
05 06 07 08 09 10 9 8 7 6 5 4 3 2 1

Dedicated to Professor Azriel Rosenfeld (1931–2004)

CONTENTS

CONTRIBUTORS

Hervé Abdi, University of Texas, Dallas, Texas

Benjamin J. Balas, Massachusetts Institute of Technology, Cambridge, Massachusetts

Marian Stewart Bartlett, University of California, San Diego, California

J. Ross Beveridge, Colorado State University, Fort Collins, Colorado

Volker Blanz, University of Siegen, Siegen, Germany

Kevin W. Bowyer, University of Notre Dame, Notre Dame, Indiana

Alexander M. Bronstein, Technion-Israel Institute of Technology, Haifa, Israel

Michael M. Bronstein, Technion-Israel Institute of Technology, Haifa, Israel

Kyong Chang, Philips Medical Systems, Seattle, Washington

Rama Chellappa, University of Maryland, College Park, Maryland

Antonio J. Colmenarez, University of Illinois, Urbana-Champaign, Urbana, Illinois

Bruce A. Draper, Colorado State University, Fort Collins, Colorado

Christopher K. Eveland, Equinox Corporation, New York, New York

Ward Fisher, Colorado State University, Fort Collins, Colorado

Patrick J. Flynn, University of Notre Dame, Notre Dame, Indiana

Geof H. Givens, Colorado State University, Fort Collins, Colorado

M. Ida Gobbini, Princeton University, Princeton, New Jersey

Himanshu Gupta, ObjectVideo, Reston, Virginia

James V. Haxby, Princeton University, Princeton, New Jersey

Bernd Heisele, Honda Research Institute USA, Inc., Boston

Jeffrey Ho, University of Florida, Gainesville, Florida

Thomas S. Huang, University of Illinois, Urbana-Champaign, Urbana, Illinois

Anil K. Jain, Michigan State University, East Lansing, Michigan

Fang Jiang, University of Texas, Dallas, Texas

Takeo Kanade, Carnegie Mellon University, Pittsburgh, Pennsylvania

Ron Kimmel, Technion-Israel Institute of Technology, Haifa, Israel

David Kriegman, University of California, San Diego, California

Zicheng Liu, Microsoft Research, Redmond, Washington

Xiaoguang Lu, Michigan State University, East Lansing, Michigan

Aleix M. Martínez, Ohio State University, Columbus, Ohio

Javier R. Movellan, University of California, San Diego, California

Karthik Nandakumar, Michigan State University, East Lansing, Michigan

Yuri Ostrovsky, Massachusetts Institute of Technology, Cambridge, Massachusetts

Alice J. O'Toole, University of Texas, Dallas, Texas

Jean-Sébastien Pierrad, University of Basel, Switzerland

Ravi Ramamoorthi, Columbia University, New Youk, New York

Dana Roark, University of Texas, Dallas, Texas

Sami Romdhani, University of Basel, Switzerland

Amit K. Roy-Chowdhury, University of California, Riverside, California

Richard Russell, Massachusetts Institute of Technology, Cambridge, Massachusetts

Terrence J. Sejnowski, University of California, San Diego, and Howard Hughes Medical Institute at the Salk Institute

Andrea Selinger, Equinox Corporation, New York, New York

Pawan Sinha, Massachusetts Institute of Technology, Cambridge, Massachusetts

Diego A. Socolinsky, Equinox Corporation, New York, New York

Matthew Turk, University of California, Santa Barbara, California

Umut Uludag, Michigan State University, East Lansing, Michigan

Thomas Vetter, University of Basel, Switzerland

Lawrence B. Wolff, Equinox Corporation, New York, New York

Ziyou Xiong, University of Illinois, Urbana-Champaign, Urban, Illinois

Akihiko Yamada, Sanyo Electric Co. Ltd., Osaka, Japan

Ruigang Yang, University of Kentucky, Lexington, Kentucky

Yongbin Zhang, Ohio State University, Columbus, Ohio

Zhengyou Zhang, Microsoft Research, Redmond, Washington

Wenyi Zhao, Sarnoff Corporation, Princeton, New Jersey

Shaohua Kevin Zhou, Siemens Corporate Research, Princeton, New Jersey

PREFACE

As one of the most important applications of image analysis and understanding, face recognition has recently received significant attention, especially during the past 10 years. There are at least two reasons for this trend: the first is the wide range of commercial and law-enforcement applications, and the second is the availability of feasible technologies.

Though machine recognition of faces has reached a certain level of maturity after more than 35 years of research, e.g., in terms of the number of subjects to be recognized, technological challenges still exist in many aspects. For example, robust face recognition in outdoor-like environments is still difficult. As a result, the problem of face recognition remains attractive to many researchers across multiple disciplines, including image processing, pattern recognition and learning, computer vision, computer graphics, neuroscience, and psychology. Research efforts are ongoing to advance the state of the art in various aspects of face processing. For example, understanding of how humans can routinely perform robust face recognition can shed some light on how to improve machine recognition of human faces. Another example is that modeling 3D face geometry and reflectance properties can help design a robust system to handle illumination and pose variations.

To further advance this important field, we believe that continuous dialog among researchers is important. It is in this spirit that we decided to edit a book on the subject of advanced methods and modeling. Because this subject is experiencing very rapid developments, we choose the book format as a collection of chapters, written by experts. Consequently, readers will have the opportunity to read the most recent advances reported directly by experts from diverse backgrounds. For those who are new to this field, we have included three introductory chapters on the general state of the art, eigen approaches, and evaluation of face recognition systems. It is our hope that this book would engage readers at various levels and, more importantly, provoke debates among all of us in order to make a big leap in performance.

The title of this book is *Face Processing: Advanced Methods and Modeling*. We have picked face processing as the main title, rather than face recognition, intentionally. We know that the meaning of *face recognition* could be interpreted differently depending on the context. For example, in the narrowest sense, face recognition means the *recognition* of facial ID. In a broad sense or from a system point of view, it implies *face detection*, *feature extraction*, and recognition of facial ID. In the broadest sense, it is essentially *face processing*, including all types of processing: detection, tracking, modeling, analysis, and synthesis. This book includes chapters on various aspects of face processing that have direct or potential impact on machine recognition of faces. Specifically, we have put an emphasis on advanced methods and modeling in this book.

This book has been divided into three parts: The first part covers all the basic aspects of face processing, and the second part deals with face modeling from both computational and psychophysical aspects. The third part includes advanced methods which can be further divided into reviews, handling of illumination, pose and expression variations, multi-modal face recognition, and miscellaneous topics.

The first introductory part of this book includes three book chapters. The chapter by Zhao and Chellappa provides a guided tour to face processing. Various aspects of face processing are covered, including development history, psychophysics/neuroscience aspects, reviews on still- and video-based face processing, and brief a introduction to advanced topics in face recognition. The chapter by Turk reviews the well-known eigenface approach for face recognition and gives a rare personal treatment on the original context and motivation, and on how it stands from a more general perspective. The chapter by Beveridge et al. provides a systematic treatment on how to scientifically evaluate face-recognition systems. It offers many insights into how complex the task of evaluation could be and how to avoid apparent pitfalls.

In the second part, we have four chapters on computational face modeling and three chapters on the neurophysiological face modeling. The chapter by Rodmhani et al. proposes a unified approach for analysis and synthesis of images based on their previous work on 3D morphable face model. The chapter by Bronstein et al. presents an expression-invariant face representation for robust recognition. The key here is to model facial expressions as isometrics of the facial surface. The chapter by Roy-Chowdhury et al. presents two algorithms for 3D face modeling from a monocular video sequence based on structure from motion and shape adaption from contour. The chapter by Bartlett et al. describes how to model face images using higher-order statistics (compared to principal-components analysis). In particular, a version of independent-component analysis derived from the principle of maximum information transfer through sigmoidal neurons is explored.

The chapter by Sinha et al. reviews four key aspects of human face perception performance. Taken together, the results provide strong constraints and guidelines for computational models of face recognition. The chapter by O'Toole et al. enumerates the factors that affect the accuracy of human face recognition and suggests that computational system should emulate the amazing human recognition system. The chapter by Gobbini and Haxby provides interesting insights into how humans perceive human faces and other objects. In summary, distributed local representations of faces and objects in the ventral temporal cortex are proposed.

In the third part of this book, we have four chapters devoted to handling the significant issues of illumination and pose variations in face recognition, and one chapter on a complete modeling system with industrial applications. They are followed by two review chapters, one on 2D+3D recognition methods and one on image-sequence recognition methods. We then have two chapters on advanced methods for real-time face detection and subset modeling of face localization errors. We conclude our book with two interesting chapters that discuss thermal-image-based face recognition and integration of other cues with faces for multimodal biometrics.

The chapter by Ho and Kriegman addresses the impact of illumination upon face recognition and proposes a method of modeling illumination based on spherical harmonics. The following chapter by Ramamoorthi focuses more on the theoretical aspects of illumination modeling based on spherical harmonics. The chapter by Kanade and Yamada proposes an algorithm based on a probabilistic approach for face recognition to address the problem of pose change, by a probabilistic approach that takes into account the pose difference between probe and gallery images. The chapter by Heisele and Blanz talks about how to apply the method of 3D morphable face model to first generate a 3D model based on three input images and then to synthesize images under various poses and illumination conditions. These images are then fit into a component-based recognition system for training. The chapter by Zhang et al. describes a complete 3D face-modeling system that constructs textured 3D animated face models from videos with minimal user interaction in the beginning. Various industrial applications are demonstrated, including interactive games, eye-gaze correction, etc.

The chapter by Bowyer et al. offers a rather unique survey on face recognition method based on three-dimensional shape models, either alone or in combination with two-dimensional intensity images. The chapter by Zhou and Chellappa investigates the additional properties manifested in image sequence and reviews approaches for video-based face recognition.

The chapter by Martinez and Zhang investigates a method for modeling disjoint subsets and its applications to localization error, occlusion, and expression. The chapter by Colmenarez et al. describes a near-time robust method for face and facial feature detection based on information-theoretic discrimination.

The chapter by Socolinsky et al. provides a nice review on face recognition based on thermal images. Finally, the chapter by Jain et al. presents a discussion on multi-modal biometrics where faces are integrated with other cues such as fingerprint and voice for enhanced performance.

PART 1

THE BASICS

CHAPTER 1

A GUIDED TOUR OF FACE PROCESSING

1.1 INTRODUCTION TO FACE PROCESSING

As one of the most successful applications of image analysis and understanding, face recognition has recently received significant attention, especially during the past few years. This is evidenced by the emergence of face-recognition conferences such as AFGR [7] and AVBPA [9] and many commercially available systems. There are at least two reasons for this trend: the first is the wide range of commercial and law-enforcement applications and the second is the availability of feasible technologies after 30 years of research. In addition, the problem of machine perception of human faces continues to attract researchers from disciplines such as image processing [8], pattern recognition [1, 3], neural networks and learning [5], computer vision [4, 6], computer graphics [2], and psychology.

Although very reliable methods of biometric personal identification exist, e.g., fingerprint analysis and retinal or iris scans, these methods rely on the cooperation of the participants, whereas a personal identification system based on analysis of frontal or profile images of the face is often effective without the participant's cooperation or knowledge. Some of the advantages/disadvantages of different biometrics are described in [16]. Table 1.1 lists some of the applications of face recognition in a broadest sense.

Commercial and law-enforcement applications of face-recognition technology (FRT) range from static, controlled-format photographs to uncontrolled video images, posing a wide range of technical challenges and requiring an equally wide range of techniques from image processing, analysis, understanding, and pattern recognition. One can broadly classify face recognition systems into two groups

Table 1.1: Typical applications of face processing, including detection and tracking, recognition of identity and expressions, and personalized realistic rendering.

Areas	Specific Applications
Entertainment	Video games, virtual reality, training programs, human–robot interaction, human–computer interaction, family photo albums
Smart Cards	Drivers' licenses, passports, voter registrations, welfare fraud/entitlement programs
Information security	TV parental control, cellphone access, desktop logon, database security, file encryption, medical records, internet access, secure trading terminals
Law enforcement and surveillance	Advanced video surveillance, CCTV control, portal control, post-event analysis, shoplifting, suspect tracking and investigation
Emerging	Tool for psychology study

depending on whether they make use of static images or video. Within these groups, significant differences exist, depending on specific applications. The differences are due to image quality, amount of background clutter (posing challenges to segmentation algorithms), variability of the images of a particular individual that must be recognized, availability of a well-defined recognition or matching criterion, and the nature, type, and amount of input from a user.

1.1.1 Problem Statement

A general statement of the problem of machine recognition of faces can be formulated as follows: Given still or video images of a scene, identify or verify one or more persons in the scene using a stored database of faces. Available collateral information such as race, age, gender, facial expression, or speech may be used in narrowing the search (enhancing recognition). The solution to the problem involves segmentation of faces (face detection) from cluttered scenes, feature extraction from the face regions, recognition, or verification (Figure 1.1). We will describe these steps in Sections 1.3 and 1.4. In identification problems, the input to the system is an unknown face, and the system reports back the determined identity from a database of known individuals, whereas in verification problems, the system needs to confirm or reject the claimed identity of the input face.

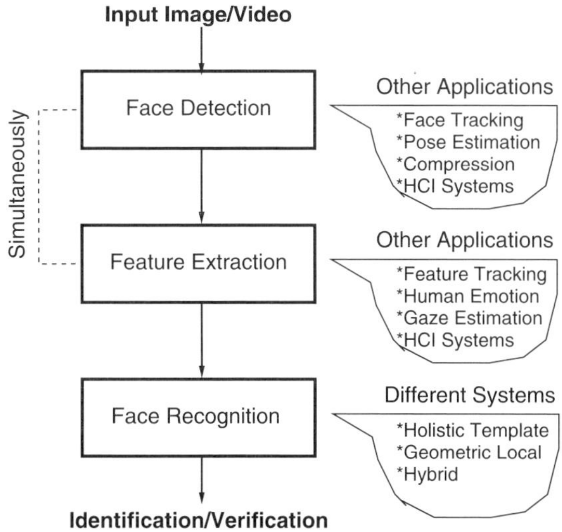

FIGURE 1.1: Configuration of a generic face-recognition/processing system.

1.1.2 Performance Evaluation

As any pattern-recognition system, the system performance is important. The measurements of system performance include, for example, the FA (false acceptance or false positive) rate and the FR (false rejection or false negative) rate. An ideal system should have very low scores on FA and FR, while a practical system often needs to make a tradeoff between these two scores. To facilitate discussions, the stored set of faces of known individuals is often referred to as a *gallery*, while a *probe* is a set of unknown faces presented for recognition.

Due to the difference in identification and verification, people often use different performance measurements. For example, the percentage of probe images that are correctly identified is often used for the task of face identification, while the equal error rate between FA scores and FR scores could be used for the task of verification. For a more complete performance report, the ROC (receiver operation characteristics) curve should be used for verification and a cumulative match score (e.g., rank-1 and rank-5 performance) for identification.

To adequately evaluate/characterize the performance of a face-recognition system, large sets of probe images are essential. In addition, it is important to keep in mind that the operation of a pattern-recognition system is statistical, with measurable distributions of success and failure. These distributions are very application-dependent. This strongly suggests that an evaluation should be based as closely as possible on a specific application scenario.

This has motivated researchers to collect several large, publicly available face databases and design corresponding testing protocols, including the FERET protocol [165], the FRVT vendor test [164], the XM2VTS protocol [161], and the BANCA protocol [160].

1.1.3 Brief Development History

The earliest work on face recognition can be traced back at least to the 1950s in psychology [29] and to the 1960s in the engineering literature [20]. (Some of the earliest studies include work on facial expression of emotions by Darwin [21] (see also Ekman [30]) and on facial profile based biometrics by Galton [22]). But research on automatic machine recognition of faces really started in the 1970s after the seminal work of Kanade [23] and Kelly [24]. Over the past thirty years, extensive research has been conducted by psychophysicists, neuroscientists, and engineers on various aspects of face recognition by humans and machines.

Psychophysicists and neuroscientists have been concerned with issues such as whether face perception is a dedicated process (this issue is still being debated in the psychology community [26, 33]) and whether it is done holistically or by local feature analysis. With the help of powerful engineering tool such as functional MRI, new theories continue to emerge [35]. Many of the hypotheses and theories put forward by researchers in these disciplines have been based on rather small sets of images. Nevertheless, many of the findings have important consequences for engineers who design algorithms and systems for machine recognition of human faces.

Until recently, most of the existing work formulate the recognition problem as recognizing 3D objects from 2D images. As a result, earlier approaches treated it as a 2D pattern-recognition problem. During the early and mid-1970s, typical pattern-classification techniques, which use measured attributes of features (e.g., the distances between important points) in faces or face profiles, were used [20, 24, 23]. During the 1980s, work on face recognition remained largely dormant. Since the early 1990s, research interest in FRT has grown significantly. One can attribute this to several reasons: an increase in interest in commercial opportunities; the availability of real-time hardware; and the emergence of surveillance-related applications.

Over the past 15 years, research has focused on how to make face-recognition systems fully automatic by tackling problems such as localization of a face in a given image or a video clip and extraction of features such as eyes, mouth, etc. Meanwhile, significant advances have been made in the design of classifiers for successful face recognition. Among appearance-based holistic approaches, eigenfaces [122, 73] and Fisherfaces [52, 56, 79] have proved to be effective in experiments with large databases. Feature-based graph-matching approaches [49] have also been quite successful. Compared to holistic approaches, feature-based methods are less sensitive to variations in illumination and viewpoint and to inaccuracy in face localization. However, the feature-extraction techniques needed for this type of approach are still not reliable or accurate enough [55]. For example, most eye-localization techniques assume some geometric and textural models and do not work if the eye is closed.

During the past five to ten years, much research has been concentrated on video-based face recognition.The still-image problem has several inherent advantages and disadvantages. For applications such as airport surveillance, automatic location and segmentation of a face could pose serious challenges to any segmentation algorithm if only a static picture of a large, crowded area is available. On the other hand, if a video sequence is available, segmentation of a moving person can be more easily accomplished using motion as a cue. In addition, a sequence of images might help to boost the recognition performance if we can effectively utilize all these images. But the small size and low image quality of faces captured from video can significantly increase the difficulty in recognition.

More recently, significant advances have been made in 3D-based face recognition. Though it is known that face recognition using 3D images has many advantages over recognition using a single or sequence of 2D images, no serious effort was made for 3D face recognition until recently. This was mainly due to the feasibility, complexity, and computational cost for acquiring 3D data in real time. Now, the availability of cheap real-time 3D sensors [90] makes it much easier to apply 3D face recognition.

Recognizing a 3D object from its 2D images poses many challenges. The illumination and pose problems are two prominent issues for appearance- or image-based approaches [19]. Many approaches have been proposed to handle these issues, and the key here is to model the 3D geometry and reflectance properties of a face. For example, 3D textured models can be built from given 2D images, and they can then be used to synthesize images under various poses and illumination conditions for recognition or animation. By restricting the image-based 3D object modeling to the domain of human faces, fairly good reconstruction results can be obtained using the state-of-the-art algorithms. Other potential applications where modeling is crucial include computerized aging, where an appropriate model needs to be built first and then a set of model parameters are used to create images simulating the aging process.

1.1.4 A Multidisciplinary Approach for Face-Recognition Research

As an application of image understanding, machine recognition of face recognition also benefits tremendously from advances in other relevant disciplines. For example, the analysis-by-synthesis approach [112] for building 3D face models from images is a classical example where techniques from computer vision and computer graphics are nicely integrated. In a brief summary, we list some related disciplines and their direct impact upon face recognition.

- **Pattern recognition.** The ultimate goal of face recognition is recognition of personal ID based on facial patterns, including 2D images, 3D structures, and any preprocessed features that are finally fed into a classifier.
- **Image processing.** Given a single raw face image, or a sequence of them, it is important to normalize the image size, enhance the image quality, and localize local features prior to recognition.
- **Computer vision.** The first step in face recognition involves the detection of face regions based on appearance, color, and motion. Computer-vision techniques also make it possible to build a 3D face model from a sequence of images by aligning them together. Finally, 3D face modeling holds great promises for robust face recognition.
- **Computer graphics.** Traditionally, computer graphics is used to render human faces with increasingly realistic appearances. Combined with computer vision, it has been applied for building 3D models from images.
- **Learning.** Learning plays a significant role for building a mathematical model. For example, given a training set (or bootstrap set by many researchers) of 2D or 3D images, a generative model can be learned and applied to other novel objects in the same class of face images.
- **Neuroscience and psychology.** Study of the amazing capability of human perception of faces can shed some light on how to improve existing systems for machine perception of faces.

Before we move to the following sections for detailed discussions of specific topics, we would like to mention a few survey papers worth reading. In 1995, a review paper [11] gave a thorough survey of FRT at that time. (An earlier survey [17] appeared in 1992.) At that time, video-based face recognition was still in a nascent stage. During the past ten years, face recognition has received increased attention and has advanced technically. Significant advances have been made in video-based face modeling/tracking, recognition, and system integration [10, 7]. To reflect these advances, a comprehensive survey paper [19] that covers a wide range of topics has recently appeared. There are also a few books on this subject [14, 13, 119]. As for other applications of face processing, there also exist review papers, for example, on face detection [50, 44] and recognition of facial expression [12, 15].

1.2 FACE PERCEPTION: THE PSYCHOPHYSICS/NEUROSCIENCE ASPECT

Face perception is an important capability of human perception system and is a routine task for humans, while building a similar computer system is still a daunting task. Human recognition processes utilize a broad spectrum of stimuli, obtained from many, if not all, of the senses (visual, auditory, olfactory, tactile, etc.). In many situations, contextual knowledge is also applied, e.g., the context plays an important role in recognizing faces in relation to where they are supposed to be located. It is futile to even attempt to develop a system using existing technology, which will mimic the remarkable face-perception ability of humans. However, the human brain has its limitations in the total number of persons that it can accurately "remember". A key advantage of a computer system is its capacity to handle large numbers of face images. In most applications the images are available only in the form of single or multiple views of 2D intensity data, so that the inputs to computer face-recognition algorithms are visual only.

Many studies in psychology and neuroscience have direct relevance to designing algorithms or systems for machine perception of faces. For example, findings in psychology [27, 38] about the relative importance of different facial features have been noted in the engineering literature [56] and could provide direct guidance for system design. On the other hand, algorithms for face processing provide tools for conducting studies in psychology and neuroscience [37, 34]. For example, a possible engineering explanation of the bottom-lighting effects studied in [36] is as follows: when the actual lighting direction is opposite to the usually assumed direction, a shape-from-shading algorithm recovers incorrect structural information and hence makes human perception of faces harder.

Due to space restriction, we will not list all the interesting results here. We briefly discuss a few published studies to illustrate that these studies indeed reveal fascinating characteristics of the human perception system. One study we would like to mention is on whether face perception is holistic or local [27, 28]. Apparently, both holistic and feature information are crucial for the perception and recognition of faces. Findings also suggest the possibility of global descriptions serving as a front end for finer, feature-based perception. If dominant features are present, holistic descriptions may not be used. For example, in face recall studies, humans quickly focus on odd features such as big ears, a crooked nose, a staring eye, etc. One of the strongest pieces of evidence to support the view that face recognition involves more configural/holistic processing than other object recognition has been the face inversion effect in which an inverted face is much harder to recognize than a normal face (first demonstrated in [40]). An excellent example is given in [25] using the "Thatcher illusion" [39]. In this illusion, the eyes and mouth of an expressing face are excised and inverted, and the result looks grotesque in an upright face; however, when shown inverted, the face looks fairly normal in appearance, and the inversion of the internal features is not readily noticed.

Another study is related to whether face recognition is a dedicated process or not [32, 26, 33]: It is traditionally believed that face recognition is a dedicated process different from other object recognition tasks. Evidence for the existence of a dedicated face-processing system comes from several sources [32]. (a) Faces are more easily remembered by humans than other objects when presented in an upright orientation; (b) prosopagnosia patients are unable to recognize previously familiar faces, but usually have no other profound agnosia. They recognize people by their voices, hair color, dress, etc. It should be noted that prosopagnosia patients recognize whether a given object is a face or not, but then have difficulty in identifying the face. There are also many differences between face recognition and object recognition [26] based on empirical evidence. However, recent neuroimaging studies in humans indicate that level of categorization and expertise interact to produce the specification for faces in the middle fusiform gyrus[1] [33]. Hence it is possible that the encoding scheme used for faces may also be employed for other classes with similar properties. More recently, distributed and overlapping representations of faces and objects in ventral temporal cortex are proposed [35]. The authors argue that the distinctiveness of the response to a given category is not due simply to the regions that respond maximally to that category, by demonstrating that the category being viewed can still be identified on the basis of the pattern of response when those regions were excluded from the analysis.

1.3 FACE DETECTION AND FEATURE EXTRACTION

As illustrated in Figure 1.1, the problem of automatic face recognition involves three key steps/subtasks: (1) detection and coarse normalization of faces, (2) feature extraction and accurate normalization of faces, (3) identification and/or verification. Sometimes, different subtasks are not totally separated. For example, the facial features (eyes, nose, mouth) are often used for both face recognition and face detection. Face detection and feature extraction can be achieved simultaneously, as indicated in Figure 1.1. Depending on the nature of the application, e.g., the sizes of the training and testing databases, clutter and variability of the background, noise, occlusion, and speed requirements, some of the subtasks can be very challenging.

A fully automatic face-recognition system must perform all three subtasks, and research on each subtask is critical. This is not only because the techniques used for the individual subtasks need to be improved, but also because they are critical in many different applications (Figure 1.1). For example, face detection is needed to initialize face tracking, and extraction of facial features is needed for recognizing human emotion, which in turn is essential in human–computer interaction (HCI) systems. Isolating the subtasks makes it easier to assess and advance the state of the art of the component techniques. Earlier face-detection techniques could only handle single or a few well-separated frontal faces in images with simple backgrounds, while state-of-the-art algorithms can detect faces and their poses

[1] The fusiform gyrus or occipitotemporal gyrus, located on the ventromedial surface of the temporal and occipital lobes, is thought to be critical for face recognition.

in cluttered backgrounds. Without considering feature locations, face detection is declared as successful if the presence and rough location of a face has been correctly identified. However, without accurate face and feature location, noticeable degradation in recognition performance is observed [127, 18].

1.3.1 Segmentation/Detection

Up to the mid 1990s, most work on segmentation was focused on single-face segmentation from a simple or complex background. These approaches included using a whole-face template, a deformable feature-based template, skin color, and a neural network.

Significant advances have been made in recent years in achieving automatic face detection under various conditions. Compared to feature-based methods and template-matching methods, appearance- or image-based methods [48, 46] that train machine systems on large numbers of samples have achieved the best results. This may not be surprising, since face objects are complicated, very similar to each other, and different from nonface objects. Through extensive training, computers can be quite good at detecting faces.

More recently, detection of faces under rotation in depth has been studied. One approach is based on training on multiple-view samples [42, 47]. Compared to invariant-feature-based methods [49], multiview-based methods of face detection and recognition seem to be able to achieve better results when the angle of out-of-plane rotation is large (35°). In the psychology community, a similar debate as to whether face recognition is viewpoint-invariant or not is ongoing. Studies in both disciplines seem to support the idea that, for small angles, face perception is view-independent, while for large angles it is view-dependent.

In any detection problem, two metrics are important: true positives (also referred to as detection rate) and false positives (reported detections in nonface regions). An ideal system would have very high true positive and very low false positive rates. In practice, these two requirements are conflicting. Treating face detection as a two-class classification problem helps to reduce false positives dramatically [48, 46] while maintaining true positives. This is achieved by retraining systems with false-positive samples that are generated by previously trained systems.

1.3.2 Feature Extraction

The importance of facial features for face recognition cannot be overstated. Many face recognition systems need facial features in addition to the holistic face, as suggested by studies in psychology. It is well known that even holistic matching methods, e.g., eigenfaces [73] and Fisherfaces [52], need accurate locations of key facial features such as eyes, nose, and mouth to normalize the detected face [127, 50].

Three types of feature-extraction method can be distinguished: (1) generic methods based on edges, lines, and curves; (2) feature-template-based methods that are used to detect facial features such as eyes; (3) structural matching methods that take into consideration geometrical constraints on the features. Early approaches focused on individual features; for example, a template-based approach is described in [43] to detect and recognize the human eye in a frontal face. These methods have difficulty when the appearances of the features change significantly, e.g., closed eyes, eyes with glasses, open mouth. To detect the features more reliably, recent approaches use structural matching methods, for example, the active-shape model (ASM) [116]. Compared to earlier methods, these recent statistical methods are much more robust in terms of handling variations in image intensity and feature shape.

An even more challenging situation for feature extraction is feature "restoration", which tries to recover features that are invisible due to large variations in head pose. The best solution here might be to hallucinate the missing features either by using the bi-lateral symmetry of the face or using learned information. For example, a view-based statistical method claims to be able to handle even profile views in which many local features are invisible [117].

Representative Methods

A template-based approach to detecting the eyes and mouth in real images is presented in [133]. This method is based on matching a predefined parametrized template to an image that contains a face region. Two templates are used for matching the eyes and mouth respectively. An energy function is defined that links edges, peaks, and valleys in the image intensity to the corresponding properties in the template, and this energy function is minimized by iteratively changing the parameters of the template to fit the image. Compared to this model, which is manually designed, the *statistical* shape model, ASM proposed in [116] offers more flexibility and robustness. The advantages of using the so-called "analysis through synthesis" approach come from the fact that the solution is *constrained* by a *flexible* statistical model. To account for texture variation, the ASM model has been expanded to statistical appearance models including a flexible-appearance model (FAM) [59] and an active-appearance model (AAM) [114]. In [114], the proposed AAM combined a model of shape variation (i.e., ASM) with a model of the appearance variation of shape-normalized (shape-free) textures. A training set of 400 images of faces, each manually labeled with 68 landmark points, and approximately 10,000 intensity values sampled from facial regions were used. The shape model (mean shape, orthogonal mapping matrix $\boldsymbol{P}_S$ and projection vector $\boldsymbol{b}_S$) is generated by representing each set of landmarks as a vector and applying principal-component analysis (PCA) to the data. Then, after each sample image is warped so that its landmarks match the mean shape, texture information can

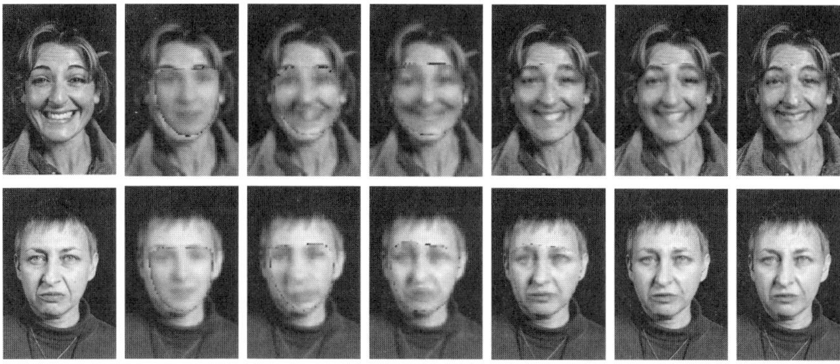

FIGURE 1.2: Multiresolution search from a displaced position using a face model. (Courtesy of T. Cootes, K. Walker, and C. Taylor.)

be sampled from this shape-free face patch. Applying PCA to this data leads to a shape-free texture model (mean texture, $\boldsymbol{P}_{\mathrm{g}}$ and $\boldsymbol{b}_{\mathrm{g}}$). To explore the correlation between shape and texture variations, a third PCA is applied to the concatenated vectors ($\boldsymbol{b}_{\mathrm{s}}$ and $\boldsymbol{b}_{\mathrm{g}}$) to obtain the combined model in which one vector $\boldsymbol{c}$ of appearance parameters controls both the shape and texture of the model. To match a given image and the model, an optimal vector of parameters (displacement parameters between the face region and the model, parameters for linear intensity adjustment, and the appearance parameters $\boldsymbol{c}$) are searched by minimizing the difference between the synthetic image and the given one. After matching, a best-fitting model is constructed that gives the locations of all the facial features so that the original image can be reconstructed. Figure 1.2 illustrates the optimization/search procedure for fitting the model to the image. To speed up the search procedure, an efficient method is proposed that exploits the similarities among optimizations. This allows the direct method to find and apply directions of rapid convergence which are learned off line.

1.4 METHODS FOR FACE RECOGNITION

1.4.1 Recognition Based on Still Intensity and Other Images

Brief Review

Many methods of face recognition have been proposed during the past 30 years. Face recognition is such a challenging problem that it has attracted researchers from different backgrounds: psychology, pattern recognition, neural networks, computer vision, and computer graphics. It is due to this fact that the literature on face recognition is vast and diverse. Often, a single system involves techniques motivated by different principles. The use of a combination of techniques makes it difficult to classify these systems based purely on what types of techniques they use for representation or classification. To have a clear and high-level categorization, we instead follow a guideline suggested by the psychological study of how humans use holistic and local features. Specifically, we have the following categorization:

1. **Holistic matching methods.** These methods use the whole face region as the raw input to a recognition system. One of the most widely used representations of the face region is the eigenpicture [122], which is based on principal-component analysis.
2. **Feature based (structural) matching methods.** Typically, in these methods, local features such as eyes, nose, and mouth are first extracted and their locations and local statistics (geometric and/or appearance) are fed into a structural classifier.
3. **Hybrid methods.** Just as the human perception system uses both local features and the whole face region to recognize a face, a machine recognition system should use both. One can argue that these methods could potentially offer the best of the two types of methods.

Within each of these categories, further classification is possible. Using subspace analysis, many face-recognition techniques have been developed: eigenfaces [73], which use a nearest-neighbor classifier; feature-line-based methods, which replace the point-to-point distance with the distance between a point and the feature line linking two stored sample points [61]; Fisherfaces [71, 52, 79], which use linear/fisher discriminant analysis (FLD/LDA) [143]; Bayesian methods, which use a probabilistic distance metric [65]; and SVM methods, which use a support-vector machine as the classifier [68]. Utilizing higher-order statistics, independent-component analysis (ICA) is argued to have more representative power than PCA, and hence in theory can provide better recognition performance than PCA [51]. Being able to offer potentially greater generalization through learning, neural-network/learning methods have also been applied to face recognition. One example is the probabilistic decision-based neural network (PDBNN) method [62] and the other is the evolution pursuit (EP) method [63].

Most earlier methods belong to the category of structural matching methods, using the width of the head, the distances between the eyes and from the eyes to the mouth, etc. [24], or the distances and angles between eye corners, mouth extrema, nostrils, and chin top [23]. More recently, a mixture–distance–based approach using manually extracted distances was reported [55]. Without finding the exact locations of facial features, methods based on the hidden Markov model (HMM) use strips of pixels that cover the forehead, eye, nose, mouth, and chin [69, 66]. [66] reported better performance than [69] by using the KL projection coefficients instead of the strips of raw pixels. One of the most successful systems in this

Table 1.2: Categorization of still-face recognition techniques

Approach	Representative work
Holistic methods	
Principal-component analysis (PCA)	
Eigenfaces	Direct application of PCA [73]
Probabilistic eigenfaces	Two-class problem with prob. measure [65]
Fisherfaces/subspace LDA	FLD on eigenspace [71, 52, 79]
SVM	Two-class problem based on SVM [68]
Evolution pursuit	Enhanced GA learning [63]
Feature lines	Based on point-to-line distance [61]
ICA	ICA-based feature analysis [51]
Other representations	
LDA/FLD	FLD/LDA on raw image [56]
PDBNN	Probabilistic decision-based NN [62]
Kernel faces	Kernel methods [78, 152, 151]
Tensorfaces	Multilinear analysis [75, 158]
Feature-based methods	
Pure geometry methods	Earlier methods [24, 23]; recent methods [64, 55]
Dynamic link architecture	Graph-matching methods [49]
Hidden Markov model	HMM methods [69, 66]
Convolution neural network	SOM learning-based CNN methods [60]
Hybrid methods	
Modular eigenfaces	Eigenfaces and eigenmodules [67]
Hybrid LFA	Local-feature method [129]
Shape-normalized	Flexible-appearance models [59]
Component-based	Face region and components [57]

category is the graph-matching system [49] which is based on the dynamic-link architecture (DLA) [123]. Using an unsupervised-learning method based on a self-organizing map (SOM), a system based on a convolutional neural network (CNN) has been developed [60].

In the hybrid-method category, we have the modular-eigenface method [67], a hybrid representation based on PCA and local-feature analysis (LFA) [129], a method based on the flexible-appearance model [59], and a recent development [57] along this direction. In [67], the use of hybrid features by combining eigenfaces and other eigenmodules is explored: eigeneyes, eigenmouth, and eigennose. Though experiments show only slight improvements over holistic eigenfaces or eigenmodules based on structural matching, we believe that these types of methods are important and deserve further investigation. Perhaps many relevant problems need to be solved before fruitful results can be expected, e.g., how to optimally arbitrate the use of holistic and local features.

Many types of system have been successfully applied to the task of face recognition, but they all have some advantages and disadvantages. Appropriate schemes should be chosen based on the specific requirements of a given task. Most of the systems reviewed here focus on the subtask of recognition, but others also include automatic face detection and feature extraction, making them fully automatic systems [65, 49, 62].

Projection/Subspace Methods Based on Intensity Images

To help readers who are new to this field, we present a class of linear projection/subspace algorithms based on image appearances. The implementation of these algorithms is straightforward, yet they are very effective under constrained situations. These algorithms helped to revive the research activities in 1990s with the introduction of eigenfaces [122, 73] and are still being actively adapted for continuous improvements.

The linear projection algorithms include PCA, ICA, and LDA, to name a few. In all these projection algorithms, classification is performed by (1) projecting the input $\boldsymbol{x}$ into a subspace via a projection/basis matrix $\boldsymbol{P}_{\text{proj}}$:[2]

$$z = \boldsymbol{P}_{\text{proj}}\boldsymbol{x}; \tag{1}$$

(2) comparing the projection coefficient vector z of the input to all the previously stored projection vectors of labeled classes to determine the input class label. The vector comparison varies in different implementations and can influence the system's performance dramatically [162]. For example, PCA algorithms can use either the angle or the Euclidean distance (weighted or unweighted) between two projection vectors. For LDA algorithms, the distance can be unweighted or weighted.

Starting from the successful low-dimensional reconstruction of faces using KL or PCA projection [122], eigenpictures have been one of the major driving forces behind face representation, detection, and recognition. It is well known that there exist significant statistical redundancies in natural images [131]. For a limited class of objects such as face images that are normalized with respect to scale, translation, and rotation, the redundancy is even more [129, 18] and one of the best global compact representations is KL/PCA that decorrelate the outputs. More specifically, mean-subtracted sample vectors $\boldsymbol{x}$ can be expressed as a linear combination of the orthogonal bases $\boldsymbol{\Phi}_i$: $\boldsymbol{x} = \sum_{i=1}^{n} a_i\boldsymbol{\Phi}_i \approx \sum_{i=1}^{m} a_i\boldsymbol{\Phi}_i$

[2]The matrix $\boldsymbol{P}_{\text{proj}}$ is $\boldsymbol{\Phi}$ for eigenfaces, $\boldsymbol{W}$ for Fisherfaces with pure LDA projection, and $\boldsymbol{W\Phi}$ for Fisherfaces with sequential PCA and LDA projections; these three bases are shown for visual comparison in Figure 1.3.

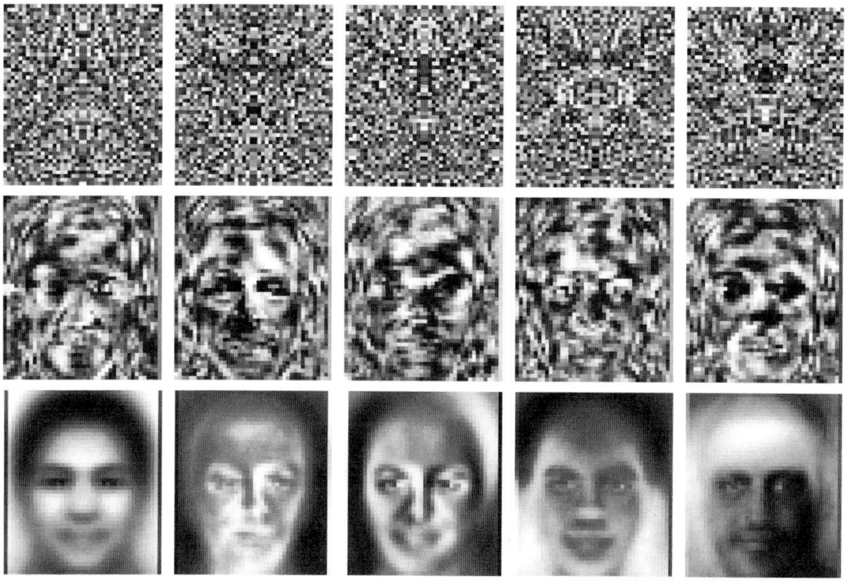

FIGURE 1.3: Different projection bases constructed from a set of 444 individuals, where the set is augmented via adding noise and mirroring. The first row shows the first five pure LDA basis images $\boldsymbol{W}$; the second row shows the first five subspace LDA basis images $\boldsymbol{W\Phi}$; the average face and first four eigenfaces $\boldsymbol{\Phi}$ are shown on the third row [79].

(typically $m \ll n$), via solving an eigenproblem

$$\boldsymbol{C\Phi} = \boldsymbol{\Phi\Lambda}, \tag{2}$$

where $\boldsymbol{C}$ is the covariance matrix for input $\boldsymbol{x}$.[3]

Another advantage of using such a compact representation is the reduced sensitivity to noise. Some of this noise could be due to small occlusions as long as the *topological structure* does *not* change. For example, good performance against blurring, partial occlusion, and changes in background has been demonstrated in many eigenpicture-based systems and was also reported in [79] (Figure 1.4). This should not come as a surprise, since the images reconstructed using PCA are much better than the original distorted images in terms of the global appearance (Figure 1.5).

[3]Equivalently, we can obtain the eigenvalues and eigenvectors through SVD decomposition of an image matrix that has $\boldsymbol{x}$ as its column.

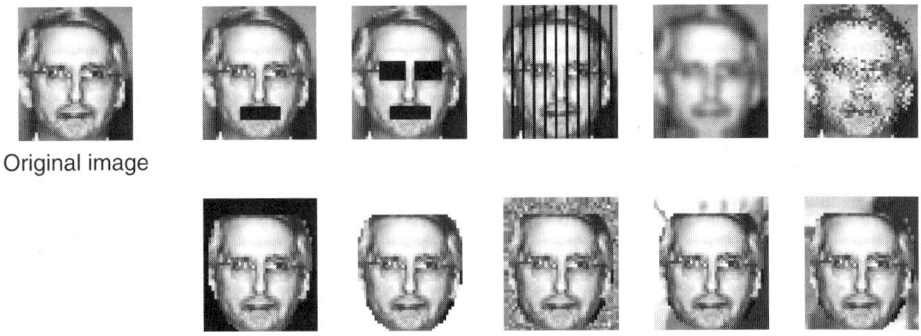

FIGURE 1.4: Electronically modified images which have been correctly identified [79].

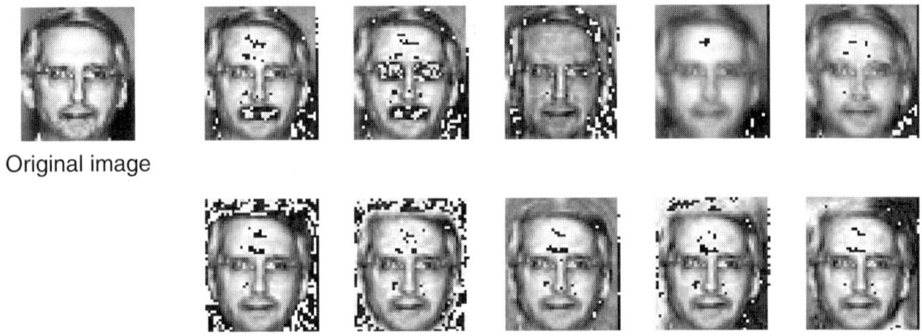

FIGURE 1.5: Reconstructed images using 300 PCA projection coefficients for electronically modified images (Figure 1.4) [18].

Improved reconstruction for face images outside the training set, using an extended training set that adds mirror-imaged faces was demonstrated in [122]. Using such an extended training set, the eigenpictures are shown to be either symmetric or antisymmetric with the most leading eigenpictures typically being symmetric.

The first real successful demonstration of machine recognition of faces was made by Turk and Pentland [73] using eigenpictures (also known as eigenfaces) for face detection and identification. Given the eigenfaces, every face in the database is represented as a vector of weights obtained by projecting the image into eigenface components by a simple inner-product operation. When a new test image whose identification is required is given, the new image is also represented by its vector of weights. The identification of the test image is done by locating the image in the database whose weights are the closest (in Euclidean distance) to the weights of the test image. By using the observation that the projection of a face image and a nonface image are quite different, a method of detecting the presence of a face in a given image is obtained. Turk and Pentland illustrate their method using a large database of 2500 face images of sixteen subjects, digitized at all combinations of three head orientations, three head sizes and three lighting conditions.

Using a probabilistic measure of similarity, instead of the simple Euclidean distance used with eigenfaces [73], the standard eigenface approach was extended in [65] using a Bayesian approach. Often, the major drawback of a Bayesian method is the need to estimate probability distributions in a high-dimensional space from very limited numbers of training samples per class. To avoid this problem, a much simpler two-class problem was created from the multiclass problem by using a similarity measure based on a Bayesian analysis of image differences. Two mutually exclusive classes were defined: Ω_I, representing *intrapersonal* variations between multiple images of the same individual, and Ω_E, representing *extrapersonal* variations due to differences in identity. Assuming that both classes are Gaussian distributed, likelihood functions $P(\Delta|\Omega_I)$ and $P(\Delta|\Omega_E)$ were estimated for a given intensity difference $\Delta = I_1 - I_2$. Given these likelihood functions and using the MAP rule, two face images are determined to belong to the same individual if $P(\Delta|\Omega_I) > P(\Delta|\Omega_E)$. A large performance improvement of this probabilistic matching technique over standard nearest-neighbor eigenspace matching was reported using large face datasets including the FERET database [165]. In [65], an efficient technique of probability density estimation was proposed by decomposing the input space into two mutually exclusive subspaces: the principal subspace F and its orthogonal subspace $\hat{F}$ (a similar idea was explored in [48]). Covariances only in the principal subspace are estimated for use in the Mahalanobis distance [145].

Face recognition systems using LDA/FLD have also been very successful [71, 52, 56]. LDA training is carried out via scatter matrix analysis [145]. For an M-class problem, the within- and between-class scatter matrices S_w, S_b are computed as follows:

$$\boldsymbol{S}_{\mathrm{w}} = \sum_{i=1}^{M} Pr(\omega_i) C_i,$$

$$\boldsymbol{S}_{\mathrm{b}} = \sum_{i=1}^{M} Pr(\omega_i)(\boldsymbol{m}_i - \boldsymbol{m}_0)(\boldsymbol{m}_i - \boldsymbol{m}_0)^{\mathrm{T}}, \tag{3}$$

where $Pr(\omega_i)$ is the prior class probability, and is usually replaced by $1/M$ in practice with the assumption of equal priors. Here $\boldsymbol{S}_{\mathrm{w}}$ is the *within-class scatter matrix*, showing the average scatter C_i of the sample vectors $\boldsymbol{x}$ of different classes ω_i around their respective means $\boldsymbol{m}_i$: $C_i = E[(\boldsymbol{x}(\omega) - \boldsymbol{m}_i)(\boldsymbol{x}(\omega) - \boldsymbol{m}_i)^{\mathrm{T}} | \omega = \omega_i]$. Similarly, $\boldsymbol{S}_{\mathrm{b}}$ is the *between-class scatter matrix*, representing the scatter of the conditional mean vectors $\boldsymbol{m}_i$ around the overall mean vector $\boldsymbol{m}_0$. A commonly used measure for quantifying the discriminatory power is the ratio of the determinant of the between-class scatter matrix of the projected samples to the determinant of the within-class scatter matrix: $\mathcal{J}(\boldsymbol{T}) = |\boldsymbol{T}^{\mathrm{T}} \boldsymbol{S}_{\mathrm{b}} \boldsymbol{T}| / |\boldsymbol{T}^{\mathrm{T}} \boldsymbol{S}_{\mathrm{w}} \boldsymbol{T}|$. The optimal projection matrix $\boldsymbol{W}$ which maximizes $\mathcal{J}(\boldsymbol{T})$ can be obtained by solving a generalized eigenvalue problem:

$$\boldsymbol{S}_{\mathrm{b}} \boldsymbol{W} = \boldsymbol{S}_{\mathrm{w}} \boldsymbol{W} \boldsymbol{\Lambda}_L \tag{4}$$

There are several ways to solve the generalized eigenproblem of Equation 4. One is to directly compute the inverse of $\boldsymbol{S}_{\mathrm{w}}$ and solve a nonsymmetric (in general) eigenproblem for matrix $\boldsymbol{S}_{\mathrm{w}}^{-1} \boldsymbol{S}_{\mathrm{b}}$. But this approach is numerically unstable since it involves the direct inversion of a potentially very large matrix which is probably close to being singular. A stable method to solve this equation is to solve the eigenproblem for $\boldsymbol{S}_{\mathrm{w}}$ first [145], i.e., removing the within-class variations (whitening). Since $\boldsymbol{S}_{\mathrm{w}}$ is a real symmetric matrix, there exist orthonormal $\boldsymbol{W}_{\mathrm{w}}$ and diagonal $\boldsymbol{\Lambda}_{\mathrm{w}}$ such that $\boldsymbol{S}_{\mathrm{w}} \boldsymbol{W}_{\mathrm{w}} = \boldsymbol{W}_{\mathrm{w}} \boldsymbol{\Lambda}_{\mathrm{w}}$. After whitening, the input $\boldsymbol{x}$ becomes $\boldsymbol{y}$:

$$\boldsymbol{y} = \boldsymbol{\Lambda}_{\mathrm{w}}^{-1/2} \boldsymbol{W}_{\mathrm{w}}^{\mathrm{T}} \boldsymbol{x}. \tag{5}$$

The between-class scatter matrix for the new variable $\boldsymbol{y}$ can be constructed as in Equation 4:

$$\boldsymbol{S}_{\mathrm{b}}^{y} = \sum_{i=1}^{M} Pr(\omega_i)(\boldsymbol{m}_i^{y} - \boldsymbol{m}_0^{y})(\boldsymbol{m}_i^{y} - \boldsymbol{m}_0^{y})^{\mathrm{T}}. \tag{6}$$

Now the purpose of FLD/LDA is to maximize the class separation of the now whitened samples $\boldsymbol{y}$, leading to another eigenproblem: $\boldsymbol{S}_{\mathrm{b}}^{y} \boldsymbol{W}_{\mathrm{b}} = \boldsymbol{W}_{\mathrm{b}} \boldsymbol{\Lambda}_{\mathrm{b}}$. Finally, we apply the change of variables to $\boldsymbol{y}$:

$$\boldsymbol{z} = \boldsymbol{W}_{\mathrm{b}}^{\mathrm{T}} \boldsymbol{y}. \tag{7}$$

Combining Equations 5 and 7, we have the relationship $\boldsymbol{z} = \boldsymbol{W}_{\mathrm{b}}^{\mathrm{T}} \boldsymbol{\Lambda}_{\mathrm{w}}^{-1/2} \boldsymbol{W}_{\mathrm{w}}^{\mathrm{T}} \boldsymbol{x}$, and $\boldsymbol{W}$ is simply

$$\boldsymbol{W} = \boldsymbol{W}_{\mathrm{w}} \boldsymbol{\Lambda}_{\mathrm{w}}^{-1/2} \boldsymbol{W}_{\mathrm{b}}. \tag{8}$$

To improve the performance of LDA-based systems, a regularized subspace LDA system that unifies PCA and LDA was proposed in [80, 18]. Good generalization ability of this system was demonstrated by experiments on new classes/individuals without retraining the PCA bases $\boldsymbol{\Phi}$, and sometimes even the LDA bases $\boldsymbol{W}$. While the reason for not retraining PCA is obvious, it is interesting to test the adaptive capability of the system by fixing the LDA bases when images from new classes are added[4]. The fixed PCA subspace of dimensionality 300 was trained from a large number of samples. An augmented set of 4056 mostly frontal-view images constructed from the original 1078 FERET images of 444 individuals by adding noisy and mirrored images was used in [80]. At least one of the following three characteristics separates this system from other LDA-based systems [71, 52]: (1) the unique selection of the universal face subspace dimension, (2) the use of a weighted distance measure, and (3) a regularized procedure that modifies the within-class scatter matrix $\boldsymbol{S}_{\mathrm{w}}$. The authors selected the dimensionality of the universal face subspace based on the characteristics of the eigenvectors (face-like or not) instead of the eigenvalues [79], as is commonly done. Later it was concluded in [130] that the global-face-subspace dimensionality is on the order of 400 for large databases of 5000 images. The modification of $\boldsymbol{S}_{w}$ into $\boldsymbol{S}_{\mathrm{w}} + \delta \boldsymbol{I}$ has two motivations: first to resolve the issue of small sample size [147] and second to prevent the significantly discriminative information[5] contained in the null space of $\boldsymbol{S}_{\mathrm{w}}$ [54] from being lost.

For comparison of the above projection/subspace face-recognition methods, a unified framework was presented in [76]. Considering the whole process of linear subspace projection and nearest-neighbor classification, the authors were able to single out the common features among PCA, LDA, and Bayesian methods: the *image differences* between each probe image and the stored gallery images $\Delta_{\mathrm{pg}} = I_{\mathrm{probe}} - I_{\mathrm{gallery}}$. A method based on the analysis was also proposed to take advantage of PCA, Bayesian and LDA methods. In practice, it is essentially a subspace LDA method where optimal parameters for the subspace dimensions (i.e., the principal bases of $\boldsymbol{\Phi}$, $\boldsymbol{W}_{\mathrm{w}}$, and $\boldsymbol{W}_{\mathrm{b}}$) are sought.

In order to handle nonlinearity caused by pose, illumination, and expression, variations presented in face images, the above linear subspace methods have been extended to kernel faces [78, 152, 151] and tensorfaces [75, 158].

Early Methods Based on Sketches and Infra-Red Images

In [74, 58], face recognition based on sketches, which is quite common in law enforcement, is described. Humans have a remarkable ability to recognize faces from sketches. This ability provides a basis for forensic investigations: an artist

[4]This makes sense because the final classification is carried out in the projection space z by comparison against previously stored projection vectors with nearest-neighbor rule.

[5]The null space of $\boldsymbol{S}_{\mathrm{w}}$ contains important discriminant information since the ratio of the determinants of the scatter matrices would be maximized in the null space.

draws a sketch based on the witness's verbal description; then a witness looks through a large database of real images to determine possible matches. Usually, the database of real images is quite large, possibly containing thousands of real photos. Therefore, building a system capable of automatically recognizing faces from sketches has practical value.

In [58], a system called PHANTOMAS (phantom automatic search) is described. This system is based on [123], where faces are stored as flexible graphs with characteristic visual features (Gabor features) attached to the nodes of the graph. The system was tested using a photo database of 103 persons (33 females and 70 males) and 13 sketches drawn by a professional forensic artist from the photo database. The results were compared with the judgments of five human subjects and were found to be comparable. For some recent work on sketch-based face recognition, please refer to [72].

[77] describes an initial study comparing the effectiveness of visible and infrared (IR) imagery for detecting and recognizing faces. One of the motivations in this work is that changes in illumination can cause significant performance degradation for visible-image-based face recognition. Hence infrared imagery, which is insensitive to illumination variation, can serve as an alternative source of information for detection and recognition. However, the inferior resolution of IR images is a drawback. Further, though IR imagery is insensitive to changes in illumination, it is sensitive to changes in temperature. Three face-recognition algorithms were applied to both visible and IR images. The recognition results on 101 subjects suggested that visible and IR imagery perform similarly across algorithms, and that by fusing IR and visible imagery one can enhance the performance compared to using either one. For more recent work on IR-based face recognition, please refer to [70].

1.4.2 Recognition Based on a Sequence of Images

A typical video-based face recognition system automatically detects face regions, extracts features from the video, and recognizes facial identity if a face is present. In surveillance, information security, and access control applications, face recognition and identification from a video sequence is an important problem. Face recognition based on video is preferable over using still images, since as demonstrated in [28], motion helps in recognition of (familiar) faces when the images are negated, inverted, or at threshold. It was also demonstrated that humans can recognize animated faces better than randomly rearranged images from the same set. Though recognition of faces from a video sequence is a direct extension of still-image-based recognition, in our opinion, *true* video-based face-recognition techniques that coherently use both spatial and temporal information started only a few years ago and still need further investigation. Significant challenges for video-based recognition still exist; we list several of them here.

- **The quality of video is low.** Usually, video acquisition occurs outdoors (or indoors but with bad conditions for video capture) and the subjects are not cooperative; hence there may be large illumination and pose variations in the face images. In addition, partial occlusion and disguises are possible.
- **Face images are small.** Again, due to the acquisition conditions, the face image sizes are smaller (sometimes much smaller) than the assumed sizes in most still-image-based face recognition systems. For example, the valid face region can be as small as 15×15 pixels[6], whereas the face image sizes used in feature-based still-image-based systems can be as large as 128×128. Small-size images not only make the recognition task more difficult, but also affect the accuracy of face segmentation, as well as the accurate detection of the fiducial points/landmarks that are often needed in recognition methods.
- **The characteristics of human faces and body parts.** During the past ten years, research on human action/behavior recognition from video has been very active and fruitful. Generic description of human behavior not particular to an individual is an interesting and useful concept. One of the main reasons for the feasibility of generic descriptions of human behavior is that the intraclass variations of human bodies, and in particular faces, is much smaller than the difference between the objects inside and outside the class. For the same reason, recognition of individuals within the class is difficult. For example, detecting and localizing faces is typically easier than recognizing a specific face.

Before we review existing video-based face-recognition algorithms, we briefly examine three closely related techniques: face segmentation and pose estimation, face tracking, and 3D face modeling. These techniques are critical for the realization of the full potential of video-based face recognition.

Basic Techniques of Video-Based Face Recognition

In [11], four computer-vision areas were mentioned as being important for video-based face recognition: segmentation of moving objects (humans) from a video sequence; structure estimation; 3D models for faces; and nonrigid-motion analysis. For example, in [120] a face-modeling system which estimates facial features and texture from a video stream is described. This system utilizes all four techniques: segmentation of the face based on skin color to initiate tracking; use of a 3D face model based on laser-scanned range data to normalize the image (by facial

[6]Notice that this is totally different from the situation where we have images with large face regions but where the final face regions fed into a classifier is 15×15.

feature alignment and texture mapping to generate a frontal view) and construction of an eigensubspace for 3D heads; use of structure from motion (SfM) at each feature point to provide depth information; and nonrigid motion analysis of the facial features based on simple 2D sum of squared differences (SSD) and tracking constrained by a global 3D model. Based on the current development of video-based face recognition, we think it is better to review three specific face-related techniques instead of the above four general areas. The three video-based face related techniques are: face segmentation and pose estimation, face tracking, and 3D face modeling.

Face segmentation and pose estimation. Early attempts [73] at segmenting moving faces from an image sequence used simple pixel-based change-detection procedures based on difference images. These techniques may run into difficulties when multiple moving objects and occlusion are present. More sophisticated methods use estimated flow fields for segmenting humans in motion [154]. More recent methods [87, 82] use motion and/or color information to speed up the process of searching for possible face regions. After candidate face regions are located, still-image-based face-detection techniques can be applied to locate the faces [50]. Given a face region, important facial features can be located. The locations of feature points can be used for pose estimation, which is important for synthesizing a virtual frontal view [82]. Newly developed segmentation methods locate the face and simultaneously estimate its pose without extracting features [42, 125]. This is achieved by learning multiview face examples which are labeled with manually determined pose angles.

Face and feature tracking. After faces are located, faces and their features can be tracked. Face and feature tracking is critical for reconstructing a face model (depth) through SfM, facial expression and gaze recognition. Tracking also plays a key role in spatiotemporal-based recognition methods [85, 86] which directly use the tracking information.

In its most general form, tracking is essentially motion estimation. However, general motion estimation has fundamental limitations, such as the aperture problem. For images like faces, some regions are too smooth to estimate flow accurately, and sometimes the change in local appearances is too large to give reliable flow estimate. Fortunately, these problems can be alleviated by face modeling, which exploits domain knowledge. In general, tracking and modeling are interdependent: tracking is constrained by a generic 3D model or a learned statistical model under deformation, and individual models are refined through tracking. Face tracking can be roughly divided into three categories: (1) head tracking, which involves tracking the motion of a rigid object that is performing rotations and translations; (2) facial-feature tracking, which involves tracking nonrigid deformations that are limited by the anatomy of the head, i.e., articulated motion due to speech or facial expressions and deformable motion due to muscle contractions and relaxations; and (3) complete tracking, which involves tracking both the head and facial features.

Early efforts focused on the first two problems: head tracking [136] and facial-feature tracking [156]. In [136], an approach to head tracking using points with high Hessian values was proposed. Several such points on the head are tracked, and the 3D motion parameters of the head are recovered by solving an overconstrained set of motion equations. Facial-feature tracking methods may make use of feature boundary or the feature region. Feature boundary tracking attempts to track and accurately delineate the shape of the facial feature, e.g., to track the contours of the lips and mouth [156]. Feature region tracking addresses the simpler problem of tracking a region such as a bounding box that surrounds the facial feature [141].

In [141], a tracking system based on local parametrized models is used to recognize facial expressions. The models include a planar model for the head, local affine models for the eyes, and local affine models and curvature for the mouth and eyebrows. A face tracking system was used in [150] to estimate the pose of the face. This system used a graph representation with about 20–40 nodes/landmarks to model the face. Knowledge about faces is used to find the landmarks in the first frame. Two tracking systems described in [120, 155] model faces completely with texture and geometry. Both systems use generic 3D models and SfM to recover the face structure. [120] tracks fixed feature points (eyes, nose tip), while [155] tracks only points with high Hessian values. Also, [120] tracks 2D features in 3D by deforming them, while [155] relies on direct comparison of a 3D model to the image. Methods are proposed in [146, 139] to solve the varying appearance (both geometry and photometry) problem in tracking. Some of the newest model-based tracking methods calculate the 3D motions and deformations directly from image intensities [142], thus eliminating the information-losing intermediate representations.

3D face modeling. Modeling of faces includes 3D shape modeling and texture modeling. For large texture variations due to changes in illumination, we will address the illumination problem in Section 1.5.1. Here we focus on 3D shape modeling. We will address the general modeling problem, including that of 2D and 3D images in Section 1.5.2.

In computer vision, one of the most widely used methods of estimating 3D shape from a video sequence is SfM, which estimates the 3D depths of interest points. The unconstrained SfM problem has been approached in two ways. In the differential approach, one computes some type of flow field (optical, image, or normal) and uses it to estimate the depths of visible points. The difficulty in this approach is reliable computation of the flow field. In the discrete approach, a set of features such as points, edges, corners, lines, or contours are tracked over a sequence of frames, and the depths of these features are computed. To overcome the difficulty of feature tracking, bundle adjustment [157] can be used to obtain better and more robust results.

Recently, multiview-based 2D methods have gained popularity. In [125], a model consists of a sparse 3D shape model learned from 2D images labeled with pose and landmarks, a shape-and-pose-free texture model, and an affine geometric model. An alternative approach is to use 3D models such as the deformable model of [118] or the linear 3D object class model of [112]. In [155], real-time 3D modeling and tracking of faces is described; a generic 3D head model is aligned to match frontal views of the face in a video sequence.

Review of Video-Based Face Recognition

Historically, video face recognition originated from still-image-based techniques. That is, the system automatically detects and segments the face from the video, and then applies still-image face-recognition techniques. Many methods reviewed in Section 4.1 belong to this category: eigenfaces [73], probabilistic eigenfaces [65], the EBGM method [49], and the PDBNN method [62]. An improvement over these methods is to apply tracking; this can help in recognition, in that a virtual frontal view can be synthesized via pose and depth estimation from video. Due to the abundance of frames in a video, another way to improve the recognition rate is the use of "voting" based on the recognition results from each frame. The voting could be deterministic, but probabilistic voting is better in general [13, 87]. One drawback of such voting schemes is the cost of computing the deterministic/probabilistic results for each frame.

The next phase of video-based face recognition dealt with the use of multimodal cues. Since humans routinely use multiple cues to recognize identities, it is expected that a multimodal system will do better than systems based on faces only. More importantly, using multimodal cues offers a comprehensive solution to the task of identification that might not be achievable by using face images alone. For example, in a totally noncooperative environment, such as a robbery, the face of the robber is typically covered, and the only way to perform faceless identification might be to analyze body motion characteristics [84]. Both face and voice have been used in many multimodal systems [81, 82, 9].

More recently, a third phase of video face recognition has begun. These methods [86, 85] simultaneously exploit both spatial information (in each frame) and temporal information (such as the trajectories of facial features). A big difference between these methods and the probabilistic voting methods [87] is the use of representations in a joint temporal and spatial space for identification.

The availability of video/image sequences gives video-based face recognition a distinct advantage over still-image-based face recognition: the abundance of temporal information. However, the typically low-quality images in video present a significant challenge: reduced spatial information. The key to building a successful

Table 1.3: Categorization of video-based face-recognition techniques

Approach	Examples
Still-image methods	Basic methods [73, 65, 49, 129, 62]
	Tracking-enhanced [87, 88, 83]
Multimodal methods	Video- and audio-based [81, 82]
Spatiotemporal methods	Feature-trajectory-based [85, 86]
	Video-to-video methods [89]

video-based system is to use temporal information to compensate for the lost spatial information. For example, a high-resolution frame can in principle be reconstructed from a sequence of low-resolution video frames and used for recognition. A further step is to use the image sequence to reconstruct the 3D shape of the tracked face object via SfM and thus enhance face recognition performance. Finally, a comprehensive approach is to use simultaneously spatial and temporal information for face recognition. This is also supported by related psychological studies.

A video-based method. While most face-recognition algorithms take still images as probe inputs, a video-based face-recognition approach that takes video sequences as inputs has recently been developed [89]. Since the detected face might be moving in the video sequence, one has to deal with uncertainty in tracking as well as in recognition. Rather than resolving these two uncertainties separately, [89] performs simultaneous tracking and recognition of human faces from a video sequence.

In still-to-video face recognition, where the gallery consists of still images, a time-series state-space model is proposed to fuse temporal information in a probe video, which simultaneously characterizes the kinematics and identity using a motion vector and an identity variable, respectively. The joint posterior distribution of the motion vector and the identity variable is first estimated at each time instant and then propagated to the next time instant. Marginalization over the motion vector yields a robust estimate of the posterior distribution of the identity variable, and marginalization over the identity variable yields a robust estimate of the posterior distribution of the motion vector, so that tracking and recognition are handled simultaneously. A computationally efficient sequential importance sampling (SIS) algorithm is used to estimate the posterior distribution. Empirical results demonstrate that, due to the propagation of the identity variable over time, *degeneracy* in the posterior probability of the identity variable is achieved to give improved recognition. The gallery is generalized to videos in order to realize video-to-video face recognition. An exemplar-based learning strategy is employed to automatically select video representatives from the gallery, serving as mixture centers in an updated likelihood measure. The SIS algorithm is used to approximate the posterior distribution of the motion vector, the identity variable, and the exemplar index. The marginal distribution of the identity variable produces the recognition result. The model formulation is very general and allows a variety of image representations and transformations. Experimental results using images/videos collected at UMD, NIST/USF, and CMU with pose/illumination variations illustrate the effectiveness of this approach in both still-to-video and video-to-video scenarios with appropriate model choices.

1.4.3 Methods Based on 3D Images

As we will discuss later in this chapter, significant 2D appearance change of a 3D face under different viewing angles and illumination conditions poses major challenges for appearance-based face-recognition systems. A good way to handle both illumination and pose problems is to use 3D face models. During the early years, 3D methods were apparently not popular due to the complexity and computational cost of acquiring 3D data. One of the exceptions is the early work reported in [93]. [93] describes a template-based recognition system involving descriptors based on curvature calculations made on range image data. The data are stored in a cylindrical coordinate system. At each point on the surface, the magnitude and direction of the minimum and maximum normal curvatures are calculated. Since the calculations involve second-order derivatives, smoothing is required to remove the effects of noise in the image. The strategy used for face recognition is as follows: (1) The nose is located; (2) Locating the nose facilitates the search for the eyes and mouth; (3) Other features such as forehead, neck, cheeks, etc. are determined by their surface smoothness (unlike hair and eye regions); (4) This information is then used for depth template comparison. Using the locations of the eyes, nose and mouth the faces are normalized into a standard position. This position is reinterpolated to a regular cylindrical grid and the volume of space between the two normalized surfaces is used as the mismatch measure. This system was tested on a dataset of 24 images of eight persons with three views of each. The data represented four male and four female faces. Adequate feature detection was achieved for 100% of these faces. 97% recognition accuracy was reported for the individual features and 100% for the whole face.

Recently, the availability of cheap real-time 3D sensors [90] makes it much easier to apply 3D face recognition. There are mainly two real-time approaches for estimating the 3D structure of a face: stereo and active projection of structured light. Based on 3D models, recognition methods have been derived that can handle illumination variation [95, 104] and facial expression variation [92]. 3D face recognition methods can be divided into two categories: (1) single-modal methods that use 3D shapes only, and (2) multimodal methods that use both 3D shapes and

2D images. Though performance evaluation based on large 3D data sets have not been reported, it was suggested that multimodal methods are the preferred choice. For more details, please refer to a recent survey paper [91].

1.5 ADVANCED TOPICS IN FACE RECOGNITION

In this section, we discuss some advanced topics that are related to face recognition. In our opinion, the best face-recognition techniques reviewed in Section 1.4 are successful in terms of recognition performance on large databases under well-controlled environment. However, face recognition in an uncontrolled environment is still very challenging. For example, the recent FRVT tests have revealed that there are at least two major challenges: the illumination variation problem and the pose variation problem. Though many existing systems build in some sort of performance invariance by applying preprocess such as histogram equalization or pose learning, significant illumination or pose change can cause serious performance degradation. In addition, face images could be partially occluded, or the system needs to recognize a person from an image in the database that was acquired some time ago. In an extreme scenario, for example, the search for missing children, the time interval could be as high as 10 years. Such scenario poses a significant challenge to building a robust system that can tolerate the variations in appearances across many years.

These problems are unavoidable when face images are acquired under uncontrolled and uncooperative environments as in surveillance applications. We first discuss two well-defined problems, the illumination problem and the pose problem, and review various approaches to solving them. We then extend our discussion to a broader perspective: mathematical modeling (or approximation in many cases). Mathematical modeling allows us to mathematically describe physical entities and hence transfer the physical phenomena into a series of numbers. The decomposition of a face image into a linear combination of eigenfaces is a classical example of mathematical modeling. Similarly, mathematical modeling is applied for the decomposition of a 3D face shape into a linear combination of 3D base shapes. As we will discuss in detail, mathematical modeling of surface reflectance and 3D geometry helps to handle the illuminant and pose issues in face recognition. Finally, mathematical modeling can be applied to handle the issues of occlusion, low-resolution, and aging.

1.5.1 Illumination and Pose Variations in Face Recognition

To facilitate the discussion and analysis, we adopt an illumination model. There are many illumination models available, which can be broadly categorized into diffuse-reflectance models and specular models. Among these models, the Lambertian model is the most popular one for diffuse reflectance and is chosen here. To approximate the real surface reflectance better, we adopt a varying-albedo Lambertian reflectance model that relates the image I of an object to its surface partial derivative (p, q) [148],

$$I = \rho_s \boldsymbol{s} \cdot \boldsymbol{n} = \rho \frac{1 + pP_s + qQ_s}{\sqrt{1 + p^2 + q^2}\sqrt{1 + P_s^2 + Q_s^2}}, \tag{9}$$

where $\boldsymbol{s}$ is the illuminant direction (a unit vector), $\boldsymbol{n}$ is the point-wise surface normal, and ρ is the pointwise albedo representing the reflectance of the surface. Alternatively, the image can be represented by (p, q), the partial derivatives of the object and $(P_s, Q_s, 1)$, the illumination direction. To simplify the notation, we replace the constant $\sqrt{1 + P_s^2 + Q_s^2}$ with K.

The Illumination Problem in Face Recognition

The illumination problem is illustrated in Figure 1.6, where the same face appears different due to a change in lighting. The changes induced by illumination are often larger than the differences between individuals, causing systems based on comparing images to misclassify input images. This was experimentally observed in [94] using a dataset of 25 individuals.

In [18], an analysis is carried out on how the illumination variation changes the eigensubspace projection coefficients for images under the assumption of Lambertian surface. For the purpose of easier analysis, we assume that frontal face objects are bilaterally symmetric about the middle vertical lines of faces. Consider the basic expression for the subspace decomposition of a face

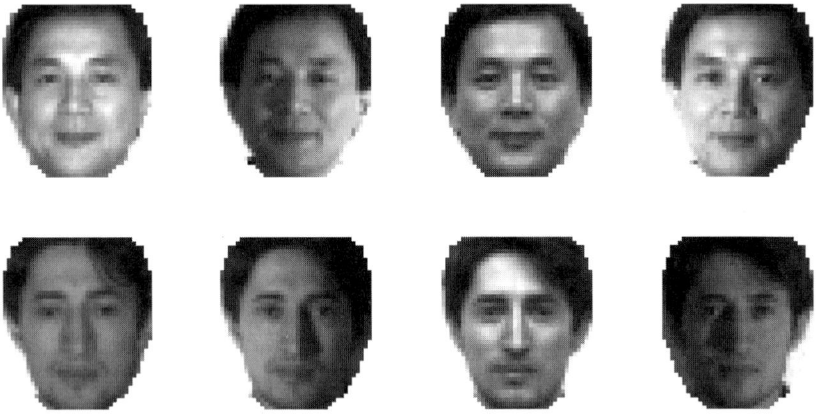

FIGURE 1.6: In each row, the same face appears differently under different illuminations (from the Yale face database).

image I: $I \simeq I_A + \sum_{i=1}^{m} a_i \Phi_i$, where I_A is the average image, Φ_i are the eigenimages and a_i are the projection coefficients. Let us assume that for a particular individual we have a prototype image I_p that is normally lighted frontal view ($P_s = 0, Q_s = 0$ in Equation 9) in the database and we want to match it against a new image $\tilde{I}$ of the same class under lighting $(P_s, Q_s, -1)$. Then their corresponding subspace-projection coefficient vectors $\boldsymbol{a} = [a_1, a_2, \cdots, a_m]^T$ (for I_p) and $\tilde{\boldsymbol{a}} = [\tilde{a}_1, \tilde{a}_2, \cdots, \tilde{a}_m]^T$ (for $\tilde{I}$) are computed as follows:

$$\begin{aligned} a_i &= I_p \odot \Phi_i - I_A \odot \Phi_i, \\ \tilde{a}_i &= \tilde{I} \odot \Phi_i - I_A \odot \Phi_i, \end{aligned} \tag{10}$$

where $\odot$ denotes the sum of all elementwise products of two matrices (vectors). If we divide the images and the eigenimages into two halves, i.e., the left and the right, then we have

$$\begin{aligned} a_i &= I_p^L \odot \Phi_i^L + I_p^R \odot \Phi_i^R - I_A \odot \Phi_i, \\ \tilde{a}_i &= \tilde{I}^L \odot \Phi_i^L + \tilde{I}^R \odot \Phi_i^R - I_A \odot \Phi_i. \end{aligned} \tag{11}$$

Based on Equation 9, the symmetric property of eigenimages and face objects, we have

$$\begin{aligned} a_i &= 2I_p^L[x,y] \odot \Phi_i^L[x,y] - I_A \odot \Phi_i, \\ \tilde{a}_i &= \left(\frac{2}{K}\right) (I_p^L[x,y] + I_p^L[x,y] q^L[x,y] Q_s) \odot \Phi_i^L[x,y] - I_A \odot \Phi_i, \end{aligned} \tag{12}$$

leading to the relation

$$\tilde{\boldsymbol{a}} = \left(\frac{1}{K}\boldsymbol{a}\right) + \frac{Q_s}{K}\left[f_1^a, f_2^a, \cdots, f_m^a\right]^T - \frac{K-1}{K}\boldsymbol{a}_A, \tag{13}$$

where $f_i^a = 2(I_p^L[x,y]q^L[x,y]) \odot \Phi_i^L[x,y]$ and $\boldsymbol{a}_A$ is the projection coefficient vector of the average image I_A: $[I_A \odot \Phi_1, \cdots, I_A \odot \Phi_m]$. Now let us assume that the training set is extended to include mirrored images as in [122]. Similar analysis can be carried out since, in such a case, the eigenimages are shown to be either symmetric (for most leading eigenimages) or antisymmetric.

In general Equation 13 suggests that a significant illumination change can seriously degrade the performance of subspace-based methods. Hence it is necessary to seek methods that compensate for these changes. Figure 1.7 plots the projection coefficients for the same face under different illuminations ($\alpha \in [0°, 40°]$, $\tau \in [0°, 180°]$) and compare them against the variations of the projection

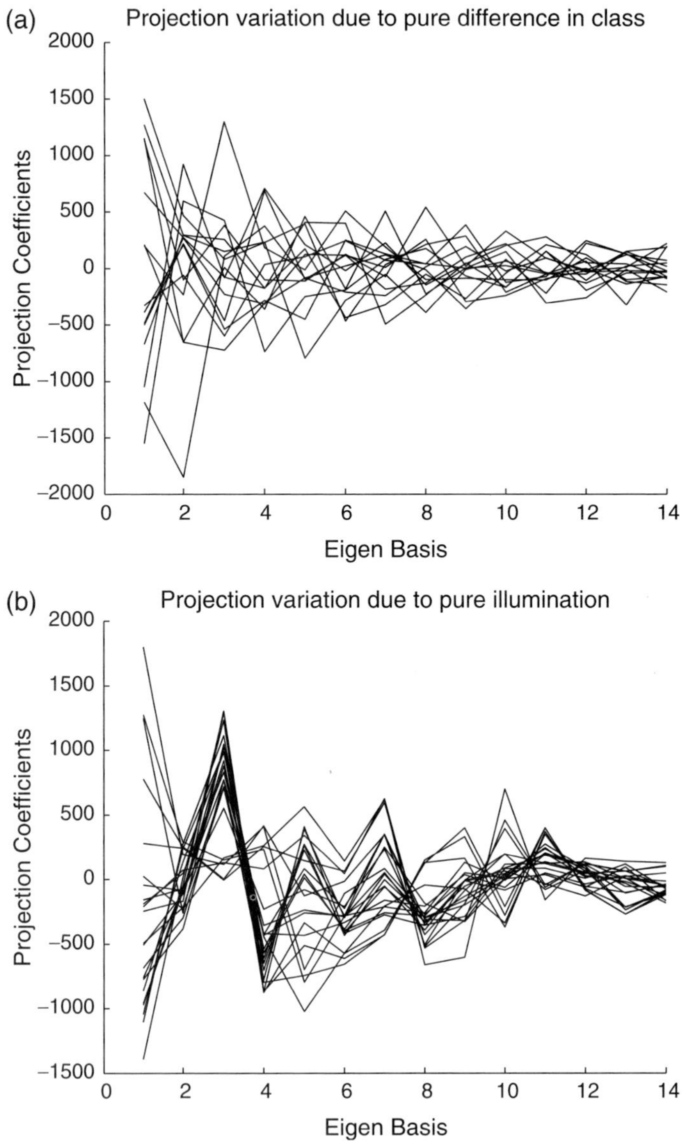

FIGURE 1.7: Change of projection vectors due to (a) class variation, (b) illumination change [18].

coefficient vectors due to pure difference in class. This confirms that significant illumination changes can severely affect system performance.

In general, the illumination problem is quite difficult and has received consistent attention in the image-understanding literature. In the case of face recognition, many approaches to this problem have been proposed that make use of the domain knowledge that all faces belong to one face class. These approaches can be divided into four types [18]: (1) heuristic methods, e.g., discarding the leading principal components, (2) image-comparison methods in which appropriate image representations and distance measures are used, (3) class-based methods using multiple images of the same face in a fixed pose but under different lighting conditions, and (4) model-based approaches in which 3D models are employed.

Heuristic approaches. Many existing systems have heuristic methods to compensate for lighting changes. For example, in [65] a simple contrast normalization is used to preprocess the detected faces, while in [48] normalization in intensity is done by first subtracting a best-fit brightness plane and then applying histogram equalization. When the face eigensubspace domain is used, it has been suggested that by discarding the three most significant principal components, variations due to lighting can be reduced. It was experimentally verified in [52] that discarding the first few principal components works reasonably well for images obtained under different lighting conditions. The plot in Figure 1.7(b) also supports this observation. However, in order to maintain system performance for normally illuminated images, while improving performance for images acquired under changes in illumination, it must be assumed that the first three principal components capture only variations due to lighting. Other heuristic methods based on frontal-face symmetry have also been proposed [18, 134]. For example, detected shadow can be replaced with texture mapped from the horizontally opposite side.

Image-comparison approaches. In [94], approaches based on image comparison using different image representations and distance measures were evaluated. The image representations used were edge maps, derivatives of the gray level, images filtered with 2D Gabor-like functions, and a representation that combines a log function of the intensity with these representations. The distance measures used were pointwise distance, regional distance, affine-GL (gray level) distance, local affine-GL distance, and log pointwise distance. For more details about these methods and about the evaluation database, see [94]. It was concluded that none of these representations alone can overcome the image variations due to illumination.

A recently proposed image-comparison method [103] used a new measure robust to illumination change. The reason for developing such a method of directly comparing images is due to the potential difficulties in building a complete representation of an object's possible images as suggested in [137]. The authors argued that it is not clear whether it is possible to construct the complete representation using a small number of training images taken under uncontrolled viewing conditions, containing multiple light sources. It was shown that, given two images of an object with unknown structure and albedo, there is always a large family of solutions. Even in the case of known light sources, only two out of three independent components of the Hessian of the surface can be determined. Instead, the authors argue that the difference between two images of the same object is smaller than the difference between images of different objects. Based on such observations, the complexity of the ratio of two aligned images was proposed as the similarity measure. The authors noticed the similarity between this measure and the measure of simply comparing the edges. It is also clear that the proposed measure is not strictly illumination-invariant, because the measure changes for a pair of images of the same object when the illumination changes. Experiments on face recognition showed the improved performance over eigenface that is somewhat worse than illumination-cone-based method [101] on the same set of data.

Class-based approaches. Under the assumptions of Lambertian surfaces and no shadowing, a 3D linear illumination subspace for a person was constructed in [153, 102, 137, 105] for a fixed viewpoint, using three aligned faces/images acquired under different lighting conditions. Under ideal assumptions, recognition based on this subspace is illumination-invariant. An illumination cone has been proposed as an effective method of handling illumination variations, including shadowing and multiple light sources [137, 101]. This method is an extension of the 3D linear-subspace method [153, 102] and also requires three aligned training images acquired under different lighting conditions. One drawback of using this method is that more than three aligned images *per* person are needed.

A method based on *quotient image* was introduced [105]. Similar to other class-based methods, this method assumes that faces of different individuals have the same shape and different textures. Given two objects $\boldsymbol{a}$ and $\boldsymbol{b}$, the quotient image Q is defined to be the ratio of their albedo functions ρ_a/ρ_b, and hence illumination invariant. Once Q is computed, the entire illumination space of object $\boldsymbol{a}$ can be generated by Q and a linear illumination subspace constructed using three images of object $\boldsymbol{b}$. To make this basic idea work in practice, a training set consisting of N objects where each object has images under various lighting conditions is needed (called the bootstrap set in the paper) and the quotient image of a novel object is defined against the average object of the bootstrap set. More specifically, the bootstrap set consists of $3N$ images taken from three fixed, linearly independent light sources s_1, s_2, and s_3 that are not known. Under this assumption, any light source s can be expressed as the linear combination of s_i: $s = x_1 s_1 + x_2 s_2 + x_3 s_3$. The authors further define the *normalized albedo* function ρ of the bootstrap set as the square sum of ρ_i where ρ_i is the albedo function of object i. Next an interesting energy/cost function is defined that is quite different from the traditional bilinear form. Let $\boldsymbol{A}_1, \boldsymbol{A}_2, \cdots, \boldsymbol{A}_N$ be $m \times 3$ matrices whose columns are the images of object i that contains the same m pixels, then the bilinear energy/cost function is $(\boldsymbol{y}_s - \sum_{i=1}^{N} \alpha_i \boldsymbol{A}_i \boldsymbol{x})^2$ which is a bilinear problem in the $N+3$ unknowns $\boldsymbol{x}$ and α_i. To compare, the proposed energy function is $\sum_{i=1}^{N} (\alpha_i \boldsymbol{y}_s - \boldsymbol{A}_i \boldsymbol{x})^2$. In our opinion, such a form of energy function is one of the major reasons why the quotient-image method works better than the "reconstructionist" methods in that it needs a smaller bootstrap set and is more tolerant on pixelwise image alignment errors. As pointed out by the authors, another factor contributing to the success of using only a small size of bootstrap set is that the albedo functions only occupy a small subspace of m.

Model-based approaches. In model-based approaches, a 3D face model is used to synthesize the virtual image under desired illumination condition from a given image. When the 3D model is unknown, *accurately* recovering the shape from images is difficult without using any priors: Shape from shading (SFS) can be used if only one image is available; stereo or SfM methods can be used when multiple images of the same object are available.

Fortunately, for face recognition, the difference in the 3D shapes of different face objects is not dramatic. This is especially true after the images are aligned and normalized. Recall that this assumption has been used in the class-based methods reviewed above. Using a statistical representation of the 3D heads, PCA was suggested as a tool for solving the parametric SFS problem [135]. An eigenhead approximation of a 3D head was obtained after training on about 300 laser-scanned range images of real human heads. The ill-posed SFS problem is thereby transformed into a parametric problem. The authors also demonstrate that such a representation helps to determine the light source. For a new face image, its 3D head can be approximated as the linear combination of eigenheads and then used to determine the light source. Using this complete 3D model, any virtual view of the face image can be generated. One major drawback of this approach is the assumption of *constant* albedo. This assumption does not hold for most real face images even though it is the most common assumption used in SFS algorithms.

To address the issue of varying albedo, a *direct* 2D-to-2D approach was proposed with the assumption that front-view faces are *symmetric* and the use of a generic 3D model [108]. Recall that a prototype image I_p is a frontal-view with $P_s = 0, Q_s = 0$. Substituting this into Equation 9, we have $I_p[x,y] = \rho \frac{1}{\sqrt{1+p^2+q^2}}$. Simple manipulation yields $I_p[x,y] = \frac{K}{2(1+qQ_s)}(I[x,y] + I[-x,y])$. This equation directly relates the prototype image I_p to $I[x,y] + I[x,-y]$ which is already available. The two advantages of this approach are: (1) there is no need to recover the varying albedo $\rho[x,y]$; (2) There is no need to recover the full shape gradients (p,q) except q, which could be approximated by a value derived from a generic 3D shape. As part of the proposed automatic method, a model-based light-source identification method is also proposed to improve existing source-from-shading algorithms.

Recently, a general method of approximating Lambertian reflectance using second-order spherical harmonics has been reported [95, 104]. Assuming Lambertian objects under distant isotropic lighting, the authors were able to show that the set of all reflectance functions can be approximated using the surface spherical-harmonic expansion. Specifically, they have proved that using a second-order (nine harmonics, i.e., 9D-space) approximation, the accuracy for any light function exceeds 97.97%. They then extend this analysis to image formation, which is a much more difficult problem due to possible occlusion, shape, and albedo variations. As indicated by the authors, worst-case image approximation can be arbitrarily bad, but most cases are good. Using their method, an image can be decomposed into so-called *harmonic images*, which are produced when the object is illuminated by harmonic functions. An interesting comparison was made between the proposed method and the 3D linear illumination subspace methods [153, 102]; the 3D linear methods are just first-order harmonic approximations without the DC components.

The Pose Problem in Face Recognition

It is not surprising that the performance of face-recognition systems drops significantly when large pose variations are present in the input images. This difficulty has been documented in the FERET test and FRVT test reports [163, 164] and suggested as a major research issue. When illumination variation is also present the task of face recognition becomes even more difficult. Here we focus on the out-of-plane rotation problem, since in-plane rotation is a pure 2D problem and can be resolved much easier.

Earlier methods focused on constructing invariant features [49] or synthesizing a prototypical view (frontal view) after a full model is extracted from the input image [59].[7] Such methods work well for small rotation angles, but they fail when the angle is large, say 60°, causing some important features to be invisible. Most proposed methods are based on using large numbers of multiview samples. This seems to concur with the findings of the psychology community; face perception is believed to be view-independent for small angles, but view-dependent for large angles.

Researchers have proposed various methods for handling the rotation problem. They can be divided into three classes [18]: (1) multiview image methods, when multiview database images of each person are available; (2) class-based methods, when multiview training images are available during training but only one database image per person is available during recognition; and (3) single image/shape-based

[7] One exception is the multiview eigenfaces of [67].

methods where no training is carried out. [96, 107, 99, 100] are examples of the first class and [132, 106, 98, 97, 117] of the second class. Up to now, the second type of approach has been the most popular. The third approach does not seem to have received much attention.

Multiview-based approaches. One of the earliest examples of the first class of approaches is the work of Beymer [96], which uses a template-based correlation matching scheme. In this work, pose estimation and face recognition are coupled in an iterative loop. For each hypothesized pose, the input image is aligned to database images corresponding to that pose. The alignment is first carried out via a 2D affine transformation based on three key feature points (eyes and nose), and optical flow is then used to refine the alignment of each template. After this step, the correlation scores of all pairs of matching templates are used for recognition. The main limitations of this method, and other related methods, are (1) many different views per person are needed in the database; (2) no lighting variations or facial expressions are allowed; and (3) the computational cost is high, since iterative searching is involved.

More recently [99], image synthesis based on illumination cones [137] has been proposed to handle both pose and illumination problems in face recognition. It handles illumination variation quite well, but not pose variation. To handle variations due to rotation, it needs to completely resolve the GBR (generalized-bas-relief) ambiguity and then reconstruct the Euclidean 3D shape. Without resolving this ambiguity, images from nonfrontal viewpoints synthesized from a GBR reconstruction will differ from a valid image by an affine warp of the image coordinates.[8] To address the GBR ambiguity, the authors propose exploiting face symmetry (to correct tilt) and the fact that the chin and the forehead are at about the same height (to correct slant), and requiring that the range of heights of the surface be about twice the distance between the eyes (to correct scale) [100]. They propose a pose- and illumination-invariant face recognition method based on building illumination cones at each pose for each person. Though conceptually this is a good idea, in practice it is too expensive to implement. The authors suggested many ways of speeding up the process, including first subsampling the illumination cone and then approximating the subsampled cone with a 11-D linear subspace. Experiments on building illumination cones and on 3D shape reconstruction based on seven training images per class were reported.

Class-based approaches. Numerous algorithms of the second type have been proposed. These methods, which make use of prior class information, are the

[8]GBR is a 3D affine transformation with three parameters: scale, slant, and tilt. A weak-perspective imaging model is assumed.

most successful and practical methods up to now. We list several representative methods here: (1) a view-based eigenface method [67], (2) a graph-matching-based method [49], (3) a linear class-based method [132, 112], (4) a vectorized image-representation-based method [98, 97], and (5) a view-based appearance model [117]. Some of these methods are very closely related — for example, methods 3, 4, and 5. Despite their popularity, these methods have two common drawbacks: (1) they need many example images to cover the range of possible views; (2) the illumination problem is not explicitly addressed, though in principle it can be handled if images captured under the same pose but different illumination conditions are available.

The popular eigenface approach [73] to face recognition has been extended to a view-based eigenface method in order to achieve pose-invariant recognition [67]. This method explicitly codes the pose information by constructing an individual eigenface for each pose. More recently, a unified framework called the bilinear model was proposed in [144] that can handle either pure-pose variation or pure-class variation.

In [49], a robust face-recognition scheme based on EBGM is proposed. The authors assume a planar surface patch at each feature point (landmark), and learn the transformations of "jets" under face rotation. Their results demonstrate substantial improvements in face recognition under rotation. Their method is also fully automatic, including face localization, landmark detection, and flexible graph matching. The drawback of this method is its requirement for accurate landmark localization, which is not an easy task, especially when illumination variations are present.

The image-synthesis method in [132] is based on the assumption of linear 3D object classes and the extension of linearity to images (both shape and texture) that are 2D projections of the 3D objects. It extends the linear shape model (which is very similar to the active shape model of [116]) from a representation based on feature points to full images of objects. To implement this method, a correspondence between images of the input object and a reference object is established using optical flow. Correspondences between the reference image and other example images having the same pose are also computed. Finally, the correspondence field for the input image is linearly decomposed into the correspondence fields for the examples. Compared to the parallel-deformation scheme in [98], this method reduces the need to compute correspondences between images of different poses. On the other hand, parallel-deformation was able to preserve some peculiarities of texture that are nonlinear and that could be "erased" by linear methods. This method is extended in [106] to include an additive error term for better synthesis. In [112], a morphable 3D face model consisting of shape and texture is directly matched to single/multiple input images.

Recently, a method that combines the light-field subspace and illumination subspace is proposed to handle both illumination and pose problems [110].

Single-image/shape-based approaches. Finally, the third class of approaches includes methods based on low-level features, invariant features, and 3D models. In [64], a Gabor wavelet-based feature-extraction method is proposed for face recognition which is robust to small-angle rotations. In 3D model-based methods, face shape is usually represented by a polygonal or mesh model which simulate tissue. In earlier days, no serious attempt to apply this approach to face recognition was made except for [93] due to the complexity and computational cost. In [109], a unified approach was proposed to solving both pose and illumination problems. This method is a natural extension of the method proposed in [108] to handle the illumination problem. Using a generic 3D model, they approximately solve the correspondence problem involved in a 3D rotation, and perform an input-to-prototype image computation. To address the varying albedo issue in the estimation of both pose and light source, the use of a *self-ratio image* is proposed. The self-ratio image $r_I[x, y]$ is defined as

$$r_I[x,y] = \frac{I[x,y] - I[-x,y]}{I[x,y] + I[-x,y]} = \frac{p[x,y]P_s}{1 + q[x,y]Q_s},$$

where $I[x, y]$ is the original image and $I[-x, y]$ is the mirrored image.

1.5.2 Mathematical Modeling for Face Recognition

In previous sections, we have discussed two problems related to 2D image-based recognition of 3D objects: appearance changes due to illumination and pose variations. Mathematical modeling of various types has been applied to handle these two problems. There are other factors that have impact on machine recognition of human faces: aging and surgery, to name two. It is our opinion that mathematical modeling is a key to overcoming many difficulties encountered in practical face recognition.

In the following, we first review the modeling of 2D and 3D face images, and then briefly discuss other modeling techniques that have potential impact upon face processing.

Modeling of 2D and 3D Face Images

Modeling of face images has been widely used in many fields for various purposes. In computer graphics, the purpose of modeling faces is for high-quality rendering under desired viewing and lighting conditions. In the compression community, it is used mainly for reducing the number of bits used for representation. In the field of image analysis and understanding, face modeling has been used for detection of faces and recognition of facial identity and facial expression.

For example, FACS (Facial Action Coding System) [31] has been used for recognition of facial expressions [15].

For the task of facial recognition, modeling of 2D face images has been used extensively: face template by Brunelli and Poggio [53], eye and mouth templates by Yullie et al. [133], eigenface by Turk and Pentland [73], ASM and AAM models by Cootes et al. [116, 114]. Modeling of 3D face images has also been very successful for both computer animation and machine recognition [132, 112].

A look into many successful face modeling methods offers the following insights:

- Methods are most successful if modeling is constrained by the fact that all face images are quite similar.
- Methods are most successful if they model both shape and texture.
- Methods are most successful if they compact/robust and process generalization capability.

A statistical method for modeling. A successful statistical method for face modeling is described in [115]. Using PCA to compactly represent key features and their spatial relationships, a statistical shape model ASM was proposed [116]. More specifically, given a set of examples each of which is represented by a set of n labeled landmark points, we align them into a common coordinate frame via applying translation, rotation, and scaling. Each shape can then be represented by a $2n$-element vector $\boldsymbol{x} = (x_1, \ldots, x_n, y_1, \ldots, y_n)$. The aligned training set forms a cloud in the $2n$ dimensional space, and can be considered to be a sample from a probability density function. This is how the phrase point-distribution model (PDM) is coined. In the simplest formulation, the cloud can be approximated with a Gaussian distribution. To improve robustness and generalization capability, PCA is applied to capture only the main axes of the cloud, and the shape model is $\boldsymbol{x} = \boldsymbol{x}_{\mathrm{m}} + \boldsymbol{Pb}$, where $\boldsymbol{x}_{\mathrm{m}}$ is the mean of the aligned training examples, $\boldsymbol{P}$ is a $2n \times t$ matrix whose columns are unit vectors along the principal axes of the cloud, and $\boldsymbol{b}$ is a t-element vector of shape parameters. Benefiting from its statistical nature that offers flexibility and robustness that constrains arbitrary deformation, the ASM has proven to be more robust and flexible than methods based on simply shape deformation, for example, the *snake* (active contour) [121]. Applications of ASM have been found in face modeling, hand modeling, and brain modeling [115]. Combining with texture information, a flexible-appearance model (FAM) was demonstrated to be effective for (1) locating facial features, (2) coding and reconstructing, and (3) face tracking and identification [124].

A method for high-quality rendering. In [112], a method was presented to synthesize high-quality textured 3D faces. As the key technique in any 2D/3D morphing algorithms, the difficult correspondence problem between a new image and

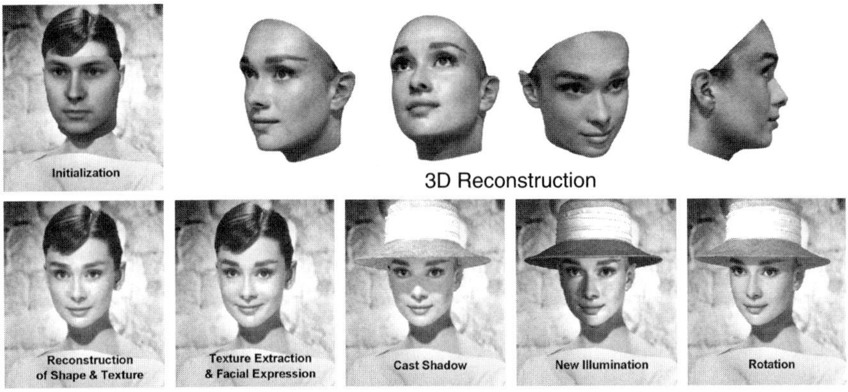

FIGURE 1.8: After manual initialization, the algorithm automatically matches a colored morphable model to the image. Rendering the inner part of the 3D face on top of the image, new shadows, facial expressions, and poses can be generated [112]. (Courtesy of V. Blanz and T. Vetter.)

the average class (3D head here) is solved by *manual* alignment plus optical flow algorithm [138, 149] coupled with statistical 3D model fitting. The basic steps for reconstructing the 3D model for a new face image is to first compute the correspondence between the image and the morphable 3D model (shape + texture) learned from a training/bootstrap set, and then synthesize the view of the face for comparison against the input face image. This is essentially an optimization procedure and needs to be iterated. During the optimization steps, head orientation, illumination conditions, and other free parameters are determined. Figure 1.8 shows that various high-quality images can be generated from just one input example.

Other Modeling Methods

In the following, we briefly discuss several interesting modeling methods that have potential impact on face processing.

Shape from shading. To obtain accurate 3D reconstruction of a face object that has very smooth regions and subtle shading, a traditional stereo or structure from motion algorithms may not be effective. Shape from shading is an effective approach complementary to stereo and SfM.

In [135], PCA was suggested as a tool for solving the parametric SFS problem. An eigenhead approximation of a 3D head was obtained after training on about 300 laser-scanned range images of real human heads. The ill-posed SFS problem is thereby transformed into a parametric problem, but *constant* albedo is still assumed. This assumption does not hold for most real face images, and it is one of the reasons why most SFS algorithms fail on real face images. The authors of [159] proposed using a varying-albedo reflectance model (Equation 9), where the albedo ρ is also a function of pixel location. They first proposed a symmetric SFS scheme based on the use of self-ratio image r_I. Unlike existing SFS algorithms, the symmetric SFS method theoretically allows pointwise shape information, the gradients (p, q), to be uniquely recovered from a single 2D image.

Modeling aging. One approach to recognizing a person using images taken many years apart is to predict his/her appearance using a computerized aging model to synthesize changes in the skin (both surface and color), muscle and tissue, and the 3D skull structure.

In [128], a standard facial-caricaturing algorithm [113] was applied to a 3D representation of human heads to study distinctiveness and the perception of facial age. Authors proposed [128] directly applying this standard algorithm to laser-scanned 3D head data to caricature the 3D heads where the lines connecting the landmark locations on individual face and the average face are now 3D instead of two-dimensional. A technical difference between this 3D caricature approach and the traditional 2D approaches is the use of PCA that yields 99 principal components as the feature vector to be exaggerated. Using about 85 scanned heads, the mean distance of the heads from the average in the PCA space was 9.9. Of those 85 heads, 30 male and 30 female heads are randomly chosen to be caricatured by changing the distance from 9.9 to 6.5 (anticaricature), 10.0 (estimate of the original head), 13.5, and 17.0 (both caricatures). To support their findings, the authors also carried out an experiment that asks 10 human observers to judge the facial age from these caricatured heads. It turns out that the perceived face age increased nearly linearly as a function of caricature level.

Instead of modeling the skull structure changes, a computer vision/graphics based approach is presented [126]. This method transfers the appearance change (aging) from a pair of young and senior faces to a new young face to obtain the aged face, or vice versa. The basic idea of this approach is based on the assumption of Lambertian reflectance (Equation 9) and the only aging factor considered here is the local surface normal (i.e., the wrinkling). The authors further assume that the change in local surface normal between the young face and senior face of one person can be transferred to another person when the two faces are properly aligned. Compared to the above approach, the 3D skull change is not considered. On the other hand, the person-specific skin information (albedo ρ) is preserved.

Modeling of misnormalization and occlusion. Ideally, an automatic system would segment a facial region and normalize it with the *right* scale, rotation angle, and translation. But the estimate of these parameters could be inaccurate, causing the recognition performance of the system to degrade [18, 134, 127]. We refer these issues collectively as the misnormalization problem.

Several methods have been proposed to address the misnormalization problem [18, 134, 111, 140]. In [140], the major concern is how to handle scale change in tracking; while the major concern in [111] is recognition under scale change and occlusion. In [18, 134], a method based on *distorted eigenspace* decomposition that is similar to [111] is proposed to handle cases of varying scale and mislocalization, including in-plane rotation, translation.

In reality, a face could be partially occluded, for example, by eyeglasses. In the image-understanding literature, there exist two traditional approaches to recognize partially occluded objects: (1) attempt to recover the occluded parts, (2) recognition based on nonoccluded parts only. The first approach has the advantage of using full information *if* the recovery of occluded parts is successful. On the other hand, the second approach uses only partial but *correct* information. The occlusion problem in face recognition has been addressed using both approaches. Within the first approach, eigenrepresentation is used to partially recover the occluded parts for recognition [80, 111] (Figure 1.5). In [59], a statistical model that combines ASM and gray-level information is used to recover the whole facial region. It is interesting to note that the algorithm also removes eyeglasses presented in the input images based on limited examples. Later, the statistical ASM model has been augmented to the view-based AAM to recover the whole facial region from a rotated view, say a profile that has half of the face being occluded [83].

Many general object recognition methods exist within the second approach that are based on local parts. In particular, the face can be divided into eye, mouth, and nose modules, each module has an output, and finally all the outputs are fused to form the final decision. For example, in the context of face detection, Burl et al. [45] introduced a principled framework for representing object deformations using probabilistic shape models. Local part detectors were used to identify candidate locations for object parts. These candidates were then grouped into object hypotheses (yes or no) and scored based on the spatial arrangement of the parts. One drawback of this approach is that the part-detection algorithm employs a hard detection strategy, that is, if the response of a part detector is above a threshold, only the position of the part is recorded; the actual response values are not retained for subsequent processing. In [41], this semiprobabilistic (local deterministic and global probabilistic) approach is expanded into a full probabilistic approach that combines both the local photometry (part match) and the global geometry (shape likelihood), which was shown to have significant improvement.

Recently, a probabilistic method that models the face with six ellipsoid-shape parts for recognition is proposed [127]. For each part/region, an eigenspace is learned from a training set of correct region images. Next to model the occlusion/localization error, a large set of synthetic samples with errors are projected onto the eigensubspace of small dimension to estimate the parameters of the assumed

Gaussian (or mixture of Gaussian) per class: covariance matrix and mean vector. To perform recognition, a face region is first warped to its final 120×170 pixel array, and then each of its six regions is projected to the above computed eigenspace to compute the probability of being correct match. Finally, all these part-based matching probabilities are summed to form the final matching score.

The same idea of using synthetic samples to estimate the probability has also been applied to recognition under mislocalization [127]. The only difference here is that the whole faces are the input samples, hence just one eigenspace is constructed.

ACKNOWLEDGMENTS

Portions of this chapter were expanded and modified, with permission from the paper by W. Zhao, R. Chellappa, P.J. Phillips, and A. Rosenfeld: "Face recognition: a literature survey", *ACM Computing Survey*, 35(4), pages 399–458, 2003.

REFERENCES

General Reference

[1] *Proceedings of International Conference on Pattern Recognition*, 1973–.
[2] *Proceedings of ACM SIGGRAPH*, 1974–.
[3] *Proceedings of IEEE Conference on Computer Vision and Pattern Recognition*, 1983–.
[4] *Proceedings of International Conference on Computer Vision*, 1987–.
[5] *Proceedings of Neural Information Processing Systems*, 1987–.
[6] *Proceedings of European Conference on Computer Vision*, 1990–.
[7] *Proceedings of International Conference on Automatic Face and Gesture Recognition*, 1995–.
[8] *Proceedings of International Conference on Image Processing*, 1994–.
[9] *Proceedings of International Conferences on Audio- and Video-Based Person Authentication*, 1997–.
[10] *Proceedings of IEEE International Workshop on Analysis and Modeling of Faces and Gestures*, 2003.
[11] R. Chellappa, C. Wilson, and S. Sirohey. Human and machine recognition of faces, a survey. *Proceedings of the IEEE*, 83:705–740, 1995.
[12] G. Donato, M. Bartlett, J. Hager, P. Ekman, and T. Sejnowski. Classifying facial actions. *IEEE Trans. on Pattern Analysis and Machine Intelligence*, 21:974–989, 1999.
[13] S. Gong, S. J. Mckenna, and A. Psarrou. *Dynamic Vision: From Images to Face Recognition*. Imperial College Press, London, 2000.
[14] S. Z. Li and A. K. Jain, editors. *Handbook of Face Recognition*. Springer, New York, 2005.
[15] M. Pantic and L. Rothkrantz. Automatic analysis of facial expressions: The state of the art. *IEEE Trans. on Pattern Analysis and Machine Intelligence*, 22:1424–1446, 2000.

[16] P. J. Phillips, R. M. McCabe, and R. Chellappa. Biometric image processing and recognition. In *Proc. of the European Signal Processing Conference*, 1998.

[17] A. Samal and P. Iyengar. Automatic recognition and analysis of human faces and facial expressions: A survey. *Pattern Recognition*, 25:65–77, 1992.

[18] W. Zhao. *Robust Image Based 3D Face Recognition*. PhD thesis, University of Maryland, 1999.

[19] W. Zhao, R. Chellappa, P. J. Phillips, and A. Rosenfeld. Face recognition: A literature survey. *ACM Computing Survey*, 35:399–458, 2003.

Early Work on Face Recognition

[20] M. Bledsoe. The model method in facial recognition. Technical Report PRI 15, Panoramic Research Inc., Palo Alto, CA, 1964.

[21] C. Darwin. *The Expression of the Emotions in Man and Animals*. John Murray, London, 1872.

[22] F. Galton. Personal identification and description. *Nature*, pages 173–188, 1888.

[23] T. Kanade. *Computer Recognition of Human Faces*. Birkhauser, 1973.

[24] M. Kelly. Visual identification of people by computer. Technical Report AI 130, Stanford, CA, 1970.

Psychology and Neuroscience

[25] J. Bartlett and J. Searcy. Inversion and configuration of faces. *Cognitive Psychology*, 25:281–316, 1993.

[26] I. Biederman and P. Kalocsai. Neural and psychophysical analysis of object and face recognition. In: H. Wechsler, P. J. Phillips, V. Bruce, F. Soulie, and T. S. Huang, editors, *Face Recognition: From Theory to Applications*, pages 3–25, 1998.

[27] V. Bruce. *Recognizing Faces*. Lawrence Erlbaum Associates, London, 1988.

[28] V. Bruce, P. Hancock, and A. Burton. Human face perception and identification. In: H. Wechsler, P. J. Phillips, V. Bruce, F. Soulie, and T. S. Huang, editors, *Face Recognition: From Theory to Applications*, pages 51–72, 1998.

[29] I. Bruner and R. Tagiuri. The perception of people. In: G. Lindzey, editor, *Handbook of Social Psychology*, volume 2, pages 634–654, Reading, MA, 1954. Addison-Wesley.

[30] P. Ekman, editor. *Charles Darwin's THE EXPRESSION OF THE EMOTIONS IN MAN AND ANIMALS*. HarperCollins and Oxford University Press, London and New York, 3rd edition, 1998. With introduction, afterwords, and commentaries by Paul Ekman.

[31] P. Ekman and W. Friesen. *Facial Action Coding System: A Technique for the Measurement of Facial Movement*. Consulting Psychologists Press, Palo Alto, CA, 1978.

[32] H. Ellis. Introduction to aspects of face processing: Ten questions in need of answers. In: H. Ellis, M. Jeeves, F. Newcombe, and A. Young, editors, *Aspects of Face Processing*, pages 3–13, 1986.

[33] I. Gauthier and N. Logothetis. Is face recognition so unique after all? *Journal of Cognitive Neuropsychology*, 17:125–142, 2000.

[34] P. Hancock, V. Bruce, and M. Burton. A comparison of two computer-based face recognition systems with human perceptions of faces. *Vision Research*, 38:2277–2288, 1998.

[35] J. Haxby, M. I. Gobbini, M. Furey, A. Ishai, J. Schouten, and P. Pietrini. Distributed and overlapping representations of faces and objects in ventral temporal cortex. *Science*, 293:425–430, 2001.

[36] A. Johnston, H. Hill, and N. Carman. Recognizing faces: Effects of lighting direction, inversion and brightness reversal. *Cognition*, 40:1–19, 1992.

[37] P. Kalocsai, W. Zhao, and E. Elagin. Face similarity space as perceived by humans and artificial systems. In: *Proc. of International Conference on Automatic Face and Gesture Recognition*, pages 177–180, 1998.

[38] J. Shepherd, G. Davies, and H. Ellis. Studies of cue saliency. In: G. Davies, H. Ellis, and J. Shepherd, editors, *Perceiving and Remembering Faces*, 1981.

[39] P. Thompson. Margaret Thatcher—a new illusion. *Perception*, 9:483–484, 1980.

[40] R. Yin. Looking at upside-down faces. *Journal of Experimental Psychology*, 81: 141–151, 1969.

Feature Extraction and Face Detection

[41] M. Burl, M. Weber, and P. Perona. A probabilistic approach to object recognition using local photometry and global geometry. In: *Proc. of European Conference on Computer Vision*, 1998.

[42] L. Gu, S. Z. Li, and H. Zhang. Learning probabilistic distribution model for multi-view face dectection. In: *Proc. of IEEE Conference on Computer Vision and Pattern Recognition*, 2001.

[43] P. Hallinan. Recognizing human eyes. In: *SPIE Proc. of Vol. 1570: Geometric Methods In Computer Vision*, pages 214–226, 1991.

[44] E. Hjelmas and B. K. Low. Face detection: A survey. *Computer Vision and Image Understanding*, 83:236–274, 2001.

[45] T. Leung, M. Burl, and P. Perona. Finding faces in cluttered scene using random labeled graph matching. In: *Proc. of International Conference on Computer Vision*, pages 637–644, 1995.

[46] H. Rowley, S. Baluja, and T. Kanade. Neural network based face detection. *IEEE Trans. on Pattern Analysis and Machine Intelligence*, 20:39–51, 1998.

[47] H. Schneiderman and T. Kanade. Probabilistic modelling of local appearance and spatial reationships for object recognition. In: *Proc. of IEEE Conference on Computer Vision and Pattern Recognition*, pages 746–751, 2000.

[48] K. Sung and T. Poggio. Example-based learning for view-based human face detection. *IEEE Trans. on Pattern Analysis and Machine Intelligence*, 20:39–51, 1997.

[49] L. Wiskott, J.-M. Fellous, and C. v. d. Malsburg. Face recognition by elastic bunch graph matching. *IEEE Trans. on Pattern Analysis and Machine Intelligence*, 19: 775–779, 1997.

[50] M. H. Yang, D. Kriegman, and N. Ahuja. Detecting faces in images: A survey. *IEEE Trans. on Pattern Analysis and Machine Intelligence*, 24:34–58, 2002.

Face Recognition Based on Still Images

[51] M. Bartlett, H. Lades, and T. Sejnowski. Independent component representation for face recognition. In: *Proc. of SPIE Symposium on Electronic Imaging: Science and Technology*, pages 528–539, 1998.

[52] P. N. Belhumeur, J. Hespanha, and D. J. Kriegman. Eigenfaces vs. Fisherfaces: Recognition using class specific linear projection. *IEEE Trans. on Pattern Analysis and Machine Intelligence*, 19:711–720, 1997.

[53] R. Brunelli and T. Poggio. Face recognition: Features versus templates. *IEEE Trans. on Pattern Analysis and Machine Intelligence*, 15:1042–1052, 1993.

[54] L. Chen, H. Liao, M. Ko, J. Lin, and G. Yu. A new LDA-based face recognition system which can solve the small sample size problem. *Pattern Recognition*, 33:1713–1726, 2000.

[55] I. Cox, J. Ghosn, and P. Yianilos. Feature-based face recognition using mixture-distance. In: *Proc. of IEEE Conference on Computer Vision and Pattern Recognition*, pages 209–216, 1996.

[56] K. Etemad and R. Chellappa. Discriminant analysis for recognition of human face images. *Journal of the Optical Society of America*, 14:1724–1733, 1997.

[57] J. Huang, B. Heisele, and V. Blanz. Component-based face recognition with 3D morphable models. In: *Proc. of International Conference on Audio- and Video-Based Person Authentication*, 2003.

[58] W. Konen. Comparing facial line drawings with gray-level images: A case study on phantomas. In: *Proc. of International Conference on Artifical Neural Networks*, pages 727–734, 1996.

[59] A. Lanitis, C. Taylor, and T. Cootes. Automatic face identification system using flexible appearance models. *Image and Vision Computing*, 13:393–401, 1995.

[60] S. Lawrence, C. Giles, A. Tsoi, and A. Back. Face recognition: A convolutional neural-network approach. *IEEE Trans. on Neural Networks*, 8:98–113, 1997.

[61] S. Z. Li and J. Lu. Face recognition using the nearest feature line method. *IEEE Trans. on Neural Networks*, 10:439–443, 1999.

[62] S. Lin, S. Kung, and L. Lin. Face recognition/detection by probabilistic decision-based neural network. *IEEE Trans. on Neural Networks*, 8:114–132, 1997.

[63] C. Liu and H. Wechsler. Evolutionary pursuit and its application to face recognition. *IEEE Trans. on Pattern Analysis and Machine Intelligence*, 22:570–582, 2000.

[64] B. Manjunath, R. Chellappa, and C. v. d. Malsburg. A feature based approach to face recognition. In: *Proc. of IEEE Conference on Computer Vision and Pattern Recognition*, pages 373–378, 1992.

[65] B. Moghaddam and A. Pentland. Probabilistic visual learning for object representation. *IEEE Trans. on Pattern Analysis and Machine Intelligence*, 19:696–710, 1997.

[66] A. Nefian and M. Hayes III. Hidden Markov models for face recognition. In: *Proc. of International Conference on Acoustics, Speech and Signal Processing*, pages 2721–2724, 1998.

[67] A. Pentland, B. Moghaddam, and T. Starner. View-based and modular eigenspaces for face recognition. In: *Proc. of IEEE Conference on Computer Vision and Pattern Recognition*, 1994.

[68] P. Phillips. Support vector machines applied to face recognition. In: *Proc. of Neural Information Processing Systems*, pages 803–809, 1998.
[69] F. Samaria and S. Young. HMM based architecture for face identification. *Image and Vision Computing*, 12:537–583, 1994.
[70] D. Socolinsky, L. Wolff, J. Neuheisel, and C. Eveland. Illumination invariant face recognition using thermal infrared imagery. In: *Proc. of IEEE Conference on Computer Vision and Pattern Recognition*, volume 1, pages 527–534, 2001.
[71] D. Swets and J. Weng. Using discriminant eigenfeatures for image retrieval. *IEEE Trans. on Pattern Analysis and Machine Intelligence*, 18:831–836, 1996.
[72] X. Tang and X. Wang. Face sketch synthesis and recognition. In: *Proc. of International Conference on Computer Vision*, 2003.
[73] M. Turk and A. Pentland. Eigenfaces for recognition. *Journal of Cognitive Neuroscience*, 3:72–86, 1991.
[74] R. Uhl and N. Lobo. A framework for recognizing a facial image from a police sketch. In: *Proc. of IEEE Conference on Computer Vision and Pattern Recognition*, pages 586–593, 1996.
[75] M. A. Vasilescu and D. Terzopoulos. Multilinear analysis of image ensembles: Tensorfaces. In: *Proc. of European Conference on Computer Vision*, pages 447–460, 2002.
[76] X. Wang and X. Tang. A unified framework for subspace face recognition. *IEEE Trans. on Pattern Analysis and Machine Intelligence*, 26:1222–1228, 2004.
[77] J. Wilder, P. Phillips, C. Jiang, and S. Wiener. Comparison of visible and infra-red imagery for face recognition. In: *Proc. of International Conference on Automatic Face and Gesture Recognition*, pages 182–187, 1996.
[78] M.-H. Yang. Kernel eigenfaces v.s. kernel fisherfaces: Face recognition using kernel methods. In: *Proc. of International Conference on Automatic Face and Gesture Recognition*, pages 215–220, 2002.
[79] W. Zhao, R. Chellappa, and A. Krishnaswamy. Discriminant analysis of principal components for face recognition. In: *Proc. of International Conference on Automatic Face and Gesture Recognition*, pages 336–341, 1998.
[80] W. Zhao, R. Chellappa, and P. Phillips. Subspace linear discriminant analysis for face recognition. Technical Report CAR-TR 914, University of Maryland, 1999.

Video Based Face Recognition

[81] J. Bigun, B. Duc, F. Smeraldi, S. Fischer, and A. Makarov. Multi-modal person authentication. In: H. Wechsler, P. J. Phillips, V. Bruce, F. Soulie, and T. S. Huang, editors, *Face Recognition: From Theory to Applications*, pages 26–50, 1998.
[82] T. Choudhury, B. Clarkson, T. Jebara, and A. Pentland. Multimodal person recognition using unconstrained audio and video. In: *Proc. of International Conference on Audio- and Video-Based Person Authentication*, pages 176–181, 1999.
[83] G. Edwards, C. Taylor, and T. Cootes. Learning to identify and track faces in image sequences. In: *Proc. of International Conference on Automatic Face and Gesture Recognition*, 1998.
[84] L. Klasen and H. Li. Faceless identification. In: H. Wechsler, P. J. Phillips, V. Bruce, F. Soulie, and T. S. Huang, editors, *Face Recognition: From Theory to Applications*, pages 513–527, 1998.

[85] B. Li and R. Chellappa. Face verification through tracking facial features. *Journal of the Optical Society of America*, 18, 2001.

[86] Y. Li, S. Gong, and H. Liddell. Constructing facial identity surfaces in a nonlinear discriminating space. In: *Proc. of IEEE Conference on Computer Vision and Pattern Recognition*, 2001.

[87] S. McKenna and S. Gong. Recognising moving faces. In H. Wechsler, P. J. Phillips, V. Bruce, F. Soulie, and T. S. Huang, editors, *Face Recognition: From Theory to Applications*, pages 578–588, 1998.

[88] J. Steffens, E. Elagin, and H. Neven. Personspotter—fast and robust system for human detection, tracking and recognition. In: *Proc. of International Conference on Automatic Face and Gesture Recognition*, pages 516–521, 1998.

[89] S. Zhou, V. Krueger, and R. Chellappa. Probabilistic recognition of human faces from video. *Computer Vision and Image Understanding*, 91:214–245, 2003.

3D Face Recognition

[90] International workshop on real time 3d sensors and their use, 2004. Associated with IEEE Conference on Computer Vision and Pattern Recognition.

[91] K. W. Bowyer, K. Chang, and P. J. Flynn. A survey of 3d and multi-modal 3d+2d face recognition. In: *Proc. of International Conference on Pattern Recognition*, 2004.

[92] A. Bronstein, M. Bronstein, E. Gordon, and R. Kimmel. 3d face recognition using geometric invariants. In: *Proc. of International Conference on Audio- and Video-Based Person Authentication*, 2003.

[93] G. Gordon. Face recognition based on depth maps and surface curvature. In: *SPIE Proc. of Volume 1570: Geometric Methods in Computer Vision*, pages 234–247, 1991.

Advanced Topics

Illumination and Pose

[94] Y. Adini, Y. Moses, and S. Ullman. Face recognition: The problem of compensating for changes in illumination direction. *IEEE Trans. on Pattern Analysis and Machine Intelligence*, 19:721–732, 1997.

[95] R. Basri and D. Jacobs. Lambertian refelectances and linear subspaces. In: *Proc. of International Conference on Computer Vision*, volume II, pages 383–390, 2001.

[96] D. Beymer. Face recognition under varying pose. AI Lab Technical Report 1461, MIT, 1993.

[97] D. Beymer. Vectorizing face images by interleaving shape and texture computations. AI Lab Technical Report 1537, MIT, 1995.

[98] D. Beymer and T. Poggio. Face recognition from one example view. In: *Proc. of International Conference on Computer Vision*, pages 500–507, 1995.

[99] A. Georghiades, P. Belhumeur, and D. Kriegman. Illumination-based image synthesis: Creating novel images of human faces under differing pose and lighting. In: *Proc. of Workshop on Multi-View Modeling and Analysis of Visual Scenes*, pages 47–54, 1999.

[100] A. Georghiades, P. Belhumeur, and D. Kriegman. From few to many: Illumination cone models for face recognition under variable lighting and pose. *IEEE Trans. on Pattern Analysis and Machine Intelligence*, 23:643–660, 2001.

[101] A. Georghiades, D. Kriegman, and P. Belhumeur. Illumination cones for recognition under variable lighting: Faces. In: *Proc. of IEEE Conference on Computer Vision and Pattern Recognition*, pages 52–58, 1998.

[102] P. Hallinan. A low-dimensional representation of human faces for arbitrary lighting conditions. In: *Proc. of IEEE Conference on Computer Vision and Pattern Recognition*, pages 995–999, 1994.

[103] D. Jacobs, P. Belhumeur, and R. Basri. Comparing images under variable illumination. In: *Proc. of IEEE Conference on Computer Vision and Pattern Recognition*, pages 610–617, 1998.

[104] R. Ramamoorthi and P. Hanrahan. On the relationship between radiance and irradiance: Determining the illumination from images of a convex lambertian object. *Journal of the Optical Society of America A*, 18:2448–2459, 2001.

[105] T. Riklin-Raviv and A. Shashua. The quotient image: Class based re-rendering and recognition with varying illuminations. In: *Proc. of IEEE Conference on Computer Vision and Pattern Recognition*, pages 566–571, 1999.

[106] E. Sali and S. Ullman. Recognizing novel 3-d objects under new illumination and viewing position using a small number of example views or even a single view. In: *Proc. of IEEE Conference on Computer Vision and Pattern Recognition*, pages 153–161, 1998.

[107] S. Ullman and R. Basri. Recognition by linear combinations of models. *IEEE Trans. on Pattern Analysis and Machine Intelligence*, 13:992–1006, 1991.

[108] W. Zhao and R. Chellappa. Illumination-insensitive face recognition using symmetric shape-from-shading. In: *Proc. of IEEE Conference on Computer Vision and Pattern Recognition*, pages 286–293, 2000.

[109] W. Zhao and R. Chellappa. SFS based view synthesis for robust face recognition. In: *Proc. of International Conference on Automatic Face and Gesture Recognition*, 2000.

[110] S. Zhou and R. Chellappa. Image-based face recognition under illumination and pose variations. *Journal of the Optical Society of America A*, 2005.

Mathematical Modeling

[111] H. Bischof and A. Leonardis. Robust recognition of scaled eigenimages through a hiearchical approach. In: *Proc. of IEEE Computer Vision and Pattern Recognition*, pages 664–670, 1998.

[112] V. Blanz and T. Vetter. A morphable model for the synthesis of 3d faces. In: *Proc. ACM SIGGRAPH*, pages 187–194, 1999.

[113] S. Brennan. The caricature generator. *Leonardo*, 18:170–178, 1985.

[114] T. Cootes, G. Edwards, and C. Taylor. Active appearance models. *IEEE Trans. on Pattern Analysis and Machine Intelligence*, 23:681–685, 2001.

[115] T. Cootes and C. Taylor. Statistical models of appearance for computer vision. Technical report, University of Manchester, 2001.

[116] T. Cootes, C. Taylor, D. Cooper, and J. Graham. Active shape models—their training and application. *Computer Vision and Image Understanding*, 61:18–23, 1995.

[117] T. Cootes, K. Walker, and C. Taylor. View-based active appearance models. In: *Proc. of International Conference on Automatic Face and Gesture Recognition*, 2000.

[118] D. DeCarlo and D. Metaxas. Optical flow constraints on deformable models with applications to face tracking. *International Journal of Computer Vision*, 38:99–127, 2000.

[119] P. L. Hallinan, G. G. Gordon, A. Yuille, P. Giblin, and D. Mumford. *Two- and Three-Dimensional Patterns of the Face*. A. K. Peters, Ltd., Natick, MA, 1999.

[120] T. Jebara, K. Russel, and A. Pentland. Mixture of eigenfeatures for real-time structure from texture. Media Lab Technical Report 440, MIT, 1998.

[121] M. Kass, A. Witkin, and D. Terzopoulos. Snakes: Active contour models. In: *Proc. of International Conference on Computer Vision*, pages 259–268, 1987.

[122] M. Kirby and L. Sirovich. Application of the Karhunen–Loeve procedure for the characterization of human faces. *IEEE Trans. on Pattern Analysis and Machine Intelligence*, 12, 1990.

[123] M. Lades, J. Vorbruggen, J. Buhmann, J. Lange, C. v. Malsburg, R. Wurtz, and W. Konen. Distortion invariant object recognition in the dynamic link architecture. *IEEE Trans. on Computers*, 42:300–311, 1993.

[124] A. Lanitis, C. Taylor, and T. Cootes. Automatic interperation and coding of face images using flexible models. *IEEE Trans. on Pattern Analysis and Machine Intelligence*, pages 743–756, 1997.

[125] Y. Li, S. Gong, and H. Liddell. Modelling face dynamics across view and over time. In: *Proc. of International Conference on Computer Vision*, 2001.

[126] Z. Liu, Y. Shan, and Z. Zhang. Image-based surface detail transfer. *IEEE Trans. on Computer Graphics and Applications*, 24:30–35, 2004.

[127] A. Martinez. Recognizing imprecisely localized, partially occluded and expression variant faces from a single sample per class. *IEEE Trans. on Pattern Analysis and Machine Intelligence*, 24:748–763, 2002.

[128] A. O'Toole, T. Vetter, H. Volz, and E. Salter. Three-dimensional caricatures of human heads: Distinctiveness and the perception of facial age. *Perception*, 26:719–732, 1997.

[129] P. Penev and J. Atick. Local feature analysis: A general statistical theory for objecct representation. *Network: Computation in Neural Systems*, 7:477–500, 1996.

[130] P. Penev and L. Sirovich. The global dimensionality of face space. In: *Proc. of International Conference on Automatic Face and Gesture Recognition*, 2000.

[131] D. Ruderman. The statistics of natural images. *Network: Computation in Neural Systems*, 5:598–605, 1994.

[132] T. Vetter and T. Poggio. Linear object classes and image ssynthesis from a single example image. *IEEE Trans. on Pattern Analysis and Machine Intelligence*, 19: 733–742, 1997.

[133] A. Yuille, D. Cohen, and P. Hallinan. Feature extractiong from faces using deformable templates. *International Journal of Computer Vision*, 8:99–112, 1992.

[134] W. Zhao and R. Chellappa. Robust image based face recognition. In: *Proc. of International Conference on Image Processing*, 2000.

Computer Vision and Statistical Learning

[135] J. Atick, P. Griffin, and N. Redlich. Statistical approach to shape from shading: Reconstruction of three-dimensional face surfaces from single two-dimensional images. *Neural Computation*, 8:1321–1340, 1996.

[136] A. Azarbayejani, T. Starner, B. Horowitz, and A. Pentland. Visually controlled graphics. *IEEE Trans. on Pattern Analysis and Machine Intelligence*, 15:602–604, 1993.

[137] P. Belhumeur and D. Kriegman. What is the set of images of an object under all possible lighting conditions? In: *Proc. of IEEE Conference on Computer Vision and Pattern Recognition*, pages 52–58, 1997.

[138] J. Bergen, P. Anadan, K. Hanna, and R. Hingorani. Hieararchical model-based motion estimation. In: *Proc. of European Conference on Computer Vision*, pages 237–252, 1992.

[139] M. Black, D. Fleet, and Y. Yacoob. A framework for modelling appearance change in image sequences. In: *Proc. of International Conference on Computer Vision*, pages 660–667, 1998.

[140] M. Black and A. Jepson. Eigentracking: Robust matching and tracking of articulated objects using a view-based representation. In: *Proc. of European Conference on Computer Vision*, pages 329–342, 1996.

[141] M. Black and Y. Yacoob. Tracking and recognizing facial expressions in image sequences using local parametrized models of image motion. Technical Report CS-TR 3401, University of Maryland, 1995.

[142] M. Brand and R. Bhotika. Flexible flow for 3d nonrigid tracking and shape recovery. In: *Proc. of IEEE Conference on Computer Vision and Pattern Recognition*, 2001.

[143] R. Fisher. The statistical utilization of multiple measurements. *Annals of Eugenics*, 8:376–386, 1938.

[144] W. Freeman and J. Tenenbaum. Learing bilinear models for two-factor problems in vision. In: *Proc. of IEEE Conference on Computer Vision and Pattern Recognition*, pages 554–560, 1997.

[145] K. Fukunaga. *Statistical Pattern Recognition*. Academic Press, New York, 1989.

[146] G. Hager and P. Belhumeur. Efficient region tracking with parametric models of geometry and illumination. *IEEE Trans. on Pattern Analysis and Machine Intelligence*, 20:1–15, 1998.

[147] Z. Hong and J. Yang. Optimal discriminant plane for a small number of samples and design method of classifier on the plane. *Pattern Recognition*, 24:317–324, 1991.

[148] B. Horn and M. Brooks. *Shape from Shading*. MIT Press, Cambridge, MA, 1989.

[149] B. Lucas and T. Kanade. An iterative image registration technique with an application to stereo vision. In: *Proc. of International Joint Conference on Artificial Intelligence*, 1981.

[150] T. Maurer and C. Malsburg. Tracking and learning graphs and pose on image sequences of faces. In: *Proc. of International Conference on Automatic Face and Gesture Recognition*, pages 176–181, 1996.

[151] S. Mika, G. Rätsch, J. Weston, B. Schölkopf, and K.-R. Muller. Fisher discriminant analysis with kernels. In: *Proc. of Neural Networks for Signal Processing IX*, pages 41–48, 1999.

[152] B. Schölkopf, A. Smola, and K.-R. Muller. Nonlinear component analysis as a kernel eigenvalue problem. *Neural Computation*, 10:1299–1319, 1998.

[153] A. Shashua. *Geometry and Photometry in 3D Visual Recognition*. PhD thesis, MIT, 1994.

[154] A. Shio and J. Sklansky. Segmentation of people in motion. In: *Proc. of IEEE Workshop on Visual Motion*, pages 325–332, 1991.

[155] J. Strom, T. Jebara, S. Basu, and A. Pentland. Real time tracking and modeling of faces: An ekf-based analysis by synthesis approach. Media Lab Technical Report 506, MIT, 1999.

[156] D. Terzopoulos and K. Waters. Analysis and synthesis of facial image sequences using physical and anatomical models. *IEEE Trans. on Pattern Analysis and Machine Intelligence*, 15:569–579, 1993.

[157] B. Triggs, P. McLauchlan, R. Hartley, and A. Fitzgibbon. Bundle adjustment — a modern synthesis. In: *Vision Algorithms: Theory and Practice*, Berlin, 2000.

[158] L. R. Tucker. Some mathematical notes on three-mode factor analysis. *Psychometrika*, 31:279–311, 1966.

[159] W. Zhao and R. Chellappa. Symmetric shape-from-shading using self-ratio image. *International Journal of Computer Vision*, 45:55–75, 2001.

Evaluation Protocols

[160] E. Bailly-Bailliere, E. Bengio, F. Bimbot, M. Hamouz, J. Kittler, J. Mariethoz, J. Matas, K. Messer, V. Popovici, F. Poree, B. Ruiz, and J. Thiran. The banca database and evaluation protocol. In: *Proc. of International Conference on Audio- and Video-Based Person Authentication*, pages 625–638, 2003.

[161] K. Messer, J. Matas, J. Kittler, J. Luettin, and G. Maitre. Xm2vtsdb: The extended m2vts database. In: *Proc. of International Conference on Audio- and Video-Based Person Authentication*, pages 72–77, 1999.

[162] H. Moon and P. Phillips. Computational and performance aspects of pca-based face recognition algorithms. *Perception*, 30:301–321, 2001.

[163] P. Phillips, P. Rauss, and S. Der. Feret (face recognition technology) recognition algorithm development and test report. Technical Report ARL 995, U.S. Army Research Laboratary, 1996.

[164] P. J. Phillips, P. Grother, R. Micheals, D. Blackburn, E. Tabassi, and J. Bone. Face recognition vendor test 2002: Evaluation report. NISTIR 6965, National Institute of Standards and Technology, 2003. http://www.frvt.org.

[165] P. J. Phillips, H. Moon, S. Rizvi, and P. Rauss. The feret evaluation methodology for face-recognition algorithms. *IEEE Trans. on Pattern Analysis and Machine Intelligence*, 22, 2000.

CHAPTER 2

EIGENFACES AND BEYOND

2.1 INTRODUCTION

The subject of visual processing of human faces has received attention from philosophers and scientists for centuries. Aristotle devoted several chapters of the *Historia Animalium* to the study of facial appearance. Physiognomy, the practice or art of inferring intellectual or character qualities of a person from outward appearance[1], particularly the face and head, has had periods of fashion in various societies [65]. Darwin considered facial expression and its identification to be a significant advantage for the survival of species [66]. Developmental studies have focused on strategies of recognition or identification and the differences between infant and adult subjects. Neurological disorders of face perception have been isolated and studied, providing insight into normal as well as abnormal face processing.

The ability of a person to recognize another person (e.g., a mate, a child, or an enemy) is important for many reasons. There is something about the perception of faces that is very fundamental to the human experience. Early in life we learn to associate faces with pleasure, fulfillment, and security. As we get older, the subtleties of facial expression enhance our communication in myriad ways. The face is our primary focus of attention in social interactions; this can be observed in interaction among animals as well as between humans and animals (and even between humans and robots [67]). The face, more than any other part of the body, communicates identity, emotion, race, and age, and is also quite useful for judging gender, size, and perhaps even character.

[1] For example, from Aristotle's *Historia Animalium*: "Straight eyebrows are a sign of softness of disposition; such as curve in towards the nose, of harshness; such as curve out towards the temples, of humour and dissimulation; such as are drawn in towards one another, of jealousy."

It is often observed that the human ability to recognize faces is remarkable. Faces are complex visual stimuli, not easily described by simple shapes or patterns; yet people have the ability to recognize familiar faces at a glance after years of separation. Many people say "I'm not very good at names, but I never forget a face." Although this quote confuses recall (generally more difficult) with recognition (generally less difficult), the point is valid that our face recognition capabilities are quite good. Lest we marvel too much at human performance, though, it should also be noted that the *inability* to recognize a face is a common experience as well. Quite often we strain to see the resemblance between a picture (e.g., a driver's license photo) and the real person; sometimes we are greeted in a friendly, familiar manner by someone we do not remember ever seeing before. Although face recognition in humans may be impressive, it is far from perfect.

Recognition is not only visual; it may occur through a variety of sensory modalities, including sound, touch, and even smell. For people, however, the most reliable and accessible modality for recognition is the sense of sight. A person may be visually recognized by face, but also by clothing, hairstyle, gait, silhouette, skin, etc. People often distinguish animals not by their faces but by characteristic markings on their bodies. Similarly, the human face is not the only, and may not even be the primary, visual characteristic used for personal identification. For example, in a home or office setting, a person's face may be used merely in verifying identity, after identity has already been established based on other factors such as clothing, hairstyle, or a distinctive moustache. Indeed, the identification of humans may be viewed as a Bayesian classification system, with prior probabilities on several relevant random variables. For example, a parent is predisposed to recognize his child if, immediately prior to contact, he sees a school bus drive by and then hears yelling and familiar light footsteps. Nevertheless, because faces are so important in human interaction, no other mode of personal identification is as compelling as face recognition.

Until rather recently, face recognition was often pointed to as one of those things that "computers can't do," even by such luminaries as Marvin Minsky [68] and David Hubel [69]. This was a motivating factor for many computer-vision students and researchers. In addition, facial processing had already interested human- and biological-vision researchers for years, and there were many interesting and curious results and theories discussed in the literature.

There has been a great deal of scientific investigation into human face-recognition performance, seeking to understand and characterize the representations and processes involved. However, a thorough understanding of how humans (and animals) represent, process, and recognize faces remains an elusive goal. Although studies of face recognition in physiology, neurology, and psychology provide insight into the problem of face recognition, they have yet to provide substantial practical guidance for computer vision systems in this area.

There are several aspects of recognizing human identity and processing facial information that make the problem of face recognition somewhat ill-defined. As mentioned above, recognition of a person's identity is not necessarily (and perhaps rarely) a function of viewing the person's face in isolation. In addition, face recognition is closely related to face (and head and body) detection, face tracking, and facial-expression analysis. There are many ways in which these "face processing" tasks may interrelate. For example, the face may be initially detected and then recognized. Alternatively, detection and recognition may be performed in tandem, so that detection is merely a successful recognition event. Or facial-feature tracking may be performed and facial expression analyzed before attempting to recognize the normalized (expressionless) face. There are, of course, many additional variations possible.

For the purposes of this chapter, "face recognition" and "face identification" describe the same task.[2] Given an image of a human face, classify that face as one of the individuals whose identity is already known by the system, or perhaps as an unknown face. "Face detection" means detecting the presence of any face, regardless of identity. "Face location" is specifying the 2D position (and perhaps orientation) of a face in the image. "Face tracking" is updating the (2D or 3D) location of the face. "Facial-feature tracking" is updating the (2D or 3D) locations, and perhaps the parametrized descriptions, of individual facial features. "Face-pose estimation" is determining the position and orientation (usually 6 degrees of freedom) of a face. "Facial-expression analysis" is computing parametric, and perhaps also symbolic, descriptions of facial deformations.

Face recognition began to be a hot topic in computer vision in the late 1980s and early 1990s. In the past two decades, the field has made substantial progress: starting with a limited set of slow techniques with questionable accuracy and applicability, there are now real-time systems installed in public spaces and sold in shrink-wrapped boxes. (Whether or not their performance is good enough for the intended applications will not be a subject of discussion in this chapter, nor will we discuss the significant policy issues that these systems generate. Clearly, however, these are topics of great importance, as argued in [62, 63, 64].)

In retrospect, the "eigenfaces" approach to face recognition popularized initially by Turk and Pentland [25, 26, 27] appears to have played a significant role in the field's emerging popularity. Maybe it was due to the part-catchy, part-awkward name given to the technique (never underestimate good PR!) or maybe it was due to the simplicity of the approach, which has been re-implemented countless

[2]The distinction may be made between classifying an object into its general category ("recognition") and labeling the object as a particular member of the category ("identification"), but we will follow the common terminology and use these terms interchangeably, with the precise meaning depending on the context.

times in graduate and undergraduate computer vision courses over the years. One could argue that it was the right approach at the right time – different enough from previous approaches to the problem that it caught the attention of researchers, and also caught (and perhaps even influenced to some small degree) the emerging trends in appearance-based vision and learning in vision.

Eigenfaces attracted people to the topic and pushed the state of the art at the time, and it has been a very useful pedagogical tool over the years, as well as a useful benchmark for comparison purposes. But it has been about fifteen years since the first publications of this work, and many other techniques have been introduced; in addition, there have been several modifications and improvements to the original eigenface technique. In this chapter we take a brief look back at face recognition research over the years, including the motivations and goals of early work in the field and the rationale for pursuing eigenfaces in the first place; we discuss some of the spinoffs and improvements over the original method; and we speculate on what may be in store for the future of automated face recognition, and for eigenfaces in particular.

2.2 ORIGINAL CONTEXT AND MOTIVATIONS OF EIGENFACES

2.2.1 Human and Biological Vision

Face recognition can be viewed as the problem of robustly identifying an image of a human face, given some database of known faces. It has all the difficulties of other vision-based recognition problems, as the image changes drastically due to several complex and confounding factors such as illumination, camera position, camera parameters (e.g., the specific lens used, the signal gain), and noise. In addition, human faces change due to aging, hairstyle, facial hair, skin changes, facial expression, and other factors. These all make the face-recognition problem quite difficult and unsolvable by direct image matching. Early researchers in the area took several different approaches attempting to deal with the complexity; these approaches were motivated by several different factors.

Much of the interest in face recognition in the mid-1980s was motivated by an interest in human and biological vision, and especially some exciting findings related to agnosia and object-selective visual recognition. Visual agnosia is a neurological impairment in the higher visual processes which leads to a defect in object recognition [70]. Agnosic patients can often "see" well, in that there is little apparent deficit in spatial vision or perception of form. The dysfunction is specific to some class of objects or shapes, such as perceiving letters or any object from an unusual viewpoint. Etcoff et al. [71] report a patient's description of his agnosia to be like "attempting to read illegible handwriting: you know that it is handwriting, you know where the words are and letters stop and start, but you have no clue as to what they signify."

Prosopagnosia, from the Greek *prosopon* (face) and *agnosia* (not knowing), refers to the inability to recognize familiar faces by visual inspection [72, 73, 74]. Prosopagnosic patients, although very few in number, have proved to be a valuable resource in probing the function of face recognition. Prosopagnosics can typically identify the separate features of a face, such as the eyes or mouth, but have no idea to whom they belong. They may recognize the sex, age, pleasantness, or expression of a face, without an awareness of the identity:

> I was sitting at the table with my father, my brother and his wife. Lunch had been served. Suddenly... something funny happened: I found myself unable to recognize anyone around me. They looked unfamiliar. I was aware that they were two men and a woman; I could see the different parts of their faces but I could not associate those faces with known persons.... Faces had normal features but I could not identify them. (Agnetti et al., p. 51, quoted in [75])

There is evidence that damage to a particular area of the right hemisphere has a predominant role in producing face-recognition difficulties. The question arises, is face recognition a special, localized, subsystem of vision? One way to approach this question, and additionally to learn about the neural mechanisms involved in face recognition and object recognition in general, is by recording the activity of brain cells while performing visual tasks including observing and recognizing faces. Through single-cell recording, a number of physiologists found what seem to be "face" neurons in monkeys, responding selectively to the presence of a face in the visual field. Perrett et al. [76, 77, 78] found cells in area STS of the rhesus monkey which were selectively responsive to faces in the visual field. Many of these cells were insensitive to transformations such as rotation. Different cells responded to different features or subsets of features, while most responded to partially obscured faces. Some cells responded to line drawings of faces. About 10% of the cells were sensitive to identity. Other researchers (e.g., [4, 79, 80]) have found cells with similar properties in monkey inferior temporal cortex, concluding that there may be specialized mechanisms for the analysis of faces in IT cortex. Kendrick and Baldwin [81] even found face-selective cells in sheep.

Psychologists have used both normal and prosopagnosic subjects to investigate models of face processing, recognition, and identification. In addition to the theoretical and clinical pursuits of neuroscience, the validity and limitations of eyewitness testimony in criminal proceedings has spurred much face-recognition research in cognitive psychology. Yin [82] presented pictures of faces in various orientations and tested subsequent recall, finding that the recall performance for inverted faces was degraded more than that of other configuration-specific stimuli such as landscapes or animals. He argued for a special face-processing mechanism to account for this effect. Others have furthered these techniques to experiment with face images which have been modified in several ways. Developmental studies (e.g., [83]) have observed the development of face recognition from infant to adult. Carey and Diamond [84] found that the effect of inversion on face recognition described by Yin increases over the first decade of life, suggesting that young children represent faces in terms of salient isolated features ("piecemeal representation"), rather than in terms of configurational properties used by older children and adults ("configurational representation"). In recent years there seems to be a growing consensus that both configurational properties and feature properties are important for face recognition [4]. Carey and Diamond [85] claimed that face recognition is not a special, unique system, and that the inversion effect may be due to a gain in the ability to exploit distinguishing "second-order relational features."

A number of experiments have explored feature saliency, attempting to discern the relative importance of different features or areas of the face. Although the early experiments generally agreed on the importance of face outline, hair, and eyes – and the relative unimportance of the nose and mouth – there is evidence that these results may be biased by the artifacts of the techniques and face presentations used [4]. Along with stored face images, a number of researchers [86, 87] have used face stimuli constructed from Identikit or Photofit2 to explore strategies of face recognition. Use of these kits may actually bias the experiments, however, since there is an underlying assumption that a face can be properly decomposed into its constituent features: eyes, ears, nose, mouth, etc. One lesson from the study of human face recognition is that approaches which treat faces as a collection of independent parts are unlikely to be relevant to the perception of real faces, where the parts themselves are difficult or impossible to delimit [4]. Consequently artists' sketches are better than face construction kits in reproducing the likeness of a target face. Faces grow and develop in a way such that features are mutually constraining. In fact these growth patterns can be expressed mathematically and used to predict the effects of aging [88]. Such techniques have already been used successfully in the location of missing children years after their disappearance [89]. Other studies have shown that expression and identity seem to be relatively independent tasks [90, 91], which is also supported by some neurological studies of prosopagnosics.

2.2.2 Compelling Applications of Face Recognition

The research that led to eigenfaces had several motivations, not the least of which was industry funding with the aim of developing television set-top boxes that could visually monitor viewers as an automated "people meter" to determine television ratings. This required a real-time system to locate, track, and identify possible viewers in a scene. Besides identifying typical members of the household, such a system should also detect unknown people (and perhaps ask them to enter relevant information, such as their sex and age) and distinguish between valid viewers and, for example, the family dog. It was also important to know when the television was on but no one was viewing it, and when people were in the vicinity but not actively watching. For this application, computation efficiency and ease of use (including training) were primary considerations.

General surveillance and security applications were also a motivating application, especially criminal mugshot identification. Whereas recognizing members of a family for use as an automated people meter required a small database of known identities, many security applications required a very large database, so memory and storage requirements were important; full images could not be stored and matched for each entry.

Other areas of interest that motivated work in face recognition at the time included image compression, film development, and human–computer interaction. In the areas of image compression for transmission of movies and television, and in general any "semantic understanding" of video signals, the presence of people in the scene is important. For example, in partitioning the spatial-temporal bandwidth for an advanced HDTV transmission, more bandwidth should be given to people than to cars, since the audience is much more likely to care about the image quality and detail of the human actors than of inanimate objects. The detection of faces in photograph negatives or originals could be quite useful in color-film development, since the effect of many enhancement or noise-reduction techniques depends on the picture content. Automated color enhancement is desirable for most parts of the scene, but may have an undesirable effect on flesh tones. (It is fine for the yellowish grass to appear greener, but not so fine for Uncle Harry to look like a Martian!) In human–computer interaction, there was a growing interest in creating machines that understand, communicate with, and react to humans in natural ways; detecting and recognizing faces is a key component of such natural interaction.

Interest in computer-based automated face recognition in the mid-1980s was motivated by several factors, including an interest in the mechanisms of biological and human vision, general object recognition, image coding, and neural networks. Today, the field is largely driven by applications in security and surveillance, perceptual interfaces [92], and content-based query of image and video. Questions of relevance to biological vision (mechanisms of human face recognition) are still of interest, but they are not a major motivation.

2.2.3 Object-Recognition Strategies

Object recognition has long been a goal of computer vision, and it has turned out to be a very difficult endeavor. The primary difficulty in attempting to recognize objects from imagery comes from the immense variability of object appearance due to several factors, which are all confounded in the image data. Shape and reflectance are intrinsic properties of an object, but an image of the object is a function of several other factors, including the illumination, the viewpoint of the camera (or, equivalently, the pose of the object), and various imaging parameters such as aperture, exposure time, lens aberrations, and sensor spectral response. Object recognition in computer vision has been dominated by attempts to infer from images information about objects that is relatively invariant to these sources of image variation. In the Marr paradigm [7], the prototype of this approach, the first stage of processing extracts intrinsic information from images, i.e., image features such as edges that are likely to be caused by surface reflectance changes or discontinuities in surface depth or orientation. The second stage continues to abstract away from the particular image values, inferring surface properties such as orientation and depth from the earlier stage. In the final stage, an object is represented as a three-dimensional shape in its own coordinate frame, completely removed from the intensity values of the original image.

This general approach to recognition can be contrasted with *appearance-based* approaches, such as correlation, which matches image data directly. These approaches tend to be much easier to implement than methods based on object shape – correlation only requires a stored image of the object, while a full 3D shape model is very difficult to compute – but they tend to be very specific to an imaging condition. If the lighting, viewpoint, or anything else of significance changes, the old image template is likely to be useless for recognition.

The idea of using pixel values, rather than features that are more invariant to changes in lighting and other variations in imaging conditions, was counterintuitive to many. After all, the whole point of the Marr paradigm of vision was to abstract away from raw pixel values to higher level, invariant representations such as 3D shape. Mumford [8] illustrated some of these objections with a toy example: recognizing a widget that comprises a one-dimensional black line with one white dot somewhere on it. He shows that, for this example, the eigenspace is no more efficient than the image space, and a feature-based approach (where the feature is the position of the white dot) is a much simpler solution. This example, however, misses the point of the eigenface approach, which can be seen in the following counterexample.

Imagine starting with images of two different people's faces. They differ in the precise location of facial features (eyes, nostrils, mouth corners, etc.) and in grayscale values throughout. Now warp one image so that all the extractable features of that face line up with those of the first face. (Warping consists of applying a two-dimensional motion vector to every pixel in the image and interpolating properly to avoid blank areas and aliasing.) The eyes line up, the noses line up, the mouth corners line up, etc. The feature-based description is now identical for both images. Do the images now look like the same person? Not at all – in many (perhaps most) cases the warped image is perceived as only slightly different from its original. Here is a case where an appearance-based approach will surely outperform a simple feature-based approach.

Soon after Mumford's toy example was introduced, Brunelli and Poggio [9] investigated generic feature-based and template-based approaches to face recognition and concluded that the template-based approach worked better, at least for their particular database of frontal view face images.

Of course, both of these examples are extreme cases. A face is nothing like a black line with a white dot. Nor is the variation in facial-feature locations and feature descriptions so small as to be insignificant. Clearly, both geometric and photometric information can be useful in recognizing faces.

In the past decade, learning has become a very significant issue in visual recognition. Rather than laboriously constructing 3D shape models or expected features manually, it would be beneficial for a system to learn the models automatically. And rather than enumerating all the conditions that require new image models or templates, it would be helpful for the system to analyze the imaging conditions to decide on optimal representations, or to learn from a collection of images what attributes of appearance will be most effective in recognition. It is likely that no recognition system of any reasonable complexity – that is, no system that solves a non-trivial recognition problem – will work without incorporating learning as a central component. For learning to be effective, enough data must be acquired to allow a system to account for the various components of the images, those intrinsic to the object and otherwise.

The concept of robustness (stability in the presence of various types of noise and a reasonable quantity of outliers) has also become very important in computer vision in the past decade or more. System performance (e.g., recognition rate) should decrease gracefully as the amount of noise or uncertainty increases. Noise can come from many sources: thermal noise in the imaging process, noise added in transmission and storage, lens distortion, unexpected markings on an object's surface, occlusions, etc. An object-recognition algorithm that requires perfect images will not work in practice, and the ability to characterize a system's performance in the presence of noise is vital.

For face-recognition systems, learning and robustness must also be balanced with practical speed requirements. Whether the task is offline, real-time, or the intermediate "interactive-time" (with a human in the loop), constraints on processing time are always an issue in face recognition.

As with most recognition tasks, the source images (face images) comprise pixel values that are influenced by several factors such as shape, reflectance, pose, occlusion, and illumination. The human face is an extremely complex object, with both rigid and non-rigid components that vary over time, sometimes quite rapidly. The object is covered with skin: a nonuniformly textured material that is difficult to model either geometrically or photometrically. Skin can change color quickly when one is embarrassed or becomes warm or cold. The reflectance properties of the skin can also change rather quickly, as perspiration level changes. The face is highly deformable, and facial expressions reveal a wide variety of possible configurations.

Other time-varying changes include the growth and removal of facial hair, wrinkles and sagging of the skin brought about by aging, skin blemishes, and changes in skin color and texture caused by exposure to sun. Add to that the many common artifact-related changes, such as cuts and scrapes, bandages, makeup, jewelry and piercings, and it is clear that the human face is much more difficult to model (and thus recognize) than most objects.

Partly because of this difficulty, face recognition has been considered a challenging problem in computer vision for some time, and the amount of effort in the research community devoted to this topic has increased significantly over the years.

2.2.4 Face Recognition through the 1980s

In general, face recognition has been viewed as both an interesting scientific direction of research (possibly helping to understand human vision) and as a useful technology to develop. The big question in the mid-1980s was how to go about solving the problem – or, perhaps more importantly – how to go about defining the problem. Face recognition was viewed as a high-level visual task, while basic areas such as stereo and motion perception were still not completely understood. However, the tasks involved in face processing are reasonably constrained; some may even have a degree of "hardwiring" in biological systems. Faces present themselves quite consistently in expected positions and orientations; their configuration (the arrangement of the components) seldom changes; they are rather symmetrical. On the other hand, human face recognition and identification is very robust in the face of external changes (e.g. hair styles, tan, facial hair, eyeglasses), so a recognition scheme cannot be rigid or overly constrained.

Should one approach face recognition via the framework of the "Marr paradigm," building a primal sketch or intrinsic images from the raw image data, then a viewer-centered 2½ D sketch revealing scene discontinuities, and finally a 3D object-centered representation on which to perform recognition? Or should face recognition be a very specialized task, somewhat separate from other general object recognition approaches, as perhaps suggested by some of the human-vision literature? Until the late 1980s, most of the work in automated face detection and recognition had focused on detecting individual features such as the eyes, nose, mouth, and head outline, and defining a face model by the position, size, and relationships among these features, with some newer methods based on neural networks, correlation-based techniques, and shape matching from range data.

Attempts to automate human face recognition by computers began in the late 1960s and early 1970s. Bledsoe [12] was the first to report semi-automated face recognition, using a hybrid human–computer system which classified faces on the basis of fiducial marks entered on photographs by hand. Parameters for the classification were normalized distances and ratios among points such as eye corners, mouth corners, nose tip, and chin point. Kelly [13] and Kanade [14] built probably the first fully automated face recognition systems, extracting feature measurements from digitized images and classifying the feature vector. At Bell Labs, Harmon, Goldstein, and their colleagues [15, 93, 94] developed an interactive system for face recognition based on a vector of up to 21 features, which were largely subjective evaluations (e.g. shade of hair, length of ears, lip thickness) made by human subjects. The system recognized known faces from this feature vector using standard pattern classification techniques. Each of these subjective features however would be quite difficult to automate.

Sakai et al. [95] described a system which locates features in a Laplacian-filtered image by template matching. This was used to find faces in images, but not to recognize them. A more sophisticated approach by Fischler and Elschlager [96] attempted to locate image features automatically. They described a linear-embedding algorithm which used local-feature template matching and a global measure to perform image matching. The technique was applied to faces, but not to recognition. The first automated system to recognize people was developed by Kelly [13]. He developed heuristic, goal-directed methods to measure distances in standardized images of the body and head, based on edge information.

Kanade's face identification system [14] was the first automated system to use a top-down control strategy directed by a generic model of expected feature characteristics of the face. His system calculated a set of facial parameters from a single face image, comprised of normalized distances, areas, and angles between fiducial points. He used a pattern-classification technique to match the face to one of a known set, a purely statistical approach depending primarily on local histogram analysis and absolute gray-scale values.

In a similar spirit, Harmon et al. [15] recognized face profile silhouettes by automatically choosing fiducial points to construct a 17-dimensional feature vector for recognition. Gordon [16] also investigated face recognition using side view facial profiles. Others have also approached automated face recognition by characterizing a face by a set of geometric parameters and performing pattern recognition based on the parameters (e.g., [97, 98, 99]).

Yuille et al. [17] and others have used deformable templates, parametrized models of features and sets of features with given spatial relations. Various approaches using neural networks (e.g., [18, 19]) have attempted to move away from purely feature-based methods. Moving beyond typical intensity images, Lapresté [20], Lee and Milios [21], Gordon [22], and others used range data to build and match models of faces and face features.

By the late 1980s, there had been several feature-based approaches to face recognition. For object recognition in general, the most common approach was to extract features from objects, build some sort of model from these features, and perform recognition by matching feature sets. Features, and the geometrical relationships among them, are stable under varying illumination conditions and pose – if they can be reliably calculated. However, it is often the case that they cannot, so the problem became more and more complex. Indexing schemes and other techniques were developed to cope with the inevitable noisy, spurious, and missing features.

A number of researchers (e.g., [100, 101]) were using faces or face features as input and training patterns to neural networks with a hidden layer, trained using backpropagation, but on small data sets. Fleming and Cottrell [19] extended these ideas using nonlinear units, training the system by back propagation. The system accurately evaluated "faceness," identity, and to a lesser degree gender, and reported a degree of robustness to partial input and brightness variations. Cottrell and Metcalfe [102] built on this work, reporting identity, gender, and facial expression evaluations by the network. The WISARD system [103] was a general-purpose binary pattern recognition device based on neural net principles. It was applied with some success to face images, recognizing both identity and expression.

Range data has the advantage of being free from many of the imaging artifacts of intensity images. Surface curvature, which is invariant with respect to viewing angle, may be quite a useful property in shape matching and object recognition. Lapresté et al. [20] presented an analysis of curvature properties of range images of faces, and propose a pattern vector comprised of distances between characteristic points. Sclaroff and Pentland [104] reported preliminary recognition results based on range data of heads.

Lee and Milios [21] explored matching range images of faces represented as extended Gaussian images. They claimed that meaningful features correspond to convex regions and are therefore easier to identify than in intensity images. Gordon [22] represented face features based on principal curvatures, calculating minimum and maximum curvature maps which are used for segmentation and feature detection. The major drawback of these approaches is the dependency on accurate, dense range data, which are currently not available using passive imaging systems, while active range systems can be very cumbersome and expensive. In addition, it is not clear that range information alone is sufficient for reliable recognition [105].

In 1988, as Sandy Pentland and I began to think about face recognition, we looked at the existing feature-based approaches and wondered if they were erring by discarding most of the image data. If extracting local features was at one extreme, what might be an effective way of experimenting with the other extreme, i.e., working with a global, holistic face representation? We began to build on work by Sirovich and Kirby [23] on coding face images using principal components analysis (PCA). Around the same time, Burt [24] was developing a system for face recognition using pyramids, multiresolution face representations. The era of appearance-based approaches to face recognition had begun.

2.3 EIGENFACES

The eigenface approach, based on PCA, was never intended to be the definitive solution to face recognition. Rather, it was an attempt to reintroduce the use of information "between the features"; that is, it was an attempt to swing back the pendulum somewhat to balance the attention to isolated features. Given the context of the problem at the time – the history of various approaches and the particular requirements of the motivating applications – we wanted to consider face-recognition methods that would meet the basic application requirements but with a different approach than what had been pursued up to that time. Feature-based approaches seemed to solve some issues but threw away too much information. Neural-network approaches at the time seemed to depend on "black box" solutions that could not be clearly analyzed. Approaches using range data were too cumbersome or expensive for our main application interests, and also did not appeal to our human-vision motivations.

Much of the previous work on automated face recognition had ignored the issue of just what aspects of the face stimulus are important for identification, by either treating the face as a uniform pattern or assuming that the positions of features are an adequate representation. It is not evident, however, that such representations are sufficient to support robust face recognition. Depending too much on features, for example, causes problems when the image is degraded by noise or features are occluded (e.g., by sunglasses). We would like to somehow allow for a system to decide what is important to encode for recognition purposes, rather than specifying that initially.

This suggested that an information-theory approach of coding and decoding face images may give insight into the information content of face images, emphasizing the significant local and global features, which may or may not be directly related to our intuitive notion of face features such as the eyes, nose, lips, and ears. This may even have important implications for the use of construction tools such as Identikit and Photofit [4], which treat faces as "jigsaws" of independent parts. Such a system motivated by information theory would seek to extract the relevant information in a face image, encode it as efficiently as possible, and compare one face encoding with a database of models encoded similarly. One approach to extracting the information contained in an image of a face is to somehow capture the variation in a collection of face images, independent of any judgment of features, and use this information to encode and compare individual face images.

2.3.1 Image Space

Appearance-based approaches to vision begin with the concept of *image space*. A two-dimensional image $I(x, y)$ may be viewed as a point (or vector) in a very high-dimensional space, called image space, where each coordinate of the space corresponds to a sample (pixel) of the image. For example, an image with 32 rows and 32 columns describes a point in a 1024-dimensional image space. In general, an image of r rows and c columns describes a point in N-dimensional image space, where $N = rc$. This representation obfuscates the neighborhood relationship (distance in the image plane) inherent in a two-dimensional image. That is, rearranging the pixels in the image (and changing neighborhood relationships) will have no practical effect on its image-space representation, as long as all other images are identically rearranged. Spatial operations such as edge detection, linear filtering, and translation are not local operations in image space. A 3×3 spatial image filter is not an efficient operation in image space; it is accomplished by multiplication with a very large, sparse $\mathrm{N} \times \mathrm{N}$ matrix. On the other hand, the image-space representation helps to clarify the relationships among collections of images.

With this image representation, the image becomes a very high-dimensional "feature," and so one can use traditional feature-based methods in recognition. So, merely by considering an image as a vector, feature-based methods can be used to accomplish appearance-based recognition; that is, operations typically performed on feature vectors, such as clustering and distance metrics, can be performed on images directly. Of course, the high dimensionality of the image space makes many feature-based operations implausible, so they cannot be applied without some thought towards efficiency. As image resolution increases, so does the dimensionality of the image space. At the limit, a continuous image maps to an infinite-dimensional image space. Fortunately, many key calculations scale with the number of sample images rather than the dimensionality of the image space, allowing for efficiency even with relatively high-resolution imagery.

If an image of an object is a point in image space, a collection of M images gives rise to M points in image space; these may be considered as samples of a probability distribution. One can imagine that all possible images of the object (under all lighting conditions, scales, etc.) define a manifold within the image space. How large is image space, and how large might a manifold be for a given object? To get an intuitive estimate of the vastness of image space, consider a tiny 8×8 binary (one bit) image. The number of image points (the number of distinct images) in this image space is 2^{64}. If a very fast computer could evaluate one billion images per second, it would take almost *600 years* to exhaustively evaluate the space. For grayscale and color images of reasonable sizes, the corresponding numbers are unfathomably large. It is clear that recognition by exhaustively enumerating or searching image space is impossible.

This representation brings up a number of questions relevant to appearance-based object recognition. What is the relationship between points in image space that correspond to all images of a particular object, such as a human face? Is it possible to efficiently characterize this subset of all possible images? Can this subset be learned from a set of sample training images? What is the "shape" of this subset of image space?

Consider an image of an object to be recognized. This image $I(r, c)$ is a point $\boldsymbol{x}$ in image space, or, equivalently, a feature in a high-dimensional feature space. The image pixel $I(r, c)$ can be mapped to the ith component of the image point $(\boldsymbol{x}_i)$ by $i = r \times width + c$. A straightforward pattern classification approach to recognition involves determining the minimal distance between a new face image $\boldsymbol{x}$ and pre-existing face classes $\tilde{\boldsymbol{x}}$. That is, given k prototype images of known objects, find the prototype $\tilde{\boldsymbol{x}_i}$ that satisfies

$$\min_i \boldsymbol{d}(\boldsymbol{x},\ \tilde{\boldsymbol{x}}_i) \quad (i = 1, \ldots, k).$$

A common distance metric is merely the Euclidean distance in the feature space:

$$d(\boldsymbol{x}_1, \boldsymbol{x}_2) = \|\boldsymbol{x}_1 - \boldsymbol{x}_2\| = \sqrt{(\boldsymbol{x}_1 - \boldsymbol{x}_2)^{\mathrm{T}}(\boldsymbol{x}_1 - \boldsymbol{x}_2)} = \sqrt{\sum_{i=1}^{rc} (\boldsymbol{x}_{1i} - \boldsymbol{x}_{2i})^2}.$$

This is the L2 norm, the mean squared difference between the images. Other metrics, such as the L1 norm, or other versions of the Minkowski metric, may also be used to define distance. However, these are relatively expensive to compute. Correlation is a more efficient operator, and under certain conditions maximizing correlation is equivalent to minimizing the Euclidean distance, so it is often used as an approximate similarity metric.

If all images of an object clustered around a point (or a small number of points) in image space, and if this cluster were well separated from other object clusters, object recognition – face recognition, in this case – would be relatively straightforward. In this case, a simple metric such as Euclidean distance or correlation would work just fine. Still, it would not be terribly efficient, especially with large images and many objects (known faces). The eigenface approach was developed in an attempt to improve on both performance and efficiency.

2.3.2 PCA

Considering the vastness of image space, it seems reasonable to begin with the following presuppositions:

- Images of a particular object (such as an individual's face), under various transformations, occupy a relatively small but distinct region of the image space.
- Different objects (different faces) occupy different regions of image space.
- Whole classes of objects (all faces under various transformations) occupy a still relatively small but distinct region of the image space.

These lead to the following questions about face images:

- What is the shape and dimensionality of an individual's "face space," and how can it be succinctly modeled and used in recognition?
- What is the shape and dimensionality of the complete face space, and how can it be succinctly modeled and used in recognition?
- Within the larger space, are the individual spaces separated enough to allow for reliable classification among individuals?
- Is the complete face space distinct enough to allow for reliable face/nonface classification?

The eigenface framework [25, 26, 27] provided a convenient start to investigating these and related issues. Let us review the basic steps in an eigenface-based recognition scheme. Principle-component analysis (PCA) [28] provides a method to efficiently represent a collection of sample points, reducing the dimensionality of the description by projecting the points onto the principal axes, an orthonormal set of axes pointing in the directions of maximum covariance in the data. PCA minimizes the mean squared projection error for a given number of dimensions (axes), and provides a measure of importance (in terms of total projection error) for each axis. Transforming a point to the new space is a linear transformation. A simple example of PCA is shown in Figure 2.1. Projecting face and non-face images into face space is shown in Figure 2.2.

Let a set of face images $\boldsymbol{x}_i$ of several people be represented as a matrix $\boldsymbol{X}$, where

$$\boldsymbol{X} = [\boldsymbol{x}_1\, \boldsymbol{x}_2\, \boldsymbol{x}_3 \cdots \boldsymbol{x}_M]$$

and $\boldsymbol{X}$ is of dimension $N \times M$, where N is the number of pixels in an image, the dimension of the image space which contains $\{\boldsymbol{x}_i\}$. The difference from the average face image (the sample mean) $\bar{x}$ is the matrix $\boldsymbol{X}'$:

$$\boldsymbol{X}' = [(\boldsymbol{x}_1 - \bar{\boldsymbol{x}})\,(\boldsymbol{x}_2 - \bar{\boldsymbol{x}})\,(\boldsymbol{x}_3 - \bar{\boldsymbol{x}}) \cdots (\boldsymbol{x}_M - \bar{\boldsymbol{x}})] = \left[\boldsymbol{x}'_1\, \boldsymbol{x}'_2\, \boldsymbol{x}'_3 \cdots \boldsymbol{x}'_M\right].$$

Principal-components analysis seeks a set of $M - 1$ orthogonal vectors, $\boldsymbol{e}_i$, which best describes the distribution of the input data in a least-squares sense, i.e., the Euclidean projection error is minimized. The typical method of computing the principal components is to find the eigenvectors of the covariance matrix $\boldsymbol{C}$, where

$$\boldsymbol{C} = \sum_{i=1}^{M} \boldsymbol{x}'_i \boldsymbol{x}'^{\mathrm{T}}_i = \boldsymbol{X}'\boldsymbol{X}'^{\mathrm{T}}$$

is $N \times N$. This will normally be a huge matrix, and a full eigenvector calculation is impractical. Fortunately, there are only $M - 1$ nonzero eigenvalues, and they can

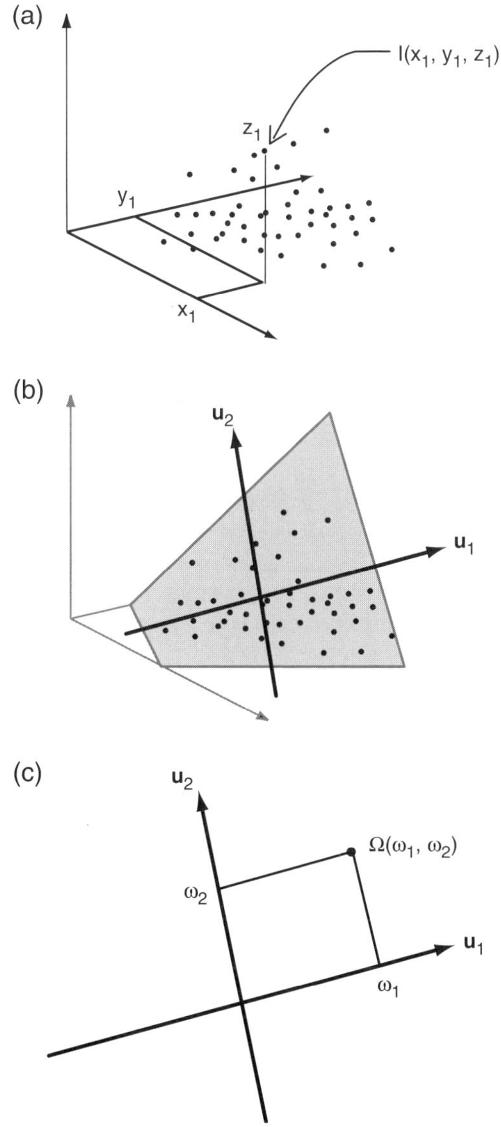

FIGURE 2.1: A simple example of principal-component analysis. (a) Images with three pixels are described as points in three-space. (b) The subspace defined by a planar collection of these images is spanned by two vectors. One choice for this pair of vectors is the eigenvectors of the covariance matrix of the ensemble, u_1 and u_2. (c) Two coordinates are now sufficient to describe the points, or images; their projections onto the eigenvectors, (ω_1, ω_2).

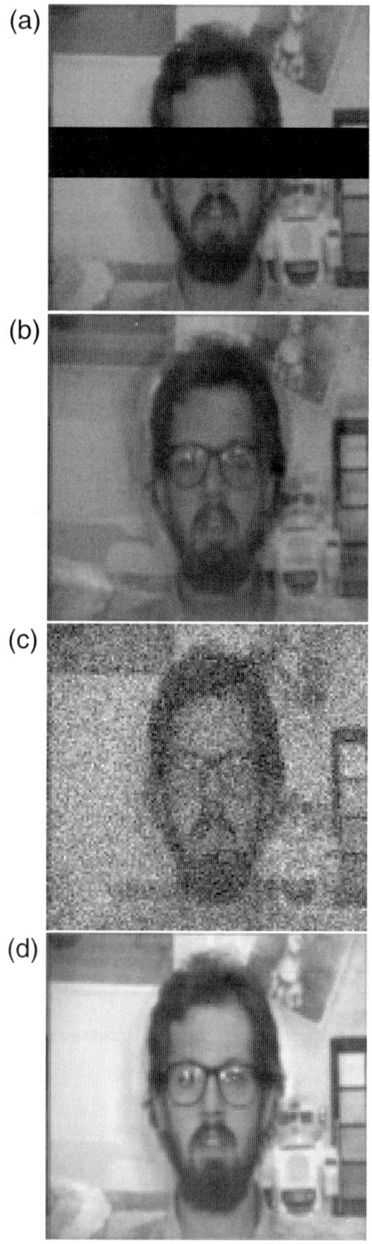

FIGURE 2.2: (a) Partially occluded face image from the test set and (b) its projection onto face space. The occluded information is encoded in the eigenfaces. (c) Noisy face image and (d) its face-space projection.

be computed more efficiently with an $M \times M$ eigenvector calculation. It is easy to show the relationship between the two. The eigenvectors $\boldsymbol{e}_i$ and eigenvalues $\boldsymbol{\lambda}_i$ of $\boldsymbol{C}$ are such that

$$\boldsymbol{C}\boldsymbol{e}_i = \boldsymbol{\lambda}_i \boldsymbol{e}_i.$$

These are related to the eigenvectors $\hat{\boldsymbol{e}}_i$ and eigenvalues $\boldsymbol{\mu}_i$ of the matrix $\boldsymbol{D} = \boldsymbol{X}'^{\mathrm{T}}\boldsymbol{X}'$ in the following way:

$$\boldsymbol{D}\hat{\boldsymbol{e}}_i = \boldsymbol{\mu}_i \hat{\boldsymbol{e}}_i, \qquad \boldsymbol{X}'^{\mathrm{T}}\boldsymbol{X}'\hat{\boldsymbol{e}}_i = \boldsymbol{\mu}_i \hat{\boldsymbol{e}}_i, \qquad \boldsymbol{X}'\boldsymbol{X}'^{\mathrm{T}}\boldsymbol{X}'\hat{\boldsymbol{e}}_i = \boldsymbol{\mu}_i \boldsymbol{X}'\hat{\boldsymbol{e}}_i,$$
$$\boldsymbol{C}\boldsymbol{X}'\hat{\boldsymbol{e}}_i = \boldsymbol{\mu}_i \boldsymbol{X}'\hat{\boldsymbol{e}}_i, \qquad \boldsymbol{C}(\boldsymbol{X}'\hat{\boldsymbol{e}}_i) = \boldsymbol{\mu}_i(\boldsymbol{X}'\hat{\boldsymbol{e}}_i), \qquad \boldsymbol{C}\boldsymbol{e}_i = \boldsymbol{\lambda}_i \boldsymbol{e}_i,$$

showing that the eigenvectors and eigenvalues of $\boldsymbol{C}$ can be computed as

$$\boldsymbol{e}_i = (\boldsymbol{X}'\hat{\boldsymbol{e}}_i), \qquad \boldsymbol{\lambda}_i = \boldsymbol{\mu}_i.$$

In other words, the eigenvectors of the (large) matrix $\boldsymbol{C}$ are equal to the eigenvectors of the much smaller matrix $\boldsymbol{D}$, premultiplied by the matrix $\boldsymbol{X}'$. The nonzero eigenvalues of $\boldsymbol{C}$ are equal to the eigenvalues of $\boldsymbol{D}$.

Once the eigenvectors of $\boldsymbol{C}$ are found, they are sorted according to their corresponding eigenvalues; a larger eigenvalue means that more of the variance in the data is captured by the eigenvector. Part of the efficiency of the eigenface approach comes from the next step, which is to eliminate all but the "best" k eigenvectors (those with the highest k eigenvalues). From there on, the "face space," spanned by the top k eigenvectors, is the feature space for recognition. The eigenvectors are merely linear combinations of the images from the original data set. Because they appear as somewhat ghostly faces, as shown in Figure 2.3, they are called eigenfaces.

PCA has been used in systems for pattern recognition and classification for decades. Sirovich and Kirby [23, 29] used PCA to form *eigenpictures* to compress face images, a task for which reproduction with low mean-squared error is important. Turk and Pentland [25] used PCA for representing, detecting, and recognizing faces. Murase and Nayar [30] used a similar eigenspace in a parametric representation that encoded pose and illumination variation, as well as identity. Finlayson et al. [31] extended grayscale eigenfaces to color images. Craw [32], Moghadam [33], Lanitis et al. [34] and others have subsequently used eigenfaces as one component of a larger system for recognizing faces.

The original eigenface recognition scheme involves two main parts, creating the eigenspace and recognition using eigenfaces. The first part (described above) is an off-line initialization procedure; that is, it is performed initially and only needs to be recomputed if the training set changes. The eigenfaces are constructed

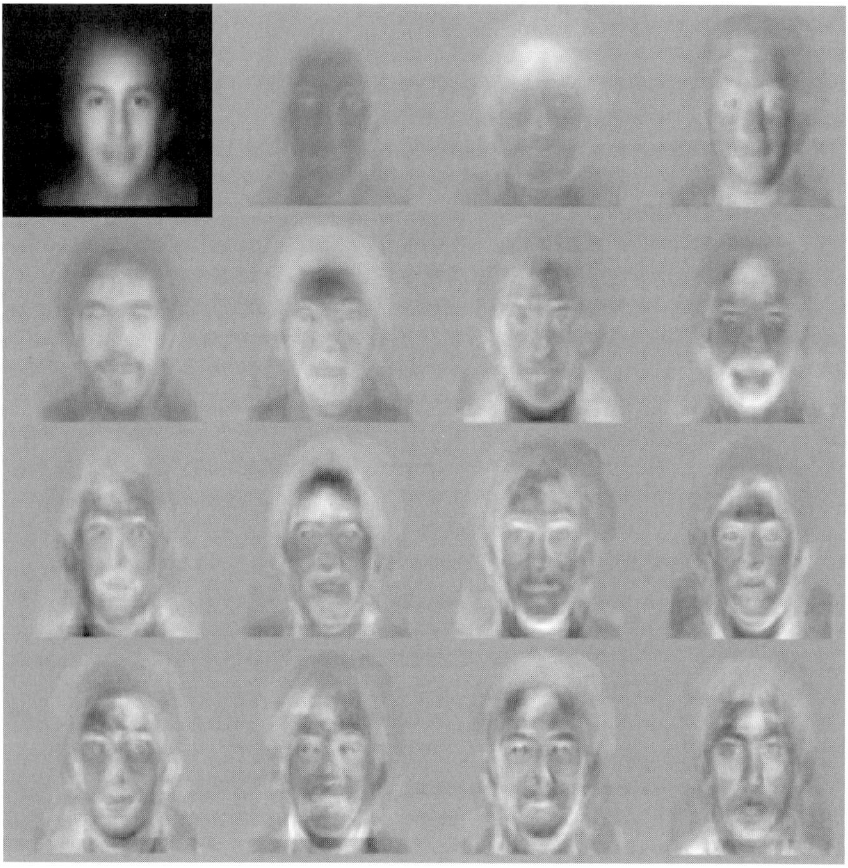

FIGURE 2.3: The average face image $\bar{x}$ and a set of eigenface images. The eigenfaces are real-valued images scaled so that a value of zero displays as a medium gray, negative values are dark, and positive values are bright.

from an initial set of face images (the training set) by applying PCA to the image ensemble, after first subtracting the mean image. The output is a set of eigenfaces and their corresponding eigenvalues. Only the eigenfaces corresponding to the top M eigenvalues are kept – these define the *face space*. For each individual in the training set, the average face image is calculated (if there is more than one instance of that individual), and this image is projected into the face space as the individual's class prototype.

The second part comprises the ongoing recognition procedure. When a new image is input to the system, the mean image is subtracted and the result is projected into the face space. This produces a value for each eigenface; together, the values comprise the image's eigenface descriptors. The Euclidean distance between the new image and its projection into face space is called the "distance from face space" (DFFS), the reconstruction error. If the DFFS is above a given threshold, the image is rejected as not a face – in other words, it is not well enough represented by the eigenfaces to be deemed a possible face of interest.

If the DFFS is sufficiently small, then the image is classified as a face. If the projection into face space is sufficiently close to one of the known face classes (by some metric such as Euclidean distance) then it is recognized as the corresponding individual. Otherwise, it is considered as an unknown face (and possibly added to the training set). Figure 2.4 shows a DFFS map corresponding to an input scene image; the face is located at the point where the DFFS map is a minimum.

Eigenfaces were originally used both for detection (via DFFS) and identification of faces. Figure 2.5 shows the overall system that was first constructed to use simple motion processing to indicate the likely area for the head; then within a smaller image segment, DFFS was used to determine the most likely face position, where face recognition was attempted. The complete tracking, detection, and recognition ran at a few frames per second on a 1990 workstation.

Although the early eigenface papers did not clearly articulate how the background was handled, Figure 2.6 shows the two standard mechanisms used to eliminate or minimize background effects. Given a particular expected scale (size of face), masking out the background around the face (possibly including most of the hair and ears) adequately removes the background from consideration.

The basic eigenface technique raises a number of issues, such as:

- How to select k, the number of eigenfaces to keep
- How to efficiently update the face space when new images are added to the data set
- How best to represent classes and perform classification within the face space
- How to separate intraclass and interclass variations in the initial calculation of face space
- How to generalize from a limited set of face images and imaging conditions.

There are obvious shortcomings of the basic eigenface technique. For example, significant variation in scale, orientation, translation, and lighting will cause it to fail. Several appearance-based recognition methods first scale the input image to match the scale of the object in a prototype template image. While this is usually an effective approximation, one must consider that scaling an image is equivalent to changing a camera's focal length, or performing an optical zoom, but it is not equivalent to moving a camera closer to the object. A translated image introduces occlusion, while a zoomed image does not. In addition, the reflectance is different for a translated image because of a slightly different angle of incidence. For an

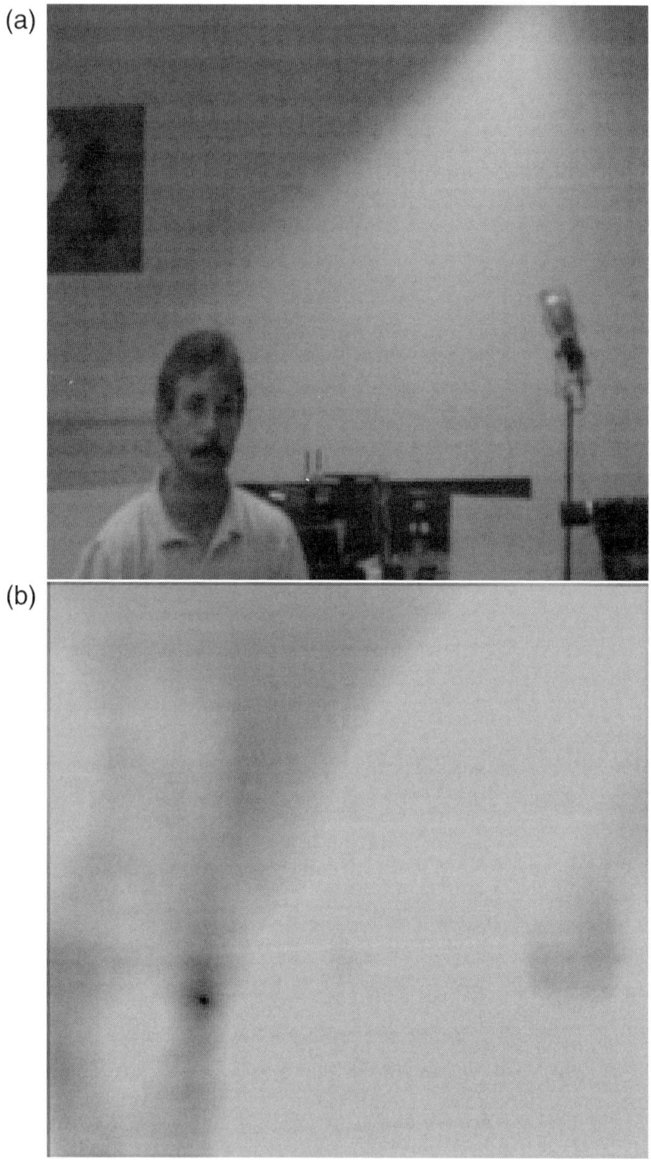

FIGURE 2.4: (a) Original image. (b) Corresponding "distance from face space" (DFFS) map, where low values (dark areas) indicate the likely presence of a face.

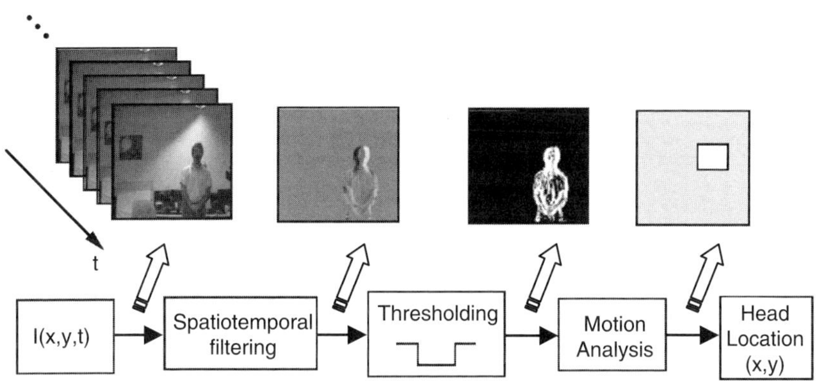

FIGURE 2.5: The initial implementation of the full almost-real-time eigenface system, using simple motion analysis to restrict the search area (for the DFFS calculation).

object with significant depth and nearby light sources, approximating translation with an image zoom may not work well. In other words, an image from the database of a face taken from one meter away will not perfectly match another image of the same face taken five meters away and zoomed in an appropriate amount.

2.3.3 Eigenobjects

An obvious question in response to approaching face recognition via eigenfaces is what about recognizing other objects and entities using a similar approach? Is this method particular to faces, or to some particular class of objects, or is it a general recognition method? What recognition problems are best tackled from a view-based perspective in general?

Since the initial eigenface work, several researchers have introduced PCA-based approaches to recognition problems. These include "eigeneyes," "eigennoses," and "eigenmouths" for the representation, detection, and recognition of facial features [27, 57, 56, 55]; "eigenears" as a biometric modality [58]; "eigenexpressions" for facial-expression analysis [27, 53]; "eigenfeatures" for image registration and retrieval [51, 50]; "eigentracking" as an approach to view-based tracking, and "eigentextures" as a 3D texture representation [59]. In addition to these vision techniques, there have been similar approaches, some inspired by the eigenface work, in other areas: for example, "eigenvoices" for speech recognition and adaptation [48, 49] and "eigengenes" and "eigenexpressions" in genetics.

Most notably, the Nayar and Murase work [30], mentioned above, utilized an eigenspace approach to represent and recognize general 3D objects at various poses, formulating object and pose recognition as parametrized appearance

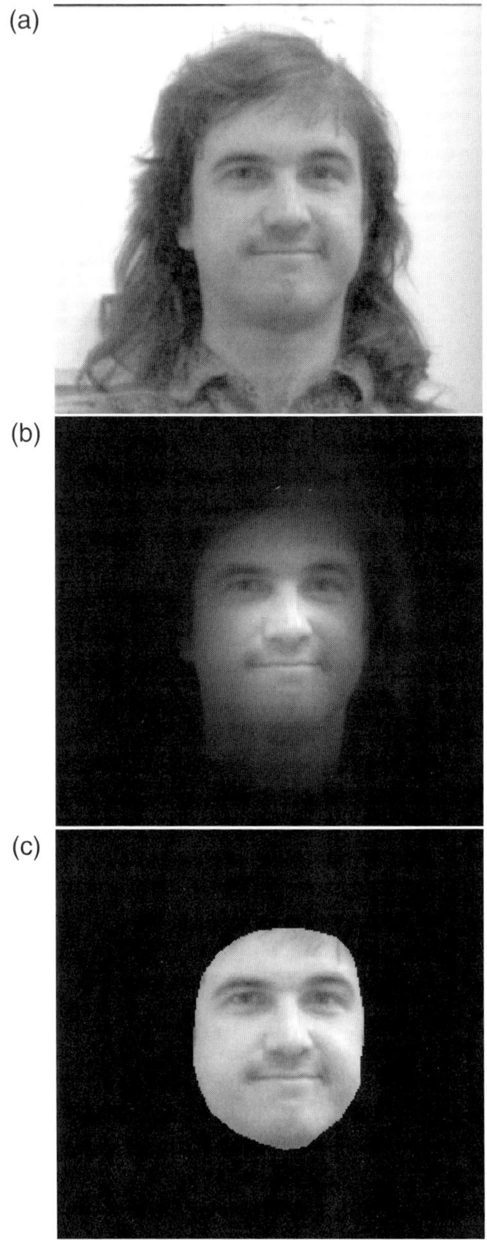

FIGURE 2.6: Two methods to reduce or eliminate the effect of background. (a) An original face image. (b) Multiplied by a Gaussian window, emphasizing the center of the face. (c) Multiplied by a binary face mask outlined by the operator (while gathering the training set).

matching. They recognized and determined the pose of 100 objects in real-time [60] by creating appearance manifolds based on the learned eigenspace.

2.4 IMPROVEMENTS TO AND EXTENSIONS OF EIGENFACES

Despite its shortcomings, there are a number of attractive aspects to eigenface methods, especially including the progress of the past decade. Since Burt [24], Turk and Pentland [26], Craw [32], and others began to use appearance-based methods in detecting and recognizing faces, there has been a voluminous amount of work on the topic, motivated by several factors. Applications of computer vision in human–computer interaction (HCI), biometrics, and image and video database systems have spurred interest in face-recognition (as well as human gesture recognition and activity analysis). There are currently several companies that market face-recognition systems for a variety of biometric applications, such as user authentication for ATM machines, door access to secure areas, and computer login, as well as a variety of HCI/entertainment applications, such as computer games, videoconferencing with computer-generated avatars, and direct control of animated characters (digital puppeteering). Conferences now exist, which are well attended, devoted to face recognition and related topics, and several good survey papers are available that track the various noteworthy results (e.g., Zhao et al. [106]). The state of the art in face recognition is exemplified both by the commercial systems, on which much effort is spent to make them work in realistic imaging situations, and by various research groups exploring new techniques and better approaches to old techniques.

The eigenface approach, as originally articulated, intentionally threw away all feature-based information in order to explore the boundaries of an appearance-based approach to recognition. Subsequent work by Moghaddam [33], Lanitis et al. [35], and others have moved toward merging the two approaches, with predictably better results than either approach alone. The original eigenface framework did not explicitly account for variations in lighting, scale, viewing angle, facial expressions, or any of the other many ways facial images of an individual may change. The expectation was that the training set would contain enough variation so that it would be modeled in the eigenface. Subsequent work has made progress in characterizing and accounting for these variations (e.g., [36] and [37]) while merging the best aspects of both feature-based and appearance-based approaches.

A few approaches in particular are significant in terms of their timing and impact. Craw et al. [32] were among the first to combine processing face shape (two-dimensional shape, as defined by feature locations) with eigenface-based recognition. They normalized the face images geometrically based on 34 face landmarks in an attempt to isolate the photometric (intensity) processing from geometric factors. Von der Malsburg and his colleagues [38, 39] introduced several systems based on elastic graph matching, which utilizes a hybrid approach where local grayscale information is combined with global feature structure. Cootes and Taylor and colleagues [40] presented a unified approach to combining local and global information, using flexible shape models to explicitly model both shape and intensity.

Recent results in appearance-based recognition applied to face recognition and other tasks include more sophisticated learning methods (e.g., [41]), warping and morphing face images [42, 43] to accommodate a wider range of face poses, including previously unseen poses, explicitly dealing with issues of robustness [44], and better methods of modeling interclass and intraclass variations and performing classification [45]. Independent-component analysis (ICA), for example, is a generalization of PCA that separates the high-order dependencies in the input, in addition to the second-order dependencies that PCA encodes [46]. The original eigenface method used a single representation and transformation for all face images, whether they originated from one individual or many; it also used the simplest techniques possible, nearest-neighbor Euclidean distance, for classification in the face space. Subsequent work has improved significantly on these first steps. Moghaddam et al. [33] developed a probabilistic matching algorithm that uses a Bayesian approach to separately model both interclass and intraclass distributions. This improves on the implicit assumption that the images of all individuals have a similar distribution. Penev and Sirovich [47] investigated the dimensionality of face space, concluding that, for very large databases, at least 200 eigenfaces are needed to sufficiently capture global variations such as lighting, small scale and pose variations, race, and sex. In addition, at least twice that many are necessary for minor, identity-distinguishing details such as exact eyebrow, nose, or eye shape.

2.5 SUMMARY

Appearance-based approaches to recognition have made a comeback from the early days of computer-vision research, and the eigenface approach to face recognition may have helped bring this about. Clearly, though, face recognition is far from being a solved problem, whether by eigenfaces or any other technique. The progress during the past decade on face recognition has been encouraging, although one must still refrain from assuming that the excellent recognition rates from any given experiment can be repeated in different circumstances. They usually cannot. Reports of dismal performance of commercial face-recognition systems in real-world scenarios [61] seem to confirm this.

Eigenface (and other appearance-based) approaches must be coupled with feature- or shape-based approaches to recognition, possibly including 3D data and models, in order to build systems that will be robust and will scale to real-world environments. Because many imaging variations (lighting, scale, orientation, etc.)

have an approximately linear effect when they are small, linear methods can work, but in very limited domains. Eigenfaces are not a general approach to recognition, but one tool out of many to be applied and evaluated in context. The ongoing challenge is to find the right set of tools to be applied at the appropriate times.

In addition to face recognition, significant progress is being made in related areas such as face detection, face tracking, face pose estimation, facial expression analysis, and facial animation. The "holy grail" of face processing is a system that can detect, track, model, recognize, analyze, and animate faces. Although we are not there yet, current progress gives us much reason to be optimistic. The future of face processing looks promising.

ACKNOWLEDGMENTS

The writing of this chapter was supported in part by NSF grant #0205740. I would like to acknowledge the vital contributions of Sandy Pentland in the original Eigenfaces work as well as the support of Tony Gochal and the Arbitron Company, and to salute everyone who has re-implemented and improved the method over the years. Portions of this chapter were expanded with permission from M. Turk, "A Random Walk through Eigenspace," IEICE Trans. Inf. & Syst., Vol. E84-D, No. 12, December 2001.

REFERENCES

[1] D. I. Perrett, E. T. Rolls, and W. Caan, "Visual neurons responsive to faces in the monkey temporal cortex," *Exp. Brain Res.*, **47**, pp. 329–342, 1982.

[2] K. M. Kendrick and B. A. Baldwin, "Cells in temporal cortex of sheep can respond preferentially to the sight of faces," *Science*, **236**, pp. 448–450, 1987.

[3] R. Desimone, "Face-selective cells in the temporal cortex of monkeys," *J. Cognitive Neuroscience* **3**, No. 1, pp. 1–8, 1991.

[4] V. Bruce, *Recognizing Faces*, Lawrence Erlbaum Associates, London, 1988.

[5] A. M. Burton, "A model of human face recognition," in *Localist Connectionist Approaches to Human Cognition*, J. Grainger and A.M. Jacobs, eds., pp. 75–100. London: Lawrence Erlbaum Associates, 1998.

[6] A. M. Burton, V. Bruce, and P. J. B. Hancock, "From pixels to people: a model of familiar face recognition,"*Cognitive Science*, **23**, pp. 1–31, 1999.

[7] D. Marr, *Vision*. W. H. Freeman, San Francisco, 1982.

[8] D. Mumford, "Parametrizing exemplars of categories," *J. Cognitive Neuroscience* **3**(1), pp. 87–88, 1991.

[9] R. Brunelli and T. Poggio, "Face recognition: Features versus templates," *IEEE Transactions on Pattern Analysis and Machine Intelligence* **15**, pp. 1042–1052, 1993.

[10] W. E. L. Grimson. Object Recognition by Computer: The Role of Geometric Constraints. The MIT Press, Cambridge, 1990.

[11] Y. Lamdan and H. J. Wolfson, Geometric Hashing: A General and Efficient Model-Based Recognition Scheme, *Proc. International Conf. Computer Vision*, pp. 238–249, Tampa FL, December 1988.

[12] W.W. Bledsoe. *Man–Machine Facial Recognition*. Technical Report PRI 22, Panoramic Research Inc., Palo Alto, CA, August 1966.

[13] M. D. Kelly, "Visual identification of people by computer," Stanford Artificial Intelligence Project Memo AI-130, July 1970.

[14] T. Kanade, "Picture processing system by computer complex and recognition of human faces," Dept. of Information Science, Kyoto University, Nov. 1973.

[15] L. D. Harmon, M. K. Khan, R. Lasch, and P. F. Ramig, "Machine identification of human faces," *Pattern Recognition* **13**, No. 2, pp. 97–110, 1981.

[16] G. G. Gordon, "Face recognition from frontal and profile views," *Proc. Intl. Workshop on Automatic Face- and Gesture-Recognition*, pp. 47–52, Zurich, 1995.

[17] A. L. Yuille, P. W. Hallinan, and D. S. Cohen, "Feature extraction from faces using deformable templates," *International Journal of Computer Vision* **8**(2), pp. 99–111, 1992.

[18] D. Valentin, H. Abdi, A. J. O'Toole, G. W. Cottrell, "Connectionist models of face processing: a survey," *Pattern Recognition* **27**, pp. 1209–1230, 1994.

[19] M. Flemming and G. Cottrell, "Face recognition using unsupervised feature extraction," *Proceedings of International Neural Network Conference*, Paris, 1990.

[20] J. T. Laprest, J. Y. Cartoux, M. Richetin, "Face recognition from range data by structural analysis," in *Syntactic and Structural Pattern Recognition*, G. Ferrat et al., eds., NATO ASI series, Vol. F45, Springer-Verlag, Berlin, Heidelberg, 1988.

[21] J. C. Lee and E. Milios, "Matching range images of human faces," *Proc. IEEE Third Intl. Conf. on Computer Vision*, pp. 722–726, Osaka, Japan, December 1990.

[22] G. G. Gordon, "Face recognition from depth and curvature," Ph.D. thesis, Harvard University, 1991.

[23] L. Sirovich and M. Kirby, "Low dimensional procedure for the characterization of human faces," *Journal of the Optical Society of America* **4**, No. 3, pp. 519–524, 1987.

[24] P. Burt, "Smart sensing within a pyramid vision machine," *Proc. of the IEEE* **76**(8), pp. 1006–1015, 1988.

[25] M. Turk and A. Pentland, "Face recognition without features," *Proc. IAPR Workshop on Machine Vision Applications*, Tokyo, pp. 267–270, November 1990.

[26] M. Turk and A. P. Pentland, "Eigenfaces for recognition," *J. Cognitive Neuroscience* **3**(1):71–96, 1991.

[27] M. Turk, "Interactive-time vision: face recognition as a visual behavior," Ph.D. Thesis, The Media Laboratory, Massachusetts Institute of Technology, September 1991.

[28] I. T. Jolliffe, *Principal Component Analysis*, Springer-Verlag, New York, 1986.

[29] M. Kirby and L. Sirovich. "Appliction of the Karhumen–Loeve procedure for the characterization of human faces," *IEEE Trans. Pattern Analysis and Machine Intelligence* **12**(1), pp. 103–108, 1990.

[30] H. Murase and S. Nayar, "Visual learning and recognition of 3D objects from appearance," *Intl. J. of Computer Vision* **14**:5–24, 1995.

[31] G. D. Finlayson, J. Dueck, B. V. Funt, and M. S. Drew, "Colour eigenfaces," *Proc. Third Intl. Workshop on Image and Signal Processing Advances in Computational Intelligence*, November 1996, Manchester, UK.

[32] I. Craw, N. Costen, T. Kato, G. Robertson, and S. Akamatsu, "Automatic face recognition: combining configuration and texture," *Proc. Intl. Workshop on Automatic Face- and Gesture-Recognition*, Zurich, 1995, pp. 53–58.

[33] B. Moghaddam, W. Wahid, and A. Pentland, "Beyond eigenfaces: probabilistic matching for face recognition," *Proc. Third Intl. Conf. on Automatic Face- and Gesture-Recognition*, Nara, Japan, pp. 30–35, 1998.

[34] A. Lanitis, C. J.Taylor, and T. F.Cootes, "A unified approach to coding and interpreting face images," *Proc. Fifth Intl. Conf. on Computer Vision*, pages 368–373, 1995.

[35] A. Lanitis, C. J. Taylor, and T. F. Cootes, "Automatic interpretation and coding of face images using flexible models," *IEEE Trans. Pattern Analysis and Machine Intelligence* **19**(7):743–756, 1997.

[36] A. S. Georghiades, D. J. Kriegman, and P. N. Belhumeur, "Illumination cones for recognition under variable lighting: faces," *IEEE Conf. on Computer Vision and Pattern Recognition*, 1998.

[37] L. Zhao and Y. H. Yang, "Theoretical analysis of illumination in PCA-based vision systems," *Pattern Recognition* **32**, pp. 547–564, 1999.

[38] M. Lades, J. Vorbruggen, J. Buhmann, J. Lange, C. V. Malsburg, R. Wurtz, and W. Konen, "Distortion invariant object recognition in the dynamic link architecture," *IEEE Trans. Computers* **42**(3):300–311, 1993.

[39] R. P. Würtz, J. C. Vorbrüggen, C. von der Malsburg, and J. Lange, "Recognition of human faces by a neuronal graph matching process," in *Applications of Neural Networks*, H. G. Schuster, ed., pp. 181–200. VCH, Weinheim, 1992.

[40] A. Lanitis, C. J. Taylor, and T. F. Cootes, "A unified approach to coding and interpretting faces," *Proc. of 5th Intl. Conf. on Computer Vision* pp. 368–373, 1995.

[41] Y. Li, S. Gong, and H. Liddell, "Support vector regression and classification based multi-view face detection and recognition," *Proc. Conf. On Automatic Face and Gesture Recognition*, Grenoble, France, pp. 300–305, March 2000.

[42] T. Ezzat and T. Poggio, "Facial analysis and synthesis using image-based models," *Proc. Second Intl. Conf. on Automatic Face and Gesture Recognition*, Killington, VT, pp. 116–121, 1996.

[43] M. Bichsel, "Automatic interpolation and recognition of face images by morphing," *Proc. Second Intl. Conf. on Automatic Face and Gesture Recognition, Killington,* VT, pp. 128–135, 1996.

[44] A. Leonardis and H. Bischof, "Robust recognition using eigenfaces," *Computer Vision and Image Understanding* **78**, pp. 99–118, 2000.

[45] P. Belhumeur, J. Hespanha, and D. Kriegman, "Eigenfaces vs. Fisherfaces: recognition using class specific linear projection," *IEEE Trans. on Pattern Analysis and Machine Intelligence* **19**(7):711–720, July 1997.

[46] T. W. Lee, *Independent Component Analysis: Theory and Applications*, Kluwer Academic Publishers, Dordrecht, 1998.

[47] P. S. Penev and L. Sirovich, "The global dimensionality of face space," *Proc. 4th Intl. Conf. Automatic Face and Gesture Recognition*, Grenoble, France, pp. 264–270, 2000.

[48] R. Kuhn, J.-C. Junqua, P. Nguyen, and N. Niedzielski, "Rapid speaker adaptation in eigenvoice space," *IEEE Transactions on Speech and Audio Processing* **8**(4):695–707, Nov 2000.
[49] J. Kwok, B. Mak, S. Ho, "Eigenvoice speaker adaptation via composite kernel principal component analysis," *Proc. NIPS-2003*, Vancouver, B. C., Dec. 9–11, 2003.
[50] D. L. Swets and J. Weng, "Using discriminant eigenfeatures for image retrieval," *IEEE Transactions on Pattern Analysis and Machine Intelligence* **18**, No. 8, pp. 831–836, 1996.
[51] H. Schweitzer. "Optimal eigenfeature selection by optimal image registration," in *Conference on Computer Vision and Pattern Recognition*, pages 219–224, 1999.
[52] O. Alter, P. Brown,and D. Botstein, "Singular value decomposition for genome-wide expression data processing and modeling," *Proc. Natl. Acad. Sci.* **97**(18), pp. 10101–10106, Aug. 2000.
[53] Cao, X. and Guo, B. "Real-time tracking and imitation of facial expression," *Proc. SPIE Intl. Conf. on Imaging and Graphics*, Hefei, PRC, Aug., 2002.
[54] M. Black and A. Jepson, "Eigentracking: robust matching and tracking of articulated objects using a view-based representation," *International Journal of Computer Vision* **26**(1), pp. 63–84, 1998.
[55] T. de Campos, R. Feris, and R. Cesar, "Eigenfaces versus eigeneyes: first steps toward performance assessment of representations for face recognition," *Lecture Notes in Artificial Intelligence*, **1793**, pp. 197–206, April 2000.
[56] E. Hjelmås and J. Wroldsen, "Recognizing faces from the eyes only," *Proceedings of the 11th Scandinavian Conference on Image Analysis*, 1999.
[57] A. Pentland, B. Moghaddam, and T. Starner, "View-based and modular eigenspaces for face recognition," *Proc. Computer Vision and Pattern Recognition*, pp. 84–91, June 1994.
[58] K. Chang, K. Bowyer, S. Sarkar, and B. Victor, "Comparison and combination of ear and face images in appearance-based biometrics," *IEEE Trans. PAMI* **25**, No. 9, pp. 1160–1165, Sept. 2003.
[59] K. Nishino, Y. Sato and K. Ikeuchi, "Eigen-texture method: appearance compression and synthesis based on a 3D model," *IEEE Transactions on Pattern Analysis and Machine Intelligence* **23**, No.11, pp. 1257–1265, Nov. 2001.
[60] S. Nayar, S. Nene, and H. Murase, "Real-Time 100 Object Recognition System," *Proceedings of ARPA Image Understanding Workshop*, San Francisco, Feb. 1996.
[61] S. Murphy and H. Bray, "Face recognition fails at Logan," *Boston Globe*, September 3, 2003.
[62] P. Agre, "Your face is not a bar code: arguments against automatic face recognition in public places," *Whole Earth* **106**, pp. 74–77, 2001.
[63] "Face Recognition," Electronic Privacy Information Center, http://www.epic.org/privacy/facerecognition/.
[64] "Privacy," American Civil Liberties Union, http://www.aclu.org/Privacy/.
[65] D. Balme (ed.), *Aristotle: Historia Animalium*, Volume 1, Books I-X, Cambridge University Press, 2002.
[66] C. Darwin, *The Expression of the Emotions in Man and Animals*, Oxford University Press, Third edition, 2002.

[67] C. Breazeal, *Designing Sociable Robots*, MIT Press, 2002.
[68] M. Minsky, *The Society of Mind*, Simon and Schuster, New York, 1986.
[69] D. Hubel, *Eye, Brain, and Vision* (Scientific American Library, No 22), W. H. Freeman and Company, 1989.
[70] M. J. Farah, *Visual Agnosia*, MIT Press, Cambridge, MA, 1990.
[71] N. L. Etcoff, R. Freeman, and K. R. Cave, "Can we lose memories of faces? Content specificity and awareness in a prosopagnosic," *Journal of Cognitive Neuroscience* **3**, No. 1, pp. 25–41, 1991.
[72] H. D. Ellis and M. Florence, "Bodamer's (1947) paper on prosopagnosia," *Cognitive Neuropsychology*, **7**(2), pp. 81–105, 1990.
[73] O. Sacks, *The Man Who Mistook His Wife For a Hat*, Harper & Row, New York, 1987.
[74] A. R. Damasio, H. Damasio, and G. W. Van Hoesen, "Prosopagnosia: anatomic basis and behavioral mechanisms," *Neurology* **32**, pp. 331–41, April 1982.
[75] G. M. Davies, H. D. Ellis, and J. W. Shepherd (eds.), *Perceiving and Remembering Faces*, Academic Press, London, 1981.
[76] D. I. Perrett, E. T. Rolls, and W. Caan, "Visual neurones responsive to faces in the monkey temporal cortex," *Exp. Brain Res.* **47** pp. 329–342, 1982.
[77] D. I. Perrett, A. J. Mistlin, and A. J. Chitty, "Visual neurones responsive to faces," *TINS* Vol. **10**, No. 9, pp. 358–364, 1987.
[78] D. I. Perrett, A. J. Mistlin, A. J. Chitty, P. A. J. Smith, D. D. Potter, R. Broennimann, and M. Harries, "Specialized face processing and hemispheric asymmetry in man and monkey: evidence from single unit and reaction time studies," *Behavioural Brain Research* **29**, pp. 245–258, 1988.
[79] R. Desimone, T. D. Albright, C. G. Gross, and C. J. Bruce, "Stimulus-selective properties of inferior temporal neurons in the macaque," *Neuroscience* **4**, pp. 2051–2068, 1984.
[80] E. T. Rolls, G. C. Baylis, M. E. Hasselmo, and V. Nalwa, "The effect of learning on the face selective responses of neurons in the cortex in the superior temporal sulcus of the monkey," *Exp. Brain. Res.* **76**, pp. 153–164, 1989.
[81] K. M. Kendrick and B. A. Baldwin, *Science* 236, pp. 448–450, 1987.
[82] R. K. Yin, "Looking at upside-down faces," *J. Exp. Psychol.* **81**, pp. 141–145, 1969.
[83] S. Leehey, "Face recognition in children: evidence for the development of right hemisphere specialization," Ph.D. Thesis, Dept. of Psychology, Massachusetts Institute of Technology, May 1976.
[84] S. Carey and R. Diamond, "From piecemeal to configurational representation of faces," *Science* **195**, pp. 312–313, Jan. 21, 1977.
[85] R. Diamond and S. Carey, "Why faces are and are not special: an effect of expertise," *J. Exp. Psych*: G **115**, No. 2, pp. 107–117, 1986.
[86] J. L. Bradshaw and G. Wallace, "Models for the processing and identification of faces," *Perception and Psychophysics* **9**(5), pp. 443–448, 1971.
[87] K. R. Laughery and M. S. Wogalter, "Forensic applications of facial memory research," in *Handbook of Research on Face Processing*, A.W. Young and H.D. Ellis (eds.), Elsevier Science Publishers B. V. (North-Holland), 1989.
[88] J. B. Pittenger and R. E. Shaw, "Ageing faces as viscal-elastic events: Implications for a theory of nonrigid shape perception," *Journal of Experimental Psychology: Human Perception and Performance* **1**, pp. 374–382, 1975.

[89] "Faces From the Future," *Newsweek*, Feb. 13, 1989, p. 62.
[90] D. C. Hay and A. W. Young, "The human face," in *Normality and Pathology in Cognitive Functions*, A.W. Ellis (ed.), Academic Press, London, 1982.
[91] A. W. Young, K. H. McWeeny, D. C. Hay, and A. W. Ellis, "Matching familiar and unfamiliar faces on identity and expression," *Psychol Res* **48**, pp. 63–68, 1986.
[92] M. Turk and M. Kölsch, "Perceptual interfaces," in *Emerging Topics in Computer Vision*, G. Medioni and S.B. Kang (eds.), Prentice Hall, 2004.
[93] A. J. Goldstein, L. D. Harmon, and A. B. Lesk, "Identification of human faces," *Proc. IEEE* **59**, pp. 748–760, 1971.
[94] A. J. Goldstein, L. D. Harmon, and A. B. Lesk, "Man–machine interaction in human-face identification," *The Bell System Technical Journal* **41**, No. 2, pp. 399–427, Feb. 1972.
[95] T. Sakai, M. Nagao, and M. Kidode, "Processing of multilevel pictures by computer – the case of photographs of human faces," *Systems, Computers, Controls* **2**, No. 3, pp. 47–53, 1971.
[96] M. A. Fischler and R. A. Elschlager, "The representation and matching of pictorial structures," *IEEE Trans. Computers* C **22**, No. 1, pp. 67–92, January 1973.
[97] S. R. Cannon, G. W. Jones, R. Campbell, and N. W. Morgan, "A computer vision system for identification of individuals," *Proc. IECON* **1**, pp. 347–351, 1986.
[98] I. Craw, H. Ellis, and J. R. Lishman, "Automatic extraction of face features," *Pattern Recognition Letters* **5**, pp. 183–187, 1987.
[99] K. Wong, H. Law, and P. Tsang, "A system for recognising human faces," *Proc. ICASSP*, pp. 1638–1642, May 1989.
[100] H. Midorikawa, "The face pattern identification by back-propagation learning procedure," *Abstracts of the First Annual INNS Meeting*, Boston, p. 515, 1988.
[101] E. I. Hines and R. A. Hutchinson, "Application of multi-layer perceptrons to facial feature location," *IEE Third International Conference on Image Processing and Its Applications*, pp. 39–43, July 1989.
[102] G. W. Cottrell and J.Metcalfe, "EMPATH: Face, gender and emotion recognition using holons," in *Advances in Neural Information Processing Systems 3*, San Mateo, CA, R.P. Lippman, J. Moody, and D.S. Touretzky (eds.), Morgan Kaufmann, 1991.
[103] T. J. Stonham, "Practical face recognition and verification with WISARD," in *Aspects of Face Processing*, H. Ellis, M. Jeeves, F. Newcombe, and A. Young (eds.), Martinus Nijhoff Publishers, Dordrecht, 1986.
[104] S. Sclaroff and A. Pentland, "Closed-form solutions for physically-based shape modeling and recognition," *Proc. CVPR*, Maui, Hawaii, pp. 238–243, June 1991.
[105] V. Bruce and A. Young, "Understanding face recognition," *British Journal of Psychology*, **77**, pp. 305–327, 1986.
[106] W. Zhao, R. Chellappa, P. J. Phillips, and A. Rosenfeld, "Face recognition: a literature survey," *ACM Computing Surveys* **35**, Issue 4, pp. 399–458, Dec. 2003.

CHAPTER 3

INTRODUCTION TO THE STATISTICAL EVALUATION OF FACE-RECOGNITION ALGORITHMS

3.1 INTRODUCTION

When is the difference between two algorithms significant? This question lies at the heart of research into face recognition, and addressing it fully and convincingly can be surprisingly tricky. To start, the term "significant" can denote two equally important but very different things. In the first case, the word may be used to mean that an author feels the difference is of some clear practical or scientific importance. While this is noteworthy, it can be a matter of expert judgment. The second sense of the word is more precise and more limited, meaning that the difference is "statistically significant". These words mean that the magnitude of an observed performance difference between algorithms is larger than could have likely occurred by chance alone.

The purpose of this chapter is to provide an introduction to some elementary methods for addressing the more narrow question of statistical significance. In other words, when may one conclude that an observed difference is statistically significant, and what set of assumptions underly such a finding? We also describe some ideas useful for assessing scientific significance through estimation of how often one algorithm might be observed to outperform another.

We first discuss simple probability models as background for the techniques described. As we formalize both the face-identification problem and some approaches to statistical evaluation, a concise mathematical nomenclature will be developed. A key to our use of symbols is provided in Appendix A To make these definitions and concepts more accessible, concrete examples are provided whenever possible.

In particular, this chapter will utilize a running example comparing a PCA [30] face recognition algorithm to a PCA followed by a LDA algorithm [33]. The algorithms are taken directly from version 5.0 of the Colorado State University (CSU) face identification evaluation system [3]. The data for the study is from the Notre Dame face data set. While mainly provided for illustration, we hope the specific findings are also of interest: the comparison of algorithms is new and constitutes a large comparative test of PCA and LDA.

The remainder of this chapter is laid out as follows. Section 3.2 reviews the basics of how face identification is defined and what performance measures are typically used, with two examples. Section 3.3 introduces the binomial approach to modeling randomness. The essential idea is to view algorithm testing as a sequence of independent success/failure trials. Section 3.4 introduces a Monte Carlo approach to measuring uncertainty, which is particularly applicable when not all trials are independent due to subject replication. Section 3.5 provides a more complete summary of the PCA versus LDA algorithm comparison. It allows us to demonstrate the practical use of the Monte Carlo approach. It also answers an important question regarding the relative value of multiple images of people to a LDA algorithm versus a PCA algorithm. Finally, Section 3.6 surveys some more advanced modeling strategies than can be quite powerful tools for detailed evaluation of recognition algorithms, particularly for assessing the effect of other variables on performance.

3.2 FACE-IDENTIFICATION DATA, ALGORITHMS, AND PERFORMANCE MEASURES

In discussing face recognition, it is important to distinguish between three distinct problems or tasks. The first task is detection and localization, i.e., determining if there is a face present and where it is [23]. The second task is identification, i.e., naming a person in a novel image [26]. The third task is verification, i.e., deciding if a person in a novel image is who they claim to be [28, 1]. The FERET evaluations [26] focused primarily on face identification.

In the FERET evaluation, a series of tests of face-identification algorithms was orchestrated. This began with the collection of the FERET data, which remains one of the larger publicly available datasets. In the FERET tests, a number of independent participants ran their algorithms on the FERET data. The results represent a well organized effort to involve multiple academic institutions in a joint empirical comparison of computer vision algorithms. The results of the FERET identification evaluation are summarized in [26] along with the associated evaluation protocol.

The FERET protocol remains relevant and important today, and researchers presenting identification performance results will do well to follow the FERET protocol where possible. However, face detection and face verification also are large research topics in their own right with their own evaluation protocols [1]. In addition, more recent tests have evaluated integrated systems, most notably the vendor tests carried out in 2000 and 2002 [5, 27]. Thus, while the FERET protocol remains important, it has limitations and researchers should keep informed regarding recent and ongoing work, for example the Face Recognition Grand Challenge[1].

3.2.1 Definitions and Fundamentals

The most common performance measure for a face-identification algorithm is the recognition rate at a fixed rank. To understand what this means, a variety of definitions are needed. Here we review these definitions, starting with the selection of data and the either implicit or explicit relationship between data and the target population to which one wants to apply algorithms.

Data

Let Ω denote a target population of images, and assume that the goal is to compare the performance of algorithms over Ω. The population Ω represents our commitment to the type of problem being solved. For example, it might be defined as all US passport photos, or all FBI mug shots. A sampling of the factors to consider when defining Ω includes demographics, the type of sensor, constraints on illumination, facial expression, etc.

While it is important to think explicitly about what constitutes the appropriate target population, it is equally important to realize that we as evaluators typically have access to only some finite sample W of Ω, i.e., $W \subseteq \Omega$. Moreover, face data collection is costly and most researchers use existing data sets both because they represent a standard of comparison and because large amounts of new data can be difficult to acquire. Thus, our choice of W has often already been made for us.

Images are elements of W and are denoted $w_{i,j}$, where the subscripts i, j indicate the jth image of the ith person. Using double subscripting highlights the distinction between images of a single person and images of different people.

Algorithms

When evaluating algorithms, three subsets of W are of particular interest:

- The training set $\mathcal{T}$
- The gallery set $\mathcal{G}$
- The probe set $\mathcal{P}$.

[1] See website http://www.frvt.com/FRGC/

Most face-identification algorithms are trained in one manner or another using a sampling of face images $\mathcal{T}$. When doing identification, algorithms match probe images in $\mathcal{P}$ to sets of stored example images, called the gallery $\mathcal{G}$. The intersection of the sets $\mathcal{T}$ and $\mathcal{G}$ may or may not be empty. In other words, in most settings it is permissible to train on the gallery data, but the training data may also be distinct from the gallery. It is generally not permissible to train an algorithm on $\mathcal{P}$. The probe image in $\mathcal{P}$ is presumed to be novel, and the algorithm should not have seen it before testing.

An empirical evaluation of one or more face-recognition algorithms typically reports results for one or several combinations of $\mathcal{T}$, $\mathcal{G}$, and $\mathcal{P}$ images. As a concise notation, for any algorithm A, this dependency is indicated by subscripting, e.g., $A_{\mathcal{T}\mathcal{G}}$ is algorithm A trained on $\mathcal{T}$ and matching to gallery $\mathcal{G}$.

Matching Images – Similarity Matrices

Most commonly used recognition algorithms may be characterized by a similarity matrix Υ that represents all the information used to perform identification. The elements of Υ are similarity measures υ which may be defined by the function

$$\upsilon : W \times W \Rightarrow \mathbb{R}. \tag{1}$$

Similarity is used to rank gallery images relative to a specific probe image. Thus, the best match to a probe image $x_{i,j} \in \mathcal{P}$ is the gallery image $y_{k,l} \in \mathcal{G}$ such that:

$$\upsilon\left(x_{i,j}, y_{k,l}\right) \geqslant \upsilon\left(x_{i,j}, y_{\kappa,\lambda}\right) \quad \forall\, y_{\kappa,\lambda} \in \mathcal{G}. \tag{2}$$

In theory, the similarity relation υ must induce a complete order on $\mathcal{G}$ for each probe image $x_{i,j} \in \mathcal{P}$. In practice, ties may arise. So long as ties are rare, it is generally safe to impose arbitrary choices and otherwise ignore the problem.

Recognition Rate and Recognition Rank

When a good identification algorithm sorts the gallery by similarity to a probe image $x_{i,j}$, images of the same person should come first or at least be near the top. More formally, for each probe image $x_{i,j} \in \mathcal{P}$, let $L\left(x_{i,j}\right) = \left\{L_1\left(x_{i,j}\right), L_2\left(x_{i,j}\right), \ldots\right\}$ represent the gallery sorted by decreasing similarity. Thus, $L_1\left(x_{i,j}\right)$ is the gallery image most similar to the probe image $x_{i,j}$, $L_2\left(x_{i,j}\right)$ is the next most similar gallery image, and in general, $L_\zeta\left(x_{i,j}\right)$ is the ζth most similar gallery image. To express when identification succeeds, the indicator function $b\left(x_{i,j}\right)$ may be defined as:

$$b\left(x_{i,j}\right) = \begin{cases} 1 & \text{if } L\left(x_{i,j}\right)_1 = y_{k,l} \quad \text{and} \quad k = i, \\ 0 & \text{otherwise.} \end{cases} \tag{3}$$

In other words, a person is identified when the subject[2] index for the probe image matches the subject index for the top ranked gallery image.

The requirement that an image of the same person appear at the top ranked position, e.g. $L_1\left(x_{i,j}\right)$, may be relaxed such that an algorithm is said to succeed if an image of the same person appears among the τ most similar gallery images. A family of indicator functions for different ranks may thus be defined:

$$b_\tau\left(x_{i,j}\right) = \begin{cases} 1 & \text{if there is an image of subject } i \text{ in the first } \tau \text{ images in } L\left(x_{i,j}\right), \\ 0 & \text{otherwise.} \end{cases} \quad (4)$$

The recognition rate for a fixed choice of τ over a probe set $\mathcal{P}$ of size n is:

$$\rho_\tau\left(\mathcal{P}\right) = \frac{c_\tau}{n} \quad \text{where} \quad c_\tau = \sum_{x_{i,j}\in\mathcal{P}} b_\tau(x_{i,j}). \quad (5)$$

The observed recognition rate $\rho_\tau\left(\mathcal{P}\right)$ should be viewed as an estimator of the unknown recognition rate $\rho_\tau\left(\Omega\right)$ that would be achieved if the algorithm were applied to the entire target population. Had we applied the algorithm to a different sample of Ω, say $\mathcal{P}'$, we would have observed a different value of the estimator, namely $\rho_\tau\left(\mathcal{P}'\right)$. Thus, we see that associated with $\rho_\tau\left(\mathcal{P}\right)$ we must have a measure of uncertainty quantifying the range of recognition rates that might reasonably have been observed from different data. Statisticians refer to the sampling distribution of $\rho_\tau\left(\mathcal{P}\right)$.

If we have two algorithms with different observed recognition rates, the magnitude of the difference may be small compared to the variability associated with the two rates, or it may be large in comparison. When the observed difference exceeds what could reasonably be expected from chance variation alone, the two algorithms have statistically significantly different recognition rates. In subsequent sections, we will introduce two simple methods for estimating the relevant sampling distributions and determining statistical significance.

Finally, the recognition rank for a probe image can also be an extremely useful indicator of performance. Using the same sorted list of gallery images $L\left(x_{i,j}\right)$ defined above, let $L_\zeta\left(x_{i,j}\right) = y_{k,l}$ be the first gallery image in the sequence that is of the same subject as the probe image $x_{i,j}$, i.e. index i equals index k. Then the recognition rank $r\left(x_{i,j}\right)$ is:

$$r\left(x_{i,j}\right) = \zeta. \quad (6)$$

[2]The term "subject" is short for "human subject" and it is a somewhat more formal way of indicating a person whose data are included in a study.

3.2.2 Example One: FERET Gallery and Probe Sets

The FERET studies made extensive use of the fact that, once an algorithm is fixed through training, the similarity υ assigned to a pair of images is a constant. Therefore, for an algorithm $A_{\mathcal{T}}$, experiments may be conducted in a "virtual" fashion. For each algorithm in the FERET evaluations, a single large similarity matrix Υ was generated and sampled for different combinations of probe and gallery sets to produce distinct comparisons. In particular, much of the study focused on one gallery and four probe sets. A training set was also distributed to the participants. However, participants were free to train their algorithms as they saw fit on their own data so long as training did not involve FERET probe images.

A short description of the standard FERET data partitions is given in Table 3.1. The four probe sets were used in conjunction with the single gallery set to compare algorithm performance. The four probe sets were designed to test algorithms under distinct conditions. The $\mathcal{P}_{\text{fb}}$ to $\mathcal{G}_{\text{fa}}$ comparison addresses how changes in facial expression influence performance, the $\mathcal{P}_{\text{dupI}}$ to $\mathcal{G}_{\text{fa}}$ and $\mathcal{P}_{\text{dupII}}$ to $\mathcal{G}_{\text{fa}}$ comparisons address how the time between images influences performance, and finally the $\mathcal{P}_{\text{fc}}$ to $\mathcal{G}_{\text{fa}}$ comparison addresses how changes in lighting influence performance.

Algorithms in the FERET evaluation were commonly compared by plotting the recognition rate at different ranks; plotting $\rho_\tau(P)$ (Equation 5) against τ. These plots are called cumulative match characteristic (CMC) curves. In a CMC curve, the horizontal axis is rank τ. The vertical axis is recognition rate $\rho_\tau(P)$. Rank values τ may range from 1 to a cutoff, such as 50, or to m, where m is the size of the gallery $\mathcal{G}$. At $\tau = m$, all algorithms achieve (trivially) a recognition rate of 1.0.

An example CMC curve is shown in Figure 3.1 for the FERET data, using the gallery $\mathcal{G}_{\text{fa}}$ and probe set $\mathcal{P}_{\text{fc}}$. The range of observed recognition rates for different algorithms on this probe set in the original FERET tests was quite large, as is the range shown here. Results for two classes of algorithm are represented in Figure 3.1:

Table 3.1: Five partitions used in most of the FERET 1996/97 evaluations.

Set	Images	Description
$\mathcal{G}_{\text{fa}}$	1196	Images taken with one of two facial expressions: neutral versus other
$\mathcal{P}_{\text{fb}}$	1195	Images taken with the other facial expression relative to $\mathcal{G}_{\text{fa}}$
$\mathcal{P}_{\text{dupI}}$	722	Subjects taken later in time
$\mathcal{P}_{\text{dupII}}$	234	Subjects taken later in time; this is a harder subset of $\mathcal{P}_{\text{dupI}}$
$\mathcal{P}_{\text{fc}}$	194	Subjects taken under different illumination
$\mathcal{T}$	501	Training Images, roughly 80 percent from $\mathcal{G}_{\text{fa}}$ and 20 percent from $\mathcal{P}_{\text{dupI}}$

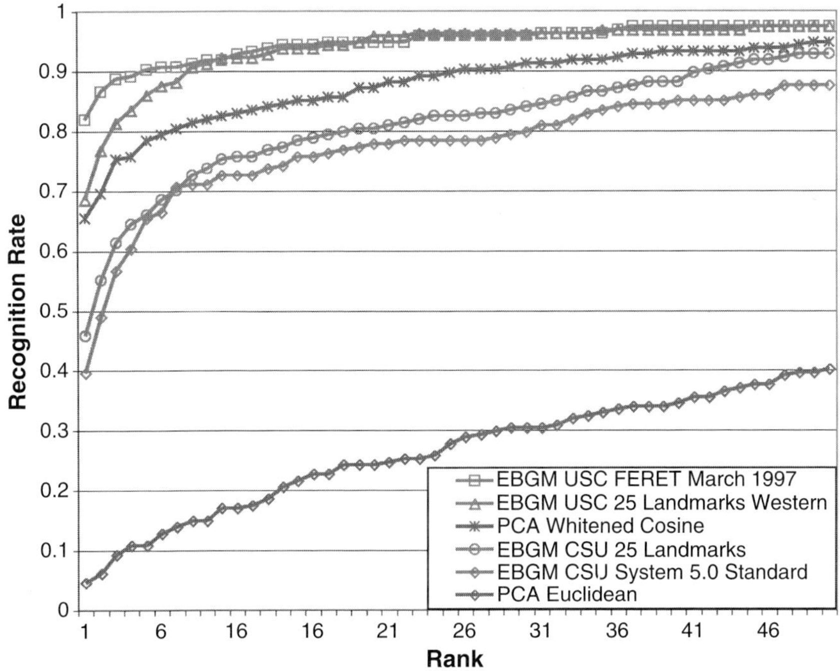

FIGURE 3.1: CMC curves for two classes of algorithms and six algorithm variants run on the gallery $\mathcal{G}_{\mathrm{fa}}$ and probe set $\mathcal{P}_{\mathrm{fc}}$.

PCA. A standard eigenface [30] algorithm that creates a a low-dimensional subspace, e.g., 100 to 500 dimensions, by applying principal-components analysis to a sample covariance matrix derived from a set of training images.

EBGM. An elastic-bunch graph-matching algorithm [25]. This algorithm uses a fixed set of positions on the face as landmarks, extracts Gabor Jets at each landmark, and then matches images relative to the quality of the match between corresponding Gabor jets [6].

The specific algorithm variants shown are:

EBGM USC FERET March 1997. The original CMC curve derived from the similarity matrix generated by University of Southern California (USC) during the FERET evaluations. Inclusion of this curve shows one of the great strengths of evaluations that use and retain similarity matrices: we reconstructed this curve using CSU analysis tools and the original similarity matrix for the EBGM algorithm distributed by NIST as part of the FERET data.

EBGM CSU 25 Landmarks Western. The best result achieved by CSU using our version of the EBGM algorithm. It uses only 25 landmarks and a modified image pre-processing procedure loosely referred to as "Western Photo". The 25 landmarks used and the pre-processing enhancements were made after Version 5.0 of the CSU Face Identification Evaluation System was released. Both are briefly described in Appendix B below.

PCA Whitened Cosine. The standard PCA algorithm using the whitened cosine distance measure [3].

EBGM CSU 25 Landmarks. The same as CSU EBGM algorithm above with the 25 landmarks, but with standard EBGM image pre-processing in place of the Western Image pre-processing.

EBGM CSU System 5.0 Standard. The CSU EBGM algorithm as distributed and run by default in Version 5.0 of the CSU Face Identification Evaluation System. These settings, along with the EBGM algorithm as a whole, are well documented in [6].

PCA Euclidean. Standard eigenfaces using the L2 norm, i.e., Euclidean distance.

Figure 3.1 illustrates how CMC curves are typically presented, and the results shown for multiple versions of two standard algorithms point out several critical concerns for anyone carrying out empirical comparisons of face recognition algorithms.

The most obvious observation is the lack of a clear ranking between algorithms. The relative performance of some algorithms depends on the recognition rank at which they are compared. This underscores the importance of end-to-end algorithm configuration. For example, the importance of image pre-processing is apparent in the difference between the CSU EBGM algorithm using the original, FERET-like pre-processing and the newer Western Photo preprocessing. It is important for authors to fully explain all the steps in the recognition process, including data preprocessing.

Another issue is the appropriate use of PCA as a standard baseline algorithm. Authors of new face recognition algorithms frequently use PCA, typically with Euclidean distance, as a standard of comparison. In Figure 3.1, the rank 1 recognition rate is $\rho_1 = 0.05$ for PCA using Euclidean distance, but $\rho_1 = 0.66$ for the same PCA algorithm using whitened cosine as the distance measure. This is a tremendous difference, improving a baseline algorithm many dismiss as no longer interesting and making it competitive in terms of performance.

3.2.3 Example Two: Comparing PCA and LDA on the Notre Dame Data

For the remainder of this chapter, a recently concluded comparison of PCA and LDA (linear discriminant analysis) on a subset of the Notre Dame face data[3] will be used for illustration. The LDA algorithm uses Fisher linear discriminants to create a subspace designed to maximally separate face images of different people. Swets and Weng [29], Belhumeur et al. [2] Etemad and Chellapa [13, 14] were early advocates of linear discriminant analysis for face recognition. The specific algorithm considered here uses PCA to perform dimensionality reduction, and then finds the $c - 1$ Fisher linear discriminants when trained with c people. It is based upon the work [33]. In our comparison, PCA with whitened cosine distance will serve as the baseline.

This is a new and previously unpublished comparison. It takes advantage of the fact that the Notre Dame dataset includes at least 20 images per person for more than 300 people. In comparison, the FERET has far fewer replicates per person. Two hypotheses motivate this study.

Hypothesis 1: LDA will outperform PCA when given many training images per person.

Hypothesis 2: LDA will not outperform PCA when trained and tested on different people.

For the study presented here, we use 20 images each of 336 people acquired over the fall of 2002. The images are all frontal face images under controlled lighting with either neutral or "other" expressions. The images are arrayed according to collection dates, earlier images first. The first 16 images per person are reserved for training. The remaining 4 images are reserved for testing. Finally, the people are divided into two disjoint groups of 168, group α and group β.

The PCA and LDA algorithms were run under a variety of configurations described more fully below. For the moment, however, let us focus our attention on the simple comparison shown in Figure 3.2 where both algorithms have been trained on three images per person from the α group. The results are for people in the β group. The probe set is the first of the four test images set aside for each person, the gallery set is the third. All images were preprocessed using the standard CSU preprocessing algorithm [3].

The recognition rates at rank one, ρ_1, are 0.857 for LDA and 0.833 for PCA. On the surface, an advocate of LDA might be encouraged by this 2.4% difference. However, the sample size is small, so let us restate the result making the sample size explicit: $\rho_1 = \frac{144}{168}$ for LDA and $\rho_1 = \frac{140}{168}$ for PCA. The difference is 4 images out of 168.

[3]There is not at this time a formal citation, but information about this data set may be obtained at http://www.nd.edu/ Ecvrl/UNDBiometricsDatabase.html

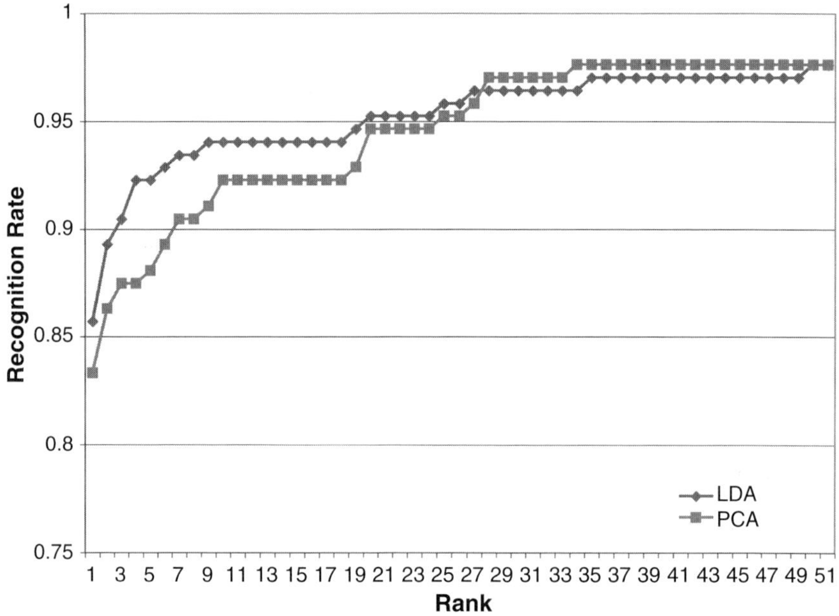

FIGURE 3.2: CMC curves for the PCA and LDA algorithms. Each algorithm was trained on 3 images for each of 168 people and then tested on a different set of 168 people.

Is a difference of 4 out of 168 statistically significant? Perhaps the observed difference is typical of differences that might have been observed from other experiments using other portions of Ω, in which case the observed difference is not statistically significant. One elementary way to model the uncertainty associated with the estimated difference is to view algorithm testing as a process of independent trials where algorithms succeed or fail. This is called a Bernoulli probability model.

3.3 A BERNOULLI MODEL FOR ALGORITHM TESTING

3.3.1 Establishing the Formal Model and Associated Assumptions

Each time an algorithm is shown a probe image, it is being given the opportunity to succeed or fail. Since the identification behavior of an algorithm is fully specified by the similarity scores in the pairwise similarity matrix Υ, all identification algorithms are deterministic given the images tested. Randomness in outcomes, and in algorithm performance differences, is introduced because the tested images are considered to be a random sample from a target population. One's confidence that a difference between two algorithms is greater than can be attributed to chance depends upon the number of times one algorithm is observed to succeed relative to the other.

We begin by considering the probability of success. Consider what happens when algorithm A trained on $\mathcal{T}$ and provided with gallery $\mathcal{G}$ (hereafter denoted $A_{\mathcal{T}\mathcal{G}}$) is applied to a collection of probe images drawn at random from a larger population of possible probe images. Let s_ℓ denote the ℓth randomly selected probe image[4]. Suppose that

$$P\left[b\left(s_\ell\right)=1\right]=p, \tag{7}$$

where s_ℓ is the random variable. In words, the probability of algorithm $A_{\mathcal{T}\mathcal{G}}$ correctly identifying a randomly selected probe image s_ℓ is p.

To tie this back into our running example, recall from above that the LDA algorithm correctly identified $\frac{144}{168}$ images at $\tau = 1$. Imagine you are asked to guess the true probability p that this algorithm would correctly identify a new probe image. With respect to our target population, the correct answer is the unknown value $\rho_1(\Omega)$. Absent of more details, $\rho_1(\mathcal{P}) = \frac{144}{168}$ is a pretty good guess. We justify this below, and develop approaches to estimate the uncertainty about $\rho_1(\Omega)$.

Algorithm Testing as Drawing Marbles from a Jar

You are given a jar containing a mix of red and green marbles. You reach into the jar and select a marble at random, record a 1 if the marble is green and a 0 otherwise. You then place the marble back into the jar. If you do this n times, you are conducting Bernoulli trials, and the number of 1s recorded is a random variable described by a binomial distribution. The distribution is parametrized by the number of marbles drawn, n, and by the fraction of all the marbles that are green, p. The equation for our marble drawing experiment that is analogous to Equation 7 above is:

$$P\left[\text{marble drawn is green}\right]=p. \tag{8}$$

It should be obvious for the jar of marbles that the probability of any given randomly selected marble being green is the ratio p of green marbles to the total number of marbles. Likewise for the randomly selected probe images. Some fraction p of the

[4]The switch from double to single subscripting of images is intentional. The single subscript emphasizes that the index refers to placement in the collection of randomly selected images. Clearly $s_\ell = x_{i,j}$ for some i and j, but random sampling means that i and j are not known.

probe images in the population of possible probe images $\mathcal{P}$ are correctly recognized by algorithm $A_{\mathcal{TG}}$.

The true fraction p dictates how the indicator function b will behave over a sequence of independently selected probe images s_ℓ. It also characterizes how the recognition rate defined in Equation 5 behaves. Restating this equation for rank $\tau = 1$ gives

$$\rho(\mathcal{P}) = \frac{c}{n} \quad \text{where} \quad c = \sum_{x_{i,j} \in \mathcal{P}} b_1(x_{i,j}). \tag{9}$$

It follows directly from the assumptions above that c is a binomially distributed random variable, in which case

$$P[c = k] = \frac{n!}{k!\,n-k!} p^k q^{n-k} \quad p + q = 1. \tag{10}$$

Suppose we observe K successes out of n trials. Then we might ask: what value of p is most likely, given these data? To answer this, we maximize $P[c = K]$ with respect to p. Setting the first derivative (with respect to p) of the log of the right side of Equation 10 to zero and solving for p yields $\hat{p} = K/n$. Thus, K/n is called the maximum-likelihood estimator for p, and it is the best guess for p in a formal statistical sense. Further, maximum-likelihood theory [7] shows that the standard error of this estimator is approximately $s_{\hat{p}} = \sqrt{[K/n(1 - K/n)/n]}$. Had we replicated the experiment, drawing a new n marbles, we probably would have found a slightly different proportion of greens, and hence a different estimate of p. The interval $(\hat{p} - 1.96s_{\hat{p}},\ \hat{p} + 1.96s_{\hat{p}})$ will include the true value of p about 95% of the time, thereby quantifying the uncertainty in our estimator.

3.3.2 Appropriateness of a Binomial Model

This binomial model is extremely simple, and there are several aspects where further consideration is warranted. The following discussion may help eliminate some areas of possible concern.

Sampling With and Without Replacement

In Bernoulli trials, the marble goes back in the jar after each draw, but in a recognition experiment, elements of Ω are sampled without replacement. Thus, the question arises as to how a model built upon the assumption of sampling with replacement may be used in experiments where sampling is done without replacement.

This concern is of little practical significance. If one assumes the target population is very large relative to the sample size, then the difference between sampling with and without replacement is negligible in terms of sampling probabilities, estimation, and hypothesis testing. While not always the case, the norm in face-identification experiments is that the number of samples (probe images) is much smaller than the population over which we are attempting to draw statistical inferences about the performance of a recognition algorithm.

Some People are Harder to Identify

Without question, some people are harder for an automated algorithm to identify than others [11, 16]. Is this a concern for the binomial model? When we have only one probe image per subject and we seek marginal inference about algorithm performance, then the fact that some people are harder to recognize than others is irrelevant. Marginal inference is when we are uninterested or unable to adjust for factors that affect recognition performance, so we effectively average out such factors.

A helpful way to illustrate when the binomial model can be applied despite some individuals being harder to recognize than others is to extend the marble example. Instead of selecting a marble from a jar, consider selecting one from a cabinet with m drawers, each containing N = 1,000,000 marbles. Consider that each person in the probe set corresponds to a drawer, and that the inherent difficulty of recognizing each person is inversely related to the number of green marbles, Np_i in his drawer. Drawing a green is inversely related to a recognition success.

A simple binomial trial from the population of potential trials corresponds to selecting a marble after combining the contents of all drawers, in which case

$$P[\text{marble drawn is green}] = \frac{1}{mN}\sum_{i=1}^{m} Np_i = \frac{1}{m}\sum_{i=1}^{m} p_i \equiv p. \tag{11}$$

Next let us keep track of the varying recognition difficulties of different people explicitly. Note that

$$P[\text{marble drawn is green} \mid \text{marble is drawn from drawer } i] = p_i. \tag{12}$$

Then the marginal probability of a green marble is

$$P[\text{marble drawn is green}] = \frac{1}{m}\sum_{i=1}^{m}\frac{Np_i}{N} = p. \tag{13}$$

Thus the staged experiment with drawer selection is equivalent to the experiment where first all the marbles are poured out of the drawers into a single jar, they are mixed, and then a marble is selected from the jar. Hence, although people have unequal probabilities (p_i) of being successfully recognized by $A_{\mathcal{TG}}$, the unconditional probability that $A_{\mathcal{TG}}$ correctly recognizes a randomly selected probe image is p. As long as $\mathcal{P}$ is representative of Ω, it is not a problem if there is only one probe image per person in $\mathcal{P}$ because this argument can equally be applied to the sampling of $\mathcal{P}$ from Ω.

Of course, when individual drawer attributes are known, elementary sampling techniques [8] allow more powerful inferences about performance over Ω. Further, when replicate images per subject are probed, more sophisticated models can be used to account for so-called extra-binomial variation and to estimate the effects of factors affecting recognition performance. See Section 3.6.

3.3.3 Comparing Algorithms Using the Binomial Model

So far we've developed the essential idea of viewing the process of testing an algorithm $A_{\mathcal{TG}}$ against a sequence of probe images as a series of success/failure trails. Recall that the recognition rate $\rho_{\tau}(\mathcal{P})$ is the maximum-likelihood estimate $\hat{p}$ for the true probability p of successfully identifying a new probe image under the assumptions set out above. Maximum-likelihood theory provides confidence intervals for the success rates of each tested algorithm, but what we really want to do is to determine whether the tested algorithms have significantly different success rates.

For the specific task of asking whether the difference between two algorithms is significant, the binomial model leads naturally to McNemar's test [21]. See also [19, 32]. We describe here a version of McNemar's test that computes exact p-values [24]. McNemar's test has the virtue of being both easy to understand and easy to apply.

Paired Success/Failure Trials: McNemar's Test

Consider the hypothetical identification success/failure data for two imaginary algorithms $A_{\mathcal{TG}}$ and $B_{\mathcal{TG}}$ summarized in Table 3.2. Note that the only difference between Table 3.2a and Table 3.2b is that 400 observations are made in Table 3.2b, while only 100 observations are made Table 3.2a. Otherwise, the relative frequency of the four outcomes are identical. The recognition rate for $A_{\mathcal{TG}}$ is $56/100 = 224/400 = 0.56$, and $48/100 = 192/400 = 0.48$ for $B_{\mathcal{TG}}$.

McNemar's test begins by discarding those cases where both algorithms have the same result. Only the cases where one succeeds and the other fails are considered. As a shorthand, we use SF to indicate the case where $A_{\mathcal{TG}}$ succeeds and $B_{\mathcal{TG}}$ fails, and FS to indicate the case where $A_{\mathcal{TG}}$ fails and $B_{\mathcal{TG}}$ succeeds. The null hypothesis, H_0, associated with McNemar's test is that, when exactly one of $A_{\mathcal{TG}}$ and $B_{\mathcal{TG}}$ succeeds, it is equally likely to be either algorithm. We hope to reject H_0 in favor of the alternative hypothesis that the two algorithms are not equally likely in such a case.

Table 3.2: Hypothetical summary of paired recognition data suitable for McNemar's test. Here 'S' means that the algorithm correctly identified a probe, and 'F' represents a failure. The relative number of successes and failures in (a) and (b) are the same, but the absolute number in (b) is quadrupled. This alters the p-value of McNemar's test.

	Outcome of $B_{\mathcal{TG}}$			Outcome of $B_{\mathcal{TG}}$	
Outcome of $A_{\mathcal{TG}}$	S	F	Outcome of $A_{\mathcal{TG}}$	S	F
S	32	24	S	128	96
F	16	28	F	64	112
	(a)			(b)	

Let n_{SF} denote the number of times SF is observed and n_{FS} denote the number of times FS is observed. The p-value of McNemar's test is the probability under H_0 of observing a value of $|n_{\mathrm{SF}} - n_{\mathrm{FS}}|$ at least as extreme as the observed value. Under H_0, counting n_{SF} and n_{FS} is equivalent to counting heads and tails when flipping a fair coin, so we may use Binomial $(n, 0.5)$ probabilities to find the p-value:

$$\text{p-value} = \sum_{i=0}^{\min\{n_{\mathrm{SF}}, n_{\mathrm{FS}}\}} \frac{n!}{i!\,(n-i)!} 0.5^n + \sum_{i=\max\{n_{\mathrm{SF}}, n_{\mathrm{FS}}\}}^{n} \frac{n!}{i!\,(n-i)!} 0.5^n \qquad (14)$$

where $n = n_{\mathrm{SF}} + n_{\mathrm{FS}}$. Now, for the data in Table 3.2a, $n_{\mathrm{SF}} = 24$, $n_{\mathrm{FS}} = 16$, and $n = 40$. The p-value is 0.268. For the data in Table 3.2b, $n_{\mathrm{SF}} = 96$, $n_{\mathrm{FS}} = 64$, and $n = 160$. The p-value is 0.014.

This illustration shows that the same difference in recognition rates can be statistically significant or not depending on sample size. Uncertainty is reduced and statistical significance is increased by collecting more observations.

McNemar's Test illustrated on the LDA versus PCA comparison

Table 3.3 summarizes success/failure outcomes for the comparison between the PCA and LDA algorithm introduced above. The SF column indicates how often the LDA algorithm correctly identified a person, while the PCA algorithm failed to identify the same person. Comparisons are shown for rank 1, 2 and 4. Note that as τ grows the separation between the algorithms grows as well; this is evident in Figure 3.2. Of the three ranks shown, the greatest separation is for $\tau = 4$. This is the only case where the observed difference between the two algorithms is found to be statistically significant (since the p-value is less than 0.05).

Table 3.3: McNemar's Test comparing PCA and LDA algorithms at three ranks τ.

τ	SS	SF	FS	FF	test p-value
1	136	8	4	20	0.388
2	143	7	2	16	0.180
4	145	10	2	11	0.039

3.4 NONPARAMETRIC RESAMPLING METHODS

Monte Carlo methods are a flexible alternative to parametric approaches like the ones discussed above, and can provide information about the scientific significance of performance differences as well as the statistical significance. Note that, when $\mathcal{P}$ contains multiple images of the same subject, McNemar's test cannot be used to test statistical significance because outcomes are not independent. We present a resampling approach that can be used in this case to draw conclusions.

We will first briefly describe the familiar resampling method known as bootstrapping and provide a simple illustration: bootstrapping the distribution of success/failure counts by sampling a set of independent success/failure outcomes from algorithm $A_{\mathcal{TG}}$ run on a single probe set $\mathcal{P}$. For this particularly simple illustration, we observe that the form of the resulting distribution is already known: it is the binomial distribution discussed above.

Next, a more general Monte Carlo approach based upon randomly resampling probe and gallery images is introduced. We first introduced this technique in [4] and note that Ross J. Micheals and Terry Boult applied a related nonparametric sampling test to face identification in [22]. This approach provides an approximation to the sampling distribution for ρ_τ. We illustrate this approach by using it to compare LDA and PCA.

3.4.1 Bootstrapping

For estimators based on independent, identically distributed (i.i.d.) data, the simple nonparametric bootstrap proceeds as follows. Let S be an i.i.d. sample dataset of size n from a larger population. Let $\widehat{\theta}$ be a statistic, namely $\widehat{\theta}(S)$, estimating a quantity of interest, θ. For example, θ might be the theoretical median and $\widehat{\theta}$ the sample median. Bootstrapping allows us to estimate the probability distribution of $\widehat{\theta}$ in cases where the analytic derivation of this distribution would be difficult or impossible.

The distribution is estimated by repeatedly drawing pseudodatasets, S^*. Each pseudodataset is formed by selecting n elements of S independently, completely at random, and with replacement. For each pseudodataset S^*, a value for the associated statistic $\widehat{\theta}(S^*)$ is computed; denote it as $\widehat{\theta}^*$. The pseudosampling is repeated many times. A normalized histogram of the resulting values for $\widehat{\theta}^*$ will often be a good approximation to the probability distribution function for $\widehat{\theta}$. Thorough introductions to bootstrapping are given by Efron and Gong [12] and Davison and Hinkley [10].

3.4.2 Bootstrapping Performance Measures for Fixed $A_{\mathcal{TG}}$

Bootstrapping can be applied to the problem of estimating the probability distribution for recognition rate given fixed training and gallery sets. Consider again the success indicator function from equation 4. Given a representative probe set $\mathcal{P}$ of size n (having only one image per subject to ensure independence), a sample of success/failure outcomes for algorithm $A_{\mathcal{TG}}$ may be expressed as:

$$S = \left\{ b\left(x_{i,j}\right) \quad : \quad x_{i,j} \in \mathcal{P} \right\}. \tag{15}$$

The recognition rate ρ for a sample S is:

$$\rho\left(\mathcal{P}\right) = \frac{c}{n} \quad \text{where} \quad c = \sum_{s_\ell \in S} b(s_\ell), \tag{16}$$

where s_ℓ is the ℓth element of S. Now, to bootstrap the statistic ρ, generate $10,000$ pseudodatasets S^* from S and, for each, compute ρ^*:

$$\rho^*\left(\mathcal{P}\right) = \frac{c^*}{n} \quad \text{where} \quad c^* = \sum_{s_i \in S^*} b(s_i). \tag{17}$$

The normalized histogram of ρ^* values is the bootstrap distribution for ρ.

Of course, applying bootstrapping to the recognition-rate problem is of pedagogical interest only, since it can be shown analytically that c^* is a binomially distributed random variable. The bootstrapping in this case amounts to nothing more than a Monte Carlo approximation to the binomial distribution. However, there are related statistics whose distribution is not so easily known. For example, consider bootstrapping to determine the sampling distribution for median recognition rank over a probe set of size n. Again, the procedure would be to draw, say, $10,000$ random pseudodatasets of size n with replacement from the set of recognition ranks determined for $\mathcal{P}$ and compute the median rank in each pseudodataset. The normalized histogram of these resulting $10,000$ values represents the bootstrap distribution for median recognition rank.

In most identification experiments, there are two images of each subject, and one is assigned to $\mathcal{P}$ and the other to $\mathcal{G}$. Our bootstrapping examples have left the choice of gallery fixed. Thus, performance variability due to random gallery assignment were not measured. This understates the amount of variation that will be seen in practical circumstances where assignment of images to $\mathcal{P}$ and $\mathcal{G}$ is random. Further, in experiments with more images of each subject, there is additional information available about within-subject image variation that contributes to the overall uncertainty about recognition performance. The next section introduces a Monte Carlo approach to resampling both gallery and probe sets that addresses these concerns.

3.4.3 Resampling Gallery and Probe Choices

The idea of the Monte Carlo approach introduced here is to generate a sampling distribution for the statistic of interest by repeatedly computing the statistic from different pseudodatasets that are somehow equivalent to the observed one. To formalize this, define a subject-replicate table U. This is just a slight refinement of our previous notation, where as before probe images $x_{i,j}$ and gallery images $y_{i,j}$ are doubly subscripted, as are the elements $u_{i,j} \in U$. So, for example, $u_{1,1}$ is the first image for the first person in U. Then the creation of different pseudodatasets relies on an exchangeability assumption: that any image $u_{i,j}$ of person i may be exchanged with another image $u_{i,k}$.

The statistic of interest is the recognition rate ρ_τ and the pseudodatasets are obtained by resampling the choice of gallery and probe images among the exchangeable options. Specifically, randomly selected probe sets $\mathcal{P}$ and galleries $\mathcal{G}$ are created by repeatedly sampling U. The exact manner in which $\mathcal{P}$ and $\mathcal{G}$ are assembled is best illustrated by example.

Recall from Section 3.2.3 that images for each of 336 people were partitioned into two groups of 168 people each, the α group and the β group. Further, the 20 replicate images per person were divided into training and test sets, with 4 replicate images per person reserved for testing. Part of the rationale for this design will become clear as we note that this test data allows us to define two 168-by-4 subject-replicate tables U_α and U_β.

Table 3.4 illustrates our sampling procedure for 168 people and 4 replicates per person. The first row indicates positions in $\mathcal{P}$ and $\mathcal{G}$ running from 1 to 168. The second row shows the subject indices i specifying the people. The remaining two rows indicate which replicate j is selected for the corresponding subject i. So, for the specific permutation shown, $\mathcal{P} = \left\{u_{82,1}, u_{39,1}, \ldots, u_{13,4}\right\}$ and $\mathcal{G} = \left\{u_{82,2}, u_{39,3}, \ldots, u_{13,3}\right\}$.

To carry out the resampling, one could simply randomly reselect probe and gallery images for each subject. However, this would likely result in some images being used more than others, when tallied across all pseudodatasets. To obtain a perfectly balanced collection of pseudodatasets, we use the following balanced resampling scheme. Fix the bottom two rows of Table 3.4. This pattern balances

Table 3.4: Permutation of probe and gallery images sampling strategy illustrated for a case of 168 people and 4 replicate images per person.

Position	1	2	3	4	5	6	7	8	9	10	11	12	13	...	168
Person index i	82	39	12	160	83	114	162	55	40	149	154	94	107	...	13
$\mathcal{P}$ Replicate index j	1	1	1	2	2	2	3	3	3	4	4	4	1	...	4
$\mathcal{G}$ Replicate index j	2	3	4	1	3	4	1	2	4	1	2	3	2	...	3

out the number of times each element from U appears. Then randomly reshuffle the person index row of Table 3.4 with respect to the bottom rows to crease a new pseudodataset. This strategy generalizes naturally to additional numbers of people and replicate images per person. The final resampling step is to repeat the process of generating random pseudodatasets many times, in our case typically $10,000$ times, and each time to record the corresponding recognition rate ρ_τ for those ranks τ of interest.

In the following subsections we consider how to use the information generated in these replicated experiments. First, we note some practical issues regarding computation. It is possible to carry out all 10,000 Monte Carlo trials and generate sampling distribution for ρ_τ in a moderately efficient manner. The software to accomplish this task need only read in the similarity matrix Υ for the images in U once. Second, it is possible to sort all images in U by similarity relative to all other images in U just once. This removes the sorting step from the individual Monte Carlo trials. Bit arrays can be used to quickly indicate which images are in or out of the current random probe set and gallery, minimizing data movement in a single Monte Carlo trial. Finally, at the end of each of the 10,000 trials, the only action taken is to increment a tally of recognition successes. All of this is done by the csuAnalyzePermute tool included in Version 5.0 of the CSU Face Identification Evaluation System [3].

Illustration with the LDA and PCA data

Table 3.5 shows a portion of the sample histogram generated by applying the permutation of probe and gallery methodology described above to the similarity matrix U_β for the LDA algorithm described in Section 3.2.3. The raw recognition counts c are shown prior to normalizing the histogram. Keep in mind that the sample probability distribution function for c is obtained by simply dividing the counts shown by $10,000$, the total number of trials.

It is possible to quickly observe a number of features of this method by examining Table 3.5. First, note that for $\tau = 168$, where recognition rank equals the size of the random galleries, the LDA algorithm correctly identifies $c = 168$ people in

Table 3.5: Portion of recognition count (c) and rank τ recognition rate (ρ_τ) histogram for the LDA algorithm. Counts are generated by 10,000 trials of the Monte Carlo sampling to generate randomized probe sets and galleries.

c	ρ_τ	1	2	3	4	5	6	7	8	9	10	...	168
168	1.00	0	0	0	0	0	0	0	1	1	3	...	10000
167	0.99	0	0	0	0	0	1	8	15	24	40	...	0
166	0.99	0	0	0	1	8	27	53	91	165	210	...	0
165	0.98	0	0	5	30	67	111	196	328	447	636	...	0
164	0.98	0	3	44	117	193	319	480	678	955	1162	...	0
163	0.97	0	12	108	284	501	712	1012	1354	1657	1874	...	0
162	0.96	1	58	301	636	905	1218	1560	1865	2024	2115	...	0
161	0.96	2	149	571	1062	1435	1758	1946	1979	1938	1838	...	0
160	0.95	10	352	991	1519	1809	1854	1708	1595	1364	1112	...	0
159	0.95	29	588	1413	1783	1683	1515	1414	1115	816	629	...	0
158	0.94	54	953	1661	1602	1413	1202	881	582	406	267	...	0
157	0.93	167	1265	1543	1276	984	736	467	272	143	79	...	0
156	0.93	309	1465	1332	873	596	332	197	92	48	31	...	0
155	0.92	537	1501	952	453	242	143	54	24	11	4	...	0
154	0.92	822	1257	541	226	110	49	17	7	1	0	...	0
153	0.91	1049	1012	333	83	35	16	6	2	0	0	...	0
152	0.90	1199	649	124	41	14	6	0	0	0	0	...	0
151	0.90	1344	399	58	8	3	0	1	0	0	0	...	0
150	0.89	1208	209	12	3	1	1	0	0	0	0	...	0
149	0.89	1063	80	7	2	1	0	0	0	0	0	...	0
148	0.88	868	26	2	1	0	0	0	0	0	0	...	0
147	0.88	606	12	2	0	0	0	0	0	0	0	...	0
146	0.87	371	10	0	0	0	0	0	0	0	0	...	0
145	0.86	190	0	0	0	0	0	0	0	0	0	...	0
144	0.86	97	0	0	0	0	0	0	0	0	0	...	0
143	0.85	39	0	0	0	0	0	0	0	0	0	...	0
142	0.85	24	0	0	0	0	0	0	0	0	0	...	0
141	0.84	10	0	0	0	0	0	0	0	0	0	...	0
140	0.83	1	0	0	0	0	0	0	0	0	0	...	0
139	0.83	0	0	0	0	0	0	0	0	0	0	...	0
	⋮						⋮						
1	0.01	0	0	0	0	0	0	0	0	0	0	...	0

all 10,000 trials: see the upper right entry. This is of course the limiting behavior associated with the upper end of a standard CMC curve. Second, a good estimator of the expected performance (say $\rho_\tau(\Omega)$) is the mean of the 10,000 rank τ recognition rate values, about 0.90 for $\tau = 1$. Note also that for $\tau = 1$, the LDA algorithm most often correctly identifies 151 out of the 168 people: it does this 1344 out of 10,000 times. This also corresponds to a recognition rate $\rho_1 = 0.90$ as shown in the second column. Thus, for $\tau = 1$, $\rho_1 = 0.90$ is the mode of the recognition rate distribution. The mode of the sample distribution is the statistic most easily picked out of the table by eye. There is also a notable visual effect if one views the pattern of values as a whole; the overall impression is somewhat that of a blurred CMC curve. Indeed, the resampling results can be used to assess uncertainty in the CMC curve.

Suppose that we wanted to predict the recognition rate achieved by LDA on a new sample. A symmetric 95% prediction interval is obtained by scanning in from both ends of the distribution until the total probability on each side sums to 0.025. Translating this procedure to the frequency values shown in Table 3.5, scan in from both ends until each sum first equals or exceeds 250. To illustrate for the $\tau = 1$ column coming down from the top, the histogram values $1+2+10+29+54+167 = 263$ at $c = 157$, and hence the upper bound on the 95% prediction interval is $c = 157$, $\rho = 0.93$. The associated lower bound for $\tau = 1$ is $c = 145$, $\rho = 0.86$. Thus we estimate using our Monte Carlo method that the probability of observing a recognition rate greater than 0.93 or less than 0.86 for the LDA algorithm under these conditions is below 0.05.

Figure 3.3 shows the sample probability distributions for ρ_1 as a plot for both the LDA and PCA algorithm. The distribution for LDA comes directly from the frequency data shown in the $\tau = 1$ column of Table 3.5. The modes of the distributions in Figure 3.3 are $\rho_1 = 0.90$ and $\rho_1 = 0.88$ for LDA and PCA respectively. Figure 3.4 shows the pointwise 95% prediction intervals for the CMC curve comparing LDA to PCA for $\tau = 1$ to $\tau = 50$. Note these intervals are all determined in the manner previously described for the LDA algorithm at $\tau = 1$. The mean values are $\rho_1 = 0.898$ and $\rho_1 = 0.874$ for LDA and PCA respectively; these are the starting points for the CMC curves shown in Figure 3.4.

Now is a good time again to raise the basic question motivating this chapter: is this difference we observe significant? McNemar's test leads us to conclude the difference at $\tau = 1$ and $\tau = 2$ is not significant, but the difference at $\tau = 4$ is significant. Contrast the McNemar's result obtained for the original single probe set and gallery with the curves and associated error bars at $\tau = 1$, $\tau = 2$ and $\tau = 4$ in Figure 3.4, and our confidence in the reproducibility of a result showing LDA clearly superior to PCA begins to erode.

It is critical to keep in mind that the resampling of probe and gallery images technique is accounting for a source of uncertainty not addressed with McNemar's test: within-subject image variation. We cannot simply include results from all

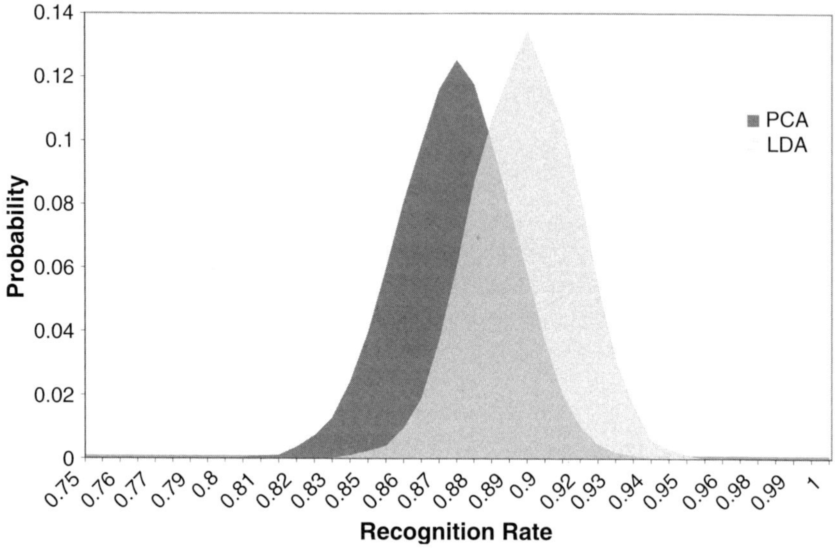

FIGURE 3.3: Sample probability distributions for ρ_τ for the LDA and PCA algorithms at $\tau = 1$.

probe and gallery assignments in a table for McNemar's test because those trials are not all independent. There is within-subject correlation. The resampling approach uses the extra possible probe and gallery assignments to estimate the variance of ρ_τ in a manner that incorporates within-subject variation.

One of the most concrete ways to think about the outcome of the permutation of probe and gallery images technique is as a way of answering the question; What happens when a single probe set and gallery experiment is repeated many times? With this interpretation in mind, again consider Figure 3.4 where the average recognition rate at rank one is $\rho_1 = 0.898$ and $\rho_1 = 0.874$ for LDA and PCA respectively. The error bars tell us to expect that 95 out of 100 times we might repeat our test, ρ_1 for the LDA algorithm will range between 0.863 and 0.935 and ρ_1 for the PCA algorithm will range between 0.828 and 0.898.

There is overlap in these intervals. It is tempting to conclude that PCA must therefore sometimes have a higher recognition rate than LDA. This may happen—or it may not—and the key is to realize that when comparing any two algorithms, what matters most is how one algorithm does relative to the other on each of the Monte Carlo trials. Therefore, in the next section we introduce one additional component to the permutation technique that explicitly measures how one algorithm does relative to another.

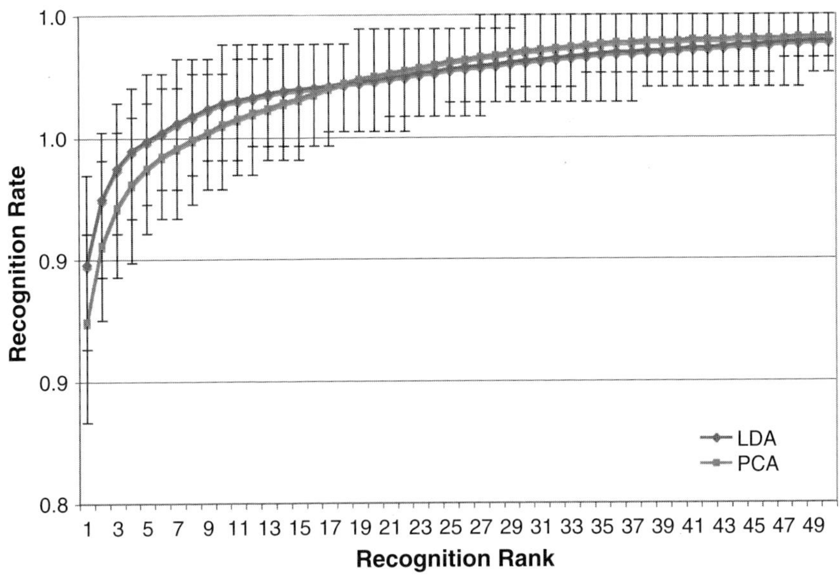

FIGURE 3.4: CMC curves for LDA and PCA with pointwise 95% prediction intervals derived from the permutation of probe and gallery images.

Finally, note from Figure 3.2 that $\rho_1 = 0.857$ and $\rho_1 = 0.833$ for the LDA and PCA algorithm respectively in the single probe set and gallery test presented in Section 3.2.3. These recognition rates are much lower than the averages for the permutation of probe and gallery images results just presented, e.g. $\rho_1 = 0.898$ and $\rho_1 = 0.874$. The obvious question is, why?

The explanation lies in the way the probe set and gallery were selected. The result in Section 3.2.3 was obtained by using the first column of test images in U_β as the probe set and the third column as the gallery, while the permutation test draws probe and gallery images equally from all columns according to the sampling pattern shown in Table 3.4. Thus, our result for the single probe set and gallery in Section 3.2.3 represented a biased choice, and one that is clearly harder than is observed on average when sampling probe sets and galleries. The most likely source of the increased difficulty is the increase in the elapsed time between when gallery images and corresponding probe images were acquired. Replicate images in U_β are ordered chronologically, and the choice of first and third columns forces all comparisons to be between probe and gallery images taken on different days.

This outcome serves to highlight the need to tailor the Monte Carlo sampling process to reflect, to the greatest extent possible, the constraints one assumes will be observed in selecting probe and gallery image pairs in practice. For example, if one is interested only in cases where probe and gallery images are taken on different days, the sampling procedure should be modified to respect this constraint. Indeed, this was done in our earliest published comparison of LDA and PCA [4] using the permutation of probe and gallery image technique. However, adding constraints to the sampling procedure complicates it somewhat, and it is then not possible to use the standard analysis tool included in the CSU Face Identification Evaluation System Version 5.0 [3].

3.4.4 Is Algorithm *A* Better than Algorithm *B* ?

The Monte Carlo procedure presented above to generate sample probability distributions for recognition rates can be used to assess both the statistical and scientific significance of a comparison between two algorithms, *A* and *B*.

To begin, we turn our attention to counting how often one algorithm does better than another while pseudosampling probe sets and galleries. Formally, for the ith experiment ($i = 1, \ldots, 10{,}000$), let $Y_i = 1$ if the recognition rate for algorithm A exceeded that for B, and $Y_i = 0$ otherwise. Let $p = P[Y_i = 1]$ where the randomness in Y_i reflects sampling variability: each experiment uses a different random sample of data. We estimate $\hat{p} = \sum_{i=1}^{10{,}000} Y_i/10{,}000$. Define A and B to perform equally over Ω if $p = 1/2$, in which case each algorithm outperforms the other on half of the possible probe/gallery pseudodatasets. Then we may use $\hat{p}$ in a standard one-sample z test of the null hypothesis $H_0 : p = 0.5$ by rejecting when

$$\frac{|\hat{p} - 0.5|}{\sqrt{\frac{1}{2}(1 - \frac{1}{2})/10000}} > 1.96.$$

This test is approximate since the independence assumption is not fully met. However, it is extremely powerful because of how many data are used. Indeed, whenever $\hat{p} < 0.49$ or $\hat{p} > 0.51$, the null hypothesis will be rejected in favor of the conclusion that one algorithm is statistically significantly better than the other.

This conclusion illustrates how misleading statistical significance can be: a tiny performance difference can yield a highly significant p-value if the sample size is large enough. Had we run 1,000,000 samples instead of 10,000, the range of $\hat{p}$ not yielding rejection of H_0 would be 10 times narrower. So far, generating all these data via resampling has provided a high degree of certainty about which algorithm performs better on average, but has not provided scientifically meaningful information about the degree of superiority in light of sample-to-sample performance variation.

Suppose from an analysis like the one above, we know with a high degree of statistical certainty that a new algorithm, *A*, performs better than *B* in the sense that

$\hat{p}$ is statistically significantly greater than 0.5. Then, two important and distinct questions arise:

1. If a researcher collected a single dataset (of the same size used here) and compared the recognition rates of A and B for these data, what is the probability of observing the misleading outcome of A performing worse than B?
2. If an algorithm developer or user is considering investing resources to implement A for similar data, how often will A actually perform better than B in practice?

Both questions are asking about the magnitude of the difference between algorithms, and may be answered using the resampling strategy described above.

To begin, it is important to restate our earlier conclusion. Namely, that so long as $\hat{p}$ is well outside the range $0.49 \ldots 0.51$, we may conclude with great confidence which algorithm is superior. To keep our discussion simple, assume that the hypothesis test determines that A is superior. In other words, $\hat{p}$ is statistically significantly greater than 0.5. If by chance B is significantly better than A, reverse the roles of A and B in the following discussion.

We want to focus attention on how often the inferior algorithm was observed doing better than the superior algorithm, so we count the instances where the recognition rate of B exceeds that of A among the 10,000 replicate experiments. Suppose B beats A in 300 cases, then there is a 300/10,000 = 3% chance that a "misleading" outcome will be observed in a future experiment. In this case, statistical significance is paired with scientific significance since A will almost always be found to beat B. On the other hand, if there are 4800 such cases, then A will outperform B only about 52% of the time. This is likely not a scientifically useful improvement. We can further quantify the degree of improvement for A relative to B using the mean or mode of the 10,000 values of the recognition rate difference

$$D_\tau(A, B) = \rho_{\tau,A} - \rho_{\tau,B}. \tag{18}$$

We suggest that the threshold for establishing scientific significance should generally be high. One way to conceptualize such a threshold is in terms of the first question posed above. Imagine we are only willing to conclude that A is really an improvement over B if the probability of being misled about which algorithm is better from a single experiment is equal to or less than 0.05. Our permutation of probe sets and galleries procedure has provided us a sample probability distribution for $D_\tau(A, B)$, and this distribution tells us directly the probability of observing B doing better than A. Thus, if $D_\tau(A, B) < 0$ in fewer than 500 of the 10,000 instances, the probability of being misled by a single comparison between A and B is less than or equal to 0.05. One might of course relax this threshold somewhat, adopting perhaps 0.10 as the acceptable upper bound on being misled by a single comparison. The exact choice here is of course a matter of judgment.

The second question, namely on what fraction of problems drawn from the sample population does one expect A to outperform B, is also directly answered from the sample probability distribution for $D_{\tau}(A, B)$. It is not hard to imagine circumstances where a more modest difference between algorithms might still provide incentive to use a new algorithm A in place of an older algorithm B. Even a modest 10% margin of victory for A, evidenced by $\hat{p} = \frac{5500}{10{,}000} = 0.55$, might be reason enough in some cases to favor A over B.

Viewed together, question 1 and question 2 above guide our interpretation of the $D_{\tau}(A, B)$ sample probability distribution. Ultimately how we threshold these outcomes is a subjective choice that should reflect the priorities of the investigator. What should be remembered in any case is that when the frequency of nonmisleading outcomes greatly exceeds 50%, the test of statistical significance will yield a very small p-value. Thus, there is essentially no uncertainty about which algorithm is actually superior on average, and our determination of scientific significance is not confounded by doubt over which outcomes are misleading.

To tie back in results from our running example with LDA and PCA, LDA correctly identified more images than did PCA 8693 times, and PCA correctly identified more images than did LDA 747 times. Note these two number do not sum to $10,000$, since the two algorithms correctly identified exactly the same number of images in 560 of the 10,000 trials. Thus, choosing LDA as algorithm A, $\hat{p} = \frac{8693}{10{,}000} \simeq 0.87$. Thus, LDA has about an 87% probability of performing better than PCA on a probe set and gallery drawn from the sample population. Whether such a difference is ultimately of practical importance is of course a question that can only be answered in some specific and practical context.

This concludes our introduction to and illustration of parametric and nonparametric techniques for comparing face identification algorithms. The next section expands upon the comparison between PCA and LDA. Recall from Section 3.2.3 where the comparison between LDA and PCA was introduced that our comparison was motivated by two hypotheses. These have not yet been adequately addressed. Whenever two general classes of algorithms are to be compared, there are issues of how each is setup, and whether such configuration factors matter.

3.5 EXPANDING THE LDA VERSUS PCA COMPARISON

The results used above for the LDA versus PCA comparison were selected from a much larger set of tests. The tests considered all combinations of five factors summarized in Table 3.6. The first factor is the subspace where images are matched: PCA or LDA. The second factor is the number of replicate images per person used to train each algorithm: 3, 4, or 16. The third factor concerns how the dimensionality of the PCA subspace is reduced. The truncate option truncates the number of PCA

Table 3.6: Summary of five factors varied in the complete LDA versus PCA comparison discussed in Section 3.5.

Factor	Level 1	Level 2	Level 3
Subspace	LDA (PCA + LDA)	PCA (only)	
Training replicates	3	4	16
PCA cutoff	truncate	energy	
Training group	α	β	
Testing group	α	β	

dimensions to match the number of LDA classes, in other words the number of people, 168, minus one. This choice is made consistently in both the PCA and the LDA algorithms. The other option truncates the PCA dimensions using an energy cutoff [31], retaining a sufficient number of PCA dimensions to account for 95% of the total energy. The energy option always retains more dimensions, ranging between around 230 for the 3-replicate training up to around 600 for the 16-replicate training. The fourth factor is whether the algorithm is trained on the α or the β group, and the fifth factor is analogous to the fourth except it indicates which group of people is used for testing.

These five factors have 2, 3, 2, 2, and 2 levels respectively. Therefore, there are a total of 48 possible combinations in a full factorial design. Hence, both and LDA and PCA were run under a total of 24 settings each. These are too many combinations to readily view in one graphic. Thus, a natural first question is whether some factors may be dismissed as unimportant.

Training on 3 replicates was included because that was the number used in our previous study [4]. Is the distinction between 3- and 4-replicate training important? Table 3.7 presents the sample probabilities with that 4-replicate training beats 3-replicate training and vice versa for all pairwise comparisons between algorithm configurations that differ only in the number of training replicates. These $\hat{p}$s were calculated using the method from Section 3.4.4.

The 16 algorithm variants are sorted from largest to smallest difference. The largest difference (0.81 versus 0.08) favors the 3-replicate configuration. While this might be considered a substantial difference in some circumstances, it does not meet the strong criterion of $\hat{p} \geq 0.95$ discussed above. Particularly for the algorithm variants further out in the table, the difference is truly negligible. For the configurations 10 through 16 we stop indicating one as superior to the other (boldface), since $\hat{p}$ is essentially at 0.5 or below for both training options. In all cases, we do not expect the two probabilities to sum to 1, since algorithms can tie.

Table 3.7: Comparison of 3-replicate (T3) versus 4-replicate (T4) training with all other factors equal. The $\hat{p}$ probabilities in both directions are shown in the last two rows.

	Configuration															
	1	2	3	4	5	6	7	8	9	10	11	12	13	14	15	16
Space	PCA	PCA	LDA	LDA	LDA	LDA	LDA	PCA	LDA	LDA	PCA	PCA	LDA	PCA	PCA	PCA
Cut	E	E	E	T	E	E	E	E	T	T	T	E	T	T	T	T
Train	α	α	α	β	β	β	α	β	β	α	β	β	α	β	α	α
Test	α	β	β	α	β	α	α	β	β	β	α	α	α	β	β	α
(T3 > T4)	**0.81**	**0.69**	0.17	0.15	0.17	0.17	0.19	**0.62**	0.20	0.28	0.33	0.44	0.33	0.34	0.39	0.38
(T4 > T3)	0.08	0.16	**0.68**	**0.68**	**0.66**	**0.64**	**0.63**	0.21	**0.60**	0.51	0.45	0.35	0.43	0.42	0.38	0.37

To leave ourselves a manageable number of comparisons, based upon the results in Table 3.7 the 3-replicate training will be dropped. Note, however, that the 16-fold multiple comparison shown in Table 3.7 could be conducted as part of a full analysis of variance [9], and while beyond the scope of what we will demonstrate here, this is preferable when trying to draw conclusions over many factors. Pointers to some examples of multifactor analysis applied to face identification will be provided in Section 3.6 below.

Focusing only on the 4-replicate training and the 16-replicate training cases, we are left with 32 algorithm variants to compare. Figure 3.5 shows the mean ρ_1 and 95% prediction intervals for each of these configurations sorted from best to worst. The names for each variant shown on the left indicate the subspace, number of training replicates, whether PCA cutoff was fixed or based upon energy, and finally the training and test groups respectively.

Several interesting trends are apparent in Figure 3.5. Perhaps most compelling, in all cases LDA appears above PCA. This leads us to revisit the two hypotheses made in Section 3.2.3.

3.5.1 Revisiting LDA Versus PCA Hypotheses

Recall the first hypothesis:

Hypothesis 1: LDA will outperform PCA when given many training images per person.

Observe that LDA trained on 16 images does outperform LDA trained on 4 images. Specifically, the best settings are "LDA 16 E α α" and "LDA 16 E β β". The mean values for ρ_1 are 0.973 and 0.972 respectively. To assess whether LDA outperforms PCA when using many training images, we explicitly compare the best LDA settings to the best PCA settings using the technique described in Section 3.4.4.

The best PCA result is for "PCA 4 T α α" with mean $\rho_1 = 88.4$. In none of the 10,000 Monte Carlo trials did this PCA variant correctly identify more people than comparable " α α": the LDA variant. The other case of training and testing on the same people, "LDA 16 E β β" versus "PCA 4 T β β", came out similarly with the PCA variant never beating the LDA variant. These results support the first hypothesis that LDA outperforms PCA when given many replicates per person and tested on the same people as it is trained to recognize.

Now recall the second hypothesis:

Hypothesis 2: LDA will not outperform PCA when trained and tested on different people.

Two relevant comparisons for this second hypothesis are "LDA 16 E α β" versus "PCA 4 T α β" and "LDA 16 E β α" versus "PCA 4 T β α". A quick look at Figure 3.5 shows that LDA is still doing much better than PCA in these two

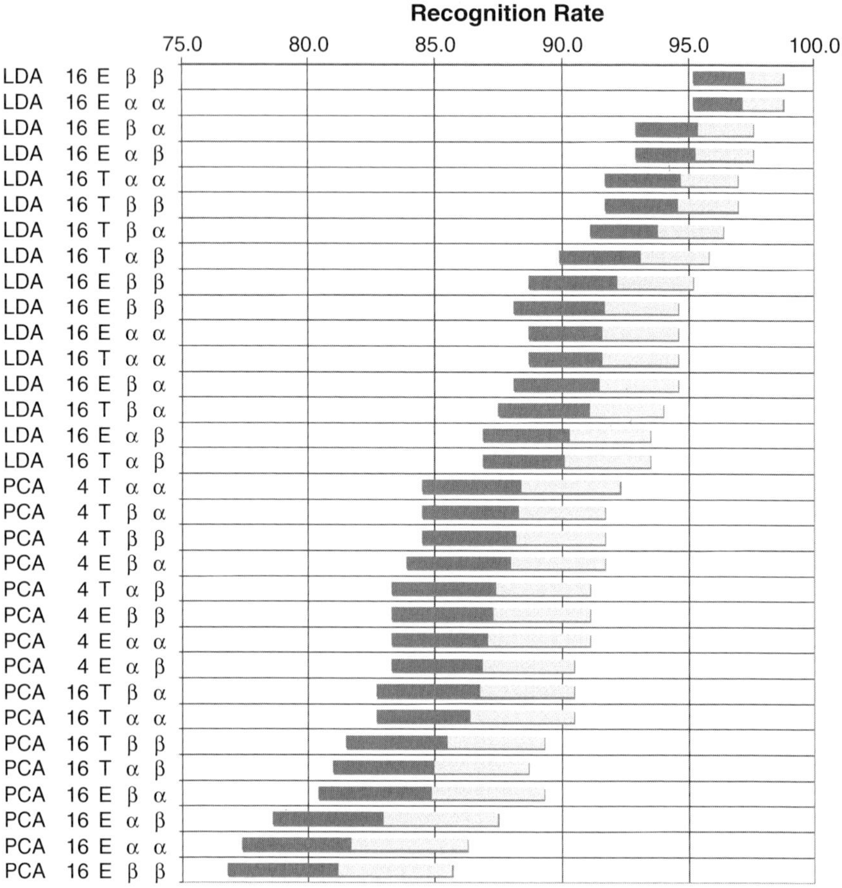

FIGURE 3.5: Recognition rate ranges for 32 distinct variants of LDA and PCA. The mean for ρ_1 is where the two bars meet, and the width of the bars indicate the 95% prediction interval on each recognition rate estimate. The 32 configurations are labeled in the left margin; see the text.

comparisons, and indeed PCA again never outperforms LDA. Thus, hypothesis 2 is not supported by the experimental data. LDA outperforms PCA even when trained on one group of people and tested on another.

This second result is worth a moment's thought, since it is unclear why LDA should behave in this manner. The theory behind the use of Fisher linear discriminants clearly rests upon a presumption that one knows apriori the classes to be recognized and is configuring a subspace to maximally separate samples into these distinct classes. Since LDA used on faces typically treats individual people as classes, it is not immediately obvious that an LDA subspace trained to maximally separate one group of 168 people should also be good when used on a completely different set of 168 people. Clearly there is some second-order transference of generalization taking place, and indeed nearly all previous applications of LDA to faces have implicitly relied on this being the case. However, it is still in our minds an open question exactly why this takes place.

3.6 ADVANCED MODELING

This chapter has focused on inferences about the recognition rate of an algorithm, and on inferences about the difference in recognition rates of two algorithms. We described methods for estimating sampling variability, which allow us to determine whether two algorithms differ by more than could be explained by chance alone and whether a proven difference is scientifically important. Another interesting question is: why do they differ?

A class of experiments to address this question collect both performance data and data on covariates, i.e., variables that might affect recognition performance. For example, subject covariates such as race, gender, and facial expression are of considerable interest; Are women easier to recognize than men? Another class of covariates are image covariates such as left/right contrast, lighting angle, indoor/outdoor, and so forth. A third class of covariates was mentioned previously: algorithm configuration and tuning variables that may affect performance. Of course, algorithm identity (PCA versus LDA) can also be a covariate.

Various classes of statistical regression models are useful for assessing the effect of covariates. Such models control for the effect of other covariates while estimating the effect of each. This is superior to marginal tabulations of results that are easily confounded by ignored variables. Further, such models allow for the estimation of interaction effects. A result like "Women are easier to recognize than men for PCA, but the reverse is true for LDA" is an interaction effect. Clearly such conclusions are of high scientific importance. Finally, some classes of models allow for careful formal modeling and estimation of the variance/covariance structure induced by testing multiple images of the same subject, possibly in different conditions.

The simplest class of models are ordinary ANOVA and regression models. These require a continuous response variable. In some of our past work with human subject covariates we have used image distances or similarities (i.e., the elements of Υ) as responses [15]. A concern with such models, however, is the extent to which similarity scores directly translate into recognition performance. To address this concern, we have used generalized linear models [20] with a binary response variable for each trial, namely $b_\tau(x_{i,j})$. These are logistic regression models with, say, rank one recognition outcome ($b_1(x_{i,j})$) as the success/failure response; and the human subject covariate work has been expanded to use these types of models [18].

With either of these model classes, there are additional complexities that arise when analyzing multiple trials arising from the same subject. Ideally, one should account for the variance/covariance structure of such data. Trials from the same person are correlated, and some people are harder to recognize than others, so there is an extra source of variation that contributes to the overall variability. In this case, one can add random effects for subjects. Ordinary ANOVA and regression models become mixed models. Generalized linear models become generalized linear mixed models. For examples of how to use generalized linear mixed models to study the configuration space of the LDA algorithm see [16], and for an example relating human subject covariates to probability of correct verification at different false alarm rates see [17].

Although the use of statistical models such as linear mixed models and generalized linear mixed models is well beyond the scope of this chapter, researchers wishing to address questions of what factors influence the difficulty of recognition should keep these more advanced techniques in mind.

3.7 CONCLUSION

We have reviewed several simple approaches for detecting statistically and scientifically significant differences between face-identification algorithms. The first, McNemar's test, is most useful in simple settings with a single trial per subject. When several images per subject are available and we wish to account explicitly for within-subject variation, the second method, based on Monte Carlo resampling, can be applied. It is our hope that future work comparing face identification algorithms will go beyond simple comparisons of CMC curves, instead relying on methods that estimate and account for the sampling variability inherent in estimated recognition rates. At a minimum, researchers should report confidence intervals, and when possible direct pairwise comparisons between new and established algorithms are best.

The running example comparing PCA to LDA has demonstrated how to construct confidence intervals as well as how to compare two algorithms. At the same time, it suggests further scientific questions not fully addressed here. In particular, what factors or covariates influence an algorithm's performance? Exploring the space of possible algorithm configurations is almost always important: note how much would be missed had we not gone beyond the first comparison in Section 3.3.3. In past work [18, 16, 17] we have demonstrated the value of statistical models to examine the effects of various factors in much greater depth, using some of the techniques described in Section 3.6.

Appendix A NOTATIONAL SUMMARY

As a guide to the notation introduced in this chapter, Table 3.8 provides short definitions for the most common terms.

Table 3.8: Notation summary.

Symbol	Usage
Ω	Target population of images to be recognized.
W	The finite set of images available for performance evaluation; the sampled population
$w_{i,j}$	The jth image of person i in W
$\mathcal{T}$	Training images, typically $\mathcal{T} \subseteq W$
$\mathcal{G}$	Gallery images, typically $\mathcal{G} \subseteq W$
$\mathcal{P}$	Probe images, typically $\mathcal{P} \subseteq W$
A	An algorithm
$A_{\mathcal{T}}$	An algorithm trained on $\mathcal{T}$
$A_{\mathcal{T}\mathcal{G}}$	An algorithm trained on $\mathcal{T}$ and using $\mathcal{G}$ as exemplars
U	Union of all gallery and probe sets used for a set of experiments, $U \subseteq W$
$u_{i,j}$	The jth image of person i in U
υ	A similarity relation between pairs of images
Υ_U	Pairwise similarity matrix for U
$x_{i,j}$	A probe image in $\mathcal{P}$
$y_{k,l}$	A gallery image in $\mathcal{G}$
$L(x)$	Gallery images sorted by decreasing similarity υ to image x
$L_\zeta(x)$	The ζth element of L for probe image x
$b_\tau\left(x_{i,j}\right)$	Success indicator function, 1 if $x_{i,j}$ matched in first τ sorted gallery images
$\rho_\tau(\mathcal{P})$	Rank τ recognition rate of an algorithm on probe set $\mathcal{P}$
$r(x)$	Rank of first correct match for probe image x
$r_\tau(x)$	Rank censored at maximum value τ
$\tilde{r}_\tau$	Median of censored rank r_τ over probe set $\mathcal{P}$
S	A random sequence of probe images or a results of an indicator function applied to them
s_ℓ	The ℓth randomly selected probe image in the sequence S
$P[e]$	The probability of some event e

Appendix B PARTICULARS FOR THE ELASTIC BUNCH GRAPH ALGORITHM

Results were presented above for a variant of the CSU EBGM algorithm not previously described. Here the essential differences between the EBGM algorithm as it is described in [6].

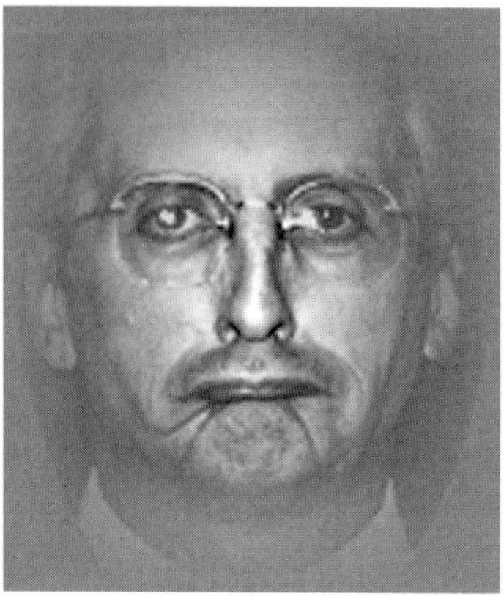

FIGURE 3.6: Illustrating the improved "Western Photo" EBGM image preprocessing procedures used to generate best results shown above in Figure 3.2.

B.1 Image Preprocessing

David Bolme [6] encountered serious problems using standard FERET style image preprocessing [3] with the EBGM algorithm and developed an alternative procedure that softened the edges created when the nonface pixels are masked out. Albert Lionelle extended this work and developed an even more refined preprocessing procedure. The preprocessing steps developed by David Bolme and supplied with Version 5.0 of the CSU Face Identification Evaluation System does the following:

1. Integer to float conversion: Converts 256 gray levels into floating-point equivalents.
2. PreFade Borders: Fades the border edges of the image, but does not overlap the face.
3. Geometric normalization: Lines up human-chosen eye coordinates. This setting is applied differently, leaving the shoulders and most the face showing.
4. PostFade Border: Fades the border of the now modified geometric image. This may overlap the face, but in the majority of cases it didn't.

The improved "Western Photo" preprocessing developed by Albert Lionelle does the following:

1. Integer to float conversion: Converts 256 gray levels into floating-point equivalents.
2. Geometric normalization: Lines up human-chosen eye coordinates.

3. Masking: Crops the image using an elliptical mask and image borders such that the face, hair, and shoulders are visible in addition to minor background areas.
4. Histogram equalization: Equalizes the histogram of the unmasked part of the image.
5. Pixel normalization: Scales the pixel values to have a mean of zero and a standard deviation of one.
6. PostFade Border: Fades the border of the modified image, providing a broken-up oval from the original harsh edge mask.

B.2 The 25 Landmark Variant of EBGM

The 25 landmarks are essentially equivalent to the 25 face features identified by David Bolme, but repositioned to account for the somewhat larger overall images size associated with the "Western Photo" preprocessing. There is one other difference. In the EBGM version developed by David Bolme, additional landmarks were placed midway between some of the facial features. This was also done by original USC EBGM algorithm. This increased the overall number of landmarks to 52. In the 25 landmark version of the EBGM algorithm, these additional landmarks are not used.

ACKNOWLEDGMENTS

This work is supported in part by the Defense Advanced Research Projects Agency under contract DABT63-00-1-0007.

REFERENCES

[1] A. J. Mansfield and J. L. Wayman. Best practices in testing and reporting of biometric devices: Version 2.01. Technical Report NPL Report CMSC 14/02, Centre for Mathematics and Scientific Computing, National Physical Laboratory, UK, August 2002.

[2] P. Belhumeur, J. Hespanha, and D. Kriegman. Eigenfaces vs. Fisherfaces: Recognition using class specific linear projection. *IEEE Transactions on Pattern Analysis and Machine Intelligence* **19**(7):771–720, 1997.

[3] J. R. Beveridge, D. Bolme, M. Teixeira, and B. A. Draper. The csu face identification evaluation system: Its purpose, features and structure. *Machine Vision and Applications*, page online, January 2005.

[4] J. R. Beveridge, K. She, B. Draper, and G. H. Givens. A nonparametric statistical comparison of principal component and linear discriminant subspaces for face recognition. In: *Proceedings of the IEEE Conference on Computer Vision and Pattern Recognition*, pages 535–542, December 2001.

[5] D. M. Blackburn, M. Bone, and P. J. Phillips. Facial recognition vendor Test 2000: Executive Overview. Technical report, Face Recognition Vendor Test (www.frvt.org), 2000.

[6] D. S. Bolme. Elastic bunch graph matching. Master's thesis, Computer Science, Colorado State University, June 2003.

[7] G. Casella and R. L. Berger. *Statistical Inference*. Brooks/Cole, Pacific Grove, CA, 2nd edition, 2001.

[8] W. G. Cochran. *Sampling Techniques*. Wiley and Sons, New York, 1953.

[9] P. Cohen. *Empirical Methods for AI*. MIT Press, 1995.

[10] A. C. Davison and D. V. Hinkley. *Bootstrap Methods and their Applications*. Cambridge University Press, Cambridge, 1997.

[11] G. Doddington, W. Ligget, and A. Martin et al. Sheep, goats, lambs, and wolves: a statistical analysis of speaker performance in the nist 1998 speaker recognition evaluation. *5th international conference on spoken language processing (ICSLP)*, page paper 608, 1998.

[12] B. Efron and G. Gong. A leisurely look at the bootstrap, the jackknife, and cross-validation. *American Statistician*, **37**:36–48, 1983.

[13] K. Etemad and R. Chellappa. Face recognition using descriminant eigenvectors. In: *Proceedings of the International Conference on Acoustics, Speech and Signal Processing*, pages 2148–2151, 1996.

[14] K. Etemad and R. Chellappa. Discriminant analysis for recognition of human fdace iages. *Journal of the Optical Society of America* **14**:1724–1733, August 1997.

[15] G. H. Givens, J. R. Beveridge, B. A. Draper, and D. Bolme. A statistical assessment of subject factors in the pca recognition of human faces. In: *CVPR 2003 Workshop on Statistical Analysis in Computer Vision Workshop*. IEEE Computer Society, June 2003.

[16] G. H. Givens J. R. Beveridge, B. A. Draper, and D. Bolme. Using a generalized linear mixed model to study the configuration space of a PCA+LDA human face recognition algorithm. In: *Proceedings of Articulated Motion and Deformable Objects, Third International Workshop (AMDO 2004)*, pages 1–11, September 2004.

[17] G. H. Givens, J. R. Beveridge, B. A. Draper, and P. J. Phillips. Repeated measures GLMM estimation of subject-related and false positive threshold effects on human face verification performance. In: *Empirical Evaluation Methods in Computer Vision Workhsop: In Conjunction with CVPR 2005*, page to appear, June 2005.

[18] G. Givens, J. R. Beveridge, B. A. Draper, P. Grother, and P. J. Phillips. How features of the human face affect recognition: a statistical comparison of three face recognition algorithms. In: *Proceedings of the IEEE Computer Vision and Pattern Recognition 2004*, pages 381–388, 2004.

[19] IFA. Statistical tests, http://fonsg3.let.uva.nl:8001/service/statistics.html). Website, 2000.

[20] P. McCullagh and J. A. Nelder. *Generalized Linear Models*. Chapman and Hall, New York, NY, 1989.

[21] Q. McNemar. Note on the sampling error of the difference between correlated proportions or percentages. *Psychometrika*, **12**:153–157, 1947.

[22] R. J. Micheals and T. Boult. Efficient evaluation of classification and recognition systems. In: *IEEE Computer Vision and Pattern Recognition 2001*, pages I:50–57, December 2001.

[23] M. H. Yang, D. Kriegman, and N. Ahuja. Detecting Faces in Images: A Survey. *IEEE Transactions on Pattern Analysis and Machine Intelligence (*PAMI*)*, **24**(1):34–58, 2002.

[24] F. Mosteller. Some statistical problems in measuring the subjective response to drugs. *Biometrics* **8**:220–226, 1952.

[25] K. Okada, J. Steffens, T. Maurer, H. Hong, E. Elagin, H. Neven, and C. von der Malsburg. The Bochum/USC face recognition system and how it fared in the FERET phase III test. In: H. Wechsler, P. J. Phillips, V. Bruce, F. Fogeman Soulié, and T. S. Huang, editors, *Face Recognition: From Theory to Applications*, pages 186–205. Springer-Verlag, 1998.

[26] P. J. Phillips, H. J. Moon, S. A. Rizvi, and P. J. Rauss. The FERET evaluation methodology for face-recognition algorithms. *T-PAMI* **22**(10):1090–1104, October 2000.

[27] P. J. Phillips, P. Grother, R. J. Micheals, D. M. Blackburn, E. Tabassi, and J. M. Bone. FRVT 2002: Overview and summary. Technical report, Face Recognition Vendor Test 2002 (www.frvt.org), 2002.

[28] S. A. Rizvi, P. J. Phillips, and H. Moon. The feret verification testing protocol for face recognition algorithms. Technical Report 6281, NIST, October 1998.

[29] D. Swets and J. Weng. Using discriminant eigenfeatures for image retrievel. *IEEE Transactions on Pattern Analysis and Machine Intelligence* **18**(8):831–836, 1996.

[30] M. A. Turk and A. P. Pentland. Face recognition using eigenfaces. In: *Proc. of IEEE Conference on Computer Vision and Pattern Recognition*, pages 586–591, June 1991.

[31] B. A. Draper, W. S. Yambor, and J. R. Beveridge. Analyzing pca-based face recognition algorithms: eigenvector selection and distance measures. In: H. Christensen and J. Phillips, editors, *Empirical Evaluation Methods in Computer Vision*. World Scientific Press, Singapore, 2002.

[32] W. S. Yambor, B. A. Draper, and J. R. Beveridge. Analyzing pca-based face recognition algorithms: eigenvector selection and distance measures. In: *Second Workshop on Empirical Evaluation in Computer Vision*, Dublin, Ireland, July 2000.

[33] W. Zhao, R. Chellappa, and A. Krishnaswamy. Discriminant analysis of principal components for face recognition. In: Wechsler, Philips, Bruce, Fogelman-Soulie, and Huang, editors, *Face Recognition: From Theory to Applications*, pages 73–85, 1998.

PART 2

FACE MODELING

COMPUTATIONAL ASPECTS

CHAPTER 4

3D MORPHABLE FACE MODEL, A UNIFIED APPROACH FOR ANALYSIS AND SYNTHESIS OF IMAGES

4.1 INTRODUCTION

The requirement of pattern synthesis for pattern analysis has often been proposed within a Bayesian framework [22, 31] or has been formulated as an alignment technique [44]. This is in contrast to pure *bottom-up* techniques which have been advocated especially in the early stages of visual signal processing [28]. Here, a standard strategy is to reduce a signal to a lower-dimensional feature vector and to compare this vector with those expected for signals in various categories. A crucial problem of these algorithms is that they cannot explicitly describe variations between or within the categories and therefore have difficulty separating unexpected noise from the variations within a particular category.

In contrast, morphable models, described in this chapter, work by actively reconstructing the signal analyzed [5, 7, 39]. In an additional *top-down* path, an estimated signal is generated and compared to the present input signal. Then, by comparing the real signal with its reconstruction, it is decided whether the analysis is sufficient to explain the signal or not. Clearly, the crucial component in such an approach is the ability of the image model function to reconstruct the input signal.

For object classes, such as faces or cars, where all objects are similar, a model function can be learned from examples. That is, the general image model for the whole class of human faces is derived by exploiting some prototypical face examples.

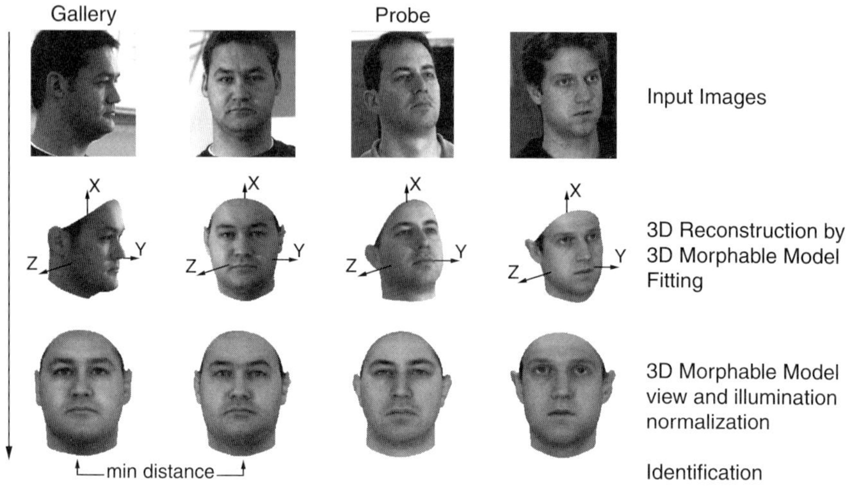

FIGURE 4.1: After a pose illumination normalization using the 3D Morphable Model, identification is performed. Identification can be performed by a comparison of either model parameters of normalized images. (See also color plate section).

Models developed for the analysis and synthesis of images of a specific class of objects must solve two problems simultaneously:

- The image model must be able to synthesize all images that cover the whole range of possible images of the class.
- It must be possible to fit the image model to a novel image. Formally, this leads to an optimization problem with all of the associated requirements that a global minimum can be found.

Solving the Full Problem

The automated analysis of images of the human face has many aspects, and the literature on the topic is enormous. Many of the proposed methods are driven by specific technical applications such as person identification or verification, usually under the aspect of real-time performance. Especially the later constraint often requires a reduction of generality, such as the requirement of cooperative subjects, or a fixed perspective or a restriction on the number of subjects.

In contrast, the method proposed here, illustrated by an example shown in Figure 4.1, tries to develop a unifying approach for all the different aspects of face image analysis, not trading generality or accuracy against speed. Simply, the system should have no requirement on the image to be analyzed. Before giving the details of our morphable-face-model approach, we would like to describe three formal aspects for the comparison or the design of an automated face analysis system. These are the most relevant aspects for an understanding of the problem.

1. What are all the parameters of variation in face images that a method can explicitly cope with?
2. What is the formal image model to represent and separate all these different parameters?
3. What is the fitting strategy for comparing a given image with the image model?

4.2 PARAMETERS OF VARIATION IN IMAGES OF HUMAN FACES

Human faces differ in shape and texture, and additionally each individual face by itself can generate a variety of different images. This huge diversity in the appearance of face images makes the analysis difficult. Besides the general differences between individual faces, the appearance variations in images of a single face can be separated into the following four sources.

- Pose changes can result in dramatic changes in images showing different views of a face. Due to occlusions, different parts become visible or invisible, and additionally the parts seen in two views change their spatial configuration relative to each other.
- Lighting changes influence the appearance of a face even if the pose of the face is fixed. Positions and distribution of light sources around a face have the effect of changing the brightness distribution in the images, the locations of attached shadows, and specular reflections. Additionally, cast shadows can generate prominent contours in facial images.
- Facial expressions, an important tool in human communication, constitute another source of variation of facial images. Only a few facial landmarks which are directly coupled with the bony structure of the skull like the interoccular distance or the general position of the ears are constant in a face. Most other features can change their spatial configuration or position due to the articulation of the jaw or to muscle action like moving eyebrows, lips, or cheeks.
- Over a longer period of time, a face changes due to aging, to a changing hairstyle or according to makeup or accessories.

Without any requirement on the image to be analyzed, none of the parameters mentioned above can be assumed constant over a set of face images. The isolation and explicit description of all these different sources of variation must be the ultimate goal of a facial-image analysis system. For example, it is desirable not to confuse the identification of a face with expression changes, or, vice versa, the recognition of the expression of a person might by eased by first identifying the person. This implies that an image model is required that accounts for each of these variations independently by explicit parameters. Image models not able to separate two of these parameters are not able to distinguish images varying along these parameters.

4.3 TWO- OR THREE-DIMENSIONAL IMAGE MODELS

In this section we discuss image representations that try to code explicitly all parameters of face variations as mentioned above.

The most elaborate techniques for modeling images of the human face have been developed in the field of computer graphics. The most general approach, in terms of modeling all parameters explicitly, consist of a 3D mesh describing the geometry of a face and bidirectional reflectance maps of the face surface that simulate the interaction of the human skin with light [33]. Geometric parameters for modeling the surface variations between individual humans as well as physical parameters for modeling the reflectance properties of human skin are derived from empirical studies on sample sets of humans. Reflectance and geometric variations are stored in an object-centered coordinate system [24, 29]. For a long time, these *object-centered 3D*-based techniques suffered by not having photorealistic image quality. In parallel, *image-based rendering* techniques have been developed to close this gap [21, 27]. Here, light-ray intensities (pixel intensities) are collected from large samples of real photographs taken from various directions around a specific object. Since all images are calibrated in 3D, according to their position and viewing direction, each pixel intensity represents the light intensity along a certain ray in 3D space. For static objects, assuming a dense sampling of all possible 'rays', novel images from novel perspectives can be synthesized with perfect quality by reassembling the light rays for the novel perspective. Since the rays are defined in 3D world coordinates and not in a coordinate system of the object depicted, modeling of the face shape for animation or for a different person is not obvious and has so far not been reported. A similar difficulty exists for modeling light variations, since all computations are done in world coordinates, modeling the surface light interaction cannot be performed. For face animation or modeling different individuals, image-based methods seem to be of little use, and in current graphics applications the object-centered three-dimensional geometry and reflectance models seem to be superior.

For image analysis in the field of computer vision, a third type of face image model has been developed known as *linear object classes* [46, 47], as active appearance models (AAMs) [26] and as an extension of a deformable model [23]. Similar to the image based methods in computer graphics, they also start from large collections of example images of human individuals. However, instead of modeling the pixel intensities in world coordinates, a two-dimensional object-centered reference system is chosen by registering all images to a single reference face image. The model parameters are then derived by statistical analysis of the variation of the pixel intensities and the variation of the correspondence fields. Since no three-dimensional information is directly available in this process, on the one hand, illumination and individual reflectance properties are mixed, and on the other hand pose and individual face-shape variations are not separable. This convolution of external imaging parameters with individual face characteristics limits the ability to explicitly model pose or illumination and can easily result in unrealistic face images and, hence, reduce the value of these models for graphics applications. This problem was recognized for image analysis, also, and different methods for separating the external parameters have been proposed. One way is to train different models, each tuned to a specific parameter setting [14, 40, 45]. That is, for each pose, a separate model is built that accounts for the head variations of different persons. Additionally, to separate illumination variations from variations of human skin color, a separate image model should be developed for each illumination condition. These methods often assume that some external parameters are either constant or known beforehand. For the general case, when no prior knowledge is available, a huge number of different models would be required. To escape from this combinatorial explosion, two directions have been pursued. Interpolation between a few discrete image models can be applied to approximate the full range of image variations. However, handling several models simultaneously is still difficult with current computing technology. The other direction is to add explicit information on the three-dimensional nature of human faces. With a three-dimensional geometric object representation, changes in viewing direction and perspective can be directly computed by artificially rotating the object. Self-occlusion can be easily treated by using a z-buffer approach. Surface normals also can be directly derived from such a representation. Assuming additionally some information on the light sources around the face, the surface normals can be used to physically simulate the results of illumination variations. How can we obtain such a convenient three-dimensional representation from a set of faces images? Implicitly, according to theory, having the positions of corresponding points in several images of one face, the three-dimensional structure of these points can be computed. However, it has proved to be difficult to extract the three-dimensional face surface from the images using techniques such as shape from shading or structure from motion, or from corresponding points visible in several images. Recently, exploiting structure from motion techniques from video streams of moving faces, a three-dimensional image model could be built [8, 17, 48]. However, the geometric representation is still coarse, and hence a detailed evaluation of the surface normals for illumination modeling was not pursued.

Morphable models, described in this chapter [2, 5, 7, 39], integrate information on the three-dimensional nature of faces into the image model by taking a different approach in the model building step, without changing the basic structure of the image-based models. The morphable models are derived from a large set of

3D-textured laser scans instead of photographs or video images. Therefore, the difficult step of model building from raw images is eased by technical means. A direct modeling of the influence of illumination must be performed, since illumination conditions in the images to be analyzed can rarely match the standard conditions for model formation. While, in image-based models, a direct illumination modeling is impossible, it is quite straightforward using 3D morphable models. In 3D, surface normals can be directly computed, and the interaction with light can be simulated. Here the morphable model incorporates the technique that was developed in computer graphics for photorealistic rendering. Morphable models combine the high quality rendering techniques developed in computer graphics with the techniques developed in computer vision for the modeling of object classes. Morphable face models constitute a unified framework for the analysis and synthesis of facial images. Current applications lie in the field of computer graphics, for photorealistic animation of face images, and in the domain of computer vision, for face-recognition applications compensating variations across pose, illumination, and expressions.

From our current view, the question "two- or three-dimensional image models?" has two aspects that should be considered separately. These are the process of model formation and the internal structure of the models for representing the three-dimensional nature of faces. For the second aspect, we do not see any simpler approach for handling self-occlusion, perspective or illumination variation than using a three-dimensional representation that includes surface normals. For the model formation, we think that the 3D laser scans, as used in our morphable model, are not essential. In the near future, improved shape-from-shading and structure-from-motion techniques could transform sample sets of face images into a three-dimensional representation.

4.4 IMAGE ANALYSIS BY MODEL FITTING

For the remainder of this chapter, we assume that the image to be analyzed can be instantiated by the model. Therefore, the analysis problem boils down to finding the internal model parameters that reconstruct the image. That is, we have to solve an inverse problem. Since the problem is ill-posed, the inverse of the image modeling function cannot be computed analytically. It is a common strategy to apply some search techniques to find the parameters that best reconstruct the image. Instead of searching the parameter space exhaustively in a brute-force approach, we would like to use some gradient-based optimization method that guides to the solution. It is clear that not all image models are equally suited. For most image models used in graphics, it is almost impossible to tune a model to a given image, as the domain of validity of the model parameters is often disconnected and, hence, many steps require manual interactions. In computer vision, on the other hand, the structures of the different approaches are quite similar, and all image models, based on 2D or on 3D examples, lead to similar least-squares problems. First- and second-order derivatives of the image model can be computed, and the parameter domain is often modeled as a convex domain from a multivariate normal distribution obtained by a principal-component analysis. The main differences are in the strategy chosen to solve the nonlinear optimization problem. As discussed later in detail, the different methods vary from linearizing the problem to applying a Newton method exploiting second-order derivatives. The computation time as well as the accuracy in terms of convergence of the different methods vary tremendously. Simpler strategies for solving the problem also tend to handle fewer model parameters. The morphable model approach presented in this chapter does not trade efficiency against accuracy or problem simplification. We demonstrate that a high-quality image model with many parameters accounting for person identity, and modeling explicitly the pose and illumination variations, can be successfully fitted to arbitrary face images.

4.5 MORPHABLE FACE MODEL

As mentioned in the previous sections, the 3D morphable model (3DMM) is based on a series of example 3D scans represented in an object-centered system and registered to a single reference scan. A detailed description of the generation of a 3D morphable model is available in Basso et al. [2]. Briefly, to construct a 3DMM, a set of M example 3D laser scans are put into correspondence with a reference laser scan (in our case $M = 200$). This introduces a consistent labeling of all N_v 3D vertices across all the scans. The shape and texture surfaces are parametrized in the (u, v) reference frame, where one pixel corresponds to one 3D vertex (Figure 4.2). The 3D position in Cartesian coordinates of the N_v vertices of a face scan are arranged in a shape matrix, $\boldsymbol{S}$; and their color in a texture matrix, $\boldsymbol{T}$.

$$\boldsymbol{S} = \begin{pmatrix} x_1 & x_2 & \cdots & x_{N_v} \\ y_1 & y_2 & \cdots & y_{N_v} \\ z_1 & z_2 & \cdots & z_{N_v} \end{pmatrix}, \qquad \boldsymbol{T} = \begin{pmatrix} r_1 & r_2 & \cdots & r_{N_v} \\ g_1 & g_2 & \cdots & g_{N_v} \\ b_1 & b_2 & \cdots & b_{N_v} \end{pmatrix} \tag{1}$$

Having constructed a linear face space, we can make linear combinations of the shapes, $\boldsymbol{S}_i$ and the textures $\boldsymbol{T}_i$ of M example individuals to produce faces of new individuals.

$$\boldsymbol{S} = \sum_{i=1}^{M} \alpha_i \boldsymbol{S}_i, \qquad \boldsymbol{T} = \sum_{i=1}^{M} \beta_i \boldsymbol{T}_i. \tag{2}$$

Equation 2 assumes a uniform distribution of the shapes and the textures. We know that this distribution yields a model that is not restrictive enough: For

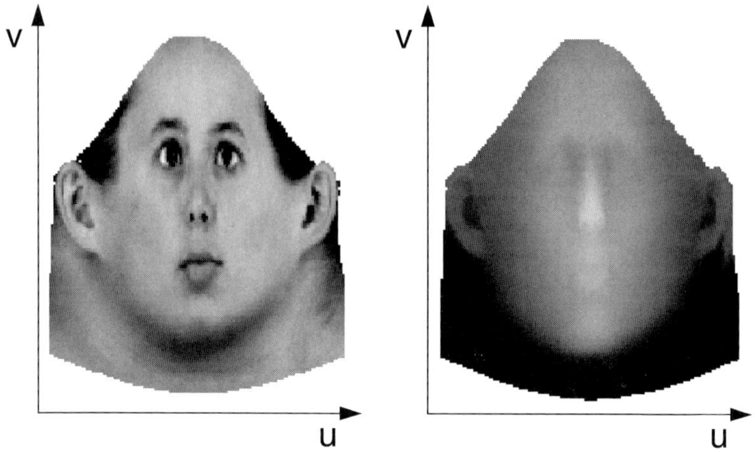

FIGURE 4.2: Texture and shape in the reference space (u, v).

instance, if some α_i or β_i is $\gg 1$, the face produced is unlikely. Therefore, we assume that the shape and the texture spaces have a Gaussian probability distribution function. Principal-component analysis (PCA) is a statistical tool that transforms the space such that the covariance matrix is diagonal (i.e., it decorrelates the data). PCA is applied separately to the shape and texture spaces. We describe the application of PCA to shapes; its application to textures is straightforward. After subtracting their average, $\overline{\boldsymbol{S}}$, the exemplars are arranged in a data matrix $\boldsymbol{A}$ and the eigenvectors of its covariance matrix $\boldsymbol{C}$ are computed using the singular-value decomposition [36] of $\boldsymbol{A}$.

$$\overline{\boldsymbol{S}} = \frac{1}{M}\sum_{i=1}^{M}\boldsymbol{S}_i, \quad \boldsymbol{a}_i = \text{vec}(\boldsymbol{S}_i - \overline{\boldsymbol{S}}), \quad \boldsymbol{A} = [\boldsymbol{a}_1, \boldsymbol{a}_2, \ldots, \boldsymbol{a}_M] = \boldsymbol{U}\boldsymbol{\Lambda}\boldsymbol{V}^{\mathrm{T}},$$
$$\boldsymbol{C}_A = \frac{1}{M}\boldsymbol{A}\boldsymbol{A}^{\mathrm{T}} = \frac{1}{M}\boldsymbol{U}\boldsymbol{\Lambda}^2\boldsymbol{U}^{\mathrm{T}}. \tag{3}$$

The component vec($\boldsymbol{S}$) vectorized $\boldsymbol{S}$ by stacking its columns. The M columns of the orthogonal matrix $\boldsymbol{U}$ are the eigenvectors of the covariance matrix $\boldsymbol{C}_A$, and $\sigma_i^2 = \lambda_i^2/M$ are its eigenvalues, where the λ_i are the elements of the diagonal matrix $\boldsymbol{\Lambda}$, arranged in decreasing order. Now, instead of representing the data matrix $\boldsymbol{A}$ in its original space, it can be projected to the space spanned by the eigenvectors of its covariance matrix. Let us denote by the matrix $\boldsymbol{B}$ this new representation of the data, and by $\boldsymbol{C_B}$ its covariance matrix.

$$\boldsymbol{B} = \boldsymbol{U}^{\mathrm{T}}\boldsymbol{A} \qquad \boldsymbol{C_B} = \frac{1}{M}\boldsymbol{B}\boldsymbol{B}^{\mathrm{T}} = \frac{1}{M}\boldsymbol{\Lambda}^2. \tag{4}$$

The second equality of the last equation is obtained because $\boldsymbol{U}^{\mathrm{T}}\boldsymbol{U} = \boldsymbol{I}$ and $\boldsymbol{V}^{\mathrm{T}}\boldsymbol{V} = \boldsymbol{I}$, where $\boldsymbol{I}$ is the identity matrix with the appropriate number of elements. Hence the projected data are decorrelated, as they have a diagonal covariance matrix. Hereafter, we denote by $\sigma_{\mathrm{S},i}$ and $\sigma_{\mathrm{T},i}$ the variances of, respectively, the shape and the texture vectors.

There is a second advantage in expressing a shape (or texture) as a linear combination of shape principal components, namely dimensionality reduction. It is demonstrated in [16] that the subspace, spanned by the columns of an orthogonal matrix $\boldsymbol{X}$, with N dimensions, that minimizes the mean squared difference between a data vector $\boldsymbol{a}$, sampled from the same population as the column vectors of the matrix $\boldsymbol{A}$, and its reconstruction, $\boldsymbol{X}\boldsymbol{X}^{\mathrm{T}}\boldsymbol{a}$, is the one formed by the N eigenvectors having largest eigenvalues: $\boldsymbol{X} = \left[\boldsymbol{U}_{\cdot,1}, \ldots, \boldsymbol{U}_{\cdot,N}\right]$, where $\boldsymbol{U}_{\cdot,i}$ denotes the ith column of $\boldsymbol{U}$.

Let us denote $\boldsymbol{U}_{\cdot,i}$, the column i of $\boldsymbol{U}$, and the principal component i, reshaped into a $3 \times N_{\mathrm{V}}$ matrix, by $\boldsymbol{S}^i = \boldsymbol{U}_{\cdot,i}^{(3)}$. The notation $\boldsymbol{a}_{m \times 1}^{(n)}$ [30] folds the $m \times 1$ vector $\boldsymbol{a}$ into an $n \times (m/n)$ matrix.

Now, instead of describing a novel shape and texture as a linear combination of examples, as in Equation 2, we express them as a linear combination of N_{S} shape and N_{T} texture principal components:

$$\boldsymbol{S} = \overline{\boldsymbol{S}} + \sum_{i=1}^{N_{\mathrm{S}}} \alpha_i \boldsymbol{S}^i, \qquad \boldsymbol{T} = \overline{\boldsymbol{T}} + \sum_{i=1}^{N_{\mathrm{T}}} \beta_i \boldsymbol{T}^i. \tag{5}$$

The third advantage of this formulation is that the probabilities of a shape and a texture are readily available from their parameters:

$$p(\boldsymbol{S}) \sim e^{-\frac{1}{2}\sum_i \frac{\alpha_i^2}{\sigma_{\mathrm{S},i}^2}}, \qquad p(\boldsymbol{T}) \sim e^{-\frac{1}{2}\sum_i \frac{\beta_i^2}{\sigma_{\mathrm{T},i}^2}}. \tag{6}$$

4.5.1 Segmented Morphable Model

As mentioned, our morphable model is derived from statistics computed on 200 example faces. As a result, the dimensions of the shape and texture spaces, N_{S} and N_{T}, are limited to 199. This might not be enough to account for the rich variation of individuals present in mankind. Naturally, one way to augment the dimension of the face space would be to use 3D scans of more persons, but they were not available in our experiments. Hence, we resort to another scheme: We segment the face into four regions (nose, eyes, mouth and the rest) and use a separate set of shape and texture coefficients to code them [7]. This method multiplies by four the expressiveness of the morphable model. We denote the shape and texture parameters by α and β when they can be used interchangeably for the global and the segmented parts of the model. When we want to distinguish them, we use, for the shape parameters, α^{g} for the global model (full face) and $\alpha^{\mathrm{S}1}$ to $\alpha^{\mathrm{S}4}$ for the segmented parts (the same notation is used for the texture parameters).

4.5.2 Morphable Model to Synthesize Images

One part of the analysis by synthesis loop is the synthesis (i.e., the generation of accurate face images viewed from any pose and illuminated by any condition). This process is explained in this section.

Shape Projection

To render the image of a face, the 3D shape is projected to the 2D image frame. This is performed in two steps. First, a 3D rotation and translation (i.e., a rigid transformation) maps the object-centered coordinates, $\boldsymbol{S}$, to a position relative to the camera in world coordinates:

$$\boldsymbol{W} = \boldsymbol{R}_{\gamma}\boldsymbol{R}_{\theta}\boldsymbol{R}_{\phi}\boldsymbol{S} + \boldsymbol{t}_{\mathrm{w}}\boldsymbol{1}_{1\times N_{\mathrm{v}}}. \tag{7}$$

The angles ϕ and θ control in-depth rotations around the vertical and horizontal axes, and γ defines a rotation around the camera axis; $\boldsymbol{t}_{\mathrm{w}}$ is a 3D translation. A perspective projection then maps a vertex i to the image plane in (x_i, y_i):

$$x_i = t_x + f\frac{w_{1,i}}{w_{3,i}}, \qquad y_i = t_y + f\frac{w_{2,i}}{w_{3,i}}, \tag{8}$$

where f is the focal length of the camera (located in the origin), (t_x, t_y) defines the image-plane position of the optical axis, and $w_{k,i}$ denotes the (k, i) element of W.

For ease of explanation, the shape-transformation parameters are denoted by the vector $\boldsymbol{\rho} = [f, \phi, \theta, \gamma, t_x, t_y, \boldsymbol{t}_{\mathrm{w}}^{\mathrm{T}}]^{\mathrm{T}}$, and $\boldsymbol{\alpha}$ is the vector whose elements are the $\boldsymbol{\alpha}_i$. The projection of the vertex i to the image frame (x, y) is denoted by the vector-valued function $\boldsymbol{p}(u_i, v_i; \boldsymbol{\alpha}, \boldsymbol{\rho})$. This function is clearly continuous in $\boldsymbol{\alpha}$, $\boldsymbol{\rho}$. To provide continuity in the (u, v) space as well, we use a triangle list and interpolate between neighboring vertices, as is common in computer graphics. Note that only N_{vv} vertices, a subset of the N_{v} vertices, are visible after the 2D projection (the remaining vertices are hidden by self-occlusion). We call this subset the domain of the shape projection $\boldsymbol{p}(u_i, v_i; \boldsymbol{\alpha}, \boldsymbol{\rho})$ and denote it by $\Omega(\boldsymbol{\alpha}, \boldsymbol{\rho}) \in (u, v)$.

In conclusion, the shape modeling and its projection provides a mapping from the parameter space $\boldsymbol{\alpha}, \boldsymbol{\rho}$ to the image frame (x, y) via the reference frame (u, v). However, to synthesize an image, we need the inverse of this mapping, detailed in the next section.

Inverse Shape Projection

The shape projection aforementioned maps a (u, v) point from the reference space to the image frame. To synthesize an image, we need the inverse mapping: An image is generated by looping on the pixels (x, y). To know which color must be drawn on that pixel, we must know where this pixel is mapped into the reference frame. This is the aim of the inverse shape mapping explained in this section.

The inverse shape projection, $\boldsymbol{p}^{-1}(x, y; \boldsymbol{\alpha}, \boldsymbol{\rho})$, maps an image point (x, y) to the reference frame (u, v). Let us denote the composition of a shape projection and its inverse by the symbol $\circ$; hence, $\boldsymbol{p}(u, v; \boldsymbol{\alpha}, \boldsymbol{\rho}) \circ \boldsymbol{p}^{-1}(x, y; \boldsymbol{\alpha}, \boldsymbol{\rho})$ is equal to $\boldsymbol{p}(\boldsymbol{p}^{-1}(x, y; \boldsymbol{\alpha}, \boldsymbol{\rho}); \boldsymbol{\alpha}, \boldsymbol{\rho})$, but we prefer the former notation for clarity. The inverse shape projection is defined by the following equation, which specifies that under the same set of parameters the shape projection composed with its inverse is equal to the identity.

$$\begin{aligned} \boldsymbol{p}(u, v; \boldsymbol{\alpha}, \boldsymbol{\rho}) \circ \boldsymbol{p}^{-1}(x, y; \boldsymbol{\alpha}, \boldsymbol{\rho}) &= (x, y), \\ \boldsymbol{p}^{-1}(x, y; \boldsymbol{\alpha}, \boldsymbol{\rho}) \circ \boldsymbol{p}(u, v; \boldsymbol{\alpha}, \boldsymbol{\rho}) &= (u, v). \end{aligned} \tag{9}$$

Because the shape is discrete, it is not easy to express $\boldsymbol{p}^{-1}(\cdot)$ analytically as a function of $\boldsymbol{p}(\cdot)$, but it can be computed using the triangle list: The domain of the plane (x, y) for which there exists an inverse under the parameters $\boldsymbol{\alpha}$ and $\boldsymbol{\rho}$, denoted by $\Psi(\boldsymbol{\alpha}, \boldsymbol{\rho})$, is the range of $\boldsymbol{p}(u, v; \boldsymbol{\alpha}, \boldsymbol{\rho})$. Such a point of (x, y) lies in a single visible triangle under the projection $\boldsymbol{p}(u, v; \boldsymbol{\alpha}, \boldsymbol{\rho})$. Therefore, the point in (u, v) under the inverse projection has the same relative position in this triangle in the (u, v) space. This process is depicted in Figure 4.3.

Illumination Modeling and Color Transformation

Ambient and directed light. We simulate the illumination of a face using an ambient light and a directed light. We use the standard Phong reflectance model that accounts for the diffuse and specular reflection on a surface [18]. The parameters of this model are the intensity of the ambient light ($L_{\mathrm{r}}^{\mathrm{a}}$, $L_{\mathrm{g}}^{\mathrm{a}}$, $L_{\mathrm{b}}^{\mathrm{a}}$), the intensity of the directed light ($L_{\mathrm{r}}^{\mathrm{d}}$, $L_{\mathrm{g}}^{\mathrm{d}}$, $L_{\mathrm{b}}^{\mathrm{d}}$), its direction ($\theta_{\mathrm{l}}$ and ϕ_{l}), the specular reflectance of human skin (k_{s}), and the angular distribution of the specular reflections of human skin (ν). For clarity, we denote by the vector $\boldsymbol{t}_i$ the ith column of the matrix $\boldsymbol{T}$, representing the RGB color of the vertex i. When illuminated, the color of this

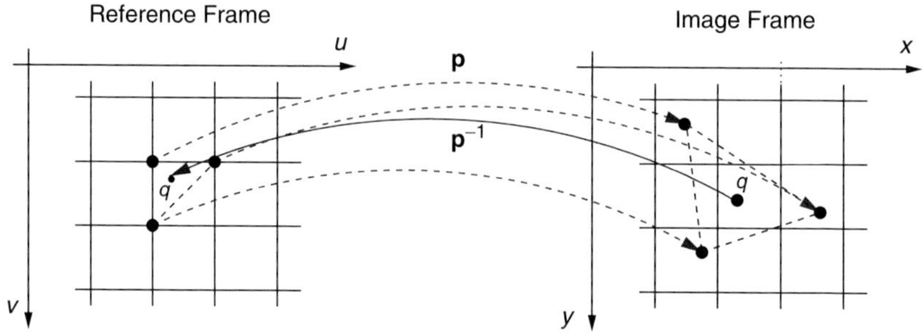

FIGURE 4.3: Inverse shape function $\boldsymbol{p}^{-1}(x, y; \boldsymbol{\alpha}, \boldsymbol{\rho})$ maps the point q (defined in the (x, y) coordinate system), onto the point q' in (u, v). This is done by recovering the triangle that would contain the pixel q under the mapping $\boldsymbol{p}(u, v; \boldsymbol{\alpha}, \boldsymbol{\rho})$. Then the relative position of q in that triangle is the same as the relative position of q' in the same triangle in the (u, v) space.

vertex is transformed to $\boldsymbol{t}_i^{\mathrm{I}}$.

$$\boldsymbol{t}_i^I = \begin{pmatrix} L_{\mathrm{r}}^{\mathrm{a}} & 0 & 0 \\ 0 & L_{\mathrm{g}}^{\mathrm{a}} & 0 \\ 0 & 0 & L_{\mathrm{b}}^{\mathrm{a}} \end{pmatrix} \boldsymbol{t}_i + \begin{pmatrix} L_{\mathrm{r}}^{\mathrm{d}} & 0 & 0 \\ 0 & L_{\mathrm{g}}^{\mathrm{d}} & 0 \\ 0 & 0 & L_{\mathrm{b}}^{\mathrm{d}} \end{pmatrix} \left(\langle \boldsymbol{n}_i^{\mathrm{v,w}}, \boldsymbol{d} \rangle \boldsymbol{t}_i + k_s \langle \boldsymbol{r}_i, \boldsymbol{v}_i \rangle^{\nu} \mathbf{1}_{3\times 1} \right). \quad (10)$$

The first term of this equation is the contribution of the ambient light. The first term of the last parenthesis is the diffuse component of the directed light, and the second term is its specular component. To take account of the attached shadows, these two scalar products are lower-bounded to zero. To take account of the cast shadows, a shadow map is computed, using standard computer-graphic techniques [18]. The vertices in shadows are illuminated by the ambient light only.

In Equation 10, $\boldsymbol{d}$ is the unit-length light direction in Cartesian coordinates, which can be computed from its spherical coordinates by

$$\boldsymbol{d} = \begin{pmatrix} \cos(\theta_l)\sin(\phi_l) \\ \sin(\theta_l) \\ \cos(\theta_l)\cos(\phi_l) \end{pmatrix}. \quad (11)$$

The normal of the vertex i, $\boldsymbol{n}_i^{\mathrm{v,w}}$, is expressed in world coordinates. World coordinates of a normal are obtained by rotating the normal from the object-centered coordinates to the world coordinates:

$$\boldsymbol{n}_i^{\mathrm{v,w}} = \boldsymbol{R}_\gamma \boldsymbol{R}_\theta \boldsymbol{R}_\phi \boldsymbol{n}_i^{\mathrm{v}}. \quad (12)$$

The normal of a vertex in object-centered coordinates, $\boldsymbol{n}_i^{\mathrm{v}}$, is defined as the unit-length mean of the normals of the triangles connected to this vertex (i.e., the triangles for which this vertex is one of the three corners):

$$\boldsymbol{n}_i^{\mathrm{v}} = \frac{\sum_{j \in \mathcal{T}_i} \boldsymbol{n}_j^{\mathrm{t}}}{\| \sum_{j \in \mathcal{T}_i} \boldsymbol{n}_j^{\mathrm{t}} \|}, \tag{13}$$

where $\mathcal{T}_i$ is the set of triangle indexes connected to the vertex i. The normal of the triangle j, denoted by $\boldsymbol{n}_j^t$, is determined by the unit-length cross product of the vectors formed by two of its edges. If $\boldsymbol{s}_{i_1}$, $\boldsymbol{s}_{i_2}$, and $\boldsymbol{s}_{i_3}$ are the Cartesian object-centered coordinates of the three corners of the triangle j (these indexes are given by the triangle list), then its normal is

$$\boldsymbol{n}_j^t = \frac{(\boldsymbol{s}_{i_1} - \boldsymbol{s}_{i_2}) \times (\boldsymbol{s}_{i_1} - \boldsymbol{s}_{i_3})}{\|(\boldsymbol{s}_{i_1} - \boldsymbol{s}_{i_2}) \times (\boldsymbol{s}_{i_1} - \boldsymbol{s}_{i_3})\|}. \tag{14}$$

In Equation (10), $\boldsymbol{v}_i$ is the viewing direction of the vertex i, which is the unit-length direction connecting the vertex i to the camera centerer. The camera center is at the origin of the world coordinates:

$$\boldsymbol{v}_i = -\frac{\boldsymbol{W}_{\cdot,i}}{\|\boldsymbol{W}_{\cdot,i}\|}. \tag{15}$$

The vector $\boldsymbol{r}_i$ in Equation 10 is the direction of the reflection of the light coming from the direction $\boldsymbol{d}$, computed as

$$\boldsymbol{r}_i = 2 \cdot \langle \boldsymbol{n}_i^{\mathrm{v,w}}, \boldsymbol{d} \rangle \boldsymbol{n}_i^{\mathrm{v,w}} - \boldsymbol{d}. \tag{16}$$

Color transformation. Input images may vary a lot with respect to the overall tone of color. To be able to handle a variety of color images as well as gray-level images and even paintings, we apply gains g_{r}, g_{g}, g_{b}, offsets o_{r}, o_{g}, o_{b}, and a color contrast c to each channel [7]. This is a linear transformation that yields the definitive color of a vertex, denoted by $\boldsymbol{t}_i^{\mathrm{C}}$. It is obtained by multiplying the RGB color of a vertex after it has been illuminated, $\boldsymbol{t}_i^{\mathrm{I}}$, by the matrix $\boldsymbol{M}$ and adding the vector $\boldsymbol{o} = [o_{\mathrm{r}}, o_{\mathrm{g}}, o_{\mathrm{b}}]^{\mathrm{T}}$.

$$\boldsymbol{t}_i^{\mathrm{C}} = \boldsymbol{M}\boldsymbol{t}_i^{\mathrm{I}} + \boldsymbol{o}, \text{ where} \tag{17}$$

$$\boldsymbol{M} = \begin{pmatrix} g_{\mathrm{r}} & 0 & 0 \\ 0 & g_{\mathrm{g}} & 0 \\ 0 & 0 & g_{\mathrm{b}} \end{pmatrix} \left[\boldsymbol{I} + (1 - c) \begin{pmatrix} 0.3 & 0.59 & 0.11 \\ 0.3 & 0.59 & 0.11 \\ 0.3 & 0.59 & 0.11 \end{pmatrix} \right]. \tag{18}$$

For brevity, the illumination and color transformation parameters are regrouped in the vector $\boldsymbol{\iota}$. Hence the illuminated and color-corrected texture depends on the coefficients of the texture linear combination regrouped in $\boldsymbol{\beta}$, on the light parameters $\boldsymbol{\iota}$, and on $\boldsymbol{\alpha}$ and $\boldsymbol{\rho}$ used to compute the normals and the viewing direction of the vertices required for the Phong illumination model.

Similar to the shape, the color of a vertex i, t_i^{C}, is represented on the (u, v) reference frame by the vector valued function $\boldsymbol{t}^{\mathrm{C}}(u_i, v_i; \boldsymbol{\beta}, \boldsymbol{\iota}, \boldsymbol{\alpha}, \boldsymbol{\rho})$, which is extended to the continuous function $\boldsymbol{t}^{\mathrm{C}}(u, v; \boldsymbol{\beta}, \boldsymbol{\iota}, \boldsymbol{\alpha}, \boldsymbol{\rho})$ by using the triangle list and interpolating.

Image Synthesis

Synthesizing the image of a face is performed by mapping a texture from the reference to the image frame using an inverse shape projection.

$$I^{\mathrm{m}}(x_j, y_j; \boldsymbol{\alpha}, \boldsymbol{\beta}, \boldsymbol{\rho}, \boldsymbol{\iota}) = \boldsymbol{t}^{\mathrm{C}}(u, v; \boldsymbol{\beta}, \boldsymbol{\iota}, \boldsymbol{\alpha}, \boldsymbol{\rho}) \circ \boldsymbol{p}^{-1}(x_j, y_j; \boldsymbol{\alpha}, \boldsymbol{\rho}), \tag{19}$$

where j runs over the pixels that belong to $\Psi(\boldsymbol{\alpha}, \boldsymbol{\rho})$ (i.e., the pixels for which a shape inverse exist, as defined in Section 4.5.2).

4.6 COMPARISON OF FITTING ALGORITHM

The previous section detailed the 3D morphable model, a mathematical formulation of the full image formation process. It takes into account most of the sources of facial image variations (flexible shape deformation, varying albedo, 3D rotation, and directed lights). A face-recognition algorithm is intrinsically a method that is inverting this image formation process, i.e., inverting Equation 19, thereby separating the identity from the imaging parameters. The first algorithm that inverted the full image formation process, and that made the least assumptions, treating the problem in its full complexity, is the stochastic Newton optimization (SNO) [4, 6, 7]. It cast the task as an optimization problem estimating all the model parameters (shape, texture, rigid transformation and illumination). The only assumption made is that the pixels are independent and identically distributed with a residual normally distributed with a variance equal to σ_I^2. This is performed by maximizing the posterior of the parameters given the image, thereby minimizing the following energy function:

$$E = \min_{\boldsymbol{\alpha}, \boldsymbol{\beta}, \boldsymbol{\rho}, \boldsymbol{\iota}} \frac{1}{\sigma_{\mathrm{I}}^2} \sum_{x,y} \|I^{\mathrm{m}}(x, y; \boldsymbol{\alpha}, \boldsymbol{\beta}, \boldsymbol{\rho}, \boldsymbol{\iota}) - I(x, y)\|^2 + \sum_i \frac{\alpha_i^2}{\sigma_{\mathrm{S},i}^2} + \sum_i \frac{\beta_i^2}{\sigma_{\mathrm{T},i}^2}. \tag{20}$$

This is a difficult and computationally expensive optimization problem, as the energy function is nonconvex. To avoid the local minima problem, a stochastic optimization algorithm is used: At each iteration of the fitting algorithm, the energy function and its derivatives are evaluated on a very small set of points (40 points) that are randomly chosen. This step introduces a perturbation on the derivatives that minimizes the risk of locking on a local minimum. The price to pay is a high computational expense, with several thousand iterations.

Over the past years, many fitting algorithms have been presented. Having computational efficiency and tractability as main goals, these methods restrain the domain of applicability and make several assumptions. It is interesting to analyze the assumptions made by each of these methods and their limitations in light of the 3DMM.

Point-Distribution Model and Active-Shape Model

Craw and Cameron [15] were the first to align some landmark pixels on a set of face images. Then they applied PCA on the textures put in coarse correspondence by sampling them on a reference frame defined by a few landmarks. Cootes et al. [9, 13] represented 2D face shapes as a sparse set of feature points that corresponded to one another across a face image ensemble. Applying PCA on this set of points yielded a point-distribution model (PDM).

Though it can be argued that there is no major conceptual difference between the PDMs of the early 1990s and the 3D shape model of the 3DMM, as both models compute statistics on a set of points in correspondence, yet there are three major differences between the two models: (1) The use of 3D rather than 2D enabling accurate out of the image plane rotation and illumination modeling. (2) The 3DMM is a dense model including several thousand vertices, whereas the PDM uses a few tens. (3) As implemented by Cootes et al., the PDM includes landmarks on the contour between the face and the background. This contour depends on the pose of the face on the image and on the 3D shape of the individual. Thus the landmarks on it do not represent the same physical points, and hence should not be put in correspondence across individuals. The authors of the PDM argue that it is capable of handling a $\pm 20°$ rotation out of the image plane. However, the 2D shape variation induced by this 3D rigid transformation is encoded in the 2D shape parameters, resulting in shape parameters not independent of the face pose. As explained in the first sections of this chapter, this reduces the identification capabilities of this model.

The first algorithm used to fit a PDM to an image is called the active-shape model (ASM), and its first version appeared in Cootes and Taylor [11]. This version used the edge evidence of the input image to adjust the 2D translation, 2D scale, image-plane rotation, and shape parameters of the model. Each iteration of this fitting algorithm includes two steps: First, each model point is displaced in the direction normal to the shape contour, toward the strongest edge, and with an amplitude proportional to the edge strength at the point. This yields a position for each model point that would fit better to the image. The second step is to update the 2D pose and shape parameters reflecting the new position of the model points. To increase the likelihood of the resulting shape, hard limits are put on the shape parameters. This method proved itself not robust enough to deal with complex objects, where the model points do not necessarily lie on strong edges. Therefore, the second version of the ASM [12] modeled the gray levels along the normal of the contour at each model point. Then, during model search, a better position for a model point was given by the image point, along the normal of the contour, that minimized the distance to its local gray-level model. An exhaustive search for this minimizer was performed that evaluated all the points along the normal within a given distance of the model point. The PCA model was used to model a gray-level profile along the normal of a contour point. Then the distance minimized during fitting is the norm of the reconstruction error of a gray-level profile obtained from a point along the normal evaluated as a potential minimizer. A predicament of the local gray-level models, as they were implemented in the ASM, is the fact that one half of the pixels modeled are outside the face area, in the background area, and hence could change randomly from image to image.

The first identification experiment on facial images fitted by an ASM was made by Lanitis et al. [25]. After fitting an image by the ASM, the gray texture enclosed within the face contour was sampled on the reference frame defined by the mean shape (and called shape-free texture) and modeled by a PCA model. The features used for identification were the shape and texture coefficients of the shape and texture models.

Active Appearance Model

Continuing in this direction, researchers started to use not only the pixels along the normal of the landmark points, but rather the full texture in the face area to drive the parameters fitting algorithm. The main motivation was that, as the algorithm would use more information, the fitting would converge faster to a more accurate minimum more robustly. First, Gleicher [20], with the image-difference-decomposition (IDD) algorithm, then Sclaroff and Isidoro [42], with the active blobs, and Cootes et al. [10], with the AAM, used the full texture error to compute, linearly, an update of the model parameters. The texture error, $\delta \boldsymbol{t}$, is the difference between the texture extracted from the input image and sampled using the shape parameters and the model texture:

$$\delta \boldsymbol{t} = I(x, y) \circ \boldsymbol{p}(u, v; \boldsymbol{\alpha}) - \boldsymbol{t}(u, v; \boldsymbol{\beta}). \tag{21}$$

As this algorithm is applied to the AAM, which neither models rotation out of the image plane nor illumination, this last equation depends neither on $\boldsymbol{\rho}$ nor on $\boldsymbol{\iota}$.

This equation defining the texture error is the same as the term of the SNO energy function, inside the norm, with the difference that the texture error is sampled in the reference frame, not in the image frame.

The aim of the fitting algorithm is to estimate the shape and texture model parameters that minimize the square of the norm of the texture error:

$$\min_{\boldsymbol{\alpha},\boldsymbol{\beta}} \|\delta \boldsymbol{t}\|^2. \tag{22}$$

For efficiency reasons, the texture difference is projected onto a constant matrix, which yields the update for the shape and texture model parameters:

$$\begin{pmatrix} \delta\boldsymbol{\alpha} \\ \delta\boldsymbol{\beta} \end{pmatrix} = \boldsymbol{A}\delta\boldsymbol{t}. \tag{23}$$

Then the next estimate is obtained by adding the update to the current estimate.

$$\boldsymbol{\alpha} \leftarrow \boldsymbol{\alpha} + \delta\boldsymbol{\alpha} \quad \text{and} \quad \beta \leftarrow \boldsymbol{\beta} + \delta\boldsymbol{\beta}. \tag{24}$$

This algorithm would be a gradient-descent optimization algorithm, if the matrix relating the texture error to the model update, the matrix $\boldsymbol{A}$, was the inverse of the Jacobi matrix. Assuming $\boldsymbol{A}$ to be constant, is equivalent to assuming that the model Jacobi matrix is constant and hence that the rendering model is linear. However, the sources of nonlinearities of the rendering model are multiple: (1) The rotation out of the image plane and the perspective effects induce a nonlinear variation of the shape points projected onto the image plane. (2) The modification of the light-source direction produce a nonlinear variation in pixel intensities. (3) The warping of the texture using the shape parameters is nonlinear as well. As the face model, on which this fitting algorithm is applied, is 2D, allows only small out of the image plane rotation, and does not model directed light sources, the first two sources of nonlinearities could be limited. Thus, the authors showed that this fitting was effective on facial images with no directed light and with small pose variations. However, it does not produce satisfactory results on the full 3D problem addressed in this chapter.

The constant matrix relating the texture error to the model update, the matrix $\boldsymbol{A}$, was first computed by a regression using texture errors generated from training images and random model-parameter displacement. Then, it was estimated by averaging Jacobian matrices obtained by numerical differentiation on typical facial images. A problem with these two approaches is that not only the pixels inside the face area, but also the ones outside, on the background area, are sampled to form the training texture error. Thus, the quality of the estimate obtained on an input image depends on the resemblance of the background of this image to one of the images used for computing the matrix $\boldsymbol{A}$.

As mentioned in the introduction of this chapter, the AAM algorithm is an instance of a fitting algorithm that favors efficiency over accuracy and generality.

To enlarge the domain of application of AAM to faces viewed from any azimuth angle, Cootes et al. [14] introduced the multiview AAM. It is constituted of five AAMs, each trained on facial images at different poses (front, left and right oblique views, and left and right profile). So, the fitting was also constituted of five constant Jacobi matrices. This is an ad-hoc solution addressing one of the limitations of a 2D model that was not pursued afterward.

Inverse Compositional Image Alignment Algorithm

As aforementioned, for efficiency reasons, the AAM treats the matrix relating the texture error to the model-parameter update as a constant. This is based on the assumption that the Jacobi matrix (which should be recomputed at each iteration) is well approximated by a constant matrix. However, this matrix is not constant owing to the warping of the texture by the shape. Baker and Matthews [1] introduced the inverse compositional image alignment (ICIA) algorithm, which also uses a constant Jacobi matrix, but, here, the matrix is shown, to a first order, to be constant. The fixedness of the updating matrix is not assumed anymore. This was achieved by a modification of the cost function to be minimized. Instead of optimizing Equation 22, the following cost function is minimized:

$$\min_{\delta\boldsymbol{\alpha}} \|\boldsymbol{t}(u,v;0) \circ \boldsymbol{p}(u,v;\delta\boldsymbol{\alpha}) - I(x,y) \circ \boldsymbol{p}(u,v;\boldsymbol{\alpha})\|^2. \tag{25}$$

To clarify the notations, in what follows, the dependency on the frame coordinates (x, y) and (u, v) is not explicit anymore; only the dependency on the model parameters is left. A first Taylor expansion of the term inside the norm of the cost function yields

$$\min_{\delta\boldsymbol{\alpha}} \|\boldsymbol{t}(0) \circ \boldsymbol{p}(0) + \boldsymbol{t}(0) \left.\frac{\partial \boldsymbol{p}}{\partial \boldsymbol{\alpha}}\right|_{\boldsymbol{\alpha}=0} \delta\boldsymbol{\alpha} - I \circ \boldsymbol{p}(\boldsymbol{\alpha})\|^2. \tag{26}$$

Differentiating this cost function with respect to the shape-parameter update, equating to zero, and rearranging the terms yields the expression of the parameter update:

$$\delta\boldsymbol{\alpha} = \left[\boldsymbol{t}(0) \left.\frac{\partial \boldsymbol{p}}{\partial \boldsymbol{\alpha}}\right|_{\boldsymbol{\alpha}=0}\right]^{\dagger} (I \circ \boldsymbol{p}(\boldsymbol{\alpha}) - \boldsymbol{t}(0) \circ \boldsymbol{p}(0)), \tag{27}$$

where the notation † denotes the pseudoinverse of a matrix. This derivation leads to a Gauss–Newton optimization [19]. In a least-squares optimization, the Hessian is a sum of two terms: the Jacobi matrix transposed and multiplied by itself and the purely second-derivative terms. In Gauss–Newton optimization, the Hessian is approximated by its first term only. The reason is that, in a least-squares optimization (such as $\min_x \sum_i e_i^2$), the second-derivative term of the Hessian of the cost function is the sum of the Hessian matrices of the elements, $\partial^2 e_i/\partial x^2$, multiplied by their residual, e_i. Near to the minimum, the residuals e_i are small, and the approximation is adequate. A Gauss–Newton optimization algorithm is more efficient than a Newton algorithm, as the second derivatives are not computed. It is not surprising that the ICIA algorithm is equivalent to a Gauss–Newton optimization, as it was derived using a first-order Taylor approximation.

The essence of the ICIA algorithm, is that the Jacobi matrix, $\boldsymbol{t}(0)\ \partial \boldsymbol{p}/\partial \boldsymbol{\alpha}|_{\boldsymbol{\alpha}=0}$ does not depend either on the current estimate of the model parameters, nor on the input image. It is then constant throughout the fitting algorithm. This constancy is not assumed but derived from a first-order expansion. The update is, then, not added to the current estimate as in the case of the AAM fitting algorithm, but composed with the current estimate:

$$\boldsymbol{p}(u, v; \boldsymbol{\alpha}) \leftarrow \boldsymbol{p}(u, v; \boldsymbol{\alpha}) \circ \boldsymbol{p}^{-1}(u, v; \delta\boldsymbol{\alpha}). \tag{28}$$

Baker et al., in [1], do not make the distinction between the reference and the image frames. A consequence of this is that they require the set of warps to be closed under inversion. This leads them to a first-order approximation of the inverse shape projection (called inverse warping in their nomenclature): $\boldsymbol{p}^{-1}(x, y; \boldsymbol{\alpha}) = \boldsymbol{p}(u, v; -\boldsymbol{\alpha})$. This does not agree with the identity defining the inverse shape projection of Equation 9: As shown in Figure 4.4, a point from (u, v), q', is mapped under $\boldsymbol{p}(u, v; \boldsymbol{\alpha})$ to q in (x, y). Hence, to agree with the identity, this point q must be warped back to q' under $\boldsymbol{p}^{-1}(x, y; \boldsymbol{\alpha})$. So, the displacement in q which should be inverted is the one from q'. However, in Baker et al. [1], the displacement function $\boldsymbol{p}$ is inverted at the point q, leading to the point b, instead of q'. This is due to the fact that the distinction between the two coordinate systems

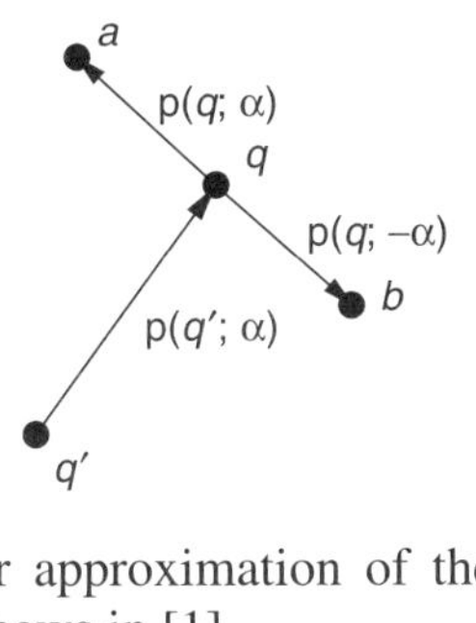

FIGURE 4.4: First-order approximation of the inverse shape projection defined by Baker and Matthews in [1].

is not made. This approximation is less problematic for a sparse-correspondence model as used by Baker for which the triangles are quite large (see plot (b) of Figure 2 of [1]), because the chances that both q and q' fall in the same triangle are much higher than in our dense-correspondence model for which the triangles are much tinier. When q and q' fall in the same triangle, then their displacements are similar to a first-order approximation, due to the linear interpolation inside triangles, and the error made during composition is small.

The improvement of ICIA over AAM is that, as the updating matrix is derived mathematically from the cost function and not learned over a finite set of examples, the algorithm is more accurate and requires fewer iterations to converge. The fact that the Jacobi matrix is constant is the first factor of the efficiency of the ICIA algorithm. The second factor is the fact that only the shape parameters are iteratively updated. The texture parameters are estimated in a single step after the recovery of the shape parameters. This is achieved by making the shape Jacobi matrix $\boldsymbol{t}(0) \left.\frac{\partial \boldsymbol{p}}{\partial \boldsymbol{\alpha}}\right|_{\boldsymbol{\alpha}=0}$ orthogonal to the texture Jacobi matrix, by projecting out the shape Jacobi matrix onto the texture Jacobi matrix. It is therefore called the *project-out* method. This induces a perturbation on the shape Jacobi matrix. However, if the texture model has few components (Baker and Matthews use fewer than ten components), then the error is small.

ICIA is an efficient algorithm. However, its domain of application is limited. It is a fitting algorithm for 2D AAM, and hence cannot handle rotation out of the image plane and directed light. After discussion with Simon Baker, it also appears that it achieves its best performance when fitting images of individuals used for training the model. On novel subjects, the performance and accuracy are comparable to the AAM fitting algorithm. The ICIA is hence a person-specific fitting algorithm.

ICIA applied to the 3DMM

The ICIA fitting algorithm was adapted to use the 3DMM in [39]. Several modifications of the original algorithm had to be made in order to obtain accurate results. The first one was to use the precise inverse shape projection defined in Section 4.5.2, which makes the distinction between the reference and image frames. The use of this inverse shape projection leads to the definition of the following cost function:

$$\|\boldsymbol{t}(u, v; \Delta\boldsymbol{\beta}) \circ \boldsymbol{p}^{-1}(x, y; \boldsymbol{\gamma}^{\mathrm{d}}) \circ \boldsymbol{p}(u, v; \boldsymbol{\gamma}^{\mathrm{d}} + \Delta\boldsymbol{\gamma}) - \boldsymbol{t}^{-1}(I(x, y) \circ \boldsymbol{p}(u, v; \boldsymbol{\gamma}); \boldsymbol{\beta})\|^2, \tag{29}$$

where $\boldsymbol{\gamma}$ is the vector formed by the concatenation of the shape parameters, $\boldsymbol{\alpha}$, and the rigid transformation parameters, $\boldsymbol{\rho}$. The derivatives are precomputed at the parameters $\boldsymbol{\gamma}^{\mathrm{d}}$. Projecting out the shape Jacobi matrix onto the texture Jacobi matrix would significantly perturb the shape update, hence the project out method was not used. Thus, the texture parameters, $\boldsymbol{\beta}$, are also iteratively updated. The texture parameter update is denoted by the vector $\Delta\boldsymbol{\beta}$. This required the definition of the inverse texture transformation:

$$t^{-1}(t(u_i, v_i); \boldsymbol{\beta}) = t(u_i, v_i) - \sum_{k=1}^{N_T} \beta_k \boldsymbol{T}^k_{\cdot,i}, \tag{30}$$

where $\boldsymbol{T}^k_{\cdot,i}$ denotes the ith column of the matrix $\boldsymbol{T}^k$, i.e., the deviation of the RGB color of vertex i along the principal component k. This definition was chosen for the texture inverse because, then a texture composed with its inverse, under the same set of parameters, is equal to the mean texture: $t^{-1}(t(u_i, v_i; \boldsymbol{\beta}); \boldsymbol{\beta}) = \overline{\boldsymbol{T}}_{\cdot,i}$; see Equation 5.

This algorithm is not as efficient as the original ICIA, but it is more accurate, and its domain of applicability is also wider: It is able to fit input facial images at any pose and of any individual. This was demonstrated in the identification experiments reported in [37]. However, this algorithm does not handle directed light sources. There is a second predicament of this algorithm. In an implementation of ICIA, the parameters at which the derivatives are computed, $\boldsymbol{\gamma}^{\mathrm{d}} = [\boldsymbol{\alpha}^{\mathrm{d}^{\mathrm{T}}} \boldsymbol{\rho}^{\mathrm{d}^{\mathrm{T}}}]^{\mathrm{T}}$ must be selected. A natural choice for the shape parameters is $\boldsymbol{\alpha}^{\mathrm{d}} = 0$. The selection of $\boldsymbol{\rho}^{\mathrm{d}}$ is not as trivial, because the derivatives of the shape projections are computed in a particular image frame set by θ^{d} and by ϕ^{d}. Therefore, the two rotation angles should be close to their optimal values (depending on the input image). Hence, a set of Jacobians is computed for a series of different directions. During the iteration, the derivatives used are the ones closest to the current estimation of the angles. Note that, at first, this approach might seem very close to the view-based approach [14, 32, 34]. The difference is, however, fundamental. In this approach, the extraneous (rotation) parameters are clearly separated from the intrinsic (identity, i.e., $\boldsymbol{\alpha}, \boldsymbol{\beta}$) parameters. They are, however, convolved with one another in the view-based approach.

2D+3D Active-Appearance Model

Recently, Xiao et al. [48] extended the ICIA algorithm, originally developed for the fitting of 2D AAM, to the fitting of a 2D+3D AAM. The aim of this fitting algorithm is to recover the 3D shape and appearance of a face very efficiently (more than 200 fps). They argued that the difference between a 2D AAM and the 3DMM is that the 3DMM codes, additionally to the (X,Y) coordinates, the Z coordinate for each shape vertex. We will see, shortly, that there is an additional major difference between these two models. Xiao et al. [48] showed that, in fact, any 2D projection of a 3D shape in the span of a 3DMM can also be instantiated by a 2D AAM, but at the expense of using more parameters. The 2D AAM requires up to 6 times more parameters than the 3DMM to model the same phenomenon

. A weak-perspective-projection model was used to demonstrate this. This property would not hold for perspective projection. A weak perspective projection is governed by the following equation:

$$x_i = t_x + f w_{1,i}, \qquad y_i = t_y + f w_{2,i}. \tag{31}$$

Xiao et al. also showed that such a 2D AAM would be capable of instantiating invalid shapes, which is natural, as 3D transformations projected to 2D are not linear in 2D. The conclusion is that it is possible to fit facial images with non-frontal poses with a 2D AAM trained on a frontal pose. To ensure the validity of the shape estimated and to increase the efficiency of the algorithm, Xiao et al. impose the constraint that the 2D shape is a legitimate projection of a 3D shape modeled by a 3DMM. Thus, 3DMM shape model parameters and weak projection parameters are required to exist such as they produce a 2D shape equal to the one estimated by the 2D fitting algorithm. This is implemented as a soft constraint by augmenting the original ICIA cost function by a term proportional to the discrepancy between the 2D AAM estimated shape and the projection of the 3DMM shape. This seems to be a rather inefficient solution as, here, two shapes have to be estimated: The 2D AAM shape as well as the 3DMM shape and the projection parameters. The 3DMM model is only used to ensure the validity of the 2D AAM shape. The 2D AAM shape is used to warp the shape-free texture onto the image frame.

Xiao et al. [48] argued that the difference between a 2D AAM and the 3DMM is that the 3DMM codes depth information for each vertex and the 2D AAM does not. There is another major difference. The 3DMM is a dense model whereas the AAM is a sparse shape model. 76,000 vertices are modeled by the 3DMM and a few tens by the AAM. This dense sampling enables the 3DMM to separate the texture from the illumination, thereby estimating a texture free of illumination effects. This is because the shading of a point depends mostly on the normal of this point and on its reflectance properties. (The self-cast shadow term of the illumination depends also on the full 3D shape of the object.) Therefore, to accurately model the illumination, it is required to accurately model the normals. The normal of a point depends on the local 3D shape in a neighborhood of this point. Hence, the 3D surface is required to be densely sampled in order to permit an accurate computation of the normals. Thus, it is not possible for a sparse 3D shape model to accurately separate the shading from the texture. It would then be difficult to relight a facial image with a different lighting configuration, as it is possible with the dense 3DMM, or to obtain high identification rates on a face recognition application across illumination.

We have just seen that it would be difficult to estimate the illumination parameters from a coarse-shape model, for the normal may not be computed accurately. It seems also unclear how to extend this algorithm to estimate the illumination while retaining the constancy of the Jacobi matrix. This might be the reason why this algorithm has never been used to estimate the lighting and to compensate for it.

As demonstrated in [48], the 2D+3D ICIA algorithm recovers accurately the correspondence between the model 3D vertices and the image. These vertices are located on edges (eyebrows, eyes, nostrils, lips, contour). The edge features are, hence, implicitly used, through the input image gradient, to drive the fitting. Although, the correspondences are recovered, it does not imply that the estimated Z values of the landmarks is close to the Z value of the corresponding physical points on the face surface. In a single 2D image, the only depth information is contained in the lighting. To estimate accurately the 3D shape of a surface, the lighting must be estimated and its 3D shape recovered using a reflectance model. For example, in a frontal image, the only clue about the distance between the nose tip and, say, the lips, is in the shading of the nose. Failing to take the shading into account in a fitting algorithm, as it is done by the 2D+3D AAM fitting algorithm, results in an imprecise 3D shape.

As the original ICIA fitting algorithm, the 2D+3D ICIA fitting is person-specific: it is able to fit accurately only individual within the training set. Its main domain of application is real-time 3D face tracking.

Linear Shape and Texture Fitting Algorithm

LiST [38] is a 3DMM fitting algorithm that addresses the same problem as the SNO algorithm in a more efficient manner by use of the linear parts of the model. (A fitting is 5 times faster than with the SNO algorithm.) It is based on the fact that, if the correspondences between the model reference frame and the input image are estimated, then fitting the shape and rigid parameters is a simple bilinear optimization that can be solved accurately and efficiently. To obtain a bilinear relationship between the 2D vertices projection and the shape and rigid parameters, a weak-perspective projection is used (Equation 31). One way of estimating these correspondences is by the use of a 2D optical-flow algorithm [3]. Hence an optical-flow algorithm is applied to a model image synthesized using the current model parameters and the input image. These correspondences are then used to update the shape and rigid parameters. The texture and illumination parameters can also be recovered efficiently using the correspondences: The input image is sampled at the location given by the correspondences and first the illumination parameters are recovered using a Levenberg–Marquardt optimization [36], while keeping the texture parameters constant. This optimization is fast, as only a few parameters need to be estimated. Then, using the estimated light parameters, the light effect of the extracted texture is inverted, yielding an illumination-free texture used to estimate the texture parameters. The texture parameters are recovered by inverting a linear system of equations. It is hence efficient and provide an accurate estimate.

A drawback of this algorithm is that the 3D shape is estimated using correspondence information only, not using the shading, as it is done in the SNO algorithm.

4.7 RESULTS

We explained in Section 4.2 that the morphable face model is a representation of human face images in which the pose and illumination parameters are separated from the shape and texture parameters. Then, in Section 4.6, we outlined SNO: a 3D morphable model fitting algorithm that estimate the 3D shape, the texture, and the imaging parameters from a single facial image. Some example images obtained after fitting and pose and illumination normalization are displayed in Figure 4.1. Examples of illumination normalization are shown on Figure 4.5. The images of the first row, illuminated from different directions, are fitted. Renderings of the fitting results are shown in the second row. The same renderings, but using the illumination parameters from the leftmost input image, appear in the third row. The last row presents the input image with illumination normalized to the illumination of the leftmost image.

4.7.1 Identification Results

In this section, the 3D morphable model and its fitting algorithm are evaluated on an identification application; these results were first published in [6]. In an identification task, an image of an unknown person is provided to the system. The unknown face image is then compared to a database of known people, called the gallery set. The ensemble of unknown images is called the probe set. It is assumed that the individual in the unknown image is in the gallery.

Dataset

We evaluate our approach on two datasets. *Set 1*: a portion of the FERET dataset [35] containing images with different poses. In the FERET nomenclature these images correspond to the series *ba* through *bk*. We omitted the images *bj* as the subjects present a smiling expression that is not accounted for by the current 3D morphable model. This dataset includes 194 individual across 9 poses at constant lighting condition except for the series *bk*: frontal view at another illumination condition than the rest of the images. *Set 2*: A portion of the CMU–PIE dataset [43] containing images of 68 individuals at 3 poses (frontal, side and profile) and illuminated by 21 different directions and by ambient light only. Among the 68 individuals, 28 wear glasses, which are not modeled and could decrease the accuracy of the fitting. None of the individuals present in these sets were used to construct the 3D morphable model. These sets cover a large ethnic variety, not present in the set of 3D scans used to build the model.

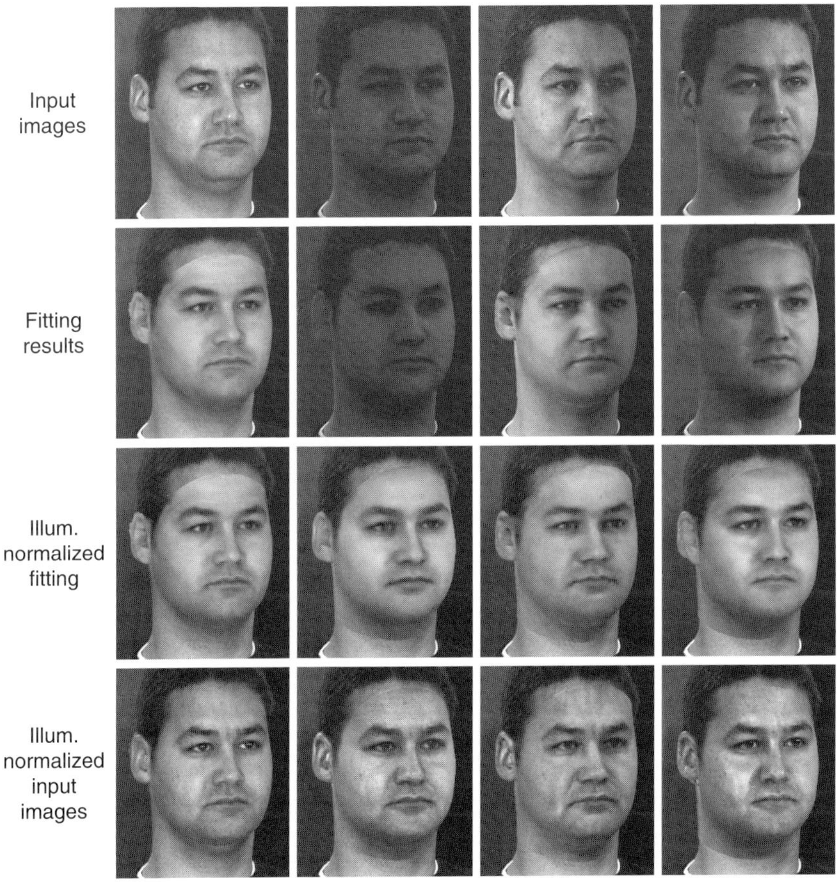

FIGURE 4.5: Demonstration of illumination normalization on a set of input images obtained using the SNO fitting algorithm. Renderings of the input image of the top row are shown on the second row. The same renderings with standard illumination, taken from the leftmost input image, are displayed on the third row. Finally, the rendering using the extracted texture taken from the original images and again with the standard illumination appear on the bottom row. (See also color plate section)

Distance Measure

Identification and verification are performed by fitting an input face image to the 3D morphable model, thereby extracting its identity parameters, α and β. Then, recognition tasks are achieved by comparing the identity parameters of the input image with those of the gallery images. We define the identity parameters of a face image, denoted by the vector $\boldsymbol{c}$, by stacking the shape and texture parameters of the global and segmented models (see Section 4.5.1) and rescaling them by their standard deviations:

$$\boldsymbol{c} = \left[\frac{\alpha_1^{\text{g}}}{\sigma_{\text{S},1}}, \ldots, \frac{\alpha_{99}^{\text{g}}}{\sigma_{\text{S},99}}, \frac{\beta_1^{\text{g}}}{\sigma_{\text{T},1}}, \ldots, \frac{\beta_{99}^{\text{g}}}{\sigma_{\text{T},99}}, \frac{\alpha_1^{\text{S1}}}{\sigma_{\text{S},1}}, \ldots, \frac{\alpha_{99}^{\text{S1}}}{\sigma_{\text{S},99}}, \ldots, \ldots, \frac{\beta_{99}^{\text{S4}}}{\sigma_{\text{T},99}} \right]^{\text{T}}. \quad (32)$$

We define a distance measure to compare two identity parameters $\boldsymbol{c}_1$ and $\boldsymbol{c}_2$. The measure, d, is based on the angle between the two vectors (it can also be seen as a normalized correlation). This measure is insensitive to the norm of both vectors. This is favorable for recognition tasks as increasing the norm of $\boldsymbol{c}$ produces a caricature which does not modify the perceived identity:

$$d = \frac{\boldsymbol{c}_1 \boldsymbol{C}_{\text{W}}^{-1} \boldsymbol{c}_2}{\sqrt{\left(\boldsymbol{c}_1^{\text{T}} \boldsymbol{C}_{\text{W}}^{-1} \boldsymbol{c}_1\right)\left(\boldsymbol{c}_2^{\text{T}} \boldsymbol{C}_{\text{W}}^{-1} \boldsymbol{c}_2\right)}}. \quad (33)$$

In this equation, $\boldsymbol{C}_{\text{W}}$ is the covariance matrix of the within-subject variations. It is computed on the FERET fitting results for the identification of the CMU–PIE dataset and vice versa.

Table 4.1 lists percentages of correct rank 1 identification obtained using the FERET dataset. The pose chosen as gallery is the one with an average azimuth angle of 11.2°, i.e., the condition *be*.

The CMU–PIE dataset is used to test the performance of this method in presence of combined pose and illumination variations. Table 4.2 presents the rank 1 identification performance averaged over all the lighting conditions corresponding to front, side and profile view galleries. Illumination 13 was selected for the galleries.

4.7.2 Improved Fitting Using an Outlier Map

The major source of fitting inaccuracies is the presence of outlying pixels in a facial image. An outlying pixel, or outlier, is a pixel inside the face area of an image whose value cannot be predicted by the model nor by the noise distribution assumed by the fitting cost function[1]. Typical examples of outliers are glasses, specular highlight due to the presence of glasses, and occluding objects such as facial hair. Naturally, a black pixel due to facial hair may be predicted by the model, but doing so would substantially modify the model parameters and deteriorate the fitting of the rest of the face. This is shown on the top row images of Figure 4.6.

[1]We assumed that the noise is independent and identically distributed over all pixels of the image with a normal distribution.

Table 4.1: Percentage of correct identification on the FERET dataset obtained using the SNO fitting algorithm. The gallery images are the view *be*. ϕ denotes the average estimated azimuth pose angle of the face. Ground truth for ϕ is not available. Condition *bk* has different illumination than the others. (Results from [6]).

Probe view	Pose ϕ	Correct identification
bb	38.9°	94.8%
bc	27.4°	95.4%
bd	18.9°	96.9%
be	11.2°	99.5
ba	1.1°	*gallery*
bf	−7.1°	97.4%
bg	−16.3°	96.4%
bh	−26.5°	95.4%
bi	−37.9°	90.7%
bk	0.1°	96.9%
Mean		*95.9%*

Table 4.2: Mean percentage of correct identification obtained on PIE images using the SNO fitting algorithm, averaged over all lighting conditions for front, oblique, and profile view galleries. In brackets are percentages for the worst and best illumination within each probe set. The overall mean of the table is 92.1%. (Results from [6]).

Gallery view	Probe view						Mean
	front		side		profile		
front	99.8%	(97.1–100)	97.8%	(82.4–100)	79.5%	(39.7–94.1)	92.3 %
side	99.5%	(94.1–100)	99.9%	(98.5–100)	85.7%	(42.6–98.5)	95.0 %
profile	83.0%	(72.1–94.1)	86.2%	(61.8–95.6)	98.3%	(83.8–100)	89.0 %

Thus, fitting an outlier is prejudicial not only because the fitted model parameter would account for an artifact such as hair but also because it would substantially decrease the overall quality of the fitting in the outlier-free region of the face. Discarding all pixels with a large residual is not a general solution to this problem: Some inliers may have a large fitting error, and discarding them would jeopardize the quality of the fitting. This may happen at the beginning of the fitting algorithm when there is an important lack of correspondence. For instance, the model pupil may overlap the skin area between the eye and the eyebrows inducing a large

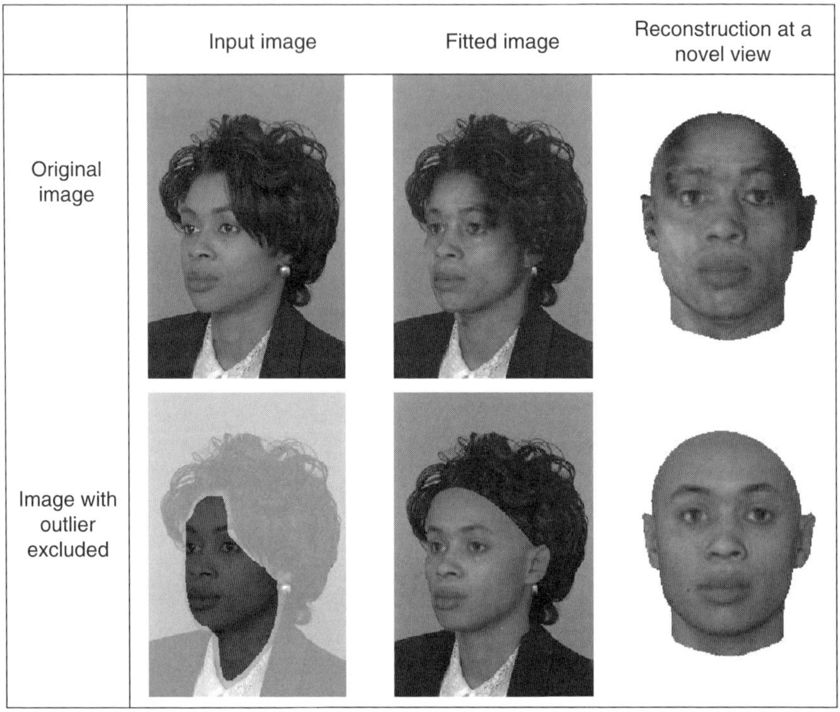

FIGURE 4.6: Example of the benefit of excluding the outlier part of a face image. Top row: The second image is a rendering of the fitting result obtained by fitting the first image. The right image is a rendering of the same fitting result at a frontal pose. Bottom row: The first image is the input image with the outlier region shown in bright. This region was excluded to produce the fitting shown on the second image. The third image is a rendering of this fitting result at a frontal pose.

residual. It is this residual which is to drive the model parameters to improve the correspondence and, hence, it should not be down-weighted.

To appreciate the importance of this problem, the fitting of the same image was performed with the outlier pixels discarded. To do so, a outlier mask was automatically generated. The mask is shown on the first image of the bottom row of Figure 4.6: The brighter pixels are treated as outliers and are not sampled in the sum of the energy function of Equation 20. The visual quality of the reconstruction yielded by this fitting, shown on the middle of the bottom row of the figure, is clearly improved. The rendering of the novel view (last column) is also superior when the outlier are excluded.

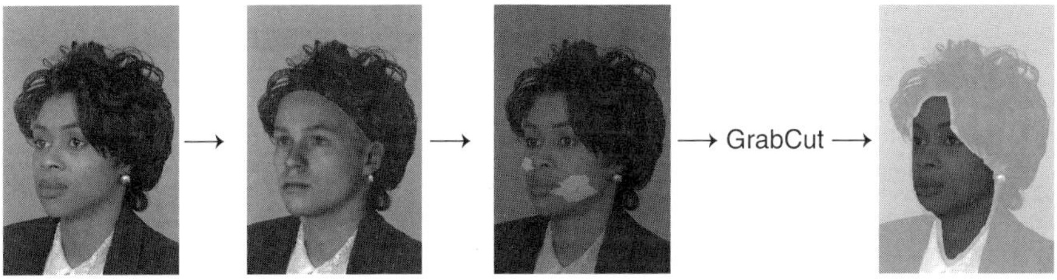

FIGURE 4.7: Automatic outlier mask generation. The outlier mask is produced by first performing a coarse fitting (second image) of the input image (first image). Then the five image patch with minimum residual error are selected (third image). The outlier mask is obtained by applying GrabCut using as foreground the five image patches selected.

To automatically produce an outlier mask, we use the following algorithm: First a coarse fitting is performed without outlier mask, but with a large weight on the prior probability (σ_I^2 is increased in Equation 20). Only a few iterations of the fitting algorithm are necessary at this stage. A rendering of the result of this fitting is shown on the first image of Figure 4.7. Then five image patches are selected in the face area. These image patches are the one with minimum residual; they are shown in the second image of Figure 4.7. These image patches are used to initialize a *GrabCut* algorithm, developed by Rother et al. [41], that segments the skin region from the nonskin part of the image, thereby producing the outlier masks. GrabCut is an image segmentation algorithm that combines color and contrast information together with a strong prior on region coherence. Foreground and background regions are estimated by solving a graph-cut algorithm.

4.8 CONCLUSION

Recognizing a signal by explaining it is a standard strategy that has been used successfully. The two requirements for this technique are to obtain a model that can account for the input signals in their whole diversity and to provide an algorithm to estimate the model parameters explaining a signal.

We showed in this chapter that an accurate and general model is one that separates the sources of variations of the signals and represents them by independent parameters. As well as identity, the sources of variations of facial images include pose changes and the lighting changes. To accurately account for pose and illumination variations, the method of computer graphics proposed a 3D object-centered representation. On the other hand, linear combinations of exemplar faces are used to produce the face of a novel individual. This produces a valid face if the faces are represented on a common reference frame on which all facial features are labeled.

We proposed, in this chapter, the 3D morphable model: a model that takes advantage of the 3D representation and of the correspondence principle, to account for any individual, viewed from any angles and under any light direction. The second ingredient of a analysis-by-synthesis loop is the model inversion, or analyzing, algorithm. The most general and accurate algorithm proposed so far is the stochastic Newton optimization whose update, at each iteration, is based on the first and second derivatives of a MAP energy function. The first derivatives are computed at each iteration, thereby favoring accuracy over efficiency. A stochastic optimization scheme was chosen to reduce the risk of locking into a local minimum.

Additionally to SNO that gives greater importance to accuracy and generality, there exists other fitting algorithms that favor efficiency at the expense of the domain of application and the precision. Using 3DMM, we outlined the principles of the major fitting algorithms, and described their advantages and predicaments. The algorithms reviewed are based on ASM, AAM, and ICIA, ICIA applied to 3DMM, 2D+3D AAM, and the linear shape and texture fitting algorithm.

REFERENCES

[1] S. Baker and I. Matthews. Equivalence and efficiency of image alignment algorithms. In: *Proceedings of the IEEE Conf. on Computer Vision and Pattern Recognition*, 2001.

[2] C. Basso, T. Vetter, and V. Blanz. Regularized 3d morphable models. In: *Proc. of the IEEE Workshop on Higher-Level Knowledge in 3D Modeling and Motion Analysis (HLK 2003)*, 2003.

[3] J. R. Bergen and R. Hingorani. Hierarchical motion-based frame rate conversion. Technical report, David Sarnoff Research Center, Princeton NJ 08540, 1990.

[4] V. Blanz. Automatische Rekonstruction der dreidimensionalen Form von Gesichtern aus einem Einzelbild. PhD thesis, Universität Tübingen, 2001.

[5] V. Blanz and T. Vetter. Face recognition based on fitting a 3d morphable model. *IEEE Transactions on Pattern Analysis and Machine Intelligence* **25**(9):1063–1074, 2003.

[6] V. Blanz and T. Vetter. Face recognition based on fitting a 3d morphable model. *IEEE Trans. Pattern Anal. Mach. Intell.*, 2003.

[7] Volker Blanz and Thomas Vetter. A morphable model for the synthesis of 3D faces. In: Alyn Rockwood, editor, *Siggraph 1999, Computer Graphics Proceedings*, pages 187–194, Los Angeles, 1999. Addison Wesley Longman.

[8] M. E. Brand. Morphable 3D models from video. In: *Proceedings IEEE Conf. on Computer Vision and Pattern Recognition*, 2001.

[9] T. Cootes, D. H. Cooper, C. J. Taylor, and J. Graham. A trainable method of parametric shape description. In: *Proc. British Machine Vison Conference*, 1991.

[10] T. Cootes, G. Edwards, and C. Taylor. Active appearance model. In: *Proc. European Conference on Computer Vision*, volume 2, pages 484–498. Springer, 1998.

[11] T. Cootes and C. Taylor. Active shape models—smart snakes. In: *Proc. British Machine Vision Conference*, 1992.

[12] T. Cootes, C. Taylor, A. Lanitis, D. Cooper, and J. Graham. Building and using flexible models incorporating grey-level information. In: *Proceedings of the 4th International Conference on Computer Vision*, pages 355–365, 1993.
[13] T. Cootes, C. J. Taylor, D. H. Cooper, and J. Graham. Training models of shape from sets of examples. In: *Proc. British Machine Vision Conference*, pages 266–275, Berlin, 1992. Springer.
[14] T. Cootes, K. Walker, and C. Taylor. View-based active appearance models. In: *Fourth International Conference on Automatic Face and Gesture Recognition*, pages 227–232, 2000.
[15] I. Craw and P. Cameron. Parameterizing images for recognition and reconstruction. In: Peter Mowforth, editor, *Proc. British Machine Vision Conference*, pages 367–370. Springer, 1991.
[16] K. I. Diamantaras and S. Y. Kung. *Principal Component Neural Networks*. Wiley, 1996.
[17] M. Dimitrijevic, S. Ilic, and P. Fua. Accurate face models from uncalibrated and ill-lit video sequences. In: *Proceedings IEEE Conf. on Computer Vision and Pattern Recognition*, 2004.
[18] J. D. Foley, A. van Dam, S. K. Feiner, and J. F. Hughes. *Computer Graphics: Principles and Practice*. Addison-Wesley, 1996.
[19] P. E. Gill, W. Murray, and M. H. Wright. *Practical Optimization*. Academic Press, 1981.
[20] M. Gleicher. Projective registration with difference decomposition. In: *Proceedings IEEE Conf. on Computer Vision and Pattern Recognition*, pages 331–337, 1997.
[21] S. J. Gortler, R. Grzeszczuk, R. Szeliski, and M. F. Cohen. The lumigraph. *Computer Graphics* **30** (Annual Conference Series):43–54, 1996.
[22] U. Grenander. *Pattern Analysis, Lectures in Pattern Theory*. Springer, New York, 1st edition, 1978.
[23] P. W. Hallinan. A deformable model for the recognition of human faces under arbitrary illumination. PhD thesis, Harvard University, Cambridge, Massachusetts, 1995.
[24] H. W. Jensen, S. R. Marschner, M. Levoy, and P. Hanrahan. A practical model for subsurface light transport. In: *Proceedings of SIGGRAPH 2001*, pages 511–518, August 2001.
[25] A. Lanitis, C. J. Taylor, and T. F. Cootes. Automatic face identification system using flexible appearance models. *Image and Vision Computing* **13**(5):393–401, June 1995.
[26] A. Lanitis, C. J. Taylor, and T. F. Cootes. Automatic interpretation and coding of face images using flexible models. *IEEE Transactions on Pattern Analysis and Machine Intelligence* **19**(7):743–756, 1997.
[27] M. Levoy and P. Hanrahan. Light field rendering. *Computer Graphics* **30**(Annual Conference Series):31–42, 1996.
[28] D. Marr. *Vision,*. W. H. Freeman, San Fancisco, 1982.
[29] S. R. Marschner, S. H. Westin, E. P. F. Lafortune, K. E. Torrance, and D. P. Greenberg. Reflectance measurements of human skin. Technical Report PCG-99-2, 1999.
[30] T. P. Minka. Old and new matrix algebra useful for statistics. http://www.stat.cmu.edu/~minka/papers/matrix.html, 2000.
[31] D. Mumford. Pattern theory: A unifying perspective. In: D.C. Knill and W. Richards, editors, *Perception as Bayesian Inference*. Cambridge University Press, 1996.

[32] H. Murase and S. K. Nayar. Visual learning and recognition of 3d objects from appearance. *International Journal of Computer Vision* **14**:5–24, 1995.

[33] F. I. Parke and K. Waters. *Computer Facial Animation.* AKPeters, Wellesley, Massachusetts, 1996.

[34] A. Pentland, B. Moghaddam, and T. Starner. View-based and modular eigenspaces for face recognition. In: *Proceedings IEEE Conf. on Computer Vision and Pattern Recognition*, 1994.

[35] P. J. Phillips, P. Rauss, and S. Der. Feret (face recognition technology) recognition algorithm development and test report.Technical report, U.S. Army Research Laboratory, 1996.

[36] Vetterling Press, Teukolsky and Flannery. *Numerical recipes in C : the Art of Scientific Computing*. Cambridge University Press, Cambridge, 1992.

[37] S. Romdhani, V. Blanz, C. Basso, and T. Vetter. Morphable models of faces. In: S. Z. Li and A. Jain, editors, *Handbook of Face Recognition*. Springer, 2005.

[38] S. Romdhani, V. Blanz, and T. Vetter. Face identification by fitting a 3d morphable model using linear shape and texture error functions. In: *Proc. European Conference on Computer Vision*, 2002.

[39] S. Romdhani and T. Vetter. Efficient, robust and accurate fitting of a 3d morphable model. In: *Proceedings of the International Conference on Computer Vision*, 2003.

[40] S. Romdhani, A. Psarrou, and S. Gong. On utilising template and feature-based correspondence in multi-view appearance models. In: *Proc. European Conference on Computer Vision*, 2000.

[41] C. Rother, V. Kolmogorov, and A. Blake. Grabcut—interactive foreground extraction using iterated graph cuts. *Proc. ACM Siggraph*, 2004.

[42] S. Sclaroff and J. Isidoro. Active blobs. In: *Proceedings of the 6th International Conference on Computer Vision*, 1998.

[43] T. Sim, S. Baker, and M. Bsat. The cmu pose, illumination and expression (pie) database of human faces. Technical report, CMU, 2000.

[44] S. Ullman. Aligning pictorial descriptions: An approach for object recognition. *Cognition* **32**:193–254, 1989.

[45] T. Vetter. Recognizing faces from a new viewpoint. In: *ICASSP97 Int. Conf. Acoustics, Speech, and Signal Processing*, volume 1, pages 139–144, IEEE Comp. Soc. Press, Los Alamitos, CA, 1997.

[46] T. Vetter and T. Poggio. Linear object classes and image synthesis from a single example image. *IEEE Trans. Pattern Anal. Mach. Intell.*, **19**(7): 733–742, 1997.

[47] T. Vetter. Synthestis of novel views from a single face image. *International Journal of Computer Vision* **28**(2):103–116, 1998.

[48] J. Xiao, S. Baker, I. Matthews, R. Gross, and T. Kanade. Real-time combined 2d+3d active appearance model. In: *Proceedings IEEE Conf. on Computer Vision and Pattern Recognition*, pages 535–542, 2004.

CHAPTER 5

EXPRESSION-INVARIANT THREE-DIMENSIONAL FACE RECOGNITION

5.1 INTRODUCTION

It is well known that some characteristics or behavior patterns of the human body are strictly individual and can be observed in two different people with a very low probability – a few such examples include the DNA code, fingerprints, structure of retinal veins and iris, individual's written signature or face. The term *biometrics* refers to a variety of methods that attempt to uniquely identify a person according to a set of such features.

While many of today's biometric technologies are based on the discoveries of the last century (like DNA, for example), some of them have been exploited from the dawn of the human civilization [17]. One of the oldest written testimonies of a biometric technology and the first identity theft dates back to biblical times, when Jacob fraudulently used the identity of his twin brother Esau to benefit from their father's blessing. The Genesis book describes a combination of hand scan and voice recognition that Isaac used to attempt to verify his son's identity, without knowing that the smooth-skinned Jacob had wrapped his hands in kidskin:

> And Jacob went near unto Isaac his father; and he felt him, and said, "The voice is Jacob's voice, but the hands are the hands of Esau." And he recognized him not, because his hands were hairy, as his brother Esau's hands.

The false acceptance which resulted from this very inaccurate biometric test had historical consequences of unmatched proportions.

Face recognition is probably the most natural biometric method. The remarkable ability of the human vision to recognize faces is widely used for biometric authentication from prehistoric times. These days, almost every identification document contains a photograph of its bearer, which allows the respective officials to verify a person's identity by comparing his actual face with the one on the photo.

Unlike many other biometrics, face recognition does not require physical contact with the individual (like fingerprint recognition) or taking samples of the body (like DNA-based identification) or observing the individual's behavior (like signature recognition). For these reasons, face recognition is considered a natural, less intimidating, and widely accepted biometric identification method [4, 46], and as such, has the potential of becoming the leading biometric technology. The great technological challenge is to perform face recognition automatically, by means of computer algorithms that work without any human intervention. This problem has been traditionally considered the realm of computer vision and pattern recognition. It is also believed to be one of the most difficult machine vision problems.

5.1.1 The problems of face recognition

The main difficulty of face recognition stems from the immense variability of the human face. Facial appearance depends heavily on the environmental factors, e.g., lighting conditions, background scene, and head pose. It also depends on facial hair, the use of cosmetics, jewelry, and piercing. Last but not least, plastic surgery or long-term processes like aging and weight gain can have a significant influence on facial appearance. Yet, much of the facial appearance variability is inherent to the face itself. Even if we hypothetically assume that external factors do not exist, e.g., that the facial image is always acquired under the same illumination, pose, and with the same haircut and make up, still, the variability in a facial image due to facial expressions may be even greater than a change in the person's identity. That is, unless the right measure is used to distinguish between individuals.

Theoretically, it is possible to recognize an individual's face reliably in different conditions, provided that the same person *has been previously observed in similar conditions*. However, the variety of images required to cover all the possible appearances of the face can be very large (see Figure 5.1). In practice, only a few observations of the face (and sometimes, even a single one) are available.

Broadly speaking, there are two basic alternatives in approaching this problem. One is to find features of the face that are not affected by the viewing conditions. Early face recognition algorithms [8, 36, 28] advocated this approach by finding a set of *fiducial points* (eyes, nose, mouth, etc.) and comparing their geometric relations (angles, lengths, and ratios). Unfortunately, there are only few such features

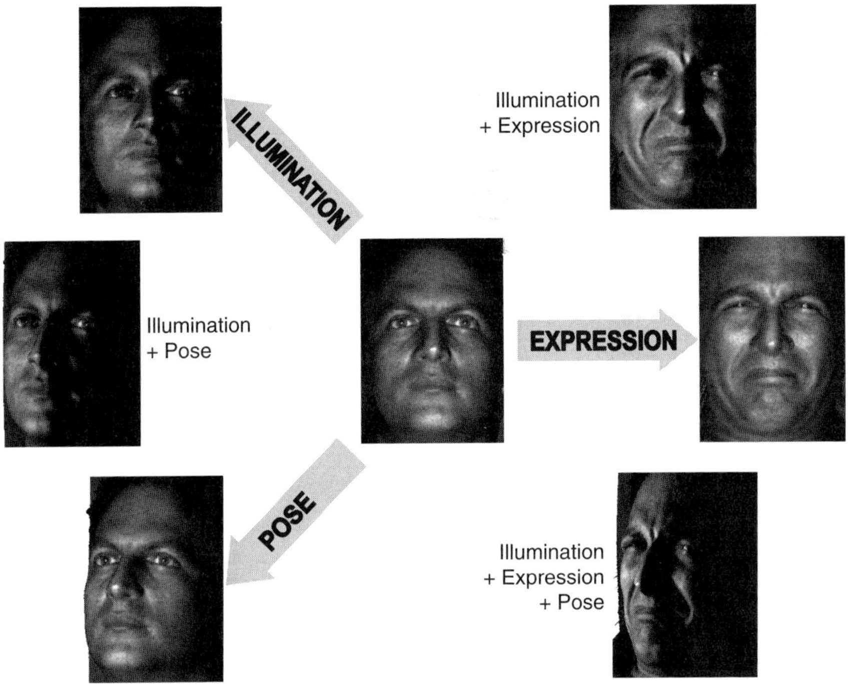

FIGURE 5.1: Illustration of some factors causing the variability of the facial image.

that can be reliably extracted from a 2D facial image and would be insensitive to illumination, pose, and expression variations [21].

The second alternative is to *generate* synthetic images of the face under new, unseen conditions. Generating facial images with new pose and illumination requires some 3D facial surface as an intermediate stage. It is possible to use a generic 3D head model [35], or estimate a rough shape of the facial surface from a set of observations (e.g., using photometric stereo [27]) in order to synthesize new facial images and then apply standard face recognition methods like eigenfaces [48, 54] to the synthetic images.

Figure 5.2 shows a simple visual experiment that demonstrates the generative approach. We created synthetic faces of Osama Bin Laden (first row, right) and George Bush (second row, right) in different poses by mapping respective textures onto the facial surface of another subject (left). The resulting images are easily recognized as the world number one terrorist and the forty-third president of the United States, though in both cases the facial geometry belongs to a completely

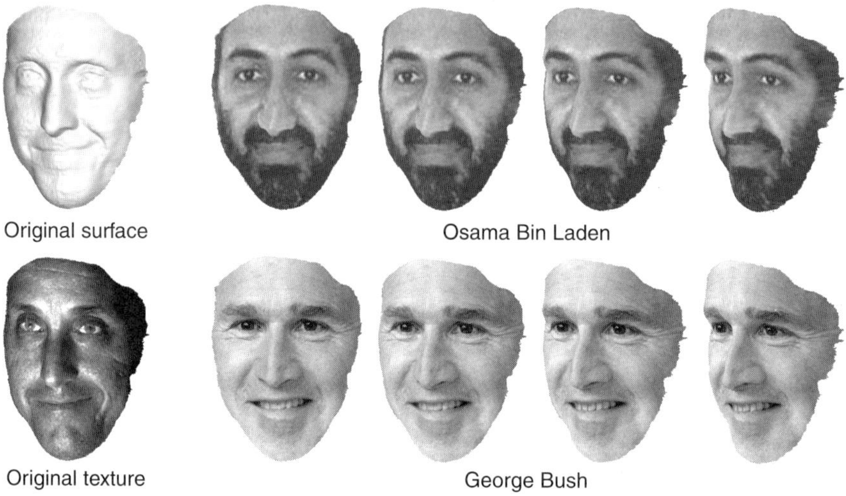

FIGURE 5.2: Simple texture mapping on the same facial surface can completely change the appearance of the 2D facial image and make the same face look like George Bush or Osama Bin Laden. This illustrates the fact that being able to change the texture of the face, e.g., by using make up, one can completely change his or her appearance as captured by a 2D image and disguise to another person. The 3D geometry of the face is the same in this example, and is more difficult to disguise.

different individual. Simple texture mapping in our experiment allowed us to create natural-looking faces, yet, the individuality of the subject concealed in the 3D geometry of his face was completely lost. This reveals the intrinsic weakness of all the 2D face recognition approaches: the face "lives" in a three-dimensional space, and using only its 2D projection can be misleading. Practically, one has the ability to draw any face on his own, so that he could essentially appear like any other person and deceive any 2D face recognition method.

5.1.2 A new dimension to face recognition

Three-dimensional face recognition is a relatively recent trend that in some sense breaks the long-term tradition of mimicking the human visual recognition system, as 2D methods attempt to do. Three-dimensional facial geometry represents the internal anatomical structure of the face rather than its external appearance influenced by environmental factors. As a result, unlike the 2D facial image, 3D facial surface is insensitive to illumination, head pose [10], and cosmetics [40].

The main problem in 3D face recognition is how to find similarity between 3D facial surfaces. Earliest works on 3D face recognition did not use the whole facial surface, but a few profiles extracted from the 3D data [19, 45, 7, 31]. Attempts were made to extend conventional dimensionality reduction techniques (e.g., PCA) to range images or combination of intensity and range images [2, 34, 40, 20, 53]. Tsalakanidou et al. applied the hidden Markov model to depth and color images of the face [52]. Many academic (e.g., [41, 1]), as well as some commercial 3D face recognition algorithms treat faces as *rigid* surfaces by employing variants of rigid surface matching algorithms.

The intrinsic flaw of these approaches is their difficulty in handling deformations of the facial surface as the result of expressions. To date, only little research has been focused on trying to make face recognition deal with facial expressions. The majority of the papers, starting from the earliest publications on face recognition [8, 36, 28] and ending with the most recent results, address mainly the external factors like illumination, head pose, etc. Moreover, though many authors mention the problem of facial expressions [30], a decent evaluation of currently available algorithms on a database of faces containing sufficiently large expression variability has never been done before [10].

In [13, 16, 14] we introduced an expression-invariant 3D face recognition algorithm, on which the 3DFACE recognition system built at the Department of Computer Science, Technion, is based. Our approach uses a geometric model of facial expressions, which allowed us to build a representation of the face insensitive to expressions. This enabled us to successfully handle even extreme facial expressions.

5.2 ISOMETRIC MODEL OF FACIAL EXPRESSIONS

In order to treat faces as deformable, nonrigid surfaces, we use the Riemannian geometry framework. We model faces as 2D smooth connected compact Riemannian surfaces (manifolds). Broadly speaking, a Riemannian surface $\mathcal{S}$ can be described by a coordinate mapping $\mathcal{S} = x : U \subset \mathbb{R}^2 \to \mathbb{R}^3$ from a domain U on a plane to the 3D Euclidean space and the *metric tensor* g, which is an intrinsic characteristic of the surface that allows us to measure local distances on $\mathcal{S}$ independently of the coordinates [39].

The deformations of the facial surface as the result of expressions can be expressed as a diffeomorphism $f : (\mathcal{S}, g) \to (\mathcal{S}', g')$ on the surface $\mathcal{S}$. Our observations show that in most parts of the human face the facial skin does not stretch significantly, and therefore we can model facial expressions as *isometries*. An experiment validating the isometric model is described in [18]. We showed that the isometric model faithfully describes most natural facial expressions, and that such a model is better than the rigid one. Isometric transformations preserve the *intrinsic geometry* of the surface. That is, the metric and consequently, the *geodesic distances* (the shortest paths between any two points on $\mathcal{S}$) remain invariant. To illustrate the idea of isometry, imagine a 2D creature that lives on the

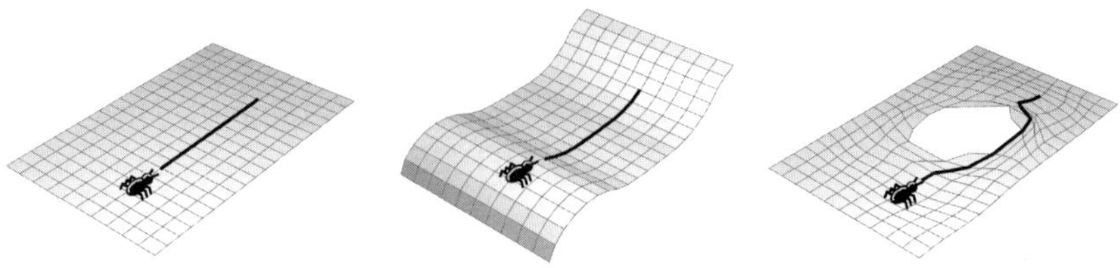

FIGURE 5.3: Illustration of isometric and nonisometric transformations of a surface. Isometries do not change the intrinsic geometry of the surface, such that an imaginary creature living on the surface does not feel the transformation.

surface (Figure 5.3). An isometry is a transformation that bends the surface such that the creature does not "feel" it. Faces in presence of facial expressions are thereby modeled as (approximately) isometric surfaces, that is, surfaces that can be obtained from some initial facial surface ("neutral expression") by means of an isometry. From the point of view of Riemannian geometry, such surfaces are indistinguishable as they have identical intrinsic geometry.

Isometry also tacitly implies that the *topology* of the facial surface is preserved. For example, expressions are not allowed to introduce "holes" in the facial surface (Figure 5.3, right). This assumption is valid for most regions of the face, yet, the *mouth* cannot be treated by the isometric model. Opening the mouth, for example, changes the topology of the facial surface by virtually creating a "hole." As a consequence, the isometric model is valid for facial expressions with the mouth either always open or always closed.

This flaw of the isometric model can be dealt with by enforcing a fixed topology on the facial surface. For example, assuming that the mouth is always closed and thereby "gluing" the lips when the mouth is open; or, alternatively, assuming the mouth to be always open, and "disconnecting" the lips by introducing a cut in the surface when the mouth is closed. This new model, which we refer to as the *topologically constrained isometric model*, is applicable to all facial expressions, including those with both open and closed mouth. The problem of open mouth in 3D face recognition is addressed in [18]. Here we assume that the mouth is always closed and thus limit our discussion to the isometric model.

5.3 EXPRESSION-INVARIANT REPRESENTATION

A cornerstone problem in three-dimensional face recognition is the ability to identify facial expressions of a subject and distinguish them from facial expressions of

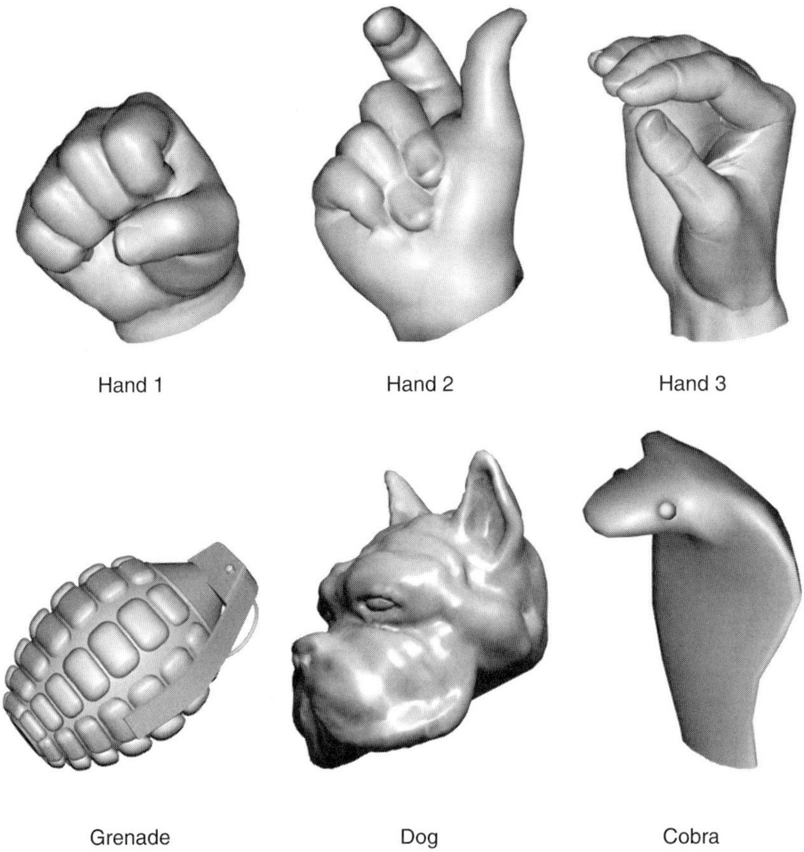

FIGURE 5.4: Illustration of the isometric surface matching problem. First row: isometries of a hand. Second row: different objects that resemble the hands if treated in a rigid way.

another subject. Under the isometric model assumption, the problem is reduced to finding similarity between isometric surfaces.

Figure 5.4 illustrates the problem of isometric surface matching. The first row shows three isometries of the human hand (assume that the fingers do not touch each other, so that the topology is preserved), which, with a bit of imagination, look like a grenade, dog, and cobra (second row). In other words, from the point of view of their *extrinsic* geometry, isometric surfaces can look different, while being just instances of the same surface. Deformations of the facial surface due to facial expressions are not as extreme as those of a human hand, yet sufficiently

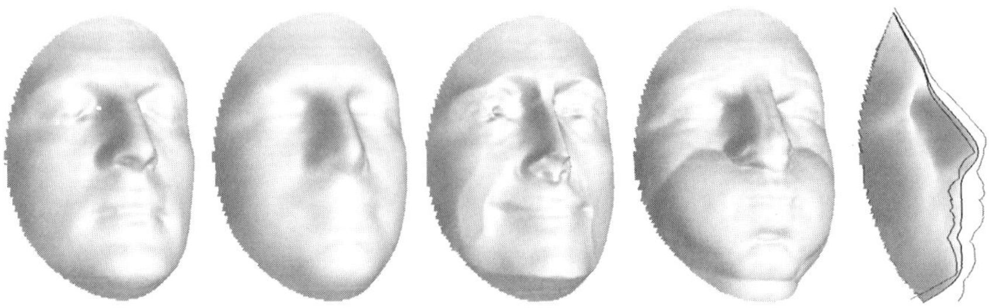

FIGURE 5.5: Variation of the facial surface due to facial expressions (left to right): neutral expression of subject Eyal, average facial expression, minimum and maximum envelopes, profile view showing the minimum and the maximum envelope.

significant to make the uncertainty region around the facial surface large enough that many other faces can fit within (see Figure 5.5).

Let us assume that we have two instances of the same face differing one from another by a facial expression, and let S and Q denote the corresponding facial surfaces. According to our isometric model, there exists an isometric transformation $f(\mathcal{S}) = \mathcal{Q}$ that maps any point $\boldsymbol{x}$ on the surface $\mathcal{S}$ to a point $\boldsymbol{y}$ on the surface $\mathcal{Q}$. Since f is isometric, the geodesic distances are preserved, i.e., $d_{\mathcal{S}}(\boldsymbol{x}_1, \boldsymbol{x}_2) = d_{\mathcal{Q}}(\boldsymbol{y}_1, \boldsymbol{y}_2)$.

Theoretically, the geodesic distances give a unique expression-invariant representation of the face. However, since in practice the surfaces $\mathcal{S}$ and $\mathcal{Q}$ are represented by a discrete set of samples $\boldsymbol{x}_1, \ldots, \boldsymbol{x}_{N_x}$ and $\boldsymbol{y}_1, \ldots, \boldsymbol{y}_{N_y}$, respectively, there is neither guarantee that the surface is sampled at the same points, nor that the number of points in two surfaces are necessarily the same ($N_x \neq N_y$ in general). Moreover, even if the samples are the same, they can be ordered arbitrarily, and thus the matrix $\boldsymbol{D} = (d_{ij}) = (d(\boldsymbol{x}_i, \boldsymbol{x}_j))$ is invariant up to some permutation of the rows and columns. Therefore, though the matrix $\boldsymbol{D}$ can be considered as an invariant, making use of it has little practicality. Nevertheless, there have been very recent efforts to establish some theory about the properties of such matrices [42, 43].

The alternative proposed in [25, 26] is to avoid dealing explicitly with the matrix of geodesic distances and find a representation to the original Riemannian surface as a submanifold of some convenient space, with an effort to preserve (at least approximately) the intrinsic geometry of the surface. Such an approximation is called *isometric embedding*. Typically, a low-dimensional Euclidean space is used as the embedding space; the embedding in this case is called *flat*.

In our discrete setting, flat embedding is a mapping

$$\varphi : (\{\boldsymbol{x}_1, \ldots, \boldsymbol{x}_N\} \subset \mathcal{S}, d) \rightarrow (\{\boldsymbol{x}'_1, \ldots, \boldsymbol{x}'_N\} \subset \mathbb{R}^m, d') \tag{1}$$

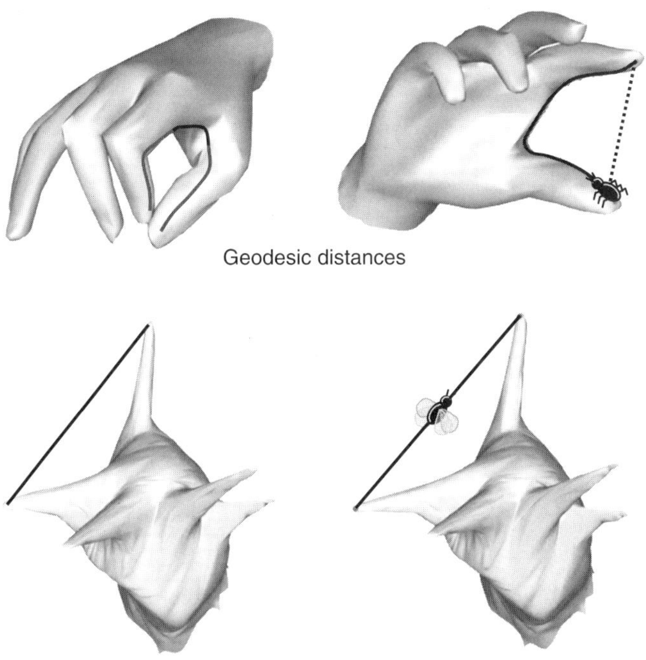

FIGURE 5.6: Illustration of the embedding problem and the canonical forms. First row: a Riemannian surface (hand) undergoing isometric transformations. The solid line shows the geodesic distance between two points on the surface, and the dotted line is the corresponding Euclidean distance. Second row: the hand surfaces embedded in a three-dimensional Euclidean space. The geodesic distances become Euclidean ones.

that maps N samples $\boldsymbol{x}_1, \ldots, \boldsymbol{x}_N$ of the surface $\mathcal{S}$ into a set of points $\boldsymbol{x}'_1, \ldots, \boldsymbol{x}'_N$ in a m-dimensional Euclidean space, such that the resulting Euclidean distances $d'_{ij} = \|\boldsymbol{x}'_i - \boldsymbol{x}'_j\|_2$ approximate the original geodesic distances d_{ij} in an optimal way (here the matrices $\boldsymbol{D}$ and $\boldsymbol{D}'(\boldsymbol{X}')$ denote the mutual geodesic and Euclidean distances, respectively). We use $\boldsymbol{X}' = (\boldsymbol{x}'_1, \ldots, \boldsymbol{x}'_N)$ to denote an $m \times N$ matrix representing the coordinates of the points in the embedding space. The resulting set of points $\boldsymbol{x}'_1, \ldots, \mathbf{x}'_N$ in the Euclidean space is called the *canonical form* of the facial surface [25, 26]. The canonical forms are defined up to a rotation, translation, and reflection, and can be therefore treated by conventional algorithms used for rigid surface matching. Figure 5.6 shows an example of a deformable surface (human hand) undergoing isometric transformations, and the corresponding canonical forms of the hand.

We would like to find such a mapping φ that deforms the geodesic distances the least. The embedding error can be measured as a discrepancy $s(\boldsymbol{X}';\boldsymbol{D})$ between the original geodesic and the resulting Euclidean distances.

5.3.1 Multidimensional Scaling

Finding the best approximate flat embedding is possible by minimization of $s(\boldsymbol{X}';\boldsymbol{D})$ with respect to $\boldsymbol{X}'$. A family of algorithms used to carry out such an approximate flat embedding is known as *multidimensional scaling* (MDS) [9]. These algorithms differ in the choice of the embedding error criterion and the numerical method used for its minimization.

One of the most straightforward possibilities is to have the metric distortion defined as a sum of squared differences

$$s(\boldsymbol{X}';\boldsymbol{D}) = \sum_{i>j}(d_{ij} - d'_{ij})^2, \tag{2}$$

and the MDS problem is posed as a least-squares problem (LS-MDS). Such an embedding-error criterion is called the *raw stress*. Since the stress is a nonconvex function in $\boldsymbol{X}'$, standard convex optimization techniques do not guarantee convergence to the global minimum. Different techniques can be employed to prevent convergence to small local minima. Examples are the iterative convex majorization algorithm (SMACOF, standing for scaling by majorization of a convex function) [9, 22] or our recent multigrid optimization approach [11A].

An alternative to LS MDS is an algebraic embedding method due to Torgerson and Gower [51, 32] based on theoretical results of Eckman, Young, and Householder [23, 55], known as *classical scaling*. Classical scaling works with the squared geodesic distances, which can be expressed as the Hadamard (coordinatewise) product $\boldsymbol{\Delta} = \boldsymbol{D} \circ \boldsymbol{D}$. The matrix $\boldsymbol{\Delta}$ is first double-centered:

$$\boldsymbol{B} = -\frac{1}{2}\boldsymbol{J}\boldsymbol{\Delta}\boldsymbol{J} \tag{3}$$

(here $\boldsymbol{J} = \boldsymbol{I} - \frac{1}{N}\boldsymbol{1}\boldsymbol{1}^{\mathrm{T}}$ and $\boldsymbol{I}$ is an $N \times N$ identity matrix). Then, the eigendecomposition $\boldsymbol{B} = \boldsymbol{V}\boldsymbol{\Lambda}\boldsymbol{V}^{\mathrm{T}}$ is computed, where $\boldsymbol{V} = (\boldsymbol{v}_1, \ldots, \boldsymbol{v}_N)$ is the matrix of eigenvectors of $\boldsymbol{B}$ corresponding to the eigenvalues $\lambda_1 \geq \lambda_2 \geq \ldots \geq \lambda_N$. Denoting by $\boldsymbol{\Lambda}_+$ the matrix of first m positive eigenvalues and by $\boldsymbol{V}_+$ the matrix of the corresponding eigenvectors, the coordinate matrix in the embedding space is given by

$$\boldsymbol{X}' = \boldsymbol{V}_+\boldsymbol{\Lambda}_+. \tag{4}$$

In practice, since we are usually interested in embedding into $\mathbb{R}^3$ or $\mathbb{R}^2$, no full eigendecomposition of $\boldsymbol{B}$ is needed – it is enough to find only the first three or even

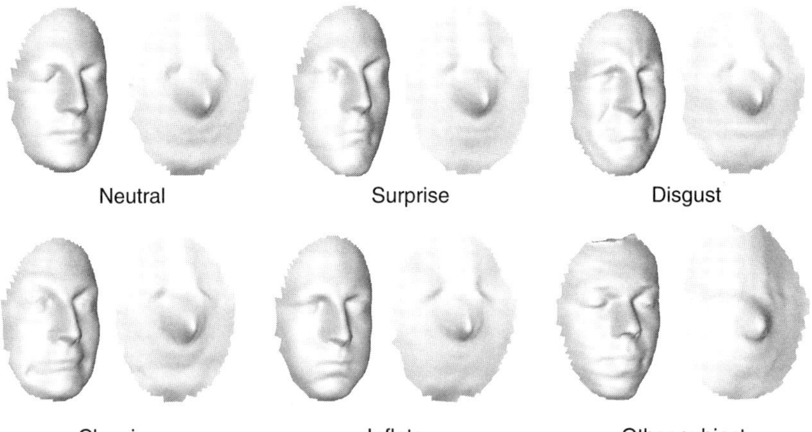

FIGURE 5.7: Examples of canonical forms of faces with strong facial expressions. For comparison, canonical form of a different subject is shown (second row, right).

two eigenvectors. The Arnoldi [3], Lanzcos, or block-Lanzcos [29, 5] algorithms can be used to performs this task efficiently.[1]

5.3.2 Canonical Forms of Facial Surfaces

When embedding is performed into a space of dimension $m = 3$, the canonical form can be plotted as a surface. Figure 5.7 depicts canonical forms of a person's face with different facial expressions. It demonstrates that, although the facial surface changes are substantial, the changes between the corresponding canonical forms are insignificant. Embedding into $\mathbb{R}^2$ is a special case – in this case, the codimension of the canonical form in the embedding space is zero. Such an embedding can be thought of as an intrinsic parametrization of the facial surface, which leads to a "warping" of the facial texture. This serves as a way of performing geometry-based registration of 2D facial images [12]. Flat embedding into $\mathbb{R}^2$ was previously used for cortical surface matching [47] in brain analysis, and adopted to texture mapping [56] in computer graphics.

[1]We thank Gene Golub and Michael Saunders (Stanford University) for their valuable comments on efficient numerical implementations of such eigendecomposition algorithms.

FIGURE 5.8: The 3DFACE prototype system and its main components: DLP projector (1), digital camera (2), monitor (3), magnetic card reader (4), mounting (5).

5.4 THE 3DFACE SYSTEM

We designed a prototype of a fully automatic 3D face recognition system based on the expression-invariant representation of facial surfaces. The 3DFACE system is shown in Figure 5.8. It can work both in one-to-one and one-to-many recognition modes. In one-to-one (verification) mode, the user swipes a magnetic card (4 in Figure 5.8) bearing his or her personal identification information. The system compares the subject's identity with the claimed one. In one-to-many (recognition) mode, the subject's identity is unknown *a priori* and it is searched for in a database of faces. In the current prototype, no automatic face detection and tracking is implemented. The monitor (3 in Figure 5.8) is used as a "virtual mirror" allowing the user to align himself relative to the camera.

Three-dimensional structure of the face is acquired by an active stereo time-multiplexed structured-light-range camera [11]. World coordinates of each point are computed by triangulation, based on the knowledge of the point location in the 2D coordinate system of the camera and the 1D coordinate system of the projector. The latter is inferred from a code which is projected in a form of light stripes onto the face of the subject using a DLP projector (1 in Figure 5.8). We use a 10-bit binary gray code, which allows us to obtain about 0.5 mm depth resolution with scan duration less than 200 msec. The scanner output is a cloud of 640×480 points.

The raw data of the scanner is first down-sampled to the resolution of 320×240, then undergoes initial cropping which roughly separates the facial contour from the background. In the topologically-constrained case, the lips are also cut off, hole filling (which removes acquired spike-like artifacts) and selective smoothing by a Beltrami-like geometric filter [49, 37, 14].

The smoothed surface is resized again about 3 times in each axis, and then the facial contour is extracted by using the *geodesic mask*. The key idea is locating invariant "source" points on the face and measuring an equidistant (in sense of the geodesic distances) contour around it. The geodesic mask is defined as the interior of this contour; all points outside the contour are removed. In the 3DFACE system, two fiducial points are used for the geodesic mask computation: the tip of the nose and the nose apex (the topmost point of the nose bone). These fiducial points are located automatically using a curvature-based 3D feature detector, similar to [44] (see details in [15]). The same feature detector is used to locate the left and the right eye.

The geodesic mask allows us to crop the facial surface in a geometrically consistent manner, insensitively to facial expressions. After cropping, the resulting surface contains between $2500 - 3000$ points. Computation of the geodesic distances is performed using the fast marching algorithm [38].

As the final stage, "canonization" of the facial surface is performed by computing the mutual geodesic distances between all the surface points and then applying MDS. We embed facial surfaces into $\mathbb{R}^3$ using LS MDS.

The 3D acquisition lasts less than 200 msec. The overall end-to-end processing time (including acquisition) is about 5 sec.

Since embedding is defined up to a Euclidean transformation (including reflection), the canonical surface must be aligned. We perform the alignment by first setting to zero the first-order moments (the center of gravity) $\mu_{100}, \mu_{010}, \mu_{001}$ of the canonical surface to resolve the translation ambiguity (here

$$\mu_{pqr} = \sum_{i=1}^{N} (x_i^1)^p (x_i^2)^q (x_i^3)^r \tag{5}$$

denotes the pqrth moment); then, the mixed second-order moments μ_{110}, μ_{011}, μ_{101} are set to zero to resolve the rotation ambiguity. Finally, using the coordinate relations of three fiducial points on the face (two eyes and the nose tip), the reflection ambiguity is resolved. The sequence of processing in 3DFACE system is illustrated in Figure 5.9

5.4.1 Surface Matching

The final stage of the face recognition algorithm is surface matching. Since the flattening compensates for the nonrigid isometries of the surface, standard

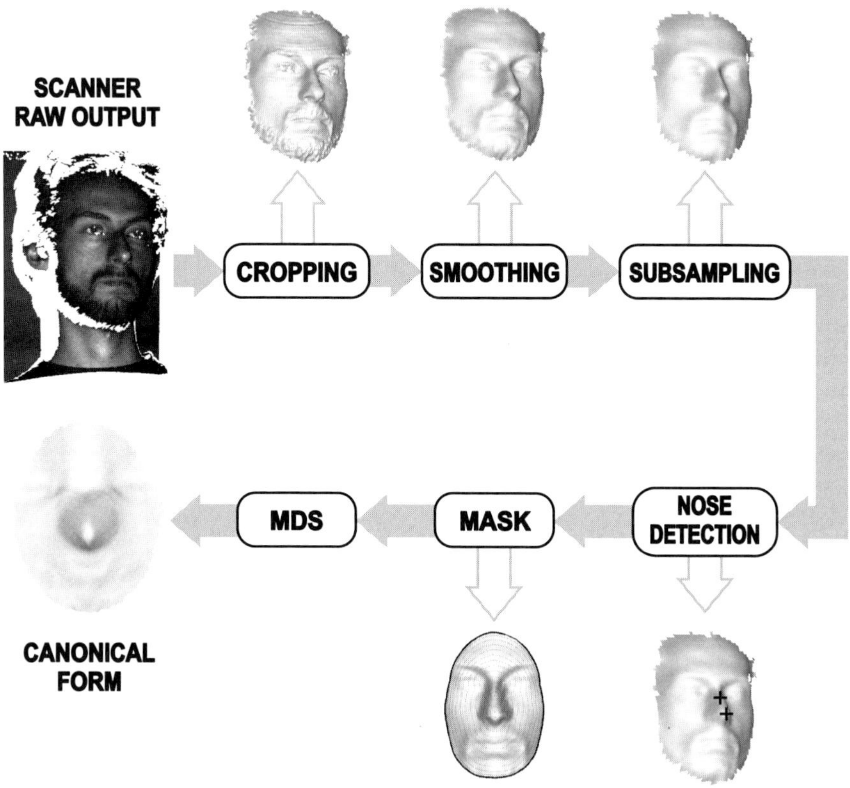

FIGURE 5.9: Scheme of preprocessing and canonization of the facial surface used in the 3DFACE system.

rigid matching (see, e.g., [33]) can be used for comparing the canonical surfaces. The standard choice in surface matching is the iterative closest-point (ICP) method and its variants [6], yet, it is disadvantageous from the point of view of computational complexity.

We use a simple and efficient surface-matching method based on high-order moments [50]. The main idea is to represent the surface by its moments μ_{pqr} up to some degree P, and compare the moments as vectors in a Euclidean space. Given two facial surface $\mathcal{S}$ and $\mathcal{Q}$ with the corresponding canonical forms $\boldsymbol{X}'$ and $\boldsymbol{Y}'$, we can define the distance between two faces as

$$d_{\text{can}}(\mathcal{S}, \mathcal{Q}) = \sum_{p+q+r \leq P} (\mu_{pqr}^{\boldsymbol{X}'} - \mu_{pqr}^{\boldsymbol{Y}'})^2. \tag{6}$$

In [13, 12] we proposed to treat canonical forms as images. After alignment, both the canonical surface and the flattened albedo are interpolated on a Cartesian grid, producing two images. These images can be compared using standard techniques, e.g., applying eigendecomposition like in eigenfaces or eigenpictures. The obtained representation was called in [13] *eigenforms*. The use of eigenforms has several advantages: First, image comparison is simpler than surface comparison, and second, the 2D texture information can be incorporated in a natural way as an additional classifier. Here, however, we focus on the 3D geometry, and in the following experiments use only the surface geometry ignoring the texture.

5.5 RESULTS

In this section, we present experimental results evaluating the 3DFACE method. First, we perform a set of experiments, the goal of which is to test how well canonical forms can handle strong facial expressions. The data sets used in these experiments contain facial expressions with closed mouth only. An evaluation of our algorithm on a data set containing expressions with both open and closed mouth can be found in [18]. Then, we provide a benchmark of 2D and 3D face recognition algorithms and compare them to our approach.

5.5.1 Sensitivity to Facial Expressions

In the first experiment (Figure 5.10), we studied the sensitivity of canonical forms to facial expressions. We used a data set containing 10 human and 3 artificial

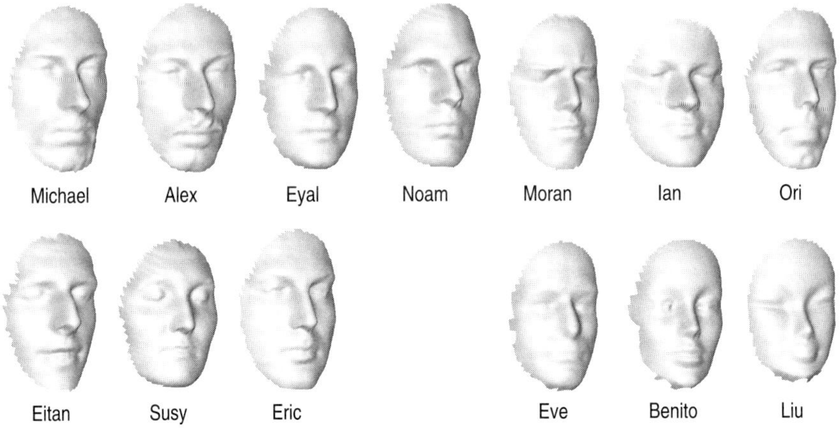

FIGURE 5.10: The subjects used in experiment I (shown with neutral expressions). Second row right: three artificial subjects.

Table 5.1: Description of the Experiment I data. Asterisk denotes artificial subjects. Double asterisk denotes identical twins.

Subject	Color	Neutral	Weak	Medium	Strong	Total
Michael**	red	6	5	6	-	17
Alex**	blue	3	1	3	1	8
Eyal	green	4	1	7	9	21
Noam	yellow	3	-	-	7	10
Moran	magenta	4	-	4	10	18
Ian	orange	5	-	16	7	28
Ori	cyan	8	-	11	10	29
Eric	d. green	5	3	-	3	11
Susy	d. magenta	6	-	9	8	23
David	l. blue	5	2	6	5	18
Eve*	black	6	-	-	-	6
Benito*	grey	7	-	-	-	7
Liu*	l. grey	8	-	-	-	8

subjects. Subjects Alex (blue) and Michael (red) are identical twins. Each face in the data set appeared with a number of instances ($6-29$ instance per subject, a total of 204 instances) in a variety of facial expressions. The database is summarized in Table 5.1.

All expressions were conducted with a closed mouth and were classified into 10 types (neutral expression + 9 expressions) and into 3 strengths (weak, medium, strong). Neutral expressions are natural postures of the face, while strong expressions are exaggerated postures rarely encountered in everyday life (Figure 5.11, second row). The group of expressions including smile, sadness, anger, surprise and disgust are basic emotions according to Ekman [24]; the group thinking, stress, grin and chewing tries to imitate facial appearance that can occur in a natural environment; finally, expressions inflate and deflate result in the most significant deformation of the facial geometry, though rarely encountered (Figure 5.11, first row). Small head rotations (up to about 10 degrees) were allowed. Since the data were acquired on a course of several months, variations in illumination conditions, facial hair, etc., are also present.

For reference, our approach was compared to rigid matching of facial surfaces. In both cases, the metric d_{can} based on moments of degree up to $P = 5$ (i.e., vectors of dimensionality 52), according to (5), was used.

The dissimilarities (distances) between different faces obtained by the metric d_{can} allow us to cluster together faces that belong to the same subjects. As a quantitative measure of the separation quality, we use the ratio of the maximum

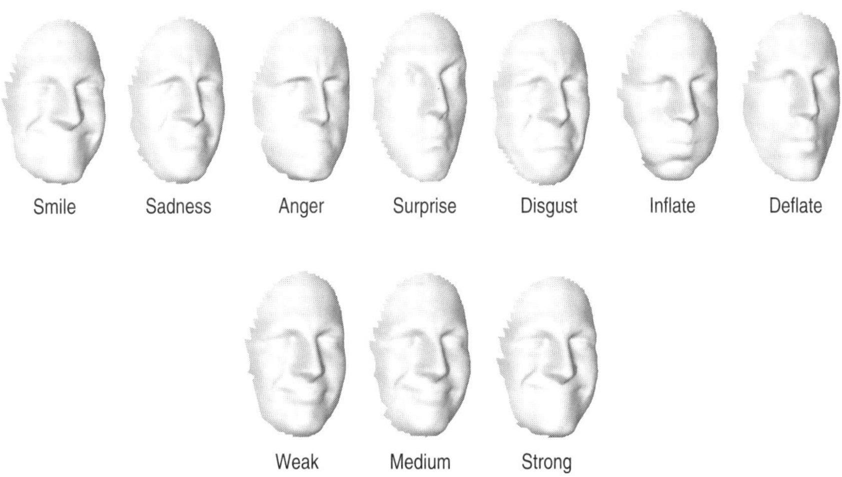

FIGURE 5.11: First row: seven representative facial expressions of subject Eyal in experiment I. Second row: three degrees of the smile expression of the same subject.

intercluster to minimum intracluster dissimilarity,

$$\varsigma_k = \frac{\max_{i,j\in C_k} \eta_{ij}}{\min_{i\notin C_k, j\in C_k} \eta_{ij}}, \tag{7}$$

and the ratio of root-mean-squared (RMS) intercluster and intracluster dissimilarities

$$\sigma_k = \sqrt{\frac{\frac{2}{|C_k|^2-|C_k|} \sum_{i,j\in C_k,\ i>j} \eta_{ij}^2}{\frac{1}{|C_k|(|C|-|C_k|)} \sum_{i\notin C_k, j\in C_k} \eta_{ij}^2}} \tag{8}$$

(C_k denotes indexes of k-th subject's faces, and η_{ij} denotes dissimilarities between faces i and j). This criterion measures how each cluster is tight and far from other clusters. Ideally, ς_k and σ_k should tend to zero.

Figure 5.12 depicts a three-dimensional visualization of the dissimilarities between faces, obtained by applying classical scaling to the dissimilarity matrix of faces. Each face on this plot is represented by a point; faces of different subjects are marked with different colors. The first row depicts the dissimilarities between faces with only neutral expressions. Faces of different subjects form tight clusters and are easily distinguishable. The advantage of canonical forms is not so apparent in this case. However, the picture changes drastically when we allow for facial expressions (Figure 5.12, second row) The clusters corresponding to canonical

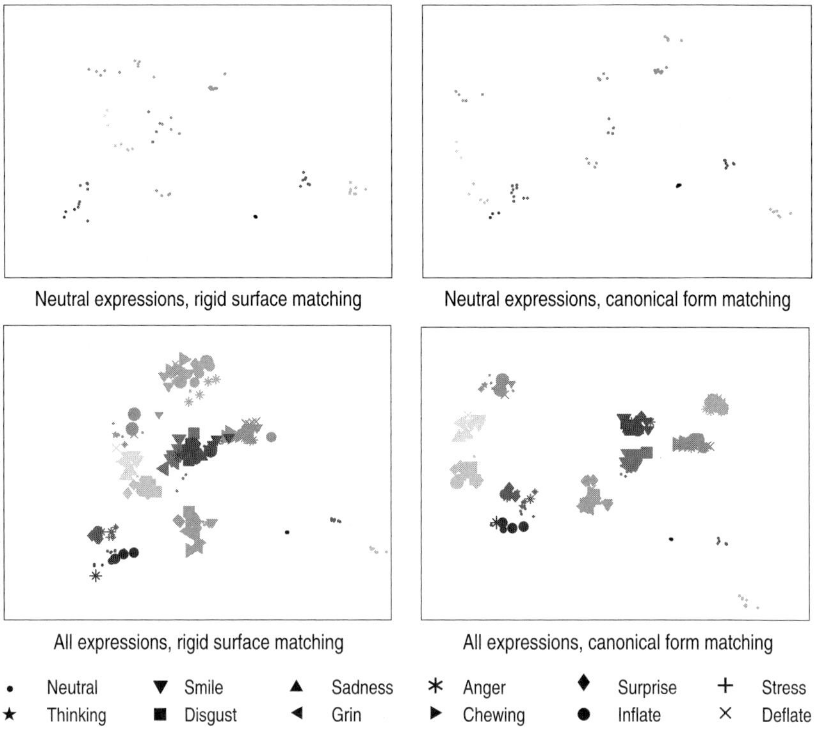

FIGURE 5.12: Low-dimensional visualization of dissimilarities between faces in Experiment I using original surface (left) and canonical form (right) matching. First row: neutral expressions only. Second row: all expressions. Colors represent different subjects. Symbols represent different facial expressions. Symbol size represents the strength of the facial expression. (See also color plate section)

surface matching are much tighter; moreover, we observe that using rigid surface matching some clusters (red and blue, dark and light magenta, light blue, yellow and green) overlap, which means that a face recognition algorithm based on rigid surface matching would confuse between these subjects.

Figure 5.13 shows the separation quality criteria (ς_k and σ_k) for rigid and canonical surface matching. When only neutral expressions are used, canonical form matching outperform rigid surface matching on most subjects in terms of ς_k and σ_k (by up to 68% in terms of ς_k and by up to 64% in terms of σ_k; slightly inferior performance in terms of ς_k is seen on artificial subject Eve and human subjects Eyal, Noam, and David). The explanation to the fact that canonical forms are better even in case when no large expression variability is present, is that "neutral

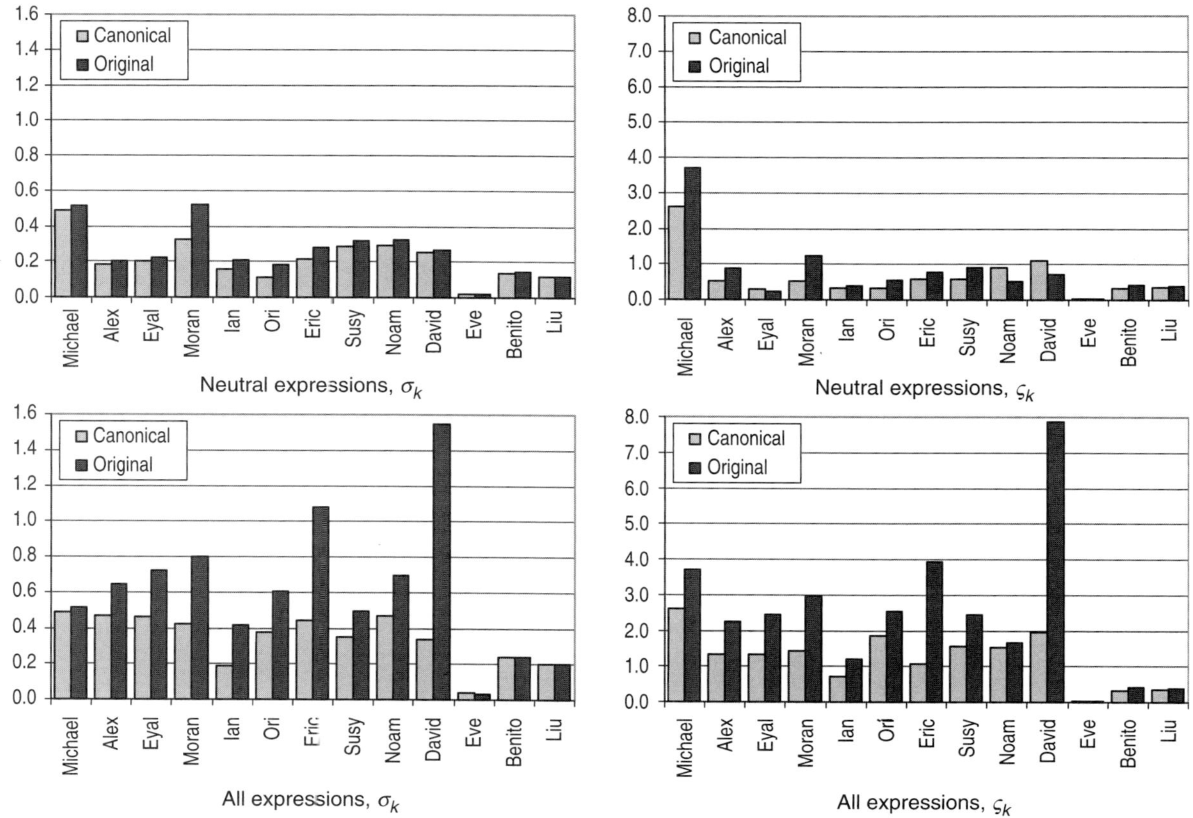

FIGURE 5.13: Separation quality criteria (ς_k and σ_k) using original (dark gray) and canonical (light gray) surface matching. The smaller ς_k and σ_k, the better is the separation quality.

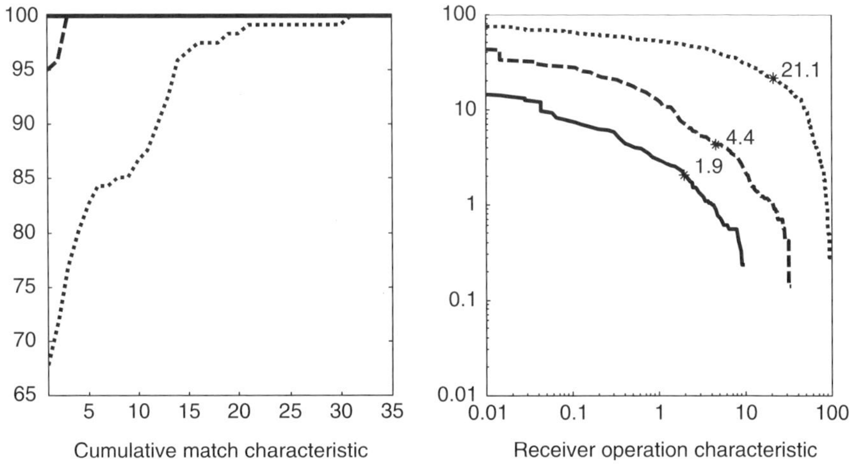

FIGURE 5.14: CMC (left) and ROC (right) curves of face recognition based on surface matching (dashed), canonical surface matching (solid), and eigenfaces (dotted). Obtained on database with all expressions. Star denotes equal error rate.

expression" as a fixed, definite expression, does not exist, and even when the face of the subject seems expressionless, its possible deformations are still sufficiently significant. When allowing for facial expressions, our approach outperforms original surface matching by up to 304% in terms of ς_k and by up to 358% in terms of σ_k.

5.5.2 Comparison of Algorithms

The goal of the second experiment is performing a benchmark of our method and comparing it to other face recognition algorithms. Faces from the probe database (30 different probe subjects in a variety of facial expression, total of 220 faces) were compared to a set of 65 gallery templates (typically, two or three templates per subject were used). Only neutral expressions were used in the gallery.

Three algorithms were tested: canonical form matching, facial surface matching, and 2D image-based eigenfaces. Eigenfaces were trained by 35 facial images that did not appear as templates; 23 eigenfaces were used for the recognition (the first two eigenfaces were excluded in order to decrease the influence of illumination variability [54]), see also Chapter 1.

Figure 5.14 (left) shows the cumulative match characteristic (CMC) curves of three algorithms compared in this experiment on full database with all facial expressions. Our approach results in rank 1 *zero* recognition error. Figure 5.14 (right) shows the receiver operation characteristic (ROC) curves. Our algorithm

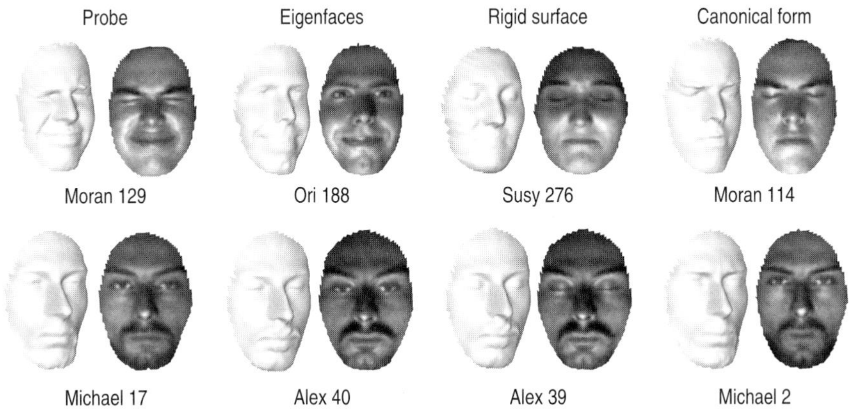

FIGURE 5.15: Example of recognition using different algorithms. The first column shows the probe subject; the second through fourth columns depict the closest (rank 1) matches found by the canonical form matching, facial surface matching and eigenfaces, respectively. Note that only the match using canonical form matching is correct. Numbers represent the subject's index in the database. Wrong matches are emphasized.

significantly outperforms both the rigid facial surface matching and the eigenface algorithm.

Figure 5.15 shows an example of rank-1 recognition on the full database (220 instances with facial expressions). The first column depicts a probe subject with extreme facial expression; columns two through four depict the rank-1 matches among the 65 templates using eigenfaces, facial surface matching, and canonical form matching.

The first row in Figure 5.15 shows results typical for the described algorithms. Eigenfaces, being image-based, finds the subject Ori 188 more similar to the reference subject Moran 129 since they have the same facial expression (strong smile), though these are different subjects. Facial surface matching is confused by 3D features (outstanding inflated cheeks) that appear on the face of subject Moran 129 due to the facial expression. These features are similar to the natural facial features (fat cheeks) of subject Susy 276. Finally, canonical surface matching finds a correct match (Moran 114), since flattening compensates for the distortion of the face of subject Moran 129 due to a smile.

The second row in Figure 5.15 shows an example of identical twins recognition – the most challenging task for a face recognition algorithm. The eigenface algorithm resulted in 29.41% incorrect matches when enrolling Michael and 25% when enrolling Alex. Facial surface matching resulted in 17.64% and 0% wrong matches,

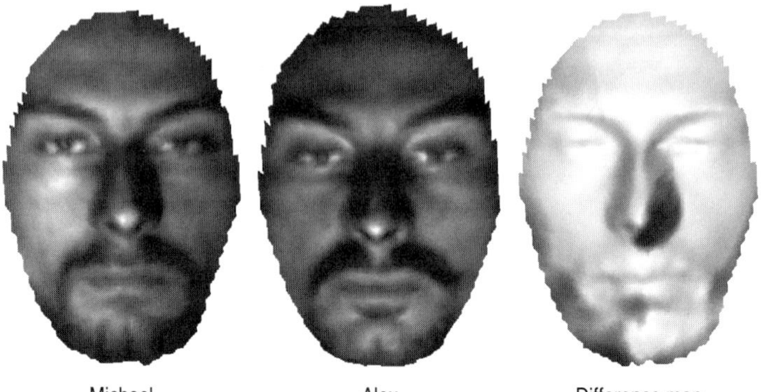

FIGURE 5.16: A pair of identical twins participating in the experiment (Alex and Michael), and the difference (blue shades represent the difference absolute value) computed in the canonical form domain and mapped onto the facial surface.

respectively. Canonical form matching resulted in 0% recognition error for both twins.

Comparing the canonical forms averaged on about 10 instances with different expressions for each of the twins, we found out a slight difference in the 3D geometry of the nose, which makes this distinction possible (Figure 5.16). Apparently, the difference is very subtle and is not distinct if using rigid surface matching, as the nose deforms quite significantly due to facial expressions.

5.6 CONCLUSIONS

The geometric framework for 3D face recognition presented here provides a solution to a major problem in face recognition: sensitivity to facial expressions. Being an internal characteristic of the human face, facial expressions are harder to deal with compared to external factors like pose or lighting. This problem is especially acute when face recognition is performed in a natural environment.

Thinking of expressions as approximated isometric transformations of a deformable facial surface allows to construct an expression-invariant representation of the face. Our approach outperforms other 3D recognition methods that treat the face as a rigid surface. The 3DFACE face recognition system prototype implementing our algorithm demonstrates high recognition accuracy and has the capability to distinguish between identical twins. It is now being evaluated for various industrial applications.

ACKNOWLEDGMENTS

We are grateful to Gene Golub and Michael Saunders (Stanford University) for valuable notes on efficient implementation of eigendecomposition algorithms, to David Donoho (Stanford University) for pointing us to Ekman's publications on facial expressions, and to everyone who contributed their faces to our database.

This research was supported by the Israel Science Foundation (ISF), Grant No. 738/04 and the Bar Nir Bergreen Software Technology Center of Excellence (STL).

REFERENCES

[1] B. Achermann and H. Bunke, Classifying range images of human faces with Hausdorff distance, *Proc. ICPR*, September 2000, pp. 809–813.

[2] B. Achermann, X. Jiang, and H. Bunke, Face recognition using range images, *Int'l Conf. Virtual Systems and Multimedia*, 1997, pp. 129–136.

[3] W. Arnoldi, The principle of minimized iterations in the solution of the matrix eigenvalue problem, *Quart. Appl. Math.* **9** (1951), 17–29.

[4] J. Ashbourn, Biometrics: advanced identity verification, Springer-Verlag, Berlin, Heidelberg, New York, 2002.

[5] Z. Bai, J. Demmel, J. Dongarra, A. Ruhe, and H. van der Vorst, *Templates for the Solution of Algebraic Eigenvalue Problems: A Practical Guide*, third ed., SIAM, Philadelphia, 2000, Online: http://www.cs.utk.edu/ dongarra/etemplates/index.html.

[6] P. J. Besl and N. D. McKay, A method for registration of 3D shapes, *IEEE Trans. PAMI* **14** (1992), 239–256.

[7] C. Beumier and M. P. Acheroy, Automatic face authentication from 3D surface, *Proc. British Machine Vision Conf.*, 1998, pp. 449–458.

[8] W. W. Bledsoe, The model method in facial recognition, Technical Report PRI 15, Panoramic Research Inc., Palo Alto (CA) USA, 1966.

[9] I. Borg and P. Groenen, *Modern Multidimensional Scaling - Theory and Applications*, Springer-Verlag, Berlin, Heidelberg, New York, 1997.

[10] K. W. Bowyer, K. Chang, and P. Flynn, A survey of 3D and multi-modal 3D+2D face recognition, Dept. of computer science and electrical engineering technical report, University of Notre Dame, January 2004.

[11A] M. M. Bronstein, A. M. Bronstein, R. Kimmel, I. Yavneh,"A multigrid approach for multi-dimensional scaling", Copper Mountain Conf. on Multigrid Methods, 2005.

[11] A. Bronstein, M. Bronstein, E. Gordon, and R. Kimmel, High-resolution structured light range scanner with automatic calibration, Tech. Report CIS-2003-06, Dept. of Computer Science, Technion, Israel, 2003.

[12] A. M. Bronstein, M. M. Bronstein, E. Gordon, and R. Kimmel, Fusion of 3D and 2D information in face recognition, *Proc. ICIP*, 2004.

[13] A. M. Bronstein, M. M. Bronstein, and R. Kimmel, Expression-invariant 3D face recognition, *Proc. Audio and Video-based Biometric Person Authentication*, 2003, pp. 62–69.

[14] ______, Three-dimensional face recognition, Tech. Report CIS-2004-04, Dept. of Computer Science, Technion, Israel, 2004.

[15] ______, Three-dimensional face recognition, *IJCV* (2005), Vol. 64, no. 1, pp. 5–30, August 2005.
[16] A. M. Bronstein, M. M. Bronstein, R. Kimmel, and A. Spira, Face recognition from facial surface metric, *Proc. ECCV*, pp. 225-237, 2004.
[17] M. M. Bronstein and A. A. Bronstein, Biometrics was no match for hair-rising tricks, *Nature* **420** (2002), 739.
[18] M. M. Bronstein, A. M. Bronstein, and R. Kimmel, Expression-invariant representations for human faces, Tech. Report CIS-2005-01, Dept. of Computer Science, Technion, Israel, 2005.
[19] J. Y. Cartoux, J. T. LaPreste, and M. Richetin, Face authentication or recognition by profile extraction from range images, *Proc. Workshop on Interpretation of 3D Scenes*, November 1989, pp. 194–199.
[20] K. Chang, K. Bowyer, and P. Flynn, Face recognition using 2D and 3D facial data, *Proc. Multimodal User Authentication Workshop*, December 2003, pp. 25–32.
[21] I. Cox, J. Ghosn, and P. Yianilos, Feature-based face recognition using mixture distance, *Proc. CVPR*, 1996, pp. 209–216.
[22] J. De Leeuw, Applications of convex analysis to multidimensional scaling, In J.R. Barra, F. Brodeau, G. Romier and B. van Custem (Eds.), Recent developments in statistics, pp. 133–145, 1977, Amsterdam, The Netherlands: North-Holland.
[23] C. Eckart and G. Young, Approximation of one matrix by another of lower rank, *Psychometrika* **1** (1936), 211–218.
[24] P. Ekman, *Darwin and Facial Expression; a Century of Research in Review*, Academic Press, New York, 1973.
[25] A. Elad and R. Kimmel, Bending invariant representations for surfaces, *Proc. CVPR*, 2001, pp. 168–174.
[26] ______, On bending invariant signatures for surfaces, *IEEE Trans. PAMI* **25** (2003), no. 10, 1285–1295.
[27] A. S. Georghiades, P. N. Belhumeur, and D.J. Kriegman, Illumination cones for recognition under variable lighting: faces, *Proc. CVPR*, 1998, pp. 52–58.
[28] A. Goldstein, L. Harmon, and A. Lesk, Identification of human faces, *Proc. IEEE* **59** (1971), no. 5, 748–760.
[29] G. H. Golub and C. F. van Loan, *Matrix Computations*, third ed., The John Hopkins University Press, 1996.
[30] G. Gordon, Face recognition based on depth and curvature features, *Proc. CVPR*, 1992, pp. 108–110.
[31] ______, Face recognition from frontal and profile views, *Proc. Int'l Workshop on Face and Gesture Recognition*, 1997, pp. 74–52.
[32] J. C. Gower, Some distance properties of latent root and vector methods used in multivariate analysis, *Biometrika* **53** (1966), pp. 325–338.
[33] A. Gruen and D. Akca, Least squares 3d surface matching, *Proc. ISPRS Working Group*, V/1: Panoramic Photogrammetry Workshop, 2004, pp. 19–22.
[34] C. Hesher, A. Srivastava, and G. Erlebacher, A novel technique for face recognition using range images, *Int'l Symp. Signal Processing and Its Applications*, 2003.
[35] J. Huang, V. Blanz, and V. Heisele, Face recognition using component-based SVM classification and morphable models, *SVM* (2002), pp. 334–341.
[36] T. Kanade, Picture processing by computer complex and recognition of human faces, Technical report, Kyoto University, Dept. of Information Science, 1973.

[37] R. Kimmel, *Numerical Geometry of Images*, Springer-Verlag, Berlin, Heidelberg, New York, 2003.

[38] R. Kimmel and J. A. Sethian, Computing geodesic on manifolds, *Proc. US National Academy of Science*, vol. 95, 1998, pp. 8431–8435.

[39] E. Kreyszig, *Differential Geometry*, Dover Publications Inc., New York, 1991.

[40] N. Mavridis, F. Tsalakanidou, D. Pantazis, S. Malassiotis, and M. G. Strintzis, The HISCORE face recognition application: Affordable desktop face recognition based on a novel 3D camera, *Proc. Euroimage Intl. conf. Augmented Virtual Environments and 3D Imaging* (ICAV 3D), May 2001.

[41] G. Medioni and R. Waupotitsch, Face recognition and modeling in 3D, Proc. AMFG, October 2003, pp. 232–233.

[42] F. Mémoli and G. Sapiro, Comparing point clouds, IMA preprint series 1978, University of Minnesota, Minneapolis, MN 55455, USA, April 2004.

[43] ———, A theoretical and computational framework for isometry invariant recognition of point cloud data, IMA preprint series 1980, University of Minnesota, Minneapolis, MN 55455, USA, June 2004.

[44] A. B. Moreno, A. Sanchez, J. Velez, and J. Diaz, Face recognition using 3D surface-extracted descriptors, *Irish Machine Vision and Image Processing Conference*, 2003.

[45] T. Nagamine, T. Uemura, and I. Masuda, 3D facial image analysis for human identification, *Proc. ICPR*, 1992, pp. 324–327.

[46] J. Ortega-Garcia, J. Bigun, D. Reynolds, and J. Gonzalez-Rodriguez, Authentication gets personal with biometrics, *IEEE Signal Processing magazine* **21** (2004), no. 2, pp. 50–62.

[47] E. L. Schwartz, A. Shaw, and E. Wolfson, A numerical solution to the generalized mapmaker's problem: flattening nonconvex polyhedral surfaces, *IEEE Trans. PAMI* **11** (1989), pp. 1005–1008.

[48] L. Sirovich and M. Kirby, Low-dimensional procedure for the characterization of human faces, *JOSA* A **2** (1987), pp. 519–524.

[49] N. Sochen, R. Kimmel, and R. Malladi, A general framework for low level vision, *IEEE Trans. Image Processing* **7** (1998), no. 3, pp. 310–318.

[50] A. Tal, M. Elad, and S. Ar, Content based retrieval of VRML objects - an iterative and interactive approach, *Eurographics Workshop in Multimedia*, 2001.

[51] W. S. Torgerson, Multidimensional scaling I - theory and methods, *Psychometrika* **17** (1952), pp. 401–419.

[52] F. Tsalakanidou, S. Malassiotis, and M. G. Strintzis, Face localization and authentication using color and depth images, *IEEE Transactions on Image Processing*, vol. 14, no. 2, pp. 152–168, February 2005.

[53] F. Tsalakanidou, D. Tzocaras, and M. Strintzis, Use of depth and colour eigenfaces for face recognition, *Pattern Recognition Letters* **24** (2003), pp. 1427–1435.

[54] M. Turk and A. Pentland, Face recognition using eigenfaces, *Proc. CVPR*, 1991, pp. 586–591.

[55] C. Eckart, G. Young and A. S. Householder, Discussion of a set of point in terms of their mutual distances, *Psychometrika* **3** (1938), pp. 19–22.

[56] G. Zigelman, R. Kimmel, and N. Kiryati, Texture mapping using surface flattening via multi-dimensional scaling, *IEEE Trans. Visualization and Computer Graphics* **9** (2002), no. 2, pp. 198–207.

CHAPTER 6

3D FACE MODELING FROM MONOCULAR VIDEO SEQUENCES

6.1 INTRODUCTION

Constructing 3D face models from a video sequence is one of the challenging problems of computer vision. Successful solution of this problem has applications in multimedia, computer graphics, and face recognition. In multimedia, 3D face models can be used in videoconferncing applications for efficient transmission. In computer-graphics applications, 3D face models form the basic building block upon which facial movements and expressions can be added. Being able to build these models automatically from video data would greatly simplify animation tasks where models are now painstakingly built with significant human intervention. By incorporating 3D models into face-recognition systems, the problems arising due to pose, illumination, and expression variations can be effectively addressed. Most current commercial systems address special cases of the 3D face-reconstruction problem that are well constrained by additional information in the form of depth estimates available from multiple cameras, projecting laser or other patterns on the face, decorating the face with special textures to make inter-frame correspondences simple, or using structured light to reveal the contours of the face. All these constraints and the special hardware required reduce the operational flexility of the system.

Various researchers have addressed the issue of 3D face modeling. In [16], the authors used an extended Kalman filter to recover the 3D structure of a face which was then used for tracking. A method for recovering nonrigid 3D shapes as a linear combination of a set of basis shapes was proposed in [4]. A factorization based method for recovering nonrigid 3D structure and motion from video was presented in [3]. In [20], the author proposes a method for self-calibration in the presence of varying internal camera parameters and reconstructs metric 3D structure. One of the common approaches to solve the problem of 3D reconstruction from a monocular video is structure from motion (SfM). Numerous SfM algorithms [15] that can reconstruct a 3D scene from two or more images exist in the literature. The basic idea behind most SfM algorithms is to recover the structure of 3D points on a rigid object from 2D point correspondences across images, or recover a dense depth map from optical flow. These methods have been adapted for face modeling in [11, 28], where the authors proposed solving the problem of 3D face modeling using a generic model. Their method of bundle-adjustment works by initializing the reconstruction algorithm with this generic model. Romdhani, Blanz and Vetter [22] came up with an impressive appearance-based approach where they showed that it is possible to recover the shape and texture parameters of a 3D morphable model from a single image. Shape from contours [2] is another promising approach for 3D reconstruction. One of the strongest cues for the 3D information contained in a 2D image is the outline of an object in the image. The occluding contour (extreme boundary) in a 2D image directly reflects the 3D shape. Shape-from-silhouette techniques have been used to reconstruct 3D shapes from multiple silhouette images of an object without assuming any previous knowledge of the object to be reconstructed [17]. However, it is impossible to recover the concavities in the shape of the object from the silhouettes or contours. But if we assume prior knowledge of the object being reconstructed, and use a generic model, contour information can be exploited as an important constraint for the exact shape of the object. Moghaddam et al [18] have developed a system to recover the 3D shape of a human face from a sequence of silhouette images. They used a downhill simplex method to estimate the model parameters, which are the coefficients of the eigenhead basis functions.

In this chapter, we present two methods that we have developed for 3D face modeling from monocular video sequences. The first method uses the multiframe SfM approach to arrive at an initial estimate of the face model [27]. This is followed by a smoothing phase where the errors in this estimate are corrected by comparison with a generic face model. The method also involves a statistical quality evaluation of the input video [24], and compensating for fluctuations in quality within the SfM algorithm framework. The second approach that we present relies on matching a generic 3D face model to the outer contours of the face to be modeled and a few of its internal features [14]. Using contours separates the geometric subtleties of the human head from the variations in shading and texture. The adaptation of the generic face model is integrated over all the frames of the video sequence, thus producing a 3D model that is accurate over a range of pose variations.

6.2 SFM-BASED 3D FACE MODELING

In this section, we present a method for face modeling from monocular video using SfM, with special emphasis on the incorporation of the generic face model and evaluating the quality of the input video sequence. The main features of our method are the following.

Reconstructing from a monocular video. This is particularly important in unregulated surveillance applications where the training data (from which the 3D model needs to be estimated) may contain a few images or a video from one view (e.g. frontal), but the probe may be another view of the person (e.g. profile). Since the motion between pairs of frames in a monocular video is usually small, we adopt the optical-flow paradigm for 3D reconstruction [19]. Also, estimating the motion between the pairs of frames accurately may be a challenge in many situations because of differences in the quality of the input video. Our method learns certain statistical parameters of the incoming video data and incorporates them in the algorithm. Quality evaluation is done by estimating the statistical error covariance of the depth estimate analytically from the optical-flow equations. The details of the statistical analysis can be found in [23, 24].

Avoid biasing the reconstruction with the generic model. Mathematically speaking, the introduction of the generic model is similar to introducing constraints on the solution of the 3D estimation problem. In our method, we introduce the generic model *after* obtaining the estimate using the SfM algorithm. The SfM algorithm reconstructs purely from the video data after computing the optical flow. This is unlike the methods in [11] and [28] where the generic model was used to initialize a bundle-adjustment approach. The difficulty with this approach is that the algorithm often converges to a solution very near this initial value, resulting in a reconstruction which has the characteristics of the generic model, rather than that of the particular face in the video which needs to be modeled. This method may give very good results when the generic model has significant similarities with the particular face being reconstructed. However, if the features of the generic model are different from those of the face being reconstructed, the solution obtained using this approach may be unsatisfactory. We provide some experimental validation for this statement later. After the initial SfM-based estimate is obtained, we use a cost function which identifies local regions in the generic face model where there are no sharp depth discontinuities, looks for deviations in the trend of the values of the 3D estimate in these regions and then corrects for the errors. The optimization of the cost function is done in a Markov-chain Monte Carlo (MCMC) framework using a Metropolis–Hastings sampler [8]. The advantage of this method is that the particular characteristics of the face that is being modeled are not lost, since the SfM algorithm does not incorporate the generic model. However, most errors (especially those with large deviations from the average representation) in the reconstruction are corrected in the energy-function minimization process by comparison with the generic model.

We start by providing a brief overview of error modeling in SfM, followed by the reconstruction algorithm, including incorporation of the generic model, and finally present the results.

6.2.1 Error Estimation in 3D Reconstruction

Since the motion between adjacent frames in a video sequence of a face is usually small, we will adopt the optical-flow framework for reconstructing the structure [19]. It is assumed that the coordinate frame is attached rigidly to the camera with the origin at the center of perspective projection and the z axis perpendicular to the image plane. The camera is moving with respect to the face being modeled (which is assumed rigid) with translational velocity $\mathbf{V} = [v_x, v_y, v_z]$ and rotational velocity $\mathbf{\Omega} = [\omega_x, \omega_y, \omega_z]$ (this can be reversed by simply changing the sign of the velocity vector). Using the small-motion approximation to the perspective-projection model for motion-field analysis, and denoting by $p(x, y)$ and $q(x, y)$ the horizontal and vertical velocity fields of a point (x, y) in the image plane, we can write the equations relating the object motion and scene depth by [19]:

$$p(x, y) = (x - fx_f)h(x, y) + \frac{1}{f}xy\omega_x - (f + \frac{1}{f}x^2)\omega_y + y\omega_z,$$

$$q(x, y) = (y - fy_f)h(x, y) + (f + \frac{1}{f}y^2)\omega_x - \frac{1}{f}xy\omega_y - x\omega_z, \tag{1}$$

where f is the focal length of the camera, $(x_f, y_f) = (\frac{v_x}{v_z}, \frac{v_y}{v_z})$ is known as the *focus of expansion* (FOE), and $h(x, y) = \frac{v_z}{z(x,y)}$ is the scaled inverse scene depth. We will assume that the FOE is known over a few frames of the video sequence. Under the assumption that the motion between adjacent frames in a video is small, we compute the FOE from the first two or three frames and then keep it constant over the next few frames [30]. For N corresponding points, using subscript i to represent the above defined quantities at the ith point, we define (similar to [30])

$$\boldsymbol{h} = (h_1, h_2, \ldots, h_N)^{\mathrm{T}}_{N\times 1},$$
$$\boldsymbol{u} = (p_1, q_1, p_2, q_2, \ldots, p_N, q_N)^{\mathrm{T}}_{2N\times 1},$$
$$\boldsymbol{r}_i = (x_i y_i, -(1+x_i^2), y_i)^{\mathrm{T}}_{3\times 1},$$
$$\boldsymbol{s}_i = (1+y_i^2, -x_i y_i, -x_i)^{\mathrm{T}}_{3\times 1},$$
$$\boldsymbol{\Omega} = (w_x, w_y, w_z)^{\mathrm{T}}_{3\times 1},$$
$$\boldsymbol{Q} = \begin{bmatrix}\boldsymbol{r}_1 & \boldsymbol{s}_1 & \boldsymbol{r}_2 & \boldsymbol{s}_2 & \ldots \boldsymbol{r}_N & \boldsymbol{s}_N\end{bmatrix}^{\mathrm{T}}_{2N\times 3}$$

$$\boldsymbol{P} = \begin{bmatrix} x_1 - x_f & 0 & \cdots & 0 \\ y_1 - y_f & 0 & \cdots & 0 \\ 0 & x_2 - x_f & \cdots & 0 \\ 0 & y_2 - y_f & \cdots & 0 \\ \vdots & \vdots & \ddots & \vdots \\ 0 & 0 & \cdots & x_N - x_f \\ 0 & 0 & \cdots & y_N - y_f \end{bmatrix}_{2N\times N}, \tag{2}$$

$$\boldsymbol{A} = [\boldsymbol{P} \quad \boldsymbol{Q}]_{2N\times(N+3)},$$

$$z = \begin{bmatrix}\boldsymbol{h} \\ \boldsymbol{\Omega}\end{bmatrix}_{(N+3)\times 1} \tag{3}$$

Then (1) can be written as

$$\boldsymbol{A}z = \boldsymbol{u}. \tag{4}$$

Our aim is to compute z from $\boldsymbol{u}$ and to obtain a quantitative idea of the accuracy of the 3D reconstruction z as a function of the uncertainty in the motion estimates $\boldsymbol{u}$. Let us denote by $\boldsymbol{R_u}$ the covariance matrix of $\boldsymbol{u}$ and by C the cost function

$$C = \frac{1}{2}||\boldsymbol{A}\, z - \boldsymbol{u}||^2 = \frac{1}{2}\sum_{i=1}^{2N} C_i^2(u_i, z). \tag{5}$$

In [24], using the implicit function theorem [33], we proved the following result.

Theorem Define

$$\begin{aligned}
\boldsymbol{A}_{\bar{i}p} &= [0 \ \cdots \ 0 \ -(x_{\bar{i}} - x_f) \ 0 \ \cdots \ 0 \ -x_{\bar{i}}y_{\bar{i}} \ (1 + x_{\bar{i}}^2) \ -y_{\bar{i}}],\\
&= [-(x_{\bar{i}} - x_f)\boldsymbol{I}_{\bar{i}}(N) | - \mathbf{r}_{\bar{i}}] = [A_{\bar{i}ph} | A_{\bar{i}pm}],\\
\boldsymbol{A}_{\bar{i}q} &= [0 \ \cdots \ 0 \ -(y_{\bar{i}} - y_f) \ 0 \ \cdots \ 0 \ -(1 + y_{\bar{i}}^2) \ x_{\bar{i}}y_{\bar{i}}(N) \ x_{\bar{i}}],\\
&= [-(y_{\bar{i}} - y_f)\boldsymbol{I}_{\bar{i}}(N) | - \mathbf{s}_{\bar{i}}] = [A_{\bar{i}qh} | A_{\bar{i}qm}],
\end{aligned} \tag{6}$$

where $\bar{i} = \lceil i/2 \rceil$ is the ceiling of i ($\bar{i}$ will then represent the number of feature points N and $i = 1, \ldots, n = 2N$) and $\boldsymbol{I}_n(N)$ denotes a 1 in the nth position of the array of length N and zeros elsewhere. The subscript p in $\boldsymbol{A}_{\bar{i}p}$ and q in $\boldsymbol{A}_{\bar{i}q}$ denotes that the elements of the respective vectors are derived from the pth and qth components of the motion in (1). Then

$$\boldsymbol{R}_z = \boldsymbol{H}^{-1} \left(\sum_i \frac{\partial C_i^{\mathrm{T}}}{\partial z} \frac{\partial C_i}{\partial \boldsymbol{u}} \boldsymbol{R}_u \frac{\partial C_i^{\mathrm{T}}}{\partial \boldsymbol{u}} \frac{\partial C_i}{\partial z} \right) \boldsymbol{H}^{-\mathrm{T}}, \tag{7}$$

$$= \boldsymbol{H}^{-1} \left(\sum_{\bar{i}=1}^{N} \left(\boldsymbol{A}_{\bar{i}p}{}^{\mathrm{T}} \boldsymbol{A}_{\bar{i}p} R_{u\bar{i}p} + \boldsymbol{A}_{\bar{i}q}{}^{\mathrm{T}} \boldsymbol{A}_{\bar{i}q} R_{u\bar{i}q} \right) \right) \boldsymbol{H}^{-\mathrm{T}}, \tag{8}$$

and

$$\boldsymbol{H} = \sum_{\bar{i}=1}^{N} \left(\boldsymbol{A}_{\bar{i}p}{}^{\mathrm{T}} \boldsymbol{A}_{\bar{i}p} + \boldsymbol{A}_{\bar{i}q}{}^{\mathrm{T}} \boldsymbol{A}_{\bar{i}q} \right), \tag{9}$$

where $\mathbf{R_u} = \mathrm{diag}\left[R_{u1p}, R_{u1q}, \ldots, R_{uNp}, R_{uNq}\right]$.
Because of the partitioning of $\mathbf{z}$ in (3), we can write

$$\boldsymbol{R}_z = \begin{bmatrix} \boldsymbol{R}_h & \boldsymbol{R}_{hm} \\ \boldsymbol{R}_{hm}^{\mathrm{T}} & \boldsymbol{R}_m \end{bmatrix}. \tag{10}$$

We can then show that, for N points and M frames, the average distortion in the reconstruction is

$$D_{\mathrm{avg}}(M, N) = \frac{1}{MN^2} \sum_{j=1}^{M} \mathrm{trace}(\boldsymbol{R}_h^j), \tag{11}$$

where the superscript is the index to the frame number. We will call (11) the multiframe SfM (MFSfM) rate-distortion function, hereafter referred to as the

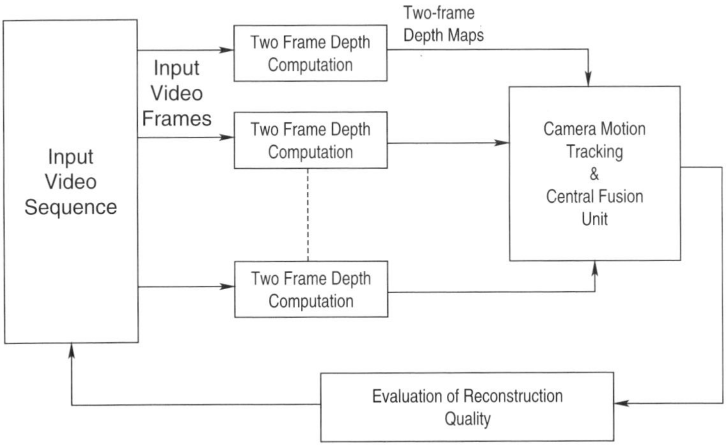

FIGURE 6.1: Block diagram of the 3D reconstruction framework.

video rate-distortion (VRD) function. Given a particular tolerable level of distortion, the VRD specifies the minimum number of frames necessary to achieve that level. In [23, 25] we proposed an alternative information-theoretic criterion for evaluating the quality of a 3D reconstruction from video and analyzed the comparative advantages and disadvantages. The above result does not require the standard assumptions of gaussianity of observations and is thus an extension of the error-covariance results presented in [34].

6.2.2 SfM Algorithm for Face Reconstruction

Figure 6.1 shows a block-diagram schematic of the complete 3D face-reconstruction framework using SfM. The input is a monocular video sequence. We choose an appropriate two-frame depth-reconstruction strategy [30]. The depth maps are aligned to a single frame of reference and the aligned depth maps are fused together using stochastic approximation.

Let $\mathbf{s}^i \in \mathbb{R}^3$ represent the structure,[1] computed for a particular point, from the ith and $(i+1)$th frame, for $i = 1, \ldots, K$, where the total number of frames is $K+1$.[2] Let the fused structure subestimate at the ith frame be denoted by $\mathbf{S}^i \in \mathbb{R}^3$. Let $\boldsymbol{\Omega}^i$ and $\mathbf{V}^i$ represent the rotation and translation of the camera between the

[1] In our description, subscripts will refer to feature points and superscripts will refer to frame numbers. Thus x_i^j refers to the variable x for the ith feature point in the jth frame.

[2] For notational simplicity, we use the ith and $(i+1)$th frames to explain our algorithm. However, the method can be applied for any two frames provided the constraints of optical flow are not violated.

ith and $(i+1)$th frames. Note that the camera motion estimates are valid for all the points in the object in that frame. The 3×3 rotation matrix $\boldsymbol{P}^i$ describes the change of coordinates between times i and $i+1$, and is orthonormal with positive determinant. When the rotational velocity $\boldsymbol{\Omega}$ is held constant between time samples, $\boldsymbol{P}$ is related to $\boldsymbol{\Omega}$ by $\boldsymbol{P} = e^{\hat{\boldsymbol{\Omega}}}$.[3] The fused subestimate $\boldsymbol{S}^i$ can now be transformed as $T^i(\boldsymbol{S}^i) = \boldsymbol{P}^i\boldsymbol{S}^i + \boldsymbol{V}^{i\mathrm{T}}$. But, in order to do this, we need to estimate the motion parameters $\boldsymbol{V}$ and Ω. Since we can determine only the direction of translational motion $(v_x/v_z, v_y/v_z)$, we will represent the motion components by the vector $\boldsymbol{m} = [\frac{v_x}{v_z}, \frac{v_y}{v_z}, \omega_x, \omega_y, \omega_z]$. Thus, the problems at stage $i+1$ will be to (i) reliably track the motion parameters obtained from the two-frame solutions, and (ii) fuse $\mathbf{s}^{i+1}$ and $T^i(\mathbf{S}^i)$. If $\{l^i\}$ is the transformed sequence of inverse depth values with respect to a common frame of reference, then the optimal value of the depth at the point under consideration is obtained as

$$u^* = \underset{u}{\operatorname{argmin}}\ \operatorname{median}_i \left(w_l^i (l^i - u)^2 \right), \tag{12}$$

where $w_l^i = (\mathbf{R}_{\boldsymbol{h}}^i(l))^{-1}$, with $\mathbf{R}_{\boldsymbol{h}}^i(l)$ representing the covariance of l^i (which can be obtained from (10)). However, since we will be using a recursive strategy, it is not necessary to align all the depth maps to a common frame of reference *a priori*. We will use a Robbins–Monro stochastic-approximation (RMSA) [21] algorithm (refer to [24] for details) where it is enough to align the fused subestimate and the two-frame depth for each pair of frames and proceed as more images become available.

For each feature point, we compute $X^i(u) = w_l^i(l^i - u)^2$, for $u \in \mathcal{U}$. At each step of the RM recursion, the fused inverse depth, $\hat{\theta}^{k+1}$, is updated according to [24]

$$\hat{\theta}^{k+1} = \hat{T}^k(\hat{\theta}^k) - a^k(p^k(\hat{\theta}^k) - 0.5), \tag{13}$$

where a^k is determined by a convergence condition, $p^k(\hat{\theta}^k) = 1_{[X^k \leq \hat{T}^k(\hat{\theta}^k)]}$, 1 represents the indicator function, $\hat{T}^k$ is the estimate of the camera motion. When $k = K$, we obtain the fused inverse depth $\hat{\theta}^{K+1}$, from which we can get the fused

[3] For any vector $\boldsymbol{a} = [a_1, a_2, a_3]$, there exists a unique skew-symmetric matrix

$$\hat{a} = \begin{bmatrix} 0 & -a_3 & a_2 \\ a_3 & 0 & -a_1 \\ -a_2 & a_1 & 0 \end{bmatrix}$$

The operator $\hat{a}$ performs the vector product on $\mathbf{R}^3$: $\hat{a}X = \mathbf{a} \times X, \forall X \in \mathbf{R}^3$. With an abuse of notation, the same variable is used for the random variable and its realization.

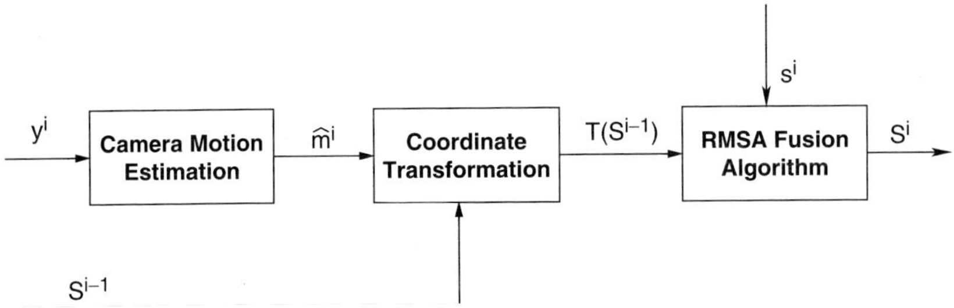

FIGURE 6.2: Block diagram of the multi-frame fusion algorithm.

depth value S^{K+1}. The camera motion $\hat{T}$ is estimated using a tracking algorithm as described in [24].

The Reconstruction Algorithm

Assume that we have the fused 3D structure S^i obtained from i frames and the two-frame depth map s^{i+1} computed from the ith and $(i + 1)$th frames. Figure 6.2 shows a block diagram of the multiframe fusion algorithm. The main steps of the algorithm are:

Track. Estimate the camera motion using the camera motion tracking algorithm [24].

Transform. Transform the previous model S^i to the new reference frame.

Update. Update the transformed model using s^{i+1} to obtain S^{i+1} from (14).

Evaluate reconstruction. Compute a performance measure for the fused reconstruction from (11).

Iterate. Decide whether to stop on the basis of the performance measure. If not, set $i \leftarrow i + 1$ and go back to Track.

6.2.3 Incorporating the Generic Face Model

The Optimization Function

The MFSfM estimate is smoothed using the generic model in an energy-minimization framework. Both the generic model and the 3D estimate have a triangular mesh representation with N vertices, and the depth at each of these vertices is known. (We will explain how this can be obtained later.)

Let $\{d_{g_i}, i = 1, \ldots, N\}$ be the set of depth values of the generic mesh for each of the N vertices of the triangles of the mesh. Let $\{d_{s_i}, i = 1, \ldots, N\}$ be the corresponding depth values from the SfM estimate. We wish to obtain a set of values $\{f_i, i = 1, \ldots, N\}$ which are a smoothed version of the SfM model, after correcting the errors on the basis of the generic mesh.

Since we want to retain the specific features of the face we are trying to model, our error-correction strategy works by comparing local regions in the two models and smoothing those parts of the SfM estimate where the *trend* of the depth values is significantly different from that in the generic model, e.g., a sudden peak on the forehead will be detected as an outlier after the comparison and smoothed. This is where our work is different from previous work [11, 28], since we do not intend to fuse the depth in the two models but to correct errors based on local geometric trends. Towards this goal, we introduce a line process on the depth values. The line process indicates the borders where the depth values have sudden changes, and is calculated on the basis of the generic mesh, since it is free from errors. For each of the N vertices, we assign a binary number indicating whether or not it is part of the line process (see Figure 6.3b). This concept of the line process is borrowed from the seminal work of Geman and Geman [12] on stochastic relaxation algorithms in image restoration.

The optimization function we propose is

$$E(f_i, l_i) = \sum_{i=1}^{N} (f_i - d_{s_i})^2 + (1 - \mu) \sum_{i=1}^{N} (f_i - d_{g_i})^2 + \mu \sum_{i=1}^{N} (1 - l_i) \sum_{j \in \mathcal{N}_i} (f_i - f_j)^2 \, 1_{d_s \neq d_g}, \tag{14}$$

where $l_i = 1$ if the ith vertex is part of a line process and μ is a combining factor which controls the extent of the smoothing, $\mathcal{N}_i$ is the set of vertices which are neighbors of the ith vertex, and $1_{d_s \neq d_g}$ represents the indicator function which is 1 if the line process of d_s is not equal to the line process of d_g, else 0. (If $d_s = d_g$ point to point, their line processes will be equal, though the converse is not true). We will now explain the performance of the optimization function in the normal operating condition when the SfM estimate is "close" to the true face, but, still contains some errors (mostly unwanted peaks and valleys). We have found experimentally that this is a usual characteristic of the SfM solution obtained from the first part of our modeling algorithm. In order to understand the importance of (14), consider the third term. When $l_i = 1$, the ith vertex is part of a line process and should not be smoothed on the basis of the values in $\mathcal{N}_i$; hence this term is switched off. Any errors in the value of this particular vertex will be corrected on the basis of the first two terms, which control how close the final smoothed mesh will be to the generic one and the SfM estimate. When $l_i = 0$, indicating that the ith vertex is not part of a line process, its final value in the smoothed mesh is determined by the neighbors as well as its corresponding values in the generic model and SfM estimate. The importance of each of these terms is controlled by the factor $0 < \mu < 1$. Note that μ is a function of d_s and d_g, as explained later when we discuss the choice of μ. This ensures that the optimization function converges to the true model in the ideal case that the SfM solution, d_s, computes the true model. For this case, the third term would be zero as the line processes would be the same, since they compare the trend in the depth values. The line process l_i has a value of 1 or 0, depending on whether there is a sudden change in the depth in the generic model at the particular vertex i. Obtaining such changes relies on derivative computations, which are known to be noise prone. Hence, in our optimization scheme, we allow the line process to be perturbed slightly around its nominal value computed from the generic model.

Besides, the normal operating condition and the ideal case discussed above, there are two other cases that may occur, though both are highly impractical. In the case where $d_s = d_g$, the second term is switched off because of μ and the third term is switched off because of the indicator function. The smoothed mesh can be either d_s or d_g as a result of (14). The other case arises if $d_s \neq d_g$, but their line processes are equal, i.e., $1_{d_s \neq d_g} = 0$. In this case, again the second and third terms are switched off, and the optimization will converge to d_s.

The design of the cost function follows conventional ideas of regularization theory [7], whereby the energy function usually consists of a data term requiring the solution to be close to the data, and a regularizer which imposes a smoothness on the solution. Various other forms of the different terms in (14) could be considered. One interesting variation would be to impose a penalty term for the discontinuities. The graduated nonconvexity algorithm of [1] is an appropriate method for solving such problems. It works by first finding the minimum of a convex approximation to the nonconvex function, followed by minimization of a sequence of functions, ending with the true cost function. However, based on experimental analysis of the solution of our reconstruction algorithm, we decided that we could work with the simpler version of (14), which does not have the penalty term. This is because, during the optimization, the line process does not move very far from its nominal value, and hence the penalty term does not have any significant contribution.

Equation 14 can be optimized in various ways. If the l_i are fixed, this is equivalent to solving a sparse linear system of equations (Chapters 9 and 10 of [13]). In the parlance of classical deterministic optimization, we then assume that we have perfect information about the loss function and that this information is used to determine the search directions in a deterministic manner in every step of the algorithm. However, as explained before, one of the major sources of noise in (14) is the estimate of l_i. If the l_i are not known perfectly, we need to optimize over this variable also. The cost function can no longer be represented as a linear system of equations. More complicated optimization schemes need to be investigated for this purpose.

We use the technique of simulated annealing built upon the Markov-chain Monte Carlo (MCMC) framework [8]. MCMC is a natural method for solving energy function minimization problems [7]. The MCMC optimizer is essentially a Monte Carlo integration procedure in which the random samples are produced by evolving a Markov chain. Let $T_1 > T_2 > \cdots > T_k > \cdots$ be a sequence of monotone decreasing temperatures in which T_1 is reasonably large and $\lim_{T_k \to \infty} = 0$. At each such T_k, we run N_k iterations of a Metropolis–Hastings (M–H) sampler [8] with the target distribution represented as $\pi_k(f, l) \propto \exp\{-E(f, l)/T_k\}$. As k increases, π_k puts more and more of its probability mass (converging to 1) in the vicinity of the global maximum of E. Since minimizing $E(f, l)$ is equivalent to maximizing $\pi(f, l)$, we will almost surely be in the vicinity of the global optimum if the number of iterations N_k of the M–H sampler is sufficiently large. The steps of the algorithm are:

- Initialize at an arbitrary configuration f_0 and, l_0 and initial temperature level T_1. Set $k = 1$.
- For each k, run N_k steps of MCMC iterations with $\pi_k(f, l)$ as the target distribution. Consider the following update strategy. For the line process, consider all the vertices (say $L < N$) for which the nominal value, $l_{i,\text{nominal}} = 1$, and their individual neighborhood sets, $N_1, \ldots, N_L$. For each $l_i = 1$, consider the neighborhood set among $N_1, \ldots, N_L$ that it lies in, randomly choose a vertex in this neighborhood set whose value is not already set to 1, and switch the values of l_i and this chosen vertex. Starting from $l_i = l_{i,\text{nominal}}$, this process ensures that the values of l_i do not move too far from the nominal values. In fact, only the vertices lying in the neighborhood sets $N_1, \ldots, N_L$ can take a value of $l_i = 1$. Next, randomly determine a new value of f, using a suitable transition function [8]. With the new values, $f_{\text{new}}, l_{\text{new}}$ of f, l, compute $\delta = E(f_{\text{new}}, l_{\text{new}}) - E(f, l)$. If $\delta < 0$, i.e., the energy decreases with this new configuration, accept $f_{\text{new}}, l_{\text{new}}$; else, accept with a probability ρ. Pass the final configuration of f, l to the next iteration.
- Increase k to $k + 1$.

Mesh Registration

The optimization procedure described above requires a one-to-one mapping of the vertices d_{s_i} and d_{g_i}. Once we obtain the estimate from the SfM algorithm, a set of corresponding points between this estimate and the generic mesh is identified. This can be done manually as in [11] or [28] or automatically as described next. This is then used to obtain a registration between the two models. Thereafter, using proper interpolation techniques, the depth values of the SfM estimate are generated corresponding to the (x, y) coordinates of the vertices of the triangles in the generic model. By this method, we obtain the meshes with the same set of N vertices, i.e., the same triangulation.

If we want to perform the registration automatically, we can follow a simple variant of our method for registering wide-baseline images [26]. In [26], we showed that it is possible to register two face images obtained from different viewing directions by considering the similarity of the shape of important facial features (e.g., eyes, nose, etc.) and compensating for the variability of the shape with viewing direction by considering prior information about it. Applying it to this problem is actually simpler because the two meshes are from the same viewing angle. We can consider the 2D projection of the generic mesh and identify the shape of some important facial features *a priori*. Once the 3D estimate from SfM is obtained, we can take its 2D projection from the same viewing direction (e.g., from the front view), automatically extract the shape of the important features (as described in [26] by using a corner-finder algorithm and k-means clustering [9]) and then register by computing the similarity of the set of two shapes.

Choice of μ

There exists substantial literature on how to optimally choose the constant μ for energy functions similar to (14) (Chapter 3 of [7]). For our problem, we decided to choose the value based on a qualitative analysis of (14). When $l_i = 1$, the third term in the optimization is switched off. These are the points where there are sharp changes in the depth (see Figure 6.4a). These changes are important characteristics particular to a person's face and need to be retained. However, any errors in these regions should also be corrected. Thus μ is computed by comparing the line process obtained from d_s with that precomputed from d_g. This gives an approximate idea about the goodness of the SfM estimate. From these considerations, the value of μ turns out to be between 0.7 and 0.8 (the correlation between the two line processes was usually in this range). When $l_i = 0$, the most important term is the third term of (14), which controls local smoothing. The errors should be corrected by comparing with neighboring values of the 3D estimate, rather than by fusing with the depth values of the vertices of the generic model. For this case, the above choice of μ is again reasonable, and the generic model is not given undue importance leading to oversmoothing. This process of choosing μ also ensures that, in the ideal case, if d_s is the true face, the optimization converges to this value as μ will be equal to 1.

The Final Algorithm

The main steps of the algorithm for incorporating the generic mesh are as follows.

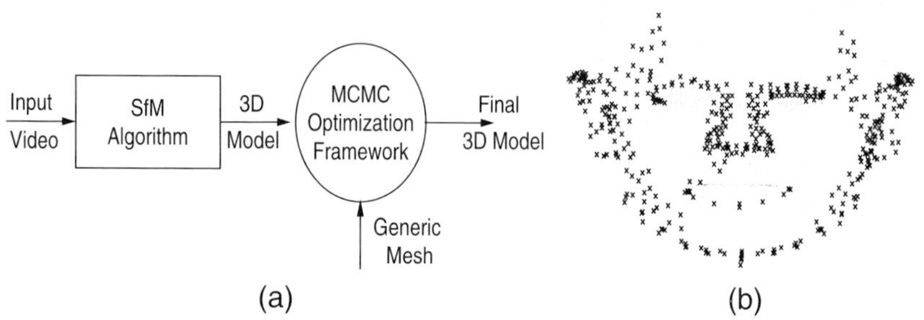

FIGURE 6.3: (a) A block diagram representation of the complete 3D modeling algorithm using the generic mesh and the SfM algorithm. (b) The vertices which form part of the line processes indicating a change in depth values are indicated with black crosses.

1. Obtain the 3D estimate from the given video sequence using SfM (output of the reconstruction algorithm of Section 6.2.1).
2. Register this 3D model with the generic mesh, and obtain the depth estimate whose vertices are in one-to-one correspondence with the vertices of the generic mesh.
3. Compute the line processes, and to each vertex i assign a binary value l_i.
4. Obtain the smoothed mesh f from the optimization function in (14).
5. Map the texture onto f from the video sequence.

The complete 3D reconstruction paradigm is composed of a sequential application of the two algorithms (3D reconstruction algorithm and the generic mesh algorithm) we have described in Sections 6.2.1 and 6.2.3 (see Figure 6.3a). Some examples of 3D reconstruction are shown in Figure 6.4.

6.2.4 Experimental Evaluation

The SfM technique computed structure from optical flow using two frames [30] and then integrated the multiple two-frame reconstructions over the video sequence using robust estimation techniques. The error covariance of the optical flow was estimated *a priori* over the first few frames of the video sequence, which were not used in the reconstruction. It was done over a sampled grid of points (rather than the dense flow) so as to simplify calculations. The technique used was similar to the gradient-based method of [31], except that, for more accurate results, it was repeated for each of these initial frames and the final estimate was obtained using bootstrapping techniques [10]. This is the stage where the quality of the video

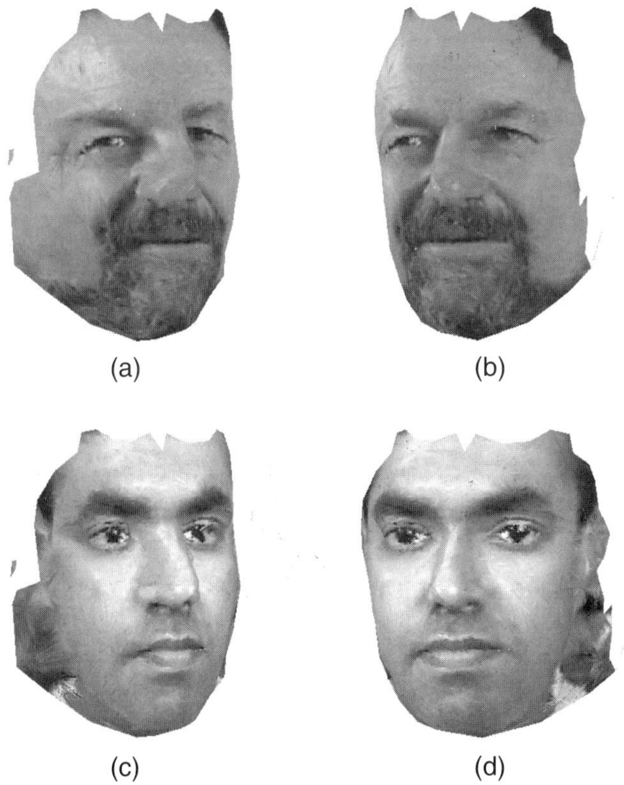

FIGURE 6.4: Different views of the two 3D models after texture mapping.

data is estimated and incorporated into the algorithm. Assuming that the statistics remain stationary over the frames used in the reconstruction, the error covariance of the 3D reconstruction, $\boldsymbol{R}_z$, in (8), was computed. The quality evaluation of the fusion algorithm was done using the rate-distortion function of (11). The model obtained from the SfM algorithm is shown in Figure 6.5b.

The combinatorial optimization function in (14) was implemented using the simulated annealing procedure based on a M-H sampler. At each temperature we carried out 100 iterations and this was repeated for a decreasing sequence of 20 temperatures. Although this is much below the optimal annealing schedule suggested by Geman and Geman [12] (whereby the temperature T_k should decrease sufficiently slowly as $\mathcal{O}(\log(\sum_{i=1}^{k} N_i)^{-1})$, N_i being the total number of iterations at temperature T_i), it does give a satisfactory result for our face-modeling example. This is because the optimization algorithm was initialized with the SfM reconstruction, which is usually a reasonable good estimate of the actual 3D model.

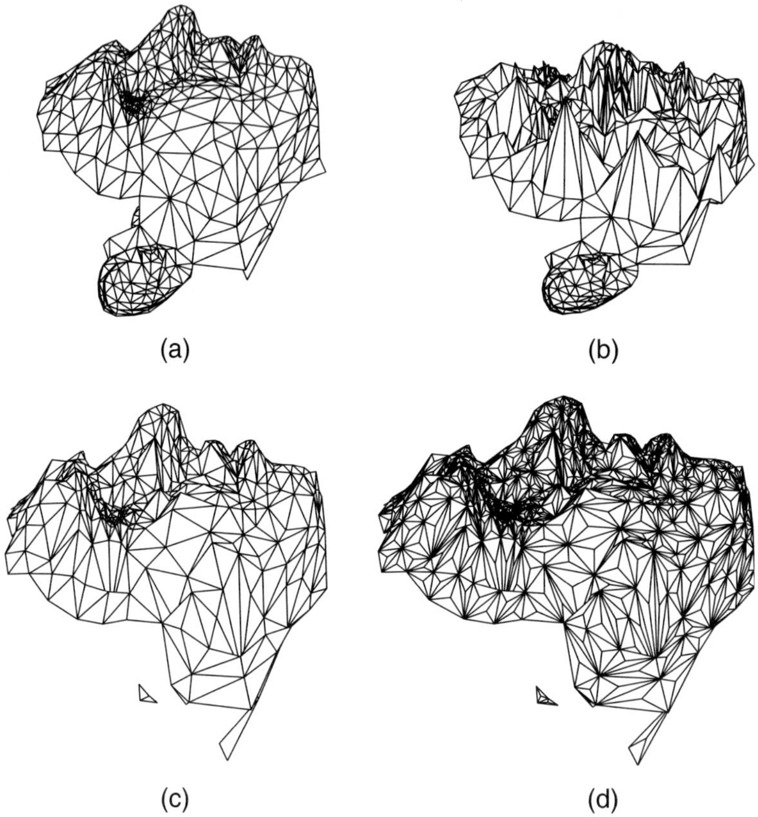

FIGURE 6.5: Mesh representations of the 3D models obtained at different stages of the algorithm. (a) A representation of the generic mesh, (b) the model obtained from the SfM algorithm (the ear region is stitched on from the generic model in order to provide an easier comparison between the different models), (c) the smoothed mesh obtained after the optimization procedure, (d) a finer version of the smoothed mesh for the purpose of texture mapping.

We used a value of $\mu = 0.7$ in (14). The final smoothed model is shown in Figure 6.5c.

Next, we map the texture onto the smoothed model in Figure 6.5c. This is done using an image from the video sequence corresponding to the front view of the face. Direct mapping of the texture from the image is not possible since the large size of the triangles smears the texture over its entire surface. In order to overcome this problem, we split each of the triangles into smaller ones. This is done only at the final texture mapping stage. The initial number of triangles is enough to obtain a good estimate of the depth values, but not to obtain a good texture mapping. This splitting at the final stage helps us save a lot of computation time, since the depth at the vertices of the smaller triangles is obtained by interpolation, not by the optimization procedure.[4] The fine mesh onto which the texture is mapped is shown in Figure 6.5d. Different views of the 3D model after the texture mapping are shown in Figure 6.4.

Computational Load

The computational complexity of the entire system can be analyzed by its individual parts. For the two-frame algorithm [30], given a $N \times N$ flow field, the complexity is $\mathcal{O}(N^2 \log N)$. For M frames, the complexity of the fusion algorithm is $\mathcal{O}(N^2 M)$. If the statistics are computed at $N' < N^2$ points, it takes computational power of the order of $\mathcal{O}(N'^2)$. The computational time for the final optimization will be determined by the actual annealing schedule chosen.

We have a running demonstration of the face-modeling software. The software runs on a 2.6 Gigahertz P4 PC with 1 Gigabyte memory. A JVC DVL9800 video camera is attached to the PC to capture the video. Using a combination of C and MATLAB implementations, the entire reconstruction (from capturing the video sequence, preprocessing it to creating of a final 3D graphics model) takes about 3–4 minutes. This can be substantially reduced by optimizing the code, converting it entirely to C and automating certain preprocessing stages (like identifying the relevant part of the input video sequence).

Accuracy Analysis of 3D Reconstructions

We have applied our algorithm to several video sequences. We present here the results on three such sequences. We will name the three people as subjects A, B, and C. Reconstructed 3D models of Subjects A and C are shown in Figure 6.4. Since the line process and the neighborhood set are calculated from the generic model, they are pre-computed. We computed the accuracy of the 3D reconstruction for these three cases by comparing the projections of the 3D model with the images from the original video sequence. Figure 6.6 plots the root mean square (RMS) projection errors as a percentage of the actual values in the original images. In order to depict the change of the error with the viewing angle, the horizontal axis of the figure represents this angle, with 0 indicating the front view. We considered all the combinations between the three subjects, i.e., A-A, A-B, A-C, B-B, B-C, and C-C. From the plots, we see that the average error in the reconstruction at the front view is about 1%. The error increases with viewing direction, as would

[4]This step may not be necessary with some of the recent computer graphics software.

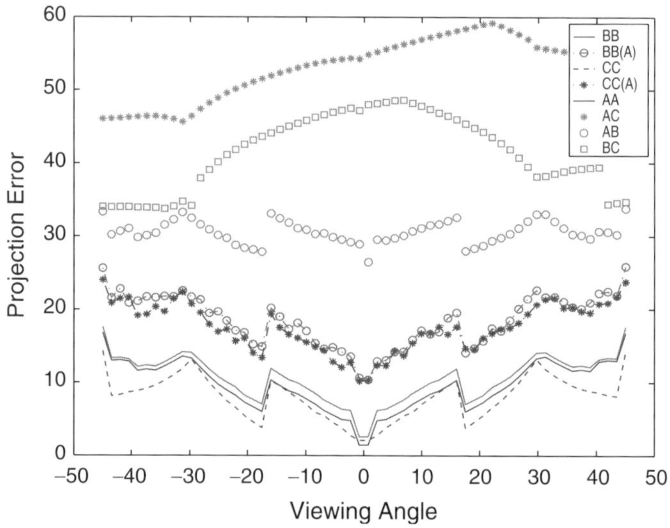

FIGURE 6.6: Percentage errors between the projections of the 3D models of subjects A, B and C and their images obtained from the video sequences, i.e., AA represents the error between the projections of 3D model of A and the images of A; similarly for AB, AC, BB, BC, and CC. BB(A) represents the projection error when B's texture is overlayed on A's model and the projections are compared to B's images; similarly for CC(A). The error is plotted as a function of the viewing angle which is represented on the x axis.

be expected. Also there are certain preferred viewing directions, a fact which has been reported before in the literature [35]. In order to obtain an understanding of the role the depth (as opposed to the texture) plays in projections, we considered the case where the texture of B and C are overlayed on the model of A and the projections are compared to the images of B and C, respectively. The experiment was also repeated for the models of B and C with similar results; however, for the sake of clarity, the plots are not shown in Figure 6.6. The separation between the error curves for the different combinations holds out hope that 3D models can be used for recognition across pose variations. For example, given the model of Subject A with its proper texture, the projection errors, at any viewing angle, with other subjects is much more than it is with itself.[5]

[5]There are many issues that contribute to the error in face recognition across pose. Accuracy of the 3D model is only one of them. Others are issues of registration of the projections of the model with the image, changes of illumination, etc. This is a separate research problem by itself and is one of our

Comparative Evaluation of the Algorithm

In order to analyze the accuracy of the 3D reconstruction directly (as opposed to comparing the 2D projections), we require the ground truth of the 3D models. We experimented with a publicly available database of 3D models obtained from a Minolta 700 range scanner. The data is available on the World Wide Web at http://sampl.eng.ohio-state.edu/sampl/data/3DDB/RID/minolta/faces-hands.1299/index.html. We will report numerical results from our algorithm on some of the data available here, though we will not publish the images or 3D models of the subjects.

In order to perform an accurate analysis of our methods, we require a video sequence of the person and the 3D depth values. This, however, is not available on this particular database or on any other that we know of. Thus we had to generate a sequence of images in order to apply our algorithm.[6] This was done using the 3D model and the texture map provided on the website. Given these images, we performed the following experiments.

- Obtain the 3D reconstruction without using the generic model (Section 6.2.1).
- Introduce the generic model at the beginning of the 3D reconstruction algorithm, by initializing (12) with the values at the vertices of the generic model.[7] Note that the statistical error analysis of the video data is still done.
- Apply the algorithm in this section, which postpones the introduction of the generic model.

We considered the error in the 3D estimate in all these methods compared to the actual 3D values. Figure 6.7 plots the percentage RMS errors (percentage taken with respect to the true value) as a function of the percentage difference of the specific 3D model (as obtained from the website) and the generic one. The percentages on the horizontal axis are calculated with respect to the generic model, while those on the vertical axis are computed with respect to the ground truth of the 3D model of the particular subject. The first five subjects on that website, referred to as "frame 001" to "frame 005", were considered. The percentage differences

---

future directions of work. For our experiments, most of these issues were taken care of manually so that the error values represented in Figure 6.6 are mostly due to errors in 3D models (though errors due to other sources cannot be completely eliminated).

[6]The optical flow computed with the generated sequences may be more accurate than in a normal setting. However, for all the methods that are compared, the images used are the same. Hence, it is reasonable to assume that the effect of the image quality would be similar in all three cases. So the comparison of the 3D reconstruction accuracy, keeping all other factors constant, should still be useful.

[7]We cannot compare it directly with [11] or [28] because of substantial differences in the input data.

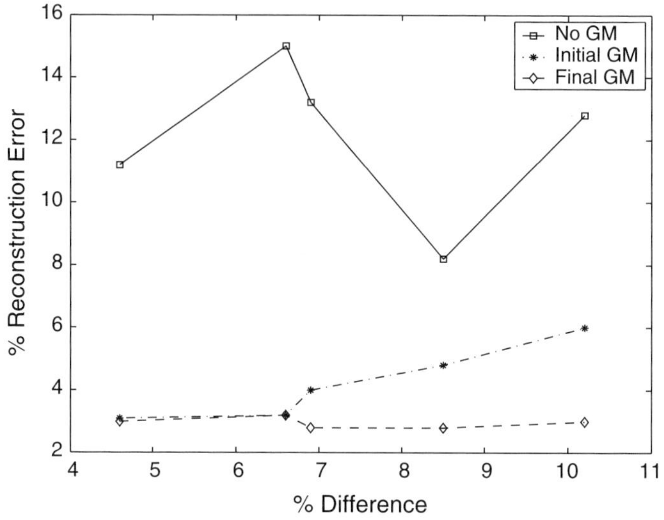

FIGURE 6.7: The error in 3D reconstruction when (-□-) no generic model (GM) is used; (·∗·) the generic model is used to initialize the reconstruction algorithm; (-◇-) the generic model is used later as described in this section. The error is plotted as a function of the difference of the specific 3D model with the generic one. Five subjects were considered in this experiment.

Table 6.1: Average percentage difference of the 3D models from the generic model.

Subject index	Percentage difference
1 (frame 001)	10.2
2 (frame 002)	8.5
3 (frame 003)	4.6
4 (frame 004)	6.6
5 (frame 005)	6.9

of the specific 3D models with the generic one are tabulated in Table 6.1. From the figure, it is clear that, if the generic model is introduced at an early stage of the algorithm, the error in the reconstruction increases as the model of the subject deviates from the generic one. On the other hand, if the generic model is introduced later (as in our algorithm), the error in the reconstruction remains approximately

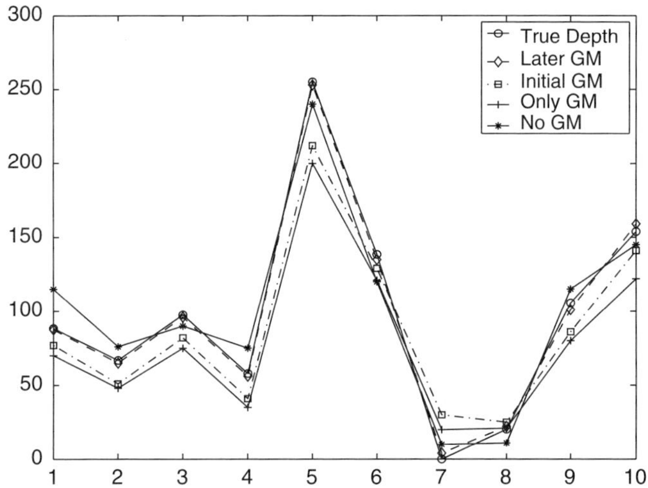

FIGURE 6.8: Plots of (-o-) the true depth, (-◇-) the depth with generic model introduced later (our algorithm), (·□·) the depth with generic model used as initial value, (-+-) the depth of the generic model and (-*-) the SfM reconstruction with no generic model. The depths are computed in local neighborhoods around a set of fiducial points on the face for Subject 1.

constant. However, the reconstructions for the case where the generic model is introduced earlier (e.g., [11] and [28]) are visually very pleasing.

An idea of the progress of the optimization can be obtained from the following experiment, the results of which are shown in Figure 6.8. The experiment was done on Subject 1 of Table 6.1, which is an interesting case since the face is very different from the generic one. We considered ten significant points on the face (similar to fiducial points). They are the left eye, the bridge between the eyes, the right eye, the left extreme part of the nose, the tip of the nose, the right extreme part of the nose, the left and right ends of the lips, and the center of the left and right cheeks. We considered a window around each of these points and computed the average depth in each of them for the various reconstructions. Figure 6.8 plots the average depth at each of these points for the following cases: true depth, depth with generic model introduced later (our algorithm), depth with generic model used as the initial value, depth of the generic model, and the SfM reconstruction with no generic model. The depth values are normalized between 0 and 255, as in a depth map image. The plots provide some very interesting insights. When the reconstruction is initialized with the generic mesh, the solution at these points do not move very far from the initial value. On the other hand, the SfM estimate

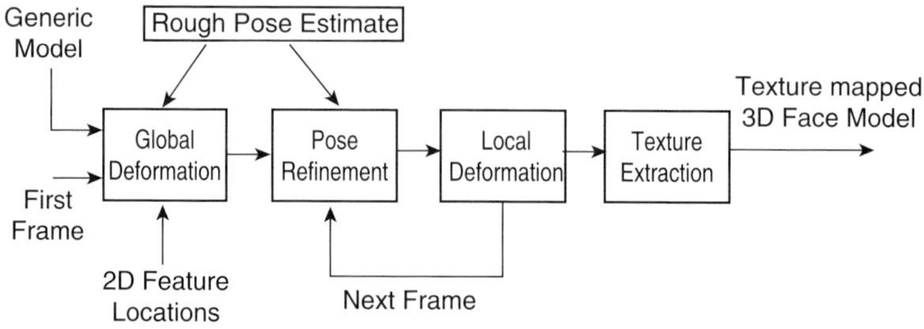

FIGURE 6.9: 3D face-modeling algorithm.

with no generic mesh gives a solution not very far from the true value, which is improved further by considering the generic model. Moreover, we find that the average error of the final reconstruction in these fiducial regions is less than the overall average error of Figure 6.7. This is very interesting since it shows that these significant regions of the face are reconstructed accurately.

6.3 CONTOUR-BASED 3D FACE MODELING

In this section, a novel 3D face-modeling approach from a monocular video captured using a conventional camera is presented. The algorithm relies on matching a generic 3D face model to the outer contours of the face to be modeled and a few of its internal features. At the first stage of the method, the head pose is estimated by comparing the edges extracted from video frames with the contours extracted from a generic face model. Next, the generic face model is adapted to the actual 3D face by global and local deformations. An affine model is used for global deformation. The 3D model is locally deformed by computing the optimal perturbations of a sparse set of control points using a stochastic-search optimization method. The deformations are integrated over a set of poses in the video sequence, leading to an accurate 3D model. Figure 6.9 shows a block diagram of the proposed algorithm.

6.3.1 Pose Estimation

In order to estimate the head pose without any prior knowledge about the 3D structure of the face, a generic 3D face model is used. Human shape variability is highly limited by both genetic and environmental constraints, and is characterized by a high degree of symmetry and approximate invariance of body lengths and ratios. Since the facial anthropometric measurements of the generic face are close to average, the algorithm can estimate the pose robustly, unless the person whose head pose is being estimated has a hugely deformed face. The problem of pose estimation along the azimuth angle is most commonly encountered in video sequences, and the system is currently limited to this case. The algorithm described in this section can be easily extended to incorporate more complex head motion (including roll and elevation). The addition of each degree of freedom of the head motion causes the search space for the face pose to grow exponentially, thus slowing down the 3D reconstruction algorithm at the pose-estimation and pose-refinement stages.

The frames extracted from the video are subsampled so that successive frames have a distinct pose variation. A simple image-difference method is used to detect the background pixels in the video. All edges in the background are removed to make sure they do not adversely affect the pose-estimation algorithm. The pose-estimation algorithm requires the coordinates of the nose tip in the image, because the edge maps obtained from the 2D projection of the 3D model and from the video frames are aligned at the nose tip. The Kanade–Lucas tracker [32] is used to automatically track the nose tip across multiple frames.

For pose estimation, an average human texture is mapped onto the generic 3D face model. An average 3D face shape and texture data was obtained from [22]. The texture-mapped generic face model is rotated along the azimuth angle, and edges are extracted using the Canny edge detector [6]. The projection of the average 3D model also has edges which result from the boundaries of the 3D model (e.g., top of forehead, bottom of neck). These edges are not the result of natural contours of the human face, and are therefore automatically removed using a modified average texture. The edge maps are computed for 3D model rotation along the azimuth angle from -90° to $+90^\circ$ in increments of 5°. These edge maps are computed only once, and stored off-line in an image array to make the procedure fast.

To estimate the head pose in a given video frame, the edges of the image are extracted using the Canny edge detector. Each of the scaled 3D model edge maps is compared to this frame edge map to determine which pose results in the best overlap of the edge maps. To compute the disparity between these edge maps, the Euclidean distance transform (DT) of the current video frame edge map is computed. For each pixel in the binary edge map, the distance transform assigns a number that is the distance between that pixel and the nearest nonzero pixel of the edge map. Figure 6.10 shows the binary edge map of a video frame, and the corresponding distance transform.

Each of the 3D model edge maps is aligned at the nose tip in the video frame, and the value of the cost function, F, is computed. The cost function, F, which measures the disparity between the 3D model edge map and the edges of the current video frame is of the form

$$F = \frac{\sum_{(i,j)\in A_{\mathrm{EM}}} \mathrm{DT}(i,j)}{N}, \tag{15}$$

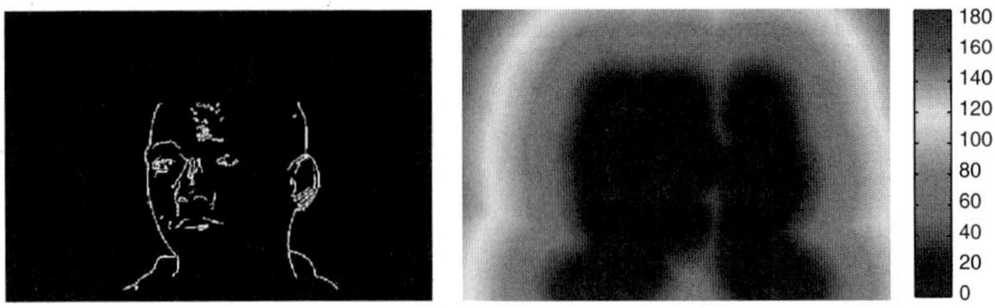

FIGURE 6.10: Left: Edge Map. Right: Distance Transform. (See also color plate section)

FIGURE 6.11: Top row: video frames. Bottom row: generic model at estimated head pose.

where $A_{\text{EM}} \triangleq \{(i,j) : \text{EM}(i,j) = 1\}$ and N is the cardinality of set A_{EM} (total number of nonzero pixels in the 3D model edge map EM). F is the average distance transform value at the nonzero pixels of the binary 3D model edge map. The pose for which the corresponding 3D model edge map results in the lowest value of F is the estimated head-pose for the current video frame. Figure 6.11 shows the head-pose estimation results for a few video frames of a subject.

6.3.2 3D Face Model Reconstruction

In this section, the contour-based algorithm for 3D face reconstruction from a monocular video is described. A generic 3D face model is assumed to be the initial estimate of the true 3D face model. The generic face model is globally and locally deformed to adapt itself to the actual 3D face of the person.

Registration and Global Deformation

Once the pose is estimated using the method described in Section 6.3.1, the next step is to perform a global deformation and register the 3D model to the 2D image. A scaled orthographic projection model for the camera, and an affine deformation model for the global deformation of the generic face model is used. The coordinates of four feature points (left eye, right eye, nose tip, and mouth center) are used to determine a solution for the affine parameters. Since these feature point locations need to be known for just the first frame, currently these are marked manually. The 3D coordinates of the corresponding feature points on the generic face model are available beforehand.

The generic 3D model is globally deformed only for the first frame. The following affine model is used for global deformation:

$$\begin{bmatrix} X_{gb} \\ Y_{gb} \\ Z_{gb} \end{bmatrix} = \begin{bmatrix} a_{11} & a_{12} & 0 \\ a_{21} & a_{22} & 0 \\ 0 & 0 & \frac{1}{2}a_{11} + \frac{1}{2}a_{22} \end{bmatrix} \begin{bmatrix} X \\ Y \\ Z \end{bmatrix} + \begin{bmatrix} b_1 \\ b_2 \\ 0 \end{bmatrix}, \tag{16}$$

where (X, Y, Z) are the 3D coordinates of the vertices of the generic model, and subscript "gb" denotes global deformation. The affine model appropriately stretches/shrinks the 3D model along the X, and Y axes and also takes into account the shearing in the X–Y plane. Considering the basic symmetry of human faces, the affine parameters contributing to shearing (a_{12}, a_{21}) are very small, and can often be neglected. Since an orthographic projection model is used, we can not have an independent affine deformation parameter for the Z-coordinate. The affine deformation parameters are obtained by minimizing the reprojection error of the 3D feature points on the rotated deformed generic model, and their corresponding 2D locations in the current frame. The 2D projection (x_f, y_f) of the 3D feature points (X_f, Y_f, Z_f) on the deformed generic face model is given by

$$\begin{bmatrix} x_f \\ y_f \end{bmatrix} = \underbrace{\begin{bmatrix} r_{11} & r_{12} & r_{13} \\ r_{21} & r_{22} & r_{23} \end{bmatrix}}_{\boldsymbol{R}_{12}} \begin{bmatrix} a_{11}X_f + a_{12}Y_f + b_1 \\ a_{21}X_f + a_{22}Y_f + b_2 \\ \frac{1}{2}(a_{11} + a_{22})Z_f \end{bmatrix}, \tag{17}$$

where $\boldsymbol{R}_{12}$ is the matrix containing the top two rows of the rotation matrix corresponding to the estimated head pose for the first frame. Using the coordinates of the four feature points, (17) can be reformulated into a linear system of equations. The affine deformation parameters $\boldsymbol{P} = [a_{11}, a_{12}, a_{21}, a_{22}, b_1, b_2]^{\mathrm{T}}$ can be determined by obtaining a least-squares (LS) solution of the system of equations. The generic mesh is globally deformed according to these parameters. This process ensures that the 3D face model matches the approximate shape of the face and the significant internal features are properly aligned.

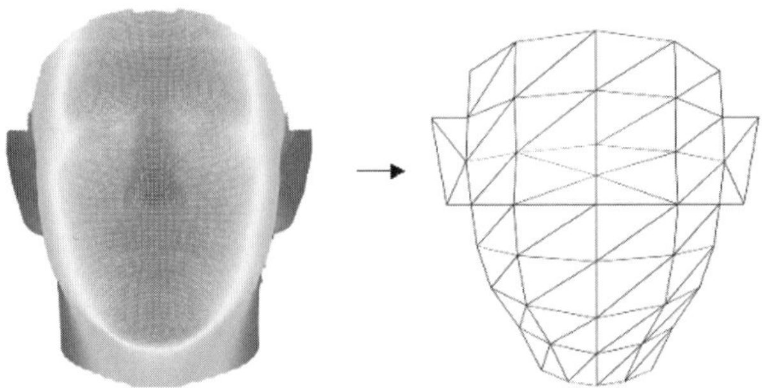

FIGURE 6.12: Sparse mesh of control points.

Local Deformation

To adapt the 3D model to a particular individual's face more accurately from the video sequence, local deformations are introduced in the globally deformed model. The algorithm for local deformation of the face model is described in this section.

Control points. The globally deformed dense face mesh is sampled at a small number of points to obtain a sparse mesh shown in Figure 6.12. Each of the vertices of this sparse mesh is a control point in the optimization procedure. Each control point is imparted a random perturbation in the X, Y, and Z direction. The perturbation for each of the vertices of the dense face mesh is computed from the random perturbations of control points using triangle-based linear interpolation. The computed perturbations are imparted to the vertices of the face mesh to obtain a locally-deformed mesh. The outer contours obtained from the locally-deformed face mesh are compared with the edges from the current video frame. The optimum local perturbations of the control points are determined using a direct random search method.

Direct random search. Direct random search methods [29] are based on exploring the domain D in a random manner to find a point that minimizes the cost function L. They are "direct" in the sense that the algorithms use minimal information about the cost function $L = L(\theta)$. The minimal information is essentially only the input–output data of the following form: input $= \boldsymbol{\theta}$, output $= L(\boldsymbol{\theta})$. In the global random search method, we repeatedly sample over D such that the current sampling for $\boldsymbol{\theta}$ does not take into account the previous samples. The domain D is a hypercube (a p-fold Cartesian product of intervals on the real line, where p is the problem dimension), and we use uniformly distributed samples. There exists a convergence proof [29] which shows that the two-step (after initialization) algorithm described below converges almost surely to the global minimum θ^*.

Global random search algorithm. *Step 0 (initialization)*: Generate an initial value of $\boldsymbol{\theta}$, say $\hat{\boldsymbol{\theta}}_0 \in D$, according to the uniform probability distribution on the domain D. Calculate $L(\hat{\theta}_0)$. Set $k = 0$.
Step 1: Generate a new independent value of $\boldsymbol{\theta} \in D$, say $\boldsymbol{\theta}_{\text{new}}(k+1)$, according to the uniform probability distribution. If $L(\boldsymbol{\theta}_{\text{new}}(k+1)) < L(\hat{\theta}_k)$, set $\hat{\theta}_{k+1} = \boldsymbol{\theta}_{\text{new}}(k+1)$. Else take $\hat{\theta}_{k+1} = \hat{\theta}_k$.
Step 2: Stop if the maximum number of L evaluations has been reached; else return to step 1 with $k \leftarrow k+1$.

In our application of the global random search algorithm to achieve the optimum local deformation of the face model, the cost function L is the disparity between the outer contours obtained from the 3D face model after applying the local deformation (θ) and the edges from the current video frame. The stochastic optimization algorithm determines the value of θ that minimizes the cost function L. The parameter vector θ whose optimum value is being sought is of the form:

$$\theta = [\Delta X_{\text{c}}\ \Delta Y_{\text{c}}\ \Delta Z_{\text{c}}]_{P\times 3}, \tag{18}$$

where ΔX_{c}, ΔY_{c}, and ΔZ_{c} are the perturbations in the X, Y, and Z directions of the P control points. The perturbations for all vertices of the dense face mesh, $[\Delta X_{\text{vt}}\ \Delta Y_{\text{vt}}\ \Delta Z_{\text{vt}}]$, are computed from the perturbations of these control points using a triangle-based linear interpolation method. The coordinates of the vertices of the locally deformed face model are given by

$$\begin{bmatrix} X_{\text{lc}} \\ Y_{\text{lc}} \\ Z_{\text{lc}} \end{bmatrix} = \begin{bmatrix} X_{\text{gb}} \\ Y_{\text{gb}} \\ Z_{\text{gb}} \end{bmatrix} + \begin{bmatrix} \Delta X_{\text{vt}} \\ \Delta Y_{\text{vt}} \\ \Delta Z_{\text{vt}} \end{bmatrix}, \tag{19}$$

where the subscript "lc" denotes local deformation. Let EM_θ be the binary edge map (outer contours) of the 2D projection of the 3D model after applying local deformation θ. The unwanted edges due to the boundaries of the 3D model are removed by the method described earlier in Section 6.3.1. Let DT be the distance transform of the edge map of the current video frame. The cost function, L, is of the form

$$L(\theta) = \frac{\sum_{(i,j)\in A_{\text{EM}_\theta}} \text{DT}(i,j)}{N}, \tag{20}$$

where $A_{\text{EM}_\theta} \triangleq \{(i,j) : \text{EM}_\theta(i,j) = 1\}$ and N is the cardinality of set A_{EM_θ}. The structure of this cost function is similar to the one used in (15). In Section 6.3.1, the edge maps were compared to determine the head pose, but here these are compared to determine the optimal local deformation for the globally deformed generic face model.

Multiresolution Search. The random perturbations are applied to the face mesh at two different resolutions. Initially, M iterations of large random perturbation are performed to scan a large search area at a coarse resolution. The final estimate at the end of M coarse-level iterations is denoted by $\hat{\theta}^{c}_{M}$. Once a coarsely deformed mesh which is close to the global minimum is available, it is used as the starting estimate for the fine resolution iterations, i.e., $\hat{\theta}^{c}_{M} = \hat{\theta}^{f}_{0}$. N small random perturbations are applied to the coarsely deformed mesh to get even closer to the actual solution. The final estimate at the end of N fine-level iterations is denoted by $\hat{\theta}^{f}_{N}$. This coarse-to-fine resolution search helps the algorithm converge to the solution much faster, compared to a search strategy where the search space is sampled at a fixed fine resolution. As the number of iterations performed for the coarse- and fine-resolution searches is increased, the performance of the algorithm gets better, but, at the cost of computational time. It was observed that the algorithm converges to a good solution in 300 iterations of large (coarse) random perturbations, and 100 iterations of small (fine) random perturbations.

The final step to fine tune the control-point perturbation estimate is an exhaustive search for the optimal perturbation, $\hat{\theta}^{sf}_{N}(i)$, for each individual control point, i, in a small neighborhood of the solution, $\hat{\theta}^{f}_{N}(i)$, obtained from fine-resolution iterations. The perturbations of all other control points are fixed to the value determined by $\hat{\theta}^{f}_{N}$. The perturbation that results in the minimum value of the cost function L is chosen to be the optimal perturbation $\hat{\theta}^{sf}_{N}(i)$ for that particular control point. The superscript "sf" denotes superfine estimate. Ideally, the exhaustive search should be performed in a combinatorial manner (instead of individually for each control point) over the entire search space, but the computational complexity for this strategy would be prohibitive.

Constraints. A few constraints are imposed on the values of perturbations that are imparted to each of the control points. Without any constraints, the algorithm will also search in the domain of unrealistic faces, thus unnecessarily slowing down the algorithm. The following constraints are imposed on the perturbations of the control points:

- Symmetry along the vertical axis passing through the nose tip, based on the fact that most human faces are symmetric about this axis.
- The maximum possible perturbation for the control points is defined.
- The perturbation of a few control points is dependent on the perturbations of their neighboring control points.
- Only control points whose movement might alter the contours of the 3D model (and hence change the cost function) are perturbed.

Pose Refinement and Model Adaptation across Time

Once an adapted 3D model from a particular frame is available, the rough pose estimate (obtained using a generic model) of the next frame is refined to obtain a more accurate head-pose estimate. The method used for this pose refinement is similar to the method described in Section 6.3.1, except that the contours to be compared to the edges of the current frame are extracted from the adapted 3D model up to that stage, instead of the generic face model. For the first video frame, the adapted 3D model used for pose refinement is just the globally deformed generic model. But, for later frames, the globally and locally adapted 3D model is used for pose refinement. This pose-refinement step is critical to the shape-estimation procedure because the described contour-based algorithm is very sensitive to the head-pose estimate. If the pose estimate is not accurate, the 3D model (at the wrong pose) will adapt itself in an inappropriate manner so that its contours conform with the edges extracted from the video frame. To refine the head-pose estimate, contours are obtained from the adapted 3D face model rotated about azimuth angles in a small neighborhood of the rough pose estimate (obtained using the method in Section 6.3.1). The angle which results in 3D model contours closest to the edges extracted from the current frame is chosen as the refined pose estimate. The similarity criterion used to compare the edge maps is the distance-transform-based measure described in Equation 15.

Once the refined pose estimate and the adapted (global and local) 3D face model are available from the first video frame, the algorithm for pose refinement and local deformation described earlier is applied recursively to all the subsequent frames to improve the quality of the reconstructed 3D face model. Integrating the algorithm across several frames makes the system robust to any noise in the contours extracted from a particular frame. The adapted 3D model from the previous frame is used as the starting estimate for the next frame. As more frames with varying head poses become available, the system gets valuable cues to model certain aspects of the face more accurately (e.g., the side view models the structure of the nose better than the front view). Figure 6.15 shows a video frame, the initial generic 3D model, and the final reconstructed 3D face model of the subject.

Texture Extraction

Once the pose refinement and the 3D model adaptation are completed for all the frames, the texture of the person's face is extracted and mapped onto the adapted 3D model for visualization. A single frame is used for texture extraction, because it was found from our experiments that texture sampling across multiple frames smears the texture slightly. The smearing occurs because of the fact that the registration (using estimated pose) across multiple frames is not exact. Now the question arises as to which frame to choose for texture extraction. Based on studies

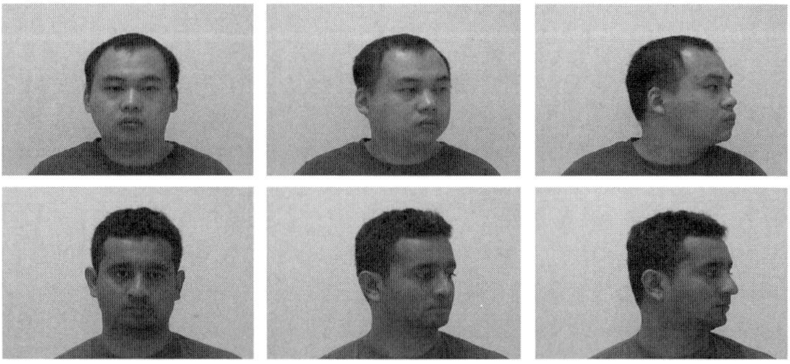

FIGURE 6.13: Video frames. Top row: Subject D. Bottom row: Subject E.

that face-recognition systems perform best on 3/4 profile view [5], we believe that this view is most representative of the person. Hence, the frame closest to 3/4 profile view ($\pm 45^\circ$) was chosen for texture extraction. In a 3/4 profile view, part of the face would be occluded for one half of the face. To get the texture for this half of face (part of which is occluded), symmetry of face texture about the vertical axis passing through the nose tip is assumed.

6.3.3 Experimental Results

In this section, the experimental results of the contour-based 3D face-modeling algorithm are presented. We have a current implementation of the algorithm in MATLAB, using video frames of size 720×480, on a 1.66 GHz Pentium 4 machine. No effort has been made yet to optimize the algorithm. Since the head motion is assumed to be only along the azimuth angle, the top and bottom pixels of the face region were marked out in the first frame, and were assumed to be constant across all video frames. All the edge maps extracted from the average 3D model were resized according to this scale. Also, only the edges in the region between the marked out top and bottom pixels of the face region are retained, thus removing the hair and shoulder edges. Figure 6.13 shows a few video frames of two subjects whose face-modeling results have been presented. Figure 6.14 shows the plot of perturbation errors (value of cost function L) at each stage of the multiresolution search across all frames of the video sequence for Subject D. It can be observed from the plot that, within each frame, the minimum perturbation error decreases as the search resolution gets finer. Figure 6.16 shows the reconstructed texture-mapped 3D face model of the subjects from different viewpoints. Extensive experimentation on a number of real video sequences has been conducted, with good results.

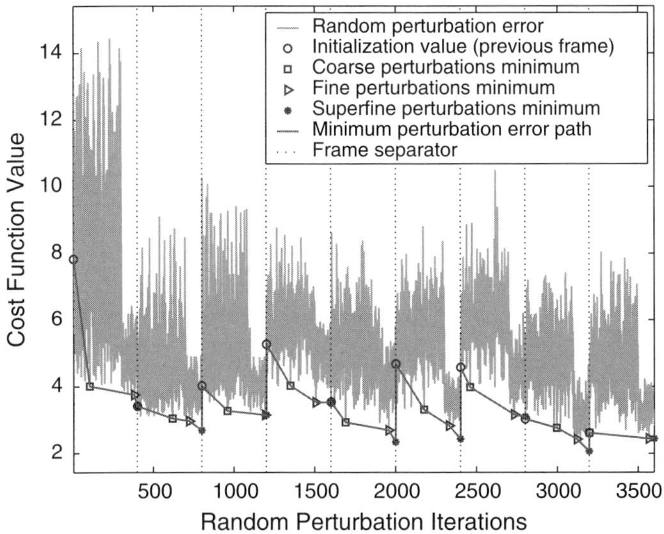

FIGURE 6.14: Perturbation errors at each stage of the multiresolution search for Subject D.

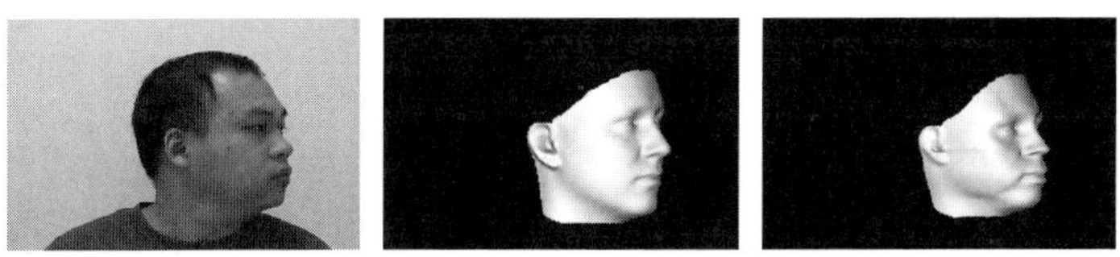

FIGURE 6.15: Left: video frame of subject. Middle: generic 3D model. Right: reconstructed 3D model.

6.4 CONCLUSIONS

In this chapter, two algorithms for 3D face modeling from a monocular video have been presented. The first method works by creating an initial estimate using multiframe SfM, which is then refined by comparing against a generic face model. The comparison is carried out using an energy-function optimization strategy. Statistical measures of the quality of the input video are evaluated and incorporated into the SfM reconstruction framework. Bias of the final reconstruction towards the generic model is avoided by incorporating it in the latter stages of the algorithm. This is validated through experimental evaluation. Results of the 3D reconstruction

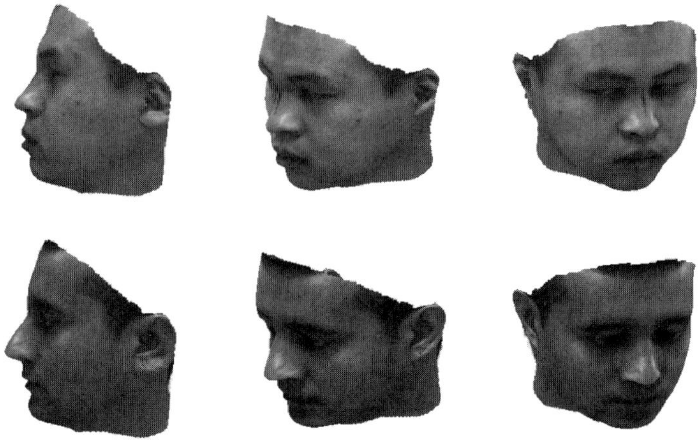

FIGURE 6.16: Reconstructed 3D face model. Top row: Subject D. Bottom row: Subject E.

algorithm, along with an analysis of the errors in the reconstruction, are presented. The second method presented in this chapter reconstructs a face model by adapting a generic model to the contours of the face over all the frames of a video sequence. The algorithm for pose estimation and 3D face reconstruction relies solely on contours, and the system does not require knowledge of rendering parameters (e.g light direction and intensity). Results and anlysis of this algorithm, which does not rely on finding accurate point correspondences across frames, is presented.

REFERENCES

[1] A. Blake and A. Zisserman. *Visual Reconstruction*. MIT Press, 1987.

[2] M. Brady and A. L. Yuille. An extremum principle for shape from contour. *IEEE Transactions on Pattern Analysis and Machine Intelligence* **6**(3):288–301, May 1984.

[3] M. Brand. Morphable 3D models from video. In: *Proc. of IEEE Computer Society Conf. on Computer Vision and Pattern Recognition*, pages II:456–463, 2001.

[4] C. Bregler, A. Hertzmann, and H. Biermann. Recovering non-rigid 3D shape from image streams. In: *Proc. of IEEE Computer Society Conf. on Computer Vision and Pattern Recognition*, pages II:690–696, 2000.

[5] V. Bruce, T. Valentine, and A. Baddeley. The basis of the 3/4 view advantage in face recognition. *Applied Cognitive Psychology* **1**:109–120, 1987.

[6] J. Canny. A computational approach to edge detection. *IEEE Trans. on Pattern Analysis and Machine Intelligence* **8**:679–698, November 1986.

[7] J. J. Clark and A. L. Yuille. *Data Fusion for Sensory Information Processing Systems*. Kluwer, 1990.

[8] A. Doucet, N. deFreitas, and N. Gordon. *Sequential Monte Carlo Methods in Practice*. Springer, 2001.

[9] R. Duda, P. Hart, and D. Stork. *Pattern Classification* (2nd edition). John Wiley and Sons, 2001.

[10] B. Efron and R. Tibshirani. *An Introduction to the Bootstrap*. Chapman and Hall, 1993.

[11] P. Fua. Regularized bundle-adjustment to model heads from image sequences without calibration data. amitrc_mmtrans*International Journal of Computer Vision* **38**(2):153–171, July 2000.

[12] S. Geman and D. Geman. Stochastic relaxation, Gibbs distributions, and the Bayesian restoration of images. *IEEE Trans. on Pattern Analysis and Machine Intelligence* **6**(6):721–741, November 1984.

[13] G. Golub and C.Van Loan. *Matrix Computations*. The Johns Hopkins University Press, Baltimore, 1989.

[14] H. Gupta, A. Roy-Chowdhury, and R. Chellappa. Contour based 3d face modeling from a monocular video. In: *British Machine Vision Conference*, 2004.

[15] R. I. Hartley and A. Zisserman. *Multiple View Geometry in Computer Vision*. Cambridge University Press, 2000.

[16] T. S. Jebara and A. P. Pentland. Parameterized structure from motion for 3d adaptive feedback tracking of faces. In: *Proc. of IEEE Computer Society Conf. on Computer Vision and Pattern Recognition*, pages 144–150, 1997.

[17] W. Matusik, C. Buehler, R. Raskar, S. J Gortler, and L. McMillan. Image-based visual hulls. In: *Computer Graphics Proceedings, Siggraph 2000, New Orleans, LA*, pages 369–374, July 2000.

[18] B. Moghaddam, J. Lee, H. Pfister, and R. Machiraju. Model-based 3-D face capture with shape-from-silhouettes. In: *IEEE International Workshop on Analysis and Modeling of Faces and Gestures, Nice, France*, pages 20–27, October 2003.

[19] V. Nalwa. *A Guided Tour of Computer Vision*. Addison-Wesley, 1993.

[20] M. Pollefeys. Self-calibration and metric 3D reconstruction from uncalibrated image sequences. PhD Thesis, ESAT-PSI, K.U.Leuven, 1999.

[21] H. Robbins and S. Monro. A stochastic approximation method. *Annals of Mathematical Statistics* **22**:400–407, 1951.

[22] S. Romdhani, V. Blanz, and T. Vetter. Face identification by fitting a 3d morphable model using linear shape and texture error functions. In: *Proc. of European Conference on Computer Vision*, page IV: 3–19, 2002.

[23] A. Roy Chowdhury. Statistical analysis of 3D modeling from monocular video streams. PhD Thesis, Univeristy of Maryland, College Park, 2002.

[24] A. Roy Chowdhury and R. Chellappa. Stochastic approximation and rate-distortion analysis for robust structure and motion estimation. *International Journal of Computer Vision* **55**(1):27–53, October 2003.

[25] A. Roy-Chowdhury and R. Chellappa. An information theoretic criterion for evaluating the quality of 3D reconstructions from video. *IEEE Trans. on Image Processing*, pages 960–973, July 2004.

[26] A. Roy Chowdhury, R. Chellappa, and T. Keaton. Wide baseline image registration with application to 3D face modeling. *IEEE Transactions on Multimedia*, pages 423–434, June 2004.

[27] A. K. Roy Chowdhury and R. Chellappa. Face reconstruction from monocular video using uncertainty analysis and a generic model. *Computer Vision and Image Understanding* **91**:188–213, July 2003.

[28] Y. Shan, Z. Liu, and Z. Zhang. Model-based bundle adjustment with application to face modeling. In: *Proc. of International Conf. on Computer Vision*, pages 644–651, 2001.

[29] J. C. Spall. *Introduction to Stochastic Search and Optimization*. Wiley, 2000.

[30] S. Srinivasan. Extracting structure from optical flow using fast error search technique. *International Journal of Computer Vision* **37**:203–230, 2000.

[31] Z. Sun, V. Ramesh, and A. M. Tekalp. Error characterization of the factorization method. *Computer Vision and Image Understanding* **82**:110–137, May 2001.

[32] C. Tomasi and J. Shi. Good features to track. In: *Proceedings IEEE Computer Society Conference on Computer Vision and Pattern Recogition, Jerusalem, Israel*, pages 593–600, October 1994.

[33] R. Walter. *Principles of Mathematical Analysis,* 3rd Edition. McGraw-Hill, 1976.

[34] G. S. Young and R. Chellappa. Statistical analysis of inherent ambiguities in recovering 3D motion from a noisy flow field. *IEEE Trans. on Pattern Analysis and Machine Intelligence* **14**:995–1013, October 1992.

[35] W. Zhao, R. Chellappa, P. J. Phillips, and A. Rosenfeld. Face recognition: A literature survey. *ACM Transactions* **35**:399–458, December 2003.

CHAPTER 7

FACE MODELING BY INFORMATION MAXIMIZATION

7.1 INTRODUCTION

Redundancy in the sensory input contains structural information about the environment. Horace Barlow has argued that such redundancy provides knowledge [5] and that the role of the sensory system is to develop factorial representations in which these dependencies are separated into independent components. Barlow argues that such representations are advantageous for encoding complex objects that are characterized by high-order dependencies. Atick and Redlich have also argued for such representations as a general coding strategy for the visual system [3].

Principal-component analysis (PCA) is a popular unsupervised statistical method to find useful image representations. Consider a set of n basis images each of which has n pixels. A standard basis set consists of a single active pixel with intensity 1, where each basis image has a different active pixel. Any given image with n pixels can be decomposed as a linear combination of the standard basis images. In fact, the pixel values of an image can then be seen as the coordinates of that image with respect to the standard basis. The goal in PCA is to find a "better" set of basis images so that, in this new basis, the image coordinates (the PCA coefficients) are uncorrelated, i.e., they cannot be linearly predicted from each other. PCA can thus be seen as partially implementing Barlow's ideas: Dependencies that show up in the joint distribution of pixels are separated out into the marginal distributions of PCA coefficients. However, PCA can only separate pairwise linear dependencies between pixels. High-order dependencies will still show in the joint distribution of PCA coefficients, and thus will not be properly separated.

Some of the most successful representations for face recognition, such as eigenfaces [70], holons [16], and "local feature analysis" [59] are based on PCA. In a task such as face recognition, much of the important information may be contained in the high-order relationships among the image pixels, and thus it is important to investigate whether generalizations of PCA which are sensitive to high-order relationships, and not just second-order relationships, are advantageous. Independent-component analysis (ICA) [15] is one such generalization. A number of algorithms for performing ICA have been proposed. See [36, 23] for reviews. Here we employ an algorithm developed by Bell and Sejnowski [12, 13] from the point of view of optimal information transfer in neural networks with sigmoidal transfer functions. This algorithm has proven successful for separating randomly mixed auditory signals (the cocktail-party problem), and for separating EEG signals [45], fMRI signals [46].

We performed ICA on the image set under two architectures. Architecture I treated the images as random variables and the pixels as outcomes, whereas architecture II treated the pixels as random variables and the images as outcomes[1] Matlab code for the ICA representations is available at http://mplab.ucsd.edu/-marni.

Face-recognition performance was tested using the FERET database [61]. Face-recognition performances using the ICA representations were benchmarked by comparing them to performances using principal-component analysis, which is equivalent to the "eigenfaces" representation [70, 60]. The two ICA representations were then combined in a single classifier.

7.2 INDEPENDENT-COMPONENT ANALYSIS

There are a number of algorithms for performing ICA [29, 15, 14, 12]. We chose the infomax algorithm proposed by Bell and Sejnowski [12], which was derived from the principle of optimal information transfer in neurons with sigmoidal transfer functions [34]. The algorithm is motivated as follows: Let $\mathbf{X}$ be an n-dimensional random vector representing a distribution of inputs in the environment. (Here boldface capitals denote random variables whereas plain capitals denote matrices.) Let W be an $n \times n$ invertible matrix, $\mathbf{U} = W\mathbf{X}$ and $\mathbf{Y} = f(\mathbf{U})$ an n-dimensional random variable representing the outputs of n-neurons. Here each component of

[1]Preliminary versions of this work appear in [9, 7]. A longer discussion of unsupervised learning for face-recognition appears in the following book [6].

$f = (f_1, \cdots, f_n)$ is an invertible squashing function, mapping real numbers into the [0, 1] interval. Typically the logistic function is used

$$f_i(u) = \frac{1}{1+e^{-u}}. \tag{1}$$

The $\mathbf{U}_1, \cdots, \mathbf{U}_n$ variables are linear combinations of inputs and can be interpreted as presynaptic activations of n neurons. The $\mathbf{Y}_1, \cdots, \mathbf{Y}_n$ variables can be interpreted as postsynaptic activation rates and are bounded by the interval [0, 1]. The goal in Bell and Sejnowski's algorithm is to maximize the mutual information between the environment $\mathbf{X}$ and the output of the neural network $\mathbf{Y}$. This is achieved by performing gradient ascent on the entropy of the output with respect to the weight matrix W. The gradient update rule for the weight matrix, W is as follows

$$\Delta W \propto \nabla_W H(\mathbf{Y}) = (W^{\mathrm{T}})^{-1} + E(\mathbf{Y}'\mathbf{X}^{\mathrm{T}}), \tag{2}$$

where $\mathbf{Y}'_i = f''_i(\mathbf{U}_i)/f'_i(\mathbf{U}_i)$, the ratio between the second and first partial derivatives of the activation function, T stands for transpose, E for expected value, $H(\mathbf{Y})$ is the entropy of the random vector $\mathbf{Y}$, and $\nabla_W H(\mathbf{Y})$ is the gradient of the entropy in matrix form, i.e., the cell in row i and column j of this matrix is the derivative of $H(\mathbf{Y})$ with respect to W_{ij}. Computation of the matrix inverse can be avoided by employing the natural gradient [1], which amounts to multiplying the absolute gradient by $W^{\mathrm{T}}W$, resulting in the following learning rule [13]

$$\Delta W \propto \nabla_W H(\mathbf{Y}) W^{\mathrm{T}} W = W + E(\mathbf{Y}'\mathbf{X}^{\mathrm{T}}) W^{\mathrm{T}} W, \tag{3}$$

where I is the identity matrix. The logistic transfer function (1) gives $\mathbf{Y}'_i = 1 - 2\mathbf{Y}_i$.

When there are multiple inputs and outputs, maximizing the joint entropy of the output $\mathbf{Y}$ encourages the individual outputs to move towards statistical independence. When the form of the nonlinear transfer function f is the same as the cumulative density functions of the underlying independent components (up to scaling and translation) it can be shown that maximizing the joint entropy of the outputs in $\mathbf{Y}$ also minimizes the mutual information between the individual outputs in $\mathbf{U}$ [49, 13]. In practice, the logistic transfer function has been found sufficient to separate mixtures of natural signals with sparse distributions including sound sources [12].

The algorithm is speeded up by including a "sphering" step prior to learning [13]. The row means of $\mathbf{X}$ are subtracted, and then $\mathbf{X}$ is passed through the whitening matrix, W_z, which is twice the inverse principal square root[2] of the covariance matrix:

$$W_z = 2(\operatorname{cov}(\mathbf{X}))^{-\frac{1}{2}}. \tag{4}$$

This removes the first and the second-order statistics of the data; both the mean and covariances are set to zero and the variances are equalized. When the inputs to ICA are the "sphered" data, the full transform matrix W_I is the product of the sphering matrix and the matrix learned by ICA,

$$W_I = WW_z. \tag{5}$$

MacKay [44] and Pearlmutter [57] showed that the ICA algorithm converges to the maximum likelihood estimate of W^{-1} for the following generative model of the data:

$$\mathbf{X} = W^{-1}\mathbf{S}, \tag{6}$$

where $\mathbf{S} = (\mathbf{S}_1, \cdots, \mathbf{S}_n)'$ is a vector of independent random variables, called the sources, with cumulative distributions equal to f_i, i.e., using logistic activation functions corresponds to assuming logistic random sources and using the standard cumulative Gaussian distribution as activation functions, corresponds to assuming Gaussian random sources. Thus W^{-1}, the inverse of the weight matrix in Bell and Sejnowski's algorithm, can be interpreted as the source-mixing matrix and the $\mathbf{U} = W\mathbf{X}$ variables can be interpreted as the maximum-likelihood estimates of the sources that generated the data.

7.2.1 ICA and Other Statistical Techniques

ICA and PCA

Principal-component analysis can be derived as a special case of ICA which uses Gaussian source models. In such a case, the mixing matrix W is unidentifiable in the sense that there is an infinite number of equally good maximum-likelihood solutions. Among all possible maximum-likelihood solutions, PCA chooses an orthogonal matrix which is optimal in the following sense: (1) Regardless of the distribution of $\mathbf{X}$, $\mathbf{U}_1$ is the linear combination of input that allows optimal linear reconstruction of the input in the mean-square sense; (2) For $\mathbf{U}_1, \cdots \mathbf{U}_k$ fixed, $\mathbf{U}_{k+1}$ allows optimal linear reconstruction among the class of linear combinations of $\mathbf{X}$ that are uncorrelated with $\mathbf{U}_1 \cdots \mathbf{U}_k$.

[2]the unique square root for which every eigenvalue has nonnegative real part.

If the sources are Gaussian, the likelihood of the data depends only on first- and second-order statistics (the covariance matrix). In PCA, the rows of W are in fact the eigenvectors of the covariance matrix of the data. In shift-invariant databases (e.g., databases of natural images) the second-order statistics capture the amplitude spectrum of images but not their phase spectrum. The high-order statistics capture the phase spectrum [22, 13]. For a given sample of natural images, we can scramble their phase spectrum while maintaining their power spectrum. This will dramatically alter the appearance of the images but will not change their second-order statistics. The phase spectrum, not the power spectrum, contains the structural information in images that drives human perception. For example, a face image synthesized from the amplitude spectrum of face A and the phase spectrum of face B will be perceived as an image of face B [54, 62]. The fact that PCA is only sensitive to the power spectrum of images suggests that it might not be particularly well suited for representing natural images.

The assumption of Gaussian sources implicit in PCA makes it inadequate when the true sources are nonGaussian. In particular it has been empirically observed that many natural signals, including speech, natural images, and EEG are better described as linear combinations of sources with long-tail distributions [22, 12]. These sources are called "high-kurtosis", "sparse", or "superGaussian" sources. Logistic random variables are a special case of sparse source models. When sparse-source models are appropriate, ICA has the following potential advantages over PCA: (1) It provides a better probabilistic model of the data, which better identifies where the data concentrate in n-dimensional space; (2) it uniquely identifies the mixing matrix W; (3) it finds a not-necessarily-orthogonal basis which may reconstruct the data better than PCA in the presence of noise; (4) It is sensitive to high-order statistics in the data, not just the covariance matrix.

Figure 7.1 illustrates these points with an example. The figure shows samples from a 3-dimensional distribution constructed by linearly mixing two high-kurtosis sources. The figure shows the basis vectors found by PCA and by ICA on this problem. Since the three ICA basis vectors are nonorthogonal, they change the relative distance between data points. This can be illustrated with the following example. Consider the three points: $x_1 = (4, 0), x_2 = (0, 10), x_3 = (10, 10)$, and the following nonorthogonal basis set: $A = [1, 1; 0, 1]$. The coordinates y under the new basis set are defined by $x = Ay$, and thus $y = A^{-1}x$ where $A^{-1} = [1, -1; 0, 1]$. Thus the coordinates under the new basis set are $y_1 = (4, 0), y_2 = (-10, 10), y_3 = (0, 10)$. Note that, in the standard coordinate system, x_1 is closer to x_2 than to x_3. However, in the new coordinate system, y_1 is closer to y_3 than to y_2. In the old coordinate system, the angle between x_1 and x_3 is the same as the angle between x_2 and x_3. However, in the new coordinate system, the angle between y_1 and y_3 is larger than the angle between y_2 and y_3.

This change in metric may be potentially useful for classification algorithms, like nearest-neighbor, that make decisions based on relative distances between points.

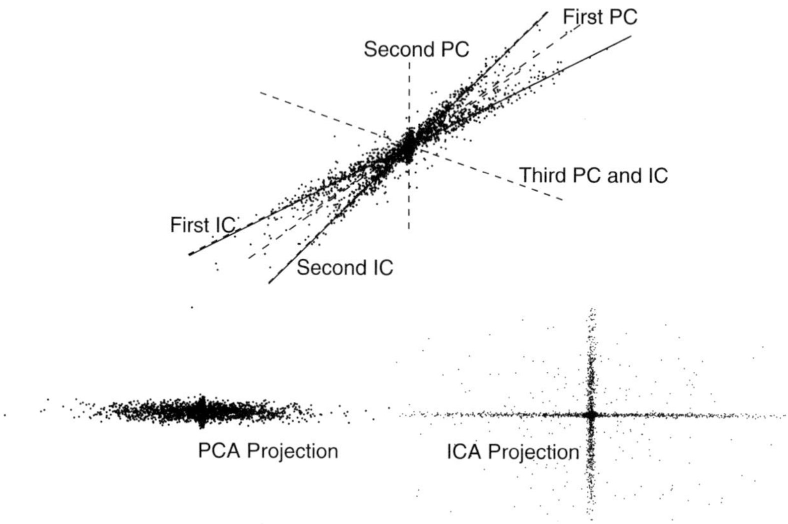

FIGURE 7.1: Top: Example 3D data distribution and corresponding principal-component and independent-component axis. Each axis is a column of the mixing matrix W^{-1} found by PCA or ICA. Note that the PC axes are orthogonal while the IC axes are not. If only two components are allowed, ICA chooses a different subspace than PCA. Bottom left: distribution of the first PCA coordinates of the data. Bottom right: distribution of the first ICA coordinates of the data. Note that, since the ICA axes are nonorthogonal, relative distances between points are different in PCA than in ICA, as are the angles between points.

The ICA basis illustrated in Figure 7.1 also alters the angles between data points, which affects similarity measures such as cosines. Moreover if an undercomplete basis set is chosen, PCA and ICA may span different subspaces. For example, in Figure 7.1, when only two dimensions are selected, PCA and ICA choose different subspaces.

The metric induced by ICA is superior to PCA in the sense that it may provide a representation more robust to the effect of noise [49]. It is therefore possible for ICA to be better than PCA for reconstruction in noisy or limited-precision environments. For example, in the problem presented in Figure 7.1 we found that, if only 12 bits are allowed to represent the PCA and ICA coefficients, linear reconstructions based on ICA are 3dB better than reconstructions based on PCA (the noise power is reduced by more than half). A similar result was obtained for PCA and ICA subspaces. If only 4 bits are allowed to represent the first 2 PCA and ICA coefficients, ICA reconstructions are 3dB better than PCA reconstructions. In some problems, one can think of the actual inputs as noisy versions of some canonical inputs. For example, variations in lighting and expressions can be seen as noisy versions of the canonical image of a person. Having input representations which are robust to noise may potentially give us representations that better reflect the data.

When the source models are sparse, ICA is closely related to the so called nonorthogonal "rotation" methods in PCA and factor analysis. The goal of these rotation methods is to find directions with high concentrations of data, something very similar to what ICA does when the sources are sparse. In such cases, ICA can be seen as a theoretically sound probabilistic method to find interesting nonorthogonal "rotations".

ICA and Cluster Analysis

Cluster analysis is a technique for finding regions in n-dimensional space with large concentrations of data. These regions are called "clusters". Typically the main statistic of interest in cluster analysis is the center of those clusters. When the source models are sparse, ICA finds directions along which significant concentrations of data points are observed. Thus, when using sparse sources, ICA can be seen as a form of cluster analysis. However, the emphasis in ICA is on finding optimal directions, rather than specific locations of high data density. Figure 7.1 illustrates this point. Note how the data concentrates along the ICA solutions, not the PCA solutions. Note also that in this case all the clusters have equal mean and thus are better characterized by their orientation rather than their position in space.

It should be noted that ICA is a very general technique. When superGaussian sources are used, ICA can be seen as doing something akin to nonorthogonal PCA and to cluster analysis, however when the source models are subGaussian, the relationship between these techniques is less clear. See [37] for a discussion of ICA in the context of subGaussian sources.

7.2.2 Two Architectures for Performing ICA on Images

Let X be a data matrix with n_r rows and n_c columns. We can think of each column of X as outcomes (independent trials) of a random experiment. We think of the ith row of X as the specific value taken by a random variable $\mathbf{X}_\mathrm{i}$ across n_c independent trials. This defines an empirical probability distribution for $\mathbf{X}_1, \cdots \mathbf{X}_{n_r}$ in which each column of X is given probability mass $1/n_\mathrm{c}$. Independence is then defined with respect to such a distribution. For example, we say that rows i and j of X are independent if it is not possible to predict the values taken by $\mathbf{X}_j$ across columns from the corresponding values taken by $\mathbf{X}_i$, i.e.,

$$P(\mathbf{X}_i = u, \mathbf{X}_j = v) = P(\mathbf{X}_i = u)P(\mathbf{X}_j = v), \quad \text{for all} \quad u, v \in \mathbb{R}, \tag{7}$$

where P is the empirical distribution as defined above.

Our goal in this chapter is to find a good set of basis images to represent a database of faces. We organize each image in the database as a long vector with as many dimensions as number of pixels in the image. There are at least two ways in which ICA can be applied to this problem:

1. We can organize our database into a matrix X where each row vector is a different image. This approach is illustrated in Figure 7.2a. In this approach, images are random variables and pixels are trials. In this approach it makes sense to talk about independence of images or functions of images. Two images i and j are independent if, when moving across pixels, it is not possible to predict the value taken by the pixel on image j based on the value taken by the same pixel on image i. A similar approach was used by Bell & Sejnowski for sound source separation [12], for EEG analysis [45], and for fMRI [46]. The image-synthesis model associated with this approach is illustrated in the top row of Figure 7.3.
2. We can transpose X and organize our data so that images are in the columns of X. This approach is illustrated in Figure 7.2b. In this approach, pixels are random variables and images are trials. Here it makes sense to talk about independence of pixels or functions of pixels. For example, pixel i and j would be independent if, when moving across the entire set of images, it is not possible to predict the value taken by pixel i based on the corresponding

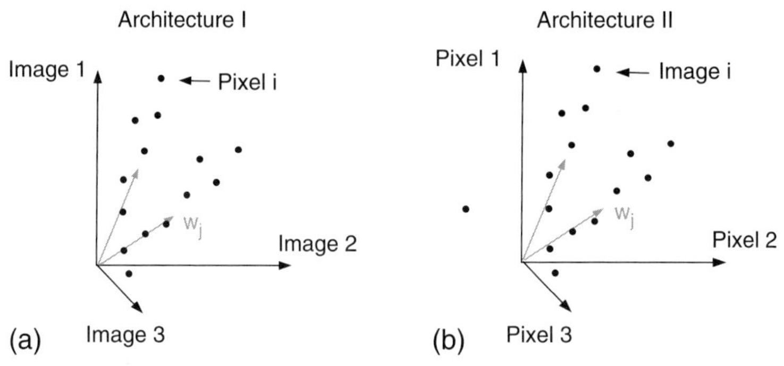

FIGURE 7.2: Two architectures for performing ICA on images. (a) Each pixel is plotted according to the grayvalue it takes on over a set of face images. ICA in Architecture I finds weight vectors in the directions of statistical dependencies among the pixel locations. This defined a set of independent basis images. (b) Here, each image is an observation in a high-dimensional space where the dimensions are the pixels. ICA in Architecture II finds weight vectors in the directions of statistical dependencies among the face images. This defined a factorial face code. (See also color plate section).

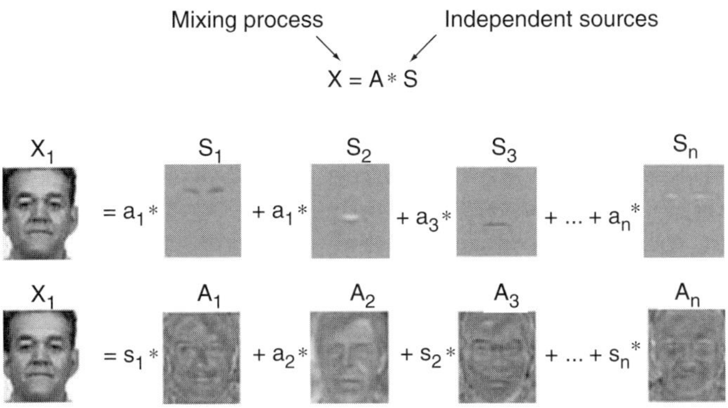

FIGURE 7.3: Image-synthesis models for the two architectures. The ICA model decomposes images as $X = AS$, where $\boldsymbol{A}$ is a mixing matrix and $\boldsymbol{S}$ is a matrix of independent sources. In Architecture I (top), $\boldsymbol{S}$ contains the basis images and $\boldsymbol{A}$ contains the coefficients, whereas in Architecture II (bottom) $\boldsymbol{S}$ contains the coefficients and $\boldsymbol{A}$ contains the basis images for constructing each face image in $\boldsymbol{X}$.

value taken by pixel j on the same image. This approach was inspired by Bell and Sejnowski's work on the independent components of natural images [13]. The image-synthesis model associated with this approach is illustrated in the bottom row of Figure 7.3.

7.3 IMAGE DATA

The face images employed for this research were a subset of the FERET face database [61]. The data set contained images of 425 individuals. There were up to four frontal views of each individual: a neutral expression and a change of expression from one session, and a neutral expression and change of expression from a second session that occurred up to two years after the first. The algorithms were trained on a single frontal view of each individual. The training set was comprised of 50% neutral-expression images and 50% change-of-expression images. The algorithms were tested for recognition under three different conditions: same session, different expression; different day, same expression; and different day, different expression (see Table 7.1).

Coordinates for eye and mouth locations were provided with the FERET database. These coordinates were used to center the face images, and then crop and scale them to 60×50 pixels. Scaling was based on the area of the triangle

Table 7.1: Image sets used for training and testing.

Image Set	Condition		No. images
Training set	Session I	50% neutral 50% expr	425
Test set 1	Same day	Different expression	421
Test set 2	Different day	Same expression	45
Test set 3	Different day	Different expression	43

defined by the eyes and mouth. The luminance was normalized by linearly rescaling each image to the interval $[0, 255]$. For the subsequent analyses, each image was represented as a 3000-dimensional vector given by the luminance value at each pixel location.

7.4 ARCHITECTURE I: STATISTICALLY INDEPENDENT BASIS IMAGES

As described earlier, the goal in this approach is to find a set of statistically independent basis images. We organize the data matrix X so that the images are in rows and the pixels are in columns, i.e., X has 425 rows and 3000 columns, and each image has zero mean.

In this approach, ICA finds a matrix W such that the rows of $U = WX$ are as statistically independent as possible. The source images estimated by the rows of U are then used as basis images to represent faces. Face image representations consist of the coordinates of these images with respect to the image basis defined by the rows of U, as shown in Figure 7.4. These coordinates are contained in the mixing matrix $A \triangleq W_I^{-1}$.

The number of independent components found by the ICA algorithm corresponds to the dimensionality of the input. Since we had 425 images in the

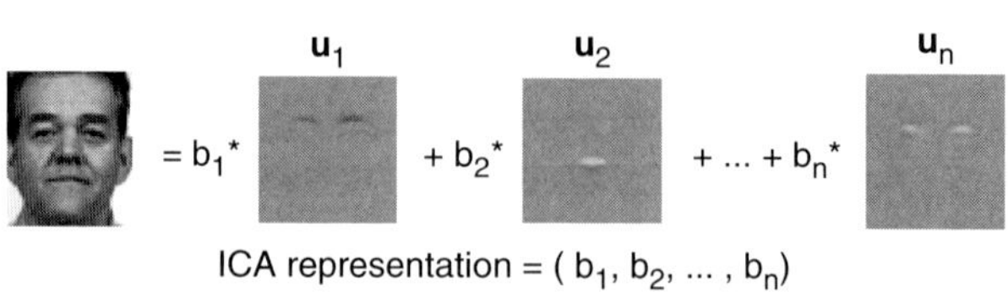

FIGURE 7.4: The independent-basis image representation consisted of the coefficients, $\boldsymbol{b}$, for the linear combination of independent basis images, U, that comprised each face image X.

training set, the algorithm would attempt to separate 425 independent components. Although we found in previous work that performance improved with the number of components separated, 425 was intractable under our present memory limitations. In order to have control over the number of independent components extracted by the algorithm, instead of performing ICA on the n_r original images, we performed ICA on a set of m linear combinations of those images, where $m < n_r$. Recall that the image synthesis model assumes that the images in X are a linear combination of a set of unknown statistically independent sources. The image synthesis model is unaffected by replacing the original images with some other linear combination of the images.

Adopting a method that has been applied to ICA of fMRI data [46], we chose for these linear combinations the first m principal component eigenvectors of the image set. Principal-component analysis on the image set in which the pixel locations are treated as observations, and each face image a measure, gives the linear combination of the parameters (images) that accounts for the maximum variability in the observations (pixels). The use of PCA vectors in the input did not throw away the high-order relationships. These relationships still existed in the data but were not separated.

Let P_m denote the matrix containing the first m principal component axes in its columns. We performed ICA on P_m^{T}, producing a matrix of m independent source images in the rows of U. In this implementation, the coefficients, $\boldsymbol{b}$, for the linear combination of basis images in U that comprised the face images in X were determined as follows:

The principal component representation of the set of zero-mean images in X based on P_m is defined as $R_m = XP_m$. A minimum-squared-error approximation of X is obtained by $\hat{X} = R_m P_m^{\mathrm{T}}$.

The ICA algorithm produced a matrix $W_I = WW_Z$ such that

$$W_I P_m^{\mathrm{T}} = U, \qquad P_m^{\mathrm{T}} = W_I^{-1} U. \tag{8}$$

Therefore

$$\hat{X} = R_m P_m^{\mathrm{T}}, \qquad \hat{X} = R_m W_I^{-1} U, \tag{9}$$

where W_Z was the sphering matrix defined in Equation 4. Hence the rows of $R_m\, W_I^{-1}$ contained the coefficients for the linear combination of statistically independent sources U that comprised $\hat{X}$, where $\hat{X}$ was a minimum-squared-error approximation of X, just as in PCA. The independent-component representation of the face images based on the set of m statistically independent feature images, U was therefore given by the rows of the matrix

$$B = R_m W_I^{-1}. \tag{10}$$

A representation for test images was obtained by using the principal-component representation based on the training images to obtain $R_{\text{test}} = X_{\text{test}} P_m$, and then computing

$$B_{\text{test}} = R_{\text{test}} W_I^{-1}. \tag{11}$$

Note that the PCA step is not required for the ICA representation of faces. It was employed to serve two purposes: (1) to reduce the number of sources to a tractable number, and (2) to provide a convenient method for calculating representations of test images. Without the PCA step, $B = W_I^{-1}$ and $B_{\text{test}} = X_{\text{test}}(U)^{\dagger}$. B_{test} can be obtained without calculating a pseudoinverse by normalizing the length of the rows of U, thereby making U approximately orthonormal, and calculating $B_{\text{test}} = X_{\text{test}} U^{\mathrm{T}}$. However, if ICA did not remove all of the second-order dependencies, then U will not be precisely orthonormal.

The principal-component axes of the training set were found by calculating the eigenvectors of the pixelwise covariance matrix over the set of face images. Independent-component analysis was then performed on the first $m = 200$ of these eigenvectors, where the first 200 principal components accounted for over 98% of the variance in the images.[3] The 1×3000 eigenvectors in P_{200} comprised the rows of the 200×3000 input matrix X. The input matrix X was sphered[4] according to Equation 4, and the weights, W, were updated according to Equation 3 for 1900 iterations. The learning rate was initialized at 0.0005 and annealed down to 0.0001. Training took 90 minutes on a Dec Alpha 2100a. Following training, a set of statistically independent source images were contained in the rows of the output matrix U.

Figure 7.4 shows a sample of basis images (i.e., rows of U) learned in this architecture. These images can be interpreted as follows: Each row of the mixing matrix W found by ICA represents a cluster of pixels that have similar behavior across images. Each row of the U matrix tell us how close each pixel is to the cluster i identified by ICA. Since we use a sparse independent source model, these basis images are expected to be sparse and independent. Sparseness in this case means that the basis images will have a large number of pixels close to zero and a few pixels with large positive or negative values. Note that the ICA images are also local (regions with nonzero pixels are nearby). This is because a majority of the statistical dependencies are in spatially proximal pixel locations. A set of principal component basis images (PCA axes), are shown in Figure 7.5 for comparison.

[3]In a pilot work, we found that face-recognition performance improved with the number of components separated. We chose 200 components as the largest number to separate within our processing limitations.

[4]Although PCA already removed the covariances in the data, the variances were not equalized. We therefore retained the sphering step.

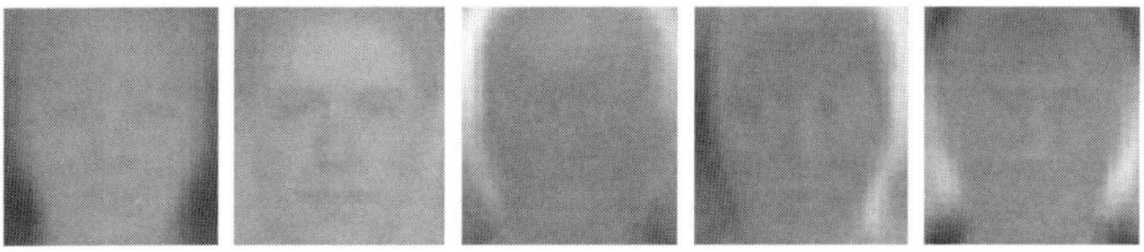

FIGURE 7.5: First 5 principal component axes of the image set (columns of P).

7.4.1 Face-Recognition Performance: Architecture I

Face recognition performance was evaluated for the coefficient vectors $\boldsymbol{b}$ by the nearest neighbor algorithm, using cosines as the similarity measure. Coefficient vectors in each test set were assigned the class label of the coefficient vector in the training set that was most similar as evaluated by the cosine of the angle between them:

$$c = \frac{\boldsymbol{b}_{\text{test}} \cdot \boldsymbol{b}_{\text{train}}}{\| \boldsymbol{b}_{\text{test}} \| \| \boldsymbol{b}_{\text{train}} \|}. \tag{12}$$

Face-recognition performance for the principal component representation was evaluated by an identical procedure, using the principal-component coefficients contained in the rows of R_{200}.

In experiments to date, ICA performs significantly better using cosines rather than Euclidean distance as the similarity measure, whereas PCA performs the same for both. A cosine similarity measure is equivalent to length-normalizing the vectors prior to measuring Euclidean distance when doing nearest neighbor:

$$\begin{aligned} d^2(\boldsymbol{x}, \boldsymbol{y}) &= \| \boldsymbol{x} \|^2 + \| \boldsymbol{y} \|^2 - 2\boldsymbol{x} \cdot \boldsymbol{y} \\ &= \| \boldsymbol{x} \|^2 + \| \boldsymbol{y} \|^2 - 2 \| \boldsymbol{x} \| \| \boldsymbol{y} \| \cos\alpha. \end{aligned} \tag{13}$$

$$\text{Thus if } \| \boldsymbol{x} \| = \| \boldsymbol{y} \| = 1, \quad \min_y d^2(x, y) \leftrightarrow \max_y \cos\alpha.$$

Such normalization is consistent with neural models of primary visual cortex [27]. Cosine similarity measures were previously found to be effective for computational models of language [28] and face processing [55].

Figure 7.6 gives face-recognition performance with both the ICA and the PCA based representations. Recognition performance is also shown for the PCA based representation using the first 20 principal component vectors, which was the eigenface representation used by Pentland, Moghaddam, and Starner [60]. Best performance for PCA was obtained using 200 coefficients. Excluding the first 1, 2, or 3 principal components did not improve PCA performance, nor did selecting

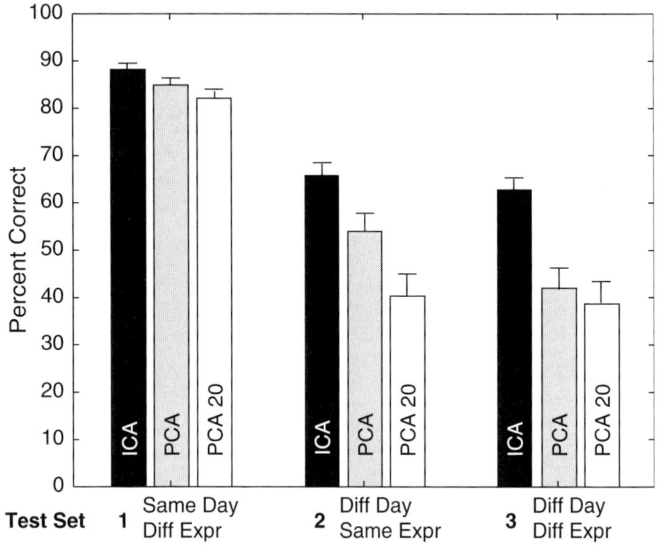

FIGURE 7.6: Percent correct face recognition for the ICA representation using 200 independent components, the PCA representation using 200 principal components, and the PCA representation using 20 principal components. Groups are performances for test set 1, test set 2, and test set 3. Error bars are one standard deviation of the estimate of the success rate for a Bernoulli distribution.

intermediate ranges of components from 20 through 200. There was a trend for the ICA representation to give superior face-recognition performance to the PCA representation with 200 components. The difference in performance was statistically significant for test set 3 ($Z = 1.94, p = 0.05$). The difference in performance between the ICA representation and the eigenface representation with 20 components was statistically significant over all three test sets ($Z = 2.5, p < 0.05$) for test sets 1 and 2, and ($Z = 2.4, p < 0.05$) for test set 3.

Recognition performance using different numbers of independent components was also examined by performing ICA on 20 to 200 image mixtures in steps of 20. Best performance was obtained by separating 200 independent components. In general, the more independent components were separated, the better the recognition performance. The basis images also became increasingly spatially local as the number of separated components increased.

7.4.2 Subspace Selection

When all 200 components were retained, PCA and ICA were working in the same subspace. However, as illustrated in Figure 7.1, when subsets of axes are selected, then ICA chooses a different subspace from PCA. The full benefit of ICA may not be tapped until ICA-defined subspaces are explored.

Face-recognition performances for the PCA and ICA representations were next compared by selecting subsets of the 200 components by class discriminability. Let $\overline{x}$ be the overall mean of a coefficient b_k across all faces, and $\overline{x}_j$ be the mean for person j. For both the PCA and ICA representations, we calculated the ratio of between-class to within-class variability, r, for each coefficient:

$$r = \frac{\sigma_{\text{between}}}{\sigma_{\text{within}}}, \tag{14}$$

where $\sigma_{\text{between}} = \sum_j (\overline{x}_j - \overline{x})^2$ is the variance of the j class means, and $\sigma_{\text{within}} = \sum_j \sum_i (x_{ij} - \overline{x}_j)^2$ is the sum of the variances within each class.

The class discriminability analysis was carried out using the 43 subjects for which four frontal view images were available. The ratios r were calculated separately for each test set, excluding the test images from the analysis. Both the PCA and ICA coefficients were then ordered by the magnitude of r. Figure 7.7 (left) compares the discriminability of the ICA coefficients to the PCA coefficients. The ICA coefficients consistently had greater class discriminability than the PCA coefficients.

Face classification performance was compared using the k most discriminable components of each representation. Figure 7.7 (right) shows the best classification performance obtained for the PCA and ICA representations, which was with the 60 most discriminable components for the ICA representation, and the 140 most discriminable components for the PCA representation. Selecting subsets of coefficients by class discriminability improved the performance of the ICA representation, but had little effect on the performance of the PCA representation. The ICA representation again outperformed the PCA representation. The difference in recognition performance between the ICA and PCA representations was significant for test set 2 and test set 3, the two conditions that required recognition of images collected on a different day from the training set ($Z = 2.9, p < .05; Z = 3.4, p < .01$), respectively. The class discriminability analysis selected subsets of bases from the PCA and ICA representations under the same criterion. Here, the ICA-defined subspace encoded more information about facial identity than PCA-defined subspace.

7.5 ARCHITECTURE II: A FACTORIAL FACE CODE

The goal in Architecture I was to use ICA to find a set of spatially independent basis images. Although the basis images obtained in that architecture are approximately independent, the coefficients that code each face are not necessarily independent. Architecture II uses ICA to find a representation in which the coefficients used to code images are statistically independent, i.e., a factorial face code. Barlow and

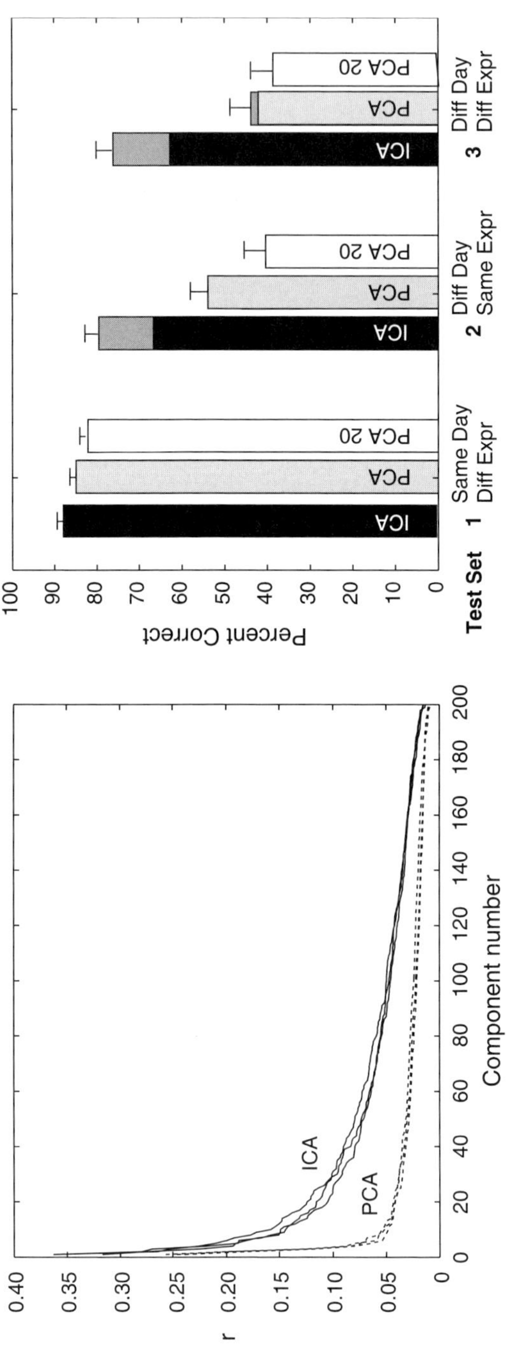

FIGURE 7.7: Selection of components by class discriminability. Left: discriminability of the ICA coefficients (solid lines) and discriminability of the PCA components (dotted lines) for the three test cases. Components were sorted by the magnitude of r. Right: improvement in face-recognition performance for the ICA and PCA representations using subsets of components selected by the class discriminability r. The improvement is indicated by the gray segments at the top of the bars.

Atick have discussed advantages of factorial codes for encoding complex objects that are characterized by high-order combinations of features [5, 2]. These include the fact that the probability of any combination of features can be obtained from their marginal probabilities.

To achieve this goal we organize the data matrix X so that rows represent different pixels and columns represent different images. (See Figure 7.2b) This corresponds to treating the columns of $A \triangleq W_I^{-1}$ as a set of basis images. (See Figure 7.3.) The ICA representations are in columns of $U = W_I X$. Each column of U contains the coefficients of the basis images in A for reconstructing each image in X (Figure 7.8). ICA attempts to make the outputs, U, as independent as possible. Hence U is a factorial code for the face images.

The representational code for test images is obtained by

$$W_I X_{\text{test}} = U_{\text{test}}, \tag{15}$$

where X_{test} is the zero-mean[5] matrix of test images, and W_I is the weight matrix found by performing ICA on the training images.

In order to reduce the dimensionality of the input, instead of performing ICA directly on the 3000 image pixels, ICA was performed on the first 200 PCA coefficients of the face images. The first 200 principal components accounted for over 98% of the variance in the images. These coefficients, ${R_{200}}^{\mathrm{T}}$, comprised the columns of the input data matrix, where each coefficient had zero mean. The Architecture II representation for the training images was therefore contained in the columns of U, where

$$W_I {R_{200}}^{\mathrm{T}} = U. \tag{16}$$

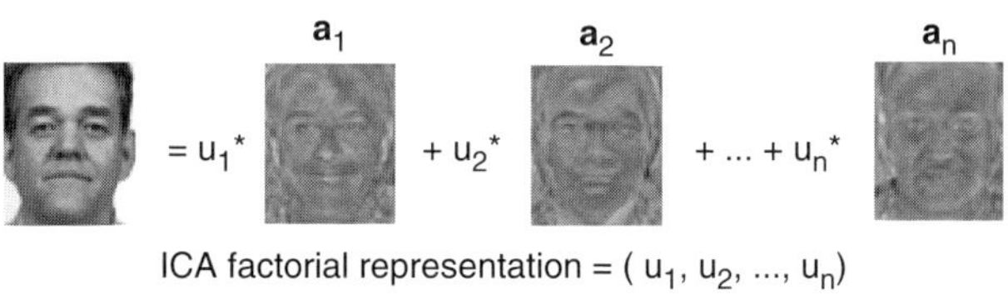

FIGURE 7.8: The factorial code representation consisted of the independent coefficients, **u**, for the linear combination of basis images in A that comprised each face image **x**.

[5]Here, each pixel has zero mean.

The ICA weight matrix W_I was 200×200, resulting in 200 coefficients in U for each face image, consisting of the outputs of each of the ICA filters.[6] The Architecture II representation for test images was obtained in the columns of U_{test} as follows:

$$W_I {R_{\text{test}}}^{\mathrm{T}} = U_{\text{test}}. \tag{17}$$

The basis images for this representation consisted of the columns of $A \triangleq W_I^{-1}$. A sample of the basis images is shown in Figure 7.8, where the principal-component reconstruction $P_{200}A$ was used to visualize them. In this approach each column of the mixing matrix W^{-1} found by ICA attempts to get close to a cluster of images that look similar across pixels. Thus this approach tends to generate basis images that look more face-like than the basis images generated by PCA, in that the bases found by ICA will average only images that look alike. Unlike the ICA output, U, the algorithm does not force the columns of A to be either sparse or independent. Indeed the basis images in A have more global properties than the basis images in the ICA output of Architecture I shown in Figure 7.4.

7.5.1 Face Recognition Performance: Architecture II

Face-recognition performance was again evaluated by the nearest-neighbor procedure using cosines as the similarity measure. Figure 7.9 compares the face-recognition performance using the ICA factorial code representation obtained with Architecture II to the independent-basis representation obtained with Architecture I and to the PCA representation, each with 200 coefficients. Again, there was a trend for the ICA-factorial representation (ICA2) to outperform the PCA representation for recognizing faces across days. The difference in performance for test set 2 is significant ($Z = 2.7, p < 0.01$). There was no significant difference in the performances of the two ICA representations.

Class discriminability of the 200 ICA factorial coefficients was calculated according to Equation 14. Unlike the coefficients in the independent-basis representation, the ICA-factorial coefficients did not differ substantially from each other according to discriminability r. Selection of subsets of components for the representation by class discriminability had little effect on the recognition performance using the ICA-factorial representation (see Figure 7.9 right). The difference in performance between ICA1 and ICA2 for test set 3 following the discriminability analysis just misses significance ($Z = 1.88, p = 0.06$).

In this implementation, we separated 200 components using 425 samples, which was a bare minimum. Test images were not used to learn the independent components, and thus our recognition results were not due to overlearning. Nevertheless,

[6]An image filter $f(\mathbf{x})$ is defined as $f(\boldsymbol{x}) = \boldsymbol{w} \cdot \boldsymbol{x}$.

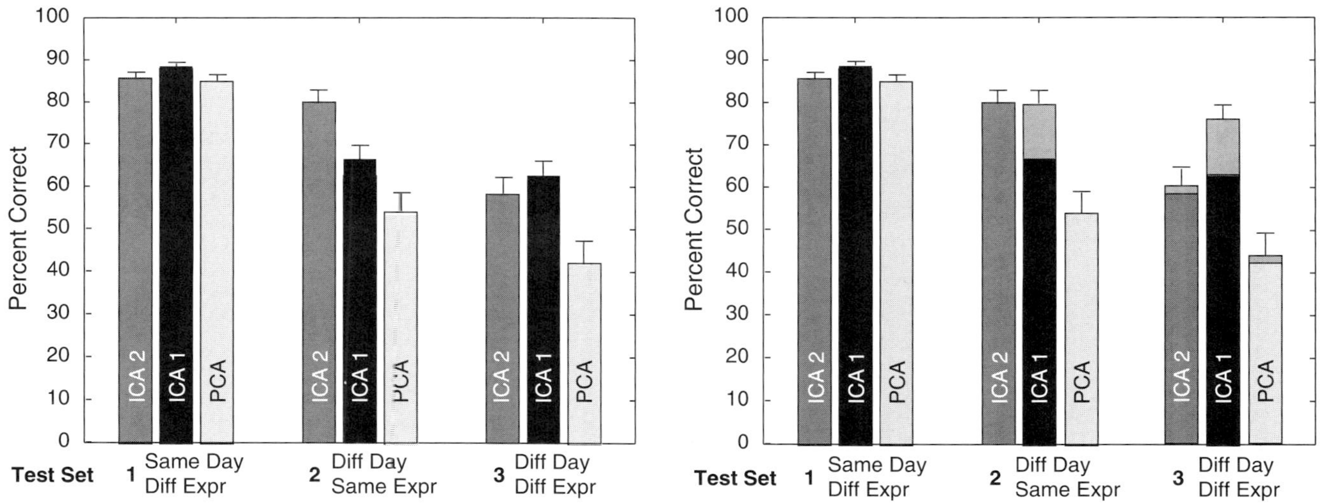

FIGURE 7.9: Left: recognition performance of the factorial-code ICA representation (ICA2) using all 200 coefficients, compared to the ICA independent-basis representation (ICA1), and the PCA representation, also with 200 coefficients. Right: improvement in recognition performance of the two ICA representations and the PCA representation by selecting subsets of components by class discriminability. Gray extensions show improvement over recognition performance using all 200 coefficients.

in order to determine whether the findings were an artifact due to small sample size, recognition performances were also tested after separating 85 rather than 200 components, and hence estimating fewer weight parameters. The same overall pattern of results was obtained when 85 components were separated. Both ICA representations significantly outperformed the PCA representation on test sets 2 and 3. With 85 independent components, ICA1 obtained 87%, 62%, and 58% correct performance respectively on test sets 1, 2, and 3, while ICA2 obtained 85%, 76%, and 56% correct performance, whereas PCA obtained 85%, 56% and 44% correct, respectively. Again, as found for 200 separated components, selection of subsets of components by class discriminability improved the performance of ICA1 to 86%, 78%, and 65%, respectively, and had little effect on the performances with the PCA and ICA2 representations. This suggests that the results were not simply an artifact due to small sample size.

7.5.2 Combined ICA Recognition System

The similarity spaces in the two ICA representations were not identical. While the two systems tended to make errors on the same images, they did not assign the same incorrect identity. In [10] we showed that an effective reliability criterion is to accept identifications only when the two systems give the same answer. Under this criterion, accuracy improved to 100%, 100%, and 97% for the three test sets.

Another way to combine the two systems is to define a new similarity measure as the sum of the similarities under Architecture I and Architecture II. In [10] we showed that this combined classifier improved performance to 91.0%, 88.9%, and 81.0% for the three test cases, which significantly outperformed PCA in all three conditions ($Z = 2.7, p < 0.01; Z = 3.7, p < .001; Z = 3.7; p < .001$).

7.6 EXAMINATION OF THE ICA REPRESENTATIONS

7.6.1 Mutual Information

A measure of the statistical dependencies of the face representations was obtained by calculating the mean mutual information between pairs of 50 basis images. Mutual information was calculated as

$$I(\mathbf{U}_1, \mathbf{U}_2) = H(\mathbf{U}_1) + H(\mathbf{U}_2) - H(\mathbf{U}_1, \mathbf{U}_2), \tag{18}$$

$$\text{where} \quad H(\mathbf{U}_i) = -E\left(\log(P_{U_i}(\mathbf{U}_i))\right).$$

Figure 7.10 (left) compares the mutual information between *basis images* for the original gray-level images, the principal-component basis images, and the ICA basis images obtained in Architecture I. Principal-component images are uncorrelated, but there are remaining high-order dependencies.

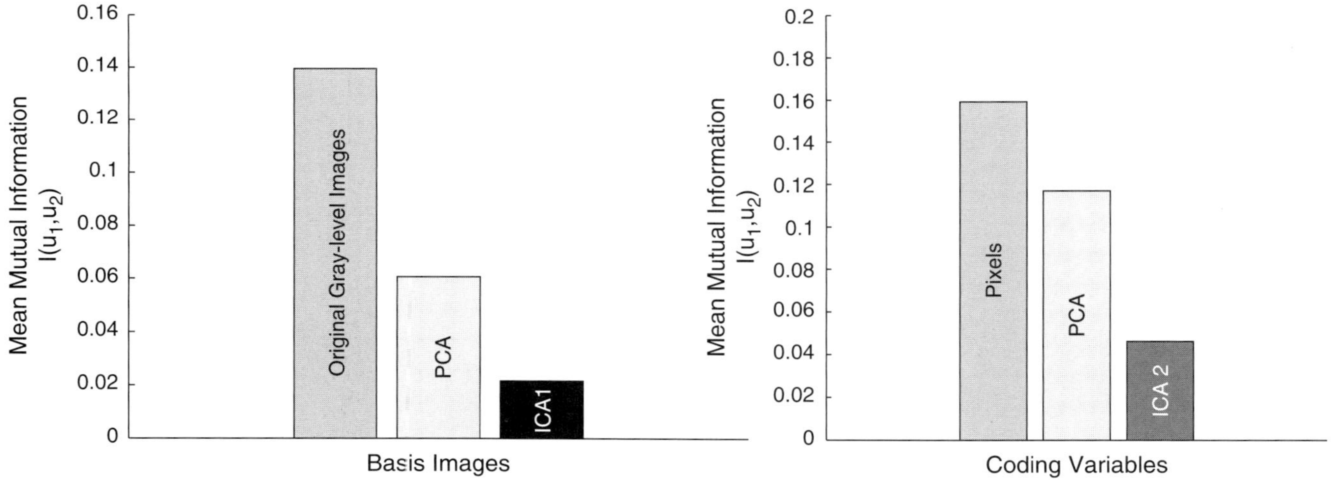

FIGURE 7.10: Pairwise mutual information. Left: mean mutual information between basis images. Mutual information was measured between pairs of gray-level images, principal component images, and independent basis images obtained by Architecture I. Right: mean mutual information between coding variables. Mutual information was measured between pairs of image pixels in gray-level images, PCA coefficients, and ICA coefficients obtained by Architecture II. (See also color plate section)

The information-maximization algorithm decreased these residual dependencies by more than 50%. The remaining dependence may be due to a mismatch between the logistic transfer function employed in the learning rule and the cumulative density function of the independent sources, the presence of subGaussian sources, or the large number of free parameters to be estimated relative to the number of training images.

Figure 7.10 (right) compares the mutual information between the *coding variables* in the ICA factorial representation obtained with Architecture II, the PCA representation, and gray-level images. For gray-level images, mutual information was calculated between pairs of pixel locations. For the PCA representation, mutual information was calculated between pairs of principal-component coefficients, and for the ICA factorial representation, mutual information was calculated between pairs of coefficients, u. Again, there were considerable high-order dependencies remaining in the PCA representation that were reduced by more than 50% by the information-maximization algorithm. The ICA representations obtained in these simulations are most accurately described not as "independent," but as "redundancy reduced," where the redundancy is less than half that in the principal component representation.

7.6.2 Sparseness

Field [22] has argued that sparse distributed representations are advantageous for coding visual stimuli. Sparse representations are characterized by highly kurtotic response distributions, in which a large concentration of values are near zero, with rare occurrences of large positive or negative values in the tails. In such a code, the redundancy of the input is transformed into the redundancy of the response patterns of the individual outputs. Maximizing sparseness without loss of information is equivalent to the minimum entropy codes discussed by Barlow [5].[7]

Given the relationship between sparse codes and minimum entropy, the advantages for sparse codes as outlined by Field [22] mirror the arguments for independence presented by Barlow [5]. Codes that minimize the number of active neurons can be useful in the detection of suspicious coincidences. Because a nonzero response of each unit is relatively rare, high-order relations become increasingly rare, and therefore more informative when they are present in the stimulus. Field contrasts this with a compact code such as principal components, in which a few units have a relatively high probability of response, and therefore high-order combinations among this group are relatively common. In a sparse distributed code, different objects are represented by indicating which units are active, rather than by how much they are active. These representations have an

[7]Information maximization is consistent with minimum-entropy coding. By maximizing the *joint* entropy of the output, the entropies of the *individual* outputs tend to be minimized.

added advantage in signal-to-noise ratio, since one need only determine which units are active without regard to the precise level of activity. An additional advantage of sparse coding for face representations is storage in associative memory systems. Networks with sparse inputs can store more memories and provide more effective retrieval with partial information [56, 11].

The probability densities for the values of the coefficients of the two ICA representations and the PCA representation are shown in Figure 7.11. The sparseness of the face representations was examined by measuring the kurtosis of the distributions. Kurtosis is defined as the ratio of the fourth moment of the distribution to the square of the second moment, normalized to zero for the Gaussian distribution by subtracting 3:

$$kurtosis = \frac{\sum_i (b_i - \overline{b})^4}{\left(\sum_i (b_i - \overline{b})^2\right)^2} - 3. \tag{19}$$

The kurtosis of the PCA representation was measured for the principal-component coefficients. The principal components of the face images had a kurtosis of 0.28. The coefficients, b, of the independent basis representation from Architecture I had a kurtosis of 1.25. Although the basis images in Architecture I had a sparse distribution of grayvalues, the face coefficients with respect to this

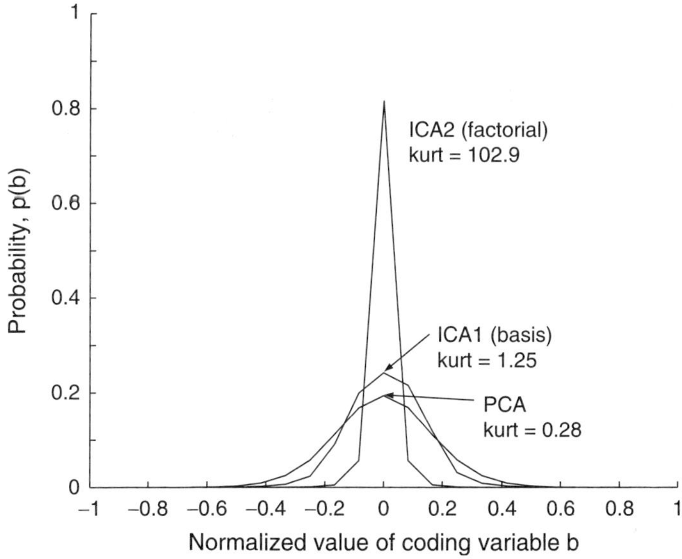

FIGURE 7.11: Kurtosis (sparseness) of ICA and PCA representations.

basis were not sparse. In contrast, the coefficients, u, of the ICA factorial code representation from Architecture II were highly kurtotic, at 102.9.

7.7 LOCAL BASIS IMAGES VERSUS FACTORIAL CODES

Draper and colleagues [20] conducted a comparison of ICA and PCA on a substantially larger set of FERET face images consisting of 1196 individuals. Performances were compared for L1 and L2 norms as well as cosines and Mahalanobis distance. This study supported the findings presented here, namely that ICA outperformed PCA, and this advantage emerged when cosines, but not Euclidean distance, was used as the similarity measure for ICA. This study included a change in lighting condition which we had not previously tested. ICA with Architecture II obtained 51% accuracy on 192 probes with changes in lighting, compared to the best PCA performance (with Mahalanobis distance) of 40% correct.

An interesting finding to emerge from the Draper study is that the ICA representation with Architecture II outperformed Architecture I for identity recognition. (See Figure 7.12). According to arguments by Barlow, Atick, and Field [5, 2, 22], the representation defined by Architecture II is a more optimal representation than the Architecture I representation given its sparse, factorial properties. Although Sections 7.4 and 7.5 showed no significant difference in recognition performance for the two architectures, Architecture II had fewer training samples to estimate the same number of free parameters as Architecture I due to the difference in the

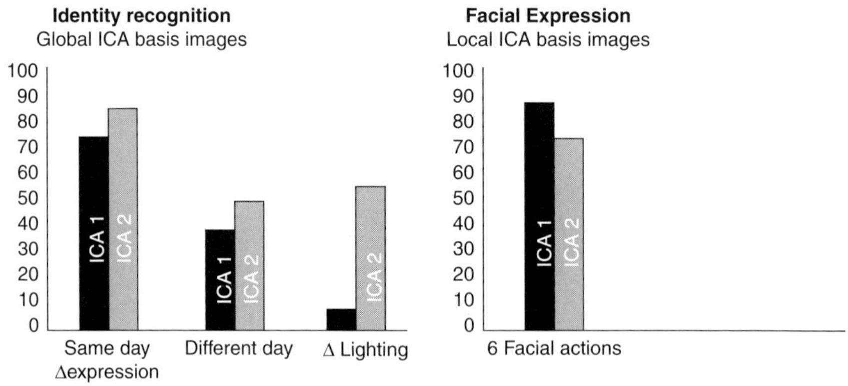

FIGURE 7.12: Face recognition performance with a larger image set from Draper et al. (2003). Architecture II outperformed Architecture I for identity recognition, whereas Architecture I outperformed Architecture II for an expression recognition task. Draper and colleagues attributed the findings to local versus global processing demands.

way the input data were defined. Figure 7.10 shows that the residual dependencies in the ICA-factorial representation were higher than in the Architecture I representation. In [6] we predicted that The ICA-factorial representation may prove to have a greater advantage given a much larger training set of images. Indeed, this prediction was born out in the Draper study [20].

When the task was changed to recognition of facial expressions, however, Draper et al. found that the ICA representation from Architecture I outperformed the ICA representation from Architecture II. The advantage for Architecture I only emerged, however, following subspace selection using the class variability ratios defined in Section 7.4.2. The task was to recognize 6 facial actions, which are individual facial movements approximately corresponding to individual facial muscles. Draper et al. attributed their pattern of results to differences in local versus global processing requirements of the two tasks. Architecture I defines local face features whereas Architecture II defines more configural face features. A large body of literature in human face processing points to the importance of configural information for identity recognition, whereas the facial-expression recognition task in this study may have greater emphasis on local information. This speaks to the issue of separate basis sets for expression and identity. The neuroscience community has been interested in this distinction, as there is evidence for separate processing of identity and expression in the brain (e.g., [30].) Here we obtain better recognition performance when we define different basis sets for identity versus expression. In the two basis sets we switch what is treated as an observation versus what is treated as an independent variable for the purposes of information maximization.

7.8 DISCUSSION

Much of the information that perceptually distinguishes faces may be contained in the higher-order statistics of the images. The basis images developed by PCA depend only on second-order image statistics, and thus it is desirable to find generalizations of PCA that are sensitive to higher-order image statistics. In this chapter we explored one such generalization: Bell and Sejnowski's infomax ICA algorithm. We explored two different architectures for developing image representations of faces using ICA. Architecture I treated images as random variables and pixels as random trials. This architecture was related to the one used by Bell and Sejnowski to separate mixtures of auditory signals into independent sound sources. Under this architecture, ICA found a basis set of statistically independent images. The images in this basis set were sparse and localized in space, resembling facial features. Architecture II treated pixels as random variables and images as random trials. Under this architecture, the image coefficients were approximately independent, resulting in a factorial face code.

Both ICA representations outperformed the "eigenface" representation [70], which was based on principal-component analysis, for recognizing images of faces sampled on a different day from the training images. A classifier that combined the two ICA representations outperformed eigenfaces on all test sets. Since ICA allows the basis images to be nonorthogonal, the angles and distances between images differ between ICA and PCA. Moreover, when subsets of axes are selected, ICA defines a different subspace than PCA. We found that when selecting axes according to the criterion of class discriminability, ICA-defined subspaces encoded more information about facial identity than PCA-defined subspaces. Moreover, with a larger training set, the factorial representation of Architecture II emerged with higher identity recognition performance than Architecture I [20], consistent with theories of the effectiveness of sparse, factorial representations for coding complex visual objects [5, 2, 22].

As discussed in Section 7.2.1, ICA representations are better optimized for transmitting information in the presence of noise than PCA, and thus they may be more robust to variations such as lighting conditions, changes in hair, makeup, and facial expression which can be considered forms of noise with respect to the main source of information in our face database: the person's identity. The robust recognition across different days is particularly encouraging, since most applications of automated face-recognition contain the noise inherent to identifying images collected on a different day from the sample images. Draper and colleagues [20], tested a specific form of noise, lighting variation in the FERET dataset, and found a considerable advantage of ICA Architecture II over PCA for robustness to changes in lighting. The Draper study also supported our finding of a substantial advantage of ICA II over PCA for images collected weeks apart. A key challenge in translating any face-recognition method into a real-world system is noise. The approach presented here would benefit from a more systematic exploration of tolerance to noise, including the effect of noise at different spatial scales. Moreover it was recently shown that flatter transfer functions than the ones learned by information maximization, proportional to the cube root of the cumulative pdf, optimize information transmission in the presence of noise since error magnification depends on the slope of the transfer function [71]. Thus another avenue of research is to explore face representations based on the optimization function in [71].

A number of research groups have independently tested the ICA representations presented here and in [9, 10]. Liu and Wechsler [42], and Yuen and Lai [76] both supported the finding that ICA outperformed PCA. Moghaddam [48] and Yang [75] employed Euclidean distance as the similarity measure instead of cosines and tested Architecture I only. No differences are expected with Euclidean distance in Architecture I, and consistent with our findings, no significant difference was found. Draper and colleagues [20] conducted a thorough comparison of ICA and PCA using four similarity measures, and supported the findings that ICA outperformed PCA, and this advantage emerged when cosines, but not Euclidean distance, was used as the similarity measure for ICA. Class-specific projections of the ICA face codes using Fisher's linear discriminant has recently been shown to be effective for face-recognition as well [32]. ICA was also shown to be effective for facial-expression recognition. The ICA representation outperformed more than eight other image representations on a task of facial expression recognition, equaled only by Gabor wavelet decomposition [19, 8], with which it has relationships discussed below.

In this chapter, the number of sources was controlled by reducing the dimensionality of the data through PCA prior to performing ICA. There are two limitations to this approach [68]. The first is the reverse-dimensionality problem. It may not be possible to linearly separate the independent sources in smaller subspaces. Since we retained 200 dimensions, this may not have been a serious limitation of this implementation. Secondly, it may not be desirable to throw away subspaces of the data with low power such as the higher principal components. Although low in power, these subspaces may contain independent components, and the property of the data we seek is independence, not amplitude. Techniques have been proposed for separating sources on projection planes without discarding any independent components of the data [68]. Techniques for estimating the number of independent components in a dataset have also been proposed [33, 47].

The information-maximization algorithm employed to perform independent-component analysis in this work assumed that the underlying "causes" of the pixel graylevels in face images had a superGaussian (peaky) response distribution. Many natural signals, such as sound sources, have been shown to have a superGaussian distribution [12]. We employed a logistic source model which has shown in practice to be sufficient to separate natural signals with superGaussian distributions [12]. The underlying "causes" of the pixel graylevels in the face images are unknown, and it is possible that better results could have been obtained with other source models. In particular, any subGaussian sources would have remained mixed. Methods for separating subGaussian sources through information maximization have been developed [37]. A future direction of this research is to examine subGaussian components of face images.

The information-maximization algorithm employed in this work also assumed that the pixel values in face images were generated from a linear mixing process. This linear approximation has been shown to hold true for the effect of lighting on face images [24]. Other influences, such as changes in pose and expression may be linearly approximated only to a limited extent. Nonlinear independent-component analysis in the absence of prior constraints is an ill-conditioned problem, but some progress has been made by assuming a linear mixing process followed by parametric nonlinear functions [38, 74]. An algorithm for nonlinear ICA based on kernel methods has also recently been presented [4]. Kernel methods have already shown to improve face-recognition performance with PCA and Fisherfaces [75], and promising results have recently been presented for face recognition with kernel-ICA [43].

Unlike PCA, ICA using Architecture I found a spatially local face representation. Local-feature analysis (LFA) [59] also finds local basis images for faces, but using second-order statistics. The LFA basis images are found by performing whitening (Equation 4) on the principal-component axes, followed by a rotation to topographic correspondence with pixel location. The LFA kernels are not sensitive to the high-order dependencies in the face image ensemble, and in tests to date, recognition performance with LFA kernels has not significantly improved upon PCA [19]. Interestingly, downsampling methods based on sequential information maximization significantly improve performance with LFA [58].

ICA outputs using Architecture I were sparse in space (within image across pixels) while the ICA outputs using Architecture II were sparse across images. Hence Architecture I produced local basis images, but the face codes were not sparse, while Architecture II produced sparse face codes, but with holistic basis images. A representation that has recently appeared in the literature, nonnegative matrix factorization (NMF) [35], produced local basis images and sparse face codes.[8] While this representation is interesting from a theoretical perspective, it has not yet proven useful for recognition. Another innovative face representation employs products of experts in restricted Boltzmann machines (RBMs). This representation also finds local features when nonnegative weight constraints are employed [69]. In experiments to date, RBM's outperformed PCA for recognizing faces across changes in expression or addition/removal of glasses, but performed more poorly for recognizing faces across different days. It is an open question as to whether sparseness and local features are desirable objectives for face-recognition in and of themselves. Here, these properties emerged from an objective of independence.

Capturing more likelihood may be a good principle for generating unsupervised representations which can be later used for classification. As mentioned in Section 7.2, PCA and ICA can be derived as generative models of the data, where PCA uses Gaussian sources, and ICA typically uses sparse sources. It has been shown that for many natural signals, ICA is a better model in that it assigns higher likelihood to the data than PCA [39]. The ICA basis dimensions presented here may have captured more likelihood of the face images than PCA, which provides a possible explanation for the superior performance of ICA in this study.

Desirable filters may be those that are adapted to the patterns of interest and capture interesting structure [40]. The more the dependencies that are encoded, the

[8]Although the NMF codes were sparse, they were not a minimum-entropy code (an independent code) as the objective function did not maximize sparseness while preserving information.

more structure that is learned. Information theory provides a means for capturing interesting structure. Information maximization leads to an efficient code of the environment, resulting in more learned structure. Such mechanisms predict neural codes in both vision [52, 13, 72] and audition [39]. The research presented here found that face representations in which high-order dependencies are separated into individual coefficients gave superior recognition performance to representations which only separate second-order redundancies.

The principle of independence may have relevance to face and object representations in the brain. Horace Barlow [5] and Joseph Atick [2] have argued for redundancy reduction as a general coding strategy in the brain. This notion is supported by the findings of Bell and Sejnowski [13] that image bases that produce independent outputs from natural scenes are local, oriented, spatially opponent filters similar to the response properties of V1 simple cells. Olshausen and Field [53, 52] obtained a similar result with a sparseness objective, where there is a close information-theoretic relationship between sparseness and independence [5, 13]. Conversely, it has also been shown that Gabor filters, which model the responses of V1 simple cells, give outputs that are sparse and independent[9] when convolved with natural scenes but not when convolved with random noise [21, 22, 66]. (See [6] for a discussion).

7.9 FACE MODELING AND INFORMATION MAXIMIZATION: A COMPUTATIONAL NEUROSCIENCE PERSPECTIVE

Dependency coding and information maximization appear to be central principles in neural coding early in the visual system. Neural systems with limited dynamic range can increase the information that the response gives about the signal by placing the more steeply sloped portions of the transfer function in the regions of highest density, and shallower slopes at regions of low density. The function that maximizes information transfer is the one that matches the cumulative probability density of the input. There is a large body of evidence that neural codes in vision and other sensory modalities match the statistical structure of the environment, and hence maximize information about environmental signals to a degree. See [67] for a review. These principles may be relevant to how we think about higher visual processes such as face recognition as well.

Here we examine algorithms for face recognition by computer from a perspective of information maximization. Principal-component solutions can be learned in neural networks with simple Hebbian learning rules [51]. Hebbian learning is a model for long-term potentiation in neurons, in which weights are increased when

[9]when response normalization is included. There is a large body of evidence for response normalization in visual cortical neurons.

the input and output are simultaneously active. The weight update rule is typically formulated as the product of the input and the output activations. This simple rule learns the covariance structure of the input (i.e., the second-order relations). In the case of a single output unit, Hebbian learning maximizes the information transfer between the input and the output [41]. For multiple inputs and outputs, however, Hebbian learning doesn't maximize information transfer unless all of the signal distributions are Gaussian. In other words, the eigenface representation performs information maximization in the case where the input distributions are Gaussian.

Independent-component analysis performs information maximization for a more general set of input distributions. (See [18] for a reference text and [6] for a review). The information-maximization learning algorithm employed here was developed from the principle of optimal information transfer in neurons with sigmoidal transfer functions. Inspection of the learning rule in Equation 2 shows that it contains a Hebbian learning term, but it is between the input and the *gradient* of the output. (In the case of the natural-gradient learning rule in Equation 3, it is between the input and the natural gradient of the output.) Here we showed that face representations derived from ICA gave better recognition performance than face representations based on PCA. This suggests that information maximization in early processing is an effective strategy for face-recognition by computer.

A number of perceptual studies support the relevance of dependency encoding to human face perception. Perceptual effects such as other-race effects are consistent with information-maximization coding. For example, face discrimination is superior for same-race than other-race faces [63], which is consistent with a perceptual transfer function that is steeper for faces in the high-density portion of the distribution in an individual's perceptual experience (i.e., same-race faces). Face-adaptation studies (e.g. [31, 50, 73]) are consistent with information maximization on short time scales. For example, after adapting to a distorted face, a neutral face appears distorted in the opposite direction. Adaptation alters the probability density on short time scales, and the aftereffect is consistent with a perceptual transfer function that has adjusted to match the new cumulative probability density. Adaptation to a nondistorted face does not make distorted faces appear more distorted, which is consistent with an infomax account, because adapting to a neutral face would not shift the population mean. There is also support from neurophysiology for information-maximization principles in face coding. The response distributions of IT face cells are sparse, and there is very little redundancy between cells [64, 65].

Unsupervised learning of second-order dependencies (PCA) successfully models a number of aspects of human face perception including similarity, typicality, recognition accuracy, and other-race effects (e.g., [17, 26, 55]. Moreover, one study found that ICA better accounts for human judgments of facial similarity than PCA, supporting the idea that the more dependencies are encoded, the better the model of human perception for some tasks [25]. The extent to which information-maximization principles account for perceptual learning and adaptation aftereffects in human face perception is an open question and an area of active research.

ACKNOWLEDGMENTS

Support for this work was provided by National Science Foundation IIS-0329287, University of California Discovery Program 10158, National Research Service Award MH-12417-02, and Howard Hughes Medical Institute. Portions of the research in this chapter use the FERET database of facial images, collected under the FERET program of the Army Research Laboratory. Early versions of this work appear in *IEEE Transactions on Neural Networks* **13**(6) pp. 1450–64, 2002, and in *Proceedings of the SPIE Symposium on Electronic Imaging: Science and Technology; Human Vision and Electronic Imaging III*, Vol. 3299 pp. 528–539, 1998.

REFERENCES

[1] S. Amari, A. Cichocki, and H. H. Yang. A new learning algorithm for blind signal separation. In: *Advances in Neural Information Processing Systems*, volume 8, Cambridge, MA, 1996. MIT Press.

[2] J. J. Atick. Could information theory provide an ecological theory of sensory processing? *Network* **3**:213–251, 1992.

[3] J. J. Atick and A. N. Redlich. What does the retina know about natural scenes? *Neural Computation* **4**:196–210, 1992.

[4] F. R. Bach and M. I. Jordan. Kernel independent component analysis. In: *Proceedings of the 3rd International Conference on Independent Component Analysis and Signal Separation*, 2001.

[5] H. B. Barlow. Unsupervised learning. *Neural Computation* **1**:295–311, 1989.

[6] M. S. Bartlett. *Face Image Analysis by Unsupervised Learning*, volume 612 of *The Kluwer International Series on Engineering and Computer Science*. Kluwer Academic Publishers, Boston, 2001.

[7] M. S. Bartlett. Face image analysis by unsupervised learning and redundancy reduction. PhD thesis, University of California, San Diego, 1998.

[8] M. S. Bartlett, G. L. Donato, J. R. Movellan, J. C. Hager, P. Ekman, and T. J. Sejnowski. Image representations for facial expression coding. In: S.A. Solla, T.K. Leen, and K.-R. Muller, editors, *Advances in Neural Information Processing Systems*, volume 12. MIT Press, 2000.

[9] M. S. Bartlett, H. M. Lades, and T. J. Sejnowski. Independent component representations for face recognition. In: B. Rogowitz and T. Pappas, editors, *Proceedings of the SPIE Symposium on Electonic Imaging: Science and Technology; Human Vision and Electronic Imaging III*, volume 3299, pages 528–539, San Jose, CA, 1998. SPIE Press.

[10] M. S. Bartlett, J. R. Movellan, and T. J. Sejnowski. Image representations for facial expression recognition. *IEEE Transactions on Neural Networks* **13**(6):1450–1464, 2002.

[11] E. B. Baum, J. Moody, and F. Wilczek. Internal representaions for associative memory. *Biological Cybernetics* **59**:217–228, 1988.

[12] A. J. Bell and T. J. Sejnowski. An information-maximization approach to blind separation and blind deconvolution. *Neural Computation* **7**(6):1129–1159, 1995.

[13] A. J. Bell and T. J. Sejnowski. The independent components of natural scenes are edge filters. *Vision Research* **37**(23):3327–3338, 1997.

[14] A. Cichocki, R. Unbehauen, and E. Rummert. Robust learning algorithm for blind separation of signals. *Electronics Letters* **30**(7):1386–1387, 1994.

[15] P. Comon. Independent component analysis - a new concept? *Signal Processing* **36**:287–314, 1994.

[16] G. Cottrell and J. Metcalfe. Face, gender and emotion recognition using holons. In: D. Touretzky, editor, *Advances in Neural Information Processing Systems*, volume 3, pages 564–571, San Mateo, CA, 1991. Morgan Kaufmann.

[17] G. W. Cottrell, M. N. Dailey, C. Padgett, and R. Adolphs. *Computational, Geometric, and Process Perspectives on Facial Cognition: Contexts and Challenges*, chapter: Is all face processing holistic? The view from UCSD. Erlbaum, 2000.

[18] T. M. Cover and J. A. Thomas. *Elements of Information Theory*. John Wiley & Sons, New York, 1991.

[19] G. Donato, M. Bartlett, J. Hager, P. Ekman, and T. Sejnowski. Classifying facial actions. *IEEE Transactions on Pattern Analysis and Machine Intelligence* **21**(10):974–989, 1999.

[20] B. A. Draper, K. Baek, Bartlett M. S., and J. R. Beveridge. Recognizing faces with pca and ica. *Computer Vision and Image Understanding*, special issue on face recognition **91**:115–137, 2003.

[21] D. J. Field. Relations between the statistics of natural images and the response properties of cortical cells. *Journal of the Optical Society of America* A **4**:2379–94, 1987.

[22] D. J. Field. What is the goal of sensory coding? *Neural Computation* **6**:559–601, 1994.

[23] M. Girolami. *Advances in Independent Component Analysis*. Springer-Verlag, Berlin, 2000.

[24] P. Hallinan. A deformable model for face recognition under arbitrary lighting conditions. PhD thesis, Harvard University, 1995.

[25] P. Hancock. Alternative representations for faces. In: *British Psychological Society, Cognitive Section*. University of Essex, September 6-8, 2000.

[26] P. J. B. Hancock, A. M. Burton, and V. Bruce. Face processing: human perception and principal components analysis. *Memory and Cognition* **24**:26–40, 1996.

[27] D. J. Heeger. Normalization of cell responses in cat striate cortex. *Visual Neuroscience* **9**:181–197, 1992.

[28] G. Hinton and T. Shallice. Lesioning an attractor network: investigations of acquired dyslexia. *Psychological Review* **98**(1):74–95, 1991.

[29] C. Jutten and J. Herault. Blind separation of sources i. an adaptive algorithm based on neuromimetic architecture. *Signal Processing* **24**(1):1–10, 1991.

[30] J. V. Haxby, E. A. Hoffman, and M. I. Gobbini. The distributed human neural system for face perception. *Trends in Cognitive Science* **4**:223–233, 2000.

[31] D. Kaping, P. Duhamel, and M. Webster. Adaptation to natural face categories. *Journal of Vision* **2**:128, 2002.

[32] J. Kim, J. Choi, and J. Yi. Face recognition based on ica combined with fld. In: *European Conference on Computer Vision*, pages 10–18, 2002.
[33] H. Lappalainen and J. W. Miskin. Ensemble learning. In: M. Girolami, editor, *Advances in Independent Component Analysis*, pages 76–92. Springer-Verlag, 2000.
[34] S. Laughlin. A simple coding procedure enhances a neuron's information capacity. *Z. Naturforsch* **36**:910–912, 1981.
[35] D. D. Lee and S. Seung. Learning the parts of objects by non-negative matrix factorization. *Nature* **401**:788–791, 1999.
[36] T. -W. Lee. *Independent Component Analysis: Theory and Applications*. Kluwer Academic Publishers, 1998.
[37] T. -W. Lee, M. Girolami, and T. J. Sejnowski. Independent component analysis using an extended infomax algorithm for mixed sub-gaussian and super-gaussian sources. *Neural Computation* **11**(2):417–41, 1999.
[38] T. -W. Lee, B. U. Koehler, and R. Orglmeister. Blind source separation of nonlinear mixing models. In: *Proceedings of the IEEE International Workshop on Neural Networks for Signal Processing*, pages 406–415, Florida, September 1997.
[39] M. Lewicki and B. Olshausen. Probabilistic framework for the adaptation and comparison of image codes. *Journal of the Optical Society of America* A **16**(7):1587–601, 1999.
[40] M. Lewicki and T. J. Sejnowski. Learning overcomplete representations. *Neural Computation* **12**(2):337–65, 2000.
[41] R. Linsker. Self-organization in a perceptual network. *Computer* **21**(3):105–117, 1988.
[42] C. Liu and H. Wechsler. Comparative assessment of independent component analysis (ica) for face recognition. In: *International Conference on Audio and Video Based Biometric Person Authentication*, 1999.
[43] Q. Liu, J. Cheng, H. Lu, and S. Ma. Modeling face appearance with nonlinear independent component analysis. In: *Sixth IEEE International Conference on Automatic Face and Gesture Recognition*, 2004.
[44] D. J. C. MacKay. Maximum likelihood and covariant algorithms for independent component analysis. personal communication, 1996.
[45] S. Makeig, A. J. Bell, T. -P. Jung, and T. J. Sejnowski. Independent component analysis of electroencephalographic data. In: D. Touretzky, M. Mozer, and M. Hasselmo, editors, *Advances in Neural Information Processing Systems*, volume 8, pages 145–151, Cambridge, MA, 1996. MIT Press.
[46] M. J. McKeown, S. Makeig, G. G. Brown, T. -P. Jung, S. S. Kindermann, A. J. Bell, and T. J. Sejnowski. Analysis of fmri by decomposition into independent spatial components. *Human Brain Mapping* **6**(3):160–88, 1998.
[47] J. W. Miskin and D. J. C. MacKay. *Ensemble Learning for Blind Source Separation ICA: Principles and Practice*. Cambridge University Press, 2001. In press.
[48] B. Moghaddam. Principal manifolds and Bayesian subspaces for visual recognition. In: *International Conference on Computer Vision*, 1999.
[49] J. -P. Nadal and N. Parga. Non-linear neurons in the low noise limit: a factorial code maximizes information transfer. *Network* **5**:565–581, 1994.
[50] M. Ng, D. Kaping, M. A. Webster, S. Anstis, and I. Fine. Selective tuning of face perception. In: *Journal of Vision* **3**:106a, 2003.

[51] E. Oja. Neural networks, principal components, and subspaces. *International Journal of Neural Systems* **1**:61–68, 1989.

[52] B. A. Olshausen and D. J. Field. Emergence of simple-cell receptive field properties by learning a sparse code for natural images. *Nature* **381**:607–609, 1996.

[53] B. A. Olshausen and D. J. Field. Natural image statistics and efficient coding. *Network: Computation in Neural Systems* **7**(2):333–340, 1996.

[54] A. V. Oppenheim and J. S. Lim. The importance of phase in signals. *Proceedings of the IEEE* **69**:529–541, 1981.

[55] A. O'Toole, K. Deffenbacher, D. Valentin, and H. Abdi. Structural aspects of face recognition and the other race effect. *Memory and Cognition* **22**(2):208–224, 1994.

[56] G. Palm. On associative memory. *Biological Cybernetics* **36**:19–31, 1980.

[57] B. A. Pearlmutter and L. C. Parra. A context-sensitive generalization of ica. In: Mozer, Jordan, and Petsche, editors, *Advances in Neural Information Processing Systems*, volume 9. MIT Press, 1996.

[58] P. S. Penev. Redundancy and dimensionality reduction in sparse-distributed representations of natural objects in terms of their local features. In: T.K. Leen, T.G. Dietterich, and V. Tresp, editors, *Advances in Neural Information Processing Systems* 13. MIT Press, 2001.

[59] P. S. Penev and J. J. Atick. Local feature analysis: a general statistical theory for object representation. *Network: Computation in Neural Systems* **7**(3):477–500, 1996.

[60] A. Pentland, B. Moghaddam, and T. Starner. View-based and modular eigenspaces for face recognition. In: *Proceedings of IEEE Conference on Computer Vision and Pattern Recognition*, 1994.

[61] P. J. Phillips, H. Wechsler, J. Juang, and P. J. Rauss. The feret database and evaluation procedure for face-recognition algorithms. *Image and Vision Computing Journal* **16**(5):295–306, 1998.

[62] L. N. Piotrowski and F. W. Campbell. A demonstration of the visual importance and flexibility of spatial-frequency, amplitude, and phase. *Perception* **11**:337–346, 1982.

[63] P. M. Walker, and J. W. Tanaka. An encoding advantage for own-race versus other-race faces. *Perception* **32**(9):1117–25, 2003.

[64] E. T. Rolls, N. C. Aggelopoulos, L. Franco, and A. Treves. Information encoding in the inferior temporal cortex: contributions of the firing rates and correlations between the firing of neurons. *Biological Cybernetics* **90**:19–32, 2004.

[65] E. T. Rolls and M. J. Tovee. Sparseness of the neuronal representation of stimuli in the primate temporal visual cortex. *Journal of Neurophysiology* **73**(2):713–726, 1995.

[66] E. P. Simoncelli. Statistical models for images: Compression, restoration and synthesis. In: *31st Asilomar Conference on Signals, Systems and Computers*, Pacific Grove, CA, November 2-5 1997.

[67] E. O. Simoncelli. Natural image statistics and neural representation. *Annual Review of Neuroscience* **24**:1193–1216, 2001.

[68] J. V. Stone and J. Porrill. Undercomplete independent component analysis for signal separation and dimension reduction. Technical Report, University of Sheffield, Department of Psychology, 1998.

[69] Y. W. Teh and G. E. Hinton. Rate-coded restricted Boltzmann machines for face recognition. In: T. Leen, T. Dietterich, and V. Tresp, editors, *Advances in Neural Information Processing Systems 13*. MIT Press, 2001.

[70] M. Turk and A. Pentland. Eigenfaces for recognition. *Journal of Cognitive Neuroscience* **3**(1):71–86, 1991.

[71] T. von der Twer and D.I.A. Macleod. Optimal nonlinear codes for the perception of natural colors. *Network: Computation in Neural Systems* **12**:395–407, 2001.

[72] T. Wachtler, T.-W. Lee, and T. J. Sejnowski. The chromatic structure of natural scenes. *Journal of the Optical Socitey of America* A **18**(1):65–77, 2001.

[73] M. M. Webster. Figural aftereffects in the perception of faces. *Psychonomic Bulletin Review* **6**(4):647–653, 1999.

[74] H. H. Yang, S.-I. Amari, and A. Cichocki. Information-theoretic approach to blind separation of sources in non-linear mixture. *Signal Processing* **64**(3):291–3000, 1998.

[75] M. Yang. Face recognition using kernel methods. In: T. Diederich, Becker S., and Z. Ghahramani, editors, *Advances in Neural Information Processing Systems*, volume 14, 2002.

[76] P. C. Yuen and J. H. Lai. Independent component analysis of face images. In: *IEEE workshop on Biologically Motivated Computer Vision*, Seoul, 2000. Springer-Verlag.

PSYCHOPHYSICAL ASPECTS

CHAPTER 8

FACE RECOGNITION BY HUMANS

8.1 INTRODUCTION

The events of September 11, 2001, in the United States compellingly highlighted the need for systems that can identify individuals with known terrorist links. In rapid succession, three major international airports—Fresno, St. Petersburg, and Logan—began testing face-recognition systems. While such deployment raises complicated issues of privacy invasion, of even greater immediate concern is whether the technology is up to the task requirements.

Real-world tests of automated face-recognition systems have not yielded encouraging results. For instance, face-recognition software at the Palm Beach International Airport, when tested on fifteen volunteers and a database of 250 pictures, had a success rate of less than fifty percent and nearly fifty false alarms per five thousand passengers (translating to two to three false alarms per hour per checkpoint). Having to respond to a terror alarm every twenty minutes would, of course, be very disruptive for airport operations. Furthermore, variations such as eyeglasses, small facial rotations and lighting changes, proved problematic for the system. Many other such tests have yielded similar results.

The primary stumbling block in creating effective face-recognition systems is that we do not know how to quantify similarity between two facial images in a perceptually meaningful manner. Figure 8.1 illustrates this issue. Images 1 and 3 show the same individual from the front and oblique viewpoints, while image 2 shows a different person from the front. Conventional measures of image similarity (such as the Minkowski metrics [26]) would rate images 1 and 2 to be more similar than images 1 and 3. In other words, they fail to generalize across important and commonplace transformations. Other transforms that lead to similar difficulties

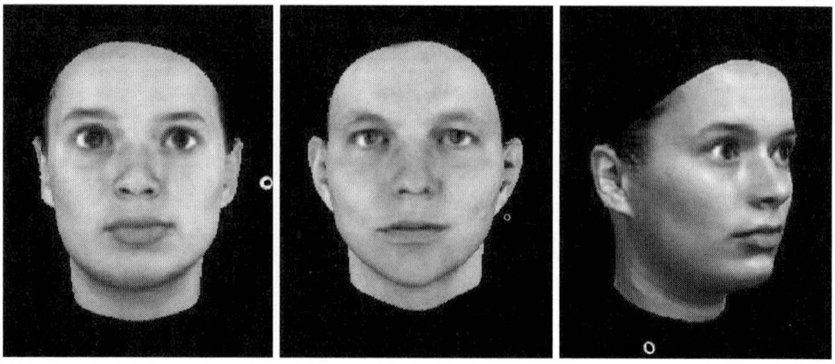

FIGURE 8.1: Most conventional measures of image similarity would declare images 1 and 2 to be more similar than images 1 and 3, even though both members of the latter pair, but not the former, are derived from the same individual. This example highlights the challenge inherent in the face-recognition task.

include lighting variations, aging, and expression changes. Clearly, similarity needs to be computed over attributes more complex than raw pixel values. To the extent that the human brain appears to have figured out which facial attributes are important for subserving robust recognition, it makes sense to turn to neuroscience for inspiration and clues.

Work in the neuroscience of face perception can influence research on machine vision systems in two ways. First, studies of the limits of human face-recognition abilities provide benchmarks against which to evaluate artificial systems. Second, studies characterizing the response properties of neurons in the early stages of the visual pathway can guide strategies for image preprocessing in the front ends of machine vision systems. For instance, many systems use a wavelet representation of the image that corresponds to the multiscale Gabor-like receptive fields found in the primary visual cortex [24, 48]. We describe these and related schemes in greater detail later. However, beyond these early stages, it has been difficult to discern any direct connections between biological and artificial face-recognition systems. This is perhaps due to the difficulty in translating psychological findings into concrete computational prescriptions.

A case in point is an idea that several psychologists have emphasized: that facial configuration plays an important role in human judgments of identity [13, 20]. However, the experiments so far have not yielded a precise specification of what is meant by "configuration" beyond the general notion that it refers to the relative placement of the different facial features. This makes it difficult to adopt this idea in the computational arena, especially when the option of using individual facial features such as eyes, noses, and mouths is so much easier to describe and implement. Thus, several current systems for face recognition, and also for the related task of facial composite generation (creating a likeness from a witness description), are based on a piecemeal approach.

As an illustration of the problems associated with the piecemeal approach, consider the facial-composite generation task. The dominant paradigm for having a witness describe a suspect's face to a police officer involves having him/her pick out the best matching features from a large collection of images of disembodied features. Putting these together yields a putative likeness of the suspect. The mismatch between this piecemeal strategy and the more holistic facial encoding scheme that may actually be used by the brain can lead to problems in the quality of reconstructions as shown in Figure 8.2. In order to create these faces, we enlisted the help of an individual who had several years of experience with the IdentiKit system, and had assisted the police department on multiple occasions for creating suspect likenesses. We gave him photographs of fourteen celebrities and requested him to put together IdentiKit reconstructions. There were no strict time constraints (the reconstructions were generated over two weeks) and the IdentiKit operator did not have to rely on verbal descriptions; he could directly consult the images we had provided him. In short, these reconstructions were generated under ideal operating conditions. The wide gulf between the face recognition performance of humans and machines suggests that there is much to be gained by improving the communication between human vision researchers on the one hand, and computer vision scientists on the other. This chapter is a small step in that direction.

FIGURE 8.2: Four facial composites generated by an IdentiKit operator at the authors' request. The individuals depicted are all famous celebrities. The operator was given photographs of the celebrities and was asked to create the best likenesses using the kit of features in the system. Most observers are unable to recognize the people depicted, highlighting the problems of using a piecemeal approach in constructing and recognizing faces. The celebrities shown are, from left to right: Bill Cosby, Tom Cruise, Ronald Reagan, and Michael Jordan.

What aspects of human face-recognition performance might be of greatest relevance to the work of computer-vision scientists? The most obvious one is simply a characterization of performance that could serve as a benchmark. In particular, we would like to know how human recognition performance changes as a function of image quality degradations that are common in everyday viewing conditions, and that machine vision systems are required to be tolerant too. Beyond characterizing performance, it would be instructive to know about the kinds of facial cues that the human visual system uses in order to achieve its impressive abilities. This provides a tentative prescription for the design of machine-based face-recognition systems. Coupled with this problem of system design is the issue of the balance between hard-wiring and on-line learning. Does the human visual system rely more on innately specified strategies for face processing, or is this a skill that emerges via learning? Finally, a computer-vision scientist would be interested in knowing whether there are any direct clues regarding face representations that could be derived from a study of the biological systems. With these considerations in mind, we shall explore the following four fundamental questions in the domain of human vision.

1. What are the limits of human face-recognition abilities, in terms of the minimum image resolution needed for a specified level of recognition performance?
2. What are some important cues that the human visual system relies upon for judgments of identity?
3. What is the timeline of development of face-recognition skills? Are these skills innate or learned?
4. What are some biologically plausible face representation strategies?

8.2 WHAT ARE THE LIMITS OF HUMAN FACE RECOGNITION SKILLS?

The human visual system (HVS) often serves as an informal standard for evaluating machine vision approaches. However, this standard is rarely applied in any systematic way. In order to be able to use the human visual system as a useful standard to strive towards, we need to first have a comprehensive characterization of its capabilities.

In order to characterize the HVS's face-recognition capabilities, we shall describe experiments that address two issues: (1) how does human performance change as a function of image resolution? and (2) what are the relative contributions of internal and external features at different resolutions? Let us briefly consider why these two questions are worthy subjects of study.

The decision to examine recognition performance in images with limited resolution is motivated by both ecological and pragmatic considerations. In the natural environment, the brain is typically required to recognize objects when they are at a distance or viewed under suboptimal conditions. In fact, the very survival of an animal may depend on its ability to use its recognition machinery as an early-warning system that can operate reliably with limited stimulus information. Therefore, by better capturing real-world viewing conditions, degraded images are well suited to help us understand the brain's recognition strategies.

Many automated vision systems too need to have the ability to interpret degraded images. For instance, images derived from present-day security equipment are often of poor resolution due both to hardware limitations and large viewing distances. Figure 8.3 is a case in point. It shows a frame from a video sequence of Mohammad Atta, a perpetrator of the World Trade Center bombing, at a Maine airport on the morning of September 11, 2001. As the inset shows, the resolution in the face region is quite poor. For the security systems to be effective, they need to be able to recognize suspected terrorists from such surveillance videos. This provides strong pragmatic motivation for our work. In order to understand how the human visual system interprets such images and how a machine-based system could do the same, it is imperative that we study face recognition with such degraded images. Furthermore, impoverished images serve as 'minimalist' stimuli, which, by dispensing with unnecessary detail, can potentially simplify our quest to identify aspects of object information that the brain preferentially encodes.

FIGURE 8.3: A frame from a surveillance video showing Mohammad Atta at an airport in Maine on the morning of the 11th of September, 2001. As the inset shows, the resolution available in the face region is very limited. Understanding the recognition of faces under such conditions remains an open challenge and motivates the work reported here.

The decision to focus on the two types of facial feature sets—internal and external—is motivated by the marked disparity that exists in their use by current machine-based face analysis systems. It is typically assumed that internal features (eyes, nose, and mouth) are the critical constituents of a face, and the external features (hair and jawline) are too variable to be practically useful. It is interesting to ask whether the human visual system also employs a similar criterion in its use of the two types of feature. Many interesting questions remain unanswered. Precisely how does face identification performance change as a function of image resolution? Does the relative importance of facial features change across different resolutions? Does featural saliency become proportional to featural size, favoring more global, external features like hair and jawline? Or, are we still better at identifying familiar faces from internal features like the eyes, nose, and mouth? Even if we prefer internal features, does additional information from external features facilitate recognition? Our experiments were designed to address these open questions by assessing face recognition performance across various resolutions and by investigating the contribution of internal and external features.

Considering the importance of these issues, it is not surprising that a rich body of research has accumulated over the past few decades. Pioneering work on face-recognition with low-resolution imagery was done by Harmon and Julesz [33, 34]. Working with block-averaged images of familiar faces of the kind shown in Figure 8.4, they found high recognition accuracies even with images containing just 16×16 blocks. However, this high level of performance could have been due at least in part to the fact that subjects were told which of a small set of people they were going to be shown in the experiment. More recent studies too have suffered from this problem. For instance, Bachmann [4] and Costen et al. (1994) used six high-resolution photographs during the 'training' session and low-resolution versions of the same during the test sessions. The prior subject priming about stimulus set and the use of the same base photographs across the training and test sessions renders these experiments somewhat unrepresentative of real-world recognition situations. Also, the studies so far have not performed some important comparisons. Specifically, it is not known how performance differs across various image resolutions when subjects are presented full faces versus when they are shown the internal features alone. Our experiments on face recognition were designed to build upon and correct some of the weaknesses of the work reviewed above. Here, we describe an experimental study with two goals: (1) assessing performance as a function of image resolution and (2) determining performance with internal features alone versus full faces.

The experimental paradigm we used required subjects to recognize celebrity facial images blurred by varying amounts (a sample set is shown in Figure 8.5). We used 36 color face images and subjected each to a series of blurs. Subjects were shown the blurred sets, beginning with the highest level of blur and proceeding on

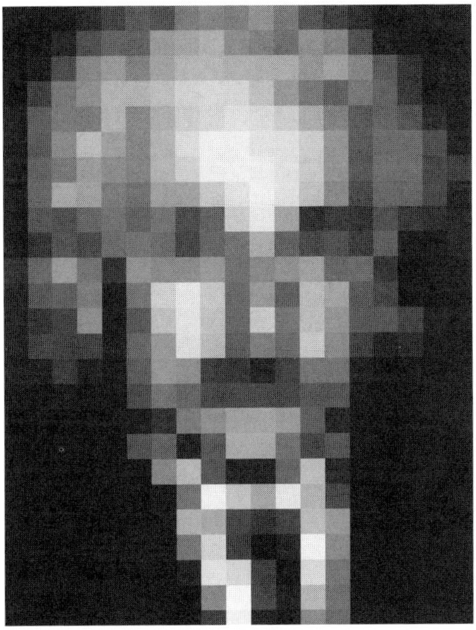

FIGURE 8.4: Images such as the one shown here have been used by several researchers to assess the limits of human face-identification processes (after [33]).

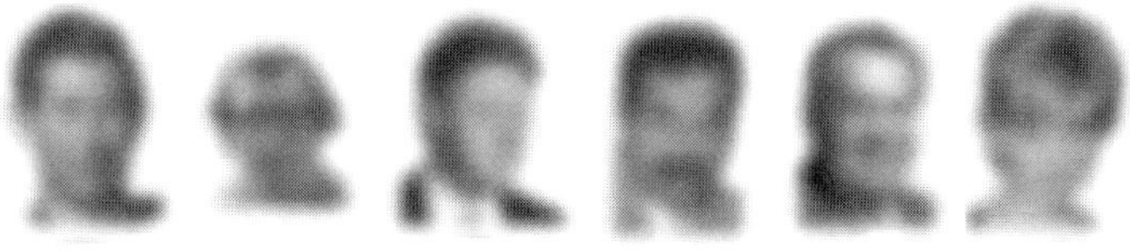

FIGURE 8.5: Unlike current machine-based systems, human observers are able to handle significant degradation in face images. For instance, subjects are able to recognize more than half of all famous faces shown to them at the resolution depicted here. The individuals shown from left to right, are: Prince Charles, Woody Allen, Bill Clinton, Saddam Hussein, Richard Nixon, and Princess Diana. (See also color plate section).

to the zero blur condition. We also created two other stimulus sets. The first of these contained the individual facial features (eyes, nose, and mouth), placed side by side while the second had the internal features in their original spatial configuration. Three mutually exclusive groups of subjects were tested on the three conditions. In all these experiments, subjects were not given any information about which celebrities they would be shown during the tests. Chance level performance was, therefore, close to zero.

8.2.1 Results

Figure 8.6 shows results from the different conditions. It is interesting to note that, in the full-face condition, subjects can recognize more than half of the faces with image resolutions of merely 7×10 pixels. Recognition reaches almost ceiling level at a resolution of 19×27 pixels. Performance of subjects with the other two stimulus sets is quite poor even with relatively small amounts of blur. This clearly demonstrates the perceptual importance of the overall head configuration for face recognition. The internal features on their own, and even their mutual configuration, are insufficient to account for the impressive recognition performance of subjects with full-face images at high blur levels. This result suggests that feature-based approaches to recognition are likely to be less robust than those based on the overall head configuration. Figure 8.7 shows an image that underscores the importance of overall head shape in determining identity.

In summary, these experimental results lead us to some very interesting inferences about face recognition:

1. A high level of face recognition performance can be obtained even with resolutions as low as 12×14 pixels. The cues to identity must necessarily

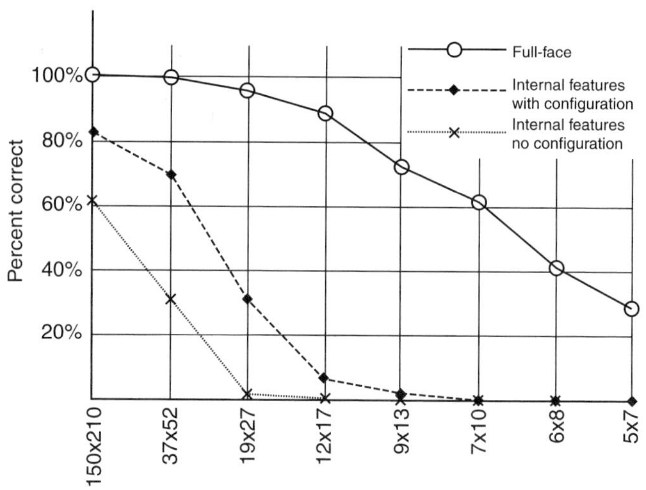

FIGURE 8.6: Recognition performance with internal features (with and without configural cues). Performance obtained with whole-head images is also included for comparison.

FIGURE 8.7: Although this image appears to be a fairly run-of-the-mill picture of Bill Clinton and Al Gore, a closer inspection reveals that both men have been digitally given identical inner face features and their mutual configuration. Only the external features are different. It appears, therefore, that the human visual system makes strong use of the overall head shape in order to determine facial identity (from [75]).

include those that can survive across massive image degradations. However, it is worth bearing in mind that the data we have reported here come from famous-face recognition tasks. The results may be somewhat different for unfamiliar faces.

2. Details of the internal features, on their own are insufficient for subserving a high-level of recognition performance.
3. Even the mutual spatial configuration of the internal features is inadequate to explain the observed recognition performance, especially when the inputs are degraded. It appears that, unlike the conventional aims in computer vision, where internal features are of paramount importance, the human visual system relies on the relationship between external head geometry and internal features. Additional experiments [39] reveal a highly nonlinear cue-combination strategy for merging information from internal and external features. In effect, even when recognition performance with only internal or only external cues is statistically indistinguishable from chance, performance with the two together is very robust.

These findings are not merely interesting facts about the human visual system; rather, they help in our quest to determine the nature of facial attributes that can subserve robust recognition. In a similar vein of understanding the recognition of "impoverished" faces, it might also be useful to analyze the work of minimalist portrait artists, especially caricaturists, who are able to capture vivid likenesses using very few strokes. Investigating which facial cues are preserved or enhanced in such simplified depictions can yield valuable insights about the significance of different facial attributes.

We next turn to an exploration of cues for recognition in more conventional settings: high-quality images. Even here, it turns out, the human visual system yields some very informative data.

8.3 WHAT CUES DO HUMANS USE FOR FACE-RECOGNITION?

Beyond the difficulties presented by suboptimal viewing conditions, as reviewed above, a key challenge for face recognition systems comes from the overall similarity of faces. Unlike many other classes of object, faces share the same overall configuration of parts and the same scale. The convolutions in depth of the face surface vary little from one face to the next, and for the most part, the reflectance distributions across different faces are also very similar. The result of this similarity is that faces seemingly contain few distinctive features that allow for easy differentiation. Despite this fundamental difficulty, human face recognition is fast and accurate. This means that we must be able to make use of some reliable cues to face identity, and in this section we will consider what these cues could be.

For a cue to be useful for recognition, it must differ between faces. Though this point is obvious, it leads us to the question, what kinds of visual differences *could* there be between faces? The objects of visual perception are surfaces. At the most basic level, there are three variables that go into determining the visual appearance of a surface: (1) the light that is incident on the surface, (2) the shape of the surface, and (3) the reflectance properties of the surface. This means that any cue that could be useful for the visual recognition of faces can be classified as a lighting cue, a shape cue, or a surface-reflectance cue. We will consider each of these three classes of cues, evaluating their relative utility for recognition.

Illumination has a large effect on the image-level appearance of a face, a fact well known to artists and machine-vision researchers. Indeed, when humans are asked to match different images of the same face, performance is worse when the two images of a face to be matched are illuminated differently [10, 36], although the decline in performance is not as sharp as for machine algorithms. However, humans are highly accurate at naming familiar faces under different illumination [59]. This finding fits our informal experience, in which we are able to recognize people under a wide variety of lighting conditions, and do not experience the identity of a person as changing when we walk with them, say, from indoor to outdoor lighting. There are certain special conditions when illumination can have a dramatic impact on assessments of identity. Lighting from below is a case in point. However, this is a rare occurrence; under natural conditions, including artificial lighting, faces are almost never lit from below. Consistent with the notion that the representation of facial identity includes this statistical regularity, faces look odd when lit from below, and several researchers have found that face recognition is impaired by bottom lighting [27, 36, 40, 53]. These findings overall are consistent with the notion that the human visual system does make some use of lighting regularities for recognizing faces.

The two other cues that could potentially be useful for face recognition are the shape and reflectance properties of the face surface. We will use the term "pigmentation" here to refer to the surface reflectance properties of the face. Shape cues include boundaries, junctions, and intersections, as well as any other cue that gives information about the location of a surface in space, such as stereo disparity, shape from shading, and motion parallax. Pigmentation cues include albedo, hue, texture, translucence, specularity, and any other property of surface reflectance. "Second-order relations", or the distances between facial features such as the eyes and mouth, are a subclass of shape. However, the relative reflectance of those features or parts of features, such as the relative darkness of parts of the eye or of the entire eye region and the cheek, are a subclass of pigmentation. This division of cues into two classes is admittedly not perfect. For example, should a border defined by a luminance gradient, such as the junction of the iris and sclera be classified as a shape or pigmentation cue? Because faces share a common configuration of parts, we classify such borders as shape cues when they are common to all faces (e.g., the iris–sclera border), but as pigmentation cues when they are unique to an individual (e.g., moles and freckles). It should also be noted that this is not an image-level description. For example, a particular luminance contour could not be classed as caused by shape or pigmentation from local contrast alone. Although the classification cannot completely separate shape and pigmentation cues, it does separate the vast majority of cues. We believe that dividing cues for recognition into shape and pigmentation is a useful distinction to draw.

Indeed, this division has been used to investigate human recognition of nonface objects, although in that literature, pigmentation has typically been referred to as "color"or "surface". Much of this work has compared human ability to recognize objects from line drawings or images with pigmentation cues, such as photographs. The assumption here is that line drawings contain shape cues, but not pigmentation cues, and hence the ability to recognize an object from a line drawing indicates reliance on shape cues. In particular, these studies have found recognition of line drawings to be as good [9, 22, 62] or almost as good [38, 69, 87] as recognition of photographs. On the basis of these and similar studies, there is a consensus that, in most cases, pigmentation is less important than shape for object recognition [64, 77, 81].

In the face recognition literature too, there is a broadly held implicit belief that shape cues are more important than pigmentation cues. Driven by this assumption, line drawings and other unpigmented stimuli are commonly used as stimuli for experimental investigation of face recognition. Similarly, many models of human face recognition use only shape cues, such as distances, as the featural inputs. This only makes sense if it is assumed that the pigmentation cues being omitted are unimportant. Also, there are many more experimental studies investigating specific aspects of shape than of pigmentation, suggesting that the research community is less aware of pigmentation as a relevant component of face representations. However, this assumption turns out to be false. In the rest of this section, we will review evidence that both shape and pigmentation cues are important for face recognition.

First, we briefly review the evidence that shape cues alone can support recognition. Specifically, we can recognize representations of faces that have no variation in pigmentation, hence no useful pigmentation cues. Many statues have no useful pigmentation cues to identity because they are composed of a single material, such as marble or bronze, yet are recognizable as representations of a specific individual. Systematic studies with 3D laser-scanned faces that similarly lack variation in pigmentation have found that recognition can proceed with shape cues only [11, 36, 50]. Clearly, shape cues are important, and sometimes sufficient, for face recognition.

However, the ability to recognize a face in the absence of pigmentation cues does not mean that such cues are not used under normal conditions. To consider a rough analogy, the fact that we can recognize a face from a view of only the left side does not imply that the right side is not also relevant to recognition. There is reason to believe that pigmentation may also be important for face recognition. Unlike other objects, faces are much more difficult to recognize from a line drawing than from a photograph [12, 23, 46, 70], suggesting that the pigmentation cues thrown away by such a representation may well be important.

Recently, some researchers have attempted to directly compare the relative importance of shape and pigmentation cues for face recognition. The experimental challenge here is to find a way to create a stimulus face that appears naturalistic, yet does not contain either useful shape or pigmentation cues. The basic approach that was first taken by O'Toole and colleagues [63] is to create a set of face representations with the shape of a particular face and the average pigmentation of a large group of faces, and a separate set of face representations with the pigmentation of an actual face and the average shape of many faces. The rationale here is that to distinguish among the faces from the set that all have the same pigmentation, subjects must use shape cues, and vice versa for the set of faces with the same shape. In O'Toole's study, the faces were created by averaging separately the shape and reflectance data from laser scans. The shape data from individual faces were combined with the averaged pigmentation to create the set of faces that differ only in terms of their shape, and vice versa for the set of faces with the same shape. Subjects were trained with one or the other set of face images, and were subsequently tested for memory. Recognition performance was about equal with both the shape and pigmentation sets, suggesting that both cues are important for recognition.

A question that arises when comparing the utility of different cues is whether the relative performance is due to differences in image similarity or differences in processing. One way to address this is to equate the similarity of a set of images that differ by only one or the other cue. If the sets of images are equated for similarity, differences in recognition performance with the sets can be attributed to differences in processing. We investigated this question in our lab [71] with sets of face images that differed only by shape or pigmentation cues and were equated for similarity using the "Gabor jet" model of facial similarity developed by von der Malsburg and colleagues [45]. We used photographic images that were manipulated with image-morphing software. Before considering the results, we can see from the stimuli (some of which are shown in Figure 8.8) that both shape and pigmentation can be used to establish identity. With the similarity of the cues equated in gray-scale images, we found slightly better performance with the shape cues. With the same images in full color (but not equated for similarity), we found slightly better performance with the pigmentation cues. Overall, the performance was approximately equal using shape or pigmentation cues. This provides evidence that both shape and pigmentation cues are important for recognition. The implication for computer vision systems is obvious: facial representations ought to encode both of these kinds of information to optimize performance.

One particular subcategory of pigmentation cues—color—has received a little extra attention, with researchers investigating whether color cues are used for face recognition. An early study reversed separately the hue and luminance channels of face images, and found that, while luminance reversal (contrast negation) caused a large decline in recognition performance, reversing the hue channel did not [44]. Thus, faces with incorrect colors can be recognized as well as those with correct colors, a point that has also been noted with respect to the use of "incorrect" colors by the early-20th-century 'Fauvist' school of art [90]. Another approach has been to investigate whether exaggerating color cues by increasing the differences in hue between an individual face image and an average face image improves recognition performance. The results suggest that this manipulation does improve performance, but only with

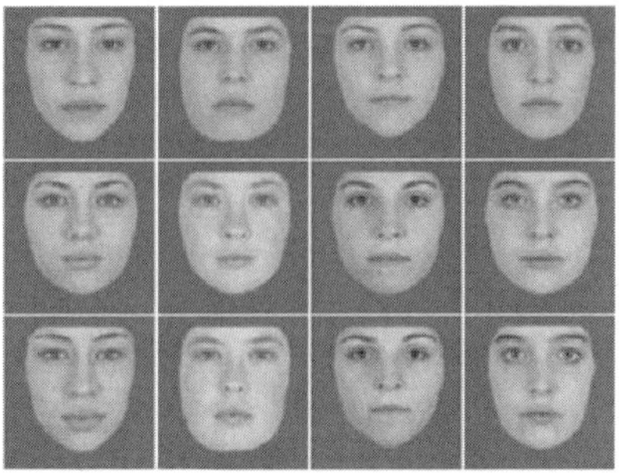

FIGURE 8.8: Some of the faces from our comparison of shape and pigmentation cues. Faces along the top row differ from one another in terms of shape cues, faces along the middle row differ in terms of pigmentation cues, and faces along the bottom row differ in terms of shape and pigmentation cues (they are actual faces). The faces along the top row all have the same pigmentation, but they do not appear to be the same person. Similarly, the faces in the middle row do not look like the same person, despite all having the same shape. This suggests that both shape and pigmentation are used for face recognition by the HVS. (See also color plate section).

fast presentation [47], and with a small bias toward color caricatures as better likenesses of the individuals than the veridical images [47]. In our laboratory, we have taken a more direct approach, comparing performance with full-color and gray-scale images. In one study investigating familiar-face recognition [88] and the study mentioned above investigating unfamiliar-face matching [71], we have found significantly better performance when subjects are viewing full-color rather than gray-scale images. Overall, color cues do seem to contribute to recognition.

To summarize the findings on the question of what cues are used in face recognition, there is evidence that all categories of cue that could potentially be used to recognize faces—illumination, shape, and pigmentation—do contribute significantly to face recognition. The human face recognition system appears to use all available cues to perform its difficult task. The system does this by making use of many weak but redundant cues. In the context of faces, the human visual system appears to be opportunistic, making use of all cues that vary across exemplars of this class to achieve its impressive face recognition skills.

An important question that arises at this juncture is whether humans are innately equipped with these skills and strategies, or have to learn them through experience. This is the issue we consider next.

8.4 WHAT IS THE TIMELINE OF DEVELOPMENT OF HUMAN FACE RECOGNITION SKILLS?

Considering the complexity of the tasks, face spotting and recognition skills develop surprisingly rapidly in the course of an infant's development. As early as a few days after birth, infants appear to be able to distinguish their mother's face from other faces. Yet, in light of the rapid progression of face recognition abilities, it may be surprising to learn that child face-processing abilities are in some ways still not adult-like, even up until the age of 10 years. This section explores the trajectory of development of face processing.

8.4.1 Bootstrapping Face Processing: Evidence from Neonates

As is the case for most visual skills, face processing must be bootstrapped with some primitive mechanism from which more advanced processes can be learned. A key question is whether infants are born with some innate abilities to process faces or are those abilities a consequence of general visual learning mechanisms?

To examine this issue, neonates (newborns) are assessed for various abilities as soon as is practical after birth. Three major findings have historically been taken as evidence for innate facial processing abilities: (1) The initial preference for faces over nonfaces, (2) The ability to distinguish the mother from strangers, and (3) imitation of facial gestures. We will look at each of these in turn.

Innate Facial Preference

Are infants prewired with a face-detection algorithm? If infants knew how to locate the faces in an image, it would be a valuable first step in the subsequent learning of face recognition processes. Morton and Johnson [58] formalized this idea into a theory called *CONSPEC and CONLERN*. CONSPEC is the structural information which guides newborn preferences for face-like patterns. CONLERN is a learning device which extracts further visual characteristics from patterns identified based on CONSPEC. CONLERN does not influence infant looking behavior until 2 months of age.

Some supporting evidence for this theory comes from the fact that newborns do indeed preferentially orient their gaze to face-like patterns. Morton and Johnson [58] (also see Simion et al. 1996; [31, 41; 83]) presented face-like stimuli as in Figure 8.9 (left) and the same display with internal features presented upside-down (right). The newborns gazed longer at the face-like pattern. Since the two patterns are largely identical except for their "faceness" and the subjects have had virtually no other visual experience prior to the presentation, the experimenters concluded that newborns have an innate preference for faces.

Recent studies, however, have called into question the structural explanation based on 'faceness' and focused more on lower-level perceptual explanations.

FIGURE 8.9: Newborns preferentially orient their gaze to the face-like pattern on the left, rather than the one shown on the right, suggesting some innately specified representation for faces (from Johnson et al., 1991).

Simion et al. (2001) showed infants displays as in Figure 8.10. These displays bore no resemblance to faces, yet newborns preferred the displays on the left. Simion and colleagues concluded that newborn preferences in this case were not for face-like stimuli per se, but rather for top-heavy displays. Similar experiments with real faces and real faces with scrambled internal features (Cassia et al. 2004) had the same pattern of results. By three months, however, genuine face preferences appear to emerge. Turati et al. (2005) performed similar experiments with 3-month-old infants and found that, by this age, infants do indeed seem to orient specifically towards face-like stimuli, and that this orientation cannot be so easily explained by the low-level attributes driving newborns' preferences.

The above pattern of experimental results shows that, although newborns seem to have some set of innate preferences, these mechanisms are not necessarily domain-specific to faces. However, even though the particular orienting heuristic identified here can match nonface stimuli, it may still be the case that the presence of this heuristic biases the orientation of attention towards faces often enough over other stimuli to target a face-specific learning process on stimuli that are usually faces. A more accurate computational characterization of these perceptual mechanisms for the orientation of attention could be applied to sample images of the world (adjusted to mimic the filter of an infant's relatively poor visual system) in order to determine whether these mechanisms do indeed orient the infant towards faces more often than to other stimuli. For now, it cannot be said whether in fact newborns prefer faces to other stimuli in their environment.

Distinguishing the Mother from Strangers

Plausible evidence for an innate mechanism for facial processing comes from the remarkable ability of newborn infants within less than 50 hours of birth [28, 15,

FIGURE 8.10: As a counterpoint to the idea of innate preferences for faces, Simion et al. (2001) has shown that newborns consistently prefer top-heavy patterns (left column) over bottom-heavy ones (right column). This may well account for their tendency to gaze longer at facial displays, without requiring an innately specified "face detector".

85] to discriminate their mothers from strangers. Newborns suck more vigorously when viewing their mother's face on a videotaped image. They also are capable of habituating to a mother's image, eventually preferring a novel image of a stranger, showing a classic novelty preference. Even with Pascalis' et al. (1995) qualification that discrimination can only be achieved when external features (mainly hair) are present, it seems that such performance can only be achieved if newborns already possess at least a rudimentary facial-processing mechanism.

A counterargument to the conclusion that an innate mechanism is necessary is given by Valentin and Abdi [82]. These researchers point out that a newborn is only asked to discriminate among a few different faces, all of which are generally very different. Given a small number of face images degraded to mimic the acuity of a newborn, Valentin and Abdi attempted to train an artificial neural network to distinguish among them using an unsupervised learning algorithm. For small numbers of images comparable to the experiments performed on newborns, the network is successful. Since this network made no initial assumptions about the content of the images (faces or otherwise), an infant's visual system also need not necessarily have a face-specific mechanism to accomplish this task.

Imitation of Facial Expressions

An oft-cited "fact" is that newborns are able to imitate the facial expressions made by others. This would entail the infant's recognizing the expression, then mapping it to its own motor functions. Many studies have shown evidence of imitation (most notably, [54, 55]). A comprehensive review by Anisfeld [1, 2], however, revealed that there were more studies with negative than positive results. Moreover, there were consistent positive results only for one expression, namely, tongue protrusion. This action might be an innate releasing mechanism, perhaps to a surprising stimulus, or an attempt at oral exploration [42]. Thus, the action is a response to the stimulus, not a recognition of it followed by imitation. The coincidence only *seems* like imitation to the observer.

On the whole, the findings so far are equivocal regarding the innateness of face-processing mechanisms. This is perhaps good news for computer vision. The machine-based systems do not necessarily have to rely on any 'magical' hardwired mechanisms, but can learn face-processing skills through experience.

8.4.2 Behavioral Progression

Although infant face recognition proficiency develops rapidly within the first few months of life, performance continues to improve up to the age of 10 years or even later. The most well-studied progression in the behavioral domain is the use of featural versus configural cues.

Adults match upright faces more easily than inverted faces. This is believed to be a consequence of the disruption in configural processing in inverted faces. Interestingly, four-month-old infants do not show this decrement in performance (the so-called inversion effect) if the images to match are otherwise identical. However, the inversion effect does appear if, for instance, pose is manipulated at each presentation of the face [73, 79]. Thus, there is evidence that configural cues start to play some role in face processing at this early age, although the processing of these cues is still rudimentary. Processing based on features appears to play the primary role in infant facial recognition [19]. Given the early onset of the usage of configural cues in child face recognition, rudimentary though it may be, one would expect that full maturation of such a fundamental system would ensue rapidly. However, numerous studies have found that, although face recognition based on features reaches near-adult levels by the age of 6 years, configural processing still lags behind until at least 10 years of age, with a gradual progression of greater accuracy and dependence on configural cues [16, 67, 52, 57, 56, 35, Bruce et al. 2000].

Is this developmental improvement in the use of configural cues an outcome of general learning processes or a maturational change? Some evidence for the latter argument comes from patients with congenital bilateral cataracts in infancy [49, 30]. Even after more than 9 years of recovery, deprivation of patterned visual input within the first 2–6 months of life leads to a substantial deficit in configural processing (see Figure 8.11). Children with a history of early visual deprivation are impaired in their ability to distinguish between faces differing in the positions of features. However, they show no such deficits when distinguishing between faces differing in their constituent features. Perhaps this early maturation is a critical period in the development of face recognition processes.

Neuroimaging

Finally, we look at the recent contributions of neuroimaging to the understanding of changes in brain activation patterns to link the development of brain processes to changes in behavioral performance.

One neural marker used to study face-specific processes is the N170, which is an event-related potential (ERP) recorded using EEG. In adults, this signal generally occurs over the occipitotemporal lobe between 140 and 170 msec after the visual onset of a face image. De Haan et al. [25] looked for the N170 in 6-month-old infants. They compared human to nonhuman primate faces, and upright to inverted faces. They report an "infant N170" (with a slower peak latency, a distribution of location more towards the medial areas of the brain, and a somewhat smaller amplitude as compared to the normal adult N170) which prefers human faces to nonhumans, but which shows no sensitivity to inversion. This is evidence of the development of a possibly face-specific process, but this particular process, at least, seems to be insensitive to configural manipulation at this early age.

Another important neural marker in face research is activation in the fusiform gyrus during functional MRI (fMRI) scans while viewing faces. Aylward et al. [3] searched for this signal in young children (8–10 years) and slightly older children (12–14 years). In a paradigm comparing the passive viewing of faces versus houses, younger children showed no significant activation in the fusiform gyrus, while older children did. Thus, although the neuroimaging data are very

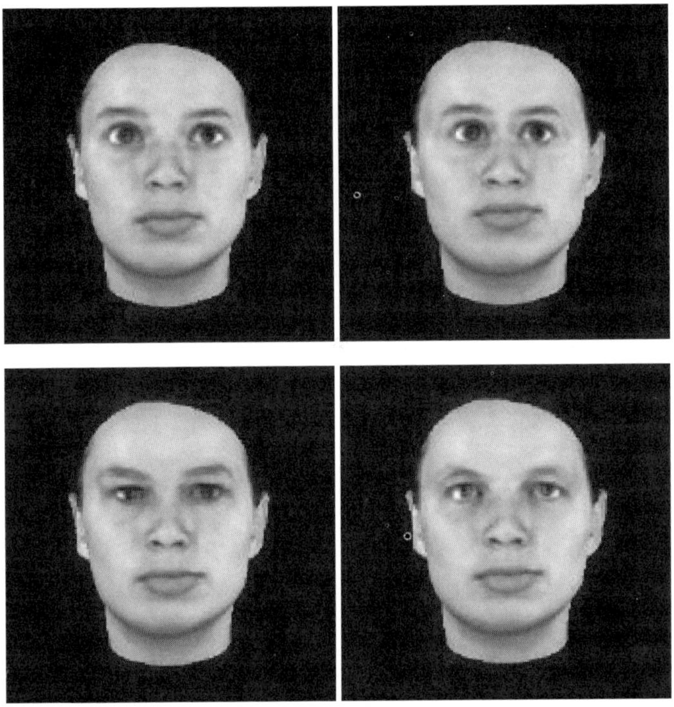

FIGURE 8.11: Children with a history of early visual deprivation are impaired in their ability to distinguish between faces differing in the positions of features (top row). However, they show no such deficits when distinguishing between faces differing in their constituent features (bottom row) (after [49]).

preliminary at the present, they do appear to be broadly consistent with the behavioral data showing continuing development of face processing well into adolescence.

8.4.3 Summary of Developmental Studies of Face Recognition

Although infant face recognition develops rapidly to the point where simple matching and recognition can take place, adult-like proficiency takes a long time to develop, at least until the age of 10. Early face processing seems to rely significantly on feature discrimination, with configural processing taking years to mature. Brain activation patterns echo this long maturational process. Interestingly, however, visual deprivation for even the first two months of life can cause possibly permanent deficits in configural face processing. The interaction of featural and configural processes may be one of the keys to fully understanding facial recognition.

8.5 WHAT ARE SOME BIOLOGICALLY PLAUSIBLE STRATEGIES FOR FACE RECOGNITION?

When developing computational methods for face recognition, it is easy to treat the recognition problem as a purely theoretical exercise, having nothing to do with faces per se. One may treat the input images as completely abstract data and use various mathematical techniques to arrive at statistically optimal decision functions. Such treatment will likely lead to the creation of a useful system for recognition, at least within the confines of the training images supplied to the algorithm.

At the same time, the machine-vision researcher is very lucky to have access to a wealth of knowledge concerning the function of the human visual system. This body of knowledge represents an incomplete blueprint of the most impressive recognition machine known. By making use of some of the same computations carried out in the visual pathway it may be possible to create extremely robust recognition systems. Also, mimicking the computations known to be carried out in the visual pathway may lead to an understanding of what is going on in higher-level cortical areas. Building a model recognition system based on our admittedly limited knowledge of the visual pathway may help us fill in the gaps, so to speak. With this goal in mind, many laboratories have developed computational models of face recognition that emphasize biologically plausible representations. We shall discuss several of those models here to demonstrate how physiological findings have informed computational studies of recognition. Also, we shall discuss possible ways in which recent computational findings might inform physiology.

8.5.1 Early Vision and Face Recognition

One of the most important findings in visual neuroscience was Hubel and Wiesel's discovery of orientation-specific cells in the feline primary visual cortex [37]. These cells were found to be tuned to stimuli along many different dimensions, including orientation, spatial frequency, motion, and color. The discovery of these cells led to the formulation of a hierarchical model of visual processing in which edges and lines were used to construct more complicated forms. An influential framework modeling how the visual pathway might combine low-level features in a cascade that leads to high-level recognition was put forth by Marr in his book *Vision* [51]. This framework has inspired many recent models of face recognition, as reviewed below.

Face Recognition: The Malsburg Model

A strategy adopted by many researchers has been to begin constructing computational models of recognition that utilize these same low-level visual primitives. In this way, the model is made to operate on data that roughly match how an image appears to the early visual cortex. The receptive fields of cells in early visual cortex are most commonly represented as *Gabor functions* in these models. A Gabor function is simply a sinusoid that has been windowed with a Gaussian. The frequency, orientation, and size of the function can be easily manipulated to produce a range of different model receptive fields.

Given these functions, there are various ways in which one can extract measurements from an image to perform recognition. One of the most useful strategies employed to date is the *Gabor jet* (also mentioned in Section 8.2, above). A jet is simply a "stack" of Gabor functions, each one with a unique orientation, frequency tuning, and scale (Figure 8.12). The construction of a jet is meant to mimic the multiscale nature of receptive field sizes as one moves upstream from the V1 neuron. It also provides a wealth of information concerning each point to which it is applied, rather than just relaying one response. Jets have been used as the basis of a very successful recognition technique known as "elastic bunch graph matching" [45, 86].

In this model, Gabor jets are applied to special landmarks on the face, referred to as *fiducial points*. The landmarks used are intuitively important structural features

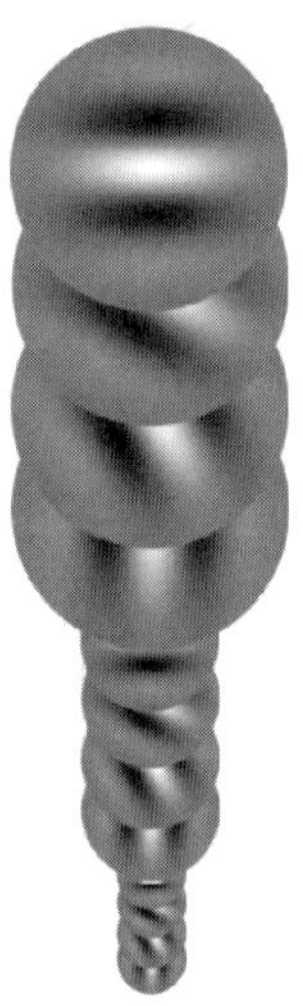

FIGURE 8.12: A Gabor jet. Linear filters are placed at a variety of scales and orientations.

such as the corners of the eyes and mouth and the outline of the face. At each of these points, the Gabor jet produces a list of numbers representing the amount of contrast energy that is present at the spatial frequencies, orientations, and scales included in the jet. These lists of numbers from each point are combined with the locations of each landmark to form the final representation of the face. Each face can then be compared with any other with a similarity metric that takes into account both the appearance and the spatial configuration of the landmarks.

The model has been shown to perform well on the FERET database benchmark for face recognition [68]. Moreover, it has also been shown that the similarity values computed by the algorithm correlate strongly with human similarity judgments [8]. This last result is particularly satisfying in that it suggests that these simple measurements capture perceptually important high-level information. The use of features that mimic visual system receptive fields appears to be a very useful strategy for representing faces when their outputs are combined in the right way.

8.5.2 Face Detection

A companion problem to face recognition (or individuation) is the challenge of face detection. In some ways, detecting a face in a cluttered scene presents deeper difficulties than determining the identity of a particular face. The main reason for this state of affairs is that when one is detecting a face in a complicated scene, one must determine a set of visual features that will always appear when a face is present and rarely appear otherwise. An additional layer of difficulty is added by the fact that one will likely have to scan a very large image to look for faces. This means that we will require fast computation at all stages of detection, as we may need to perform those measurements hundreds of times across a single image.

In this realm, as in face individuation, employing features that mimic those found in primary visual cortex has been found to be useful. We shall discuss two models that rely upon simple spatial features for face detection. The first makes use of box-like features that provide for exceptional speed and accuracy in large images containing faces. The second provides for impressive invariance to different lighting conditions by incorporating ordinal encoding: a nonlinear computation that more closely resembles the response properties of V1 cells.

Viola and Jones

Our first example of a face-detection model based on early visual processing was developed by Viola and Jones [84]. As in the Malsburg model, the selection of primitive features was motivated by the finding that Gabor-like receptive fields are found in primary visual cortex. However, rather than using true Gabor filters to process the image, this model utilizes an interesting abstraction of these functions to buy more speed for their algorithm. Box-like features are used as a proxy for Gabors because they are much faster to compute across an image and they can roughly approximate many of the spatial characteristics of the original functions. By constructing an extremely large set of these box features, the experimenters were able to determine which computations were the most useful at discriminating faces from background. It should be noted that no individual feature was particularly useful, but the combined judgments of a large family of features provides for very high accuracy. The notion that many 'weak' classifiers can be 'smart' when put together is an instance of 'boosting.' Examples of the best features for face detection are shown in Figure 8.13.

We see in Viola and Jones' model that very simple image measurements are capable of performing both detection and recognition tasks. This indicates that the early visual system may be capable of contributing a great deal to very complex visual tasks. Though this model is very good at detecting faces in cluttered backgrounds, it does break down when large variations in facial pose, expression or illumination are introduced. Human observers are capable of recognizing and detecting faces despite large variations of these kinds, of course. The challenge then is to produce a system equally capable of detecting faces despite these sources of variation.

We present in the next section a model from our laboratory which aims to uncover what processing strategies might support invariance to varying face illumination in particular. Our model overcomes the problem of varying illumination through a very simple modification of the basic edge-based representations mentioned previously. This modification, ordinal encoding, is particularly compelling in that it more closely models the behavior of V1 cells. This further bolsters the idea that biological systems may implicitly make use of computational principles that can inform our modeling.

FIGURE 8.13: Examples of the box filters evaluated by Viola and Jones for face detection showing a selection of the most useful features for face detection. Note the resemblance to the receptive fields of V1 neuron (after [84]).

8.5.3 Ordinal Encoding and Ratio Templates

The computations carried out by this model are very similar to those already discussed. We begin by noting a troubling difficulty associated with these simple measurements, however. As the lighting of a particular face changes, we note that the values produced by a simple box-like filter comparing two neighboring image regions may change a great deal (Figure 8.14). This is deeply problematic if we wish to be able to detect this face consistently. We must either hope that we can find other measurements that will not behave this way, or find a new way of using these features to represent a face.

We shall adopt the latter strategy, motivated by the observation that though the numerical values of the two regions may change, the sign of their relative magnitudes stays constant. This is precisely the kind of stable response we can use to perform face detection. Rather than maintaining precise image measurements when comparing two regions, we will only retain information about which region was the brightest. We are throwing away much quantitative information, but in a way that benefits the system's performance. We also note that this is much more representative of the behavior of a typical neuron in the visual system. Such cells are generally not capable of representing a wide range of contrasts, but rather saturate rapidly (Figure 8.15), providing an essentially binary signal (either "different" or "not different").

This kind of measurement constitutes an ordinal encoding, and we can build up a representation of what we expect faces to look like under this coding scheme for use in detection tasks. We call this representation a "ratio template" because it makes explicit only coarse luminance ratios between different face regions [76]. The model performs very well in detection tasks (Figure 8.16 shows some results), and we have also shown that relatively high-fidelity reconstructions of images are possible from only local ordinal measurements [72]. This means that, despite the coarse way we encode via ordinal measurements, the original image can be recovered very accurately.

We have seen in these three models that using computations similar to those carried out early in the visual system can provide good performance in complex visual tasks. At the heart of all three of these models is the Gabor function, which captures much of the tuning properties of neurons in the striate cortex. In the Malsburg model, these functions are applied in *jets* over the surface of a face to provide information about spatial features and configuration for individuation. In Viola and Jones' detection algorithm, box filters that approximate these functions are rapidly computed across a complex scene and combine their outputs to determine if a face is present. Finally, by incorporating the nonlinear nature of V1 cells into an ordinal code of image structure, our own algorithm provides for useful invariance to lighting changes. Taken together, these three models suggest that turning to biologically plausible mechanisms for recognition can help enhance the performance of computational systems.

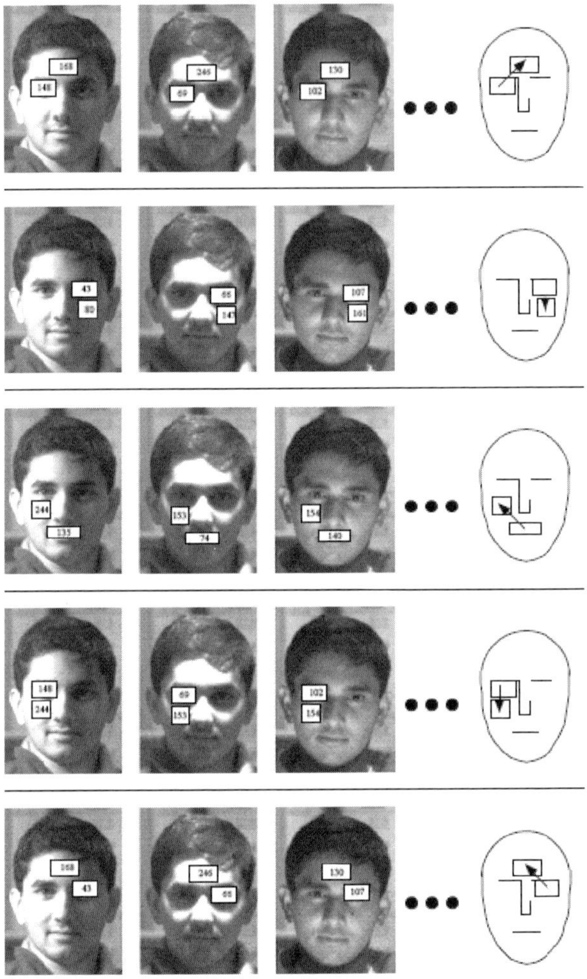

FIGURE 8.14: An example of how changes in ambient lighting can greatly alter the brightness values in neighboring regions of a face. We point out that, although the absolute difference changes, the direction of the brightness gradient between some regions stays constant (from [76]).

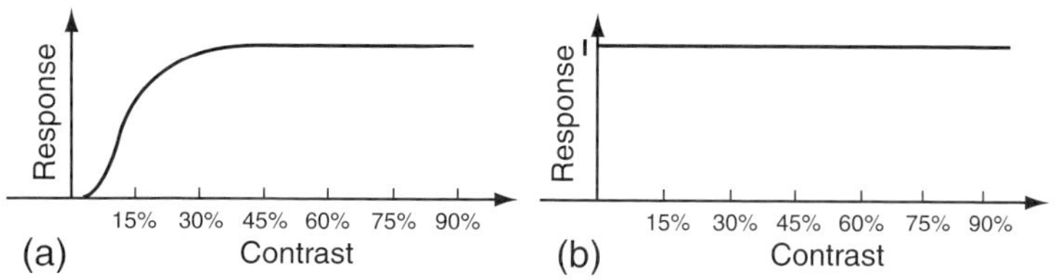

FIGURE 8.15: A demonstration of how the response of a typical neuron in the primary visual cortex flattens rapidly with increasing contrast. An idealized version of this neuron which is a pure step function is presented in (b) (from [72]).

FIGURE 8.16: Results of using a qualitative representation scheme to encode and detect faces in images. Each box corresponds to an image fragment that the system believes is a face. The representation scheme is able to tolerate wide appearance and resolution variations. The lower detection in the central image is a false positive. (See also color plate section).

8.5.4 Looking Downstream: Can Computational Models Inspire Physiological Research?

Up to this point, we have discussed how our knowledge of early visual cortex has driven the creation of various models for face recognition and detection. In this section, we shall speculate on how computational studies of face recognition might help drive physiologists to look for particular structures in higher-level visual areas.

It has been known for some time that the stimuli to which visual neurons are tuned increase in complexity as we move into higher stages of the visual pathway. While edge-like features are found in V1, cells that respond selectively to hands and faces have been found in the primate inferotemporal cortex [32]. Likewise, functional MRI studies of the human visual pathway reveal a possible "face area" in the fusiform gyrus [43]. Given the impressive selectivity in these areas, the obvious challenge is to determine what features or processes produce such specificity for complex objects. This is a very difficult endeavor, and many research groups have attempted to describe the tuning properties of various populations of high-level neurons. The chief difficulty is that the space of possible image features to choose from is very large, and one must limit oneself to some corner of that space in order to make progress. In so doing, however, it is impossible to know if there is another image feature outside of that space that would be truly "optimal" for a given neuron. The current picture of what stimuli higher-level neurons are tuned to is very complex, with reports of tuning for various forms of checkerboard grating [29] and curvature [66] in V4 through parametric studies. Nonparametric stimulus-reduction techniques have also led to a possible characterization of IT (Infero-Temporal cortex) into complex object "columns" [78].

We suggest that determining what complex features are useful for high-level tasks such as face detection and recognition may help physiologists determine what kind of selectivities to look for in downstream visual areas. In this way, computational models can contribute to physiology in much the same way as physiology has enhanced our computational understanding of these processes. We present here one very recent result from our lab that may suggest an interesting modification of our current models of cortical visual analysis based on computational results.

8.5.5 Dissociated Dipoles: Image Representation via Nonlocal Operators

We have noted thus far that many computational models of face recognition rely on edge-like analyses to create representations of images. However, one of the primary computational principles that appears to underlie the emergence of such cells in the visual cortex is that of image reconstruction. That is, edge-like cells appear to be an optimal solution to the problem of perfectly reconstructing image information with a small set of features [7, 60, 61]. However, it is not clear that one needs to perfectly reconstruct an image in order to recognize it. This led us to ask if such features are truly the most useful for a task like face recognition. We set out to determine a useful vocabulary of image features for face recognition through an exhaustive consideration of two-lobed box filters. In our analysis, we allowed lobes to be overlapping, neighboring, or completely separate. Each feature was then evaluated according to its ability to accurately identify faces from a large database [6].

Under the modified criterion of recognition, rather than reconstruction, two distinct families of model neurons appeared as good features. These "optimal" features were cells with a center-surround organization and those with two spatially disjoint lobes (Figure 8.17). The former are akin to retinal ganglion cells, but the latter do not resemble anything reported to date in the primate or human visual pathway. We have since found that despite the computational oddities they present, these nonlocal operators (which we call "dissociated dipoles")

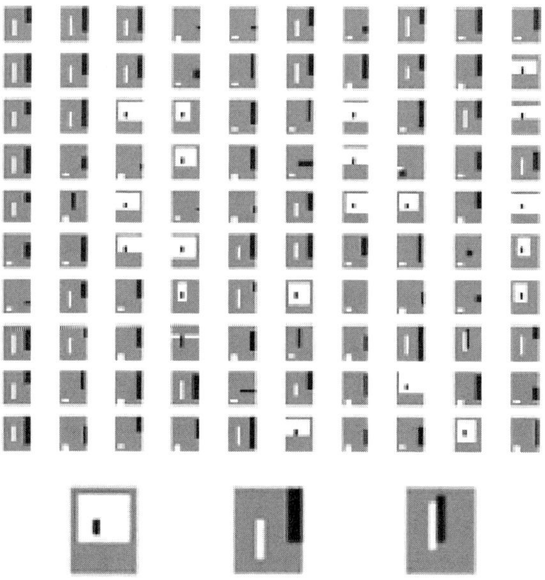

FIGURE 8.17: Examples of the best differential operators for face recognition discovered by Balas and Sinha. Note the prevalence of center-surround and non-local receptive fields. We call the non-local operators "dissociated dipoles" in reference to their spatially disjoint receptive fields.

are indeed very useful tools for performing recognition tasks with various face databases [5].

Could such computations be carried out in the visual processing stream? They appear to be found in other sensory modalities, such as audition and somatosensation [18, 89], meaning that they are certainly not beyond the capabilities of our biological machinery. We suggest that it might be useful to look for such operators in higher-level visual areas. One way in which complexity might be built up from edges to objects may be through incorporating information from widely separated image regions. While there is still no evidence of these kinds of receptive fields to date, this may be one example of how a computational result can motivate physiological inquiry.

8.6 CONCLUSION

Face recognition is one of the most active and exciting areas in neuroscience, psychology, and computer vision. While significant progress has been made on the issue of low-level image representation, the fundamental question of how to encode overall facial structure remains largely open. Machine-based systems stand to benefit from well-designed perceptual studies that can allow precise inferences to be drawn about the encoding schemes used by the human visual system. We have reviewed here a few results from the domain of human perception that provide benchmarks and guidelines for our efforts to create robust machine-based face recognition systems. It is important to stress that the limits of human performance do not necessarily define upper bounds on what is achievable. Specialized identification systems (say those based on novel sensors, such as close-range infrared cameras) may well exceed human performance in particular settings. However, in many real-world scenarios using conventional sensors, matching human performance remains an elusive goal. Data from human experiments can not only give us a better sense of what this goal is, but also what computational strategies we could employ to move towards it and, eventually, past it.

REFERENCES

[1] M. Anisfeld. Neonatal imitation. *Developmental Review* **11**, 60–97, 1991.

[2] M. Anisfeld. Only tongue protrusion modeling is matched by neonates. *Developmental Review* **16**, 149–161, 1996.

[3] E. H. Aylward, J. E. Park, K. M. Field, A. C. Parsons, T. L. Richards, S. C. Cramer, et al. Brain activation during face perception: Evidence of a developmental change. *Journal of Cognitive Neuroscience* **17**(2), 308–319, 2005.

[4] T. Bachmann. Identification of spatially quantized tachistoscopic images of faces: How many pixels does it take to carry identity? *European Journal of Cognitive Psychology* **3**, 85–103, 1991.

[5] B. J. Balas, and P. Sinha. Dissociated dipoles: image representation via non-local operators (AIM-2003-018). Cambridge, MA: MIT AI Lab, 2003.

[6] B. Balas, and P. Sinha. Receptive field structures for recognition. MIT CSAIL Memo, AIM-2005-006, CBCL-246, 2005.

[7] A. J. Bell, and T. J. Sejnowski. The "independent components" of natural scenes are edge filters. *Vision Research* **37**(23), 3327–3338, 1997.

[8] I. Biederman, and P. Kalocsai. Neurocomputational bases of object and face-recognition. *Philosophical Transactions of the Royal Society of London, Series B.* **352**(1358), 1203–1219, 1997.

[9] I. Biederman, and G. Ju. Surface versus edge-based determinants of Visual Recognition. *Cognitive Psychology* **20**, 38–64, 1988.

[10] W. L. Braje, D. Kersten. M. J. Tarr, and N. F. Troje. Illumination effects in face recognition. *Psychobiology* **26**, 371–380, 1998.

[11] V. Bruce, P. Healey, M. Burton, T. Doyle, A. Coombes, and A. Linney. Recognising facial surfaces. *Perception* **20**, 755–769, 1991.

[12] V. Bruce, E. Hanna, N. Dench, P. Healey, and M. Burton. The importance of "mass" in line drawings of faces. *Applied Cognitive Psychology* **6**, 619–628, 1992.

[13] V. Bruce, and A. Young. *In the Eye of the Beholder: The Science of Face Perception.* Oxford: Oxford University Press, 1998.

[14] V. Bruce, R. N. Campbell, G. Doherty-Sneddon, A. Import, S. Langton, S. McAuley, et al. Testing face processing skills in children. *British Journal of Developmental Psychology* **18**, 319–333, 2000.

[15] I. W. R. Bushnell, F. Sai, and J. T. Mullin. Neonatal recognition of the mothers face. *British Journal of Developmental Psychology* **7**, 3–15, 1989.

[16] S. Carey, and R. Diamond. From piecemeal to configurational representation of faces. *Science* **195**(4275), 312–314, 1977.

[17] V. M. Cassia, C. Turati, and F. Simion. Can a nonspecific bias toward top-heavy patterns explain newborn's face preference? *Psychological Science* **15**(6), 379–383, 2004.

[18] J. K. Chapin. Laminar differences in sizes, shapes, and response profiles of cutaneous receptive fields in rat SI cortex. *Experimental Brain Research* **62,** 549–559, 1986.

[19] L. B. Cohen, and C. H. Cashon. Do 7-month-old infants process independent features or facial configurations? *Infant and Child Development* **10**(1–2), 83–92, 2001.

[20] S. M. Collishaw, and G. J. Hole. Featural and configurational processes in the recognition of faces of different familiarity. *Perception* **29**(8), 893–909, 2000.

[21] N. P. Costen, D. M. Parker, and I. Craw. Spatial content and spatial quantization effects in face-recognition. *Perception* **23**, 129–146, 1994.

[22] J. Davidoff, and A. Ostergaard. The role of color in categorical judgments. *Quarterly Journal of Experimental Psychology* **40**, 533–544, 1988.

[23] G. M. Davies, H. D. Ellis, and J. W. Sheperd. Face recognition accuracy as a function of mode of representation. *Journal of Applied Psychology* **63**, 180–187, 1978.

[24] G. DeAngelis. I. Ohzawa., and R. D. Freeman. Spatiotemporal organization of simple-cell receptive fields in the cat's striate cortex. I. General characteristics and postnatal development. *Journal of Neurophysiology* **69**(4): 1091–1117, 1993.

[25] M. de Haan, O. Pascalis., and M. H. Johnson. Specialization of neural mechanisms underlying face recognition in human infants. *Journal of Cognitive Neuroscience* **14**(2), 199–209, 2002.

[26] R. Duda, and P. Hart. *Pattern Classification and Scene Analysis*. Wiley: NY, 1973.
[27] J. T. Enns, and D. I. Shore. Separate influences of orientation and lighting in the inverted-face effect. *Perception Psychophysics* **59**, 23–31, 1997.
[28] T. M. Field, D. Cohen. R. Garcia., and R. Greenberg. Mother-stranger face discrimination by the newborn. *Infant Behavior and Development* **7**(1), 19–25, 1984.
[29] J. L. Gallant, C. E. Connor, S. Rakshit. J. W. Lewis, and D. C. Van Essen. Neural responses to polar, hyperbolic, and Cartesian gratings in area V4 of the macaque monkey. *Journal of Neuroscience* **76**(4), 2718–2739, 1996.
[30] S. Geldart. C. J. Mondloch, D. Maurer. S. de Schonen, and H. P. Brent. The effect of early visual deprivation on the development of face processing. *Developmental Science* **5**(4), 490–501, 2002.
[31] C. C. Goren, M. Sarty, and P. Y. K. Wu. Visual following and pattern-discrimination of face-like stimuli by newborn infants. *Pediatrics* **56**(4), 544–549, 1975.
[32] C. G. Gross, C. E. Rocha- Miranda, and D. B. Bender. Visual properties of neurons in inferotemporal cortex of the Macaque. *Journal of Neurophysiology* **35**(1), 96–111, 1972.
[33] L. D. Harmon, and B. Julesz. Masking in visual recognition: Effects of two-dimensional noise. *Science* **180**, 1194–1197, 1973.
[34] L. D. Harmon. The recognition of faces. *Scientific American* **229**(5), 70–83, 1973.
[35] D. C. Hay, and R. Cox. Developmental changes in the recognition of faces and facial features. *Infant and Child Development* **9**, 199–212, 2000.
[36] H. Hill, and V. Bruce. Effects of lighting on the perception of facial surfaces. *Journal of Experimental Psychology: Human Perception and Performance* **22**, 986–1004, 1996.
[37] D. Hubel, and T. Wiesel. Receptive fields of single neurons in the cat's striate cortex. *Journal of Physiology* **148**, 574–591, 1959.
[38] G. Humphrey. The role of surface information in object recognition: studies of visual form agnosic and normal subjects. *Perception* **23**, 1457–1481, 1994.
[39] I. Jarudi, and P. Sinha. Contribution of internal and external features to face recognition. (Submitted), 2005.
[40] A. Johnston, H. Hill, and N. Carman. Recognising faces: effects of lighting direction, inversion, and brightness reversal. *Perception* **21**, 365–375, 1992.
[41] M. H. Johnson, S. Dziurawiec, H. Ellis, and J. Morton. Newborns, preferential tracking of face-like stimuli and its subsequent decline. *Cognition* **40**(1–2), 1–19, 1991.
[42] S. S. Jones. Imitation or exploration? Young infants' matching of adults' oral gestures. *Child Development* **67**(5), 1952–1969, 1996.
[43] N. Kanwisher, J. McDermott, and M. M. Chun. The fusiform face area: a module in human extrastriate cortex specialized for face perception. *Journal of Neuroscience* **17**(11), 4302–4311, 1997.
[44] R. Kemp, G. Pike, P. White, and A. Musselman. Perception and recognition of normal and negative faces: the role of shape from shading and pigmentation cues. *Perception* **25**, 37–52, 1996.
[45] M. Lades, J. C. Vortbruggen, J. Buhmann, J. Lange, C. von der Malsburg, R. P. Wurtz, and W. Konen. Distortion invariant object recognition in the dynamic link architecture. *IEEE Transactions on Computers* **42**, 300–311, 1993.

[46] H. Leder. Matching person identity from facial line drawings. *Perception* **28**, 1171–1175, 1999.
[47] K. J. Lee, and D. Perrett. Presentation-time measures of the effects of manipulations in colour space on discrimination of famous faces. *Perception* **26**, 733–752, 1997.
[48] T. S. Lee. Image representation using 2D Gabor wavelets. *IEEE Transactions on Pattern Analysis and Machine Intelligence* **18**(10): 959–971, 1996.
[49] R. Le Grand, C. J. Mondloch, D. Maurer, and H. P. Brent. Neuroperception—early visual experience and face processing. *Nature* **410**(6831), 890–890, 2001.
[50] C. H. Liu, C. A. Collin, A. M. Burton, and A. Chaurdhuri. Lighting direction affects recognition of untextured faces in photographic positive and negative. *Vision Research* **39**, 4003–4009, 1999.
[51] Marr, D. *Vision*. New York: W.H. Freeman and Company, 1982.
[52] D. Maurer, R. Le Grand, and C. J. Mondloch. The many faces of configural processing. *Trends in Cognitive Sciences* **6**(6), 255–260, 2002.
[53] P. A. McMullen, D. I. Shore, and R. B. Henderson. Testing a two-component model of face identification: effects of inversion, contrast reversal, and direction of lighting. *Perception* **29**, 609–619, 2000.
[54] A. N. Meltzoff, and M. K. Moore. Imitation of facial and manual gestures by human neonates. *Science* **198**(4312), 75–78, 1977.
[55] A. N. Meltzoff, and M. K. Moore. Newborn infants imitate adult facial gestures. *Child Development* **54**(3), 702–709, 1983.
[56] C. J. Mondloch, S. Geldart, D. Maurer, and R. Le Grand, Developmental changes in face processing skills. *Journal of Experimental Child Psychology* **86**(1), 67–84, 2003.
[57] C. J. Mondloch, R. Le Grand, and D. Maurer. Configural face processing develops more slowly than featural face processing. *Perception* **31**(5), 553–566, 2002.
[58] J. Morton, and M. H. Johnson. Conspec and conlern—a 2-process theory of infant face recognition. *Psychological Review* **98**(2), 164–181, 1991.
[59] Y. Moses, Y. Adini, and S. Ullman. Face recognition: the problem of compensating for illumination changes. *Proceedings of the European Conference on Computer Vision*. pp. 286–296, 1994.
[60] B. A. Olshausen. Emergence of simple-cell receptive field properties by learning a sparse code for natural images. *Nature* **381**(607–609), 1996.
[61] B. A. Olshausen, and D. J. Field. Sparse coding with an overcomplete basis set: A strategy employed by V1? *Vision Research* **37**(23), 3311–3325, 1997.
[62] A. Ostergaard, and J. Davidoff. Some effects of color on naming and recognition of objects. *Journal of Experimental Psychology: Learning, Memory, and Cognition* **11**, 579–587, 1985.
[63] A. J. O'Toole, T. Vetter, and V. Blanz. Three-dimensional shape and two-dimensional surface reflectance contributions to face-recognition: an application of three-dimensional morphing. *Vision Research* **39**, 3145–3155, 1999.
[64] S. E. Palmer. *Vision Science: Photons to Phenomenology*. Cambridge, Massachusetts: MIT Press. 1999.
[65] O. Pascalis, S. Deschonen, J. Morton, C. Deruelle, and M. Fabregrenet. Mother's face recognition by neonates—a replication and an extension. *Infant Behavior and Development* **18**(1), 79–85, 1995.

[66] A. Pasupathy, and C. E. Connor. Population coding of shape in area V4. *Nature Neuroscience* **5**(12), 1332–1338, 2002.
[67] E. Pellicano, and G. Rhodes. Holistic processing of faces in preschool children and adults. *Psychological Science* **14**(6), 618–622, 2003.
[68] P. J. Phillips, H. Moon, S. Rizvi, and P. Rauss. The FERET evaluation methodology for face recognition algorithms. *IEEE Transactions on Pattern Analysis and Machine Intelligence* **22**, 1090–1104, 2000.
[69] C. J. Price, and G. W. Humphreys. The effects of surface detail on object categorization and naming. *Quarterly Journal of Experimental Psychology* **A41**, 797–828, 1989.
[70] G. Rhodes, S. E. Brennan, and S. Carey. Recognition and ratings of caricatures: implications for mental representations of faces. *Cognitive Psychology* **19**, 473–497, 1987.
[71] R. Russell, P. Sinha, I. Biederman, and M. Nederhouser. The importance of pigmentation for face recognition. *Journal of Vision* **4**, 418a, 2004.
[72] J. Sadr, S. Mukherjee, K. Thoresz, and P. Sinha (Eds.). The fidelity of local ordinal encoding. *Advances in Neural Information Processing Systems* (Vol. 14): MIT Press. 2002.
[73] G. Schwarzer. Development of face processing: The effect of face inversion. *Child Development* **71**(2), 391–401, 2000.
[74] F. Simion, V. M. Cassia, C. Turati, and E. Valenza. The origins of face perception: specific versus non-specific mechanisms. *Infant and Child Development* **10**(1–2), 59–65, 2001.
[75] P. Sinha, and T. Poggio. I think I know that face.... *Nature* **384**, 404, 1996.
[76] P. Sinha. Qualitative representations for recognition. *Lecture Notes in Computer Science* (Vol. LNCS 2525, pp. 249–262). Springer-Verlag, 2002.
[77] J. Tanaka, D. Weiskopf, and P. Williams. The role of color in high-level vision. *Trends in Cognitive Sciences* **5**, 211–215, 2001.
[78] K. Tanaka. Columns for complex visual object features in the inferotemporal cortex: clustering of cells with similar but slightly different stimulus selectivites. *Cerebral Cortex* **13**(1), 90–99, 2003.
[79] C. Turati, S. Sangrigoli, J. Ruel, and S. de Schonen. Evidence of the face inversion effect in 4-month-old infants. *Infancy* **6**(2), 275–297, 2004.
[80] C. Turati, E. Valenza, I. Leo, and F. Simion. Three-month-old's visual preference for faces and its underlying visual processing mechanisms. *Journal of Experimental Child Psychology* **90**(3), 255–273, 2005.
[81] S. Ullman. *High-Level Vision: Object Recognition and Visual Cognition*. Cambridge, Massachusetts: MIT Press, 1996.
[82] D. Valentin, and H. Abdi. Face recognition by myopic baby neural networks. *Infant and Child Development* **10**(1–2), 19–20, 2001.
[83] E. Valenza, F. Simion, V. M. Cassia, and C. Umilta. Face preference at birth. *Journal of Experimental Psychology: Human Perception and Performance* **22**(4), 892–903, 1996.
[84] P. Viola, and M. Jones. Rapid object detection using a boosted cascade of simple features. In: *Proceedings of 2001 IEEE Computer Society Conference on Computer Vision and Pattern Recognition* (*CVPR 2001*), Jauai, Hawaii, December 8–14. IEEE Computer Society Press, 2001.

[85] G. E. Walton, N. J. A. Bower, and T. G. R. Bower. Recognition of familiar faces by newborns. *Infant Behavior and Development* **15**(2), 265–269, 1992.

[86] L. Wiskott, J. M. Fellous, N. Kruger, and C. von der Malsburg. Face recognition by elastic bunch graph matching. *IEEE Transactions on Pattern Analysis and Machine Intelligence* **19**(7), 775–779, 1997.

[87] L. H. Wurm, G. E. Legge, L. M. Isenberg, and A. Luebker. Color improves object recognition in normal and low vision. *Journal of Experimental Psychology: Human Perception and Performance* **19**, 899–911, 1993.

[88] A. Yip, and P. Sinha. Contribution of color to face recognition. *Perception* **31**, 995–1005, 2002.

[89] E. D. Young. Response characteristics of neurons of the cochlear nucleus. In: C. I. Berlin (Ed.), *Hearing Science Recent Advances*. San Diego: College Hill Press, 1984.

[90] S. Zeki. *Inner Vision*. New York, New York: Oxford University Press, 2000.

CHAPTER 9

PREDICTING HUMAN PERFORMANCE FOR FACE RECOGNITION

9.1 INTRODUCTION

Human face-recognition abilities are impressive by comparison with the performance of many current automatic face-recognition systems. Such is the belief of many psychologists and computer scientists. This belief, however, is supported more by anecdotal impression than by scientific evidence. In fact, there are relatively few systematic and direct comparisons between the performance of humans and automatic face-recognition algorithms (though see [84], for exceptions). Comparisons of these sorts can be carried out both at a quantitative and qualitative level. On the qualitative side, as the field of automatic face recognition develops, the assessment of working systems has expanded to consider, not only overall levels of performance, but performance under more restricted and targeted conditions. It may thus be important to know, for example, whether an algorithm performs accurately recognizing "male faces of a certain age", "female faces of a certain ethnicity", or faces that are "typical" in the context of the database to be searched. How secure is a security system that was developed to operate on Caucasian faces when it is put to the test in a context filled with diverse faces of many ethnicities?

By contrast to the study of automated face-recognition, the study of human face-recognition abilities has long focused on a range of qualitative factors that affect performance. The rationale for this endeavor has been to consider these factors as indices into the nature of human representations of faces. Do we encode a face in terms of its three-dimensional structure? Do we encode a face in terms of its similarity relationship to the other faces we know? In fact, much is known about the qualitative factors that affect human accuracy in recognizing faces. There is ample evidence to suggest that not all faces are equally recognizable. There is evidence also that the experiential history of individuals affects their ability to recognize faces.

Using the range of qualitative factors that are known to affect human performance, we can begin to tag individual faces with a probability of successful recognition. In addition to the properties that affect the recognizability of individual faces, much is known also about the viewing conditions that support good performance and about those that make humans prone to error. More recently, the study of human face recognition has begun to consider the factors that affect face and person recognition in more natural contexts. Accordingly, studies in recent years have considered the question of how humans recognize faces in motion. These studies open a window into understanding the mechanics of face recognition in the real world, where there is little or no control of illumination and where the relationship between the viewer and the person-to-be-recognized changes continuously. All of these factors impact the likelihood of successful human recognition in ways that are more or less knowable through the application of experimental studies of human memory.

In this chapter, we review the factors that affect human accuracy for face recognition. These factors can be classified into three categories: (1) face-based constraints, including the effects of typicality, gender, age, and ethnicity; (2) viewing constraints, including the effects of illumination, viewpoint, pose, and facial motion; and (3) experiential constraints, including how experience with individual faces and groups of faces affects recognition performance. Our purpose in each case is to define and discuss the nature of the predictors and the implications each has for understanding how humans represent and process faces. Our secondary goal is to begin to sketch out, for the field of automatic face recognition, a series of factors that might prove useful in deriving a more detailed and refined set of measures for the performance of face-recognition algorithms. When all is said and done, any given face-recognition algorithm must compete, not only against other automatic recognition algorithms, but against humans, who are currently performing the task in most applied situations.

9.2 FACE-BASED FACTORS AND THE FACE-SPACE MODEL

A face is a complex three-dimensional object, with overlying surface pigmentation that specifies the reflectance properties of the skin, eyes, mouth and other features. To remember a face as an individual, and to distinguish it from other known and unknown faces, we must encode information that makes the face unique.

We begin by introducing the concept of a face space in psychological models of face processing [121]. The model has just a few components, but is ubiquitous in psychological and computational theorizing about face recognition. In a face-space model of recognition, faces can be considered as points in a multidimensional space. The axes of the space represent the "features" with which a face is encoded. A face can be represented, therefore, by its coordinates on these axes, which specify the face's value on each of the feature dimensions. At a psychological level, it is not necessary to specify the nature of features that form the axes of a space. It is generally enough to know that faces can be described using a set of features dimensions (e.g., width of face, distance between eyes, eye color, etc.) and that different faces vary in the combination of feature values they need to be re-created.

At the computational level, physical face spaces form the core of most automatic face-recognition algorithms. The feature axes in computational models represent real physical features, often extracted statistically using principal- or independent-component analysis. These kinds of analysis have been applied to face images (e.g., [115]), separated face shapes and image data (e.g., [34]), or three-dimensional data from laser scans (e.g., [12]). Face recognition then becomes a problem of projecting a "target" face into a space and comparing its location to other faces in the space. If the target face is sufficiently close to one of these other faces, then it is recognized. Otherwise, the face is rejected as unknown.

In the context of a face-space model, it is relatively easy to see conceptually how the properties of individual faces and categories of faces may affect both human and machine recognition accuracy. We present the effects of each of these categories on human face-recognition performance using the analogy of a face-space as the supporting theoretical construct.

9.2.1 Typicality

A long-standing focus of research in human perception and memory centers on the importance of the "average" or "prototype" in guiding recognition and categorization of visual stimuli. The theory is that categories of objects, including faces, are organized around a prototype or average. The prototype is the best or most typical item in the category. Typical items, those similar to the prototype, are easier to categorize as exemplars of the category than are unusual items. For example, human subjects categorize sparrows as "birds" faster than they categorize penguins as "birds", because sparrows are "typical" birds and penguins are "unusual" birds [101]. In general, the idea is that the closer an item is to the category prototype, the easier it is to "recognize" as an exemplar of the category.

The problem for faces, however, is to "recognize" each person as an *individual*. Thus, it is not sufficient to say that something is a face. Rather, we must determine: (a) if the face is known to us and, if so, (b) whose face it is. Here, the inverse relationship holds between typicality and recognizability. Typical faces, which are similar to the prototype, are more difficult to recognize. This occurs, presumably, because typical faces are plentiful and so there are more faces that could be falsely mistaken for any given typical face than for any given distinctive face [121].

Face typicality is one of the best known and most robust predictors of human face-recognition performance. In a classic study, Light et al. [73] found that faces rated as "typical" were recognized less accurately than faces rated as "unusual". In a follow-up experiment, Light et al. showed convincingly that the disadvantage found for typical faces was due to their high inter-item similarity. Faces rated by subjects as "typical" were found to be generally more similar to other faces, than faces rated by subjects as "unusual". These findings have been replicated many times, with a variety of different measures of typicality (e.g., [122]).

The relationship between typicality and recognizability accounts well for the psychological findings suggesting the enhanced recognizability of facial caricatures relative to veridical faces. In general, artists draw caricatures in a way that exaggerates or enhances facial features that are "unusual" for the person. As such, the lips of Mick Jagger become thicker than they are, and the eyes of Prince Charles are even closer together than they actually are. Despite the fact that caricatures are grotesque distrortions of a face, they are often recognized more accurately and efficiently than actual images of the faces [9, 80, 102]. Computer generated caricatures likewise operate by comparing a face to the "average face", and then by exaggerating facial dimensions that deviate from the average [18]. The enhanced recognizability of caricatures by comparison to veridical faces may be due to the fact that the exaggeration of unusual features in these faces makes the person less confusable with other faces, and somehow or other "more like themselves".

It is worth noting that face typicality correlates well with other ratings of faces, including perceived familiarity (i.e., "this person looks like someone I know") [123] and surprisingly, facial attractiveness [68]. The odd, but replicable, finding that typical faces are attractive indicates that faces similar to the "prototype", which lack distinctive features, are aesthetically preferred to more memorable, albeit less attractive, faces.

Metaphorically, the same face-space analogy is applicable to understanding the effects of face typicality on recognition. For automatic face-recognition systems, a number of findings about the relationship between typicality and recognizability may be useful. For example, we would expect that the recognizability of individual faces should be predicted by the density of faces in the neighboring "face space". We might also expect that the face space should be most dense in the center near the average. The space should become progressively less dense as we move away from the average. If a computationally-based face space approximates the similarity space humans employ for face processing, we might expect that "typical" faces would be near the center of the space and that unusual or distinctive faces be far from the center. It follows, therefore, that computational models of face recognition will not perform equally well for all faces. These systems should, like humans, make more errors on typical faces than on ususual faces.

9.2.2 Sex

The assumptions we make about the density of face space being highest near the center of the space and progressively less dense as we move away from the average face are somewhat simplistic. A more likely version of human face space is one in which there are a number of local averages at the centers of clusters of faces that belong to different natural categories, e.g., sex, race, age. In fact, when we consider that faces can be grouped into natural categories based on their sex, race/ethnicity, and possibly age, there is little reason to expect that "typical faces" are close to the grand mean of the space. Rather, the average face at the center of a space that is based on a set of diverse faces is likely to be androgenous and unrecognizable as a member of a racial or ethnic group. The dynamics of face recognition in this space are rather different than those discussed previously, though we can expect that, within any given face category, many of the same principles of typicality and confusability should apply. This brings us to the question of how natural categories of faces such as sex, race, and age might affect human face-recognition accuracy.

We begin with the sex of a face, which is arguably the largest source of variance in facial appearance. Indeed, faces are readily identifiable as male or female by both humans and machines (cf. [1, 27, 41, 48, 89]). Computationally-based contrasts between male and female faces are easily derived with analyses like PCA. Applied to the three-dimensional laser-scan data obtained from a large number of faces [89], this contrast is illustrated in Figure 9.1. Here we see that the information specifying the sex of a face can be localized to a few eigenvectors. In Figure 9.1, we show the contrast in face shape that can be achieved by varying the weights of two individual eigenvectors (the first and sixth) in combination with the average head. The contrasts clearly change the gender appearance of the face in a perceptible way.

How does the sex of a face affect face recognition accuracy for humans? A number of studies, which posed this question in the 1960s and 1970s found effects of the sex of a face on recognition accuracy that were minimal and inconsistent (see review [105]). Shepherd (1981) concluded that the most consistent finding in these studies was a small recognition advantage for female faces. A second intriguing finding was an interaction between the sex of the subject and the sex of the face. This interaction generally indicated that women are particularly accurate at recognizing female faces. There has been little or no work since these early studies that alters their main conclusions.

More recently the focus of the relationship between the sex of the face and recognition has been to consider whether the identity-specific and sexually dimorphic information in faces are independent. As noted previously, in order to recognize a face, we must encode the information that makes it unique or different from all

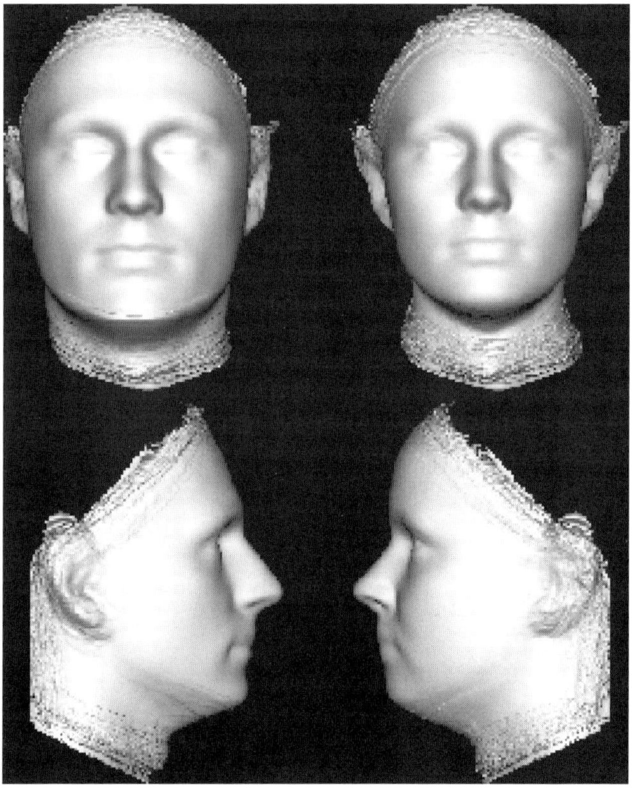

FIGURE 9.1: Illustration of the computationally derived information about the sex of a face from three-dimensional head models from laser scans [89]. The first row of the figure shows the average face plus the first eigenvector (left) and minus the second eigenvector (right). The second row of the figure shows an analgous display for the sixth eigenvector. The coordinates of a face on both the first and sixth eigenvectors predict the sex of the face.

other faces in the world. To categorize a face, for example by sex, we must encode information that the face shares with an entire group of faces. To know that a face is male, we must find features common to male faces, but unlikely to be found in female faces.

According to Bruce and Young's classic model of face recognition, the sex determining information about a face is accessed independently of the information about identity [25]. Computational assessments of the information necessary for sex classification of faces versus identification support the claim that there is reasonably good separability of these two kinds of information in faces [84].

A recent study of "speeded classification" of faces by sex and familiarity, however, is at odds with the conclusion of independent processing of these kinds of information in faces [44]. In this kind of study, subjects are asked to classify faces by sex (the target feature), for example, as quickly as possible. In some cases, other "distractor" features of the faces in the sequence change during the task (for example, facial expression or familiarity). If the change in the other irrelevant feature interferes with the processing of the target feature, one can conclude that the processing of the target and distractor features is not independent. In a speeded-up classification task of this sort, interference between familiarity and the sex of faces was found when subjects classified faces either by sex or by familiarity [44]. Specifically, irrelevant variations in familiarity slowed sex classifications and vice versa. They concluded that humans are unable to attend selectively to either identity or sex information in faces.

However, a more direct method for testing the codependence of sex and identity information in faces was used by Wild and colleagues [125]. They compared recognition and sex classification of children's and adults' faces. The rationale behind this method is based on the assumption that although children's faces have reliable information for determining their sex, this information is less salient and more difficult to access than the information that specifies the sex of adult faces. In this study, the authors began by trying to establish a more factual basis for the above assumption. Using simple morphing methods, they created a prototype "boy face" and a prototype "girl face" (see Figure 9.2). For example, the construction

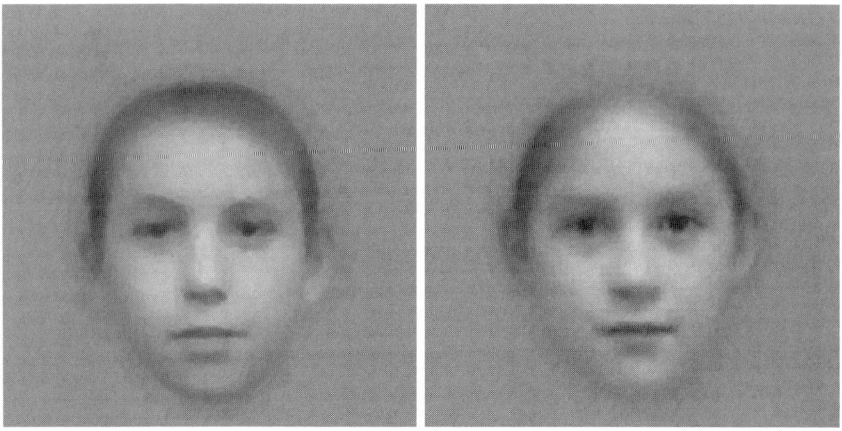

FIGURE 9.2: Computationally-derived gender information in children's faces. Left is the male prototype, made by morphing boys together (see text for details), and right is the female prototype.

of a prototype boy face was done by morphing pairs of boy faces together, and then morphing pairs of the morphs together, etc., until the resulting morph stopped changing. The prototypes indicate that the faces of boys and girls should be physically distinguishable, but that the differences are indeed subtle and difficult to describe [125].

Next, Wild and colleagues asked children and adults to classify the faces of children and adults by sex. After the sex classification task, the participants were asked to recognize the faces they had just classified. The results indicated that although children's faces were more difficult to classify by sex, they were recognized as accurately as adult faces.

In short, one can conclude that the quality of categorical information in faces is not related to the quality of the identity information. This finding is consistent with principal-component analysis (PCA) models of face recognition and categorization, which localize categorical information about faces in eigenvectors with large eigenvalues and identity information in eigenvectors with smaller eigenvalues [84]. Combined, the computational and human results reemphasize the importance of the local structure of face space in predicting recognizability.

For computational models, the best current data come from recent findings in the Face Recognition Vendor Test 2002 [95]. This test was conducted by DARPA on the performance of ten commercial, automated, face-recognition systems. The results of this analysis indicate a quite consistent advantage for recognizing male faces over female faces. The results showed that the performance advantage for male faces declined with face age. Although highly consistent, it is difficult to pinpoint the source of this advantage to idiosyncratic properties of the set of faces or to the algorithms tested (for which implementation details are proprietary). More details about the algorithms and a verification of the finding with a different set of faces is needed to be certain of the advantage. The properties of both the set of faces and the types of algorithm employed may play a role in performance, but this may need to be evaluated on a case-by-case basis.

In summary, for humans there are only minimal differences in accuracy recognizing male and female faces. For machines, there can be quite large differences, though at present it is unclear if these are differences that generalize across different sets of faces and algorithms. There is some evidence from both human and machine studies that the information specifying the sex of a face can be effectively isolated from the information useful for recognizing the face.

9.2.3 Ethnicity

Another set of natural face categories can be found in the racial or ethnic differences between human faces. Large-scale structural and reflectance differences exist between faces of different races. Psychologically, the categorical dimension of race differs from sex in that humans tend to have more experience with one race of faces, generally, their own race, than with other races. By contrast, most people have roughly equal amounts of contact and experience with males and females.

How does face race affect face recognition accuracy? Malpass (1969) was the first to demonstrate the "other-race effect" empirically [78]. He showed that human subjects recognize faces of their own race more accurately than faces of other races. This result has remained remarkably consistent over the years, with replications that have tested the effect with subjects and faces from a variety of races. In general, no effects of the race of the subject or the race of the face, per se, are found. The other-race effect consists solely of an interaction between the race of the face and the race of the subject.

The primary hypothesis for the cause of this effect is based on the differential experience individuals have with faces of their own race versus faces of other races. The so-called "contact" hypothesis predicts a relationship between the amount of experience we have with other-race faces and the size of the other-race effect. Several studies have tested this hypothesis, defining contact variously from simple questionnaires assessing previous exposure to members of other races [78] to the experience of living in an integrated neighborhood [40]. As noted previously [72], these studies have yielded inconsistent results, with some finding support for the contact hypothesis [30, 36, 33, 40, 104] and other studies failing to find support for this hypothesis [19, 70, 78, 83]. One reason for the lack of consistency among these studies might be linked to the diversity of the methods employed, and consequently, to the kinds of experience each may be measuring [42].

In reviewing evidence for the contact hypothesis many years ago, Shepherd [105] noted that, among the few studies testing children and/or defining "contact" with other-race faces developmentally [36, 40], more consistent evidence for the contact hypothesis is found. These developmental studies examined children's contact with other races in the form of integrated versus segregated schools and/or neighborhoods [40, 36]. Shepherd suggests that this kind of early contact may be critical for developing an other-race effect for face recognition [105].

Complementing these studies, the developmental course of the other-race effect was tested further with Caucasian participants between the ages of 6 and 20 years old on a memory task for Caucasian and Asian faces [31]. In this study, the youngest participants, 6-year-olds, recognized faces of both races equally well. By 10 years of age, however, there was a recognition-accuracy advantage for Caucasian faces, which became successively larger for the older participants. Combined, these studies suggest the possibility that not all "contact" is equally effective in reducing/preventing an other-race effect. Contact early in life may be related to the magnitude of the other-race effect, whereas contact later on appears to be less consistently related to recognition skills for other-race faces.

The other-race effect for humans, therefore, represents an interaction between the properties of different categories of faces as well as the experiential history of the subject. In particular, the age at which initial contact occurs may be particularly important in establishing the other-race effect.

At a glance, it is not obvious that this effect is relevant for face-recognition algorithms. However, most recognition algorithms rely on computational and statistical descriptive tools, like PCA. These analyses are, by definition, sensitive to the statistical structure of the training set of faces. The basis vectors that define the face space are extracted from sets of faces that may vary in how representative they are of any particular population (e.g., the population for which one hopes to optimize a working automatic face-recognition system). When face sets contain more faces of one race than of other races, it is reasonable to expect recognition accuracy differences for faces from the "majority" and "minority" race.

Furl et al. (2002) examined 13 face-recognition algorithms for the presence of an other-race effect [42]. In all cases, the training set of faces was biased for the inclusion of Caucasian versus Asian faces. The algorithms fell into one of three categories in terms of their computational strategies and the extent to which a training set of faces could influence their performance. Two control algorithms based on image similarity comparisons (i.e., without low-dimensional representations or complex distance measures), were not sensitive to the statistical structure of the training set. These algorithms showed no consistent difference for Asian versus Caucasian faces. Eight algorithms were based on PCA and varied in the distance measures used to assess matches. These algorithms performed best on the *minority* face race! In other words, they showed an "own-race effect". The final three algorithms used a combination of PCA with a preselected biased training set and a second analysis that warped the space to optimize the separability of faces within the space (e.g., Fischer discriminant analysis, FDA). These algorithms performed most like humans, showing an other-race effect that favored recognition of the Caucasian faces.

Furl et al. (2002) speculated on the difference between the other-race advantage showed by the pure PCA models and the own-race advantage showed by the algorithms that mixed PCA with a space-warping algorithm like FDA. The authors suggest that the former are likely to perform better on the minority-race faces due to the fact that these faces are "distinct" by comparison to the majority of faces in the face space. In other words, the minority-race faces that were included in the PCA have representations in the sparsest part of the space and so are less likely to be confused with the minority distractor faces. By contrast, the combined models have the effect of warping the face space to provide the best and most distinctive representations for the majority faces. Thus, the learning of new minority-race faces is limited to using a rather constrained and impoverished representation of these faces.

In summary, the interaction between experience and face race can affect performance for humans, and under more complicated metrics, for machines as well.

9.3 VIEWING CONSTRAINTS

In addition to the properties of faces themselves, nearly all components of the viewing parameters that have been tested with human subjects have been shown to affect face-recognition accuracy. Most prominant are the effects of changes in viewpoint/pose and changes in the direction and intensity of the illuminant. These factors are also highly relevant for face-recognition algorithms. At present, it would not be an exaggeration to say that viewing conditions constitute the single largest challenge to putting algorithms into the field. A common theme throughout this discussion is that as humans become familiar with faces, the apparent and strong effects of viewing parameters are lost or at least minimized. Common sense tells us that, when we know someone well, it is possible to recognize them under the poorest and least optimal of viewing conditions. We will return to the issue of familiarity after summarizing findings of viewing parameters.

9.3.1 Pose

To recognize a face from a novel view, we must be able to encode something unique about the face that distinguishes it from all other faces in the world and, further, we must be able to access this unique information from the novel view. Studying the representations and processes that humans use to accomplish this task is difficult, due to the complexity of the visual information observers experience in viewing faces from different viewpoints and due to the multitude of ways that such information can be encoded and represented. In general, humans can recognize faces from different views. However, they do this accurately only when the faces are well known to them.

The nature of human representations of faces and objects is a long-standing issue that is still actively under debate. Two rather divergent theories of these representations have been put forth and tested. Structure-based theories suggest that the visual system constructs a three-dimensional representation of faces and objects from the two-dimensional retinal images on the left and right retinas [79]. Accordingly, objects can be recognized in a view-independent manner by their components (i.e., volumes or parts [11]).

The second type of theory is based on the direct analysis of images. These image-based theories assume a view-dependent representation of faces and objects. As such, experience with multiple views of an object or face is essential to be able to recognize the object or face from a novel viewpoint [97].

Both structural and image-based theories have been supported by psychophysical studies [110, 10]. It has been suggested, therefore, that the most viable model of object/face recognition should incorporate the most appealing aspects of both accounts [110].

To be able to recognize a face from a new view, we need to extract invariant information that can be used subsequently to identify the face. Because we know the structure of familiar faces, we can automatically encode them, and therefore it is unlikely that we will find an effect of pose for recognition of familiar faces. Indeed, subjects do not benefit from seeing familiar faces in the same view at learning and at test [20, 117]. By contrast, subjects find it easier to recognize unfamiliar faces if the viewpoint of the face is the same at learning and at test. Thus, it seems that unfamiliar faces are sensitive to the change in viewpoint between the view of the face that was learned and the view that was tested.

Notwithstanding, it is also likely that some views of an unfamiliar face, those seen either at learning or at test, will ease the extraction of the invariant information important for face recognition. Therefore, recognition of unfamiliar faces is likely to be influenced by the change in viewpoint between learning and test and by the information provided by the viewpoint of the face.

Rotation Effects

In general, human observers' recognition performance is impaired when faces are rotated between learning and test. When the viewpoint changes between learning and test less than 30°, performance declines in a roughly linear fashion as a function of the rotation offset. For changes of more than 30°, the performance cost reaches a plateau. On the whole, a 45° rotation seems to impair performance significantly [5, 20, 57, 61, 76, 85]. Recent fMRI studies [4] seem to confirm this effect by showing that activity in the face fusiform area is sensitive to change in viewpoints. It is worth noting however that the fusiform responds to faces across viewpoints, as do other face sensitive brain regions like the superior temporal sulcus [3]. It has been possible as well to differentiate subareas of fusiform cortex and the lateral occipital complex that show varying degrees of invariance to changes in viewpoint and illumination versus changes in the size and retinal position of objects and faces [50].

In human recognition studies, subjects could recognize unfamiliar faces as long as the change in viewpoint was smaller that 30° [118, 119]. Interestingly, image-based simulations of face recognition such as PCA replicate these results. The strong correlation between simulations and human performance suggests that the deleterious effect of rotation on human performance is due to the decline of image similarity that occurs with rotation.

Contrary to the simulation results, however, human performance is still better than chance even after a 90° rotation. This could be due to the abstraction of a 3D model of the face (i.e., like one suggested by structural theories [79, 85] or by the abstraction of invariant features from the faces, e.g., moles, and other surface markings [120], or by both mechanisms [110]).

Even if the theoretical interpretation is still open to discussion, the empirical effect is clear. Unfamiliar faces are more difficult to recognize after a change of pose. Within limits, the larger the change the larger the decrement in performance. In addition, more distinctive faces are less sensitive to this effect; more typical faces are more sensitive to it [87].

Recognition from the Three-Quarter Pose

In early work, several studies suggested that the three-quarter view of faces was "special" because it seemed to be recognized better than other viewpoints [5, 61, 20]. This finding was replicated several times [76, 21] and has been found also for young infants [39]. A neural network simulation of face recognition with an image-based code input to a radial-basis-function network spontaneously yields a $\frac{3}{4}$-view advantage [118]. This suggests that the $\frac{3}{4}$-view advantage can be attributed to the "more informative" nature of the view for recognition. In other words, there may be more unique information in face images taken from the $\frac{3}{4}$ view than in face images taken from other views. This information-based explanation may also explain why the $\frac{3}{4}$ view is favored for western portraits [6] as well as for cartoons [92].

Despite the fact that the $\frac{3}{4}$-view advantage is replicable, the size of this effect is rather small, and there are cases where the effect is not actually found [75]. In recent review of 14 studies testing for the $\frac{3}{4}$-view advantage, it was reported that six of these failed to detect the effect. The authors concluded that the $\frac{3}{4}$-view advantage was not due to the $\frac{3}{4}$ view being more informative per se (i.e., being a "canonical view" or special view [91]). Rather they suggest that the $\frac{3}{4}$ view is closer to most other views (i.e., profiles or full faces) and that it is therefore easier to transfer information to the $\frac{3}{4}$ views (because the rotation angle to this view is smaller on average than to other views).

One counter-argument against this explanation is that the $\frac{3}{4}$-view advantage has been found even when the view of the test stimulus matches the view that was learned [87]. Thus, learning and testing with $\frac{3}{4}$-view faces produces better recognition performance than learning and testing with frontal or full-profile views.

In summary, there is good evidence that both the viewpoint, per se, and the change in viewpoint between learning and test affect human recognition performance.

9.3.2 Illumination

The effects of illumination variation on face and object recognition have been studied extensively and mirror the findings for viewpoint. The effects of illumination on the appearance of an object in an image can be every bit as dramatic as the effects of large changes in viewpoint. Illumination parameters can change the overall magnitude of light intensity reflected back from an object, as well as the pattern of shading and shadows visible in an image [110]. Both shading and shadows may provide cues about the three-dimensional shape of a face. Indeed, varying the direction of illumination can result in larger image differences than varying the identity [2] or the viewpoint [110] of a face.

Although face and object recognition have been shown to be sensitive to illumination variations [15, 17, 55, 110, 114], considerable illumination invariance can be achieved under some novel illumination conditions [82, 16, 17]. In combination, these findings suggest that there are processes that attempt to discount the effects of illumination when recognizing faces, but that they are not perfect. Again, as for viewpoint, the effects of illumination seem to depend on the magnitude of change in illumination conditions between learning and test. The effects of illumination have been tested systematically only for unfamiliar faces. One might assume more robust processes for recognizing familiar faces over changes in illumination.

There is further evidence to suggest that humans prefer illumination that comes from above the face. At its extreme, this preference impacts the recognition of faces quite dramatically. For example, it is notoriously difficult to recognize faces in the photographic negative [43]. Photographic negatives of faces look like faces that are illuminated from below. It is difficult to recognize people, even those we know well, from photographic negatives. These data suggest that human recognition might be based on internal neural representations that are more like images than like structural descriptions of the face. Assuming the latter, recognition of photographic negatives should not be as difficult as it is.

The effects of illumination have been considered also at the level of neural encoding. These results are consistent with the findings from psychophysical studies. For example, although it is believed that illumination information can be extracted by lower visual areas, the results from a recent functional magnetic-resonance imaging study suggest that the sensitivity to the direction of illumination is retained even in higher levels of the visual hierarchy [50]. In that study, the technique of functional magnetic-resonance imaging adaptation was employed to test "invariance". In general, the procedure works by assuming that the neural response adapts to repreated presentations of stimuli that are perceived identically. If an area of cortex is processing faces in a way that is invariant to illumination, then the neural response to a face varying in illumination, presented repeatedly, will continue to adapt. If the cortex response is sensitive to illumination, then adaptation will not occur with multiple presentations. Using this technique, Grill-Spector and colleagues found two subdivisions in the lateral occipital complex, one that showed recovery from adaptation under all transformations (size, position, illumination, and viewpoint) and another that showed recovery only for illumination and viewpoint [50]. This indicates the existence of a complex hierarchy of visual processes that ultimately contribute to human face recognition.

9.4 MOVING FACES

Recognition memory for moving faces is a new and growing area of research [88, 98, 128]. From an ecological perspective, the use of moving faces as stimuli in face-recognition studies is useful for approximating the way in which people typically encounter faces in the real world—as dynamic objects. Moving faces provide the viewer with a wealth of social information [3] and, potentially, with a unique source of identity information.

9.4.1 Social Signals and Motion

We begin with a brief taxonomy of facial movements. At the highest level of the taxonomy, facial movements are either rigid or nonrigid. Rigid motions of the head include nodding, shaking, tilting, and rotating about the vertical axis. It is worth noting that all of these movements change the view of the face available to a stationary observer. It is also important to note that each of these movements can convey a social signal or connotation. Nodding the head can convey agreement, shaking back and forth can convey disagreement, and turns of the head, either toward or away from another person, can be used to initiate or break off a communication.

Nonrigid movements of the face occur during facial speech, facial expressions, and eye-gaze changes. Again, these movements produce highly variable images of the person that can distort many of the identifying "features" of faces, like the relative distances of the features. The difference between the image of a person smiling and one of the person with a surprised expression can be quite strong. The relative position of the eyebrows, for example, with respect to the mouth and the other features, changes radically between these images. Facial speech, expression, and eye-gaze movements can also convey a social message. When a person speaks, they rarely do so with a static or neutral expression. Eye-gaze changes can signal boredom or interest, and facial expression can yield a nearly limitless amount of information about a person's internal state.

Thus, we must bear in mind that facial movements may be challenging to the perceptual system. This is because they alter the nature of the "invariants" available for recognizing a face and because they must be monitored and interpreted constantly for the social information they convey.

9.4.2 Recognizing Moving Faces

How do facial movements affect face-recognition accuracy? To begin to answer this question, we must first ask, "in what ways might motion help face recognition?" and "in what ways might motion make face recognition more difficult?" For the former, there are two theories about how motion might improve recognition [88, 98]. The first theory posits that motion might provide facial identity signatures in the idiosyncratic patterns of dynamic information they undergo. If repeated regularly (e.g., as with characteristic gestures or facial expressions), these idiosyncratic patterns of movement might provide a reliable cue for the identity of the face. A second theory posits that motion could help face recognition by providing additional structure-from-motion information about the face. This can enhance the quality of the perceptual representation of a face [88, 98].

For the dynamic-identity-signature theory, the role of characteristic motions for learning to identify faces has been studied recently using animated synthetic three-dimensional head models [56, 59]. These studies have focused on the "learnability" of a dynamic signature when it is the most, or only, reliable cue to identity. For example, Hill and Johnston projected facial animations generated by human actors onto a computer-generated average head [56] (see Figure 9.3 for an example of their stimulus). Their participants learned to discriminate among four individuals based solely on the facial-motion information. It is worth noting that nonrigid motion was less useful than rigid motion in this identity learning.

In another similar study, Knappmeyer and her colleagues trained participants to discriminate two synthetic faces that were animated with different characteristic facial motions [59]. When later viewing morphs between the two head models, the subjects' identity judgments about the intermediate morphed heads were biased by the animated-motion information they learned to associate with the faces originally. Both studies support the notion that inherently dynamic information about face

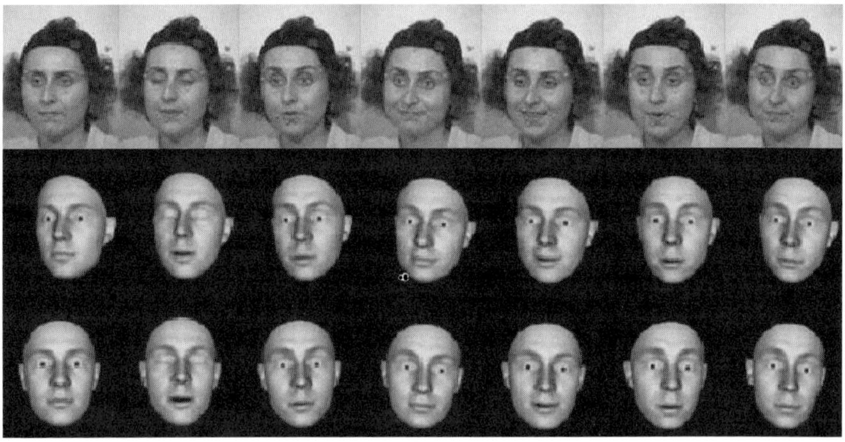

FIGURE 9.3: Illustration of stimulus creation from Hill and Johnston's [56] study. The motions of a human actor are projected onto the synthetic head. Subjects learn to identify the head by the motions.

movements can support recognition and can form a part of our representation of the identity of an individual.

More relevant for everyday face recognition, dynamic identity signatures are likely to be most helpful when faces are familiar. This is because it may take time, for any given individual, to learn the difference between characteristic movements and movements that are generated randomly. Indeed, the current literature suggests that face familiarity mediates the usefulness of facial motion as a recognition cue. The beneficial effects of facial motion are more robust and easier to demonstrate in recognition tasks with familiar/famous faces than with unfamiliar faces. For example, participants can recognize the faces of well-known politicians and celebrities more accurately from videotaped images than from static images. This finding is especially salient when the faces are presented in suboptimal viewing formats (e.g., blurred, inverted, or pixilated displays; [60, 64, 66, 67]). Thus it seems that motion becomes more important as a cue to identity when the viewing conditions are suboptimal. This probably occurs because the static features are less-reliable cues to identity in these conditions, and so subjects are more likely to require additional information available in the identity signature to successfully recognize the person.

The second hypothesis about how motion might benefit face recognition posits that structure-from-motion processes can contribute to the quality of the face representation. This should apply most clearly in studying how motion affects our ability to learn new faces (i.e., to create new face representations). In the case of newly learned or unfamiliar faces, the data are not clear as to whether motion improves face recognition. Some studies report a motion benefit [65, 96, 98], whereas other studies find no benefit of motion [23, 24, 35, 54]. A closer inspection of these results suggest that the benefits of facial motion for unfamiliar face recognition tasks may be tied to the specific parameters of the learning and test conditions used across these different studies. Differences in the type of recognition tasks implemented and variations in the kinds of stimuli employed likely account in large part for the disparity in the results.

In a recent study, we attempted to control some of the confounding factors in previous experiments in order to assess the role of motion in face recognition [107]. In particular, we wanted to control for the extra views we see of a face when it is in motion. As noted, motion almost always provides subjects with extra views of a face in addition to whatever benefit may come from the motion, per se. In a single experiment, we compared recognition performance for subjects who learned faces in four conditions. One set of subjects learned each face from a static frontal image. A second set of subjects learned each face from nine images taken from different viewpoints and presented in an ordered sequence. A third set of subjects learned from the same nine images, but presented in random order to eliminate any implied motion signal. Finally, a fouth set of subjects learned faces from a video clip of the face rotating systematically through the nine viewpoints.

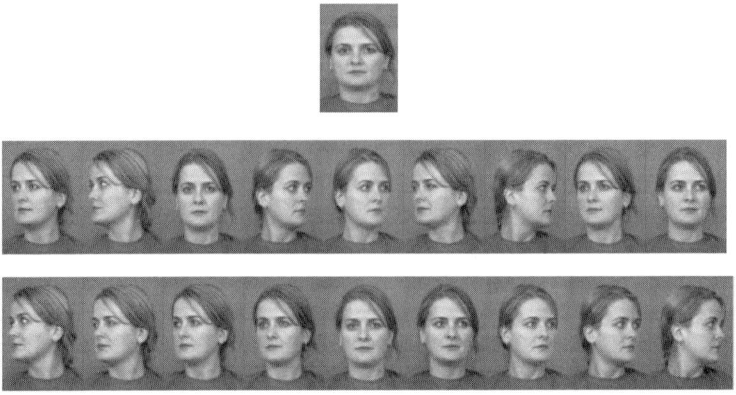

FIGURE 9.4: Stimuli for a study comparing static and dynamic learning conditions controlling for the extra views seen in the dynamic condition [107]. Subjects learned a single static image (row 1), or a random sequence of nine views (row 2), or an ordered sequence of nine views (row 3), or a video clip of the head rotating systematically through nine views.

This video looked like a person searching through a crowded room for someone. An example of the stimuli appears in Figure 9.4. Subjects were tested using the same kind of stimulus they learned.

The results of the experiment revealed absolutely no difference in recognition performance as a function of the learning condition (see Figure 9.5). This indicates that motion does not seem to benefit the learning of unfamiliar faces. To be sure that this effect was not based on the test conditions, which were matched between the learn conditions in the first experiment, we repeated the study. This time, however, we tested with a single frontal image in all conditions. Again, we found the same result. Motion provided no additional benefit for recognition.

9.5 MOTION AND FAMILIARITY

The disparity of findings concerning the role of motion in face recognition hinges primarily around the familiarity of faces. Most studies have employed the faces of celebrities as "familiar" faces or the faces of people already well known to the subjects (e.g., professors at their university). The major shortcoming of this method is that the researcher has no control over the learning conditions any given subject undergoes along the path to becoming familiar with a face. How many previous exposures have they had to the celebrity? Which views have they seen? Have the views occurred over years (e.g., Harrison Ford) or over months (e.g., some newly popular celebrity)?

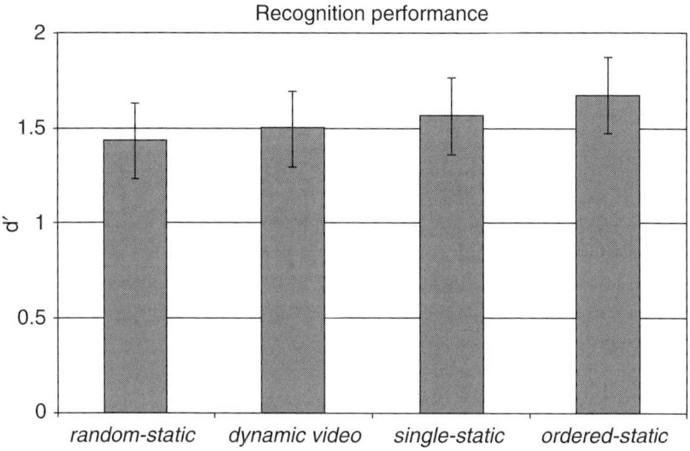

FIGURE 9.5: The results of a study comparing static and dynamic learning conditions, controlling for the extra views seen in the dynamic condition [107]. No significant differences were found among the learning conditions, indicating that motion does not provide any benefit for face learning.

One recent experiment employs a technique by which subjects gain familiarity with a face under controlled conditions during the course of the experiment [99]. In that study, subjects learned people from a variable number of exposures of the same stimulus, either a close-up of the face (moving or static) or a distant view of the person walking. They were tested with the "other" stimulus. Thus people learned from facial close-ups were tested with distanced videos and vice versa. The findings indicated, first, that pure repetition of the same stimulus was remarkably effective in improving recognition over these substantial changes in viewing conditions.

Second, recognition was more accurate when participants viewed faces in motion *both* at learning and test than when participants viewed moving images at learning on at test only. Again, this held even when the two stimuli were presented in different viewing formats (i.e., high-quality facial images versus whole-body surveillance-like videos). This finding, in light of other studies that have likewise reported inconsistent recognition benefits with moving faces (e.g., [24]), may suggest that unfamiliar-face recognition tasks benefit from facial motion primarily in learning–test conditions that involve "motion-to-motion" transfers.

Tentative converging support for this motion-match hypothesis comes from a recently advanced neural theory of face processing proposed by Haxby and colleagues [53]. In this theory, the variant and invariant aspects of faces are processed independently in the brain ([53]). In Haxby et al.'s model, the moving aspects of faces are processed in the superior temporal sulcus region (STS) of the dorsal visual stream, whereas the static aspects of faces (i.e., facial features) are processed in the fusiform face area of the ventral visual stream (FFA, [58]). The proposed separation of face processing between the two streams leaves open the possibility that recognition tasks that permit processing to remain in the same stream are more likely to support successful recognition performance. For now, however, it is clear that more research is needed to generate a coherent picture of how facial motion affects recognition.

9.6 FAMILIARITY AND EXPERIENCE

We close this chapter by recalling our beginning claim that humans are the best available face-recognition systems – superior to many, if not all, algorithms. It should be clear by now that this claim must be qualified in two ways. First, on the task of learning and remembering faces from a limited number of exposures, humans encounter the same kinds of difficulty that algorithms do, and likewise reflect these difficulties with impaired recognition performance. Second, although we generally assume that experience with faces we know well allows us to recognize them in suboptimal viewing conditions (poor illumination, short glances, bad viewing angles), this problem has been little studied by psychologists. The reason for this is quite practical. In most cases, our performance in recognizing faces we know well is generally so good that few variables are likely to affect performance substantially.

Ultimately, we think that the issues surrounding how we *become* familiar with faces, and how the internal neural representation of faces changes with experience, are much neglected topics for psychologists studying human memory for faces. They are, however, the most valuable open questions for psychologists to study when they are interested in providing data for computational modelers of face recognition. Being able to recognize a friend, at a distance, in a dimly lit train station, from an odd angle, is currently a unique accomplishment of the human visual system. Certainly, algorithms can achieve this if enough is known about how we do it.

ACKNOWLEDGMENTS

This work was supported by a contract from TSWG to A.J. O'Toole and H. Abdi.

REFERENCES

[1] H. Abdi, D. Valentine, B. Edelman, and A. J. O'Toole. More about the difference between men and women: evidence from linear neural networks and the principal-component approach. *Perception* **24**:539–562, 1995.

[2] Y. Adini, Y. Moses, and S. Ullman. Face recognition: the problem of compensating for changes in illumination direction. Report No. CS93-21. The Weizmann Institute of Science, 1995.

[3] T. Allison, A. Puce, and G. McCarthy. Social perception from visual cues: role of the STS region. *Trends in Cognitive Sciences* **4**:267–278, 2000.

[4] T. J. Andrews and M. P. Ewbank. Distinct representations for facial identity and changeable aspects of faces in the human temporal lobe. *Neuroimage*, **23**(3), 905–913, 2004.

[5] A. Baddeley and M. Woodhead. Techniques for improving eyewitness identification skills. Paper presented at the *SSRC Law and Psychology Conference*, Trinity College, Oxford, 1981.

[6] A. Baddeley and M. Woodhead. Improving face recognition ability. In: S. Llyood-Bostock and B. Clifford (Eds.) *Evaluating Witness Evidence*. Chichester: Wiley, 1983.

[7] J. C. Bartlett and J. Searcy. Inversion and configuration of faces. *Cognitive Psychology* **25**:281–316, 1993.

[8] J. C. Bartlett and A. Fulton. Familiarity and recognition of faces in old age. *Memory and Cognition* **19**:229–238, 1991.

[9] P. J. Benson and D. I. Perrett. Perception and recognition of photographic quality caricatures: implications for the recognition of natural images. *European Journal of Cognitive Psychology* **3**:105–135, 1993.

[10] I. Biederman and M. Bar. One-shift viewpoint invariance in matching novel objects. *Vision Research* **39**:2885–2889, 1998.

[11] I. Biederman and P. Gerhardstein. Recognizing depth-rotated objects: evidence and conditions for three-dimensional viewpoint invariance. *Journal of Experimental Psychology: Human Perception and Performance* **19**:1162–1183, 1993.

[12] V. Blanz and T. Vetter. A morphable model for the synthesis of 3D faces. In: *SIGGRAPH'99 Proceedings*, ACM. Computer Society Press, 187–194, 1999.

[13] V. Blanz, A. J. O'Toole, T. Vetter, and H. A. Wild. On the other side of the mean: the perception of dissimilarity in human faces. *Perception* **29**:885–891, 2000.

[14] G. H. Bower and M. B. Karlin. Depth of processing pictures of faces and recognition memory. *Journal of Experimental Psychology* **103**:751–757, 1974.

[15] W. L. Braje, D. J. Kersten, M. J. Tarr, and N. F. Trojc. Illumination effects in face recognition. *Psychobiology* **26**:371–380, 1999.

[16] W. L. Braje, G. E. Legge, and D. Kersten. Invariant recognition of natural objects in the presences of shadows. *Perception* **29**:383–398, 2000.

[17] W. L. Braje. Illumination encoding in face recognition: effect of position shift. *Journal of Vision* **3**:161–170, 2003.

[18] S. E. Brennan. The caricature generator. *Leonardo* **18**:170–178, 1985.

[19] J. C. Brigham and P. Barkowitz. Do "They all look alike?" The effects of race, sex, experience and attitudes on the ability to recognize faces. *Journal of Applied Social Psychology* **8**:306–318, 1978.

[20] V. Bruce. Changing faces: visual and non-visual coding processes in face recognition. *British Journal of Psychology* **73**:105–116, 1982.

[21] V. Bruce, T. Valentine, A. Baddeley. The basis of the 3/4 view advantage in face recognition. *Applied Cognitive Psychology* **1**:109–120, 1987.

[22] R. Brunelli and T. Poggio. Caricatural effects in automated face perception. *Biological Cybernetics* **69**:235–241, 1993.

[23] V. Bruce, Z. Henderson, K. Greenwood, P.J.B. Hancock, A.M. Burton and P. Miller. Verification of face identities from images captured on video. *Journal of Experimental Psychology—Applied*, **5**:339–360, 1999.

[24] V. Bruce, Z. Henderson, C. Newman and A.M. Burton. Matching identities of familiar and unfamiliar faces caught on CCTV images. *Journal of Experimental Psychology—Applied* **7**:207–218, 2001.

[25] V. Bruce and A. W. Young. Understanding face recognition. *British Journal of Psychology* **77**(3):305–327, 1986.

[26] D. M. Burt and D. I. Perrett. Perception of age in adult Caucasian male faces: Computer graphic manipulation of shape and colour information. *Proceedings of the Royal Society of London* **259**:137–143, 1995.

[27] A. M. Burton, V. Bruce, and N. Dench. What's the difference between men and women: evidence from facial measurement. *Perception* **22**(2):153–176, 1993.

[28] A. M. Burton, V. Bruce, and P. J. B. Hancock. From pixels to people: a model of familiar face recognition. *Cognitive Science* **23**:1–31, 1999.

[29] A. M. Burton, S. Wilson, M. Cowan, and V. Bruce. Face recognition in poor-quality video. *Psychological Science* **10**:243–248, 1999.

[30] A. W. Carroo. Other-race face recognition: A comparison of Black American and African subjects. *Perceptual and Motor Skills* **62**:135–138, 1986.

[31] J. E. Chance, A. L. Turner, and A. G. Goldstein. Development of differential recognition for own- and other-race faces. *Journal of Psychology* **112**:29–37, 1982.

[32] Y. Cheng, A. J. O'Toole, and H. Abdi. Sex classification of adults' and children's faces: computational investigations of subcategorical feature encoding. *Cognitive Science* **25**:819–838., 2001.

[33] P. Chiroro and T. Valentine. An investigation of the contact hypothesis of the own-race bias in face recognition. *Quarterly Journal of Experimental Psychology, A, Human Experimental Psychology* A**48**:879–894, 1995.

[34] I. Craw and P. Cameron. Parametrizing images for recognition and reconstruction. In: P. Mowforth, editor, *Proceedings of the British Machine Vision Conference*. Springer, London, 1991.

[35] F. Christie and V. Bruce. The role of dynamic information in the recognition of unfamiliar faces. *Memory & Cognition* **26**:780–790, 1998.

[36] J. F. Cross, J. Cross, and J. Daly. Sex, race, age, and beauty as factors in recognition of faces. *Perception & Psychophysics* **10**:393–396, 1971.

[37] S. Edelman and H. H. Bülthoff. Orientation dependence in the recognition of familiar and novel views of three-dimensional objects. *Vision Research* **32**(12):2385–2400, 1992.

[38] P. J. Ekman and W. V. Friesen. *The Facial Action Coding System: A Technique for the Measurement of Facial Movement.* Consulting Psychology Press, San Francisco, 1978.

[39] J. Fagan. The origin of facial pattern recognition. In: M. Bornstein and W. Keesen (Eds.) *Psychological Development from Infancy: Image to Intention.* 1979.

[40] S. Feinman and D. R. Entwisle. Children's ability to recognize other children's faces. *Child Development* **47**(2):506–510, 1976.

[41] M. Fleming and G. W. Cottrell. Categorization of faces using unsupervised feature extraction. *Proceedings of IJCNN-90*, Vol 2, Ann Arbor, MI:IEEE Neural Networks Council, 65–70, 1990.
[42] D. S. Furl, P. J. Phillips, and A. J. O'Toole. Face recognition algorithms as models of the other-race effect. *Cognitive Science* **96**:1–19, 2002.
[43] R. E. Galper and J. Hochberg. Recognition memory for photographs of faces. *American Journal of Psychology* **84**:351–354, 1971.
[44] T. Ganel and Y. Goshen-Gottstein. Perceptual integrality of sex and identity of faces. *Journal of Experimental Psychology: Human Perception and Performance* **28**:854–867, 2002.
[45] I. Gauthier, M. J. Tarr, A. W. Anderson, P. Skudlarski, and J. C. Gore. Activation of the middle fusiform face area increases with expertise recognizing novel objects. *Nature Neuroscience* **2**:568–573, 1999.
[46] M. S. Gazzaniga (Ed.), 1995. *The Cognitive Neurosciences.* MIT Press, Cambridge.
[47] G. Givens, J. R. Beveridge, B. A. Draper, P. Grother, and P/J. Phillips. How features of the human face affect recognition: A statistical comparison of three face recognition algorithms. *IEEE Computer Society Conference on Computer Vision and Pattern Recognition*, 2004.
[48] M. Gray, D. T. Lawrence, B. A. Golomb, T. J. Sejnowski. A perceptron reveals the face of sex. *Neural Computation* **7**:1160–1164, 1995.
[49] D. M. Green and J. A. Swets. *Signal Detection Theory and Psychophysics.* Wiley, New York, 1966.
[50] K. Grill-Spector, T. Kushnir, S. Edelman, G. Avidan, Y. Itzchak, and R. Malach. Differential processing of objects under various viewing conditions in the human lateral occipital complex. *Neuron* **24**:187–203, 1999.
[51] P. J. B. Hancock, V. Bruce, and A. M. Burton. Recognition of unfamiliar faces. *Trends in Cognitive Sciences* **4**(9):263–266, 1991.
[52] J. V. Haxby, M. I. Gobbini, M. L. Furey, A. Ishai, J. L. Shouten and J. L. Pietrini. Distributed and overlapping representations of faces and objects in ventral temporal cortex. *Science* **293**:2425–2430, 2001.
[53] J. V. Haxby, E. A. Hoffman, and M. I. Gobbini. The distributed human neural system for face perception. *Trends in Cognitive Sciences* **20**(6):223–233, 2000.
[54] Z. Henderson, V. Bruce, and A. M. Burton. Matching the faces of robbers captured on video. *Applied Cognitive Psychology* **15**:445–464, 2001.
[55] H. Hill and V. Bruce. Effects of lighting on the perception of facial surface. *Journal of Experimental Psychology: Human Perception and Performance* **4**(9):263–266, 1991.
[56] H. Hill and A. Johnston. Categorizing sex and identity from the biological motion of faces. *Current Biology* **11**:880–885, 2001.
[57] H. Hill, P. Schyns, and S. Akamatsu. Information and viewpoint dependence in face recognition. *Cognition* **62**:201–202, 1997.
[58] N. Kanwisher, J. McDermott, and M. Chun. The fusiform face area: a module in human extrastriate cortex specialized for face perception. *Journal of Neuroscience* **17**:4302–4311, 1997.
[59] B. Knappmeyer, I. M. Thornton, and H.H. Bulthoff. The use of facial motion and facial form during the processing of identity. *Vision Research* **43**:1921–1936, 2003.

[60] B. Knight and A. Johnston. The role of movement in face recognition. *Visual Cognition*, **4**:265–273, 1997.

[61] F. L. Krouse. Effects of pose, pose change, and delay on face recognition performance. *Journal of Applied Psychology* **66**:201–222, 1981.

[62] P. K. Kuhl, K. A. Williams, and F. Lacerdo. Linguistic experience alters phonetic perception in infants by 6 months of age. *Science* **255**:606–608, 1992.

[63] J. Kurucz and J. Feldmar. Prosopo-affective agnosia as a symptom of cerebral organic brain disease. *Journal of the American Geriatrics Society* **27**:91–95, 1979.

[64] K. Lander and V. Bruce. Recognizing famous faces: exploring the benefits of facial motion. *Ecological Psychology* **12**:259–272, 2000.

[65] K. Lander and V. Bruce. The role of motion in learning new faces. *Visual Cognition* **10**:897–912, 2003.

[66] K. Lander, V. Bruce, and H. Hill. Evaluating the effectiveness of pixelation and blurring on masking the identity of familiar faces. *Applied Cognitive Psychology* **15**:101–116, 2001.

[67] K. Lander, K. Christie, and V. Bruce. The role of movement in the recognition of famous faces. *Memory and Cognition* **27**:974–985, 1999.

[68] J. H. Langlois, L. A. Roggman, and L. Mussleman. What is average and what is not average about attractive faces? *Psychological Science* **5**:214–220, 1994.

[69] A. Lanitis, C. J. Taylor, and T. F. Cootes. Automatic interpretation and coding of face imaging using flexible models. *IEEE Transactions on Pattern Analysis and Machine Intelligence* **19**:743, 1997.

[70] P. J. Lavarkas, J. R. Buri, and M. S. Mayzner. A perspective on the recognition of other-race faces. *Perception and Psychophysics* **20**:475–481, 1976.

[71] D. Leopold, A. J. O'Toole, T. Vetter, and V. Blanz. Prototype-referenced shape encoding revealed by high-level aftereffects. *Nature Neuroscience* **4**:89–94, 2001.

[72] D. Levin. Race as a visual feature: using visual search and perceptual discrimination tasks to understand face categories and the and the cross-race recognition deficit. *Journal of Experimental Psychology—General* **129**:559–574, 2000.

[73] L. Light, F. Kayra-Stuart, and S. Hollander. Recognition memory for typical and unusual faces. *Journal of Experimental Psychology—Human Learning and Memory* **5**:212–228, 1979.

[74] D. S. Lindsay, P. C. Jack, and M. A. Christian. Other-race face perception. *Journal of Applied Psychology* **76**:587–589, 1991.

[75] C. H. Liu, and A. Chaudhuri. Reassessing the 3/4 view effect in face recognition. *Cognition* **83**:31–48, 2002.

[76] R. Loggie, A. Baddeley, and M. Woodhead. Face recognition, pose and ecological validity. *Applied Cognitive Psychology* **1**:53–69, 1987.

[77] N. K. Logothetis, J. Pauls, H. H. Bülthoff, and T. Poggio. Shape representation in the inferior temporal cortex of monkeys. *Current Biology* **5**:552–563, 1991.

[78] R. S. Malpass and J. Kravitz. Recognition for faces of own and other race faces. *Journal of Personality and Social Psychology* **13**:330–334, 1969.

[79] D. Marr. *Vision*. Freeman, San Francisco, 1982.

[80] R. Mauro and M. Kubovy. Caricature and face recognition. *Memory & Cognition* **20**:433–440, 1992.

[81] W. H. Merigan. P and M pathway specialization in the macaque. In: A. Valberg and B. B. Lee, editors, *From Pigments to Perception*. Plenum, New York. Pages 117–125, 1991.

[82] Y. Moses, S. Edelman, and S. Ullman. Generalization to novel images in upright and inverted faces. *Perception* **25**(4):443–461, 1996.

[83] W. Ng and R. C. L. Lindsay. Cross-race facial recognition: Failure of the contact hypothesis. *Journal of Cross-Cultural Psychology* **25**:217–232, 1994.

[84] A. J. O'Toole, H., Abdi, K. A. Deffenbacher, and D. Valentin. Low dimensional representation of faces in high dimensions of the space. *Journal of the Optical Society of America* A**10**, 405–410, 1993.

[85] A. J. O'Toole, H. Bülthoff, N. Troje, and T. Vetter. Face recognition across large viewpoint changes. In: *Proceedings if the International Workshop on Automatic Face and Gesture Recognition*. Zurich, 1995.

[86] A. J. O'Toole, K. A. Deffenbacher, and D. Valentine. Structural aspects of face recognition and the other-race. *Memory and Cognition* **22**:208–224, 1994.

[87] A. J. O'Toole, S. E. Edelman, and H. H. Bülthoff. Stimulus-specific effects in face recognition over changes in viewpoint. *Vision Research* **38**:2351–2363, 1998.

[88] A. J. O'Toole, D. Roark, and H. Abdi. Recognition of moving faces: a psychological and neural perspective. *Trends in Cognitive Sciences* **6**:261–266, 2002.

[89] A. J. O'Toole, T. Vetter, N. F. Troje, and H. H. Buelthoff. Sex classification is better with three-dimensional head structure than with image intensity information. *Perception* **26**:75–84, 1997.

[90] A. J. O'Toole, T. Vetter, H. Volz, and E. M. Salter. Three-dimensional caricatures of human heads: distinctiveness and the perception of facial age. *Perception* **26**:719–732, 1997.

[91] S. Palmer, E. Rosch, and P. Chase. Canonical perspective and the perception of objects. In: J. Long, and A.D. Baddeley (Eds.) *Attention and Performance IX*. Hillsdale: Erlbaum, 1981.

[92] D. N. Perkins. A definition of caricature and recognition. *Studies in the Anthropology of Visual Communication* **2**:1–24. 1975.

[93] D. Perrett, J. Hietanen, M. Oram, and P. Benson. Organization and function of cells responsive to faces in temporal cortex. *Philosophical Transactions of the Royal Society of London B—Biolological Science*, **335**:23–30, 1992.

[94] P. J. Phillips, H. J. Moon, S. A. Rivzi and P. J. Rauss. The FERET Evaluation methodology for face-recognition algorithms. *Transactions on Pattern Analysis and Machine Intelligence* **22**(10):1090–1104, 2003.

[95] P. J. Phillips, P. Grother, R. J. Michaels, D. M. Blackburn, E. Tabassi, and M. Bone. Face recognition vendor test 2002. *NISTIR*, 2003.

[96] G. E. Pike, R. I. Kemp, N. A. Towell, and K. C. Phillips. Recognizing moving faces: The relative contribution of motion and perspective view information. *Visual Cognition* **4**:409–437, 1997.

[97] T. Poggio and S. Edelman. A network that learns to recognize 3D objects. *Nature* **343**:263–266, 1991.

[98] D. A. Roark, S. E. Barrett, M. A. Spence, H. Abdi, and A. J. O'Toole. Psychological and neural perspectives on the role of motion in face recognition. *Behavioral and Cognitive Neuroscience Reviews* **2**:15–46, 2003.

[99] D. A. Roark, S. E. Barrett, H. Abdi, and A. J. O'Toole. Learning the moves: the effect of facial motion and familiarity on recognition across large changes in viewing format. *Perception* **36**.

[100] D. A. Roark, H. Abdi, and A. J. O'Toole. Human recognition of familiar and unfamiliar people in naturalistic videos. *Proceedings of the IEEE International Workshop on Analysis and Modeling of Faces and Gestures*, 36–43, 2003.

[101] E. Rosch and C. Mervis. Family resemblances: Studies in the internal structure of categories. *Cognitive Psychology* **7**:573–605, 1975.

[102] G. Rhodes. *Superportraits: Caricatures and Recognition*. Psychology Press, Hove, UK, 1997.

[103] M. Riesenhuber and T. Poggio. Models of object recognition. *Nature Neuroscience Supplement* **3**:1199–1204, 2000.

[104] J. W. Shepherd, J. B. Deregowski and H. D. Ellis. A cross-cultural study of recognition memory for faces. *International Journal of Psychology* **9**:205–212, 1974.

[105] J. Shepherd. Social factors in face recognition. In: G. Davies, H. Ellis, and J. Shepherd Eds., *Perceiving and Remembering Faces*. Academic Press, London, 55–78, 1981.

[106] M. Spiridon and N. Kanwisher. How distributed is visual category information in human occipito-temporal cortex? An FMRI study. *Neuron* **35**:1157, 1991.

[107] S. Snow, G. Lannen, A. J. O'Toole, and H. Abdi. Memory for moving faces: Effects of rigid and non-rigid motion. *Journal of Vision*, **2**:7, Abstract 600, 2002.

[108] J. W. Tanaka and M. J. Farah. Parts and wholes in face recognition. *Quarterly Journal of Psychology* A**46**(2):225–245, 1993.

[109] K. Tanaka. Neuronal mechanisms of object recognition. *Science* **262**:685–688, 1991.

[110] M. Tarr and H. Bülthoff. Image-based object recognition in man, monkey and machine. *Cognition* **67**:1–20, 1998.

[111] M. J. Tarr, D. Kersten, and H. H. Bulthoff. Why the visual recognition system might encode the effects of illumination. *Vision Research* **38**:2259–2275, 1998.

[112] P. Thompson. Margaret Thatcher: a new illusion. *Perception* **9**:483–484, 1980.

[113] N. F. Troje and H. H. Bülthoff. Face recognition under varying pose: the role of texture and shape. *Vision Research* **36**:1761–1771, 1996.

[114] N. F. Troje and H. H. Bulthoff. How is bilateral symmetry of human faces used for recognition of novel views? *Vision Research* **38**:79–89, 1998.

[115] M. Turk and A. Pentland. Eigenfaces for recognition. *Journal of Cognitive Neuroscience* **3**:71–86, 1991.

[116] S. Ullman. *High-Level Vision*. MIT Press, Cambridge, MA, 1996.

[117] D. Valentin. How come when you turn your head I still know who you are: evidence from computational simulations and human behavior. Ph.D. dissertation, The University of Texas at Dallas. 1996.

[118] D. Valentin, H. Abdi, and B. Edelman. What represents a face: A computational approach for the integration of physiological and psychological data. *Perception* **26**:1271–1288, 1997.

[119] D. Valentin, H. Abdi, and B. Edelman. From rotation to disfiguration: Testing a dual-strategy model for recognition of faces across view angles. *Perception* **28**:817–824 1999.

[120] D. Valentin, H. Abdi, B. Edelman, and M. Posamentier. 2D or not 2D? that is the question: What can we learn from computational models operating on 2D representations of faces? In: M. Wenger and J. Townsend (Eds.), *Computational, Geometric, and Process Perspectives on Facial Cognition*. Mahwah (NJ): Erlbaum. 2001.

[121] T. Valentine. A unified account of the effects of distinctiveness, inversion, and race in face recognition. *Quarterly Journal of Experimental Psychology* A**43**:161–204, 1991.

[122] T. Valentine and V. Bruce. The effects of distinctiveness in recognising and classifying faces. *Perception* **15**:525–536, 1986.

[123] J. Vokey and D. Read. Familiarity, memorability, and the effect of typicality on the recognition of faces. *Memory and Cognition* **22**:208–224, 1992.

[124] J. F. Werker, J. H. Gilbert, K. Humphrey, and R. C. Tees. Developmental aspects of cross-language speech perception. *Child Development* **52**:349–355, 1981.

[125] H. A. Wild, S. E. Barrett, M. J. Spence, A. J. O'Toole, Y. Cheng, and J. Brooke. Recognition and categorization of adults' and children's faces: examining performance in the absence of sex-stereotyped cues. *Journal of Experimental Child Psychology* **77**:269–291, 2000.

[126] R. K. Yin. Looking at upside-down faces. *Journal of Experimental Psychology* **81**:141–145, 1969.

[127] A. W. Young, D. Hellawell, and D. C. Hay. Configurational information in face perception. *Perception* **16**:747–759, 1987.

[128] W. Zhao, R. Chellappa, P.J. Phillips and A. Rosefeld. Face recognition: a literature survey. *ACM Computing Surveys* **35**:399–459, 2003.

CHAPTER 10

SPATIAL DISTRIBUTION OF FACE AND OBJECT REPRESENTATIONS IN THE HUMAN BRAIN

10.1 THE VENTRAL OBJECT-VISION PATHWAY

Human and nonhuman visual cortices have a macroscopic organization into two processing pathways: a dorsal spatial-vision pathway for perception of spatial location, motion, and the guidance of movements, and a ventral object-vision pathway for the perception of color and form and for face and object recognition. The human ventral object-vision pathway was first mapped out in early functional imaging studies that showed that the perception of faces, color, and shape evoked neural activity in regions of the inferior occipital, fusiform, and lingual gyri [52, 69, 16, 17, 65]

Subsequent imaging studies showed that different regions in inferior occipital and ventral temporal cortex were selective for objects and faces. Malach et al. (1995) found a region in lateral inferior occipital cortex that responded more to intact, meaningful objects than to nonsense images. Moreover, object-responsive regions in inferior occipital and ventral temporal cortex showed differential responses while viewing different categories of objects and faces [62, 45, 55, 21, 1, 35, 42, 13, 14, 19]. In particular, regions were discovered that respond maximally during face perception. The fusiform face area (FFA: [45, 55], and during perception of scenes, interior spaces, and buildings, the parahippocampal place area (PPA: [21, 1]).

These studies that found that subregions of object-responsive cortex respond more to one category than to others revealed an organizational structure that was not discovered from invasive studies in nonhuman primates. Inspired by these fMRI (functional magnetic resonance imaging) findings in humans, fMRI studies in awake, behaving monkeys have revealed that a similar segregation into category-selective regions is also found in nonhuman primates [51, 66, 61]. The discrepancy between earlier, single-unit recording studies of face and object perception in monkeys and these fMRI studies may be due to the difference between the types of neural activity that are measured by these techniques [51] and differences between the spatial resolution and spatial coverage of these techniques.

The existence of regions with category-related responses in the extrastriate cortices of the ventral object-vision pathway has led to a prolonged debate over the functional significance of these regions. One view holds that part of the object-vision pathway is organized into "modules" for specific categories – faces, places, and body parts – that have dedicated processors because of their biological significance and that the rest of object-responsive cortex is a more general-purpose object perception system [45, 21, 19]. A second view holds that the object-vision pathway is organized by perceptual processes rather than by object categories, and that processes that subserve expert visual perception are located in the fusiform region that responds more to faces. Because nearly everyone is an expert in face perception, this region shows stronger responses to faces, but people who are experts for other categories, such as birds or cars, also show stronger responses in the FFA when viewing the objects of their expertise [25]. A third view attributes at least part of the category-related patterns of responsiveness to a coarse retinotopy in object-responsive cortex [50, 33, 54]. The FFA is in cortex that is more responsive to foveal stimuli, whereas the PPA is in cortex that is more responsive to peripheral stimuli. Thus, the location of some category-related regions may be related to eccentricity biases insofar as the representations of categories that rely more on finer details, such as faces and letters, are stronger in cortex with a more foveal bias whereas representations of categories that are less likely to be foveated, such as buildings and landscapes, are stronger in cortex with a more peripheral bias.

We have proposed a different hypothesis for the functional organization of object-responsive cortex, which posits that the representation of a viewed face or object is not restricted to neural activity in cortex that responds maximally to that face or object. Thus, we have proposed that the representations of faces and different object categories in face- and object-responsive cortices are distributed and overlapping. Our hypothesis is not necessarily inconsistent with other principles of organization, particularly the roles of expertise and retinotopy, but that allows more direct investigation of *how* information about the visual appearance of faces and objects is encoded by neural activity in these regions [38, 39]. Our hypothesis stems from our demonstration that information about the face or object being viewed is carried not only by activity in regions that respond maximally to that face or object, but is also by activity in regions that respond more strongly to categories other than that of the currently viewed stimulus (Figure 10.1). With this demonstration we introduced a new method for fMRI data analysis: multivoxel pattern analysis (MVPA). With MVPA, we showed that information about the face or object being viewed is carried by distributed patterns of response in which both strong and weak levels of activity play a role (Figure 10.2). The patterns that are detected by MVPA are related more directly to the distributed population responses that encode visual information [44] and, thus, provide a means to investigate how the neural codes for faces and objects are organized, not just where they may exist.

10.2 LOCALLY DISTRIBUTED REPRESENTATIONS OF FACES AND OBJECTS IN VENTRAL TEMPORAL CORTEX

Although models of the functional architecture of ventral temporal cortex that are based on regional category preferences have provided a useful schema, they have

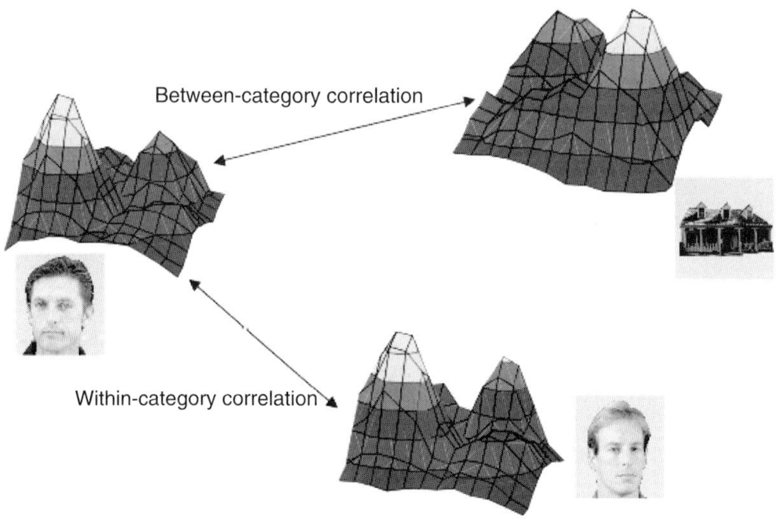

FIGURE 10.1: The schematic colored landscapes illustrate how patterns of response for different categories can be distinct and reliable. According to our hypothesis, strong and weak responses to a specific category both play a central role in the identification of that category. These pictures are only for illustration and are not based on real data. (See also color plate section)

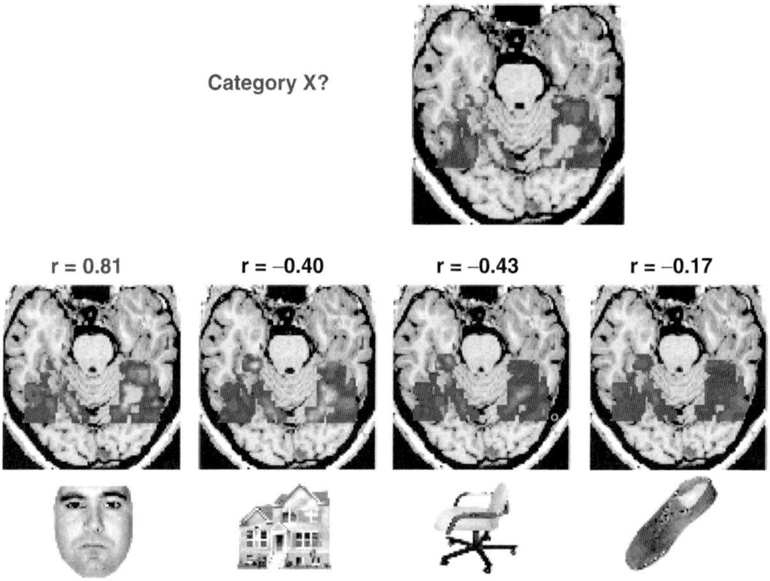

FIGURE 10.2: Example of patterns of response in the ventral temporal cortex based on data from one subject. The category that the subject was viewing can be identified based on the pattern of response. The data were split in two halves, and one half of the data were used to predict what the participant was looking at when the second half of the data were obtained. Overall accuracy of identifying the category being viewed in pairwise comparisons was 96%. The correct identification was based on a higher correlation within category than between category. (See also color plate section).

limited explanatory power. First, these models imply that submaximal responses are discarded. If a neural response represents the viewed face or object with a rate code, low response rates indicate that the viewed stimulus is simply something else, not the preferred face or object. Second, an architecture based on regional category preferences has insufficient capacity to represent all conceivable categories of faces and objects. Kanwisher and colleagues (Downing et al. 2001a; [46]) have suggested that the number of regions dedicated to a specific category is limited to a small number of categories that have special biological significance, namely faces, places, and body parts. Third, regional models provide no account for how these special categories are represented, only where those representations reside in cortex. Moreover, regional models with a small number of category-specific modules provides no account for how other categories may be represented, suggesting only that a system for general object perception exists in the remaining ventral temporal cortex.

Although we also have consistently observed regions of ventral temporal cortex that respond maximally to faces, houses and other object categories, such as tools and chairs [35, 42, 13, 14], we also observed that the responses to nonpreferred categories in these regions were significant and varied by category. We focused on these nonmaximal responses to faces and objects to address a fundamental problem for representation in the ventral object-vision pathway, namely that the category-selectivity of regions cannot afford a comprehensive account for how the unlimited variety of faces and objects can be represented in ventral temporal cortex. Faces may have specialized processors that have evolved because of their biological significance, but the neural code for faces must have the capacity to produce unique representations for a virtually unlimited variety of individual faces. The "grandmother cell" hypothesis that each individual face is encoded by the activity of a specialized cell or small set of cells was discarded long ago as insufficient. Moreover, the representation framework in the ventral pathway must also distinguish between categories of uncertain biological significance, such as furniture versus clothing, and even among finer category distinctions – easy chairs versus desk chairs, dress shoes versus sport shoes. Clearly, there is not enough cortical real estate in the ventral temporal lobe to provide a new, specialized region for every conceivable category, and the neural code must have the generative power to produce unique representations for an unbounded set of faces and objects, and the structure to embody the similarities and dissimilarities among individual faces and various objects. If, on the other hand, different patterns of strong and weak responses in the same cortical space can represent different faces and objects, the combinatorial possibilities are virtually unlimited.

In order to test our hypothesis that patterns of strong and weak responses can represent different faces and houses in the same cortical space, we introduced a new approach for analyzing functional brain imaging data, multivoxel pattern analysis [38, 39, 32, 59]. Rather than searching for regions that respond more strongly to one experimental condition as compared to others, multivoxel pattern analysis examines whether the patterns or landscapes of neural activity within a region are distinctive for different cognitive or perceptual states. Briefly, multivoxel pattern-analysis methods build models of the patterns of response for each experimental condition based on part of the data for an individual subject and test the classification performance of these models on data from the same individual that were not used to build the models. Accurate classification of patterns of response in the out-of-sample data indicates that these patterns are unique.

In our initial experiment using multivoxel pattern analysis, we measured cortical patterns of response using fMRI while subjects viewed eight different categories of objects: faces, houses, cats, chairs, scissors, shoes, bottles, and phase-scrambled images [38]. We used a split-half correlation method to test whether the patterns of response to these categories in ventral temporal cortex were reliably distinct. The data for each individual subject were split into two halves, and the patterns of response to each category in one half of the data were correlated with the patterns of response to each category in the other half of the data. If the correlation of a pattern of response to a given category with itself (a within-category correlation) was higher than the correlations of the pattern of response to that category with the response to other categories (between-category correlations), the patterns were deemed reliable and distinct. The results showed that each of the eight categories evoked reliable patterns of response that could be distinguished from the patterns of response to all of the other categories. Moreover, the distinctiveness of these patterns was not due to voxels that responded maximally to a category. When the patterns of response to two categories were compared with the voxels that responded maximally to either category removed from the analysis, the patterns were still found to be reliably distinct, with the within-category correlation being larger in magnitude than the between-category correlation for 94% of the pairwise comparisons. These results show that the patterns of nonmaximal responses carry as much information about the category being viewed as do the maximal responses. The same expanse of ventral temporal cortex can produce a different, distinctive pattern of response for each category, suggesting that a combinatorial code at the relatively coarse spatial resolution of fMRI is sufficient to account for the capacity of this cortex to produce unique responses for an unlimited variety of object categories.

Multivoxel pattern analysis detects a fundamentally different type of information in a functional imaging dataset, as compared to the information that is extracted by previous standard methods. Previous methods identify individual voxels that respond differently to different stimulus or cognitive conditions, and the results of such an analysis are interpreted as an indication that the cell populations in a given voxel are tuned to represent the stimulus or process that elicits the maximal response. Patterns of response, by contrast, are vectors that reflect the activity in numerous voxels. Information about stimulus properties or cognitive state, therefore, is carried in these patterns by differences between responses in different parts of a distributed cortical representation, not by the strength of the response in one piece of cortex.

Since our initial demonstration of the sensitivity of multivoxel pattern analysis, better methods for building classifiers of these patterns have been introduced. Linear-discriminant analysis [10, 59], support-vector machines [18, 44], and neural-net classifiers [32] have all been used successfully. Moreover, these analyses have shown another important property of the patterns, namely that the similarities among stimulus conditions, in terms of either semantic similarity or physical stimulus similarity, are reflected by the similarities of the neural patterns of response [32, 59, 44].

A recent paper by Kamitani and Tong [44] provides a clearer example of how multivoxel pattern analysis can detect distributed population codes for visual stimuli. In their study, Kamitani and Tong had subjects view oriented gratings. Multivoxel pattern analysis, using a support-vector machine, could detect the angle of orientation of the grating that was being viewed based on the pattern of response in V1 cortex. When the classifier made an error, the incorrect orientation indicated by the classifier usually differed from the correct orientation by only 22.5°. This study is especially instructive because the neural encoding of orientation in V1 neurons is already well described as a topographic arrangement of orientation-selective columns. Functional imaging, with a resolution of 3 mm, presumably can detect this columnar topography, which cycles through all orientations in less than a millimeter, because the topography is not evenly distributed. These results suggest, therefore, that an unevenly distributed high-spatial-frequency topography of differentially tuned neurons can be analyzed by a lower-spatial-frequency grid of voxels. The irregularities in the topography result in subtle biases in each voxel, and such subtle biases, pooled over a large sample of voxels, are sufficient to specify the stimulus encoding.

The tuning functions for cell populations in ventral temporal cortex that reflect differences in face and object appearance, unlike the tuning functions of orientation-selective cells in early visual cortex, are poorly understood. Nontheless, functional imaging and multivoxel pattern analysis appear to be able to detect variations in the responses of these cells that are specific to face and object categories. Presumably, detection of these variations in pattern is due to uneven distribution of cells that are differentially tuned to visual attributes of faces and objects. Future work will determine whether the patterns of activity that are evoked by faces and objects can provide a key for understanding how to decipher the neural code that underlies face and object recognition.

The basis functions that underlie the tuning functions for different face-responsive and object-responsive cell populations are unknown, as are the principles of organization that underlie their topographic arrangement. Our results clearly show that the complete pattern of strong and weak responses carries information about face and object appearance. Whether the information carried by weak responses is used for object identification is unknown, but it seems unlikely that such information would be discarded. Moreover, a population code in which nonmaximal responses contribute information has a greater capacity to produce unique representations for an unlimited variety of faces and objects.

Representation of a simpler visual property, namely color, illustrates how weak responses can play an integral role in representation by a population response. In color vision, the perceptual quality of a hue that evokes a maximal red response in red–green neurons is also dependent on small responses in yellow–blue neurons that determine whether that hue is perceived as being more orange or violet. Population response representations that use the pattern of strong and weak responses, rather than just the strong responses, may also be used for even simpler visual properties, such as orientation, because these patterns can be more accurate and can enable finer discriminations than can winner-take-all representations.

Within ventral temporal cortex, the representations of faces and objects are distributed and overlapping. The same cortical area can produce representations for an unlimited variety of individual faces and objects by virtue of a spatially distributed combinatorial code. Faces and objects also evoke neural activity in numerous regions other than the ventral temporal cortex, suggesting that the representations of faces and objects also are distributed across cortical areas.

10.3 EXTENDED DISTRIBUTION OF FACE AND OBJECT REPRESENTATIONS

The extended distribution of brain regions that show activity during face and object recognition includes extrastriate visual areas in both ventral and lateral occipitotemporal cortices and other brain regions that are not strictly visual. The extended distribution of activity across extrastriate visual and nonvisual cortices, as compared to the local distribution of activity within a cortical area, is related to a different dimension of this information in the representation of faces and objects. Whereas the local distribution of activity within a single area, such as ventral temporal cortex, reflects a population response that represents a combinatorial code for one aspect of a face or object, such as visual appearance of form, activity in other areas represents other aspects of a face or object, such as how it can move, what sounds might it make, or what social message might it convey.

Category-related responses to faces and objects also are seen in inferior occipital (including inferior and midoccipital gyri), dorsal occipital (in the dorsal occipital gyrus and intraoccipital sulcus), and lateral temporal (posterior superior temporal sulcus and middle temporal gyrus) cortices. We found the highest accuracy for identification of the category being viewed in ventral temporal cortex (96%), but we also found high accuracies for patterns of response in inferior occipital (92%), and lateral temporal (92%) cortices, with a lower but still significant level of accuracy for patterns of response in dorsal occipital (87%) cortices.

Neural activity in lateral temporal cortex appears to be related to more dynamic aspects of faces and objects as compared to the information about face and object form that is represented by activity in ventral temporal cortex (Figure 10.3). Cortex in the posterior STS and the temporal parietal junction contain regions that respond more to faces than other object categories [35, 36] and appear to play a more general role in interpretation of the intent of others and in social communication [2]. In the realm of face perception, cortex in the posterior STS plays a role in representing eye gaze direction, facial expression, and facial movement

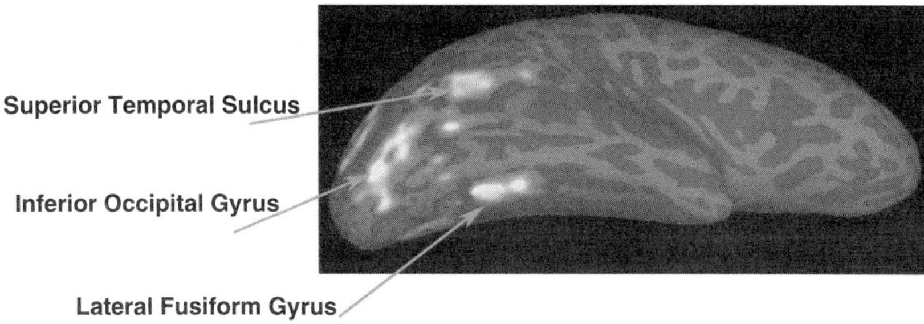

FIGURE 10.3: Location of the areas of the core system in an "inflated" image of the right cerebral cortex. The areas in red to yellow responded more to faces than to houses. The areas in shades of blue responded more to houses than to faces. The inflated cortical view allows illustration of areas in activity in sulci (indicated by a darker shade of gray) that would be hidden from view otherwise. (See also color plate section)

[41, 63, 68], suggesting that this region is central for extracting the social meaning conveyed by how the facial configuration is changed by movement, not the invariant structure that distinguishes the identity of that face [36]. Posterior STS cortices also are activated during perception of whole body movement [6, 4, 30, 31], during perception of patterns of motion that imply intentional actions [12], and while making judgments about a person's character [67]. Activity that is evoked in this region by whole bodies and animals appears to be associated with the representation of how bodies move [6, 4, 5, 30, 31]. The posterior STS region that responds to biological movement is superior to an adjacent region in the middle temporal gyrus that responds more strongly to patterns of movement associated with inanimate artifacts, such as a broom, a saw, a hammer, or a pen [4, 5]. Thus, the lateral temporal cortex appears to contain a locally distributed representation of how faces and objects move.

The face-responsive regions in ventral and lateral temporal extrastriate visual cortices are involved differentially in the representation of the invariant structure of a face, which plays a role in the recognition of identity, and in the representation of how the facial configuration can change with movement, which plays a role in social communication. We have called this part of the distributed human neural system for face perception the "core system" for the visual analysis of faces. Nonvisual areas that are associated with other types of information that is gleaned from faces are also activated during face perception. We have called the nonvisual areas that also participate in extracting information from faces the "extended system" [36]: Figure 10.4).

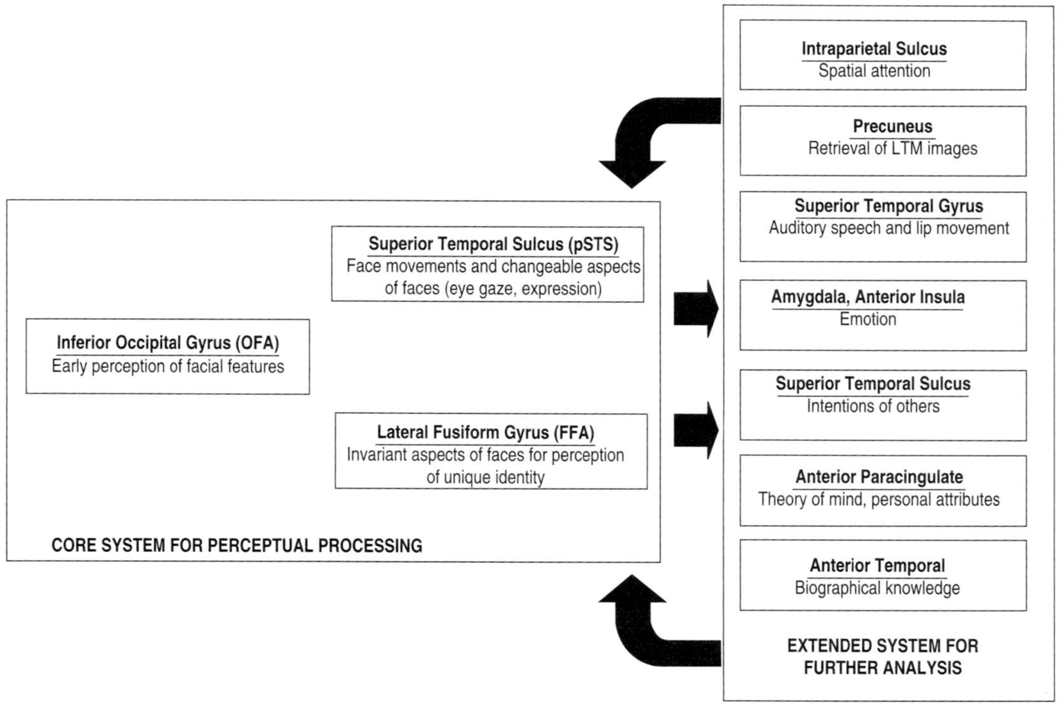

FIGURE 10.4: Model of the distributed human neural system for face perception. The cortical areas in the "core system" perform a perceptual analysis of faces, while the areas in the "extended system" are nonvisual areas recruited to gather information related to person knowledge, emotion, spatial attention, and speech (readapted from [36]). Face identification is the result of feedforward and feedback interactions between areas of the core and extended systems.

The activity that is evoked in nonvisual neural systems during face perception appears to be related to a role played by the simulation of nonvisual sensory, motor, and emotional representations in inferring the meaning of facial appearance. Lip reading is associated with the activation of a cortical area in the superior temporal gyrus that is associated with auditory speech perception [8, 9], suggesting that the perception of lip movement is augmented by simulating the auditory representations of the speech sounds that those movements can produce. Perception of expressions is associated with activation of premotor cortex in the frontal operculum that is associated with the production of expressions [11, 56], suggesting that representations of the visual appearance of expressions are augmented by activating the motor representations that could produce those expressions. Perception of emotional expressions also is associated with the activation of regions in the amygdala and in the anterior insular cortex that are associated with emotional states [60, 57, 7], suggesting that the perception of expression is augmented by simulating the emotions that might produce those expressions. Perception of eye gaze is associated with activation of the oculomotor system [41, 28] suggesting that the representation of the eye gaze of others is augmented by activating the representations of how one might move one's own eyes to direct attention to the same location or object that is the target of gaze. Perception of averted eye gaze elicits an automatic shift of spatial attention [22, 20, 40, 47] and the representation of oculomotor control and spatial shifts of attention appear to be largely coextensive [3].

Recognition of a familiar face is associated with neural activity in a widely distributed set of brain areas that presumably reflects the spontaneous activation of personal knowledge and emotions associated with that person. The perception of familiar faces is associated with activation of a region in the anterior middle temporal gyrus and the temporal poles that may reflect the spontaneous activation of biographical information [27, 49, 58]. The recognition of a personally familiar face activates regions in the anterior paracingulate cortex and the posterior superior temporal sulcus that have been associated with the representation of the mental states of others – the theory of mind [23, 24, 64] suggesting that seeing a close relative or friend leads to the spontaneous activation of one's understanding of that person's intentions, personality traits, and other mental states [26]. Recognition of familiar faces also modulates activity in the amygdala, with a reduced response while looking at personally familiar faces [26], possibly reflecting how people feel less guarded around friends, but an increased response in mothers looking at their own child [48], perhaps related to maternal protectiveness.

10.4 SPATIALLY DISTRIBUTED FACE AND OBJECT REPRESENTATIONS

In the human brain, face and object perception are mediated by concerted activity in a widely distributed neural systems. The information that is represented by this activity is distributed spatially both locally and globally. Within a cortical area, information about a specific aspect of a face or object, such as its visual form or how it moves, is encoded by a population code. Distinct locally distributed representations for different stimulus categories can be detected with fMRI measures of patterns of response. Both strong and weak levels of activity within these patterns carry information about the stimulus category. Patterns of response that carry information about different categories are overlapping, suggesting that the same cortical area can play a role in the representation of multiple stimulus categories. The extended distribution of representations across cortical areas can be divided into regions that are involved in the analysis of the visual appearance of faces and objects and regions that extract the meaning of faces that is not strictly visual. Different regions within extrastriate visual cortices participate differentially in the representation of the appearance of object form and the invariant structure of faces and in the representation of how faces and objects move and how movement changes appearance. Brain regions that are not strictly visual represent other kinds of information that can be extracted from faces – speech content, emotional state, direction of attention, and person knowledge. Many of these representations involve the simulation of motor representations, nonvisual perceptual representations, and emotional representations that could be associated with generating the visually-perceived facial configuration.

REFERENCES

[1] G. K. Aguirre, E. Zarahn, and M. D'Esposito. An area within human ventral cortex sensitive to "building" stimuli: evidence and implications. *Neuron* **21**, 373–83, 1998.

[2] T. Allison, A. Puce, and McCarthy, G. Social perception from visual cues: role of the STS region. *Trends in Cognitive Science* **4**, 267–278, 2000.

[3] M. S. Beauchamp, L. Petit, T. M. Ellmore, J. Ingeholm, and J. V. Haxby. A parametric fMRI study of overt and covert shifts of visuospatial attention. *Neuroimage* **14**, 310–321, 2001.

[4] M. S. Beauchamp, K. E. Lee, J. V. Haxby, and A. Martin. Parallel visual motion processing streams for manipulable objects and human movements. *Neuron* **34**, 149–59, 2002.

[5] M. S. Beauchamp, K. E. Lee, J. V. Haxby, and A. Martin. fMRI responses to video and point-light displays of moving humans and manipulable objects. *Journal of Cognitive Neuroscience* **15**(7):991–1001, 2003.

[6] E. Bonda, M. Petrides, D. Ostry, and A. Evans. Specific involvement of human parietal systems and the amygdala in the perception of biological motion. *Journal of Neuroscience* **16**, 3737–44, 1996.

[7] H. C. Breiter, N. L. Etcoff, P. J. Whalen, W. A. Kennedy, Rauch, S. L., Buckner, R. L., Strauss, M. M., Hyman, S. E., and B. R. Rosen. Response and habituation of the human amygdala during visual processing of facial expression. *Neuron* **17**, 875–87, 1996.

[8] G. A. Calvert, E. T. Bulmore, M. J. Brammer, R. Campbell, S. C. R. Williams, P. K. Mc Guire, P. W. R. Woodruff, S. D. Iverson, and A. S. David. Activation of auditory cortex during silent lipreading. *Science* **276**, 593–596, 1997.
[9] G. A. Calvert, and R. Campbell. Reading speech from still and moving faces: The neural substrates of visible speech. *Journal of Cognitive Neuroscience* **15**, 57–70, 2003.
[10] T. A. Carlson, P. Schrater, and S. He. Patterns of activity in the categorical representations of objects. *Journal of Cognitive Neuroscience* **15**(5):704–17, 2003.
[11] L. Carr, M. Iacoboni, M. C. Dubeau, J. C. Mazziotta, and G. L. Lenzi. Neural mechanisms of empathy in humans: a relay from neural systems for imitation to limbic areas. *Proceedings of the National Academy of Sciences of the U.S.* **100**(9): 5497–502, 2003.
[12] F. Castelli, F. Happe, U. Frith, and C. Frith. Movement and mind: a functional imaging study of perception and interpretation of complex intentional movement patterns. *Neuroimage* **12**, 314–25, 2000.
[13] L. L. Chao, J. V. Haxby, and A. Martin. Attribute-based neural substrates in temporal cortex for perceiving and knowing about objects. *Nature Neuroscience* **2**,913–9, 1999a.
[14] L. L. Chao, A. Martin, and J. V. Haxby. Are face-responsive regions selective only for faces? *Neuroreport* **10**, 2945–50, 1999b.
[15] L. L. Chao, and A. Martin. Representation of manipulable man-made objects in the dorsal stream. *Neuroimage* **12**, 478–84, 2000.
[16] M. Corbetta, F. M. Miezin, S. M. Dobmeyer, and S. E. Petersen. Attentional modulation of neural processing of shape, color, and velocity in humans. *Science* **248**, 1556–1559, 1990.
[17] M. Corbetta, F. M. Miezin, S. M. Dobmeyer, G. L. Shulman, and S. E. Petersen. Selective and divided attention during visual discriminations of shape, color, and speed: functional anatomy by positron emission tomography. *Journal of Neuroscience* **11**, 2383–2402, 1991.
[18] D. D. Cox, and R. L. Savoy. Functional magnetic resonance imaging (fMRI) "brain reading": detecting and classifying distributed patterns of fMRI activity in human visual cortex. *Neuroimage* **19**:261–70, 2003.
[19] P. E. Downing, Y. Jiang, M. Shuman, and N. Kanwisher. A cortical area selective for visual processing of the human body. *Science* **293**, 2470–3, 2001.
[20] J. Driver, G. Davis, P. Ricciardelli, P. Kidd, E. Maxwell, and S. Baron-Cohen. Gaze perception triggers reflexive visuospatial orienting. *Visual Cognition* **6**, 509–540, 1999.
[21] R. Epstein, and N. Kanwisher. A cortical representation of the local visual environment. *Nature* **392**, 598–601, 1998.
[22] C. Friesen, and A. Kingstone. The eyes have it! Reflexive orienting is triggered by nonpredictive gaze. *Psychonomic Bulletin and Review* **5**, 490–495, 1998.
[23] C. D. Frith, and U. Frith. Interacting minds—a biological basis. *Science* **286**, 1692–1695, 1999.
[24] H. L. Gallagher, F. Happe, N. Brunswick, P. C. Fletcher, U. Frith, and C. D. Frith. Reading the mind in cartoons and stories: an fMRI study of "theory of mind" in verbal and nonverbal tasks. *Neuropsychologia* **38**(1):11–21, 2000.

[25] I. Gauthier, P. Skudlarski, J. C. Gore, and A. W. Anderson. Expertise for cars and birds recruits brain areas involved in face recognition. *Nature Neuroscience* **3**, 191–197, 2000.

[26] M. I. Gobbini, E. Leibenluft, N. Santiago, J. V. Haxby. Social and emotional attachment in the neural representation of faces. *Neuroimage* **22**(4):1628–35, 2004.

[27] M. L. Gorno- Tempini, C. J. Price, O. Josephs, R. Vandenberghe, S. F. Cappa, N. Kapur, R. S. Frackowiak, and M. L. Tempini. The neural systems sustaining face and proper-name processing. *Brain* **121**, 2103–18, 1998.

[28] M.-H. Grosbras, A. R. Laird, and T. Paus. Cortical regions involved in eye movements, shifts of attention,and gaze perception. *Human Brain Mapping*, **25**, 140–154, 2005.

[29] C. G. Gross, C. E. Rocha-Miranda, and D. B. Bender. Visual properties of neurons in inferotemporal cortex of the Macaque. *Journal of Neurophysiology* **35**, 96–111, 1972

[30] E. Grossman, M. Donnelly, R. Price, D. Pickens, V. Morgan, G. Neighbor, and R. Blake. Brain areas involved in perception of biological motion. *Journal of Cognitive Neuroscience* **12**, 711–20, 2000.

[31] E. D. Grossman, and R. Blake. Brain Areas Active during Visual Perception of Biological Motion. *Neuron* **35**, 1167–75, 2002.

[32] S. J. Hanson, T. Matsuka, and J. V. Haxby. Combinatorial codes in ventral temporal lobe for object recognition: Haxby (2001) revisited: is there a "face" area? *Neuroimage* **23**(1):156–66, 2004.

[33] U. Hasson, I. Levy, M. Behrmann, T. Hendler, and R. Malach. Eccentricity bias as an organizing principle for human high-order object areas. *Neuron* **34**, 479–90, 2002.

[34] U. Hasson, M. Harel, I. Levy, and R. Malach. Large-scale mirror-symmetry organization of human occipito-temporal object areas. *Neuron* **37**, 1027–41, 2003.

[35] J. V. Haxby, L. G. Ungerleider, V. P. Clark, J. L. Schouten, E. A. Hoffman, and A. Martin. The effect of face inversion on activity in human neural systems for face and object perception. *Neuron* **22**, 189–99, 1999.

[36] J. V. Haxby, E. A. Hoffman, and M. I. Gobbini. The distributed human neural system for face perception. *Trends in Cognitive Science* **4**, 223–233, 2000a.

[37] J. V. Haxby, A. Ishai, L. L. Chao, L. G. Ungerleider, and A. Martin. Object form topology in the ventral temporal lobe. *Trends in Cognitive Sciences* **4**, 3–4, 2000b.

[38] J. V. Haxby, M. I. Gobbini, M. L. Furey, A. Ishai, J. L. Schouten, and P. Pietrini. Distributed and overlapping representations of faces and objects in ventral temporal cortex. *Science* **293**, 2425–30, 2001.

[39] J. V. Haxby. Analysis of topographically-organized patterns of response in fMRI data: Distributed representations of objects in the ventral temporal cortex. In: N. Kanwisher and J. Duncan (Eds),*Functional Neuroimaging of Visual Cognition—Attention and Performance XX*, Oxford, New York, 83–98, 2004.

[40] J. K. Hietanen. Does your gaze direction and head orientation shift my visual attention? *Neuroreport* **10**, 3443–3447, 1999.

[41] E. A. Hoffman, and J. V. Haxby. Distinct representations of eye gaze and identity in the distributed human neural system for face perception. *Nature Neuroscience* **3**, 80–4, 2000.

[42] A. Ishai, L. G. Ungerleider, A. Martin, J. L. Schouten, and J. V. Haxby. Distributed representation of objects in the human ventral visual pathway. *Proceedings of the National Academy of Sciences of the U.S.A* **96**, 9379–84, 1999.

[43] A. Ishai, L. G. Ungerleider, A. Martin, and J. V. Haxby. The representation of objects in the human occipital and temporal cortex. *Journal Of Cognitive Neuroscience* **12**, 35–51, 2000.

[44] Y. Kamitani, and F. Tong. Decoding the visual and subjective contents of the human brain. *Nature Neuroscience* **8**(5):679–85, 2005.

[45] N. Kanwisher, J. McDermott, and M. M. Chun. The fusiform face area: a module in human extrastriate cortex specialized for face perception. *Journal of Neuroscience* **17**, 4302–11, 1997.

[46] N. Kanwisher. Domain specificity in face perception. *Nature Neuroscience* **3**, 759–63, 2000.

[47] S. R. H. Langton, and V. Bruce. Reflexive visual orienting in response to the social attention of others. *Visual Cognition* **6**, 541–568, 1999.

[48] E. Leibenluft, M. I. Gobbini, T. Harrison, and J. V. Haxby. Mothers' neural activation in response to pictures of their children and other children. Biological Psychiatry **56**(4):225–32, 2004.

[49] C. L. Leveroni, M. Seidenberg, A. R. Mayer, L. A. Mead, J. R. Binder, and S. M. Rao. Neural systems underlying the recognition of familiar and newly learned faces. *Journal of Neuroscience* **20**, 878–86, 2000.

[50] I. Levy, U. Hasson, G. Avidan, T. Hendler, and R. Malach. Center–periphery organization of human object areas. *Nature Neuroscience* **4**, 533–9, 2001.

[51] N. K. Logothetis, H. Guggenberger, S. Peled, J. Pauls. Functional imaging of the monkey brain. *Natural Neuroscience* **2**, 555–62, 1999.

[52] C. J. Lueck, S. Zeki, K. J. Friston, M.-P. Deiber, P. Cope, V. J. Cunningham, A. A. Lammertsma, C. Kennard, and R. S. J. Frackowiak. The color centre in the cerebral cortex of man. *Nature* **340**, 386–389, 1989.

[53] R. Malach, J. B. Reppas, R. R. Benson, K. K. Kwong, H. Jiang, W. A. Kennedy, P. J. Ledden, T. J. Brady, B. R. Rosen, and R. B. Tooltell. Object-related activity revealed by functional magnetic resonance imaging in human occipital cortex. *Proceedings of the National Academy of Sciences of the U.S.A.* **92**, 8135–8139, 1995.

[54] R. Malach, I. Levy, and U. Hasson. The topography of high-order human object areas. *Trends in Cognitive Science* **6**, 176–184, 2002.

[55] G. McCarthy, A. Puce, J. C. Gore, and T. Allison. Face-specific processing in the human fusiform gyrus. *Journal of Cognitive Neuroscience* **9**, 605–610, 1997.

[56] K. J. Montgomery, M. I. Gobbini, J. V. Haxby. Imitation, production and viewing of social communication: an fMRI study. *Society for Neuroscience Abstracts*, Program Number 128.10, 2003.

[57] J. Morris, C. D. Frith, D. I. Perrett, D. Rowland, A. W. Young, A. J. Calder, and R. J. Dolan. A differential neural the human amygdala to fearful and happy facial expressions. *Nature* **383**, 812–815, 1996.

[58] K. Nakamura, R. Kawashima, N. Sato, A. Nakamura, M. Sugiura, T. Kato, K. Hatano, K. Ito, H. Fukuda, T. Schormann, and K. Zilles. Functional delineation of the human occipito-temporal areas related to face and scene processing: a PET study. *Brain* **123**, 1903–12, 2000.

[59] A. J. O'Toole, F. Jiang, H. Abdi, and J. V. Haxby. Partially distributed representations of objects and faces in ventral temporal cortex. *Journal of Cognitive Neuroscience* **17**(4):580–90, 2005.

[60] M. L. Phillips, A. W. Young, C. Senior, M. Brammer, C. Andrew, A. J. Calder, E. T. Bullmore, D. I. Perrett, D. Rowland, S. C Williams, J. A. Gray, and A. S. David. A specific neural substrate for perceiving facial expressions of disgust. *Nature* **389**, 495–8, 1997.

[61] M. A. Pinsk, K. Desimone, T. Moore, C. G. Gross, and S. Kastner. Representations of faces and body parts in macaque temporal cortex: a functional MRI study. *proceeding pf the National Academy of Sciences of the U.S.A* **102**(19):6996–7001, 2005.

[62] A. Puce, T. Allison, M. Asgari, J. C. Gore, and G. McCarthy. Differential sensitivity of human visual cortex to faces, letterstrings, and textures: a functional magnetic resonance imaging study. *Journal of Neuroscience* **16**, 5205–15, 1996.

[63] A. Puce, T. Allison, S. Bentin, J. C. Gore, and G. McCarthy. Temporal cortex activation in humans viewing eye and mouth movements. *Journal of Neuroscience* **18**, 2188–99, 1998.

[64] R. Saxe, and N. Kanwisher. People thinking about thinking people. The role of the temporo-parietal junction in "theory of mind". *Neuroimage* **19**(4):1835–42, 2003.

[65] J. Sergent, S. Ohta, and B. MacDonald. Functional neuroanatomy of face and object processing: A positron emission tomography study. *Brain* **115**, 15–36, 1992.

[66] D. Y. Tsao, W. A. Freiwald, T. A. Knutsen, J. B. Mandeville, and R. B. H. Tootell. Faces and Objects in Macaque Cerebral Cortex. *Nature Neuroscience* **6** 989–95, 2003.

[67] J.S. Winston, B. A. Strange, J. O'Doherty, and R. J. Dolan. Automatic and intentional brain responses during evaluation of trustworthiness of faces. *Nature Neuroscience* **5**, 277–83, 2002.

[68] J. S. Winston, R. N. Henson, M. R. Fine-Goulden, and R. J. Dolan. fMRI-adaptation reveals dissociable neural representations of identity and expression in face perception. *Journal of Nwurophysiology* **92**(3):1830–9, 2004.

[69] S. Zeki, J. D. G. Watson, C. J. Lueck, K. J. Friston, C. Kennard, and R. S. J. Frackowiak. A direct demonstration of functional specialization in human visual cortex. *Journal of of Neuroscience* **11**, 641–649, 1991.

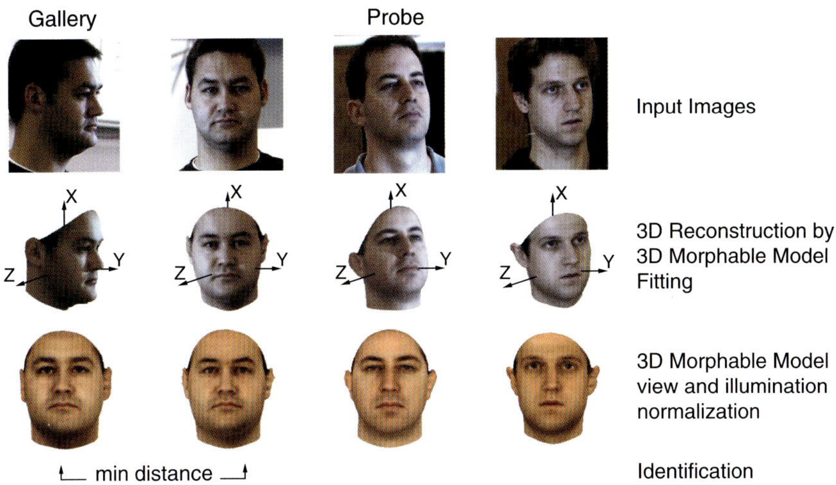

FIGURE 4.1

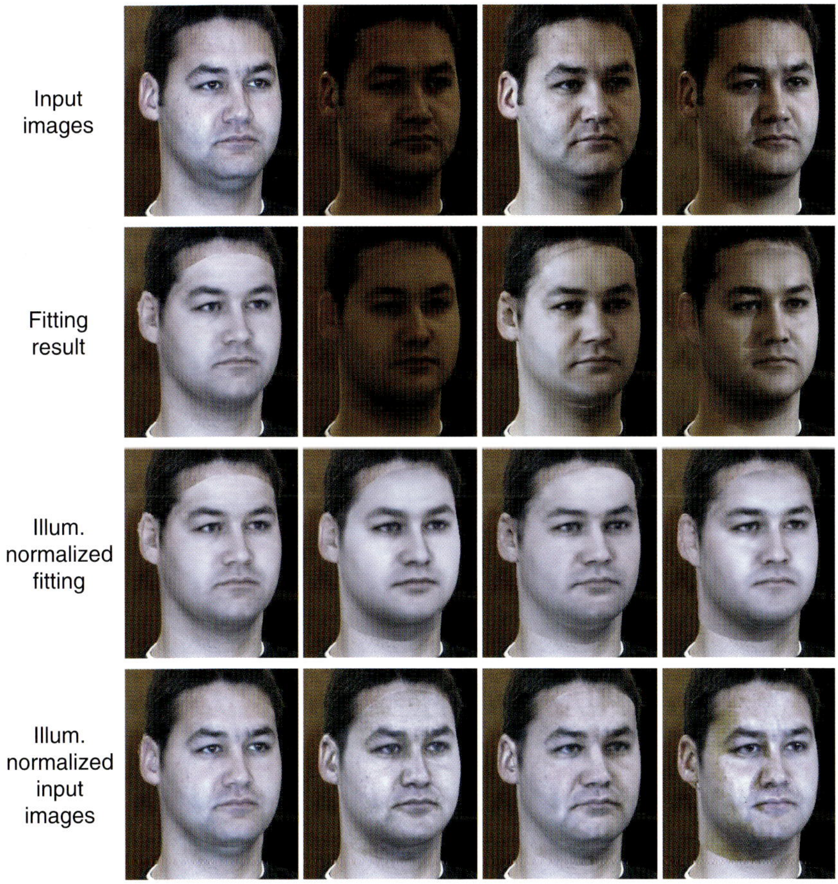

FIGURE 4.5

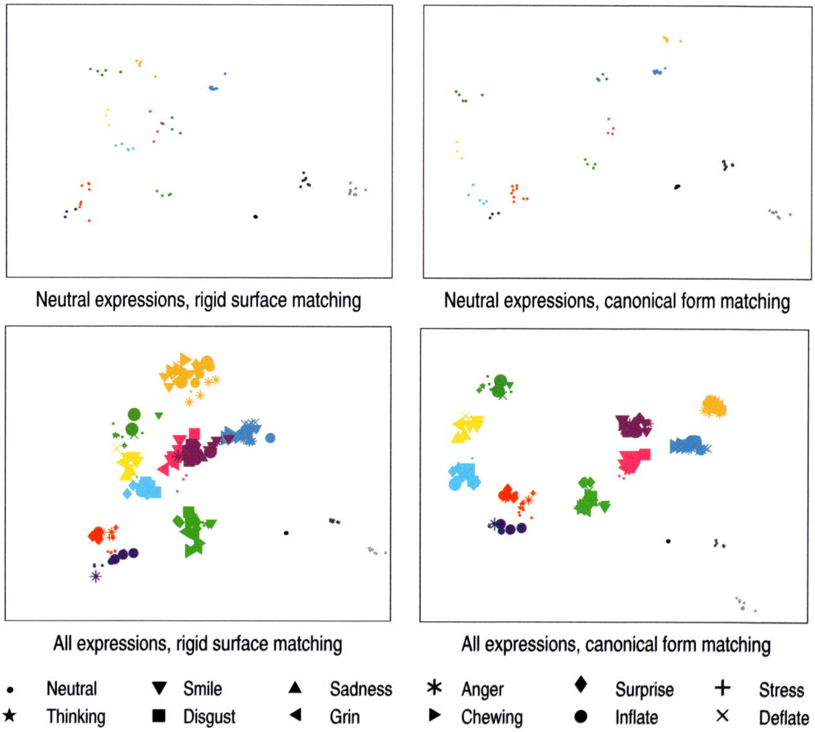

FIGURE 5.12

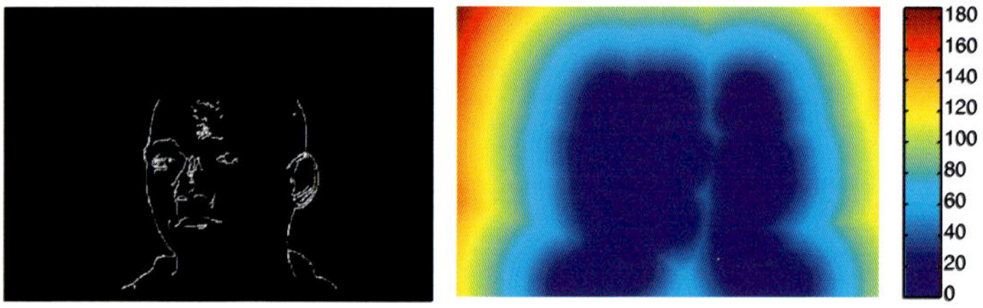

FIGURE 6.10

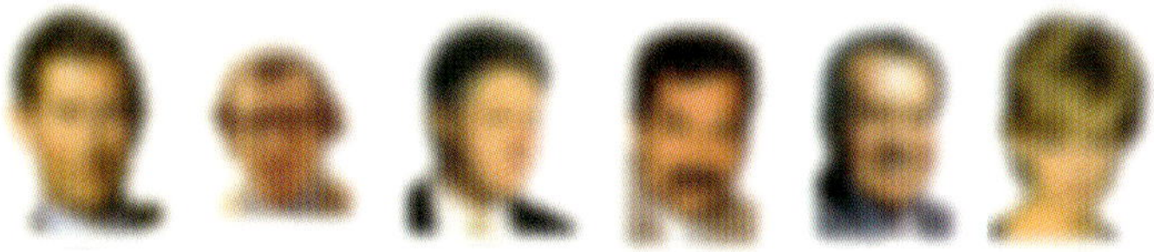

FIGURE 8.5

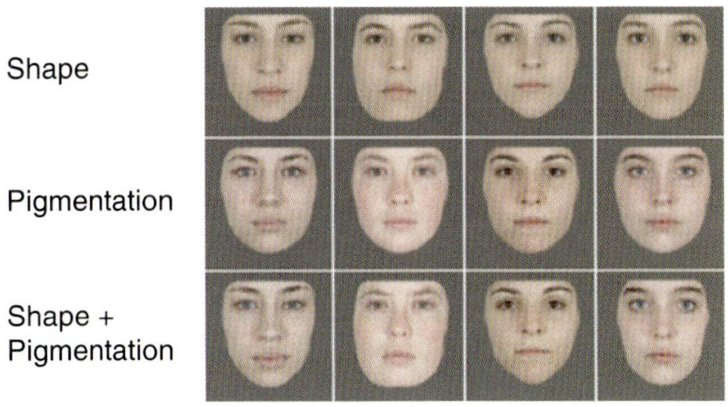

FIGURE 8.8

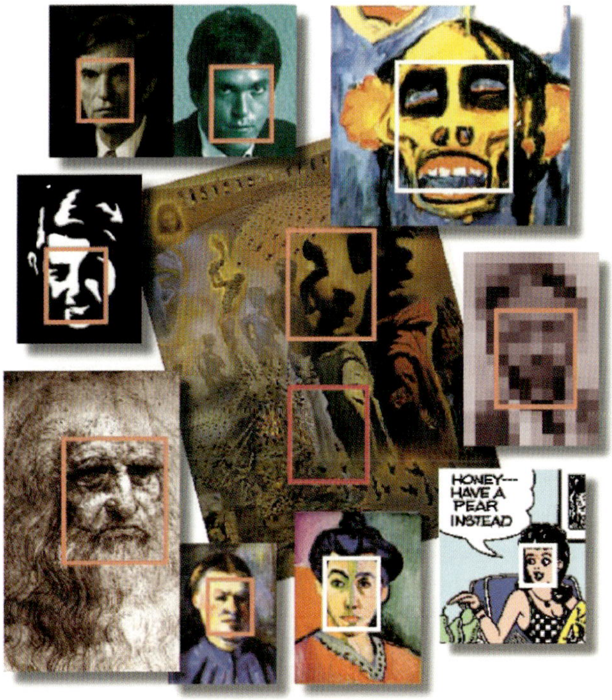

FIGURE 8.16

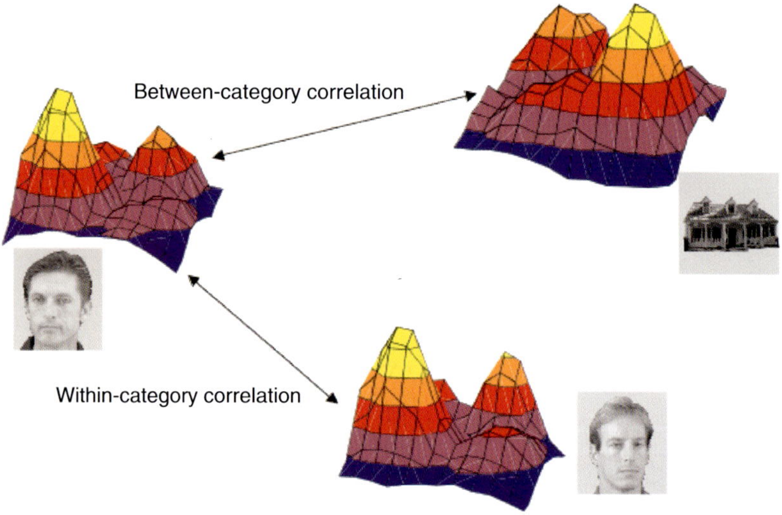

FIGURE 10.1

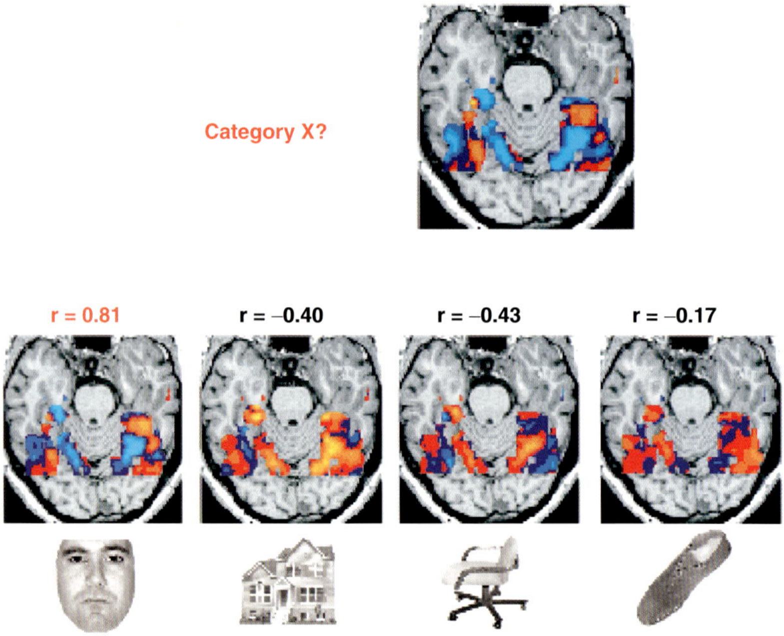

FIGURE 10.2

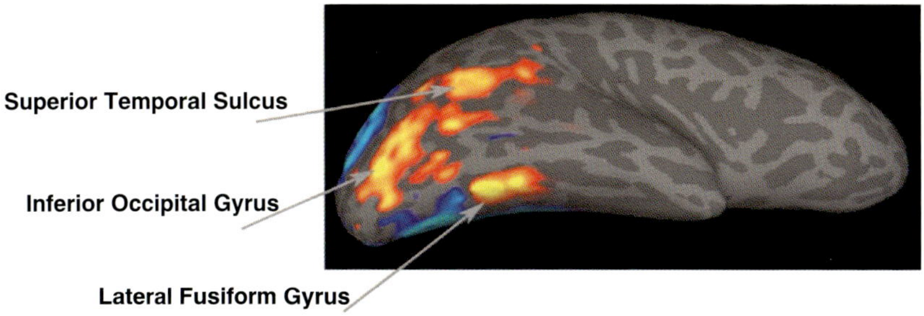

FIGURE 10.3

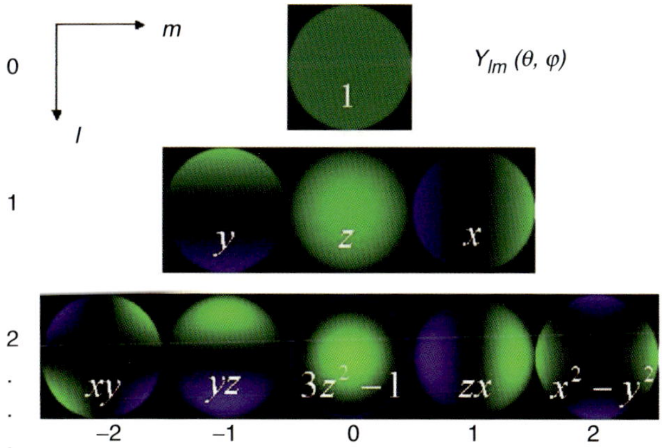

FIGURE 12.4

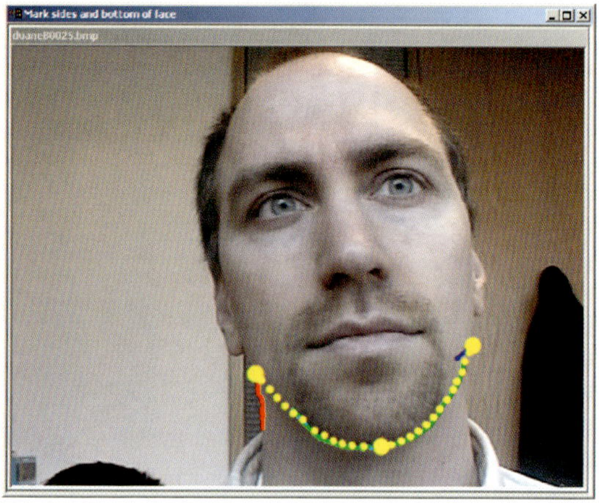

FIGURE 15.6

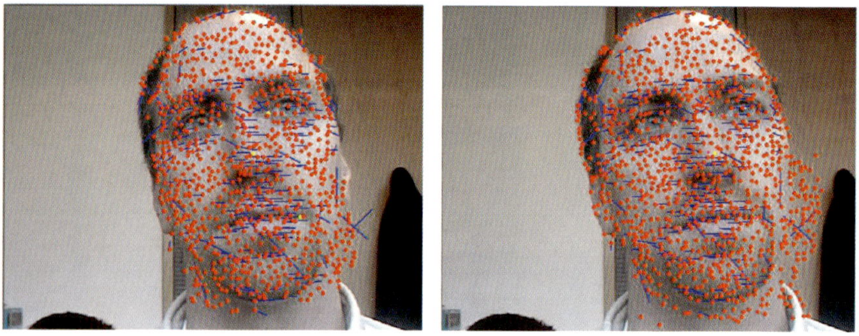

FIGURE 15.7

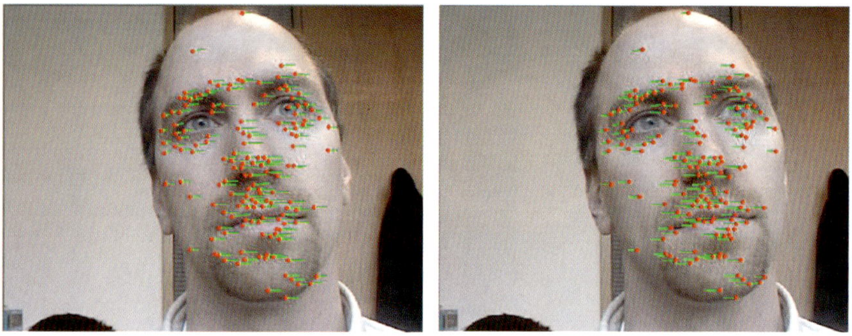

FIGURE 15.8

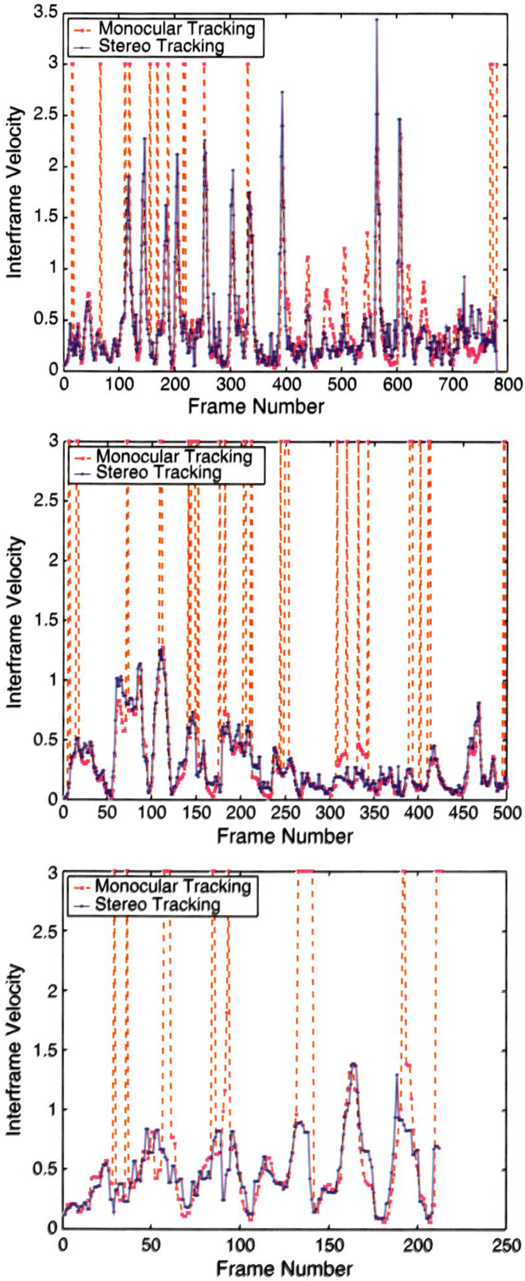
Monocular Tracking
Stereo Tracking
Interframe Velocity
Frame Number
0 100 200 300 400 500 600 700 800
0 0.5 1 1.5 2 2.5 3 3.5
Monocular Tracking
Stereo Tracking
Interframe Velocity
Frame Number
0 50 100 150 200 250 300 350 400 450 500
0 0.5 1 1.5 2 2.5 3
Monocular Tracking
Stereo Tracking
Interframe Velocity
Frame Number
0 50 100 150 200 250
0 0.5 1 1.5 2 2.5 3

FIGURE 15.26

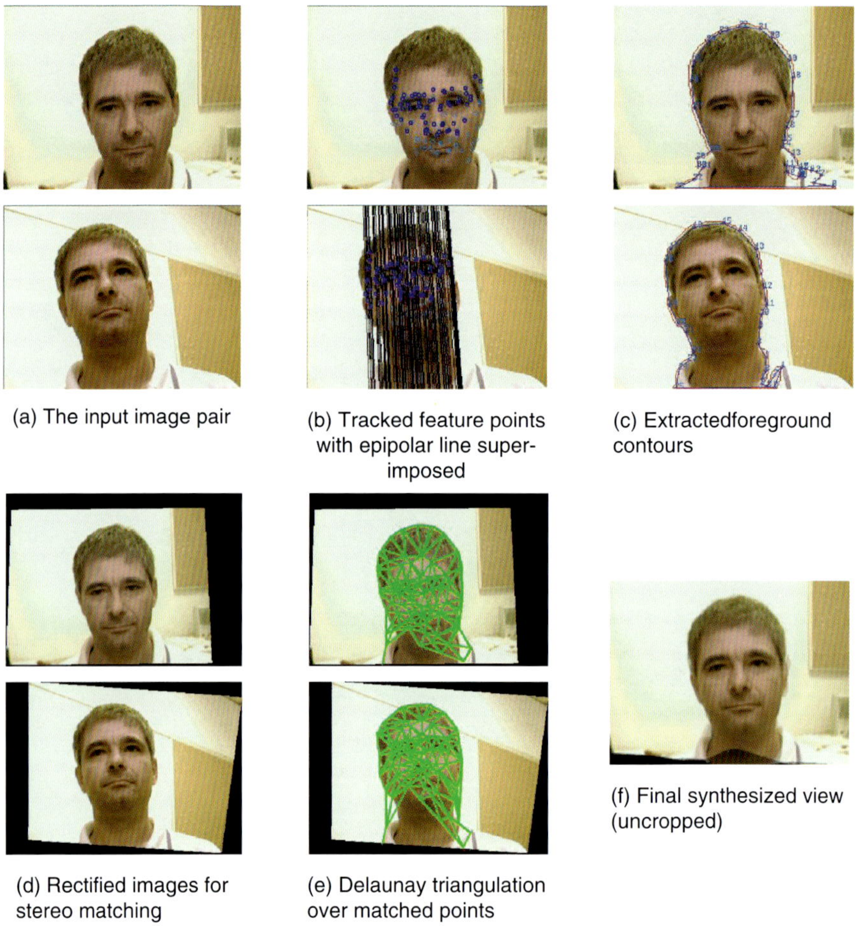

(a) The input image pair

(b) Tracked feature points with epipolar line superimposed

(c) Extractedforeground contours

(d) Rectified images for stereo matching

(e) Delaunay triangulation over matched points

(f) Final synthesized view (uncropped)

FIGURE 15.31

PART 3

ADVANCED METHODS

CHAPTER 11

ON THE EFFECT OF ILLUMINATION AND FACE RECOGNITION

11.1 INTRODUCTION

Our experience in this predominantly visual world is enriched greatly by the diverse ways the world can be illuminated. While this diversity makes our world fascinating, it also makes recognition from images, face recognition in particular, difficult. As is evident in Figure 11.1, the effect of illumination on the appearance of a human face can be striking. The four images in the top row are images of an individual taken with the same viewpoint but under different external illumination conditions. The four images in the bottom, on the other hand, are images of four individuals taken under the same viewpoint and lighting. Using the most common measure of similarity between pairs of images, the L^2 difference,[1] it is not surprising to learn that the L^2 difference between any pair of images in the bottom is always less than the L^2 difference between any pair of images from the top row. In other words, simple face-recognition algorithms based purely on L^2 similarity are doomed to fail for these images. This result corroborates well the sentiment echoed through the often-quoted observation made more than a decade ago that "the variations between the images of the same face due to illumination... are almost always larger than image variations due to change in face identity"[26].

Needless to say, a robust recognition system must be able to, among other things, identify an individual across variable illumination conditions. For decades,

[1]For images with same number of pixels n, the L^2 difference between a pair of images is simply the usual Euclidean distance between the two corresponding vectors in $\mathbf{R}^n$.

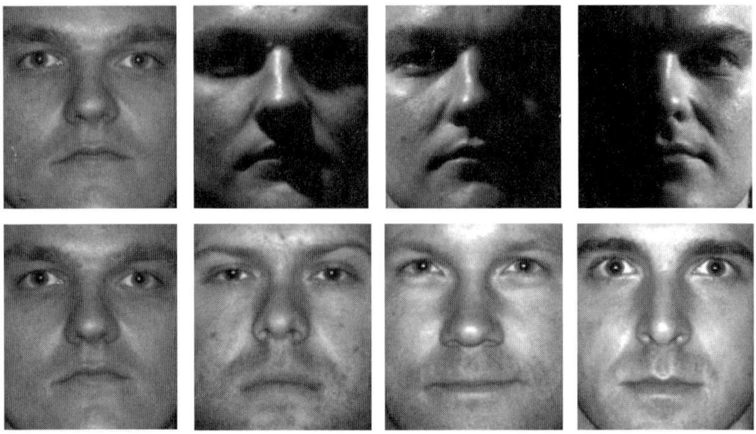

FIGURE 11.1: Striking effects of illumination on the appearances of a human face. Top row: images are taken with the same viewpoint but under different illumination conditions. Bottom row: images of four different individuals taken under the same viewpoint and illumination condition.

feature-based methods such as [15, 21] (see surveys in, e.g., [9] and references in [14]) have used properties and relations (e.g., distances and angles) between facial features such as eyes, mouth, nose, and chin to perform recognition. However, reliable and consistent extraction of these features can be problematic under difficult illumination conditions, as images in Figure 11.1 clearly indicate. In fact, it has been claimed that methods for face recognition based on finding local image features and inferring identity by the geometric relations of these features are ineffective [7]. Image-based, or appearance-based, techniques (e.g., [27, 36, 39]) offer a different approach. For this type of algorithm, local image features no longer play significant roles. Instead, image-based techniques strive to construct low-dimensional representations of images of objects that, in the least-squares sense, faithfully represent the original images. For the particular problem of face-recognition under varying illumination, some of the most successful methods (e.g., [2, 10, 14, 24, 35, 43]) are image-based. For each person to be recognized, these algorithms use a small number of (training) images to construct a low-dimensional representation of images of a given face under a wide range of illumination conditions. The low-dimensional representation is (except [10]) invariably some linear subspace in the image space, and the linearity makes the recognition algorithms efficient and easy to implement. The actual recognition process is straightforward and somewhat trivial: each query image is compared to each subspace in turn, by computing the usual L^2 distance between a linear subspace and an image (in vector form). The recognition result is the person in the database whose linear subspace has the minimal L^2 distance to the query image. What is nontrivial, however, is the discovery of the correct language and mathematical framework to model the effects of illumination [2, 4, 30], and the application of these illumination models to the designs of efficient face-recognition algorithms [2, 10, 14, 24, 35, 43] that explicitly model lighting variability using only a small number of training images. While pairwise L^2 comparisons between images would have failed miserably, L^2 comparisons between a query image and suitably chosen (according to the illumination model) subspaces are robust against illumination variation.

Figure 11.1 illustrates the two main elements in modelling illumination effects: the variation in pixel intensity and the formation of shadows. As lighting varies, the radiance at each point on the object's surface also varies according to its *reflectance*. In general, surface reflectance can be described by a 4D function $f_r(\theta_i, \phi_i, \theta_o, \phi_o)$, the bidirectional reflectance distribution function (BRDF), which gives the reflectance of a point on a surface as a function of the incident illumination direction $\omega_i = (\theta_i, \phi_i)$ and the emitted direction $\omega_o = (\theta_o, \phi_o)$. (See Figure 11.2). A full BRDF with four independent variables is difficult to model and work with, e.g., [41]. Fortunately for face-recognition, the much simpler Lambertian model [23] has been shown to be both sufficient and effective [4, 14] for modelling the reflectance of human faces:[2] the radiance (pixel intensity) I at each surface point is given by the inner product between the unit normal vector $\vec{n}$ scaled by the *albedo* value ρ and the light vector $\vec{L}$, which encodes the direction and magnitude of incident light coming from a distance,

$$I(\vec{L}) = \rho \max\{\vec{L} \cdot \vec{n}, 0\}. \tag{1}$$

The Lambertian model effectively collapses the usual 4D BRDF f_r into a constant function with value ρ. In particular, a Lambertian object appears equally bright from all viewing directions. We note that Equation 1 is linear on the image level in the sense that the image of an object produced by two light sources is simply the sum of the two images produced by the sources individually. This, of course, is the familiar superposition principle of illumination, and it is the source of linearity appearing in all illumination models discussed below. However, because of the presence of the max term, Equation 1 is only quasi-linear in $\vec{L}$: $I(\vec{L}_1 + \vec{L}_2) \neq I(\vec{L}_1) + I(\vec{L}_2)$ in general. This quasi-linearity in $\vec{L}$ is responsible for several tricky

[2]To model more fully the interaction of light with a surface such as human skin, one might also want to model the effects of subsurface scattering [20] and the effects of interreflections between say the nose and the cheek; while these effects lead to greater realism in rendering, they have not yet been considered significant enough for face recognition.

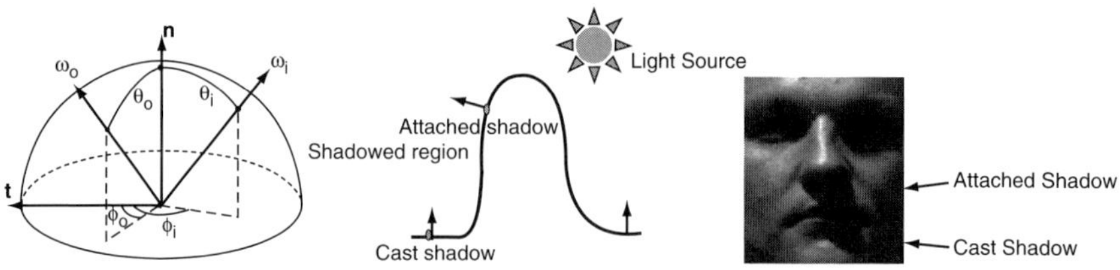

FIGURE 11.2: Left: coordinate system used in defining the bidirectional reflectance distribution function (BRDF); $\omega_i = (\theta_i, \phi_i)$ parametrizes the incident lighting direction; $\omega_o = (\theta_o, \phi_o)$ represents the viewing direction (emitted direction); **n** is the normal vector. Center: Attached and cast shadows. Right: The formation of shadows on a human face. Attached shadows are in the upper region of the eye socket. Cast shadows appear in the lower region of the eye socket and the lower part of the face.

points in analyzing illumination effects, and it is related to the formation of attached shadows.

Shadows naturally account for a significant portion of the variation in appearances. On a surface, two types of shadow can appear: attached shadows and cast shadows (Figure 11.2). An attached shadow is formed when there is no irradiance at a point on the surface: in other words, when the light source is on the "back side" of the point. This condition can be summarized concisely as $\vec{n} \cdot \vec{L} < 0$, with $\vec{n}$ the normal vector. We note that the equality $I(\vec{L}_1 + \vec{L}_2) = I(\vec{L}_1) + I(\vec{L}_2)$ fails precisely at pixels such that $\vec{n} \cdot \vec{L}_1 < 0$ or $\vec{n} \cdot \vec{L}_2 < 0$, i.e., most pixels in attached shadow. Cast shadows, on the other hand, simply shadows the object casts on itself. Clearly, cast shadows are related to the object's global geometry, and local information such as normals does not determine their formation. Consequently, they are considerably more difficult to analyze (see [22], however).

In this chapter, we discuss recent advances in modelling illumination effects [2, 4, 30] and various face-recognition algorithms based on these foundational results [2, 10, 14, 24, 35, 43]. At first glance, modelling the variability in appearance of a human face under all lighting conditions may seem to be intractable since, after all, the space of all lighting conditions is, in principle, infinite-dimensional.[3] Nonetheless, it turns out that the variability caused by illumination can be effectively captured using low-dimensional linear models. This can be

[3]In general, lighting (radiances) can be modelled as a positive function over the 4D space of rays. For face recognition, where the face is usually distant from light sources, it is reasonable to model the light as being distant and so giving rise to a positive function on the sphere. Clearly, this does not account for, say, a candlelit dinner.

largely attributed to the reflectance and geometry of human faces. While human faces are generally not Lambertian (as the often oily and specular forehead demonstrates), they can nevertheless be approximated well by a Lambertian object. Also, intensity variation and attached shadows can be succinctly modelled using only Equation 1. In fact, without the presence of cast shadows, illumination modelling for a Lambertian object can be formulated under an elegant framework using spherical-harmonic functions [2, 30], and precise results concerning the dimensionality of the approximating subspaces and the faithfulness of the approximations can be given. Furthermore, image variation due to illumination can be completely characterized by enumerating a finite number of basis images [4]. Although these foundational results only deal with the ideal case of convex Lambertian objects (where cast shadows are absent), they nevertheless form the basis for the subsequent developments. It is largely due to the geometry of human faces that the successful applications of these ideas to face recognition can be made possible. While human faces are generally not convex, considerable parts of them are, as the smoothly-curved forehead and lightly rounded cheeks attest to (Figure 11.2). Consequently, the effects of cast shadows, which might appear to be formidable, are usually manageable. Several empirical results [11, 14] have shown that even with cast shadows included, the appearance model is still low-dimensional, and its dimension is only slightly larger than the dimension predicted by the theory [2, 30].

This chapter is organized as follows. In the following section, we discuss an important result that first appeared in [10] concerning the nonexistence of illumination invariants. This interesting and somewhat unexpected result raises several subtle issues regarding illumination and face recognition. In the third section, the foundational results on illumination modeling are discussed. Section 11.4 contains a brief survey on several recently published face-recognition algorithms based on these foundational results. Their performance and other experimental results are discussed in Section 11.5. We conclude this chapter with a short summary and remark on future work.

11.2 NON-EXISTENCE OF ILLUMINATION INVARIANTS

Before delving into the details of various face-recognition algorithms, we briefly discuss the important and interesting result of [10] on the nonexistence of illumination invariants. Specifically, [10] demonstrates that for any two images, whether they are of the same object or not, there is always a family of Lambertian surfaces, albedo patterns, and light sources that could have produced them. One consequence of this surprising result is quite counterintuitive: given two images, it is not possible with absolute certainty to determine whether they were created by the same or by different objects. More specifically, [10] contains a proof of the following:[4]

PROPOSITION 11.2.1 Given two images I and J, and any two linearly independent vectors $\vec{s}, \vec{l} \in \mathbb{R}^3$, there exists a Lambertian surface $\mathcal{S}$ such that the images of $\mathcal{S}$ taken under lighting conditions (point sources at infinity) specified by $\vec{s}$ and $\vec{l}$ are I and J, respectively.

A direct consequence of this proposition is the nonexistence of (nontrivial) illumination invariants. Abstractly, an illumination invariant is a function μ of images that is constant on images of an object taken under different illumination conditions. That is, if I_1, I_2 are two images of an object $\mathcal{O}$ taken under the same viewpoint but different illumination conditions, then, $\mu(I_1) = \mu(I_2)$. Note that this implies that, if we have $\mu(I) \neq \mu(J)$ for a pair of images I, J, then I and J cannot originate from the same object. Unfortunately, the proposition above implies that any such function μ must be a (trivial) constant function, i.e., $\mu(I) = \mu(J)$ for any two images I, J. Clearly, these invariants are not discriminative, and they cannot be exploited for face recognition. The analogous nonexistence result for view invariants is presented in [8].

While the consequences of this proposition can be surprising, the motivation behind its proof, however, is straightforward. We remark that there is no analogous result for three or more images. Let's assume that the Lambertian surface $\mathcal{S}$ is viewed from the direction $(0, 0, 1)$, and $\mathcal{S}$ can be written as $(x, y, z = f(x, y))$, with $z \geq 0$ over some bounded rectangular region $\mathcal{R}$ in the xy plane. The images, I, J, are then considered as some nonnegative functions on $\mathcal{R}$, i.e., $I, J \in \mathcal{P}$, where $\mathcal{P}$ denotes the space of nonnegative functions on $\mathcal{R}$. The space of Lambertian objects is then precisely $\mathcal{P} \times \mathcal{P}$, with one factor for the geometry $(z = f(x, y))$ and the other for the albedo values. However, the variability offered by pairs of images is also $\mathcal{P} \times \mathcal{P}$. Therefore, given two fixed lighting conditions and a pair of images, we expect (heuristically) that at least one Lambertian surface can be responsible for the images. When the number of images is greater than two, we see that the variability in images is much larger than the space of Lambertian surfaces. Therefore, for a generic triplet of images, one generally does not expect to find a Lambertian surface that accounts for these images. Essentially as we will see soon, the proof exploits the following underdetermined system of linear equations (in components of $\vec{n}$):

$$I(x, y) = \alpha(x, y)\vec{s} \cdot \hat{n}(x, y) = \vec{s} \cdot \vec{n}(x, y), \tag{2}$$

$$J(x, y) = \alpha(x, y)\vec{l} \cdot \hat{n}(x, y) = \vec{l} \cdot \vec{n}(x, y). \tag{3}$$

---

[4]In this discussion, we ignore the regularity assumptions. All surfaces and images are assumed to be infinitely differentiable (C^∞).

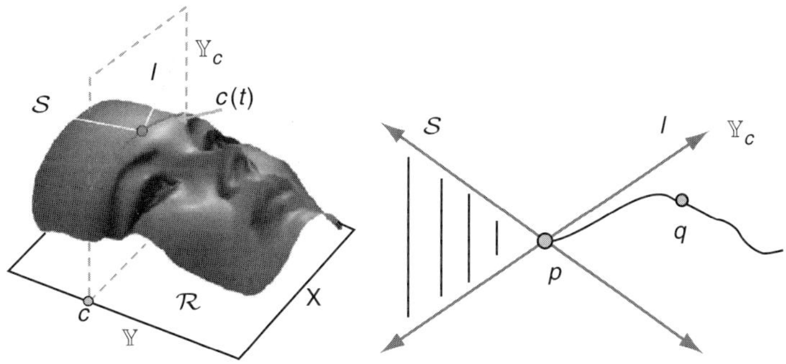

FIGURE 11.3: Left: the surface $\mathcal{S}$ is defined over a rectangular domain $\mathcal{R}$ in the xy plane. The intersection $\mathbb{Y}_c \cap \mathcal{S}$ defines the curve $\vec{c}(t)$. Right: On each plane $\mathbb{Y}_c$ and for any point $\vec{q} \in \mathbb{Y}_c \cap \mathcal{S}$ that is to the right of $\vec{p}$, because $\vec{q} = \vec{p} + a\vec{l} - b\vec{s}$ for some nonnegative a, b, the point $\vec{q}$ has to lie inside the cone generated by $\vec{l}$ and $-\vec{s}$. Similarly, any point $\vec{q}$ that is to the left of $\vec{p}$ has to lie in the cone generated by $\vec{s}$ and $-\vec{l}$.

Here, $\hat{n}(x,y)$ is the unit normal and $\vec{n}(x,y) = \alpha(x,y)\hat{n}(x,y)$ is the albedo-scaled normal vector. For the case of three images, the analogous system will generally be invertible; therefore, a normal vector can be determined uniquely at any point (x,y). However, the resulting normal vector field formed by these normals is determined pointwise and will not be integrable [5] in general. So the inconsistency among a triplet of images can be detected. For a pair of images and linearly independent $\vec{s}, \vec{l}$, because there is always a family of normal vectors satisfying the above equations at any point, we can produce an integrable normal field by choosing the normal at each point carefully.

To see how the proof works, we assume that the images I, J do not vanish simultaneously, $I(x,y) + J(x,y) > 0$ for all (x,y). This implies in particular that the albedo $\alpha(x,y)$ is also nonvanishing. The general case is only slightly more complicated. In addition, we also assume that $\vec{s} = (-1,0,1)$ and $\vec{l} = (1,0,1)$. The extension to arbitrary $\vec{s}$ and $\vec{l}$ will become clear later. Under these assumptions, let's consider $\mathcal{S}$ along a scanline, $y = c$ for some constant c (see Figure 11.3). Let $\mathbb{Y}_c$ denote the plane $y = c$. The intersection between $\mathcal{S}$ and $\mathbb{Y}_c$ defines a curve $\vec{c}(t) = (x(t), y(t), z(t))$. If $\vec{c}(t)$ satisfies the differential equation

$$\frac{d\vec{c}}{dt} = I\vec{l} - J\vec{s} \equiv \begin{Bmatrix} dx/dt = I + J \\ dy/dt = 0 \\ dz/dt = I - J \end{Bmatrix}, \tag{4}$$

then, because $d\vec{c}/dt$ is a tangent vector of $\mathcal{S}$, $0 = d\vec{c}/dt \cdot \vec{n} = (I\vec{l} - J\vec{s}) \cdot \vec{n}$ on $\vec{c}(t)$. For the system of ODEs above, a unique solution $\vec{c}$ can be found by integration provided that an initial condition (a point in $\mathbb{Y}_c$ of the form (x, c, y)) is also given. If such an initial point is given, $\vec{c}(t)$ does indeed to stay on the plane $\mathbb{Y}_c$ because $dy/dt = 0$. Furthermore, because $dx/dt > 0$ by our assumption, $x(t)$ is a strictly monotone function. This implies that $z(t)$ is a function of $x(t)$ on the "slice" $\mathbb{Y}_c \cap \mathcal{S}$. It follows that we can construct one particular $\mathcal{S}$ by specifying initial points along the left edge of the rectangular region $\mathcal{R}$ and integrating across all scanlines. If the initial points are chosen to be a smooth curve, it follows (e.g., [1]) that $\mathcal{S}$ will indeed be in the form $(x, y, z = f(x, y))$ for some smooth function f, and more importantly, $(I\vec{l} - J\vec{s}) \cdot \vec{n} = 0$ everywhere on $\mathcal{S}$. Because $\vec{l} \cdot \vec{n}$ and $\vec{s} \cdot \vec{n}$ cannot vanish simultaneously (since $\vec{n}$ is a multiple of $(\partial f/\partial x, \partial f/\partial y, -1)$), we have $I(x, y) = \alpha(x, y)\vec{s} \cdot \hat{n}$ and $J(x, y) = \alpha(x, y)\vec{l} \cdot \hat{n}$ at all (x, y) for some positive function α, the albedo.

This almost completes the proof, except, alas, that Equations 2 and 3 are not quite the same as Equation 1. They will be so only if we can show that there is no (x, y) such that $\vec{s} \cdot \vec{n} < 0$ or $\vec{l} \cdot \vec{n} < 0$. Also, we have to show that the lights $\vec{l}$ and $\vec{s}$ do not cast shadows on $\mathcal{S}$; for otherwise, Equation 2 or 3 is not valid. Both can be easily demonstrated. Since

$$\vec{n} = \left(\frac{\partial z}{\partial x}, \frac{\partial z}{\partial y}, 1\right)$$

and

$$\frac{\partial z}{\partial x} = \frac{I - J}{I + J}$$

(from Equation 4), a quick calculation gives

$$\vec{l} \cdot \vec{n} = 1 + \frac{I - J}{I + J}$$

and

$$\vec{s} \cdot \vec{n} = 1 - \frac{I - J}{I + J},$$

which are both nonnegative everywhere on $\mathcal{S}$. Next, we show that there are no cast shadows. We note that for $\vec{p}$ to be in cast shadow under $\vec{l}$ (similarly for $\vec{s}$), the ray $\vec{p} + t\vec{l}$, where $t \geq 0$, must intersect $\mathcal{S}$ transversally (Figure 11.3), i.e., $\mathcal{S}$ must be on both sides of the ray. Since $\vec{l}$ has zero y component, the ray and $\vec{p}$ are on the plane $\mathbb{Y}_c$; i.e., we are over a scanline. So points of $\mathcal{S}$ that can cast a shadow on $\vec{p}$

must be on the right of $\vec{p}$ (and for $\vec{s}$, they must be on the left). Points on the right of $\vec{p}$ are of the form

$$\vec{q} = \vec{p} + \int_0^{t=w} (I\vec{l} - J\vec{s})\, dt = \vec{p} + \int_0^{t=w} I dt\, \vec{l} - \int_0^{t=w} J dt\, \vec{s} = \vec{p} + a\vec{l} - b\vec{s},$$

for some $w > 0$ and nonnegative numbers a, b. Because b is nonnegative, this immediately shows that $\mathcal{S}$ cannot intersect the ray transversally, and hence $\vec{p}$ is not in shadow.

This completes the proof of the proposition for $\vec{s} = (-1, 0, 1)$ and $\vec{l} = (1, 0, 1)$. For general $\vec{s}, \vec{l}$, the proof above can be modified by defining the planes $\mathbb{Y}_c$ to be the affine planes $\mathbb{Y}_p$ generated by $\vec{s}$ and $\vec{l}$: $\mathbb{Y}_p = \{x | x = \vec{p} + a\vec{s} + b\vec{l}, \vec{p} \in \mathbb{R}^3, a, b \in \mathbb{R}\}$. This will ensure that each solution $\vec{c}(t)$ to Equation 4 stays in one such plane. The rest of the proof carries through without much change.

11.2.1 Image Gradients as Illumination-Insensitive Measures

The negative result on the existence of illumination invariants is not as devastating as one may have thought. As far as face recognition is concerned, there are at least two ways out of this apparent quandary. While determining whether two images are of the same object is impossible in principle, nothing prevents us from doing so for three or more images. That is, we can increase the number of training images for each person in the database, and if qualitatively and quantitatively sufficient training images are available, indeterminacy can generally be avoided. We will discuss this type of approach in the next section.

On the other hand, Proposition 11.2.1 can be largely attributed to the unrestricted access to the space of Lambertian objects, since we can always find some Lambertian surface, however bizarre and strange it may be, to account for any two images. For example, the above theorem implies that given an image of Marilyn Monroe and one of Cary Grant, along with the light source directions, there exists a Lambertian surface that could produce these images. However, it is unlikely to be face-like. Therefore, it makes sense to limit the space of available Lambertian objects, e.g., to face-like objects. Alternatively, let's consider only *planar* Lambertian objects. It follows directly from Equation 1 that the image gradient is a discriminative illumination invariant: given two images I and J of some planar Lambertian object taken under the same viewpoint, their image gradients $\nabla I, \nabla J$ must be parallel at every pixel where they are defined. This is obvious because for a planar object, there is only one surface normal, and each image is simply a constant multiple of the albedo values, with the constant determined by the illumination. While the pixel intensity can be any allowable value, given appropriate illumination, its derivative—the image gradient—cannot. Probabilistically, the distribution of pixel values under varying illumination may be random, but the distribution of image gradients is not.

Unfortunately, the dependence of image gradient on albedo is only part of the story. For general nonplanar surfaces, the image gradient (for a given light source $\vec{s} = (s_u, s_v)$) is related to both the albedo (reflectance) and surface geometry:

$$\nabla I = \overbrace{\kappa_u s_u + \hat{v}\kappa_v s_v \hat{u}}^{\text{geometric}} + \underbrace{(\vec{s} \cdot \hat{n})\nabla\alpha}_{\text{reflectance}} . \tag{5}$$

In the above, κ_v, κ_u are the two principal curvatures, and $\hat{u}, \hat{v}$ are the corresponding principal directions at a given surface point.[5] For a planar object, $\kappa_v/\kappa_u = 0$ and $\nabla I = (\nabla\alpha)\vec{s} \cdot \hat{n}$. The geometric term in the equation above destroys the simple relation between image gradient and albedo gradient that we have for planar objects. However, for the case of uniform albedo (i.e., $\nabla\alpha = 0$), a deeper analysis using only the geometric term above reveals that the image gradient distribution is still not random. More specifically, for light sources with a directionally uniform distribution given by

$$\rho_s(\vec{s} = (s_u, s_v, s_n)) = \frac{1}{(\sqrt{2\pi}\sigma)^3} e^{-\frac{1}{2\sigma^2}(s_u^2+s_v^2+s_n^2)}, \quad s_n \in [0, \infty),$$

the image gradient distribution is

$$\rho(u, v) = \frac{1}{\pi^{\frac{3}{2}}\sigma^2\kappa_u\kappa_v} e^{\frac{1}{2\sigma^2}((\frac{u}{\kappa_u})^2+(\frac{v}{\kappa_v})^2)}.$$

This result strongly suggests that the joint distribution of two image gradients from two different images under two random lighting should not be random either. In its most general form, the probability density function for this joint distribution can be written as [10]:

$$\rho(r_1, \varphi_1, r_2, \varphi_2) = \int \rho(r_1, \varphi_1 - \gamma|\kappa, \vec{\alpha})\rho(r_2, \varphi_2 - \gamma|\kappa, \vec{\alpha})dP(\gamma, \kappa, \vec{\alpha}), \tag{6}$$

where $P(\gamma, \kappa, \alpha)$ is the probability measure on the nonobservable random variables that include the surface geometry (κ), albedo (α) and camera viewpoints (γ). In the expression above, r_i and φ_i are the magnitude and orientation of the image gradient, respectively. This expression can be simplified slightly by observing that when

[5] Equation 5 is really an equation in terms of a coordinate system $\hat{u}, \hat{v}$ at the tangent space of the surface, not the image plane. Following [10], we will ignore the effects of projection, and treat $\hat{u}, \hat{v}$ as directions in the image.

comparing two images, the absolute azimuth values (φ_i) are not important since we have no information on the relative orientation between the camera and the object. Instead, their difference, $\varphi = \varphi_1 - \varphi_2$, is. This "azimuthal symmetry" [10] allows us to write ρ as a function of three variables:

$$\rho(r_1, \varphi = \varphi_1 - \varphi_2, r_2) = \int \rho(r_1, (\varphi_1 - \varphi_2) - \gamma|\kappa, \vec{\alpha})\rho(r_2, -\gamma|\kappa, \vec{\alpha})dP(\gamma, \kappa, \vec{\alpha}). \tag{7}$$

Equation 7 is only of theoretical interest since the probability measure P on the nonobservables is unknown. However, we can try to reconstruct the distribution empirically using images of objects under varying illumination. In [10], 1280 images of 20 objects under 64 different illumination conditions were gathered. The objects included folded cloth, a computer keyboard, cups, a styrofoam mannequin, etc. The values $\rho(r_1, \varphi, r_2)$ were estimated directly from a histogram of image gradients. A slice of the joint probability density ρ is shown in Figure 11.4(left). Note that for a planar (or piecewise planar) Lambertian object, ρ is a delta function at $\varphi = 0$ (angular difference is 0). It is expected that, with contributions from surface geometry and other factors, ρ should be considerably more complicated for general objects. However, the shape of ρ with its prominent ridge at $\varphi = 0$ does resemble that of a delta function. Surface geometry accounts for most of the "spread" of the density from the line $\varphi = 0$. This shows that the statistical regularity of scene radiance gradient does reflect the intrinsic geometry and reflectance properties of surfaces, and this regularity can then be exploited for face recognition as detailed in Section 11.5.

11.3 THEORY AND FOUNDATIONAL RESULTS

Let $\mathcal{C}$ denote the set of images of an object $\mathcal{O}$ under all possible illumination conditions. One of the main goals of illumination modeling is to say something about $\mathcal{C}$. We assume that the images were taken from the same viewpoint, i.e., no pose variation, and the images are all of the same size. By the usual rasterization, we can regard $\mathcal{C}$ as a subset of the image space $\mathbb{R}^n$, with n the number of pixels in the image. In this section, we discuss the important results of [2, 4, 30], which give various characterizations of the set $\mathcal{C}$ when the object $\mathcal{O}$ is Lambertian and convex. There are two main themes, the effective low dimensionality of $\mathcal{C}$ and its linearity.

Before moving on, we fix a few conventions and notations. Interreflections will be ignored throughout all subsequent discussions, and all illumination will be assumed to be generated by distant sources. In particular, if the distant source l is a point source, then l can be represented as a 3-vector $\vec{l}$ such that $|\vec{l}|$ encodes the strength of the source and the unit vector $\vec{l}/|\vec{l}|$ represents its direction. Note

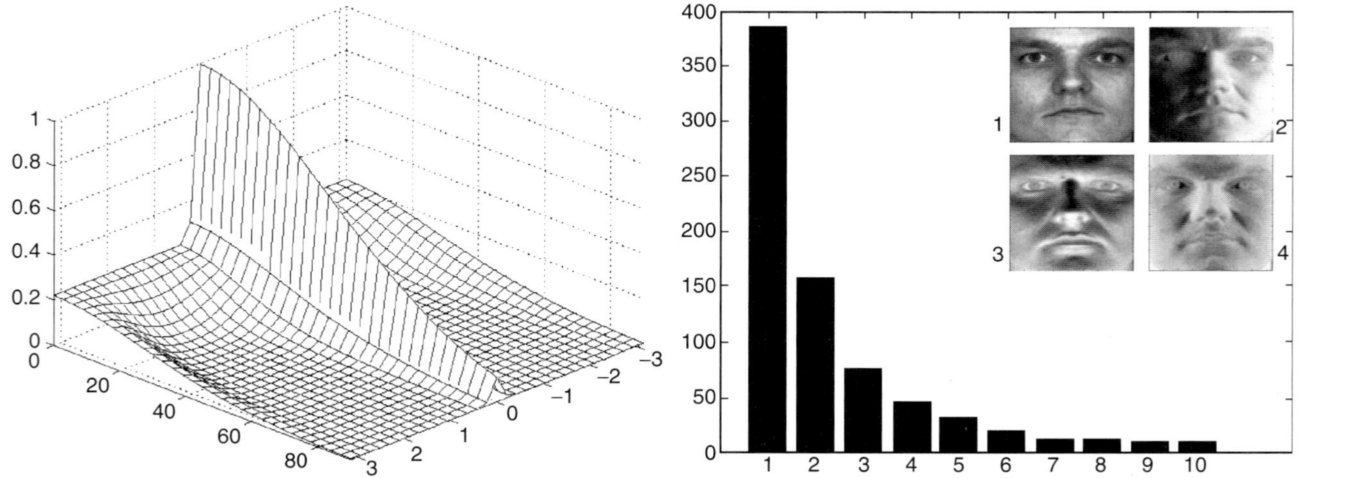

FIGURE 11.4: Left: (courtesy of [10]) empirical joint probability density of two image gradients $\rho(r_1, \varphi, r_2 = 50)$ under two random lighting conditions. Right: The magnitudes of the first ten singular values for the images in Figure 11.12. In this example, the first three eigenvalues account for more than 97% of the energy. The four eigenimages corresponding to the largest four eigenvalues are also displayed.

that the unit vector is a point on the sphere S^2, and conversely, every point $\vec{p}$ on S^2 can represent some distant point source with direction $\vec{p}$. More generally, any illumination condition can be represented as a nonnegative function on S^2. For an image, we will use the same symbol I to denote both the image *and* its associated vector in the n-pixel image space $\mathbb{R}^n$.

11.3.1 Early Empirical Observations

Convexity and low dimensionality are two important properties of $\mathcal{C}$. Convexity is a simple consequence of the superposition principle for illumination. For if I_1 and I_2 are two images taken under two different illumination conditions l_1 and l_2, any convex combination of these two images,

$$J = aI_1 + bI_2, \quad \text{where } a, b \geq 0 \quad \text{and} \quad a + b = 1,$$

is also an image of the same object under a new illumination condition specified by $al_1 \cup bl_2$, i.e., l_1 and l_2 are "turned on" simultaneously with attenuation factors a, b, respectively. This should not be confused with the illumination given by the point distant source $a\vec{l_1} + b\vec{l_2}$, when l_1, l_2 are distant point sources.

The fact that for objects with diffuse, Lambertian-like reflectance, the effective dimension of $\mathcal{C}$ is small and was also noticed quite early [11], [17]. This can be demonstrated by collecting images of an object taken under a number of different illumination conditions. If $\{I_1, \cdots, I_m\}$ are m such images, we can stack them horizontally to form the intensity matrix $I = [I_1 \cdots I_m]$. A singular-value decomposition (SVD) of I [16],

$$I = U\Sigma V^{\mathrm{T}}, \tag{8}$$

gives the singular vectors as the columns of the matrix U, and the diagonal elements of Σ as the singular values. Let $\{\sigma_1, \cdots, \sigma_m\}$ denote the singular values in descending order. These singular vectors are usually called eigenimages since they can be converted to matrix/image forms. The important fact implied by the factorization in Equation 8 is that the eigenimages can be used as a basis of a linear subspace which approximates the original images $\{I_1, \cdots, I_m\}$. If R denotes the subspace spanned by the k eigenimages associated with the k largest singular values, the L^2 reconstruction error,

$$\sum_{i=1}^{m} \mathrm{dist}^2_{L^2}(I_i, R) = \sum_{i=k+1}^{m} \sigma_i^2, \tag{9}$$

can be computed directly from the singular values. If σ_i turns out to be negligible for $i > k$, then the entire collection of images can be effective approximated using the subspace R. In particular, the effective dimension of $\{I_1, \cdots, I_m\}$ is simply the dimension of R, which is k.

Figure 11.4(right) displays the magnitude of the first ten singular values obtained by applying SVD to a collection of 45 images of a human face (in frontal pose, and under 45 different point light sources) shown in Figure 11.12 (Section 11.5). The magnitude of the singular values decreases rapidly after the first three singular values. In fact, the first three eigenvalues account for more than 97% of the entire energy. Here, the energy is defined as $\sum_{i=1}^{m} \sigma_i^2$. For a pure Lambertian object with simple geometry, this observation can be explained easily. Assuming $\{I_1, \cdots, I_m\}$ contains no shadows, then, the intensity matrix I can be factored as

$$I = B \cdot S = [\vec{n}_1 \cdots \vec{n}_n]^{\mathrm{T}} \cdot [\vec{s}_1 \cdots \vec{s}_k], \tag{10}$$

where B is a n-by-3 matrix containing the normals and albedos at each pixel, and s_i is the product of a light source's strength and direction. Since S can have a rank of at most 3, I also has rank of at most 3, and hence there are at most three nonzero singular values in Σ. For a general collection of images, the object is no longer Lambertian with simple geometry, and the lighting conditions are not describable by point sources. This means that there will be more than three nonzero singular values, and the extent of this "spread of singular values" depends on how many of the idealized assumptions have been violated.

Instead of SVD, the principal component analysis (PCA), in the spirit of [41], was used in [11] to give an analysis similar to the one above for images of non-Lambertian objects. These include objects with specular spikes, small cast shadows, and some other irregularities such as partial occlusions. The conclusion from this empirical study is surprising in that from 3 to 7 eigenimages are sufficient to model objects with wide range of reflectance properties. As mentioned earlier, at the lower end, 3 can be used to model Lambertian objects with simple geometry. In their conclusion, the first few eigenimages describe the Lambertian component, and the succeeding eigenimages describe the specular component and specular spikes, shadows, and so forth. This result is particularly encouraging because human faces are generally non-Lambertian. Still, a low-dimensional linear representation is already sufficient to capture a large portion of the possible image variations due to illumination.

Finally, we remark that the matrix factorization in the form similar to Equations 8 and 10 appears frequently in computer vision literature. Besides the example we discussed above, arguably the most well-known example in the literature is the structure from motion (SFM) algorithm of Tomasi and Kanade [38]. In both cases, matrix factorization can serve as the starting point for extracting important information from the images. In the SFM case, they are the 3D positions and camera parameters. In our example above, with extra work, we can recover the normal vectors of the object's surface as well as the lighting directions. This is photometric stereo in a nutshell [44]. The formulation we described above can be expanded in several directions. For example, the data intensity matrix I can contain images of different objects, or it can contain images taken with different poses. It should not be too surprising to learn that there are corresponding factorization algorithms for these more general formulations. For the former, [46] provides a factorization algorithm with rank constraints, and TensorFaces [40] has been applied to image collections containing both illumination and pose variations.

11.3.2 Modeling Reflectance and Illumination using Spherical Harmonics

The effective low dimensionality of $\mathcal{C}$ that we have just discussed clearly begs for explanation. Somewhat surprisingly, this empirical observation can be elegantly explained via a signal-processing framework using spherical harmonics [2, 29, 30]. The key conceptual advance is to treat a Lambertian object as a "low-pass filter" that turns complex external illumination into a smoothly shaded image. In the context of illumination, the signals are functions defined on the sphere, and spherical harmonics are the analogs of the Fourier basis functions.

First, we fix a local (x, y, z) coordinate system F_p at a point p on a convex Lambertian object such that the z axis coincides with the surface normal at p. Let (r, θ, ϕ) denote[6] the spherical coordinates centered at p. Under the assumption of homogeneous light sources, the configuration of lights that illuminates the object can be expressed as a nonnegative function $L(\theta, \phi)$ defined on S^2. The reflected radiance at p is then given by

$$r(p) = \rho \iint_{S^2} k(\theta)L(\theta, \phi)dA = \rho \int_0^{2\pi} \int_0^{\pi} k(\theta)L(\theta, \phi)\sin\theta d\theta d\phi, \tag{11}$$

where ρ is the albedo, and $k(\theta) = \max(\cos\theta, 0)$ is called the Lambertian kernel. Note that this equation is simply the integral form of Equation 1, in which we integrate over all possible incident directions at p. Because the normal at p coincides with the z axis, the Lambertian kernel is precisely the max term in Equation 1. For any other point q on the surface, the reflectance is computed by a similar integral as above. The only difference between the integrals at p and q is the lighting function L: at each point, L is expressed in a local coordinates system at that point. Therefore, considered as a function on the unit sphere, L_p and L_q differ by a rotation $g \in SO(3)$ that rotates the frame F_p to F_q. That is, $L_p(\theta, \phi) = L_q(g(\theta, \phi))$.

Since $k(\theta)$ and $L(\theta, \phi)$ are now functions on S^2, the natural thing to do next is to expand these functions in terms of some canonical basis functions, and spherical harmonics offer a convenient choice. Spherical harmonics, Y_{lm}, are a set

[6] To conform with the notation used in the spherical harmonics literature, θ denotes the elevation angle and ϕ denotes the azimuth angle.

of functions that form an orthonormal basis for the set of all square-integrable (L^2) functions defined on the unit sphere. They are the analog on the sphere to the Fourier basis on the line or circle. The function Y_{lm}, indexed by two integers l (degree) and m (order) obeying $l \geq 0$ and $-l \leq m \leq l$, has the following form:

$$Y_{lm}(\theta, \phi) = \begin{cases} N_{lm} P_l^{|m|}(\cos\theta)\cos(|m|\phi) & \text{if } m > 0, \\ N_{lm} P_l^{|m|}(\cos\theta) & \text{if } m = 0, \\ N_{lm} P_l^{|m|}(\cos\theta)\sin(|m|\phi) & \text{if } m < 0, \end{cases} \tag{12}$$

where N_{lm} is a normalization constant that guarantees the functions Y_{lm} are orthonormal in the L^2 sense:

$$\int_{S^2} Y_{lm}\, Y_{l'm'} dA = \delta_{mm'} \delta_{ll'}.$$

Here $P_l^{|m|}$ are the associated Legendre functions whose precise definition is not important here (however, see [37]). The formal definition of Y_{lm} using spherical coordinates above is somewhat awkward to work with for this purpose. Instead, it is usually more convenient to write Y_{lm} as a function of x, y, z rather than angles. Each $Y_{lm}(x, y, z)$ expressed in terms of (x, y, z) is a polynomial in (x, y, z) of degree l:

$$Y_{00} = \sqrt{\frac{1}{4\pi}}, \tag{13}$$

$$(Y_{11}; Y_{1-1}; Y_{10}) = \sqrt{\frac{3}{4\pi}}(x; y; z), \tag{14}$$

$$\begin{aligned} (Y_{21}; Y_{2-1}; Y_{2-2}) &= \sqrt{\frac{15}{4\pi}}(xz;\ yz;\ xy),\ (Y_{20}; Y_{22}) \\ &= \sqrt{\frac{5}{16\pi}}(3z^2 - 1;\ \sqrt{3}(x^2 - y^2)). \end{aligned} \tag{15}$$

In other words, spherical harmonics are just the restrictions of some homogeneous polynomials (in x, y, z) of degree l to S^2. While degree-two polynomials in x, y, z are six-dimensional, because $x^2 + y^2 + z^2 - 1 = 0$ on S^2, spherical harmonics of degree two are only five-dimensional. Using polynomials, it is straightforward to see that a rotated spherical harmonic is the linear superposition of spherical harmonics of the same degree: for a 3D rotation $g \in SO(3)$,

$$Y_{lm}(g(\theta, \phi)) = \sum_{n=-l}^{l} g^l_{mn} Y_{ln}(\theta, \phi). \tag{16}$$

The coefficients g^l_{nm} are real numbers and are determined by the rotation g. This is because a rotated homogeneous polynomial of degree l is a polynomial of the same degree, and hence it has to be a linear combination of the spherical harmonics of the same degree (see Proposition II 5.10 in [6]).

With these basic properties of spherical harmonics in hand, the idea of viewing a Lambertian object as a "low-pass filter" can be made precise by expanding the Lambertian kernel $k(\theta)$ in terms of Y_{lm}. Because $k(\theta)$ has no ϕ-dependency, its expansion, $k(\theta) = \sum_{l=0}^{\infty} k_l Y_{l0}$, has no Y_{lm} components with $m \neq 0$ (Equation 12). It can be shown [2, 30] that k_l vanishes for odd values of $l > 1$, and the even terms fall to zero rapidly; in addition, more than 99% of the L^2 energy of $k(\theta)$ is captured by its first three terms,[7] those with $l < 3$. See Figure 11.5(left). Because of these numerical properties of k_l and the orthogonality of the spherical harmonics, any high-frequency ($l > 2$) component of the lighting function $L(\theta, \phi)$ will be severely attenuated in evaluating the integral in Equation 11, and in this sense, the Lambertian kernel acts as a low-pass filter. Therefore, the reflected radiance computed using Equation 11 can be accurately approximated by the same integral with L replaced by L', obtained by truncating the harmonic expansion of L at $l > 2$; i.e., the spherical-harmonic expansion of L' contains no Y_{lm} with $l > 2$. Since rotations preserve the l-degree of the spherical harmonics (Equation 16), the same truncated L' will work at every point. Let $L'(\theta, \phi) = \sum_{i=1}^{9} l_i Y_i$ denote the expansion of L' and Y_i the nine spherical harmonics with degree less than three.[8] At any point q, we have

$$r(q) \approx \rho_q \iint_{S^2} k_q(g(\theta, \phi)) L'(\theta, \phi) dA = \rho_q \sum_{i=1}^{9} l_i \int_0^{2\pi} \int_0^{\pi} k_q(g(\theta, \phi)) Y_i dA, \tag{17}$$

where g is the rotation that rotates the local frame F_q at q to the frame F_p. We can define the nine *harmonic images* I_i whose intensity at each point (pixel) is $I_i(q) = \rho_q \int_{S^2} k_q(g(\theta, \phi)) Y_i dA$: images taken under the *virtual* lighting conditions

[7] The energy of $k(\theta)$ is, $\int_{S^2} k^2(\theta) dA$, which is the convergent infinite sum $\sum_{i=0}^{\infty} k_l^2$.

[8] $Y_1 = Y_{00}$, $Y_2 = Y_{11}$, $Y_3 = Y_{1-1}$, $Y_4 = Y_{10}$, $Y_5 = Y_{21}$, $Y_6 = Y_{2-1}$, $Y_7 = Y_{2-2}$, $Y_8 = Y_{20}$, $Y_9 = Y_{22}$.

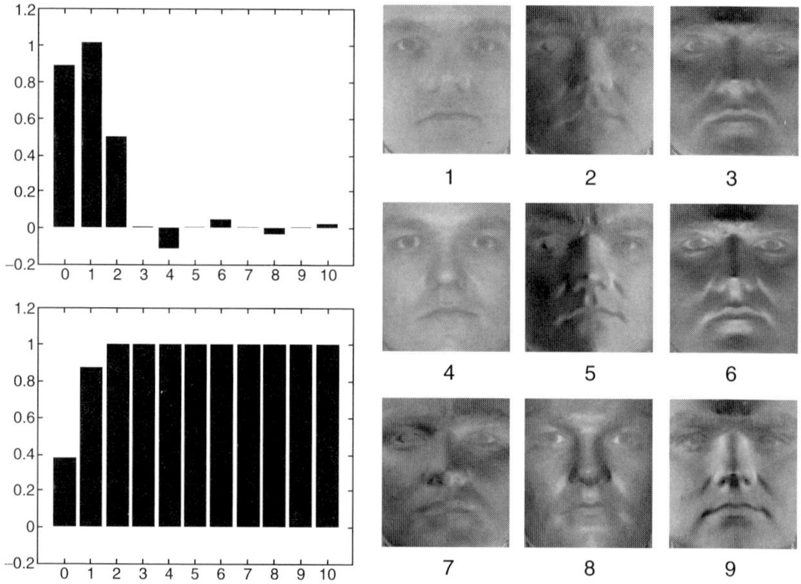

FIGURE 11.5: Analysis using spherical harmonics. Left (courtesy of [2]) top: A graph representation of the first eleven coefficients in the spherical harmonics expansion of the Lambertian kernel $k(\theta)$. The abscissa represents l, the degree of the spherical harmonics and the ordinate gives the coefficient of Y_{l0} in the expansion of $k(\theta)$. Left bottom: the cumulative energy. Here, the abscissa represents the degree l as before but the ordinate gives the cumulative energy. Note that 99% of the energy is captured for $l = 3$. Right: The nine harmonic images rendered by ray tracing. For examples of computing harmonic images without using ray tracing (Equation 19), see [29].

specified by the nine spherical harmonics. Hence, the pixelwise approximation above translates into the approximation for images.

$$I \approx \sum_{i=1}^{9} l_i I_i. \tag{18}$$

If I is an image taken under some illumination condition L with l_i as the nine coefficients in L's truncated spherical-harmonic expansion, I can be approximated by a linear combination of the nine harmonic images using the same coefficients. The far-reaching consequence of this fact is that, although lighting conditions are infinite-dimensional (the function space for $L(\theta, \phi)$), the illumination effects on a Lambertian object can be approximated by a nine-dimensional linear subspace $\mathcal{H}$, the harmonic subspace spanned by the harmonic images I_i; i.e., $\mathcal{C}$ can be approximated well by $\mathcal{H}$.

Harmonic Images

Equation 18 indicates the great importance of computing the harmonic images. Except for the first spherical harmonic (which is a constant), all others have negative values and therefore do not correspond to real lighting conditions. Hence, the corresponding harmonic images are not real images, as pointed out in [2], "they are abstractions." Nevertheless, they can be computed quickly if the object's surface normals and albedos are known.

Using the polynomial definition of spherical harmonics, the recipe for computing the nine harmonic images I_i $(1 \leq i \leq 9)$ is particularly simple: for each pixel p, let $\vec{n}_p = (x, y, z)$ denote the unit surface normal at p and ρ_p the albedo. The intensity value of I_i at p is given by

$$I_i(p) = \rho_p Y_i(x, y, z). \tag{19}$$

Another way to compute the harmonic images is to simulate the images under harmonic lightings by explicitly evaluating the integral $\rho_p \int_{S^2} k_p(g(\theta, \phi))Y_i dA$ at every point p and taking into account the cast shadows:

$$\rho_p \int_{S^2} k_p(g(\theta, \phi))\nu_p(\theta, \phi)Y_i dA, \tag{20}$$

where $\nu_p(\theta, \phi) = 1$ if the ray coming from direction (θ, ϕ) is not occluded by another point on the surface. Otherwise, $\nu_p(\theta, \phi) = 0$. Figure 11.5 shows the rendered harmonic images for a face taken from the Yale Face Database B. These synthetic images are rendered by sampling 1000 directions on a hemisphere, and the final images are the weighted sum of 1000 ray-traced images. By using the 3D information, these harmonic images also include the effects of cast shadows arising from the nonconvex human face.

11.3.3 Illumination Cones and Beyond

So far,we have shown that $\mathcal{C}$, the set of images of a convex Lambertian object under all possible illumination conditions, is a convex set in the image space, and $\mathcal{C}$ can be effectively approximated by a nine-dimensional linear subspace. However, we still do not know what $\mathcal{C}$ is. An explicit characterization of $\mathcal{C}$ was first studied and determined in [4]. This chapter shows that, for a convex Lambertian object, $\mathcal{C}$ is a polyhedral cone in the image space, and its complexity (i.e., number of generators) is quadratic in the number of distinct surface normals. In this subsection, we discuss this result and some of its implications. First, we recall that a cone in $\mathbb{R}^n$ is simply a convex subset of $\mathbb{R}^n$ that is invariant under nonnegative scalings: if x is in the cone, then λx is also in the cone for any nonnegative λ. A polyhedral cone is simply a cone with a *finite* number of generators $\{e_1, \cdots, e_l\}$: points of the cone are vectors $x \in \mathbb{R}^n$ that can be expressed as some nonnegative linear combination of the generators, $x = a_1e_1 + \cdots + a_le_l$ with $a_1, \cdots, a_l \geq 0$.

For a general object (without assumptions on reflectance and geometry), it is straightforward to establish that the set $\mathcal{C}$ is a cone, which is a direct consequence of the superposition property of illumination. This is the simplest and also the only characterization of $\mathcal{C}$ without any limiting assumptions on reflectance or geometry. Unfortunately, this result is not strong enough to be useful because it does not furnish an explicit scheme for computing $\mathcal{C}$. However, for a convex and Lambertian object, a much more precise analysis of $\mathcal{C}$ is available. Clearly, because of Equation 1, $\mathcal{C}$ will be intimately related to the distribution of the object's surface normals. The simplest starting point is the planar surface with varying albedo but only one normal vector across the entire surface. For such a surface, $\mathcal{C}$ is a ray, which is a cone with just one generator. Figure 11.6 illustrates the case for a piecewise planar surface with three distinct normals. From these two examples, one naturally wonders whether there exists a relation between the number of surface normals and the generators of $\mathcal{C}$.

Indeed, there is. It turns out that the number of generators of $\mathcal{C}$ is bounded by a quadratic expression in the number of surface normals. The main idea is to use the pseudolinear nature of Equation 1, which is both an image-formation as well as a shadow-formation equation, to explicitly enumerate a finite set of

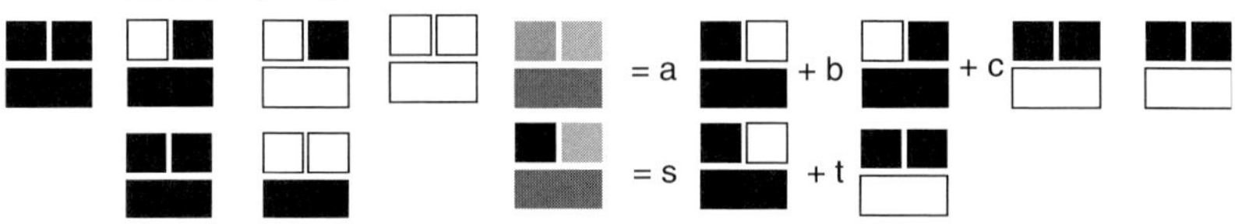

FIGURE 11.6: For a piecewise planar Lambertian surface with three distinct surface normals, there are at most three different intensity values for any given illumination. Schematically, we represent this by a three-pixel image. Assume that the surface normals are linearly independent. Left: eight shadowing configurations. Note that the three images in the second column provide a basis such that any image can be written as a nonnegative linear combination of these three images. Right: Examples of such linear combinations with $a, b, c, s, t > 0$.

"basis images" that can be used to compute $\mathcal{C}$. For two single distant lighting conditions, $\vec{L}_1$ and $\vec{L}_2$, it is easy to see that $I(\vec{L}_1 + \vec{L}_2) = I(\vec{L}_1) + I(\vec{L}_2)$ if $\vec{L}_1 \cdot \vec{n} \geq 0$ and $\vec{L}_2 \cdot \vec{n} \geq 0$ for all normal vectors $\vec{n}$ on the surface. This linearity breaks down when the two images $I(\vec{L}_1), I(\vec{L}_2)$ have different attached shadow configurations: there exists a surface normal $\vec{n}$, such that $\vec{L}_1 \cdot \vec{n}$ and $\vec{L}_2 \cdot \vec{n}$ have different signs. This indicates the possibility that we can construct basis images for each shadowing configuration, and for any given images (under single distance light source), it can then be generated by the basis images with the same shadow configuration (because linearity exists between such images). Figure 11.6 shows how a simplified form of such arithmetic can be applied to computing $\mathcal{C}$ for a surface with three distinct normals. From this simple illustration, one may come away with the impression that the number of different shadowing configurations depends exponentially on the number of surface normals. Fortunately, [4] shows that this is not so, and the actual dependence is only quadratic. This translates, after some work, into a bound on the number of generators of $\mathcal{C}$.

Therefore, at least for surfaces with finite number of different normals, we know the entire geometry of $\mathcal{C}$ from a finite number of images. For a general non-Lambertian object, there is no analogous result, since, depending on the complexity of the BRDF, the dependence of the number of generators on the number of surface normals can be more complicated than quadratic, or there may not exist a finite number of basis images at all (assuming the intensity values are not quantized). For a smooth convex Lambertian surface, as long as the image resolution is sufficiently high,[9] we can effectively approximate smooth surface normals for the set of points projecting to the same image pixels as identical, and associate an effective surface normal with each pixel in the image. This reduces the treatment for smooth surfaces to that of for piecewise planar surfaces.

PROPOSITION 11.3.1 For a convex Lambertian object, the set $\mathcal{C}$ of images of an object under all possible illumination conditions is a superior multilication sign polyhedral cone.

Below, we discuss briefly the proof of this proposition. Let I denote an image of a convex object with n pixels. Let $B \in \mathbb{R}^{3\times n}$ be a matrix where each row of B is the product of the albedo with the inward pointing unit normal for a point on the surface projecting to a particular pixel. Although the proposition deals with all possible illumination conditions, we can actually focus our analysis on single distant sources, thanks to the superposition principle. Thus, we need to examine the set $\mathcal{U}$ of images of a convex Lambertian surface created by varying the direction

[9]This implies that the radiance contribution from the surface patch that got projected to a pixel can be effectively approximated by one normal vector and albedo value.

and strength of a *single* point light source at infinity. It will turn out that $\mathcal{U}$ can be decomposed into a collection of polyhedral subcones $\mathcal{U}_i$ indexed by the set $\mathcal{S}$ of *shadowing configurations*:

$$\mathcal{U} = \{I|I = \max(Bs, 0),\ s \in \mathbb{R}^3\} = \bigcup_{i \in \mathcal{S}} \mathcal{U}_i, \tag{21}$$

As its name suggests, each element $i \in \mathcal{S}$ indexes a particular shadowing configuration, and $\mathcal{U}_i$ indexed by i is a set of images, each with the same pixels in shadow and the same pixels illuminated (images with the same "shadowing configuration"). Since the object is assumed to be convex, all shadows are attached shadows. The technical part of the proof for the above proposition is then to give a bound on the size of the set $\mathcal{S}$ as given in the following lemma:

LEMMA 11.3.2 The number of shadowing configurations is at most $m(m-1)+2$, where $m \leq n$ is the number of distinct surface normals.

To see how to count the elements in $\mathcal{S}$, we begin with a definition. As usual, we first ignore the max term in Equation 21. The product of B with all possible light source directions and magnitudes sweeps out a subspace in $\mathbb{R}^3$, and this has been called the illumination subspace $\mathcal{L}$, where

$$\mathcal{L} = \{I|x = Bs,\ s \in \mathbb{R}^3\}.$$

Note that the dimension of $\mathcal{L}$ equals the rank of B. Since B is an $n \times 3$ matrix, $\mathcal{L}$ will in general be a 3D subspace, and we will assume it to be so in the following discussion. The important point now is to observe that $\mathcal{L}$ slices through different orthants of $\mathbb{R}^n$. The most conspicuous one is the intersection of $\mathcal{L}$ with the nonnegative orthant of $\mathbb{R}^n$, and the intersection is nonempty because, when a single light source is parallel with the camera's optical axis, all visible points on the surfaces are illuminated, and consequently, all pixels in the image have nonzero values, and the image has no shadow. What can be said about the intersections of $\mathcal{L}$ with other orthants? Let $\mathcal{L}_i$ denote the intersection of $\mathcal{L}$ with an orthant i. Certain components of $x \in \mathcal{L}_i$ are always negative and others always greater than or equal to zero. To turn x into a real image, we have to apply the max term above and this leaves the nonnegative components of $x \in \mathcal{L}_i$ untouched, while the negative components of x go to zero. Note that this operation is a linear projection P_i on $\mathcal{L}_i$ that maps $\mathcal{L}_i$ to the closure of the nonnegative orthant. We then clearly have the decomposition of $\mathcal{U}$ into subcones:

$$\mathcal{U} = \bigcup_i P_i(\mathcal{L}_i),$$

$P_i(\mathcal{L}_i)$ is a cone because P_i is linear and L_i is a cone. In fact, we can identify each $P_i(\mathcal{L}_i)$ with $\mathcal{U}_i$ in Equation 21 because we can identify the set $\mathcal{S}$ with the set of orthants having nonempty intersection with $\mathcal{L}$, according to the discussion above. Although there are 2^n orthants in $\mathbb{R}^n$, we will see below that $\mathcal{L}$ can only intersect at most $m(m-1)+2$ orthants.

Representing all possible light-source directions by the usual two-sphere S^2, we see that, for a convex object, the set of light-source directions for which a given pixel in the image is illuminated corresponds to an open hemisphere; the set of light-source directions for which the pixel is shadowed corresponds to the other hemisphere of points. The boundary is the great circle defined by $\vec{n} \cdot s = 0$, where $\vec{n}$ is the normal at a point on the surface projecting to the given pixel. Each of the n pixels in the image has a corresponding great circle on the illumination sphere, and there are m distinct great circles in total, where m is the number of distinct surface normals. The collection of great circles carves up the surface of S^2 into a collection of cells S_i. See Figure 11.7. The collection of light source directions contained within a cell S_i on the sphere produces a set of images, each with the same pixels in shadow and the same pixels illuminated. This, again, immediately identifies the set $\mathcal{S}$ with the set of cells S_i, and hence, the size of $\mathcal{S}$ is the number of such cells on S^2. It is then a simple inductive argument, using the fact that two great circles intersect at two different points, to show that the number of cells S_i cannot exceed $m(m-1)+2$. Furthermore, the cone's generators are given by the images produced by light-sources at the intersection to two great circles. This then

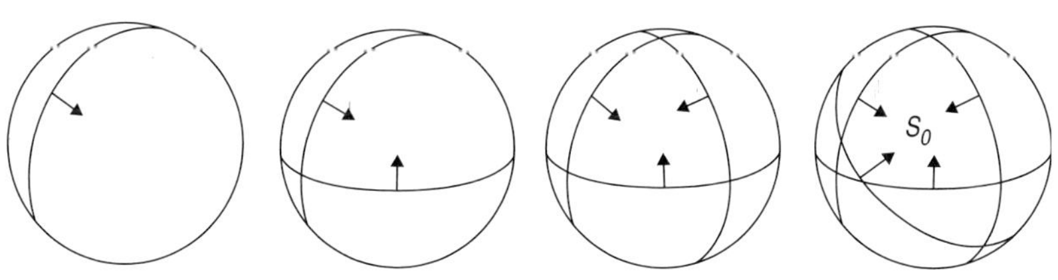

FIGURE 11.7: Great circles corresponding to individual pixels divide the sphere into cells of different shadowing configurations. The arrows indicate the hemisphere of light directions for which the particular pixel is illuminated. The generators (extreme rays) of the cone are given by the images produced by light-sources at the intersection of two circles.

immediately implies that the number of generators of $\mathcal{C}$ is quadratic in the number of distinct surface normal m.

With $\mathcal{U}$ understood, we can now construct the set $\mathcal{C}$ of all possible images of a convex Lambertian surface created by varying the direction and strength of an *arbitrary number* of point light-sources at infinity,

$$\mathcal{C} = \left\{ I | I = \sum_{i=1}^{k} \max(Bs_i, 0), s_i \in \mathbb{R}^3, k \in \mathbb{Z}^+ \right\},$$

where $\mathbb{Z}^+$ is the set of positive integers. The above discussion on $\mathcal{U}$ then immediately shows that $\mathcal{C}$ is a polyhedral cone.

Some Properties of an Illumination Cone

Since the illumination cone $\mathcal{C}$ is completely determined by the illumination subspace $\mathcal{L}$, the cone $\mathcal{C}$ can be determined uniquely if the surface normals scaled by albedo B were known. The method of uncalibrated photometric stereo ([42]) allows us to recover B up to an invertible 3×3 linear transformation $A \in GL(3)$,

$$Bs = (BA)(A^{-1}s) = B^*s^*,$$

by using as few as three images taken with unknown lighting since $\mathcal{L}$ is a 3D subspace. Although B is not uniquely determined, nevertheless, it is easy to see that B and B^* determine the same illumination subspace, and hence, the same illumination cone.

Another interesting result proven in [4] is that the actual dimension of $\mathcal{C}$ is equal to the number of distinct surface normals. For images with n pixels, this indicates that the dimension of the illumination cone is one for a planar object, is roughly $\sqrt{n}$ for a cylindrical object, and is n for a spherical object. It is to be noted, however, that having a cone span n dimensions does not mean that it covers $\mathbb{R}^n$. It is conceivable that an illumination cone could completely cover the positive orthant of $\mathbb{R}^n$. However, the existence of an object geometry that would produce this is unlikely, since for such an object, it must be possible to choose n light-source directions such that each of the n facets (pixels) are illuminated independently. On the other hand, a cone that covers the entire positive orthant cannot be approximated by a low-dimensional linear subspace, and this would contradict our analysis in the previous section using spherical harmonics. In particular, the result from the previous section indicates that the shape of the cone is "flat" with most of its volume concentrated near a low-dimensional subspace. From a face-recognition viewpoint, this is encouraging because it indicates the possibility that the illumination cones for different faces are small (compared with the ambient space $\mathbb{R}^n$) and well separated. Recognition using illumination cones should then be possible, even under extreme lighting conditions.

To compute an illumination cone, we need to obtain the illumination basis (generators of the illumination cone) first. However, these basis images all belong to the boundary of the illumination cone and therefore, compared to images in the interior, these boundary images are closer to images in other illumination cones (from other individuals). From a face-recognition viewpoint, they are the difficult images to recognize correctly. Conversely, images in cone's interior are relatively easy provided that different illumination cones do not have significant intersections. They typically include images taken under diffused ambient lighting conditions and with little or no shadowing on them. Reference [25] contains some preliminary experiment results supporting this observation. Combining these two observations, we have an explanation for the obvious fact that it is harder to recognize faces that are shadowed than those that are not.

Illumination Bases Are Not Equal

While the illumination cone provides a satisfying characterization of $\mathcal{C}$, its exact computation is, in principle, not feasible for most objects. This is because the number of basis images (generators) for an illumination cone is quadratic in the number of distinct surface normals, and for many objects, this number is on the same order as the number of pixels. Both time and space requirements for enumerating all generators can be formidable. For instance, for a typical 200×200 image, there are roughly 1.6 billion generators. Each generator is stored as a 200×200 image, and hence it requires at least 64,000 gigabytes to store all generators—indeed a formidable requirement for just one illumination cone. However, from a face-recognition viewpoint, knowing the entire cone is not really necessary. An illumination cone can contain images with unusual appearances taken under some uncommon illumination conditions, such as images with only a few bright pixels. What we would like to know is the part of the illumination cone that contains images under common lighting conditions, such as under smooth ambient illumination.

This idea can be made more precise as follows. While the harmonic subspace $\mathcal{H}$ is a nine-dimensional subspace approximating the illumination cone $\mathcal{C}$, we would like to find a subspace $\mathcal{R}$ (of the same or different dimension), with a *basis* formed by the generators of $\mathcal{C}$, such that $\mathcal{R}$ also approximates $\mathcal{C}$ well. The benefit of replacing $\mathcal{H}$ with $\mathcal{R}$ is that a basis of $\mathcal{R}$ now consists of *real* images taken under *real* lighting conditions. Taking these images as training images, a linear subspace can be immediately computed as the span of these images without recourse to estimating surface normals and albedos and to rendering images.

The discussion above can be formulated as a computational problem [24]. Let $\mathbb{D}$ be a collection of lighting conditions, and we want to determine a subset

$\{s_1, \cdots, s_n\}$ of $\mathbb{D}$ such that images $\{I_{s_1}, \cdots, I_{s_n}\}$ taken under $\{s_1, \cdots, s_n\}$ span a subspace $\mathcal{R}$ that approximates $\mathcal{C}$ well. $\mathbb{D}$ can be, for instance, the set of generators of an illumination cone, or a set of points sampled on S^2. An algorithm for computing the subset $\{s_1, \cdots, s_n\}$ is presented in [24, 25]. One possible way to solve the problem is to enumerate all possible subspaces "generated" by points in $\mathbb{D}$ and compute how good of a fit it has to the original cone. However, in practice, it is not possible to do so, and therefore [24, 25] chose a different solution. Instead, a nested sequence of linear subspaces, $R_0 \subseteq R_1 \subseteq \ldots \subseteq R_i \cdots \subseteq R_9 = R$, with R_i an i-dimensional subspace and $i \geq 0$, is computed. The idea is to make sure that all subspaces are as close to the harmonic subspace as possible. Since $\mathcal{H}$ approximates $\mathcal{C}$, and if R_i is close to $\mathcal{H}$, then R_i should also approximate $\mathcal{C}$ well. In [24, 25], the nested sequence of linear subspaces is computed iteratively by finding $s \in \mathbb{D}$ at each iteration that satisfies

$$s_i = \underset{s \in \mathbb{D}_{i-1}}{\operatorname{argmax}} \frac{\operatorname{dist}(I(s), R_{i-1})}{\operatorname{dist}(I(s), \mathcal{H})}. \tag{22}$$

Here dist is the usual L^2 distance between a subspace and a vector; $I(s)$ denotes the image formed under a light-source in direction s; and $\mathbb{D}_{i-1}$ denotes the set obtained by deleting i elements from $\mathbb{D}$. The next subspace R_i in the sequence is the subspace spanned by R_{i-1} and s_i.

To actually solve the optimization problem above, we have to know the images under lighting conditions in $\mathbb{D}$. Assuming human faces are Lambertian, this can be accomplished as before by rendering images under lighting conditions $s \in \mathbb{D}$ if face geometry and reflectance are known. In [24], a collection of 1005 sampled points on S^2 is used to define the domain $\mathbb{D}$ for the optimization problem posed in Equation 22 using face images and 3D models for ten individuals from the Yale Face Database B.

For the five faces shown in Figure 11.8, the results of computing the 9-dimensional linear subspace R are shown beneath their respective images. Since all lights are sampled from S^2, spherical coordinates are used to denote the light positions. The coordinates frame used in the computation is defined such that the center of the face is located at the origin, and the nose is facing toward the positive z axis. The x and y axes are parallel to the horizontal and vertical axes of the image plane, respectively. The spherical coordinates are expressed as (ϕ, θ) (in degrees), where $0 \leq \phi \leq 180$ is the elevation angle (angle between the polar axis and the z axis), and $-180 \leq \theta \leq 180$ is the azimuth angle. It is worthwhile to note that the set of nine lighting directions chosen by the algorithm has a particular type of configuration. The first two directions chosen are frontal directions (with small values of ϕ). The first direction chosen, by definition, is always the one whose image is closest to H, and in most cases, it is the direct frontal light given by $\phi = 0$. Second, after the frontal images are chosen, the next five directions are from the

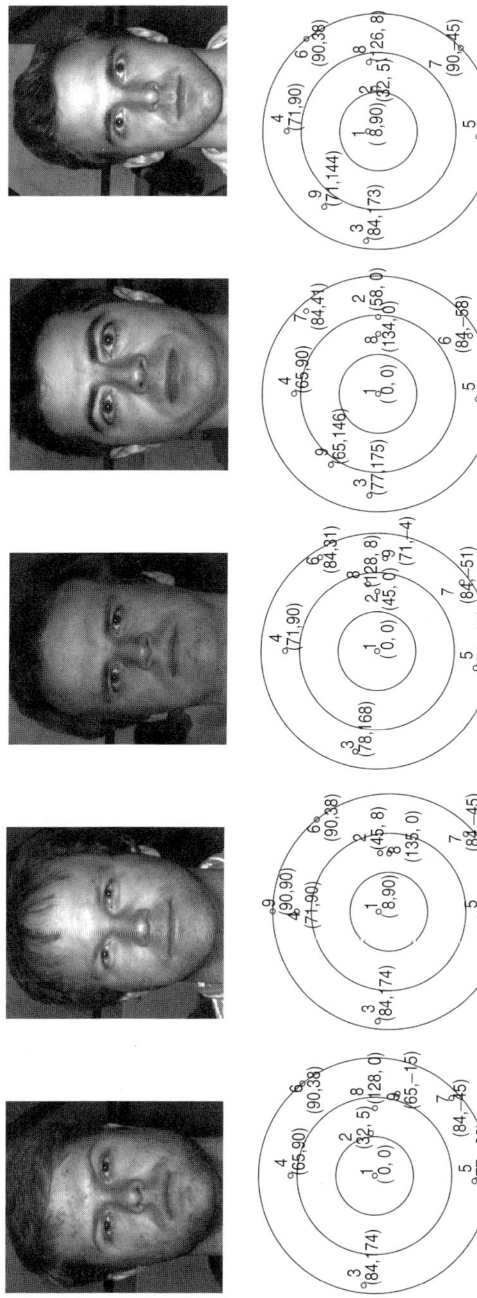

FIGURE 11.8: Experiment results for selecting basis images. Top row: five of the ten faces in the Yale database used in the experiment. Bottom row: the nine lighting directions found by maximizing Equation 22 for the five faces above. The directions are represented in spherical coordinates (ϕ, θ) centered at the face; see [24].

sides (with $\phi \approx 90$). By examining the θ values of these directions, we see that these directions spread in a quasi-uniformly manner around the lateral rim. And finally, the last chosen direction seems to be random. It is important to note that it is by no means clear *a priori* that our algorithm based on maximizing Equation 22 will favor such type of configurations. Furthermore and most importantly, the resulting configurations across all individuals are very similar.

This similarity strongly suggests that we can generalize the results here to all individuals. That is, there may exist a configuration of nine (or fewer) lighting directions such that, for any individual, the subspace spanned by images taken under these lighting conditions approximates the illumination cones $\mathcal{C}$. As we will discuss in the following section, the face-recognition algorithm proposed in [24, 25] is simply an algorithm for computing one such configuration of lighting directions. For face recognition, this configuration is particularly beneficial because it tells that, under these prespecified lighting conditions, we can then gather training of any individual, and the subspace spanned by these images will form an effective representation.

11.4 MENAGERIE

In this section, we discuss several recently published algorithms [2, 10, 14, 24, 35, 43] for face-recognition under varying illumination. All of these used an image-based metric, and except for [10], the common feature among them being the ability to produce a low-dimensional linear representation that models the illumination effect using only a handful of training images. The justification for such a generalization is provided by the discussions in the previous sections.

The six algorithms proposed in [2, 10, 24, 14, 35, 43] can roughly be categorized into two types: algorithms that explicitly estimate face geometry (shape and/or surface normals as well as albedos [2, 14, 35, 43] and algorithms that do not [10, 24]. Knowing surface normals allows one to recover the 3D structure by integration and therefore, it is possible to render images of a human face under different illumination conditions. In particular, 3D information allows the modeling of cast shadows by using, for example, ray tracing [14]. The dimensionality of the datum (images) can be reduced, for example, using principal-component analysis [35]. On the reduced space, various classifiers (such as support-vector machine and nearest-neighbors classifier) can be brought to bear on these simulated images. The approach of harmonic subspaces [2, 43] provides one way of utilizing surface normals without explicitly computing the 3D structure. Since the basis images are simply polynomials of the surface normals and albedos, they can be easily computed if the normals and albedos are known. Obviously, a part of these algorithms is the recovery of surface normals and albedos using a few training images. This can be accomplished either using photometric stereo techniques [14] or employing probabilistic methods using some learned prior distributions of normals and albedos [35, 43]. Though not discussed here, other shape reconstruction techniques such as stereo and laser range finders, could be used to acquire geometry and registered reflectance.

Papers [10] and [24] offer algorithms that do not require normal and albedo information. Reference [10] proposes an algorithm based on their joint probability density distribution (pdf) (see Equation 6). The joint pdf is obtained empirically, and recognition is performed by calculating the maximum likelihood using this pdf. Reference [24] is perhaps the simplest algorithm. The paper shows that a set of training images (as few as five) taken under prescribed lighting conditions is sufficient to yield good recognition results. The key here is to obtain a configuration of lighting conditions such that the (training) images taken under these lighting conditions form a basis of a linear subspace that approximates the illumination cone well.

Georghiades et al. [14]

In this algorithm, surface normals and albedos of the face are recovered using photometric stereo techniques [5, 42], and the 3D shape of the face is obtained by integrating the normal vector fields. Once the normals and albedos are known, it is possible to render synthetic images under new lighting conditions by applying Equation 1 directly. However, to account for the cast shadows, a simple ray tracer is employed. In both cases, the simulated images are all under distant point sources, and they can be interpreted as generators of the illumination cone. After sufficiently many images have been sampled, there are two ways to produce appearance models. One can apply principal-component analysis (PCA) to produce a low-dimensional linear representation, or one can use the cone generated by the sampled images directly. The difference between subspace and cone models is how the projection (closest image in the representation) is computed when matching a query image x: in both cases, the projection is defined by minimizing the reconstruction error:

$$\min \|x - (a_1 e_1 + \cdots , a_s e_s)\|^2.$$

In the subspace model, the e_i are the basis vectors, and the coefficients a_i are real numbers. In the cone case, the e_i are the generators (extreme rays), and the coefficients are subjected to the nonnegativity constraint, $a_i \geq 0$. Because of this constraint, determining the a_i becomes a convex programming problem, which, fortunately, can be solved efficiently.

Next, we mention briefly how 3D reconstruction is accomplished in this chapter. Using photometric stereo [42], the problem is the following: given a collection of training images $\{I_1, \cdots , I_k\}$, we want to find matrices B and S that minimize the "reconstruction error":

$$\min_{B,S} \|X - BS\|^2 \tag{23}$$

where $X = [I_1, \cdots, I_k]$ is the intensity matrix of k images (in vector form), and S is a $3 \times k$ matrix whose columns s_i are the light-source directions scaled by their intensities for all k images. B is a $n \times 3$ matrix whose rows are the normal vectors. Given X, the matrices B and S can be estimated using SVD [16]. However, there are three complications. First, a straightforward application of SVD is not robust since minimizing Equation 23 at shadowed pixels (both attached and cast shadows) is incorrect. The solution is to consider entries of X corresponding to shadowed pixels as missing values, and SVD with missing values [19, 32] is used instead of regular SVD. The second complication arises from the fact the the normal vector field B estimated by SVD is in general not integrable, i.e., it is not the normal vector field of a smooth surface. However, it is possible to efficiently compute an integrable normal vector field that has minimal L^2 distance to B using the discrete cosine transform [12]. The overall strategy to minimize Equation 23 is to minimize B and S separately and iteratively: each time B has been estimated, we find an integral normal vector field that is closest to B in the L^2 sense, and then S is estimated using this integrable field. The third complication arises from the fact that the pair B, S estimated from the factorization $X = BS$ is not unique. In fact, for any nonsingular 3×3 matrix G, the product of BG and $G^{-1}S$ is also X. Integrability of BG and B requires that G belongs to a three-dimensional subgroup of $GL(3)$, the generalized bas-relief transformations (a GBR transformation scales the surface and introduces an additive plane) [5]. Therefore, the reconstruction of the surface geometry outlined above is only up to some (unknown) GBR transformation. In [14], symmetry and face-specific information are exploited to resolve this ambiguity. Some reconstruction results from this chapter are shown in Figure 11.9. A more sophisticated reconstruction algorithm using non-Lambertian reflectance functions has been proposed recently [13].

The method proposed in [14] is a generative algorithm in that images under new illumination *and* pose conditions can be simulated. The single-pose recognition algorithm we just discussed can be generalized immediately to multiple poses by sampling the pose space, and constructing an illumination cone for each one. Each query image is then tested against all these illumination cones to determine the recognition result.

Basri and Jacobs [2]

The face-recognition algorithm proposed by the authors is a straightforward application of their illumination model based on spherical harmonics. Similar to the

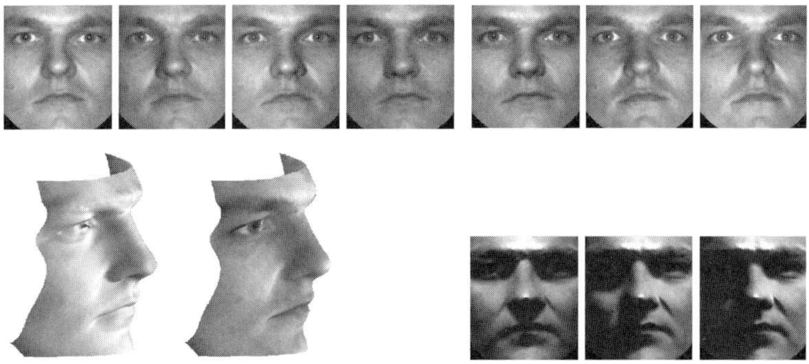

FIGURE 11.9: 3D reconstruction of a human face. Top: the seven training images. Bottom left: reconstruction results, (left) the surface is rendered with flat shading (constant albedo), (right) rendered using estimated albedos. Bottom right: three synthesized images with new lighting conditions. Note the large variations in shading and shadowing as compared to the seven training images above.

preceding algorithm, it is also a subspace-based algorithm in that the appearance model for each individual in the database is a nine-dimensional linear subspace spanned by the nine harmonic images. Assuming Lambertian reflectance, this subspace will capture more than 99% of the variance in pixel intensities. Since a harmonic image is simply a product of albedos and a polynomial (with degree less than three) in the components of the normal vectors, the nine basis images can be immediately obtained once the normals and albedos are known. The analytic description of the subspace is the strength of this algorithm, and it enables us to compute the subspace without simulating any images.

Let $B = [b_1, \cdots, b_9]$ be the matrix whose columns are harmonic images (of an individual). The face-recognition algorithm is based on computing the L^2 reconstruction error, and for a query image x, it is given by

$$\min_a \|Ba - x\|^2, \qquad (24)$$

where a can be any 9-by-1 vector. Experiment results reported in [2] have shown that the recognition algorithm based on this minimal L^2 reconstruction error has good performance. However, without any constraint on a, it is possible that the illumination condition implied by a is not physically realizable, i.e., the function $l = a_1 Y_1 + \cdots + a_9 Y_9$ has negative values somewhere on S^2. The constrained version of Equation 24 in this context is slightly harder to formulate. We start with a lighting configuration given by a collection of J point lights represented by the delta function $\delta_{\theta_j \phi_j}$,

$$l = \sum_{j=1}^{J} a_j \delta_{\theta_j \phi_j} = \sum_{j=1}^{J} a_j \sum_{n=0}^{\infty} \sum_{m=-n}^{n} Y_{nm}(\theta_j, \phi_j) Y_{nm}.$$

As before, any physically realizable lighting conditions can be approximated to an arbitrary precision using sufficiently large J and appropriate delta functions. The point, of course, is that a_i in the above equation are all nonzero, and we can rewrite equations Equation 24 above using l. Specifically, we need a matrix H that relates the delta functions $\delta_{\theta_i \phi_j}$ and the spherical harmonics. Let H be a matrix that contains a sampling of the harmonic functions, with its rows containing the transforms of the delta functions. Equation 24 can be rewritten as

$$\min_{a} \|BH^{\mathrm{T}}a - x\| \text{ s.t. } a \geq 0.$$

This gives the constrained version of the linear problem, and it guarantees that the resulting lighting configuration is physically realizable. However, the experimental results reported in [2] do not indicate any visible difference in performance between the two variations described above.

Surface normals and albedos are unexpendable components in the previous two algorithms, and photometric stereo is a commonly used technique for estimating normals and albedos. However, photometric stereo generally requires more than three images (under different lighting conditions) in order to unambiguously estimate the surface normals at every pixel. What is needed is an algorithm that estimates the normals and albedos from as few training images as possible, and Sim and Kanade's algorithm below does that for just one image.

The nonexistence of illumination invariants discussed previously has shown that it is impossible to recover the normals and albedos from one image directly. However, as we pointed out earlier, the results presented in Section 11.2 were derived without any assumption on the geometry and reflectance of the object. It is possible, however, to estimate the normals and albedos reasonably accurately if some useful prior has been given, and this is precisely what the following two algorithms strive to accomplish: given one image of an individual and some learned priors, normals and albedos are estimated based on some maximal-likelihood estimation process.

Sim and Kanade [35]

In this method, the illumination model is the usual Lambertian model augmented with an additional term $e(x, s)$:

$$i(x) = b(x)^{\mathrm{T}} s + e(x, s). \tag{25}$$

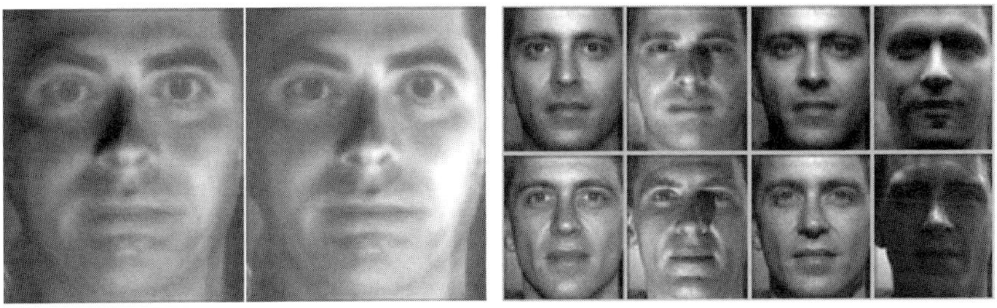

FIGURE 11.10: Hallucinated images (courtesy of [35]). Left: image rendered using strict Lambertian equation (without e in Equation 25) and one that uses the error term e, where the specular reflection on the left cheek is more accurately rendered. Right: four synthetic images using estimated normals $n(x)$ and $e(x, s)$ (top row) and actual images under the same illumination (bottom row).

Here, as before, $i(x)$ stands for the intensity at pixel x, $b(x)$ the albedo-scaled normal, and s is the direction of some single distant lighting. The extra e term models the effective ambient illumination and it depends both on x and s. With aligned images, it is assumed that the normals of human faces at pixel x forms a Gaussian distribution with some mean $\mu_b(x)$ and covariance matrix $C_b(x)$. Similarly, $e(x, s)$ is also assumed to form a Gaussian distribution, with mean $\mu_e(x, s)$ and variance $\sigma_e^2(x, s)$. All these parameters can be estimated from a collection of images with known normals and lighting directions. In addition, normals at different pixels are assumed to be independent, and this assumption makes the following MAP procedure much simpler.

Once the distributions for $b(x)$ and $e(x, s)$ have been obtained, we can estimate the normals at each pixel of a given image using Equation 25. Specifically, for an given image, we first estimate the unknown illumination s ([44, 45]). This allows $\mu_e(x, s)$ and $\sigma_e^2(x, s)$ to be computed. Then $b(x)$ can be recovered as a *maximum a posteriori* (MAP) estimate, $b_{\text{MAP}}(x) = \arg\max_{b(x)} \Pr(b(x)|i(x))$, where $i(x)$ is given by Equation 25. Simulated images under new illumination can be rendered using the estimated normals and error term e. See Figure 11.10. Face recognition proceeds exactly as before: for each individual in the database with one training image, we first estimate the normals and the error term e. Images under novel illumination conditions are simulated, and a linear subspace is computed by applying PCA to the simulated images.

Zhang and Samaras [43]

Note that using the estimated normals above, we might as well compute the nine-dimensional harmonic subspaces using the normals, and therefore, avoid simulating images. In fact, we can do better by directly estimating the nine harmonic images from just one training image, and an algorithm for doing this appeared in [43]. The starting point is an equation that is similar to Equation 25:

$$i(x) = b(x)^{\mathrm{T}}\alpha + e(x, \alpha). \tag{26}$$

Here the new $b(x)$ is a 9-by-1 vector that encodes the pixel values of the nine harmonic images at x, and e is the error term exactly as before. In place of s, we have a 9-by-1 vector α that represents the nine coefficients in the truncated spherical harmonics expansion of s. We can assume that $b(x)$ forms a Gaussian distribution at each pixel x, with some mean $\mu_b(x)$ and covariance matrix $C_b(x)$. As before, these parameters can be estimated from a collection of training images. Once they have been computed, for any given image, we can estimate $b(x)$, and hence the nine harmonic images, at each pixel.

Chen et al. [10]

Unlike the previous algorithms, this one does not estimate surface normals and albedos, and it requires only one single training image. It is essentially probabilistic, similar to the algorithms of Zhang and Samaras and Sim and Kanade, in the sense that the algorithm depends critically on a prior distribution. In this case, the distribution is on the angles between image gradients, and it is obtained empirically, instead of analytically. As we discussed in Section 11.2, the joint density distribution ρ for two image gradients can be used as an illumination insensitive measure. If we treat each pixel independently, the joint probability of observing the images gradients, $\nabla I, \nabla J$, of two images I and J of the same object is

$$P(\nabla I, \nabla J) = \prod_{i \in \text{Pixels}} \rho(\nabla I_i, \nabla J_i) = \prod_{i \in \text{Pixels}} \rho(r_1(i), \phi(i), r_2(i)), \tag{27}$$

where $r_1(i) = |\nabla I(i)|, r_2(i) = |\nabla J(i)|$, and ϕ is the angle between the two gradient vectors. With this probability value, it is then quite straightforward to come up with a face-recognition algorithm. Given a query image I, we compute $P(\nabla I, \nabla J)$ for every training image J using the empirically determined probability distribution ρ. The one training image having the largest P value is considered to be the likeliest to have come from the same face as the query image I. Therefore, no subspace is involved for this algorithm, and the computation is exceptionally fast and efficient. The obvious drawback is that we need to know how to evaluate (at least empirically) the joint density function ρ, and determining ρ accurately may require great effort.

Lee et al. [25]

Implementation-wise, this is perhaps the simplest algorithm. In this algorithm, surface normals and albedos are not needed and there is no need to simulate images under novel lighting conditions. The main insight here is to use certain configuration of lighting positions such that images taken under these lighting positions can serve as basis vectors for a subspace of the image space for recognition. This particular configuration, named the "universal configuration" in [24], can be computed for a small number of models (faces), and then it can be applied to all faces.

Suppose l models (faces) are available with sufficient information (normals and albedos) that we can simulate these faces under any new lighting condition. Given a set of sampled directions $\mathbb{D}$, we seek a fixed configuration of nine lighting directions for all l faces such that, for each face, on average, the linear space spanned by the images taken under these lighting conditions is a good linear approximation to the illumination cone. To find such a configuration, [24] tries to obtain a nested sequence of linear subspaces, $R_0 \subseteq R_1 \subseteq \ldots \subseteq R_i \cdots \subseteq R_9 = R$, by iteratively maximizing the average of the quotient in Equation 22 over all the available faces:

$$x_i = \arg \max_{x \in \mathbb{D}_{i-1}} \sum_{k=1}^{l} \frac{\text{dist}(x^k, R^k_{i-1})}{\text{dist}(x^k, H^k)}. \tag{28}$$

Since we are computing Equation 22 for all the available face models (indexed by k) simultaneously, for each $x \in \mathbb{D}_{i-1}$, x^{k}-denotes the image of model k taken under a single light source with direction x. $\mathbb{D}_{i-1}$ denotes the set obtained by deleting i elements from Ω. With k indexing the available face models, H^k denotes the harmonic subspace of model k, and R^k_{i-1} represents the linear subspace spanned by the images $\{x^k_1, \cdots, x^k_i\}$ of model k under light-source directions $\{x_1, \cdots, x_i\}$. [24] computes a R using a set of 200 uniformly sampled points on the "frontal hemisphere" (the hemisphere in front of the face). The resulting configuration as well as the 200 samples on the hemisphere are plotted in Figure 11.11. In the next section, this set of nested linear subspaces $R_0 \subseteq R_1 \subseteq \ldots \subseteq R_i \cdots \subseteq R_9 = R$ will be applied in face-recognition experiments.

11.5 EXPERIMENTS AND RESULTS

In this section, we discuss the performance of the face-recognition algorithms summarized in the previous section. With one exception ([10]), all algorithms are subspace-based algorithms. Since they are all image-based algorithms, low-level imaging processing such as edge and feature detections are unnecessary. Experiment results reported below will demonstrate that these recognition algorithms are quite robust against illumination variation. In addition, because L^2 differences can be quickly computed using a small number of matrix operations, they are efficient and easy to implement as well. However, the algorithms differ from each other

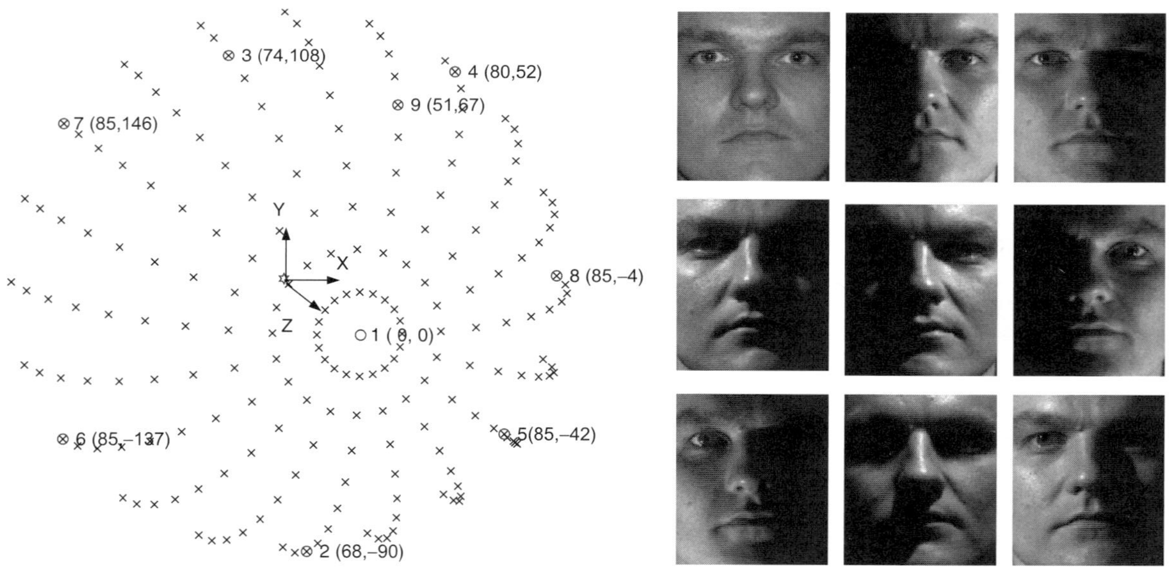

FIGURE 11.11: Left: the universal configuration of nine light-source directions with all 200 sample points plots on a hemisphere.Spherical coordinates (ϕ, θ) (on S^2) are used here, and the nine directions in spherical coordinates are $\{(0, 0), (68, -90), (74, 108), (80, 52), (85, -42), (85, -137), (85, 146), (85, -4), (51, 67)$. Right: nine images of a person illuminated by lights from the universal configuration.

in two fundamental ways: by the number of training images they require and by ways the subspaces are computed from the training images.

The algorithms are tested below using two face databases, Yale face database B [14] and the PIE (pose, illumination, and expression) database from CMU [33]. In the past few years, they have become the *de facto* standards for researchers working on illumination effect and face recognition. Both databases contain many images of different individuals taken under various illumination and viewing (pose) conditions. In the case of CMU PIE database, expression variation is also included. For the experiments below, only the illumination part of the databases will be used. The PIE database (see [34] for more information) contains 1587 images of 69 individuals and 23 different illumination conditions. The original (older) Yale database contains 10 individuals, and each individual has 45 different illumination conditions (a sample of Yale database is shown in Figure 11.12; see [14] for more details). Later, the number of individuals had increased to 38 in the extended Yale database. The images in the Yale database are grouped into four subsets according to the lighting angle with respect to the camera axis. The first two subsets cover the angular range 0° to 25°, the third subset covers 25° to 50°, and the fourth subset covers 50° to 77°. As the lighting direction moves from the frontal to the lateral positions, both attached and cast shadows develop prominently in the resulting images. These heavily shadowed images (subset four) are the most challenging for face recognition.

11.5.1 Results

Table 11.1 summarizes the experiment results reported in the literature for all the algorithms (except that of Sim and Kanade) discussed in the previous section. The original Yale face database (10 individuals, 450 images) is used in this experiment. The first five rows contain the results of using "quick-fix" algorithms without significant illumination modeling. The next eight rows display the results of using more sophisticated illumination modeling. The difference in performances between these two categories of algorithm is apparent: while the total error rates for the former category hover above 20%, algorithms in the later category can achieve less than 1% in error rates. Note that different algorithms require different numbers of training images and, in evaluating algorithm performance, we have tried to use the same number of training images whenever possible.

Before going further, we briefly describe these six "quick-fix" algorithms [14]. Correlation is a nearest-neighbor classifier in the image space [7] in which all of the images are normalized to have zero mean and unit variance. In this experiment, we take frontally illuminated images from subset 1 as training images and calculate the correlations between these normalized training images and each (normalized) query image. "9NN" is a straightforward implementation of the nearest neighbor classifier using nine training images for each individual. The nine training images

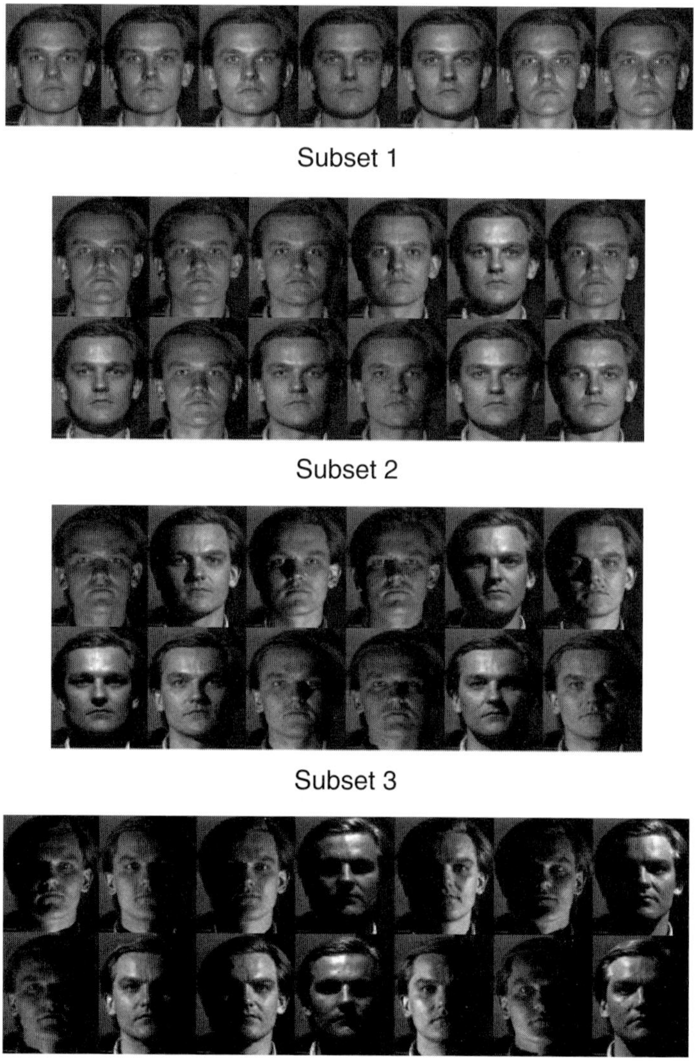

FIGURE 11.12: Example images of a single individual in frontal pose from the Yale Face Database B showing the variability due to illumination. The images have been divided into four subsets according to the angle the light-source direction makes with the camera axis: subset 1 (up to 12°), subset 2 (up to 25°), subset 3 (up to 50°), subset 4 (up to 77°).

are images taken under the lighting conditions specified in the universal lighting configuration discussed in the previous section. Therefore, unlike "correlation," the nine training images contain both frontally and laterally illuminated images. "Eigenfaces" uses PCA to obtain a subspace from training images. One proposed method for handling illumination variation using PCA is to discard the first three most significant principal components, which, in practice, yields better recognition algorithm [3]. The linear-subspace method is a simple subspace-based method. The subspace is a three-dimensional subspace built on the x, y, z components of the surface normals. This is a variant of the photometric alignment method proposed in [31] and is related to [18, 27]. While this method models the variation in shading when the surface is completely illuminated, it does not model shadowing, neither attached nor cast shadows.

Table 11.1: The error rates for various recognition methods on subsets of the Yale Face Database B. Some of the Each entry is taken directly from a published source indicated by citation.

Comparison of Recognition Methods						
Method	Number of training images	Estimate normals	Error rate (%) vs. illum.			
			Subset 1&2	Subset 3	Subset 4	Total
Correlation [14]	6–7	No	0.0	23.3	73.6	29.1
Eigenfaces [14]	6–7	No	0.0	25.8	75.7	30.4
Eigenfaces w/o 1st 3 [14]	6–7	No	0.0	19.2	66.4	25.8
9NN [25]	9	No	13.8	54.6	7.0	22.6
Linear subspace [14]	6–7	Yes	0.0	0.0	15.0	4.6
Cones attached [14]	6–7	Yes	0.0	0.0	8.6	2.7
Harmonic exemplars [43]	1	Yes	0.0	0.3	3.1	1.0
9PL (simulated images) [25]	9	No	0.0	0.0	2.8	0.87
Harmonic subspace attached (no cast shadow) [25]	6–7	Yes	0.0	0.0	3.571	1.1
Harmonic subspace cast (with cast shadow) [25]	6–7	Yes	0.0	0.0	2.7	0.85
Gradient angle [10]	1	No	0.0	0.0	1.4	0.44
Cones cast [14]	6–7	Yes	0.0	0.0	0.0	0.0
5PL (real images) [25]	5	No	0.0	0.0	0.0	0.0
9PL (real images) [25]	9	No	0.0	0.0	0.0	0.0

In Table 11.1, there are two slightly different versions of the harmonic subspace method [2] and the illumination-cone method [14]. In "harmonic subspace attached", the nine harmonic images that form the basis of the linear subspace are rendered directly according to the formulas in Equations 13–15. "Harmonic subspace cast" uses a simple ray tracer to simulate the harmonic images of a 3D face under harmonic lightings including the effects of cast shadows. Similarly, in "cone attached" and "cone cast" methods, we use images without and with cast shadows to compute the illumination cones, respectively. Harmonic exemplars is the method proposed in [43] and the result here is taken directly from that paper. "Gradient angle" comes from [10], and finally, "9PL" is the algorithm first proposed in [24] that uses nine training images taken under the nine lighting conditions specified by their universal configuration. (5PL and 9PL Real are related and will be discussed below). In this experiment, the 3D structure of the face is used to render the nine images under these nine lighting conditions (which are not included in the Yale database).

There are several ways to understand the results in Table 11.1. First, recognition is generally easier in images taken under frontal illumination. As expected, the laterally-illuminated images (those from subsets 3 and 4) are the main challenges. As the first five "quick-fix" algorithms clearly demonstrated, it is difficult to robustly perform recognition for these images without any significant illumination modeling. Second, linear-subspace models are indeed the right tool to use for modeling illumination. This, of course, is the main result we discussed in Section 11.3, and here we observe it empirically by comparing the recognition results using "9PL" and "9NN". While they use the same training images and compute the same norm (L^2 norm), the ability of the subspace to correctly extrapolate images under novel illumination conditions is the only explanation for the discrepancy in performance between "9PL" and "9NN". Finally, we see that the difference in performance between "Harmonic subspace attached" and "harmonic subspace cast" (and likewise for "cone attached" and "cone cast") are not too significant. This implies, as we mentioned in the introduction, that the degree of nonconvexity of human faces is not so severe as to render the effect of cast shadows on human faces unmanageable.

While the on-line recognition processes for algorithms listed in Table 11.1 are pretty much the same, they differ significantly, however, in their off-line training processes. For algorithms that required surface normals, at least three training images are needed in order to determine the normals and albedos. In this experiment, we require typically six frontally illuminated images to estimate the surface normals and albedos using photometric stereo techniques. Although "harmonic exemplar" can get by with just one training image, it requires the priors on harmonic images that can only be obtained using an off-line training process that typically requires a large number of training images. Same goes for "gradient angle" in which a prior on the angles between image gradients has to be estimated empirically. Perhaps, the simplest algorithm among the bunch, both implementationally and conceptually, is "9PL". Since there is practically no training involved here, the work is almost minimal: we simply need to obtain images of a person taken under nine specified lighting conditions. Further experiments have also shown that a five-dimensional subspace ("5PL") is already sufficient for robust face recognition. While "9PL" is sufficient for frontal-view face recognition, without 3D reconstructions, it cannot handle multiview face recognition with variable lighting-considered by Georghiades et al. [14].

Experiments have also been reported in the literature using the CMU PIE database. In [25], it has been demonstrated that using only a five-dimensional subspace for each individual (i.e., five training images per person), the overall recognition error rate of 3.5% can be achieved for the CMU PIE dataset using the algorithm of Lee et al. In [35], Sim and Kanade have compared the performance of two different algorithms, between the nearest neighbor (NN) classifier, and the classifier based on individual PCA subspaces using their algorithm (as discussed in the previous section). The result reported for NN has a recognition error rate of 61% while it is just 5% for their method in [35]. The dimension of the PCA subspaces used in this experiment ranged from 35 to 45.

11.5.2 Further Dimensionality Reduction

While subspace-based algorithms have performed well in the preceding experiments, they all have used subspaces with dimension greater than or equal to nine. While the numerological fixation on nine has its origin (or justification) in spherical harmonics, it is desirable to have subspaces with still lower dimension without suffering from significant degradation in recognition performance.

Further dimensionality reduction is particularly straightforward for the method of Lee et al. [25]. Here, the subspace is determined through a nested sequence of linear subspaces with increasing dimension: $R_0 \subseteq R_1 \subseteq \ldots \subseteq R_i \ldots \subseteq R_9 = R$, with R_i an i-dimensional subspace and $i \geq 0$. Any of these subspaces R_i can be used for recognition, and surprisingly, the experiments reported in [25] demonstrate that a five-dimensional subspace R_5 may well be sufficient for face recognition under large illumination variation. Figure 11.13(left) shows that the recognition error rate is negligible when R_i, $i \geq 5$, is used as the subspace. Specifically, they have tested their algorithm (with R_5 as the subspace) on the extended Yale face database (1710 images of 38 individuals). Using real images as training images this time, they have reported an error rate of 0.2%. Considering the lighting distribution specified by R_5 (the first five directions in the universal configuration), this result corroborates well with our discussion in Section 11.3.1, where the empirical observation was that using 3 to 7 eigenimages is sufficient to provide a good representation of the images of a human face under variable lighting.

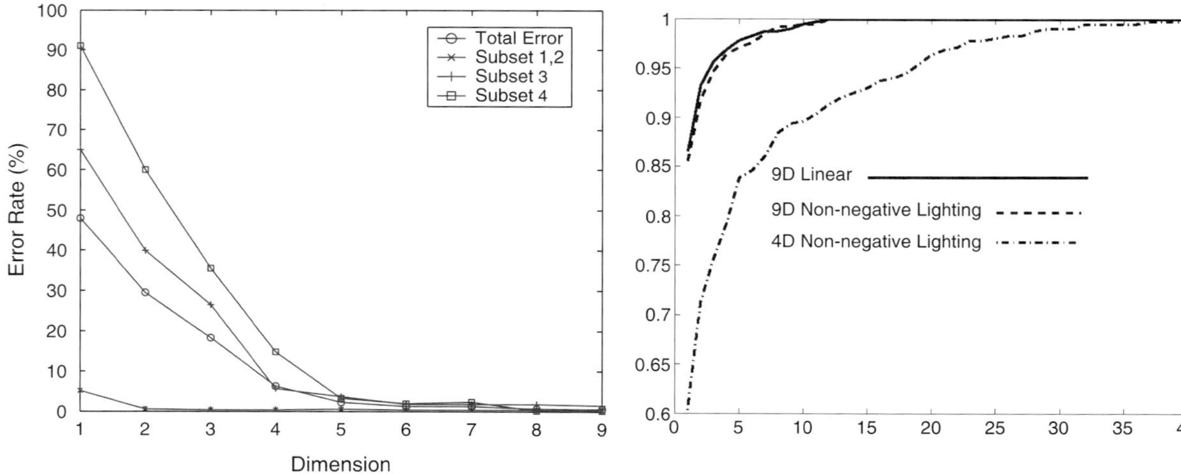

FIGURE 11.13: Further dimensionality reduction. Left: (courtesy of [25]) the error rates for face recognition using successively smaller linear subspaces R_i. The abscissa represents the dimension of the linear subspace while the ordinate gives the error rate. Right: (courtesy of [2]) ROC curve for using nine-dimensional and four-dimensional harmonic subspaces. The ordinate represents the percentage of query images for which the correct model is classified among the top k models, with k represented by the abscissa.

Under the spherical-harmonics framework, dimensionality reduction is less straightforward. For example, to define a seven-dimensional subspace R_7, presumably, we can find a basis for R_7 using some linear combinations of the nine harmonic images. Ramamoorthi [28] has proposed a method of determining such linear combinations. However, the simplest way is to use the first four spherical harmonics (i.e., ignoring spherical harmonics with degrees greater than one). By themselves, these four spherical harmonics have accounted for at least 83% of the reflected energy, and the four corresponding harmonic images encode already the albedos and surface normals. Figure 11.13(right) [2] displays the receiver operating characteristic (ROC) curves for the 9D harmonic subspace method, and 9D and 4D harmonic subspace methods with nonnegative light conditions. The experiment in [2] used a database of faces collected at NEC, Japan, which contains 42 faces with seven different poses and six different lighting conditions. The ROC curves show the fraction of query images for which the correct model (person) is classified among the top k closest models (persons) as k varies from 1 to 40. As expected, the 4D positive lighting method performs worse than the other two methods employing the full 9D subspace. However, it is much faster, and seems to be quite effective under simpler pose and lighting conditions [2].

11.6 CONCLUSION

Reexamining the images in Figure 11.1 in the introduction, we now have at our disposal a number of face-recognition algorithms that can comfortably handle these formidable-looking images. Barely a decade ago, these images would have been problematic for face-recognition algorithms of the time. The new concepts and insights introduced in studying illumination modeling in the past decade has bore many fruits in the form of face-recognition algorithms that are robust against illumination variation. In many ways, we are fortunate that human faces do not have more complicated geometry and reflectance. Coupled with the superposition nature of illumination, this allows us to utilize low-dimensional *linear* appearance models to capture a large portion of image variation due to illumination. Linearity makes the algorithms efficient and easy to implement, and the appearance models make the algorithms robust.

While great strides have been made, many problems are still awaiting formulation and solution. From the face-recognition perspective, there is the important problem of detection and alignment, which has been completely ignored in our discussion. How to make these processes robust under illumination variation is a difficult problem, and a solution to this problem would have significant impacts on other related research areas such as video face recognition. Because face tracking is an integral and indispensable part of video face recognition, it is also a challenging problem to develop a tracker that is robust against illumination variation. Pose, expression, occlusion, aging and other factors must also be considered in concert with work on illumination. Other important and interesting problems include photo-realistic simulation of human faces as well as face recognition using lighting priors.

ACKNOWLEDGMENT

Many thanks to Josh Wills for his careful reading of an earlier draft of this chapter and for his comments.

REFERENCES

[1] V. Arnold. *Ordinary Differential Geometry*. MIT press, 1973.

[2] R. Basri and D. Jacobs. Lambertian reflectance and linear subspaces. *IEEE Transactions on Pattern Analysis and Machine Intelligence* **25**(6):383–390, 2003.

[3] P. Belhumeur, J. Hespanha, and D. Kriegman. Eigenfaces vs. Fisherfaces: Recognition using class specific linear projection. *IEEE Transactions on Pattern Analysis and Machine Intelligence* **19**(7):711–720, 1997.

[4] P. Belhumeur and D. Kriegman. What is the set of images of an object under all possible lighting conditions. *Int. J. Computer Vision* **28**:245–260, 1998.

[5] P. Belhumeur, D. Kriegman, and A. Yuille. The bas-relief ambiguity. *Int. J. Computer Vision* **35**(1):33–44, 1999.

[6] T. Brocker and T. tom Dieck. *Representations of Compact Lie Groups*. Graduate Texts in Mathematics 98, Springer-Verlag, 1985.

[7] R. Brunelli and T. Poggio. Face recognition: features vs. templates. *IEEE Transactions on Pattern Analysis and Machine Intelligence* **15**(10):1042–1053, 1993.

[8] J. Burns, R. Weiss, and E. Riseman. The non-existence of general-case view-invariants. In: *Geometric Invariance in Computer Vision*. Edited by J. Mundy and A. Zisserman, MIT Press, 1991.

[9] R. Chellappa, C. Wilson, and S. Sirohey. Human and machine recognition of faces: a survey. *Proc. IEEE* **83**(5):705–740, 1995.

[10] H. Chen, P. Belhumeur, and D. Jacobs. In search of illumination invariants. In: *Proc. IEEE Conf. on Comp. Vision and Patt. Recog.*, pages 1–8, 2000.

[11] R. Epstein, P. Hallinan, and A. Yuille. 5+/-2 eigenimages suffice: an empirical investigation of low-dimensional lighting models. In *PBMCV*, 1995.

[12] R. Frankot and R. Chellapa. A method for enforcing integrability in shape from shading algorithm. *IEEE Transactions on Pattern Analysis and Machine Intelligence* **10**(4):439–451, 1988.

[13] A. Georghiades. Incorporating the Torrance and Sparrow model of reflectance in uncalibrated photometric stereo. In: *Proc. Int. Conf. on Computer Vision*, pages 816–825, 2003.

[14] A. Georghiades, D. Kriegman, and P. Belhumeur. From few to many: Generative models for recognition under variable pose and illumination. *IEEE Transactions on Pattern Analysis and Machine Intelligence* **23**(6):643–660, 2001.

[15] A. Goldstein, L. Harmon, and A. Lesk. Identification of human faces. *Proc. IEEE* **59**(5):748–760, 1971.

[16] G. Golub and C. van Loan. *Matrix Computation*. The John Hopkins Univ. Press, 1989.

[17] P. Hallinan. A low-dimensional representation of human faces for arbitrary lighting conditions. In: *Proc. IEEE Conf. on Comp. Vision and Patt. Recog.*, pages 995–999, 1994.

[18] P. Hallinan. A deformable model for face recognition under arbitrary lighting conditions. Ph.D. Thesis, Harvard Univ., 1995.

[19] D. Jacobs. Linear fitting with missing data: Applications to structure from motion and characterizing intensity images. In: *Proc. IEEE Conf. on Comp. Vision and Patt. Recog.*, 1997.

[20] H. Jensen, S. Marschner, M. Levoy, and P. Hanrahan. A practical model for subsurface light transport. In: *Proceedings of SIGGRAPH*, pages 511–518, 2001.

[21] T. Kanade. *Ph.D Thesis*. Kyoto Univ., 1973.

[22] D. Kriegman and P. Belheumer. What shadows reveal about object structure. *Journal of the Optical Society of America*, pages 1804–1813, 2001.

[23] J. H. Lambert. Photometria sive de mensure de gratibus luminis, colorum umbrae. *Eberhard Klett*, 1760.

[24] K. Lee, J. Ho, and D. Kriegman. Nine points of lights: acquiring subspaces for face recognition under variable lighting. In: *Proc. IEEE Conf. on Comp. Vision and Patt. Recog.*, pages 519–526, 2001.

[25] K.-C. Lee, J. Ho, and D. Kriegman. acquiring linear subspaces for face recognition under variable lighting. *IEEE Transactions on Pattern Analysis and Machine Intelligence*, **27**(5):684–698, May 2005.

[26] Y. Moses, Y. Adini, and S. Ullman. Face recognition: The problem of compensating for changes in illumination direction,. In: *Proc. European Conf. on Computer Vision*, pages 286–296, 1994.

[27] S. Nayar and H. Murase. Dimensionality of illumination in appearance matching. *Proc. IEEE Conf. on Robotics and Automation*, 1996.

[28] R. Ramamoorthi. Analytic PCA construction for theoretical analysis of lighting variability in images of a lambertian object. *IEEE Transactions on Pattern Analysis and Machine Intelligence* **24**:1322–1333, 2002.

[29] R. Ramamoorthi and P. Hanrahan. An efficient representation for irradiance environment. In: *Proceedings of SIGGRAPH*, pages 497–500, 2001.

[30] R. Ramamoorthi and P. Hanrahan. A signal-processing framework for inverse rendering. In: *Proceedings of SIGGRAPH*, pages 117–228, 2001.

[31] A. Shashua. On photometric issues in 3D visual recognition form a single image. *Int. J. Computer Vision* **21**:99–122, 1997.

[32] H. Shum, K. Ikeuchi, and R. Reddy. Principal component analysis with mising data and its application to polyhedral object modeling. *IEEE Transactions on Pattern Analysis and Machine Intelligence* **17**(9):854–867, 1995.

[33] T. Sim, S. Baker, and M. Bsat. The CMU pose, illumination, and expression (pie) database. In: *Proceedings of the IEEE International Conference on Automatic Face and Gesture Recognition*, pages 53–58, 2002.

[34] T. Sim, S. Baker, and M. Bsat. The CMU pose, illumination and expression (pie) database. In: *Proc. IEEE Conf. on Auto. Facial and Gesture Recog.*, pages 53–58, 2002.

[35] T. Sim and T. Kanade. Combining models and exemplars for face recognition: An illuminating example. In: *Proceedings of Workshop on Models versus Exemplars in Computer Vision*, 2001.

[36] L. Sirovitch and M. Kirby. Low-dimensional procedure for the characterization of human faces. *J. of Optical Soc. Am.* A **2**:519–524, 1987.

[37] W. Strauss. *Partial Differential Equations*. John Wiley & Sons, Inc, 1992.

[38] C. Tomasi and T. Kanade. Shape and motion from image streams under orthography: a factorization method. *Int. J. Computer Vision*, **9**(2):137–154, 1992.

[39] M. Turk and A. Pentland. Eigenfaces for recognition. *J. of Cognitive Neuroscience* **3**(1):71–96, 1991.

[40] M. Vasilescu and D. Terzopoulos. Multilinear analysis of image ensembles: Tensorfaces. In: *Proc. European Conf. on Computer Vision*, pages 447–460, 2002.

[41] S. Westin, J. Arvo, and K. Torrance. Predicting reflectance functions from complex surfaces. In: *Proceedings of SIGGRAPH*, pages 255–264, 1992.

[42] A. Yuille and D. Snow. Shape and albedo from multiple images using integrability. In: *Proc. IEEE Conf. on Comp. Vision and Patt. Recog.*, pages 158–164, 1997.

[43] L. Zhang and D. Samaras. Face recognition under variable lighting using harmonic image examplars. In: *Proc. IEEE Conf. on Comp. Vision and Patt. Recog.*, volume 1, pages 19–25, 2003.

[44] R. Zhang, P. Tsai, and J. Cryer. Shape from shading: a survey. *IEEE Transactions on Pattern Analysis and Machine Intelligence* **21**(8):690–706, 1999.

[45] Q. Zheng and R. Chellappa. Estimation of illuminant direction, albedo and shape from shading. *IEEE Transactions on Pattern Analysis and Machine Intelligence* **13**(7):680–702, 1991.

[46] S. Zhou and D. J. R. Chellappa. Characterization of human faces under illumination variations using rank, integrability, and symmetry constraints. In: *Proc. European Conf. on Computer Vision*, Volume 1, pages 588–601, 2004.

CHAPTER 12

MODELING ILLUMINATION VARIATION WITH SPHERICAL HARMONICS

12.1 INTRODUCTION

Illumination can have a significant impact on the appearance of surfaces, as the patterns of shading, specularities, and shadows change. For instance, some images of a face under different lighting conditions are shown in Figure 12.1. Differences in lighting can often play a much greater role in image variability of human faces than differences between individual people. Lighting designers in movies can often set the mood of a scene with carefully chosen lighting. To achieve a sinister effect, for instance, one can use illumination from below the subject—a sharp contrast to most natural indoor or outdoor scenes where the dominant light sources are above the person.

Characterizing the variability in appearance with lighting is a fundamental problem in many areas of computer vision, face modeling, and computer graphics. One of the great challenges of computer vision is to produce systems that can work in uncontrolled environments. To be robust, recognition systems must be able to work oudoors in a lighting-insensitive manner. In computer graphics, the challenge is to be able to efficiently create the visual appearance of a scene under realistic, possibly changing, illumination.

At first glance, modeling the variation with lighting may seem intractable. For instance, a video projector can illuminate an object like a face with essentially any pattern. In this chapter, we will stay away from such extreme examples, making a set of assumptions that are approximately true in many common situations. One assumption we make is that illumination is distant. By this, we mean that the direction to, and intensity of, the light sources is approximately the same throughout the region of interest. This explicitly rules out cases like slide projectors. This is

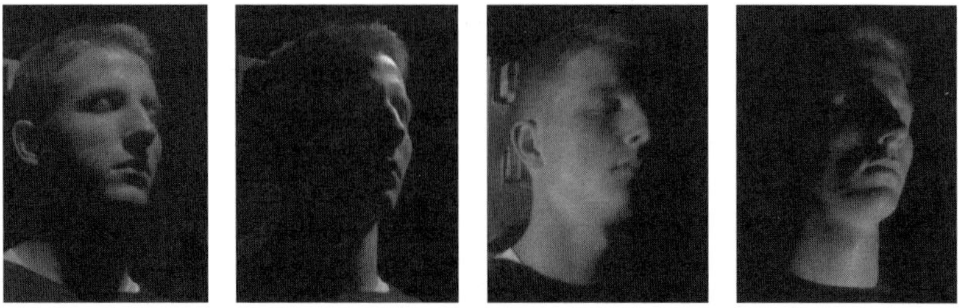

FIGURE 12.1: Images of a face, lit from a number of different directions. Note the vast variation in appearance due to illumination. Images courtesy of Debevec et al. [13].

a reasonably good approximation in outdoor scenes, where the sky can be assumed to be far away. It is also fairly accurate in many indoor environments, where the light sources can be considered much further away relative to the size of the object.

Even under the assumption of distant lighting, the variations may seem intractable. The illumination can come from any incident direction, and can be composed of multiple illuminants including localized light sources like sunlight and broad-area distributions like skylight. In the general case, we would need to model the intensity from each of infinitely many incident lighting directions. Thus, the space we are dealing with appears to be infinite-dimensional. By contrast, a number of other causes of appearance variation are low-dimensional. For instance, appearance varies with viewing direction as well. However, unlike lighting, there can only be a single view direction. In general, variation because of pose and translation can be described using only six degrees of freedom.

Given the daunting nature of this problem, it is not surprising that most previous analytic models have been restricted to the case of a single-directional (distant) light source, usually without considering shadows. In computer graphics, this model is sufficient (but not necessarily efficient) for numerical Monte Carlo simulation of appearance. In computer vision, such models can be reasonable approximations under some circumstances, such as controlled laboratory conditions. However, they do not suffice for modeling illumination variation in uncontrolled conditions like the outdoors.

Fortunately, there is an obvious empirical method of analyzing illumination variation. One can simply record a number of images of an object under light sources from all different directions. In practice, this usually corresponds to moving a light source in a sphere around the object or person, while keeping camera and pose fixed. Because of the linearity of light transport—the image under multiple lights is the sum of that under the individual sources—an image under arbitrary distant illumination can be written as a linear combination of these source images.

This observation in itself provides an attractive direct approach for relighting in computer graphics. Furthermore, instead of using the source images, we can try to find linear combinations or basis images that best explain the variability due to illumination. This is the standard technique of finding low-dimensional subspaces or principal components. The surprising result in these experiments is that most of the variability due to lighting is modeled using a very low-dimensional subspace—usually only five basis functions.

Given the discussion about the infinite dimensionality of illumination, this observation seems strikingly counterintuitive. However, some insight may be obtained by looking at an (untextured) matte or diffuse surface. One will notice that, even if the lighting is complicated, the surface has a smooth appearance. In effect, it is blurring or low-pass-filtering the illumination. In this chapter, we will make these ideas formal, explaining previous empirical results. A diffuse or matte surface (technically Lambertian) can be viewed as a low-pass filter acting on the incident illumination signal, with the output image obtained by convolving the lighting with the surface reflectance. This leads to a frequency-domain view of reflection, which has not previously been explored and which leads to many interesting insights. In particular, we can derive a product convolution formula using spherical-harmonic basis functions. The theoretical results have practical implications for computer-graphics rendering, illuminant estimation, and recognition and reconstruction in computer vision.

12.2 BACKGROUND AND PREVIOUS WORK

Lighting and appearance have been studied in many forms almost since the beginning of research in computer vision and graphics, as well as in a number of other areas. Horn's classic text [27] provides background on work in vision. In this section, we only survey the most relevant previous work, which relates to the theoretical developments in this chapter. This will also be a vehicle to introduce some of the basic concepts on which we will build.

12.2.1 Environment Maps in Computer Graphics

A common approximation in computer graphics, especially for interactive hardware rendering, is to assume distant illumination. The lighting can then be represented as a function of direction, known in the literature as an environment map. Practically, environment maps can be acquired by photographing a chrome-steel or mirror sphere, which simply reflects the incident lighting. Environment mapping corresponds exactly to the distant-illumination assumption we make in this chapter. Another common assumption, which we also make, is to neglect *cast shadows* from one part of the object on another. These should be distinguished from *attached shadows* when a point is in shadow because the light source is behind the surface (below the horizon). We will explicitly take attached shadows into account.

In terms of previous work, Blinn and Newell [7] first used environment maps to efficiently find reflections of distant objects. This technique, known as reflection mapping, is still widely in use today for interactive computer-graphic rendering. The method was greatly generalized by Miller and Hoffman [46] (and later Greene [21]). They introduced the idea that one could acquire a real environment by photographing a mirror sphere. They also precomputed diffuse and specular reflection maps, for the corresponding components of surface reflection. Cabral et al. [9] later extended this general method to computing reflections from bump-mapped surfaces, and to computing environment-mapped images with more general reflective properties [10] (technically, the bidirectional reflectance distribution function or BRDF [52]). Similar results for more general materials were also obtained by Kautz et al. [32, 33], building on previous work by Heidrich and Seidel [25]. It should be noted that both Miller and Hoffman [46], and Cabral et al. [9, 10] qualitatively described the reflection maps as obtained by convolving the lighting with the reflective properties of the surface. Kautz et al. [33] actually used convolution to implement reflection, but on somewhat distorted planar projections of the environment map, and without full theoretical justification. In this chapter, we will formalize these ideas, making the notion of convolution precise, and derive analytic formulae.

12.2.2 Lighting-Insensitive Recognition in Computer Vision

Another important area of research is in computer vision, where there has been much work on modeling the variation of appearance with lighting for robust lighting-insensitive recognition algorithms. Some of this prior work is discussed in detail in the excellent Chapter 11, *on the effect of illumination and face recognition*, by Ho and Kriegman in this volume. Illumination modeling is also important in many other vision problems, such as photometric stereo and structure from motion.

The work in this area has taken two directions. One approach is to apply an image-processing operator that is relatively insensitive to the lighting. Work in this domain includes image gradients [8], the direction of the gradient [11], and Gabor jets [37]. While these methods reduce the effects of lighting variation, they operate without knowledge of the scene geometry, and so are inherently limited in the ability to model illumination variation. A number of empirical studies have suggested the difficulties of pure image-processing methods, in terms of achieving lighting insensitivity [47]. Our approach is more related to a second line of work that seeks to explicitly model illumination variation, analyzing the effects of the lighting on a 3D model of the object (here, a face).

Linear 3D Lighting Model Without Shadows

Within this category, a first important result (Shashua [69], Murase and Nayar [48], and others) is that, for Lambertian objects, in the absence of *both* attached and cast shadows, the images under all possible illumination conditions lie in a three-dimensional subspace. To obtain this result, let us first consider a model for the illumination from a single directional source on a Lambertian object.

$$B = \rho L(\omega \cdot \boldsymbol{n}) = \boldsymbol{L} \cdot \boldsymbol{N}, \tag{1}$$

where L in the first relation corresponds to the intensity of the illumination in direction ω, $\boldsymbol{n}$ is the surface normal at the point, ρ is the surface albedo and B is the outgoing radiosity (radiant exitance) or reflected light. For a nomenclature of radiometric quantities, see a textbook, like McCluney [45] or Chapter 2 of Cohen and Wallance [12].[1] It is sometimes useful to incorporate the albedo into the surface normal, defining vectors $\boldsymbol{N} = \rho \boldsymbol{n}$ and correspondingly $\boldsymbol{L} = L\omega$, so we can simply write (for Lambertian surfaces in the absence of all shadows, attached or cast) $B = \boldsymbol{L} \cdot \boldsymbol{N}$.

Now, consider illumination from a variety of light sources $\boldsymbol{L}_1, \boldsymbol{L}_2$ and so on. It is straighforward to use the linearity of light transport to write the net reflected light as

$$B = \sum_i \boldsymbol{L}_i \cdot \boldsymbol{N} = \left(\sum_i \boldsymbol{L}_i\right) \cdot \boldsymbol{N} = \hat{\boldsymbol{L}} \cdot \boldsymbol{N}, \tag{2}$$

where $\hat{\boldsymbol{L}} = \sum_i \boldsymbol{L}_i$. But this has essentially the same form as Equation 1. Thus, in the absence of all shadows, there is a very simple linear model for lighting in Lambertian surfaces, where we can replace a complex lighting distribution by the weighted sum of the individual light sources. We can then treat the object as if lit by a *single* effective light source $\hat{\boldsymbol{L}}$.

Finally, it is easy to see that images under all possible lighting conditions lie in a 3D subspace, being linear combinations of the Cartesian components N_x, N_y, N_z, with $B = \hat{L}_x N_x + \hat{L}_y N_y + \hat{L}_z N_z$. Note that this entire analysis applies to all points on the object. The basis images of the 3D lighting subspace are simply the Cartesian

[1] In general, we will use B for the reflected radiance, and L for the incident radiance, while ρ will denote the BRDF. For Lambertian surfaces, it is conventional to use ρ to denote the albedo (technically, the surface reflectance that lies between 0 and 1), which is π times the BRDF. Interpreting B as the radiosity accounts for this factor of π.

components of the surface normals over the objects, scaled by the albedos at those surface locations.[2]

The 3D linear subspace has been used in a number of works on recognition, as well as other areas. For instance, Hayakawa [24] used factorization based on the 3D subspace to build models using photometric stereo. Koenderink and van Doorn [35] added an extra ambient term to make this a 4D subspace. The extra term corresponds to perfectly diffuse lighting over the whole sphere of incoming directions. This corresponds to adding the albedos over the surface themselves to the previous 3D subspace, i.e., adding $\rho = \| \boldsymbol{N} \| = \sqrt{N_x^2 + N_y^2 + N_z^2}$.

Empirical Low-Dimensional Models

These theoretical results have inspired a number of authors to develop empirical models for lighting variability. As described in the introduction, one takes a number of images with different lighting directions, and then uses standard dimensionality-reduction methods like principal-component analysis. PCA-based methods were pioneered for faces by Kirby and Sirovich [34, 72], and for face recognition by Turk and Pentland [78], but these authors did not account for variations due to illumination.

The effects of lighting alone were studied in a series of experiments by Hallinan [23], Epstein et al. [15], and Yuille et al. [80]. They found that for human faces, and other diffuse objects like basketballs, a 5D subspace sufficed to approximate lighting variability very well. That is, with a linear combination of a mean and 5 basis images, we can accurately predict appearance under arbitrary illumination. Furthermore, the form of these basis functions, and even the amount of variance accounted for, were largely consistent across human faces. For instance, Epstein et al. [15] report that for images of a human face, three basis images capture 90% of image variance, while five basis images account for 94%.

At the time however, these results had no complete theoretical explanation. Furthermore, they indicate that the 3D subspace given above is inadequate. This is not difficult to understand. If we consider the appearance of a face in outdoor lighting from the entire sky, there will often be attached shadows or regions in the environment that are not visible to a given surface point (these correspond to lighting below the horizon for that point, where $\omega \cdot \boldsymbol{n} < 0$).

Theoretical Models

The above discussion indicates the value of developing an analytic model to account for lighting variability. Theoretical results can give new insights, and also lead to simpler and more efficient and robust algorithms.

[2]Color is largely ignored in this chapter; we assume each color band, such as red, green, and blue, is treated separately.

Belhumeur and Kriegman [6] have taken a first step in this direction, developing the illumination-cone representation. Under very mild assumptions, the images of an object under arbitrary lighting form a convex cone in image space. In a sense, this follows directly from the linearity of light transport. Any image can be scaled by a positive value simply by scaling the illumination. The convexity of the cone is because one can add two images, simply by adding the corresponding lighting. Formally, even for a Lambertian object with only attached shadows, the dimension of the illumination cone can be infinite (this grows as $O(n^2)$, where n is the number of distinct surface normals visible in the image). Georghiades et al. [19, 20] have developed recognition algorithms based on the illumination cone. One approach is to sample the cone using extremal rays, corresponding to rendering or imaging the face using directional light sources. It should be noted that exact recognition using the illumination cone involves a slow complex optimization (constrained optimization must be used to enforce a nonnegative linear combination of the basis images), and methods using low-dimensional linear subspaces and unconstrained optimization (which essentially reduces to a simple linear system) are more efficient and usually required for practical applications.

Another approach is to try to analytically construct the principal-component decomposition, analogous to what was done experimentally. Numerical PCA techniques could be biased by the specific lighting conditions used, so an explicit analytic method is helpful. It is only recently that there has been progress in analytic methods for extending PCA from a discrete set of images to a continuous sampling [40, 83]. These approaches demonstrate better generalization properties than purely empirical techniques. In fact, Zhao and Yang [83] have analytically constructed the covariance matrix for PCA of lighting variability, but under the assumption of no shadows.

Summary

To summarize, prior to the work reported in this chapter, the illumination variability could be described theoretically by the illumination cone. It was known from numerical and real experiments that the illumination cone lay close to a linear low-dimensional space for Lambertian objects with attached shadows. However, a theoretical explanation of these results was not available. In this chapter, we develop a simple linear lighting model using spherical harmonics. These theoretical results were first introduced for Lambertian surfaces simultaneously by Basri and Jacobs [2, 4] and Ramamoorthi and Hanrahan [64]. Much of the work of Basri and Jacobs is also summarized in an excellent book chapter on illumination modeling for face recognition [5].

12.2.3 Frequency-Space Representations: Spherical Harmonics

We show reflection to be a convolution and analyze it in frequency space. We will primarily be concerned with analyzing quantities like the BRDF and distant lighting, which can be parametrized as functions on the unit sphere. Hence, the appropriate frequency-space representations are spherical harmonics [28, 29, 42]. Spherical harmonics can be thought of as signal-processing tools on the unit sphere, analogous to the Fourier series or sines and cosines on the line or circle. They can be written either as trigonometric functions of the spherical coordinates, or as simple polynomials of the Cartesian components. They form an orthonormal basis on the unit sphere, in terms of which quantities like the lighting or BRDF can be expanded and analyzed.

The use of spherical harmonics to represent the illumination and BRDF was pioneered in computer graphics by Cabral et al. [9]. In perception, D'Zmura [14] analyzed reflection as a linear operator in terms of spherical harmonics, and discussed some resulting ambiguities between reflectance and illumination. Our use of spherical harmonics to represent the lighting is also similar in some respects to previous methods such as that of Nimeroff et al. [56] that use steerable linear basis functions. Spherical harmonics have also been used before in computer graphics for representing BRDFs by a number of other authors [70, 79].

The results described in this chapter are based on a number of papers by us. This includes theoretical work in the planar 2D case or flatland [62], on the analysis of the appearance of a Lambertian surface using spherical harmonics [64], the theory for the general 3D case with isotropic BRDFs [65], and a comprehensive account including a unified view of 2D and 3D cases including anisotropic materials [67]. More details can also be found in the PhD thesis of the author [61]. Recently, we have also linked the convolution approach using spherical harmonics with principal component analysis, quantitatively explaining previous empirical results on lighting variability [60].

12.3 ANALYZING LAMBERTIAN REFLECTION USING SPHERICAL HARMONICS

In this section, we analyze the important case of Lambertian surfaces under distant illumination [64], using spherical harmonics. We will derive a simple linear relationship, where the coefficients of the reflected light are simply filtered versions of the incident illumination. Furthermore, a low-frequency approximation with only nine spherical-harmonic terms is shown to be accurate.

Assumptions

Mathematically, we are simply considering the relationship between the irradiance (a measure of the intensity of incident light, reflected equally in all directions by diffuse objects), expressed as a function of surface orientation, and the incoming radiance or incident illumination, expressed as a function of incident angle. The corresponding physical system is a curved convex homogeneous

Lambertian surface reflecting a distant illumination field. For the physical system, we will assume that the *surfaces under consideration are convex*, so they may be parametrized by the surface orientation, as described by the surface normal, and so that *interreflection and cast shadowing (but not attached shadows) can be ignored*. Also, surfaces will be assumed to be *Lambertian and homogeneous*, so the reflectivity can be characterized by a constant albedo. We will further assume here that the *illuminating light sources are distant*, so the illumination or incoming radiance can be represented as a function of direction only.

Notation used in the chapter (some of which pertains to later sections) is listed in Table 12.1. A diagram of the local geometry of the situation is shown in Figure 12.2. We will use two types of coordinates. Unprimed global coordinates denote angles with respect to a global reference frame. On the other hand, primed local coordinates denote angles with respect to the local reference frame, defined by the

Table 12.1: Notation.

B	Reflected radiance
$B_{lmn,pq}, B_{lmpq}$	Coefficients of basis-function expansion of B
L	Incoming radiance
L_{lm}	Coefficients of spherical-harmonic expansion of L
E	Incident irradiance (for Lambertian surfaces)
E_{lm}	Coefficients of spherical-harmonic expansion of E
ρ	Surface BRDF
$\hat{\rho}$	BRDF multiplied by cosine of incident angle
$\hat{\rho}_{ln,pq}, \hat{\rho}_{lpq}$	Coefficients of spherical-harmonic expansion of $\hat{\rho}$
$\theta_{\rm i}', \theta_{\rm i}$	Incident elevation angle in *local*, *global* coordinates
$\phi_{\rm i}', \phi_{\rm i}$	Incident azimuthal angle in *local*, *global* coordinates
$\theta_{\rm o}', \theta_{\rm o}$	Outgoing elevation angle in *local*, *global* coordinates
$\phi_{\rm o}', \phi_{\rm o}$	Outgoing azimuthal angle in *local*, *global* coordinates
$A(\theta_{\rm i}')$	Half-cosine Lambertian transfer function $A(\theta_{\rm i}') = \max(\cos(\theta_{\rm i}'), 0)$
A_l	Spherical-harmonic coefficients of Lambertian transfer function
X	Surface position
α	Surface normal parametrization—elevation angle
β	Surface normal parametrization—azimuthal angle
γ	Orientation of tangent frame for anisotropic surfaces
$R_{\alpha,\beta,\gamma}$	Rotation operator for tangent frame orientation (α, β, γ)
Y_{lm}	Spherical harmonic
Y_{lm}^*	Complex conjugate of spherical harmonic
$D_{mm'}^l$	Representation matrix of dimension $2l+1$ for rotation group $SO(3)$
Λ_l	Normalization constant, $\sqrt{4\pi/(2l+1)}$
I	$\sqrt{-1}$

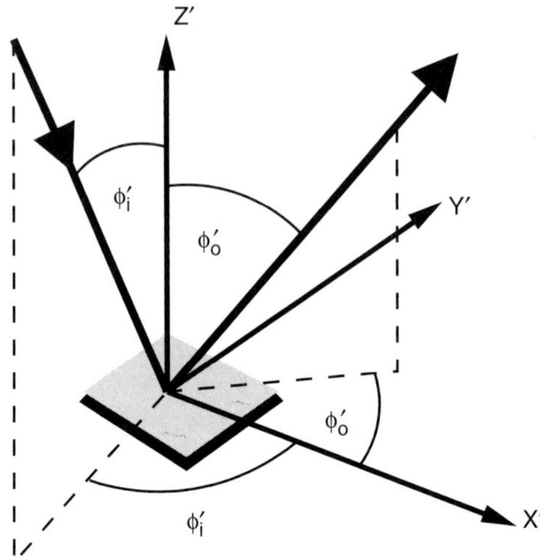

FIGURE 12.2: Diagram showing the local geometry. Quantities are primed because they are all in local coordinates.

local surface normal and an arbitrarily chosen tangent vector. These two coordinate systems are related simply by a rotation, and this relationship will be detailed shortly.

12.3.1 Reflection Equation

In local coordinates, we can relate the irradiance to the incoming radiance by

$$E(\mathrm{x}) = \int_{\Omega_{\mathrm{i}}'} L(\mathrm{x}, \theta_{\mathrm{i}}', \phi_{\mathrm{i}}') \cos\theta_{\mathrm{i}}' \, d\Omega_{\mathrm{i}}', \tag{3}$$

where E is the irradiance, as a function of position x on the object surface, and L is the radiance of the incident light field. As noted earlier, primes denote quantities in local coordinates. The integral is over the upper hemisphere with respect to the local surface normal. For the purposes of this derivation, we will be interested in the relationship of the irradiance to the radiance. In practice, the reflected light (here, radiant exitance or radiosity) can be related to the irradiance using $B(\mathrm{x}) = \rho E(\mathrm{x})$, where ρ is the surface reflectance or albedo, which lies between 0 and 1. This last relation also holds when we interpret ρ as the BRDF (obtained by scaling the reflectance by $1/\pi$) and B as the reflected radiance (obtained by scaling the radiant exitance by $1/\pi$).

We now manipulate Equation 3 by performing a number of substitutions. First, the assumption of distant illumination means the illumination field is homogeneous over the surface, i.e., independent of surface position x, and depends only on the *global* incident angle (θ_i, ϕ_i). This allows us to replace $L(x, \theta_i', \phi_i')$ by $L(\theta_i, \phi_i)$. Second, consider the assumption of a curved convex surface. This ensures that there is no shadowing or interreflection, so that the irradiance is only because of the distant illumination field L. This fact is implicitly assumed in Equation 3. Furthermore, since the illumination is distant, we may reparametrize the surface simply by the surface normal. Equation 3 now becomes

$$E(\mathrm{n}) = \int_{\Omega_i'} L(\theta_i, \phi_i) \cos\theta_i' \, d\Omega_i'. \tag{4}$$

To proceed further, we will parametrize by its spherical angular coordinates (α, β, γ). Here, (α, β) define the angular coordinates of the local normal vector, i.e.,

$$\mathrm{n} = [\sin\alpha \cos\beta, \sin\alpha \sin\beta, \cos\alpha]\,. \tag{5}$$

γ defines the local tangent frame, i.e., rotation of the coordinate axes about the normal vector. For isotropic surfaces—those where there is no preferred tangential direction, i.e., where rotation of the tangent frame about the surface normal has no physical effect—the parameter γ has no physical significance, and we have therefore not explicitly considered it in Equation 5. We will include γ for completeness in the ensuing discussion on rotations, but will eventually eliminate it from our equations after showing mathematically that it does in fact have no effect on the final results. Finally, for convenience, we will define a transfer function $A(\theta_i') = \cos\theta_i'$. With these modifications, Equation 4 becomes

$$E(\alpha, \beta, \gamma) = \int_{\Omega_i'} L(\theta_i, \phi_i) A(\theta_i') \, d\,\Omega_i'. \tag{6}$$

Note that local and global coordinates are mixed. The lighting is expressed in global coordinates, since it is constant over the object surface when viewed with respect to a global reference frame, while the transfer function $A = \cos\theta_i'$ is expressed naturally in local coordinates. Integration can be conveniently done over either local or global coordinates, but the upper hemisphere is easier to keep track of local coordinates.

Rotations: Converting between Local and Global coordinates

To do the integral in Equation 6, we must relate local and global coordinates. The *North Pole* $(0', 0')$ or +Z axis in local coordinates is the surface normal, and the corresponding global coordinates are (α, β). It can be verified that a rotation of the form $R_z(\beta)R_y(\alpha)$ correctly performs this transformation, where the subscript z denotes rotation about the Z axis and the subscript y denotes rotation about the Y axis. For full generality, the rotation between local and global coordinates should also specify the transformation of the local tangent frame, so the general rotation operator is given by $R_{\alpha,\beta,\gamma} = R_z(\beta)R_y(\alpha)R_z(\gamma)$. This is essentially the Euler-angle representation of rotations in 3D. Refer to Figure 12.3 for an illustration. The relevant transformations are given below,

$$(\theta_{\rm i}, \phi_{\rm i}) = R_{\alpha,\beta,\gamma}(\theta'_{\rm i}, \phi'_{\rm i}) = R_z(\beta)R_y(\alpha)R_z(\gamma)\left\{\theta'_{\rm i}, \phi'_{\rm i}\right\}$$

$$(\theta'_{\rm i}, \phi'_{\rm i}) = R^{-1}_{\alpha,\beta,\gamma}(\theta_{\rm i}, \phi_{\rm i}) = R_z(-\gamma)R_y(-\alpha)R_z(-\beta)\left\{\theta_{\rm i}, \phi_{\rm i}\right\}. \quad (7)$$

Note that the angular parameters are rotated as if they were a unit vector pointing in the appropriate direction. It should also be noted that this rotation of parameters

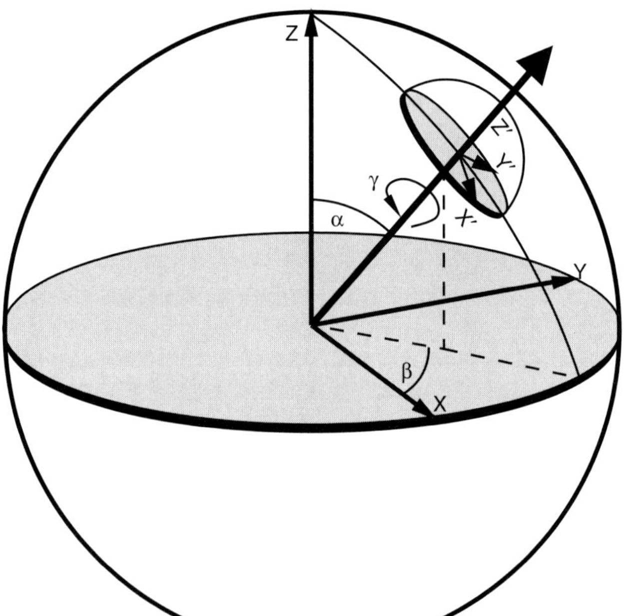

FIGURE 12.3: Diagram showing how the rotation corresponding to (α, β, γ) transforms between local (primed) and global (unprimed) coordinates.

is equivalent to an inverse rotation of the function, with R^{-1} being given by $R_z(-\gamma)R_y(-\alpha)R_z(-\beta)$.

Finally, we can substitute Equation 7 into Equation 6 to derive

$$E(\alpha, \beta, \gamma) = \int_{\Omega'_i} L\left(R_{\alpha,\beta,\gamma}\left(\theta'_i, \phi'_i\right)\right) A(\theta'_i)\, d\,\Omega'_i. \tag{8}$$

As we have written it, this equation depends on spherical coordinates. It might clarify matters somewhat to also present an alternate form in terms of rotations and unit vectors in a coordinate-independent way. We simply use R for the rotation, which could be written as a 3×3 rotation matrix, while ω_i is a unit vector (3×1 column vector) corresponding to the incident direction (with primes added for local coordinates). Equation 8 may then be written as

$$E(R) = \int_{\Omega'_i} L\left(R\omega'_i\right) A(\omega'_i)\, d\omega'_i, \tag{9}$$

where $R\omega'_i$ is simply a matrix–vector multiplication.

Interpretation as Convolution

In the spatial domain, convolution is the result generated when a filter is *translated* over an input signal. However, we can generalize the notion of convolution to other transformations T_a, where T_a is a function of a, and write

$$(f \otimes g)(a) = \int_t f\left(T_a(t)\right) g(t)\, dt. \tag{10}$$

When T_a is a translation by a, we obtain the standard expression for spatial convolution. When T_a is a rotation by the angle a, the above formula defines convolution in the angular domain, and is a slightly simplified version of Equations 8 and 9.

The irradiance can therefore be viewed as obtained by taking the incident illumination signal L and filtering it using the transfer function $A = \cos\theta'_i$. Different observations of the irradiance E, at points on the object surface with different orientations, correspond to different rotations of the transfer function—since the local upper hemisphere is rotated—which can also be thought of as different rotations of the incident light field. We will see that this integral becomes a simple product when transformed to spherical harmonics, further stressing the analogy with convolution.

12.3.2 Spherical-Harmonic Representation

We now proceed to construct a closed-form description of the irradiance. Since we are dealing with convolutions, it is natural to analyze them in the frequency domain. Since unit vectors corresponding to the surface normal or the incident direction lie on a sphere of unit magnitude, the appropriate signal-processing tools are spherical harmonics, which are the equivalent for that domain to the Fourier series in 2D (on a circle). These basis functions arise in connection with many physical systems such as those found in quantum mechanics and electrodynamics. A summary of the properties of spherical harmonics can therefore be found in many standard physics textbooks [28, 29, 42].

Key properties of Spherical Harmonics

The spherical harmonic Y_{lm} is given by (see Figure 12.4 for an illustration)

$$N_{lm} = \sqrt{\frac{2l+1}{4\pi}\frac{(l-m)!}{(l+m)!}}$$

$$Y_{lm}(\theta,\phi) = N_{lm}P_{lm}(\cos\theta)e^{Im\phi}, \tag{11}$$

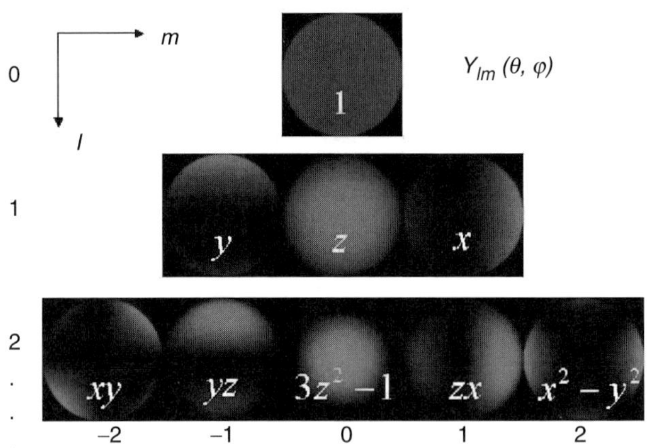

FIGURE 12.4: The first 3 orders of real spherical harmonics ($l = 0, 1, 2$) corresponding to a total of 9 basis functions. In these images, we show only the front of the sphere, with green denoting positive values and blue denoting negative values. Also note that these images show the real form of the spherical harmonics. The complex forms are given in Equation 12. (See also color plate section)

where N_{lm} is a normalization factor. In the above equation, the azimuthal dependence is expanded in terms of Fourier basis functions. The θ dependence is expanded in terms of the associated Legendre functions P_{lm}. The indices obey $l \geq 0$ and $-l \leq m \leq l$. Thus, there are $2l + 1$ basis functions for given order l. They may be written either as trigonometric functions of the spherical coordinates θ and ϕ or as polynomials of the Cartesian components x, y and z, with $x^2 + y^2 + z^2 = 1$. In general, a spherical harmonic Y_{lm} is a polynomial of maximum degree l. Another useful relation is that $Y_{l-m} = (-1)^m Y^*_{lm}$, where Y^*_{lm} denotes the complex conjugate. The first 3 orders (we give only terms with $m \geq 0$) may be written as

$$
\begin{aligned}
Y_{00} &= \sqrt{\frac{1}{4\pi}}, \\
Y_{10} &= \sqrt{\frac{3}{4\pi}} \cos\theta = \sqrt{\frac{3}{4\pi}} z, \\
Y_{11} &= -\sqrt{\frac{3}{8\pi}} (\sin\theta)\, e^{I\phi} = -\sqrt{\frac{3}{8\pi}} (x + Iy), \\
Y_{20} &= \frac{1}{2}\sqrt{\frac{5}{4\pi}} \left(3\cos^2\theta - 1\right) = \frac{1}{2}\sqrt{\frac{5}{4\pi}} \left(3z^2 - 1\right), \\
Y_{21} &= -\sqrt{\frac{15}{8\pi}} (\sin\theta\cos\theta)\, e^{I\phi} = -\sqrt{\frac{15}{8\pi}} z\,(x + Iy), \\
Y_{22} &= \frac{1}{2}\sqrt{\frac{15}{8\pi}} \left(\sin^2\theta\right) e^{2I\phi} = \frac{1}{2}\sqrt{\frac{15}{8\pi}} (x + Iy)^2 .
\end{aligned}
\tag{12}
$$

The spherical harmonics form an orthonormal and complete basis for functions on the unit sphere:

$$
\int_{\phi=0}^{2\pi} \int_{\theta=0}^{\pi} Y^*_{lm}(\theta,\phi) Y_{l'm'}(\theta,\phi) \sin\theta \, d\theta d\phi = \delta_{ll'} \delta_{mm'} . \tag{13}
$$

To find the spherical-harmonic coefficients of an arbitrary function f, we simply integrate against the complex conjugate, as with any orthonormal basis set. That is,

$$
\begin{aligned}
f(\theta,\phi) &= \sum_{l=0}^{\infty} \sum_{m=-l}^{l} f_{lm} Y_{lm}(\theta,\phi) \\
f_{lm} &= \int_{\phi=0}^{2\pi} \int_{\theta=0}^{\pi} f(\theta,\phi) Y^*_{lm}(\theta,\phi) \sin\theta \, d\theta d\phi .
\end{aligned}
\tag{14}
$$

Let us now build up the rotation operator on the spherical harmonics. Rotation about the z axis is simple, with

$$Y_{lm}\left(R_z(\beta)\{\theta,\phi\}\right) = Y_{lm}(\theta,\phi+\beta) = \exp(Im\beta)Y_{lm}(\theta,\phi). \tag{15}$$

Rotation about other axes is more complicated, and the general rotation formula can be written as

$$Y_{lm}\left(R_{\alpha,\beta,\gamma}(\theta,\phi)\right) = \sum_{m'=-l}^{l} D^{l}_{mm'}(\alpha,\beta,\gamma)Y_{lm'}(\theta,\phi). \tag{16}$$

The important thing to note here is that the m indices are *mixed*: a spherical harmonic after rotation must be expressed as a combination of other spherical harmonics with different m indices. However, the l indices are not mixed: rotations of spherical harmonics with order l are composed entirely of other spherical harmonics with order l. For given order l, D^l is a matrix that tells us how a spherical harmonic transforms under rotation, i.e., how to rewrite a rotated spherical harmonic as a linear combination of all the spherical harmonics of the same order. In terms of group theory, the matrix D^l is the $(2l+1)$-dimensional representation of the rotation group $SO(3)$. A pictorial depiction of Equation 16 as a matrix multiplication is found in Figure 12.5. An analytic form for the matrices D^l can be found in standard references, such as Inui et al. [28].

$$\begin{pmatrix} Y_{00} \\ Y_{1-1} \\ Y_{10} \\ Y_{11} \\ Y_{2-2} \\ Y_{2-1} \\ Y_{20} \\ Y_{21} \\ Y_{22} \\ \cdots \end{pmatrix} = \begin{pmatrix} D^0_{00} & & & & & & & & & \\ & D^1_{-1-1} & D^1_{-10} & D^1_{-11} & & & & & & \\ & D^1_{0-1} & D^1_{00} & D^1_{01} & & & & & & \\ & D^1_{1-1} & D^1_{10} & D^1_{11} & & & & & & \\ & & & & D^2_{-2-2} & D^2_{-2-1} & D^2_{-20} & D^2_{-21} & D^2_{-22} & \\ & & & & D^2_{-1-2} & D^2_{-1-1} & D^2_{-10} & D^2_{-11} & D^2_{-12} & \\ & & & & D^2_{0-2} & D^2_{0-1} & D^2_{00} & D^2_{01} & D^2_{02} & \\ & & & & D^2_{1-2} & D^2_{1-1} & D^2_{10} & D^2_{11} & D^2_{12} & \\ & & & & D^2_{2-2} & D^2_{2-1} & D^2_{20} & D^2_{21} & D^2_{22} & \\ & & & & & & & & & \cdots \end{pmatrix} \begin{pmatrix} Y_{00} \\ Y_{1-1} \\ Y_{10} \\ Y_{11} \\ Y_{2-2} \\ Y_{2-1} \\ Y_{20} \\ Y_{21} \\ Y_{22} \\ \cdots \end{pmatrix}$$

FIGURE 12.5: Depiction of Equation 16 as a matrix equation. Note the block-diagonal nature of the matrix, with only spherical harmonics of the same order (i.e., $l = 0, 1, 2$) mixed. The matrix elements D are functions of the angles of rotation.

In particular, since $R_{\alpha,\beta,\gamma} = R_z(\beta)R_y(\alpha)R_z(\gamma)$, the dependence of D^l on β and γ is simple, since rotation of the spherical harmonics about the z axis is straightforward, i.e.,

$$D^l_{mm'}(\alpha,\beta,\gamma) = d^l_{mm'}(\alpha)e^{Im\beta}e^{Im'\gamma}, \tag{17}$$

where d^l is a matrix that defines how a spherical harmonic transforms under rotation about the y axis. For the purposes of the exposition, we will not generally need to be concerned with the precise formula for the matrix d^l. The analytic formula is rather complicated, and is derived in Equation 7.48 of Inui et al. [28],

To derive some of the quantitative results, we will require one important property of the representation matrices D^l (see for instance the appendix in [67]),

$$D^l_{m0}(\alpha,\beta,\gamma) = d^l_{m0}(\alpha)e^{Im\beta} = \sqrt{\frac{4\pi}{2l+1}}Y_{lm}(\alpha,\beta). \tag{18}$$

Decomposition into Spherical Harmonics

We now have the tools to expand Equation 8 in spherical harmonics. We first expand the lighting in global coordinates:

$$L(\theta_{\rm i},\phi_{\rm i}) = \sum_{l=0}^{\infty}\sum_{m=-l}^{l} L_{lm}Y_{lm}(\theta_{\rm i},\phi_{\rm i}). \tag{19}$$

To obtain the lighting in local coordinates, we must rotate the above expression. Using Equation 16, we get

$$L(\theta_{\rm i},\phi_{\rm i}) = L\left(R_{\alpha,\beta,\gamma}(\theta'_{\rm i},\phi'_{\rm i})\right) = \sum_{l=0}^{\infty}\sum_{m=-l}^{+l}\sum_{m'=-l}^{l} L_{lm}D^l_{mm'}(\alpha,\beta,\gamma)Y_{lm'}(\theta'_{\rm i},\phi'_{\rm i}). \tag{20}$$

Since the transfer function $A(\theta'_{\rm i}) = \cos\theta'_{\rm i}$ has no azimuthal dependence, terms with $m' \neq 0$ will vanish when we perform the integral in Equation 8. Therefore, we will be most interested in the coefficient of the term with $m' = 0$. We have already seen (Equation 18) that, in this case, $D^l_{m0}(\alpha,\beta,\gamma) = \left(\sqrt{4\pi/(2l+1)}\right)Y_{lm}(\alpha,\beta)$.

We now expand the transfer function in terms of spherical harmonics. Since we are expanding over the full spherical domain, we should set $A(\theta'_{\rm i}) = 0$ in the invisible lower hemisphere. Thus, we define $A(\theta_i')$ as the *half-cosine function*, $A(\theta_i') = \max(\cos\theta_i', 0)$. Since this has no azimuthal dependence, terms A_{ln} with $n \neq 0$ will vanish. Therefore, we can write

$$A(\theta_i') = \max(\cos\theta_i', 0) = \sum_{l=0}^{\infty} A_l Y_{l0}(\theta_i'). \tag{21}$$

Note that the modes Y_{l0} depend only on θ_i' and have no azimuthal dependence.

Finally, we can also expand the irradiance in terms of spherical harmonics. In order to do so, we ignore the tangential rotation γ, which has no physical significance, and write

$$E(\alpha, \beta) = \sum_{l=0}^{\infty} \sum_{m=-l}^{l} E_{lm} Y_{lm}(\alpha, \beta). \tag{22}$$

Spherical-Harmonic Reflection Equation

We can now write down the reflection equation, as given by Equation 8, in terms of the expansions just defined. To do so, we multiply the expansions for the lighting and BRDF, and integrate. By orthonormality of the spherical harmonics, we require $m' = 0$. Hence, we obtain

$$E(\alpha, \beta, \gamma) = \sum_{l=0}^{\infty} \sum_{m=-l}^{l} A_l L_{lm} D^l_{m0}(\alpha, \beta, \gamma),$$

$$E(\alpha, \beta) = \sum_{l=0}^{\infty} \sum_{m=-l}^{l} \sqrt{\frac{4\pi}{2l+1}} A_l L_{lm} Y_{lm}(\alpha, \beta). \tag{23}$$

Note that, as expected, the tangential rotation γ, which has no physical meaning here, has vanished from the equations.

Finally, we can equate spherical-harmonic coefficients of the irradiance, and use a symbol for the normalization, $\Lambda_l = \sqrt{4\pi/(2l+1)}$, to obtain the key equation for Lambertian reflection:

$$E_{lm} = \Lambda_l A_l L_{lm}. \tag{24}$$

This equation states that the standard direct illumination integral in Equation 8 can be viewed as a simple product in terms of spherical-harmonic coefficients.

This is not really surprising,[3] considering that equation 8 can be interpreted as showing that the irradiance is a *convolution* of the incident illumination and the transfer function. We have simply derived the product formula for convolution in the frequency domain, analogous to the standard formula in terms of Fourier coefficients.

Representing the Transfer Function

The one remaining component is to explicitly find the spherical-harmonic coefficients of the half-cosine or clamped-cosine transfer function A_l. The coefficients are given by

$$A_l = 2\pi \int_0^{\frac{\pi}{2}} Y_{l0}(\theta_{\mathrm{i}}') \cos\theta_{\mathrm{i}}' \sin\theta_{\mathrm{i}}' \, d\theta_{\mathrm{i}}', \tag{25}$$

where the factor of 2π comes from integrating 1 over the azimuthal dependence. It is important to note that the limits of the integral range from 0 to $\pi/2$ and not π, because we are considering only the upper hemisphere ($A(\theta_{\mathrm{i}}') = 0$ in the lower hemisphere). The expression above may be simplified by writing in terms of Legendre polynomials $P(\cos\theta_{\mathrm{i}}')$. Putting $u = \cos\theta_{\mathrm{i}}'$ in the above integral and noting that $P_1(u) = u$ and that $Y_{l0}(\theta_{\mathrm{i}}') = \sqrt{(2l+1)/(4\pi)} P_l(\cos\theta_{\mathrm{i}}')$, we obtain

$$A_l = 2\pi \, \Lambda_l^{-1} \int_0^1 P_l(u) P_1(u) \, du. \tag{26}$$

To gain further insight, we need some facts regarding the Legendre polynomials. The polynomial P_l is odd if l is odd, and even if l is even. The Legendre polynomials are orthogonal over the domain $[-1, 1]$ with the orthogonality relationship being given by

$$\int_{-1}^1 P_a(u) P_b(u) = \frac{2}{2a+1} \delta_{a,b} \tag{27}$$

From this, we can establish some results about Equation 26. When l is equal to 1, the integral evaluates to half the norm above, i.e., 1/3. When l is odd but greater than 1, the integral in Equation 26 vanishes. This is because, for $a = l$ and $b = 1$, we can break the left-hand side of Equation 27, using the oddness of a and b, into two equal integrals from $[-1, 0]$ and $[0, 1]$. Therefore, both of these integrals must

[3]Basri and Jacobs [2, 4] have noticed that this result follows directly from Equation 8 by the Funk–Hecke theorem (as stated, for instance, in Groemer [22], page 98). However, that theorem does not generalize to more complex materials (where the BRDF lobe is not radially symmetric, as the half-cosine function is). The derivation above enables us to easily generalize the results to arbitrary materials, as discussed later in the chapter.

vanish, and the latter integral is the right-hand integral in Equation 26. When l is even, the required formula is given by manipulating Equation 20 in Chapter 5 of MacRobert [42]. Putting it all together, we have

$$
\begin{aligned}
l = 1: \quad & A_l = \sqrt{\frac{\pi}{3}}, \\
l > 1, \text{odd}: \quad & A_l = 0, \\
l \ \text{even}: \quad & A_l = 2\pi\sqrt{\frac{2l+1}{4\pi}}\frac{(-1)^{\frac{l}{2}-1}}{(l+2)(l-1)}\left[\frac{l!}{2^l(\frac{l}{2}!)^2}\right].
\end{aligned} \tag{28}
$$

We can determine the asymptotic behavior of A_l for large even l by using Stirling's formula. The bracketed term goes as $l^{-1/2}$, which cancels the term in the square root. Therefore, the asymptotic behavior for even terms is $A_l \sim l^{-2}$. A plot of A_l for the first few terms is shown in Figure 12.6, and approximation of the clamped cosine by spherical-harmonic terms as l increases is shown in Figure 12.7.

Since A_l vanishes for odd values of $l > 1$, as seen in Equation 28, the irradiance (per Equation 24) has zero projection onto odd-order modes, i.e., $E_{lm} = 0$ when $l > 1$ and odd. In terms of the filtering analogy in signal processing, since the filter corresponding to $A(\theta_i') = \cos\theta_i'$ destroys high-frequency odd terms in the

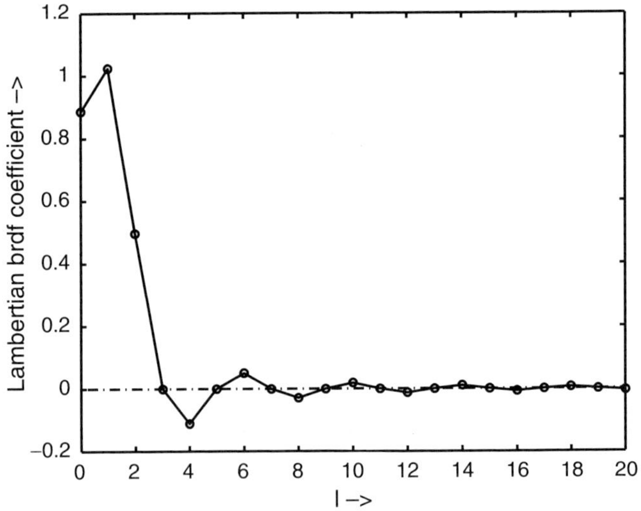

FIGURE 12.6: The solid line is a plot of A_l versus l. It can be seen that odd terms with $l > 1$ have $A_l = 0$. Also, as l increases, the BRDF coefficient or transfer function rapidly decays.

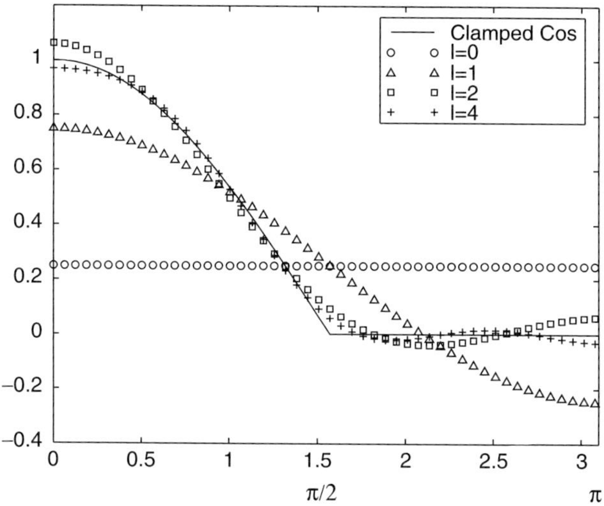

FIGURE 12.7: Successive approximations to the *clamped cosine* function by adding more spherical harmonic terms. For $l = 2$, we already get a very good approximation.

spherical-harmonic expansion of the lighting, the corresponding terms are not found in the irradiance. Further, for large even l, the asymptotic behavior of $E_{lm} \sim l^{-5/2}$ since $A_l \sim l^{-2}$. The transfer function A acts as a low-pass filter, causing the rapid decay of high-frequency terms in the lighting.

It is instructive to explicitly write out numerically the first few terms for the irradiance per Equations 24 and 28:

$$\begin{aligned}
E_{00} &= \pi L_{00} \approx 3.142 L_{00}, \\
E_{1m} &= \frac{2\pi}{3} L_{1m} \approx 2.094 L_{1m}, \\
E_{2m} &= \frac{\pi}{4} L_{2m} \approx 0.785 L_{2m}, \\
E_{3m} &= 0 \\
E_{4m} &= -\frac{\pi}{24} L_{4m} \approx -0.131 L_{4m}, \\
E_{5m} &= 0 \\
E_{6m} &= \frac{\pi}{64} L_{6m} \approx 0.049 L_{6m}.
\end{aligned} \tag{29}$$

We see that, already for E_{4m}, the coefficient is only about 1% of what it is for E_{00}. Therefore, in real applications, where surfaces are only approximately Lambertian, and there are errors in measurement, we can obtain good results using an order 2 approximation of the illumination and irradiance. Since there are $2l + 1$ indices (values of m, which ranges from $-l$ to $+l$) for order l, this corresponds to nine coefficients for $l \leq 2$: 1 term with order 0, three terms with order 1, and five terms with order 2. Note that the single order 0 mode Y_{00} is a constant, and the three order 1 modes are linear functions of the Cartesian coordinates—in real form, they are simply x, y, and z—while the five order 2 modes are quadratic functions of the Cartesian coordinates.

Therefore, the irradiance—or equivalently, the reflected light field from a convex Lambertian object—can be well approximated using spherical harmonics up to order 2 (a 9-term representation), i.e., as a quadratic polynomial of the Cartesian coordinates of the surface normal vector. This is one of the key results of this chapter. By a careful formal analysis, Basri and Jacobs [2, 4] have shown that over 99.2% of the energy of the filter is captured in orders 0, 1 and 2, and 87.5% is captured even in an order 1 four-term approximation. Furthermore, by enforcing nonnegativity of the illumination, one can show that, for *any* lighting conditions, the average accuracy of an order 2 approximation is at least 98%.

12.3.3 Explanation of Experimental Results

Having obtained the theoretical result connecting irradiance and radiance, we can now address the empirical results on low dimensionality we discussed in the background section. First, consider images under all possible illuminations. Since any image can be well described by a linear combination of 9 terms—the images corresponding to the 9 lowest-frequency spherical-harmonic lights—we expect the illumination variation to lie close to a 9-dimensional subspace. In fact, a 4D subspace consisting of the linear and constant terms itself accounts for almost 90% of the energy. This 4D subspace is exactly that of Koenderink and van Doorn [35]. Similarly, the 3D subspace of Shashua [69] can be thought of as corresponding to the lighting from order 1 spherical harmonics only. However, this neglects the ambient or order 0 term, which is quite important from our analytic results.

There remain at this point two interesting theoretical questions. First, how does the spherical-harmonic convolution result connect to principal component analysis? Second, the 9D subspace predicted by the above analysis does not completely agree with the 5D subspace observed in a number of experiments [15, 23]. Recently, we have shown [60] that, under appropriate assumptions, the principal components or eigenvectors are equal to the spherical-harmonic basis functions, and the eigenvalues, corresponding to how important a principal component is in explaining image variability, are equal to the spherical-harmonic coefficients $\Lambda_l A_l$. Furthermore, if we see only a single image, as in previous experimental tests, only the

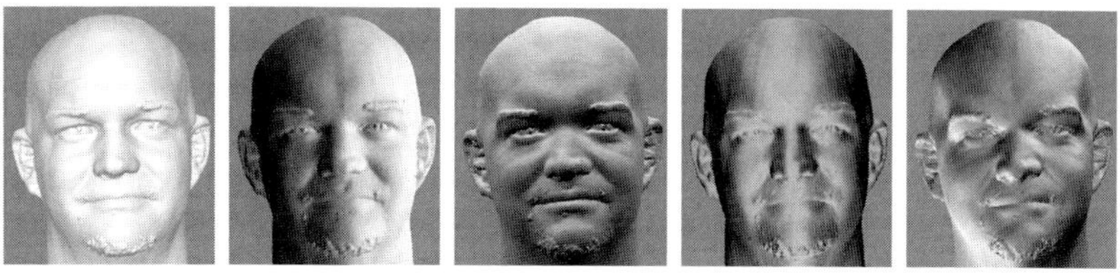

FIGURE 12.8: The first 5 principal components of a face, computed by our method [60] for analytically constructing the PCA for lighting variability using spherical harmonics. The form of these eigenmodes is strikingly similar to those derived empirically by previous researchers (see Figure 1 in Hallinan [23]). The corresponding eigenvalues are in the ratio of .42, .33, .16, .035, .021 and are in agreement with empirical observation. The first 3 and 5 eigenmodes respectively account for 91% and 97% of the variance, compared to empirical values of 90% and 94% respectively [15]. The slight discrepancy is likely due to specularities, cast shadows, and noisy or biased measurements. The principal components contain both positive values (bright regions) and negative values (dark regions), with zero set to the neutral gray of the background. The face is an actual range scan courtesy of Cyberware.

front-facing normals, and not the entire sphere of surface normals is visible. This allows approximation by a lower-dimensional subspace. We have shown how to analyze this effect with spherical harmonics, deriving analytic forms for the principal components of canonical shapes. The results for a human face are shown in Figure 12.8, and are in qualitative and quantitative agreement with previous empirical work, providing the first full theoretical explanation.

12.4 APPLICATIONS OF LAMBERTIAN 9-TERM SPHERICAL-HARMONIC MODEL

In the previous section, we have derived an analytic formula for the irradiance, or reflected light from a Lambertian surface, as a convolution of the illumination and a low-pass filter corresponding to the clamped-cosine function. Though the mathematical derivation is rather involved, the final result is very simple, as given in Equations 24 and 28. This simple 9-term Lambertian model allows one to reduce the effects of very complex lighting distributions, involving multiple point sources and extended sources, to a simple quadratic formula. It therefore enables computer-vision applications, which until now had to rely on the simple point-source or directional light-source model, to handle arbitrary distant lighting with attached shadows. In computer graphics, lighting calculations that had previously required expensive finite-element or Monte Carlo integration, can now be done with a simple analytic formula. Because of the simplicity and effectiveness of the final model, it has been widely adopted in both graphics and vision for a number of practical applications.

12.4.1 Computer Vision: Recognition

In computer vision, a critical problem is being able to recognize objects in a way that is lighting insensitive. This is especially important to build robust face-detection and face-recognition systems. In fact, face recognition was the first application of the Lambertian 9-term model (Basri and Jacobs [2]). To apply the results of the previous section, we assume the availability of 3D geometric models and albedos of the faces of interest. With the ready availability of range scanners that give detailed 3D models, this is an increasingly reasonable assumption.

Once the position of the object and pose have been estimated, the lighting-insensitive recognition problem can be stated as follows. Given a 3D geometric model and albedos for an object, determine if the query image observed can correspond to the object under some lighting condition. To solve this problem, we note that images of a Lambertian object lie close to a 9D subspace. We simply need to take the query image and project it onto this subspace, to see if it lies close enough.[4] Furthermore, our spherical harmonic formula gives an easy way of analytically constructing the subspace, simply by analytically rendering images of the object under the 9 spherical harmonic basis lighting conditions, using Equation 24. This approach is efficient and robust, compared with numerical PCA methods. The results are excellent (recognition rates $> 95\%$ on small datasets), indicating that human faces are reasonably well approximated as Lambertian convex objects for the purpose of modeling illumination variation in recognition. Hence, the Lambertian 9-term model has significant practical relevance for robust modeling and recognition systems for faces.

Recent work has addressed a number of subproblems in recognition. Lee et al. [39] show how to construct the 9D Lambertian subspace using nine images taken with point light sources, while Zhang and Samaras [82] have shown how to learn spherical-harmonic subspaces for recognition. Ho et al. [26] have used the harmonic subspace for image clustering. We (Osadchy et al. [57]) have also built on the spherical-harmonic representation, combining it with compact models for light sources to tackle the problem of recognizing specular and transparent objects (Figure 12.9).

[4] In practice, the object that lies closest to the subspace can be chosen—we recognize the best-matching face.

FIGURE 12.9: Some of the objects which we have been able to recognize [57], using a combination of the Lambertian 9-term model, and compact light source models to account for specularities.

12.4.2 Modeling

An understanding of lighting is also critical for many modeling problems. For instance, in computer vision, we can obtain the shape of an object to build 3D models using well-known and well-studied techniques like photometric stereo, stereo, or structure from motion. However, all of these methods usually make very simple assumptions about the lighting conditions, such as point light sources without shadows. Our Lambertian results enable solving these problems under realistic complex lighting, which has implications for shape modeling in uncontrolled environments.

In photometric stereo, we have a number of images under different lighting conditions (previously, these were always point sources). We seek to determine the shape (surface normals). Basri and Jacobs [3] have shown how to solve this problem for the first time under unknown complex (but distant) illumination, under the assumption of Lambertian reflectance and convex objects.

The mathematics of their solution is conceptually simple. Consider taking f images of p pixels each. One can create a large matrix M of size $f \times p$ holding all the image information. The low-dimensional Lambertian results indicate that this matrix can be approximated as $M \approx LS$, where L is a $f \times 9$ lighting matrix, that gives the lowest-frequency lighting coefficients for the f images, and S is a $9 \times p$ matrix, which corresponds to the spherical-harmonic basis images of the object under the 9 lighting conditions. Of course, once the basis images are recovered, the surface normals and albedos can be trivially estimated from the order 0 and order 1 terms, which are simply the albedos and scaled surface normals themselves. The computation of $M \approx LS$ can be done using singular-value decomposition, but is not generally unique. Indeed, there is a 9×9 ambiguity matrix (4×4 if using a linear 4D subspace approximation). Basri and Jacobs [3] analyze these ambiguities in some detail. It should be noted that their general approach is quite similar to previous work on photometric stereo with point sources, such as the factorization method proposed by Hayakawa [24]. The Lambertian low-dimensional lighting model enables much of the same machinery to be used, now taking complex illumination and attached shadows into account.

In photometric stereo, we assume the scene is static, and only the lighting changes. This is not always practical, for instance when building models of human faces. A far less restrictive acquisition scenario is to directly model objects from video, where the subject is moving normally. We would like to take your home videos, and construct a 3D geometric model of your face. This problem also relates to structure from motion, a classic computer-vision method. Simakov et al. [71] have recently addressed this problem. They assume the motion is known, such as from tracking features like the eyes or nose. They then seek to recover the 3D shape and albedos of the rigid moving object, as well as a spherical-harmonic description of the illumination in the scene. They use the Lambertian spherical-harmonic model to create a consistency measure within a stereo algorithm. Zhang et al. [81] do not assume known motion, but assume infinitesimal movement between frames. They use the 4-term or first-order spherical-harmonic model for representing the illumination.

12.4.3 Inverse Problems: Estimating Illumination

An important theoretical and practical problem is to estimate the illumination from observations. Once we have done this, we can add synthetic objects to a real scene with realistic lighting for computer graphics, or use knowledge of the lighting to solve computer-vision problems. The study of inverse problems can be seen in terms of our results as a deconvolution.

For estimating the illumination from a convex Lambertian object, our theoretical results give a simple practical solution. We just need to use Equation 24, solving for lighting coefficients using $L_{lm} = \Lambda_l^{-1} E_{lm}/A_l$. The form of the Lambertian filter coefficients indicates that inverse lighting is ill-posed (numerator and denominator vanish) for odd illumination frequencies greater than one. These frequencies of the signal are completely annihilated by the Lambertian filter, and do not appear in the output irradiance at all. Furthermore, inverse lighting is an ill-conditioned inverse problem, because the Lambertian filter is low-pass, and the irradiance cannot easily be deconvolved. In fact, we will be able to reliably estimate only the first 9 low-order spherical-harmonic terms of the incident lighting from images of a Lambertian surface.

These results settle a theoretical question on whether radiance and irradiance are equivalent [64], by showing formally that irradiance is not invertible, since the Lambertian kernel has zero entries. This corrects a long-standing (but incorrect) conjecture (Preisendorfer [58], vol. 2, pp 143–151) in the radiative-transfer community. It also explains the ill-conditioning in inverse lighting calculations observed by previous authors like Marschner and Greenberg [44]. Finally, the results are in accord with important perceptual observations like the retinex theory [38]. Since lighting cannot produce high-frequency patterns on Lambertian surfaces, such patterns must be attributable to texture, allowing perceptual separation of illumination effects and those due to texture or albedo.

12.4.4 Computer Graphics: Irradiance Environment Maps for Rendering

The OpenGL standard for graphics hardware currently has native support only for point or directional light sources. The main reason is the lack of simple procedural formulas for general lighting distributions. Instead, the lighting from all sources must be summed or integrated. Our Lambertian 9-term model makes it straightforward to compute, represent, and render with the irradiance environment map $E(\alpha, \beta)$. This was our first practical application of the Lambertian 9-term model in computer graphics [63].

First, we *prefilter* the environment map, computing the lighting coefficients L_{lm}. From this, we can compute the 9 irradiance coefficients E_{lm} as per Equation 24. This computation is very efficient, being linear in the size of the environment map. We can now adopt one of two approaches. The first is to explicitly compute the irradiance map $E(\alpha, \beta)$ by evaluating it from the irradiance coefficients. Then, for any surface normal (α, β), we can simply look up the irradiance value and multiply by the albedo to shade the Lambertian surface. This was the approach taken in previous work on environment mapping [21, 46]. Those methods needed to explicitly compute $E(\alpha, \beta)$ with a hemispherical integral of all the incident illumination for each surface normal. This was a slow procedure (it is $O(n^2)$, where n is the size of the environment map in pixels). By contrast, our spherical-harmonic prefiltering method is linear time $O(n)$. Optimized implementations can run in near real time, enabling our method to potentially be used for dynamically changing illumination. Software for preconvolving environment maps is available on our website and widely used by industry.

However, we can go much further than simply computing an explicit representation of $E(\alpha, \beta)$ quickly. In fact, we can use a simple procedural formula for shading. This allows us to avoid using textures for irradiance calculations. Further, the computations can be done at each vertex, instead of at each pixel, since irradiance varies slowly. They are very simple to implement in either vertex shaders on the graphics card, or in software. All we need to do is explicitly compute

$$E(\alpha, \beta) \approx \sum_{l=0}^{2} \sum_{m=-l}^{l} E_{lm} Y_{lm}(\alpha, \beta). \tag{30}$$

This is straightforward, because the spherical harmonics up to order 2 are simply constant, linear, and quadratic polynomials. In fact, we can write an alternative simple quadratic form,

$$E(n) = n^{\mathrm{T}} M n, \tag{31}$$

where n is the 4×1 surface normal in homogeneous Cartesian coordinates and M is a 4×4 matrix. Since this involves only a matrix–vector multiplication and dot product, and hardware is optimized for these operations with 4×4 matrices and vectors, the calculation is very efficient. An explicit form for the matrix M is given in [63]. The image in Figure 12.10 was rendered using this method, which is now widely adopted for real-time rendering in video games.

Note the duality between forward and inverse problems. The inverse problem, inverse lighting, is ill-conditioned, with high frequencies of the lighting not easy

FIGURE 12.10: The diffuse shading on all the objects is computed procedurally in real time using the quadratic formula for Lambertian reflection [63]. The middle sphere, armadillo and table are diffuse reflectors. Our method can also be combined with standard texture mapping, used to modulate the albedo of the pool ball on the right, and reflection mapping, used for specular highlights on the pool ball and mirror sphere on the left. The environment is a light probe of the Grace Cathedral in San Francisco, courtesy of Paul Debevec.

to estimate. Conversely, the forward problem, irradiance environment maps, can be computed very efficiently, by approximating only the lowest frequencies of the illumination and irradiance. This duality between forward and inverse problems has rarely been used before, and we believe it can lead to many new applications.

12.5 SPECULARITIES: CONVOLUTION FORMULA FOR GENERAL MATERIALS

In this section, we discuss the extension to general materials, including specularities, and briefly explore the implications and practical applications of these general convolution results. We first briefly derive the fully general case with anisotropic materials [61, 67]. An illustrative diagram is found in Figure 12.11. For practical applications, it is often convenient to assume isotropic materials. The convolution result for isotropic BRDFs, along with implications are developed in [65, 67], and we will quickly derive it here as a special case of the anisotropic theory.

12.5.1 General Convolution Formula

The reflection equation, analogous to Equation 8 in the Lambertian case, now becomes

$$B(\alpha, \beta, \gamma, \theta'_o, \phi'_o) = \int_{\Omega'_i} L\left(R_{\alpha,\beta,\gamma}\left(\theta'_i, \phi'_i\right)\right) \hat{\rho}(\theta'_i, \phi'_i, \theta'_o, \phi'_o)\, d\omega'_i, \tag{32}$$

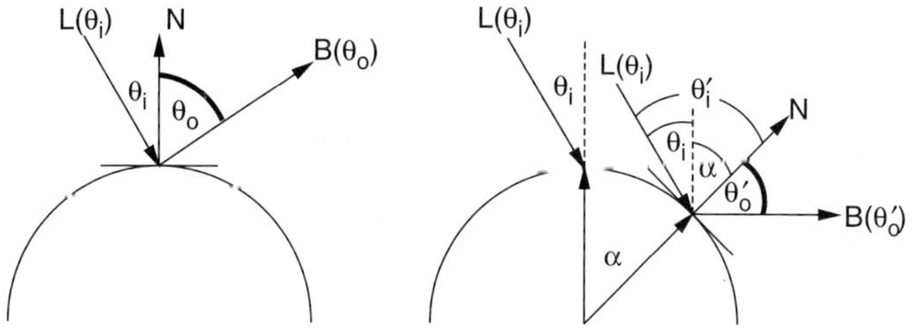

FIGURE 12.11: Schematic of reflection in 2D. On the left, we show the situation with respect to one point on the surface (the North Pole or 0° location, where global and local coordinates are the same). The right figure shows the effect of the surface orientation α. Different orientations of the surface correspond to rotations of the upper hemisphere and BRDF, with the global incident direction θ_i corresponding to a rotation by α of the local incident direction θ'_i. Note that we also keep the local outgoing angle (between N and B) fixed between the two figures.

where the reflected light field B now also depends on the (local) outgoing angles θ'_o, ϕ'_o. Here $\hat{\rho}$ is a transfer function (the BRDF multiplied by the cosine of the incident angle $\hat{\rho} = \rho \max(\cos\theta'_i, 0)$), which depends on local incident and outgoing angles.

The lighting can be expanded as per Equations 19 and 20 in the Lambertian derivation. The expansion of the transfer function is now more general than Equation 21, being given in spherical harmonics by

$$\hat{\rho}(\theta'_i, \phi'_i, \theta'_o, \phi'_o) = \sum_{l=0}^{\infty} \sum_{n=-l}^{l} \sum_{p=0}^{\infty} \sum_{q=-p}^{p} \hat{\rho}_{ln,pq} Y^*_{ln}(\theta'_i, \phi'_i) Y_{pq}(\theta'_o, \phi'_o), \tag{33}$$

where the complex conjugate for the first factor is to simplify the final results.

Finally, we need to expand the reflected light field in basis functions. For the Lambertian case, we simply used spherical harmonics over (α, β). In the general case, we must consider the full tangent frame (α, β, γ). Thus, the expansion is over mixed basis functions, $D^l_{mn}(\alpha, \beta, \gamma) Y_{pq}(\theta'_o, \phi'_o)$. With the correct normalization (Equation 7.73 of [28]),

$$B(\alpha, \beta, \gamma, \theta'_o, \phi'_o) = \sum_{l=0}^{\infty} \sum_{m=-l}^{l} \sum_{n=-l}^{l} \sum_{p=0}^{\infty} \sum_{q=-p}^{p} B_{lmn,pq} C_{lmn,pq}(\alpha, \beta, \gamma, \theta'_o, \phi'_o),$$

$$C_{lmn,pq}(\alpha, \beta, \gamma, \theta'_o, \phi'_o) = \sqrt{\frac{2l+1}{8\pi^2}} D^l_{mn}(\alpha, \beta, \gamma) Y_{pq}(\theta'_o, \phi'_o). \tag{34}$$

We can now write down the reflection equation, as given by Equation 32, in terms of the expansions just defined. As in the Lambertian case, we multiply the expansions for the lighting (Equation 20) and BRDF (Equation 33) and integrate. To avoid confusion between the indices in this intermediate step, we will use L_{lm} and $\hat{\rho}_{l'n,pq}$ to obtain

$$\begin{aligned} B(\alpha, \beta, \gamma, \theta'_o, \phi'_o) &= \sum_{l=0}^{\infty} \sum_{m=-l}^{l} \sum_{m'=-l}^{l} \sum_{l'=0}^{\infty} \sum_{n=-l'}^{l'} \sum_{p=0}^{\infty} \sum_{q=-p}^{p} L_{lm} \hat{\rho}_{l'n,pq} D^l_{mm'}(\alpha, \beta, \gamma) \\ &\quad \times Y_{pq}(\theta'_o, \phi'_o) T_{lm'l'n}, \\ T_{lm'l'n} &= \int_{\phi'_i=0}^{2\pi} \int_{\theta'_i=0}^{\pi} Y_{lm'}(\theta'_i, \phi'_i) Y^*_{l'n}(\theta'_i, \phi'_i) \sin\theta'_i \, d\theta'_i d\phi'_i \\ &= \delta_{ll'} \delta_{m'n}. \end{aligned} \tag{35}$$

The last line follows from orthonormality of the spherical harmonics. Therefore, we may set $l' = l$ and $n = m'$ since terms not satisfying these conditions vanish. We then obtain

$$B(\alpha, \beta, \gamma, \theta'_o, \phi'_o) = \sum_{l=0}^{\infty} \sum_{m=-l}^{l} \sum_{n=-l}^{l} \sum_{p=0}^{\infty} \sum_{q=-p}^{p} L_{lm} \hat{\rho}_{ln,pq} \left(D^l_{mn}(\alpha, \beta, \gamma) Y_{pq}(\theta'_o, \phi'_o) \right). \tag{36}$$

Finally, equating coefficients, we obtain the spherical-harmonic reflection equation or convolution formula, analogous to Equation 24. The reflected light field can be viewed as taking the incident illumination signal, and filtering it with the material properties or BRDF of the surface:

$$B_{lmn,pq} = \sqrt{\frac{8\pi^2}{2l+1}} L_{lm} \hat{\rho}_{ln,pq}. \tag{37}$$

Isotropic BRDFs

An important special case are isotropic BRDFs, where the orientation of the local tangent frame γ does not matter. To consider the simplification that results from isotropy, we first analyze the BRDF coefficients $\hat{\rho}_{ln,pq}$. In the BRDF expansion of Equation 33, only terms that satisfy isotropy, i.e., are invariant with respect to adding an angle $\Delta\phi$ to both incident and outgoing azimuthal angles, are nonzero. From the form of the spherical harmonics, this requires that $n = q$, and we write the now 3D BRDF coefficients as $\hat{\rho}_{lpq} = \hat{\rho}_{lq,pq}$.

Next, we remove the dependence of the reflected light field on γ by arbitrarily setting $\gamma = 0$. The reflected light field can now be expanded in coefficients [65]:

$$B(\alpha, \beta, \theta'_o, \phi'_o) = \sum_{l=0}^{\infty} \sum_{m=-l}^{l} \sum_{p=0}^{\infty} \sum_{q=-\min(l,p)}^{\min(l,p)} B_{lmpq} C_{lmpq}(\alpha, \beta, \theta'_o, \phi'_o),$$

$$C_{lmpq}(\alpha, \beta, \theta'_o, \phi'_o) = \sqrt{\frac{2l+1}{4\pi}} D^l_{mq}(\alpha, \beta, 0) Y_{pq}(\theta'_o, \phi'_o). \tag{38}$$

Then, the convolution formula for isotropic BRDFs becomes

$$B_{lmpq} = \Lambda_l L_{lm} \hat{\rho}_{lpq}. \tag{39}$$

Finally, it is instructive to derive the Lambertian results as a special case of the above formulation. In this case, $p = q = 0$ (no exitant angular dependence), and Equation 24 is effectively obtained simply by dropping the indices p and q in the above equation.

12.5.2 Implications

We now briefly discuss the theoretical implications of these results. A more detailed discussion can be found in [65, 67]. First, we consider inverse problems like illuminant and BRDF estimation. From the formula above, we can solve them simply by dividing the reflected-light field coefficients by the known lighting or BRDF coefficients, i.e., $L_{lm} = \Lambda_l^{-1} B_{lmpq}/\hat{\rho}_{lpq}$ and $\hat{\rho}_{lpq} = \Lambda_l^{-1} B_{lmpq}/L_{lm}$. These problems will be well conditioned when the denominators do not vanish and contain high-frequency elements. In other words, for inverse lighting to be well conditioned, the BRDF should be an all-pass filter, ideally a mirror surface. Mirrors are delta functions in the spatial or angular domain, and include all frequencies in the spherical-harmonic domain. Estimating the illumination from a chrome-steel or mirror sphere is in fact a common approach. On the other hand, if the filter is low-pass, like a Lambertian surface, inverse lighting will be ill-conditioned. Similarly, for BRDF estimation to be well conditioned, the lighting should include high frequencies like sharp edges. The ideal is a directional or point light source. In fact, Lu et al. [41] and Marschner et al. [43] have developed image-based BRDF measurement methods from images of a homogeneous sphere lit by a point source. More recently, Gardner et al. [17] have used a linear light source. On the other hand, BRDF estimation is ill-conditioned under soft diffuse lighting. On a cloudy day, we will not be able to estimate the widths of specular highlights accurately, since they will be blurred out.

Interesting special cases are Lambertian, and Phong and Torrance-Sparrow [77] BRDFs. We have already discussed the Lambertian case in some detail. The frequency-domain filters corresponding to Phong or microfacet models are approximately Gaussian [65]. In fact, the width of the Gaussian in the frequency domain is inversely related to with the width of the filter in the spatial or angular domain. The extreme cases are a mirror surface that is completely localized (delta function) in the angular domain and passes through all frequencies (infinite frequency width), and a very rough or near-Lambertian surface that has full width in the angular domain, but is a very compact low-pass filter in the frequency domain. This duality between angular and frequency-space representations is similar to that for Fourier basis functions. In some ways, it is also analogous to the Heisenberg uncertainty principle—we cannot simultaneously localize a function in both angular and frequency domains. For representation and rendering however, we can turn this to our advantage, using *either* the frequency or angular domain, wherever the different components of lighting and material properties can be more efficiently and compactly represented.

Another interesting theoretical question concerns *factorization* of the 4D reflected light field, simultaneously estimating the illumination (2D) and isotropic BRDF (3D). Since the lighting and BRDF are lower-dimensional entities, it seems that we have redundant information in the reflected light field. We show [65] that an analytic formula for both illumination and BRDF can be derived from reflected-light field coefficients. Thus, up to a global scale and possible ill-conditioning, the reflected-light field can in theory be factored into illumination and BRDF components.

12.5.3 Applications

Inverse Rendering

Inverse rendering refers to inverting the normal rendering process to estimate illumination and materials from photographs. Many previous inverse rendering algorithms have been limited to controlled laboratory conditions using point light sources. The insights from our convolution analysis have enabled the development of a general theoretical framework for handling complex illumination. We [61, 65] have developed a number of frequency-domain and dual angular and frequency-domain algorithms for inverse rendering under complex illumination, including simultaneous estimation of the lighting and BRDFs from a small number of images.

Forward Rendering (Environment Maps with Arbitrary BRDFs)

In the forward problem, all of the information on lighting and BRDFs is available, so we can directly apply the convolution formula [66]. We have shown how this can be applied to rendering scenes with arbitrary distant illumination (environment maps) and BRDFs. As opposed to the Lambertian case, we no longer have a simple 9-term formula. However, just as in the Lambertian case, convolution makes prefiltering the environment extremely efficient, three to four orders of magnitude faster than previous work. Furthermore, it allows us to use a new compact and efficient representation for rendering, the spherical-harmonic reflection map or SHRM. The frequency-domain analysis helps to precisely set the number of terms and sampling rates and frequencies used for a given accuracy.

Material Representation and Recognition

Nillius and Eklundh [54, 55] derive a low-dimensional representation using PCA for images with complex lighting and materials. This can be seen as a generalization of the PCA results for diffuse surfaces (to which they have also made an important contribution [53]). They can use these results to recognize or classify the materials in an image, based on their reflectance properties.

12.6 RELAXING AND BROADENING THE ASSUMPTIONS: RECENT WORK

So far, we have seen how to model illumination, and the resulting variation in appearance of objects using spherical harmonics. There remain a number of assumptions as far as the theoretical convolution analysis is concerned. Specifically, we assume distant lighting and convex objects without cast shadows or interreflections. Furthermore, we are dealing only with opaque objects in free space, and are not handling translucency or participating media. It has not so far been possible to extend the theoretical analysis to relax these assumptions. However, a large body of recent theoretical and practical work has indicated that the key concepts from the earlier analysis—such as convolution, filtering, and spherical harmonics—can be used to analyze and derive new algorithms for all of the effects noted above. The convolution formulas presented earlier in the chapter will of course no longer hold, but the insights can be seen to be much more broadly applicable.

Cast Shadows

For certain special cases, it is possible to derive a convolution result for cast shadows (though not in terms of spherical harmonics). For parallel planes, Soler and Sillion [75] have derived a convolution result, using it as a basis for a fast algorithm for soft shadows. We [68] have developed a similar result for V-grooves to explain lighting variability in natural 3D textures.

An alternative approach is to incorporate cast shadows in the integrand of the reflection equation. The integral then includes three terms, instead of two—the illumination, the BRDF (or Lambertian half-cosine function), and a binary term for cast shadows. Since cast shadows vary spatially, reflection is no longer a convolution in the general case. However, it is still possible to use spherical harmonics to represent the terms of the integrand. Three-term or triple-product integrals can be analyzed using the Clebsch–Gordan series for spherical harmonics [28], which is common for analyzing angular momentum in quantum mechanics. A start at working out the theory is made by Thornber and Jacobs [76]. Most recently, we [51] have investigated generalizations of the Clebsch–Gordan series for other bases, like Haar wavelets, in the context of real-time rendering with cast shadows.

Precomputed Light Transport:

Sloan et al. [74] recently introduced a new approach to real-time rendering called precomputed radiance transfer. Their method uses low-frequency spherical-harmonic representations (usually order 4 or 25 terms) of the illumination and BRDF, and allows for a number of complex light-transport effects including soft shadows, interreflections, global illumination effects like caustics, and dynamically changing lighting and viewpoint. Because of the high level of realism, these approaches provide compared to standard hardware rendering, spherical-harmonic lighting is now widely used in interactive applications like video games, and is incorporated into Microsoft's DirectX API.

We introduce the theory only for Lambertian surfaces—relatively straightforward extensions are possible for complex materials, including view-dependence.

The basic idea is to define a transport operator T that includes half-cosine, visibility and interreflection effects. We can then write, in analogy with Equations 3 and 6,

$$E(\mathrm{x}) = \int_{\Omega_\mathrm{i}} L(\omega_\mathrm{i}) T(\mathrm{x}, \omega_\mathrm{i})\, d\omega_\mathrm{i}, \tag{40}$$

where T denotes the irradiance at x due to illumination from direction ω_i, and incorporates all light-transport effects including cast shadows and interreflections. An offline precomputation step (that may involve ray-tracing for instance) is used to determine T for a static scene. We can expand all quantities in spherical harmonics, to write

$$E(\mathrm{x}) \approx \sum_{l=0}^{l^*} \sum_{m=-l}^{l} T_{lm}(\mathrm{x}) L_{lm}, \tag{41}$$

where l^* denotes the maximum order used for calculations (2 would be sufficient in the absence of cast shadows). We repeat this calculation for each vertex x for rendering. The expression above is simply a dot product which can be implemented efficiently in current programmable graphics hardware, allowing real-time manipulation of illumination. Somewhat more complicated, but conceptually similar expressions, are available for glossy materials. For real-time rendering with varying lighting and viewpoint, Sloan et al. [73] precompute a view-dependent operator (a 25×25 matrix of spherical-harmonic coefficients at each vertex) and compress the information over the surface using clustered principal-component analysis.

New Basis Functions

Spherical harmonics are the ideal signal-processing representation when we have a convolution formula or want to analyze functions in the frequency domain. However, they are not the best representation for "all-frequency" effects—an infinite number of spherical harmonics will be needed to accurately represent a point source or delta function. Furthermore, some quantities like the BRDF are better defined over the hemisphere rather than the full sphere. In cases where cast shadows and interreflections are involved, we no longer have a convolution formula, so spherical harmonics are sometimes not the preferred representation. Recently, we [50, 51] have introduced all-frequency relighting methods using Haar wavelets. These methods can be orders of magnitude more efficient than previous precomputed light-transport techniques using spherical harmonics. Hemispherical basis functions for BRDF representation have been proposed by Gautron et al. [18] and Koenderink and van Doorn [36].

Near-Field Illumination

Recent theoretical work [16] has tried to remove the limitation of distant illumination, by showing a convolution result (with a different kernel) to also hold for near-field lighting under specific assumptions for Lambertian surfaces. A practical solution using spherical-harmonic gradients for rendering is proposed by Annen et al. [1].

Translucency and Participating Media

Like most of computer vision and graphics, we have so far assumed clear-day conditions, without translucent materials like marble and skin, or participating media like milk, smoke, clouds, or fog. Both spherical harmonics and convolution are popular tools for subsurface scattering, volumetric rendering and radiative transport problems. Practical methods that conceptually use convolution of the incident irradiance with a point-spread function have been employed successfully for subsurface scattering [30, 31], and we have recently developed such methods for rendering multiple scattering effects in volumetric participating media [49, 59].

12.7 CONCLUSION

Much of the richness of the visual appearance of our world arises from the effects of illumination, and the varying appearance of surfaces with lighting. In this chapter, we have seen that there are many situations in the real world, such as for diffuse Lambertian surfaces, where the appearance of an object can be relatively simple to describe, even if the lighting is complicated. We have introduced new tools to model illumination and appearance using spherical harmonics. With these tools, we can derive a signal-processing approach to reflection, where the reflected light can be thought of as obtained by filtering or convolving the incident illumination signal by the material properties of the surface. This insight leads to a number of new results, and explanations of previous experimental work, such as the 9-parameter low-dimensional subspace for Lambertian objects. There are implications for face modeling, recognition and rendering, as well as a number of other problems in computer vision, graphics and perception. Already, a number of new algorithms have been developed in computer vision for recognition and modeling, and in computer graphics for both forward and inverse problems in rendering, with the promise of many further developments in the near future.

ACKNOWLEDGMENTS

Special thanks to all my coauthors and collaborators, with particular gratitude to Pat Hanrahan, David Jacobs, Ronen Basri, and Ren Ng. This work was supported in part by grants #0305322, #0430258 and #0446916 from the National Science Foundation and from Intel Corporation.

REFERENCES

[1] T. Annen, J. Kautz, F. Durand, and H. Seidel. Spherical harmonic gradients for mid-range illumination. In: *EuroGraphics Symposium on Rendering*, 2004.

[2] R. Basri and D. Jacobs. Lambertian reflectance and linear subspaces. In: *International Conference on Computer Vision*, pages 383–390, 2001.

[3] R. Basri and D. Jacobs. Photometric stereo with general, unknown lighting. In: *CVPR 01*, pages II–374–II–381, 2001.

[4] R. Basri and D. Jacobs. Lambertian reflectance and linear subspaces. *PAMI* **25**(2): 218–233, 2003.

[5] R. Basri and D. Jacobs. Illumination modeling for face recognition. Chapter 5 in *Face Recognition Handbook*. Springer Verlag, 2004.

[6] P. Belhumeur and D. Kriegman. What is the set of images of an object under all possible illumination conditions? *IJCV* **28**(3):245–260, 1998.

[7] J. Blinn and M. Newell. Texture and reflection in computer generated images. *Communications of the ACM* **19**:542–546, 1976.

[8] R. Brunelli and T. Poggio. Face recognition: Features versus templates. *IEEE Transactions on Pattern Analysis and Machine Intelligence* **15**(10):1042–1062, 1993.

[9] B. Cabral, N. Max, and R. Springmeyer. Bidirectional reflection functions from surface bump maps. In: *SIGGRAPH 87*, pages 273–281, 1987.

[10] B. Cabral, M. Olano, and P. Nemec. Reflection space image based rendering. In: *SIGGRAPH 99*, pages 165–170, 1999.

[11] H. Chen, P. Belhumeur, and D. Jacobs. In search of illumination invariants. In: *CVPR 00*, pages 254–261, 2000.

[12] M. Cohen and J. Wallace. *Radiosity and Realistic Image Synthesis*. Academic Press, 1993.

[13] P. Debevec, T. Hawkins, C. Tchou, H.P. Duiker, W. Sarokin, and M. Sagar. Acquiring the reflectance field of a human face. In: *SIGGRAPH 00*, pages 145–156, 2000.

[14] M. D'Zmura. Shading ambiguity: reflectance and illumination. In: *Computational Models of Visual Processing*, pages 187–207. MIT Press, 1991.

[15] R. Epstein, P.W. Hallinan, and A. Yuille. 5 plus or minus 2 eigenimages suffice: An empirical investigation of low-dimensional lighting models. In: *IEEE Workshop on Physics-Based Modeling in Computer Vision*, pages 108–116, 1995.

[16] D. Frolova, D. Simakov, and R. Basri. Accuracy of spherical harmonic approximations for images of lambertian objects under far and near lighting. In: *ECCV*, pages I–574–I–587, 2004.

[17] A. Gardner, C. Tchou, T. Hawkins, and P. Debevec. Linear light source reflectometry. *ACM TOG (SIGGRAPH 2003)* **22**(3):749–758, 2003.

[18] P. Gautron, J. Krivanek, S. Pattanaik, and K. Bouatouch. A novel hemispherical basis for accurate and efficient rendering. In: *EuroGraphics Symposium on Rendering*, 2004.

[19] A. Georghiades, P. Belhumeur, and D. Kriegman. From few to many: Generative models for recognition under variable pose and illumination. In: *Fourth International Conference on Automatic Face and Guesture Recognition*, pages 277–284, 2000.

[20] A. Georghiades, D. Kriegman, and P. Belhumeur. Illumination cones for recognition under variable lighting: Faces. In: *CVPR 98*, pages 52–59, 1998.

[21] N. Greene. Environment mapping and other applications of world projections. *IEEE Computer Graphics and Applications* **6**(11):21–29, 1986.

[22] H. Groemer. *Geometric Applications of Fourier Series and Spherical Harmonics*. Cambridge University Press, 1996.

[23] P.W. Hallinan. A low-dimensional representation of human faces for arbitrary lighting conditions. In: *CVPR 94*, pages 995–999, 1994.

[24] H. Hayakawa. Photometric stereo under a light source with arbitrary motion. *Journal of the Optical Society of America* A **11**(11):3079–3089, Nov 1994.

[25] W. Heidrich and H. P. Seidel. Realistic, hardware-accelerated shading and lighting. In: *SIGGRAPH 99*, pages 171–178, 1999.

[26] J. Ho, M. Yang, J. Lim, K. Lee, and D. Kriegman. Clustering Appearances of Objects Under Varying Illumination Conditions. In: *CVPR*, volume 1, pages 11–18, 2003.

[27] B. Horn. *Robot Vision*. MIT Press, 1986.

[28] T. Inui, Y. Tanabe, and Y. Onodera. *Group Theory and its Applications in Physics*. Springer Verlag, 1990.

[29] J. Jackson. *Classical Electrodynamics*. John Wiley, 1975.

[30] H. Jensen and J. Buhler. A rapid hierarchical rendering technique for translucent materials. *ACM Transactions on Graphics (SIGGRAPH 2002)* **21**(3):576–581, 2002.

[31] H. Jensen, S. Marschner, M. Levoy, and P. Hanrahan. A practical model for subsurface light transport. In: *SIGGRAPH 2001*, pages 511–518, 2001.

[32] J. Kautz and M. McCool. Approximation of glossy reflection with prefiltered environment maps. In: *Graphics Interface*, pages 119–126, 2000.

[33] J. Kautz, P. Vázquez, W. Heidrich, and H.P. Seidel. A unified approach to prefiltered environment maps. In: *Eurographics Rendering Workshop 00*, pages 185–196, 2000.

[34] M. Kirby and L. Sirovich. Application of the Karhunen–Loeve procedure for the characterization of human faces. *PAMI* **12**(1):103–108, Jan 1990.

[35] J. Koenderink and A. van Doorn. The generic bilinear calibration-estimation problem. *IJCV* **23**(3):217–234, 1997.

[36] J. Koenderink and A. van Doorn. Phenomenological description of bidirectional surface reflection. *Journal of the Optical Society of America* A **15**(11):2903–2912, 1998.

[37] M. Lades, J. Vorbruggen, J. Buhmann, J. Lange, C. von der Malsburg, R. Wurtz, and W. Konen. Distortion invariant object recognition in the dynamic link architecture. *IEEE Transactions on Computers* **42**(3):300–311, 1992.

[38] E. Land and J. McCann. Lightness and retinex theory. *Journal of the Optical Society of America* **61**(1):1–11, 1971.

[39] K. Lee, J. Ho, and D. Kriegman. Nine points of light: Acquiring subspaces for face recognition under variable lighting. In: *CVPR*, pages 519–526, 2001.

[40] A. Levin and A. Shashua. Principal component analysis over continuous subspaces and intersection of half-spaces. In: *ECCV 02*, 2002.

[41] R. Lu, J.J. Koenderink, and A.M.L. Kappers. Optical properties (bidirectional reflection distribution functions) of velvet. *Applied Optics* **37**(25):5974–5984, 1998.

[42] T. MacRobert. *Spherical Harmonics; an Elementary Treatise on Harmonic Functions, with Applications*. Dover Publications, 1948.

[43] S. Marschner, S. Westin, E. Lafortune, and K. Torrance. Image-Based BRDF measurement. *Applied Optics* **39**(16):2592–2600, 2000.

[44] S.R. Marschner and D.P. Greenberg. Inverse lighting for photography. In: *Fifth Color Imaging Conference*, pages 262–265, 1997.

[45] R. McCluney. *Introduction to Radiometry and Photometry*. Artech House, 1994.

[46] G. Miller and C. Hoffman. Illumination and reflection maps: Simulated objects in simulated and real environments. *SIGGRAPH 84 Advanced Computer Graphics Animation, Seminar Notes*, 1984.

[47] Y. Moses, Y. Adini, and S. Ullman. Face recognition: the problem of compensating for changes in illumination direction. In: *ECCV 94*, pages 286–296, 1994.

[48] H. Murase and S. Nayar. Visual learning and recognition of 3-D objects from appearance. *IJCV* **14**(1):5–24, 1995.

[49] S. Narasimhan, R. Ramamoorthi, and S. Nayar. Analytic rendering of multiple scattering in participating media. *Submitted to ACM Transactions on Graphics*, 2004.

[50] R. Ng, R. Ramamoorthi, and P. Hanrahan. All-frequency shadows using non-linear wavelet lighting approximation. *ACM Transactions on Graphics* (*SIGGRAPH 2003*) **22**(3), 2003.

[51] R. Ng, R. Ramamoorthi, and P. Hanrahan. Triple product wavelet integrals for all-frequency relighting. *ACM Transactions on Graphics* (*SIGGRAPH 2004*) **23**(3): 475–485, 2004.

[52] F. E. Nicodemus, J. C. Richmond, J. J. Hsia, I. W. Ginsberg, and T. Limperis. *Geometric Considerations and Nomenclature for Reflectance*. National Bureau of Standards (US), 1977.

[53] P. Nillius and J. Eklundh. Low-dimensional representations of shaded surfaces under varying illumination. In: *CVPR 03*, pages II:185–II:192, 2003.

[54] P. Nillius and J. Eklundh. Phenomenological eigenfunctions for irradiance. In: *ICCV 03*, pages I:568–I:575, 2003.

[55] P. Nillius and J. Eklundh. Classifying materials from their reflectance properties. In:*ECCV 04*, pages IV–366–IV–376, 2004.

[56] J. Nimeroff, E. Simoncelli, and J. Dorsey. Efficient re-rendering of naturally illuminated environments. In: *Eurographics Workshop on Rendering 94*, pages 359–373, June 1994.

[57] M. Osadchy, D. Jacobs, and R. Ramamoorthi. Using specularities for recognition. In: *ICCV*, pages 1512–1519, 2003.

[58] R.W. Preisendorfer. *Hydrologic Optics*. US Dept Commerce, 1976.

[59] S. Premoze, M. Ashikhmin, R. Ramamoorthi, and S. Nayar. Practical rendering of multiple scattering effects in participating media. In: *EuroGraphics Symposium on Rendering*, 2004.

[60] R. Ramamoorthi. Analytic PCA construction for theoretical analysis of lighting variability in images of a Lambertian object. *IEEE Transactions on Pattern Analysis and Machine Intelligence* (*PAMI*) **24**(10):1322–1333, Oct 2002.

[61] R. Ramamoorthi. A signal-processing framework for forward and inverse rendering. PhD thesis, Stanford University, 2002.

[62] R. Ramamoorthi and P. Hanrahan. Analysis of planar light fields from homogeneous convex curved surfaces under distant illumination. In: *SPIE Photonics West: Human Vision and Electronic Imaging VI*, pages 185–198, 2001.

[63] R. Ramamoorthi and P. Hanrahan. An efficient representation for irradiance environment maps. In: *SIGGRAPH 01*, pages 497–500, 2001.

[64] R. Ramamoorthi and P. Hanrahan. On the relationship between radiance and irradiance: Determining the illumination from images of a convex lambertian object. *Journal of the Optical Society of America* A **18**(10):2448–2459, 2001.

[65] R. Ramamoorthi and P. Hanrahan. A signal-processing framework for inverse rendering. In: *SIGGRAPH 01*, pages 117–128, 2001.

[66] R. Ramamoorthi and P. Hanrahan. Frequency space environment map rendering. *ACM Transactions on Graphics* (*SIGGRAPH 02 proceedings*) **21**(3):517–526, 2002.

[67] R. Ramamoorthi and P. Hanrahan. A signal-processing framework for reflection. *ACM Transactions on Graphics* **23**(4):1004–1042, 2004.

[68] R. Ramamoorthi, M. Koudelka, and P. Belhumeur. A Fourier theory for cast shadows. *IEEE Transactions on Pattern Analysis and Machine Intelligence* **27**(2):288–295, 2005.

[69] A. Shashua. On photometric issues in 3D visual recognition from a single 2D image. *IJCV* **21**:99–122, 1997.

[70] F. Sillion, J. Arvo, S. Westin, and D. Greenberg. A global illumination solution for general reflectance distributions. In: *SIGGRAPH 91*, pages 187–196, 1991.

[71] D. Simakov, D. Frolova, and R. Basri. Dense shape reconstruction of a moving object under arbitrary, unknown lighting. In: *ICCV 03*, pages 1202–1209, 2003.

[72] L. Sirovich and M. Kirby. Low-dimensional procedure for the characterization of human faces. *JOSA* A **4**(3):519–524, Mar 1987.

[73] P. Sloan, J. Hall, J. Hart, and J. Snyder. Clustered principal components for precomputed radiance transfer. *ACM Transactions on Graphics* (*SIGGRAPH 03 proceedings*): **22**(3), 2002.

[74] P. Sloan, J. Kautz, and J. Snyder. Precomputed radiance transfer for real-time rendering in dynamic, low-frequency lighting environments. *ACM Transactions on Graphics* (*SIGGRAPH 02 proceedings*) **21**(3):527–536, 2002.

[75] C. Soler and F. Sillion. Fast calculation of soft shadow textures using convolution. In: *SIGGRAPH 98* pages 321–332, 1998.

[76] K. Thornber and D. Jacobs. Broadened, specular reflection and linear subspaces. Technical Report TR#2001-033, NEC, 2001.

[77] K. Torrance and E. Sparrow. Theory for off-specular reflection from roughened surfaces. *JOSA* **57**(9):1105–1114, 1967.

[78] M. Turk and A. Pentland. Eigenfaces for recognition. *Journal of Cognitive Neuroscience* **3**(1):71–96, 1991.

[79] S. Westin, J. Arvo, and K. Torrance. Predicting reflectance functions from complex surfaces. In: *SIGGRAPH 92*, pages 255–264, 1992.

[80] A. Yuille, D. Snow, R. Epstein, and P. Belhumeur. Determining generative models of objects under varying illumination: Shape and albedo from multiple images using SVD and integrability. *IJCV* **35**(3):203–222, 1999.

[81] L. Zhang, B. Curless, A. Hertzmann, and S. Seitz. Shape and motion under varying illumination: Unifying multiview stereo, photometric stereo, and structure from motion. In: *International Conference on Computer Vision*, pages 618–625, 2003.

[82] L. Zhang and D. Samaras. Face recognition under variable lighting using harmonic image exemplars. In: *CVPR*, pages I:19–I:25, 2003.

[83] L. Zhao and Y. Yang. Theoretical analysis of illumination in PCA-based vision systems. *Pattern Recognition* **32**:547–564, 1999.

CHAPTER **13**

A MULTISUBREGION-BASED PROBABILISTIC APPROACH TOWARD POSE-INVARIANT FACE RECOGNITION

13.1 INTRODUCTION

Many face-recognition algorithms have been developed, and some have been commercialized for applications such as access control and surveillance. Several studies have been reported in recent years [1, 2, 5] which compare those algorithms and evaluate the state of the art of face-recognition technology. These studies show that current algorithms are not robust against changes in illumination, pose, facial expression, and occlusion.

Of these, pose change is one of the most important and difficult issues for the practical use of automatic face recognition. For example, the Face Recognition Vender Test 2000 [1], sponsored by the Department of Defense and the National Institute of Justice, reports that the recognition rate by representative face-recognition programs drops by 20 percent under different illumination conditions, and as much as 75 percent for different poses.

Most algorithms [3, 7, 10, 11] proposed so far for pose-invariant face recognition need several images of each subject. We propose an approach that can recognize faces in a variety of poses even if a gallery database includes images of only one pose per person.

The proposed method works as follows. When a probe image is given, the face region in the image is detected, and its landmarks (such as the eyes) are localized. The resulting probe face region is registered with that of the face in the gallery. The face region is divided into a set of small subregions, and each subregion is compared with the corresponding subregion of the face in the gallery. To compare the two, a similarity value for each subregion, defined by the sum of squared differences (SSD) after image normalization (so effectively the same as normalized correlation), is computed; this is done after finer alignment is done in order to compensate for the potential error in registration and the local deformation due to pose and other variations. The total similarity value between the probe face and the gallery face is then obtained by combining the similarity values of all subregions.

The key idea of our approach is that in combining those similarity values of subregions we take into account how the similarity value of each subregion, and thus its utility, changes as the pose of the face changes.

We have developed a probabilistic model of that change by using a large set of training images from the CMU PIE database [4], which consists of face images of a set of people from many viewing angles. In a face-recognition task of different poses, it was shown that our algorithm outperformed a baseline algorithm (PCA) and a commercial product for face recognition.

13.2 MODELING CHANGE OF LOCAL APPEARANCE ACROSS POSES

Our approach is categorically that of appearance-based template matching. In template matching, if we use the whole face region for comparison, it is not easy to take into account changes in appearance due to the pose differences, because the appearance in a different part of a face changes in a different manner due to the face's complicated three-dimensional shape. Instead, one can compare subregions of the face separately, such as the eyes, nose, and mouth [7, 9]. It is not understood, however, which subregions provide stable and discriminative information, in particular, with respect to pose changes.

Generally, the similarity value (such as SSD) of a subregion varies by three factors: the differences in the identity, the poses, and the location in the face. We will perform a systematic study by computing similarity values of various subregions of a face for a large number of combinations of the same and different identities and poses.

13.2.1 The CMU PIE Database

The CMU PIE database [4] consists of face images of 68 subjects, each in 13 poses, each under 21 different illumination conditions each on 2 occasions. We will use

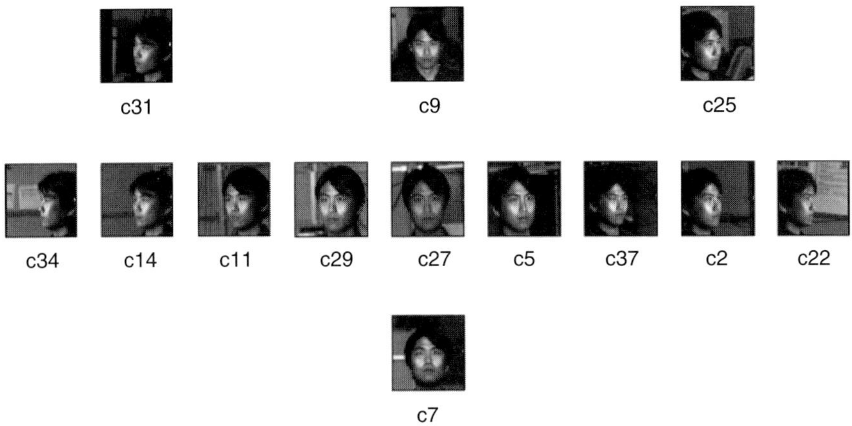

FIGURE 13.1: An example set of face images in the CMU PIE database. The database has 68 subjects with 13 poses per person, taken almost simultaneously [4]. The 13 poses cover from left profile (c34) to right profile (c22), and slightly up or down with c27 as the frontal view.

part of this database in this chapter. We will use only those images with frontal illumination; thus 13 images per person by 13 poses for 68 people. So, each image $\boldsymbol{I}$ in the study is labeled by (i, ϕ), where i is identity of the person ($|\{i\}| = 68$), and ϕ is pose ($|\{\phi\}| = 13$). Figure 13.1 shows a sample set of images of different poses for one person i=Yamada. Poses ϕ are denoted by symbols, like c34, c14, etc., where c27 is the frontal view and c37 and c11 are the views of about 45°.

13.2.2 Local Subregions in a Face and Their Similarity Value

For our study, three facial landmark points (the pupils of both eyes and the midpoint of the mouth) are manually located. The image is rotated and resized in plane so that the line that connects left and right pupils is horizontal and its length is nominal. The face region is then cropped and resized to 128 × 128 pixels.

As shown in Figure 13.2, a 7-by-3 lattice is placed on the face, whose position and orientation are defined by the three landmarks. Finally, we create a 9 × 15 pixel subregion centering at each of the lattice points, resulting in 21 subregions in total. For each subregion the intensity values are normalized to have zero mean and a unit variance.

As the similarity measure, the SSD (sum of squared differences) values $s_j(i_k, \phi_k; i_m, \phi_m)$ between corresponding jth subregions for all the pair of images $\boldsymbol{I}_k = (i_k, \phi_k)$ vs. $\boldsymbol{I}_m = (i_m, \phi_m)$ in the training dataset were calculated. Note that since we compute the SSD after image intensity normalization for each subregion, the SSD value contains effectively the same information as normalized correlation.

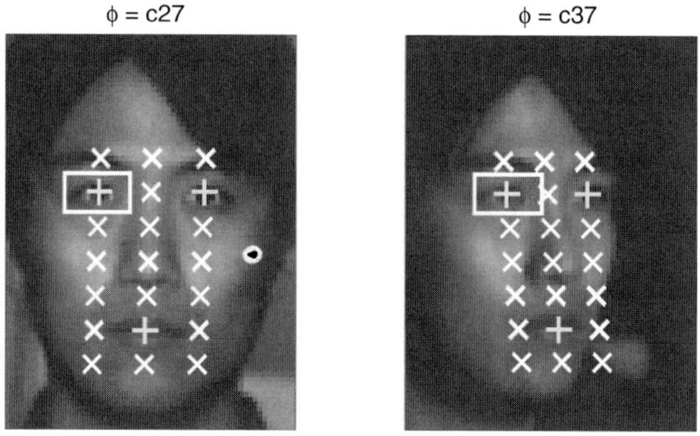

FIGURE 13.2: The facial landmark points are hand-labeled and the 7× 3 lattice points are placed on the face based on their positions. The size of each subregion at the lattice points is 9 × 15 pixels.

13.2.3 Map of Similarity Values

In order to comprehend how these similarity values $s_j(i_k, \phi_k; i_m, \phi_m)$ vary with identity and pose of the face, we plot them in two-dimensional maps. The leftmost map in Figure 13.3 shows $s_j(i_k, \text{c27}; i_m, \text{c5})$; that is, the similarity values of the subregion at the right eye plotted as a two-dimensional image with gallery's identity i_k (all with $\phi_k = \text{c27}$) as the horizontal axis and probe's identity i_m

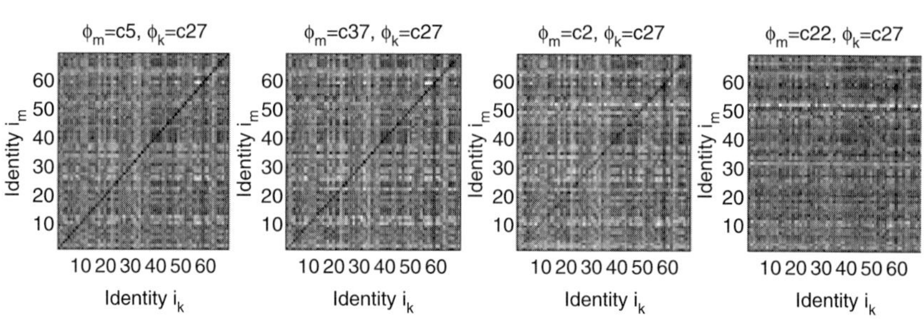

FIGURE 13.3: The two-dimensional maps of similarity values $s_j(i_k, \phi_k; i_m, \phi_m)$ of the subregion around an eye in Figure 13.2. The horizontal and vertical axes are the subjects' identities of $\boldsymbol{I}_k$ and $\boldsymbol{I}_m$, respectively. The four maps from left to right correspond to the case where the pose ϕ_m is c5, c37, c2, and c22, respectively, whereas the pose ϕ_k remains c27. The darker the pixel in the maps is, the more similar the corresponding subregion is in the two images.

(all with ϕ_m = c5) as the vertical axis. Pose c27 is the frontal and pose c5 is slightly left. The darker (the smaller SSD value) the "pixel" is, the more similar are the two corresponding regions in the gallery face and probe face. Naturally, along the diagonal of the map, that is, when $i_k = i_m$, the map is dark, meaning that the similarity is high.

The other similarity maps are for (i_k, ϕ_k = c27) versus. (i_m, ϕ_m = c37), ($i_k\phi_k$ = c27) versus (i_m, ϕ_m = c2), (i_k, ϕ_k = c27) versus. (i_m, ϕ_m = c22). They correspond to cases where, while the gallery remains to consist of frontal faces, the probe poses move from gradually left to all the way to the left profile (c22). It is clear that the similarity decreases even for the same identity (i.e., at the diagonal) as the pose moves away from frontal. These maps illustrate the difficulty of face recognition across poses.

13.2.4 Prior Distributions of Similarity Values

From each similarity map like in Figure 13.3, we compute two histograms of similarity values. One is for the diagonal part. It represents the distribution of similarity values between face images of the same person. The other is for the nondiagonal part, which is the distribution of similarity values between faces of different people. Figure 13.4 shows these histograms, each one for the corresponding map in Figure 13.3. The histograms of the first type are shown by solid curves, and the second type by broken curves.

The favorable situation is for the two histograms to be as separate as possible, because that means that the similarity values of that subregion have the

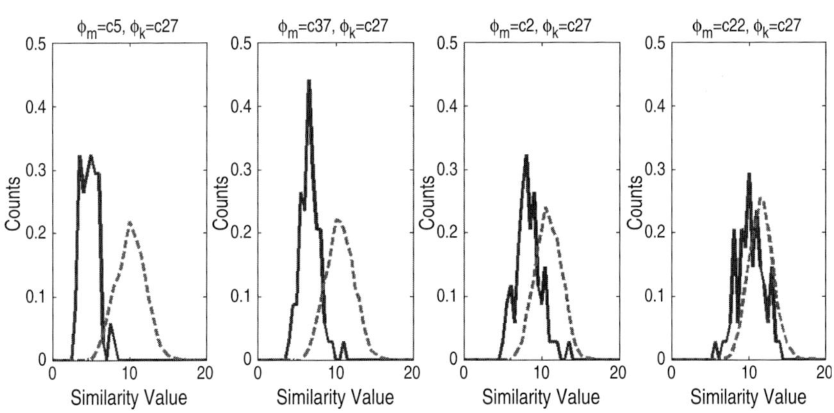

FIGURE 13.4: Each graph contains the two histograms of similarity values: their distribution for the same identity (solid curves) and that for the different identity (broken curves). The four graphs are for the combinations that correspond to those in Figure 13.3.

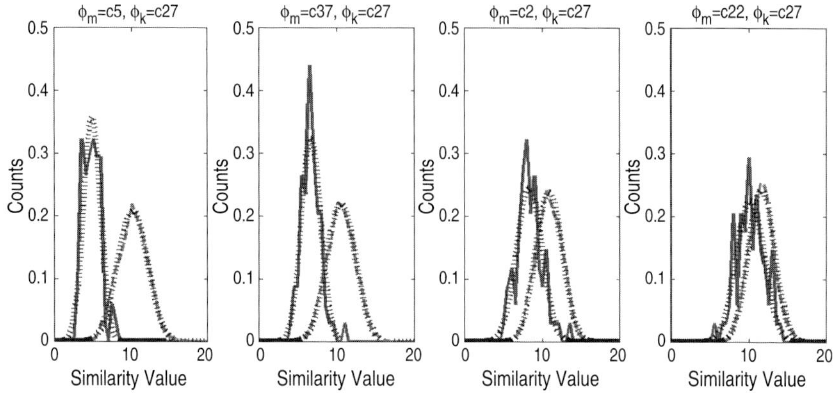

FIGURE 13.5: Histograms (solid and broken curves) and Gaussian fits (dotted curves) to them.

discriminative power to tell whether two faces are of the same person or not. It is clear that, for the frontal (c27) gallery, the discriminative power of the eye subregion decreases as the pose of the probe moves from slightly left (c5), more left (c37), further left (c2), and all the way to profile (c22).

From the histograms, we create $P(s_j|same, \phi_k, \phi_m)$, the conditional probability density of the jth similarity value s_j given that the images are of the class *same* identity and the poses of the two images are ϕ_k and ϕ_m, respectively. Likewise we also create $P(s_j|diff, \phi_k, \phi_m)$ for the class of *different*.

In this chapter, we approximate these distributions by a Gaussian distribution:

$$
\begin{aligned}
P(s_j|same, \phi_k, \phi_m) &= \frac{1}{\sqrt{2\pi}\sigma_j^{same}} \exp\left[-\frac{1}{2}\left(\frac{s_j - \mu_j^{same}}{\sigma_j^{same}}\right)^2\right], \\
P(s_j|diff, \phi_k, \phi_m) &= \frac{1}{\sqrt{2\pi}\sigma_j^{diff}} \exp\left[-\frac{1}{2}\left(\frac{s_j - \mu_j^{diff}}{\sigma_j^{diff}}\right)^2\right],
\end{aligned} \tag{1}
$$

where μ_j^{same} and μ_j^{diff}, with σ_j^{same} and σ_j^{dif}, are the means and standard deviations of class *same* and *diff*, respectively, which are obtained from the histograms. Figure 13.5 shows how these Gaussian models fit to the histograms.

13.3 RECOGNITION

Imagine that we are given a probe image $\boldsymbol{I}_{\mathrm{p}} = (i_{\mathrm{p}}, \phi_{\mathrm{p}})$ to be recognized with respect to a gallery of images $\{\boldsymbol{I}_{\mathrm{g}} = (i_{\mathrm{g}}, \phi_{\mathrm{g}})\}$. What we want to know is the probe identity i_{p}.

13.3.1 Posterior Probability

For the probe image $\boldsymbol{I}_{\mathrm{p}}$ and an image $\boldsymbol{I}_{\mathrm{g}}$ in the gallery we compute the similarity values of all subregions, i.e., $\{s_1, s_2, \ldots, s_J\}$. Once we have developed probabilistic models of similarity values of each subregion, we want to properly combine these similarity values in order to reach to the total decision for recognition: whether the two faces are from the same or different identity.

The posterior probability that the probe image and the gallery image are of the same identity, given the jth similarity value and their poses, is

$$P(same|s_j, \phi_{\mathrm{p}}, \phi_{\mathrm{g}}) = \frac{P(s_j|same, \phi_{\mathrm{p}}, \phi_{\mathrm{g}})P(same)}{P(s_j|same, \phi_{\mathrm{p}}, \phi_{\mathrm{g}})P(same) + P(s_j|diff, \phi_{\mathrm{p}}, \phi_{\mathrm{g}})P(diff).} \quad (2)$$

The values $P(same)$ and $P(diff)$ are the *a priori* probabilities of identify and nonidentity, respectively, and the conditional densities, $P(s_j|same, \phi_{\mathrm{p}}, \phi_{\mathrm{g}})$ and $P(s_j|diff, \phi_{\mathrm{g}}, \phi_{\mathrm{p}})$, are from the models obtained beforehand in (1).

13.3.2 Marginal Distribution for Unknown Pose of Probe

It is reasonable to assume that we have good knowledge of gallery pose ϕ_{g}, but we may not have that for the probe pose ϕ_{p}. In that case we cannot use (2).

One of the ways to deal with this is to determine the probe pose ϕ_{p} by using some pose estimation algorithms. However, the pose estimation may not be as easy.

Another way is to compute the marginal distributions of (1) over ϕ_{p} as

$$P(s_j|same, \phi_{\mathrm{g}}) = \sum_p P(\phi_{\mathrm{p}})P(s_j|same, \phi_{\mathrm{g}}, \phi_{\mathrm{p}}),$$

$$P(s_j|diff, \phi_{\mathrm{g}}) = \sum_p P(\phi_{\mathrm{p}})P(s_j|diff, \phi_{\mathrm{g}}, \phi_{\mathrm{p}}). \quad (3)$$

Then, we can develop a posterior probability, given the jth similarity values and the pose of the gallery image (but not that of the probe):

$$P(same|s_j, \phi_{\mathrm{g}}) = \frac{P(s_j|same, \phi_{\mathrm{g}})P(same)}{P(s_j|same, \phi_{\mathrm{g}})P(same) + P(s_j|diff, \phi_{\mathrm{g}})P(diff)}. \quad (4)$$

13.3.3 Combining Similarity Values for Identification

Given a probe image $\boldsymbol{I}_{\mathrm{p}}$, we compute for each image $\boldsymbol{I}_{\mathrm{g}}$ in the gallery (whose pose ϕ_{g} is known) the similarity values of all subregions, i.e., $\boldsymbol{S}_{\mathrm{pg}} = \{s_1, s_2, \ldots, s_J\}$. Then, for each of the similarity values in $\boldsymbol{S}_{\mathrm{pg}}$, we compute (2) or (4), depending on whether the pose information of the probe is available or not. Let's denote the resultant value as $h(same|s_j, \boldsymbol{I}_g; \boldsymbol{I}_p,)$.

The total similarity between $\boldsymbol{I}_p$ and $\boldsymbol{I}_g$ is now ready to be computed. Since we have not yet modeled the probabilistic dependency among s_js, we chose to use the sum rule [6] in order to obtain the total similarity value,

$$H(same|S_{pg}, I_g; I_p) = \sum_{j \in S_{pg}} h(same|s_j, I_g; I_p) \quad (5)$$

The identity i_p is determined to be the identity i_g of the gallery image that gives the highest value of H above.

13.4 RECOGNITION EXPERIMENTS

We have evaluated our algorithm by comparing its performance with a standard PCA-based method and a commercial product.

13.4.1 Training and Test Dataset

We used half of the subjects (34 subjects) in the CMU PIE database as a training dataset and obtained the statistical model described above. The images of the remaining 34 subjects were used as probe images in the recognition test. The test dataset, therefore, consists of the images of 13 poses for each of the 34 subjects. As the gallery images, we use frontal (c27) images of all 68 subjects. This makes this recognition task a little more difficult than otherwise since the gallery includes images of 34 subjects that are not included in the test dataset, with which no probe image should match. In this experiment below, we use $P(same) = 1/68$, $P(diff) = 1 - P(same) = 67/68$.

13.4.2 Recognition Results

Assume that we do not have any prior knowledge of the pose of the probe, and that we do not have means to compute reliably the pose of the given probe. Then we have to set $P(\phi_k) = 1/13$ (we have 13 different poses in our experiments) and we have to use Equation 4 for recognition. The dotted curve in Figure 13.6(a) plots the recognition scores with respect to the pose of the probe when the gallery pose is frontal (c27). When a probe is at a frontal pose (c27), the scores are 100%, since exactly the same images are included in the gallery for the frontal pose. As the probe pose moves away from the frontal, the scores deteriorate. The greater the width at which the scores remain high, the more pose invariant the algorithm is. Figure 13.6(b) is a two-dimensional plot where the gallery's pose is also varied. The width of the region with high scores along the diagonal indicates the degree

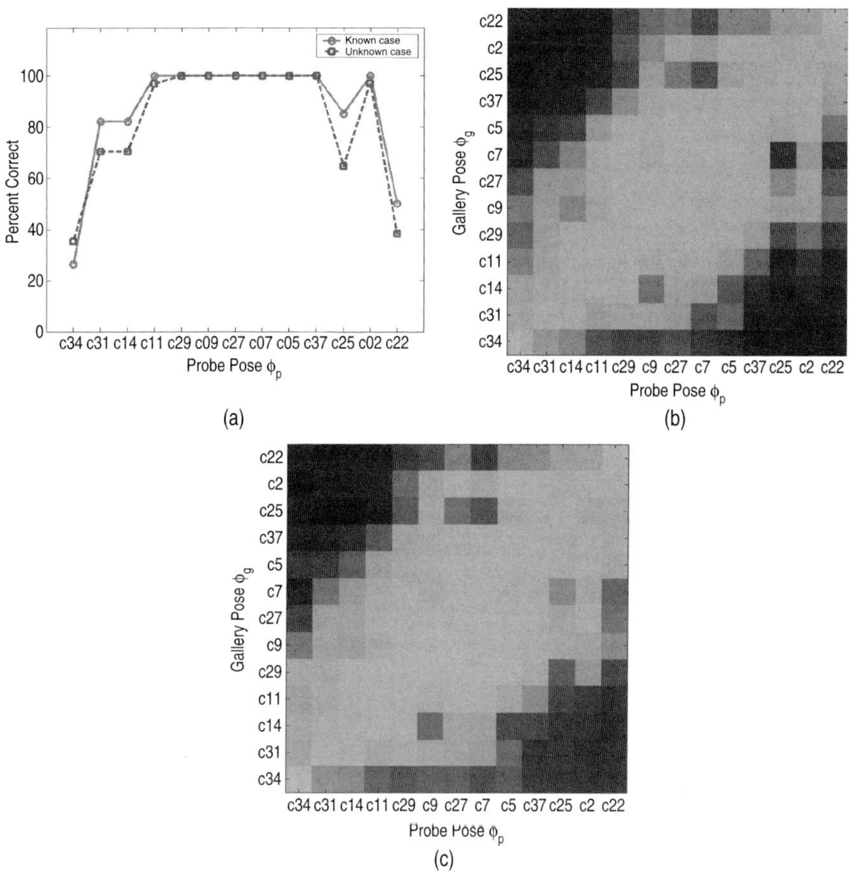

FIGURE 13.6: (a) Recognition score by our algorithm for different probe poses when the gallery pose is frontal (c27). The dotted curve is the case when the probe pose is not known, and the solid curve is for the case when the probe pose is known. (b) Recognition scores by our algorithm for all combinations of probe and gallery poses when the probe pose is unknown. (c) Recognition scores by our algorithm for all combinations of probe and gallery poses *if* the probe pose *were* known. Comparison of (b) and (c) tells that the recognition scores for unknown pose do not deteriorate much from those of known pose.

with which the algorithm can accommodate the difference of the pose between the gallery and the pose. One can see that the algorithm sustain a high recognition rate up to 45° of rotation, left or right.

One may wonder how much recognition rates improve if we knew the probe pose and could use Equation 2 instead of 4. The two dimensional plot of Figure 13.6(c), shows the result. By comparing Figures 13.6(b and c), it is interesting to note that ignorance of the probe pose does not significantly worsen the recognition score.

13.4.3 Comparison with Other Algorithms

We compare our recognition scores with those by a PCA algorithm and a commercial face-recognition program. Figure 13.7(a) shows the comparison for the case when the gallery pose is frontal (C27). The scores by the PCA algorithm drop as soon as the probe pose moves 15° away from frontal. The commercial program maintains its high performance till 30°. Our algorithm's score stays high till 45° $\{c34, c31, c14, c25, c02, c22\}$. The differences in the scores by algorithms become larger as the probe poses pull away from the gallery pose c27, especially, at poses such as $\phi_{\rm p} = \{c34, c31, c14, c25, c02, cc22\}$.

Figures 13.7(b and c) are two-dimensional plots of recognition rates of a PCA algorithm and a commercial face-recognition program, respectively. Comparison with Figure 13.6(c) shows the difference in generalization, namely the width of diagonal band of high recognition rate; our algorithm (Figure 13.6(c)) clearly outperforms the PCA algorithm (Figure 13.7(b)) and a commercial program (Figure 13.7(c)).

13.4.4 Discriminativeness of Subregions

An interesting question is which subregion (or which part) of a face has the most discriminating power for recognizing faces across poses. We performed the recognition task for the gallery at the frontal c27 pose and the probe at one pose by using only one subregion at a time. This gives us 7×3 recognition scores, which we can think of as indicating the discriminating power of each subregion for that particular combination of gallery and probe poses. We repeat this for all the probe poses.

Figure 13.8 shows the results as a 7×3 "image" for each pose of probe; the brighter the "pixel" is, the more powerful is the corresponding subregion. As the probe pose changes from central to left, as shown in {c29, c11, c14,...}, the right side of the face becomes more discriminating than the left side, which is very natural.

On the contrary, the left side of the face becomes more discriminating at {c5, c37, c2, c22}. For the nodding faces, such as {c9, c7}, the right and left sides of the face have almost the same discriminating power.

It is interesting to notice that the nose subregion and cheek subregion become less discriminating rather quickly, probably because the former is three-dimensional

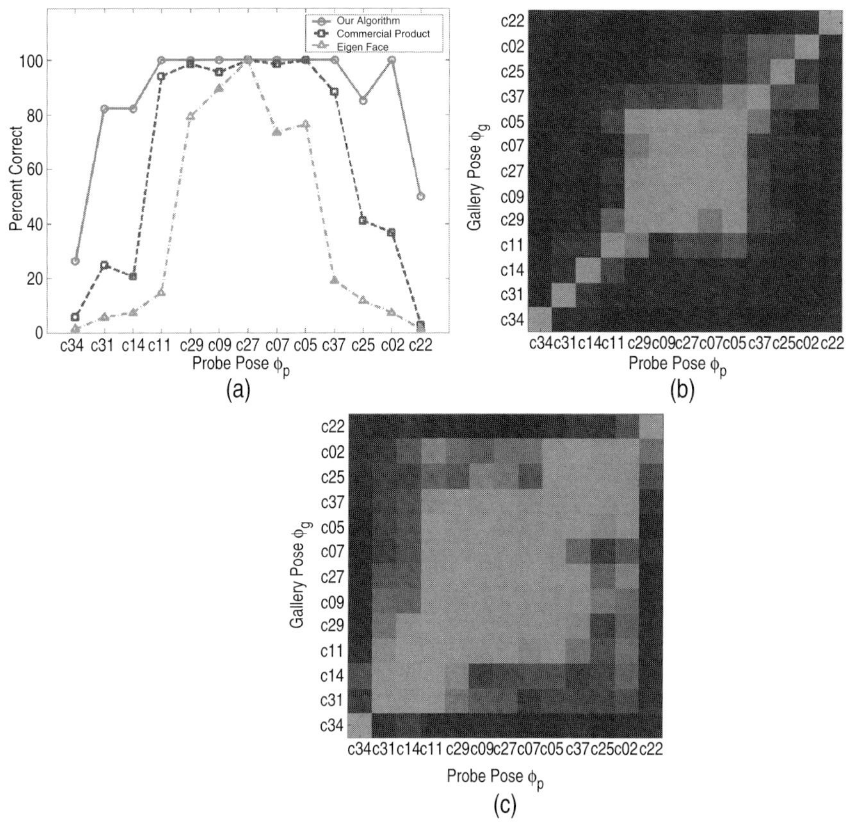

FIGURE 13.7: (a) Recognition scores by our algorithm (solid curve), an eigenface (PCA) method (dashed curve), and a commercial product (dotted curve) measured by using the CMU PIE database. (b) Recognition scores for all the combinations of gallery pose and probe by an eigenface (PCA) method. (c) Recognition scores for all the combinations of gallery pose and probe by a commercial product.

and thus changes its appearance quickly (note the proposed algorithm has no ability to compensate that effect), and the latter is uniform and thus is less useful from the beginning.

13.5 CONCLUSION

We have proposed a face-recognition algorithm based on a probabilistic modeling of how the utility of each local subregion of a face changes for the face-recognition

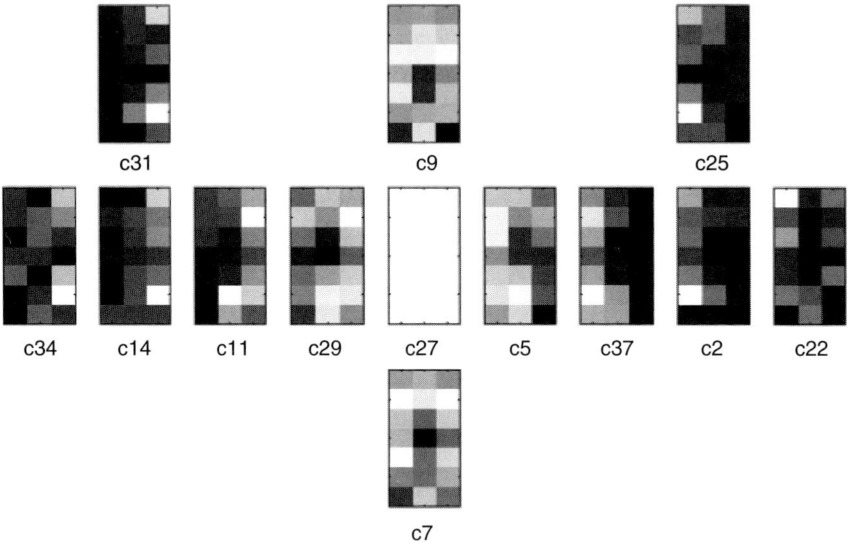

FIGURE 13.8: Maps of discriminating powers of subregions for various probe poses when the gallery pose is frontal (c27).

task as the pose varies. The algorithm outperformed a PCA method and a commercial product in the experiment using CMU PIE face image database that includes faces of 68 people with 13 different poses. While a larger-scale and more systematic experiment with face images over time is necessary to draw a decisive conclusion, the proposed method has demonstrated the importance and usefulness of correctly modeling how the face appearance changes as the pose changes.

REFERENCES

[1] D. Blackburn, M. Bone, and P. Phillips. Facial Recognition Vendor Test 2000: Evaluation Report, National Institute of Standards and Technology, 2000.

[2] P. J. Phillips, H. Moon, S. Rizvi, and P. Rauss. The FERET evaluation methodology for face recognition algorithms. *IEEE Trans. on PAMI* **22**(10): 1090–1103, 2000.

[3] Ralph Gross, Iain Matthews, and Simon Baker. Eigen light-fields and face recognition across pose. In: *Proceedings of the 5th International Conference on Automatic Face, and Gesture Recognition*, 2002.

[4] T. Sim, S. Baker, and M. Bsat. The CMU pose illumination and expression (PIE) database. In: *Proceedings of the 5th International Conference on Automatic Face and Gesture Recognition*, 2002.

[5] R. Gross, J. Shi, and J. Cohn. Quo Vadis face recognition? *Third Workshop on Empirical Evaluation Methods in Computer Vision*, December, 2001.

[6] J. Kittler, M. Hatef, R. Duin, and J. Matas. on combining classifiers: *IEEE Trans. On PAMI* **20**(3): 226–239, 1998.

[7] A. Pentland, B. Moghaddam, T. Starner. View-based and modular eigenspaces for face recognition. In *Proceedings of IEEE Conference on Computer Vision and Pattern Recognition*, 1994.

[8] P. S. Penev and J. J. Atick. Local feature analysis: a general statistical theory for object representation. *Network: Computation in Neural Systems* **7**(3), 477–500, 1996.

[9] R. Brunelli and T. Poggio. Face recognition: features versus templates. *IEEE Transactions on Pattern Analysis and Machine Intelligence*, **15**(10), October 1993.

[10] D. Beymer. Pose-invariant face recognition using real and virtual views. M.I.T., A.I. Technical Report No.1574, March 1996.

[11] T. Vetter and T. Poggio. Linear object classes and image synthesis from a single example image. A.I. Memo No.1531 (C.B.C.L. Paper No.119), March 1995.

CHAPTER 14

MORPHABLE MODELS FOR TRAINING A COMPONENT-BASED FACE-RECOGNITION SYSTEM

14.1 INTRODUCTION

The need for a robust, accurate, and easily trainable face-recognition system becomes more pressing as real-world applications in the areas of law enforcement, surveillance, access control, and human–machine interfaces continue to develop. However, extrinsic imaging parameters such as pose, illumination, and facial expression still contribute to degradation in recognition performance [48]. Another challenge is that training databases in real-world applications usually contain only a small number of face images per person.

By combining morphable models and component-based face recognition, we address the problems of pose and illumination invariance given a small number of training images. The morphable model is employed during training only, where slow speed and manual interaction is not as problematic as during classification. Based on three images of a person's face, the morphable model computes a 3D face model using an analysis-by-synthesis method. Once the 3D face models of all the subjects in the training database are computed, we generate a large number of synthetic face images under varying pose and illumination to train the component-based recognition system. The face-recognition module is preceded by a hierarchical face-detection system which performs a rough localization of the face in the image. Following the hierarchical system is a component-based face detector, which precisely localizes the face and extracts the components for face recognition.

In the following, we divide face-recognition techniques into global approaches and component-based approaches.

In the global approach, a single feature vector that represents the whole face image is used as input to a classifier. Several classification techniques have been proposed in the literature, e.g., minimum-distance classification in the eigenspace [42, 43], Fisher's discriminant analysis [2], and neural networks [19]. A comparison between state-of-the-art global techniques including eigenfaces, Fisher's discriminant analysis, and kernel PCA can be found in [32, 47]. Global techniques have been successfully applied to recognizing frontal faces. However, they are not robust against pose changes, since global features are highly sensitive to translation and rotation of the face. To overcome this problem, an alignment stage can be added before classifying the face. Aligning an input face image with a reference frontal face image requires computing correspondences between the two face images. These correspondences are usually determined for a small number of prominent points in the face, like the centers of the eyes or the nostrils. Based on these correspondences, the input face image can be warped to the reference face image. An affine transformation is computed to perform the warping in [33]. Active-shape models are used in [28] to align input faces with model faces. A semiautomatic alignment step in combination with support-vector machine (SVM) classification was proposed in [26]. Due to self-occlusion, automatic alignment procedures will eventually fail to compute the correct correspondences for large pose deviations between input and reference faces. An alternative, which allows a larger range of views, is to combine a set of view-tuned classifiers, originally proposed in a biological context in [38]. In [37], an eigenface approach was used to recognize faces under variable pose by grouping the training images into several separate eigenspaces, one for each combination of scale and orientation. Combining view-tuned classifiers has also been applied to face detection. The system presented in [41] was able to detect faces rotated in depth up to $\pm 90°$ with two naïve Bayesian classifiers, one trained on frontal views, the other one trained on profiles.

Unlike global approaches, in which the whole pattern of the face is classified, component-based methods[1] perform classification on components of the face. Component-based methods have the following two processing steps in common: In a first step, the face is scanned for a set of characteristic features. For example, a canonical gray-value template of an eye is cross-correlated with the face image to localize the eyes. We will refer to these local features as components. In a second step, the detected components are fed to a face classifier. There are three main advantages over the global approach:

- The flexible positioning of the components can compensate for changes in the pose of the face. An additional alignment step is not necessary.
- Parts of the face which don't contain relevant information for face identification can be easily omitted by selecting a proper set of components.

[1] In the literature, these methods are also referred to as local, part-based, or patch-based methods.

- Global techniques use a mask of fixed size and shape to extract the face from the background. In the component-based approach, the components can be freely arranged in the face image in order to capture variations in the shape of faces and to perform an accurate face/backround separation.

In the following, we focus on component-based techniques for face recognition. It should be mentioned, though, that classification based on local image features has also been applied to other object-recognition tasks; see, e.g., [15, 18, 30, 34, 44].

In [12], face recognition was performed by independently matching templates of three facial regions: both eyes, the nose, and the mouth. The configuration of the components during classification was unconstrained, since the system did not include a geometrical model of the face. A similar approach with an additional alignment stage was proposed in [6]. In an effort to enhance the robustness against pose changes, the global eigenface method has been further developed into a component-based system in which PCA is applied to local facial components [37]. The elastic grid matching technique described in [46] uses Gabor wavelets to extract features at grid points and graph matching for the proper positioning of the grid. In [35], a window of fixed size was shifted across the face image, and the DCT coefficients computed within the window were fed to a 2D Hidden Markov Model. A probabilistic approach using part-based matching has been proposed in [31] for expression invariant and occlusion tolerant recognition of frontal faces.

The above described global and component-based face-recognition techniques are 2D, or image-based, in a sense that they don't use 3D measurements and 3D models for identifying the face. A well-known problem with 2D systems is that they require a large number of training images which capture variations in viewpoint and lighting in order to achieve invariance to changes in pose and illumination. In most applications, however, these data are not available. An elegant solution to this problem is provided by 3D morphable models, since these models can be used to estimate the 3D geometry of a face based on few images [10]. Once the 3D model is available, synthetic training images for the 2D recognition system can be generated for arbitrary poses and illuminations. The morphable 3D face model is a consequent extension of the interpolation technique between face geometries that was introduced by Parke [36]. The approach is also rooted in the image-based work on eigenfaces [42, 43] and subsequent methods that have encoded 2D shape and gray values in images of faces separately after establishing correspondence between structures in different faces [4, 5, 17, 20, 25, 28]. Unlike eigenfaces, the separation of shape and texture defines a representation that specifically parametrizes the set of face-like images, as opposed to nonfaces, since any set of model coefficients defines a realistic face. However, each 2D model is restricted to a single viewpoint or a small range of angles around a standard view. In order to cover a number of different, discrete viewing directions, separate 2D morphable models for all views are required. If these are constructed from the same individual heads in a consistent way, model coefficients can be transferred from one model to another in order to predict new views of an individual [14, 45]. Still, this approach would be inefficient for producing the large variety of novel viewing conditions required for training the component-based system described in this article.

With a 3D representation of faces, rendering algorithms from computer graphics provide a simple, straightforward way of producing novel views. The challenge, then, is to estimate 3D shape from the input images of the new faces. In addition to the images of the individuals to be detected by the system, our approach relies on a dataset of 3D scans of other individuals. From this 3D dataset, we form the 3D morphable model that defines the natural range of face shapes and textures and plays a key role in the 3D shape reconstruction from images. Previous methods for model-based reconstruction used articulated 3D objects such as tools [29] or deformable models of 3D faces [13]. Others have used untextured 3D scans of faces in a representation that did not consider correspondence between facial structures across individuals [1]. In contrast to these methods, our system uses a correspondence-based, high-resolution morphable model for 3D shape and texture that has been created in an automated process from raw 3D scans. On a more abstract level, this process can be interpreted as learning class-specific information from samples of the class of human faces.

In the 3D morphable-model approach, faces are represented in a viewpoint-invariant way in terms of model coefficients. These coefficients are explicitly separated from the imaging parameters, such as head orientation or the direction of the light in the input image. In previous work, the model coefficients have been used for face recognition directly [9, 11, 40]. Currently, these methods have two shortcomings: The fitting process requires initialization, such as locating a set of facial features in the images, which have to be provided by the user or a feature detection algorithm, and the fitting process is computationally expensive.

The chapter is organized as follows: In Section 14.2 we review the basics of morphable models and explain how to fit a 3D model to an image. Section 14.3 describes the systems for face detection and recognition. Section 14.4 includes the experimental results. An algorithm for automatically learning relevant components for face recognition is presented in Section 14.5. The chapter is concluded with a summary and an outlook.

14.2 MORPHABLE MODELS

Generating synthetic views of a face from new camera directions, given only a single image, involves explicit or implicit inferences on the 3D structure of the facial surface. Shape-from-shading has intrinsic ambiguities [27], and the nonuniform, unknown albedo of the face makes it even more difficult to estimate the true shape of faces from images: If a region in the image is dark, there is no way of finding out if this is due to shading or to a low reflectance of the surface.

However, class-specific knowledge about human faces restricts the set of plausible shapes considerably, leaving only a relatively small number of degrees of freedom. The 3D morphable model [10, 11] captures these degrees of freedom of natural faces, representing the manifold of realistic faces as a vector space spanned by a set of examples [1, 13, 45]. In this vector space, any linear combination of shape and texture vectors $\boldsymbol{S}_i$ and $\boldsymbol{T}_i$ of the examples describes a realistic human face:

$$\boldsymbol{S} = \sum_{i=1}^{m} a_i \boldsymbol{S}_i, \quad \boldsymbol{T} = \sum_{i=1}^{m} b_i \boldsymbol{T}_i, \tag{1}$$

given that the coefficients a_i and b_i are within a certain range that will be defined later in this section. In the 3D morphable model, shape vectors $\boldsymbol{S}_i$ are formed from the 3D coordinates of surface points, and texture vectors $\boldsymbol{T}_i$ contain the red, green, and blue values of these points:

$$\boldsymbol{S}_i = (x_1, y_1, z_1, x_2, \ldots, x_n, y_n, z_n)^{\mathrm{T}}, \tag{2}$$

$$\boldsymbol{T}_i = (R_1, G_1, B_1, R_2, \ldots, R_n, G_n, B_n)^{\mathrm{T}}. \tag{3}$$

Continuous changes in the model parameters a_i generate a smooth transition such that each point of the initial surface moves towards a point on the final surface.

A crucial step in building a morphable model is to establish a point-to-point mapping between corresponding structures, such as the tips of the noses or the corners of the mouths, for all faces in the dataset. This mapping is used to encode each structure by the same vector coordinate $k(k = 1, \ldots n)$ in Equations 2, 3 in all vectors $\boldsymbol{S}_i$, $\boldsymbol{T}_i$. Errors in the correspondence cause artefacts when morphing between two faces according to the equations

$$\boldsymbol{S}(\lambda) = \lambda \boldsymbol{S}_1 + (1 - \lambda)\boldsymbol{S}_2, \tag{4}$$

$$\boldsymbol{T}(\lambda) = \lambda \boldsymbol{T}_1 + (1 - \lambda)\boldsymbol{T}_2, \quad 0 \le \lambda \le 1. \tag{5}$$

For example, an eyebrow of the first face may morph into a region of the forehead in the other face, producing double structures for $\lambda = \frac{1}{2}$.

In the morphable model described in this article, correspondence is established automatically from a reference face to a dataset of 3D face scans of 100 male and 100 female persons, aged between 18 and 45 years. One person is Asian, all others are Caucasian. The scans are recorded in a cylindrical surface parametrization

$$\boldsymbol{I}(h, \phi) = (r(h, \phi), R(h, \phi), G(h, \phi), B(h, \phi))^{\mathrm{T}}, \quad h, \phi \in \{0, \ldots, 511\}. \tag{6}$$

On these data, correspondence is computed with a modification of an optical-flow algorithm [3] that takes into account shape and texture at the same time [10, 11], essentially minimizing the difference between corresponding structures in a norm

$$\|\boldsymbol{I}\|^2 = w_{\mathrm{r}} r^2 + w_{\mathrm{R}} R^2 + w_{\mathrm{G}} G^2 + w_{\mathrm{B}} B^2 \tag{7}$$

with weights w_{r}, w_{R}, w_{G}, w_{B} that compensate for different variations within the radius data and the red, green, and blue texture components, and control the overall weighting of shape versus texture information. The optical-flow algorithm uses a coarse-to-fine approach for finding corresponding structures [3]. For details, see [11].

A principal-component analysis (PCA) of the statistical distribution of data within the vector space captures additional information about the object class. PCA defines a basis transformation from the set of examples to a basis that has two properties relevant to our algorithm: First, the basis vectors are ordered according to the variances in the dataset along each vector, which allows us to use a coarse-to-fine strategy in shape reconstruction. Second, assuming that the data have a normal distribution in face space, the probability density takes a simple form. Other properties of PCA, such as the orthogonality of the basis, are not crucial for our shape-reconstruction algorithm. We perform PCA on shape and texture separately, ignoring possible correlations between the two. For shape, we subtract the average $\bar{s} = \frac{1}{m}\sum_{i=1}^{m} \boldsymbol{S}_i$ from each shape vector, $\boldsymbol{a}_i = \boldsymbol{S}_i - \bar{s}$, and define a data matrix $\boldsymbol{A} = (\boldsymbol{a}_1, \boldsymbol{a}_2, \ldots, \boldsymbol{a}_m)$. The core step of PCA is to compute the eigenvectors $s_1, s_2, \ldots$ of the covariance matrix $\boldsymbol{C} = \frac{1}{m}\boldsymbol{A}\boldsymbol{A}^{\mathrm{T}} = \frac{1}{m}\sum_{i=1}^{m} \boldsymbol{a}_i \boldsymbol{a}_i^{\mathrm{T}}$, which can be achieved by a singular-value decomposition [39] of $\boldsymbol{A}$. The eigenvalues of $\boldsymbol{C}$, $\sigma_{S,1}^2 \geq \sigma_{S,2}^2 \geq \ldots$, are the variances of the data along each eigenvector. By the same procedure, we obtain texture eigenvectors $\boldsymbol{t}_i$ and variances $\sigma_{T,i}^2$. The two most dominant principal components are visualized in Figure 14.1. The eigenvectors form an orthogonal basis,

$$\boldsymbol{S} = \bar{s} + \sum_{i=1}^{m-1} \alpha_i s_i, \qquad \boldsymbol{T} = \bar{t} + \sum_{i=1}^{m-1} \beta_i \boldsymbol{t}_i, \tag{8}$$

and PCA provides an estimate of the probability density within face space:

$$p_S(\boldsymbol{S}) \sim e^{-\frac{1}{2}\sum_i \frac{\alpha_i^2}{\sigma_{S,i}^2}}, \qquad p_T(\boldsymbol{T}) \sim e^{-\frac{1}{2}\sum_i \frac{\beta_i^2}{\sigma_{T,i}^2}}. \tag{9}$$

14.2.1 3D Shape Reconstruction from Images

The main idea of the 3D shape-reconstruction algorithm is to find model coefficients α_i and β_i such that the linear combination (Equation 8) and subsequent rendering of the face give an image that is as similar as possible to the input image.

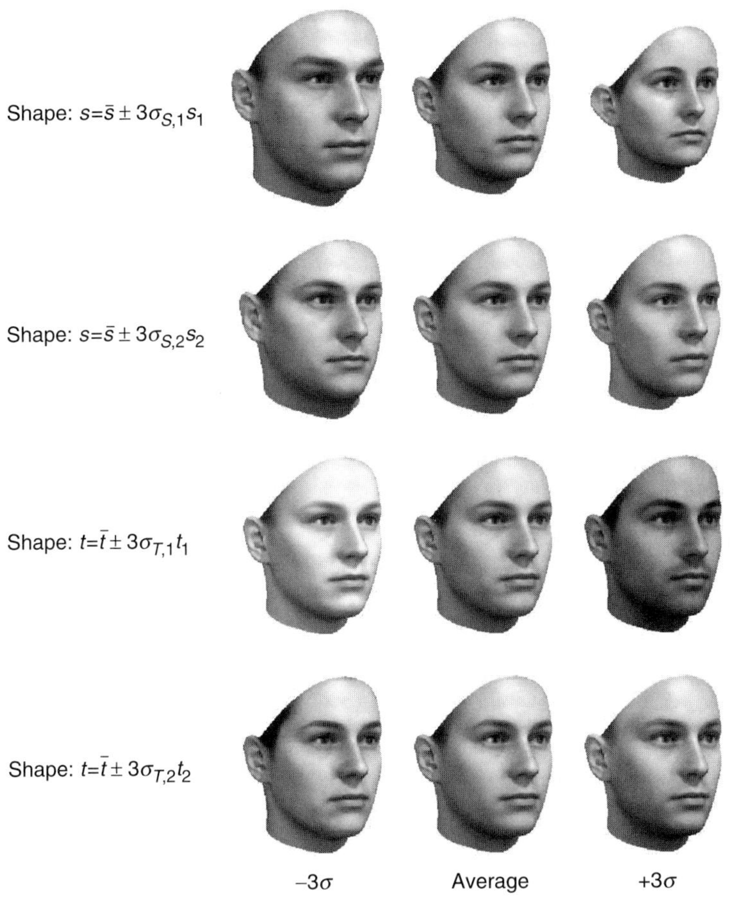

FIGURE 14.1: The average and the first two principal components of a dataset of 200 3D face scans.

In an analysis-by-synthesis loop, the fit is refined iteratively by rendering portions of an image, computing image difference along with its partial derivatives, and updating α_i, β_i, and the parameters of the rendering operation.

Rendering a 3D face model involves the following steps (in brackets, we list the parameters ρ_i that are optimized automatically):

- Rigid transformation (three angles and a translation vector)
- Perspective projection (focal length)
- Computing surface normals

- Phong illumination (intensity and direction of parallel light, intensity of ambient light in each color channel)
- Visibility and cast shadows, implemented with depth buffers
- Color transformation (color contrast and gains and offsets in the color channels).

Given an input image $\boldsymbol{I}_{\text{input}}(x, y) = (I_{\text{r}}(x, y), I_{\text{g}}(x, y), I_{\text{b}}(x, y))^{\text{T}}$, the sum-of-squares differences over all color channels and all pixels between this image and the synthetic reconstruction is

$$E_I = \sum_{x,y} \left\| \boldsymbol{I}_{\text{input}}(x, y) - \boldsymbol{I}_{\text{model}}(x, y) \right\|^2 . \tag{10}$$

E_I is the most important part of the cost function to be minimized in the reconstruction procedure. Additional terms help to achieve reliable and plausible results: For initialization, the first iterations exploit the manually defined feature points $(q_{x,j}, q_{y,j})$ and the positions (p_{x,k_j}, p_{y,k_j}) of the corresponding vertices k_j in an additional function

$$E_F = \sum_j \left\| \begin{pmatrix} q_{x,j} \\ q_{x,j} \end{pmatrix} - \begin{pmatrix} p_{x,k_j} \\ p_{y,k_j} \end{pmatrix} \right\|^2 . \tag{11}$$

Minimization of these functions with respect to α, η, ρ may cause overfitting effects similar to those observed in regression problems (see for example [16]). We therefore employ a maximum *a posteriori* estimator (MAP): Given the input image $\boldsymbol{I}_{\text{input}}$ and the feature points F, the task is to find model parameters α, β, ρ with maximum posterior probability $p(\alpha, \beta, \rho \,|\, \boldsymbol{I}_{\text{input}}, F)$. According to Bayes' rule,

$$p\left(\alpha, \beta, \rho \mid \boldsymbol{I}_{\text{input}}, F\right) \sim p\left(\boldsymbol{I}_{\text{input}}, F \mid \alpha, \beta, \rho\right) P\left(\alpha, \beta, \rho\right). \tag{12}$$

If we neglect correlations between some of the variables, the right-hand side is

$$p\left(\boldsymbol{I}_{input} \mid \alpha, \beta, \rho\right) p\left(F \mid \alpha, \beta, \rho\right) P\left(\alpha\right) P\left(\beta\right) P\left(\rho\right) \cdot \tag{13}$$

The prior probabilities $P(\alpha)$ and $P(\beta)$ were estimated with PCA (Equation 9). We assume that $P(\rho)$ is a normal distribution, and use the starting values for $\overline{\rho}_i$, and ad hoc values for $\sigma_{R,i}$. The starting condition is a frontal view of the average face in the center of the image, and frontal, white illumination.

For Gaussian pixel noise with a standard deviation σ_I, the likelihood of observing $\boldsymbol{I}_{\text{input}}$, given α, β, ρ, is a product of one-dimensional normal distributions, with one distribution for each pixel and each color channel. This product can be rewritten as $p(\boldsymbol{I}_{\text{input}}|\alpha, \beta, \rho) \sim \exp(\frac{-1}{2\sigma_I^2} E_I)$. In the same way, feature point coordinates may be subject to noise, which gives rise to a probability density $p(F|\alpha, \beta, \rho) \sim \exp(\frac{-1}{2\sigma_F^2} E_F)$.

In order to simplify computations, we maximize the overall posterior probability by minimizing

$$E = -2\log p\left(\alpha, \beta, \rho \mid \boldsymbol{I}_{\text{input}}, F\right),$$

$$E = \frac{1}{\sigma_I^2} E_I + \frac{1}{\sigma_F^2} E_F + \sum_i \frac{\alpha_i^2}{\sigma_{S,i}^2} + \sum_i \frac{\beta_i^2}{\sigma_{T,i}^2} + \sum_i \frac{(\rho_i - \overline{\rho}_i)^2}{\sigma_{R,i}^2}. \tag{14}$$

Ad hoc choices of σ_I and σ_F are used to control the relative weights of E_I, E_F, and the prior probability terms in Equation 14. At the beginning, the prior probability and E_F are weighted high. As iterations continue, more weight is given to E_I and dependance on E_F is reduced. Since a high weight on prior probability penalizes those principal components that have small standard deviations σ_i the most, during initial iterations only the most dominant principal components are considered. As the weight on prior probability is reduced, more and more dimensions of face space are taken into account. This coarse-to-fine approach helps to avoid local minima of E_I.

The evaluation of E_I would require the time-consuming process of rendering a complete image of the face in each iteration. A significant speed-up is achieved by selecting a random subset of triangles (in our case 40) for computing the cost function and its gradients. The random choice of different triangles adds noise to the optimization process that helps to avoid local minima, very much as in simulated annealing and other stochastic optimization methods. In order to make the expectation value of the approximated cost function equal to the full cost function E_I, the probability of selecting a triangle k is set to be equal to the area a_k that it covers in the image. The areas of triangles, along with occlusions and cast shadows, are computed at the beginning and updated several times in the optimization process.

In each iteration, partial derivatives $\partial E/\partial \alpha_i$, $\partial E/\partial \beta_i$, $\partial E/\partial \rho_i$ are computed analytically from the formulas of the rendering function. In a stochastic Newton optimization algorithm [11], these derivatives and the numerically computed Hessian give an update on α_i, β_i, ρ_i.

Simultaneously fitting the model to multiple images $\boldsymbol{I}_{\text{input},j}$, for $j = 1, 2, ..., m$, is achieved by a straightforward extension of the algorithm: Since we are looking for a unique face that fits each image, we need to fit one common set ofcoefficients

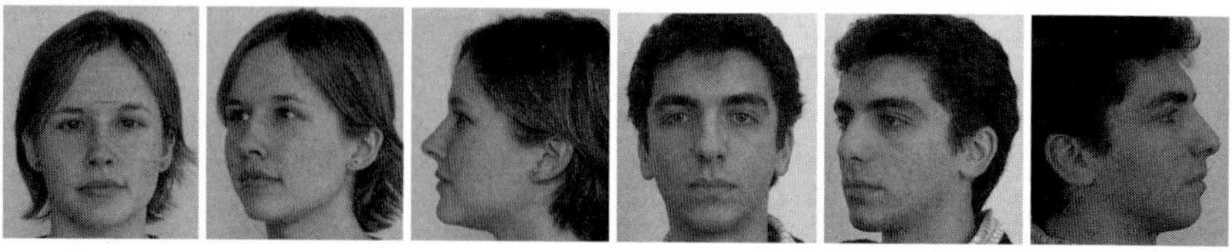

FIGURE 14.2: Examples of the image triplets used for generating the 3D models.

α_i, β_i, but use separate parameters $\rho_{i,j}$ for each image. The cost function is replaced by a sum

$$E_{\text{multiple}} = \frac{1}{m\sigma_I^2}\sum_{j=1}^{m} E_{I,j} + \frac{1}{m\sigma_F^2}\sum_{j=1}^{m} E_{F,j} + \sum_i \frac{\alpha_i^2}{\sigma_{S,i}^2} + \sum_i \frac{\beta_i^2}{\sigma_{T,i}^2} + \frac{1}{m}\sum_{i,j} \frac{\left(\rho_{i,j} - \overline{\rho}_i\right)^2}{\sigma_{R,i}^2}. \tag{15}$$

The face reconstruction algorithm is not fully automated, due to the initialization: the user has to manually define a small number of about 7–15 feature points $(q_{x,j}, q_{y,j})$. The remaining procedure is fully automated and takes about 4.5 minutes on a 2GHz Pentium 4 processor. Results of the 3D reconstruction process are shown in Figure 14.3. Some examples of the image triplets used for 3D reconstruction are shown in Figure 14.2.

14.3 FACE DETECTION AND RECOGNITION

14.3.1 System Overview

The overview of the system is shown in Figure 14.4. First the image is scanned for faces at multiple resolutions by a hierarchical face detector. Following the hierarchical system is a component-based face detector [24] which precisely localizes the face and extracts the components. The extracted components are combined into a single feature vector and fed to the face recognizer which consists of a set of nonlinear SVM classifiers.

14.3.2 Face Detection

As shown in Figure 14.4, the detection of the face is split into two modules. The first module is a hierarchical face detector similar to the one described in [23], consisting of a hierarchy of SVM classifiers which were trained on faces at different resolutions. Low-resolution classifiers remove large parts of the background at the bottom of the hierarchy; the most accurate and slowest classifier performs the final detection on the top level. In our experiments we used the following hierarchy

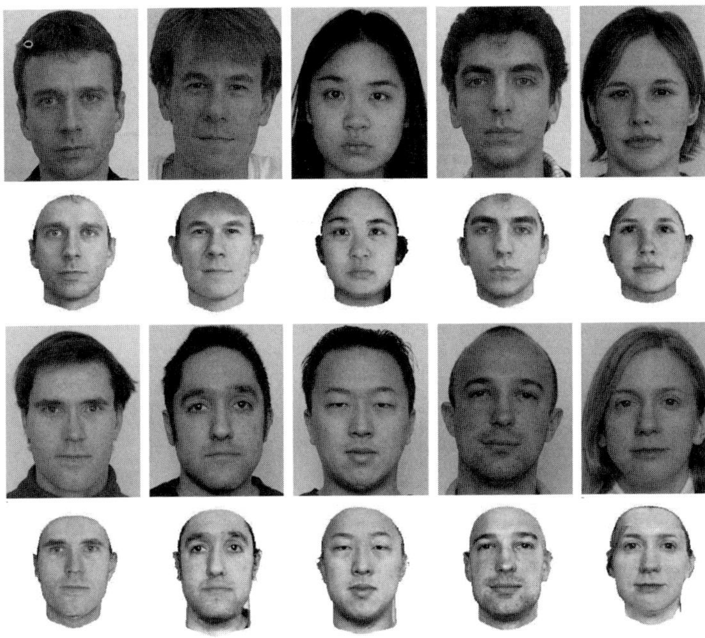

FIGURE 14.3: Original images and synthetic images generated from 3D models for all ten subjects in the training database.

of SVM classifiers: 3×3 linear, 11×11 linear, 17×17 linear, and 17×17 second-degree polynomial.[2] The positive training data for all classifiers was generated from 3D head models, with a pose range of ±45° rotation in depth and ±10° rotation in the image plane. The negative training set initially consisted of randomly selected nonface patterns which was enlarged by difficult patterns in several bootstrapping iterations.

Once the face is roughly localized by the hierarchical detector we apply a two-level, component-based face detector [24] to an image part which is slightly bigger than the detection box computed by the hierarchical classifier. The component-based classifier performs a fine search on the given image part, detects the face, and extracts the facial components. The first level of the detector consists of 14 component classifiers which independently search for facial components within the extracted image patch. The component classifiers were linear SVMs, each of which was trained on a set of extracted facial components and on a set of randomly

[2]Here $n \times n$ means that the classifier has been trained on face images of size $n \times n$ pixels.

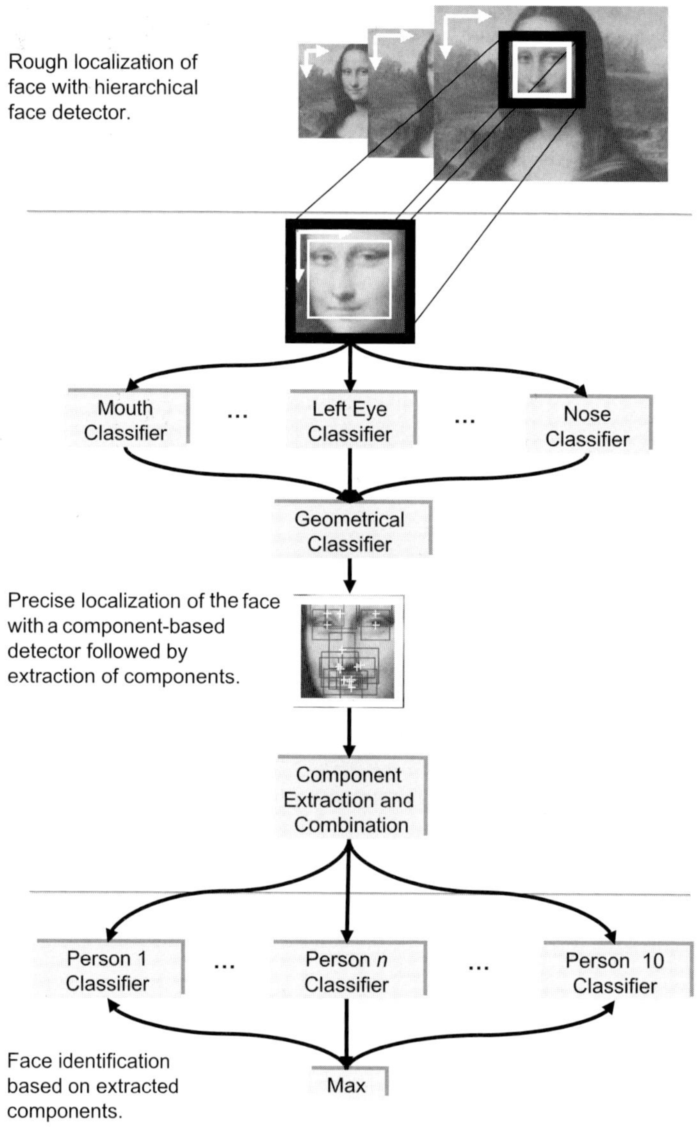

FIGURE 14.4: System overview of the component-based face-recognition system. The face is roughly localized by a hierarchical face detector. A fine localization of the face and its components is performed with a component-based face detector. The final step is the classification of the face based on the extracted facial components.

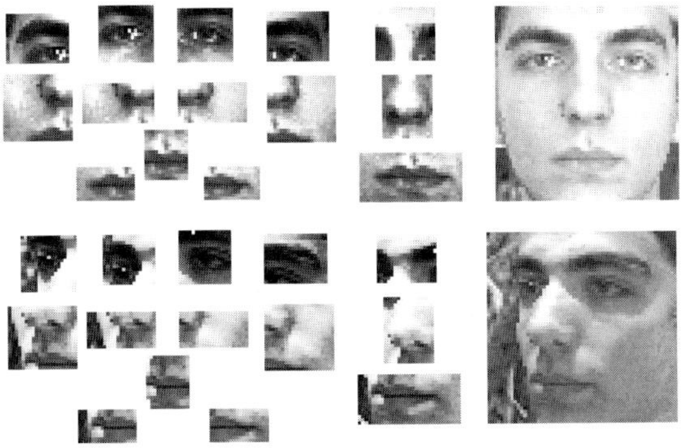

FIGURE 14.5: Examples of the 14 components extracted from frontal and half profile views of a face.

selected nonface patterns. The facial components for training were automatically extracted from synthetic face images for which the point-to-point correspondence was known.[3] Figure 14.5 shows examples of the 14 components for two training images. On the second level, the maximum continuous outputs of the component classifiers within rectangular search regions around the expected positions of the components are used as inputs to the geometrical classifier (linear SVM) which performs the final detection of the face.

14.3.3 Face Recognition

To train the system we first generate synthetic faces at a resolution of 58×58 for the ten subjects by rendering the 3D face models under varying pose and illumination. Specifically, the faces were rotated in depth from $-34°$ to $34°$ in $2°$ increments and rendered with two illumination models at each pose. The first model consisted of ambient light alone, the second model included ambient light and a directed light source, which was pointed at the center of the face and positioned between $-90°$ and $90°$ in azimuth and $0°$ and $75°$ in elevation. The angular position of directed light was incremented by $15°$ in both directions. Some example images from the training set are shown in Figure 14.6.

From the originally 14 components extracted by the face detector, only nine components were used for face recognition. Five components were eliminated because they strongly overlapped with other components or contained few gray-value structure. A global component was added. The location of this component

[3]The database for training the component-based face detector was different from the one used for face recognition; for details about training the face detector, see [24].

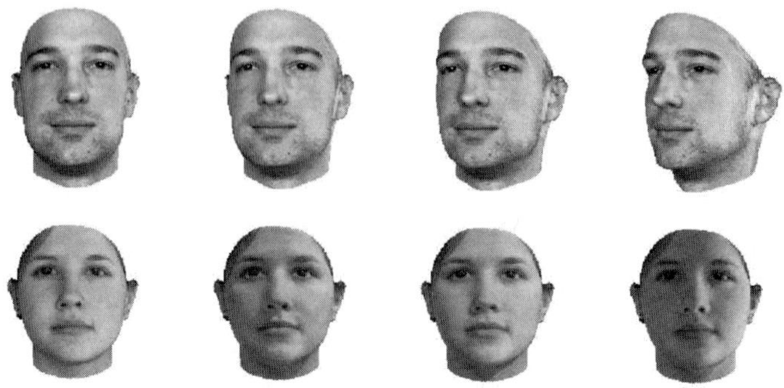

FIGURE 14.6: Effects of Pose and illumination variations.

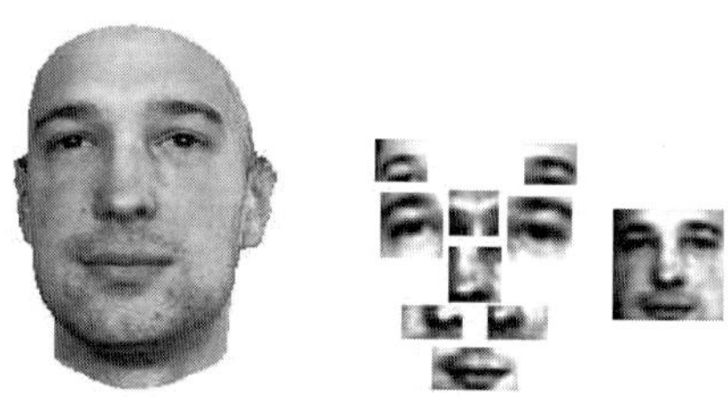

FIGURE 14.7: The ten components used for face recognition.

was computed by taking the circumscribing square around the bounding box of the other nine components. After extraction, the squared image patch was normalized to 40×40 pixels. The component-based face detector was applied to each synthetic face image in the training set to extract the ten components. Histogram equalization was then performed on each component individually. Figure 14.7 shows the histogram-equalized components for an image from the training data. The gray pixel values of each component were then combined into a single feature vector. A set of ten second-degree-polynomial SVM classifiers was trained on these feature vectors in a one-versus-all approach. To determine the identity of a person at runtime, we compared the continuous outputs of the SVM classifiers. The identity associated with the SVM classifier with the highest output value was taken to be the identity of the face.

14.4 EXPERIMENTAL RESULTS

A test set was created by taking images of the ten people in the database with a digital video camera. The subjects were asked to rotate their faces in depth and the lighting conditions were changed by moving a light source around the subject. The final test set consisted of 200 images, some of which are shown in Figure 14.8.

The component-based face-recognition system was compared to a global face-recognition system—both systems were trained and tested on the same images. The input vector to the global face recognizer [21] consisted of the histogram equalized gray values from the entire 40×40 facial region as extracted by the hierarchical face detector. The resulting ROC curves for global and component-based recognition can be seen in Figure 14.9. Each point on the ROC curve corresponds to a different rejection threshold. A test image was rejected if the maximum output of the ten SVM classifiers was below the given rejection threshold. The rejection threshold

FIGURE 14.8: Example images from the test set. Note the variety of poses and illumination conditions.

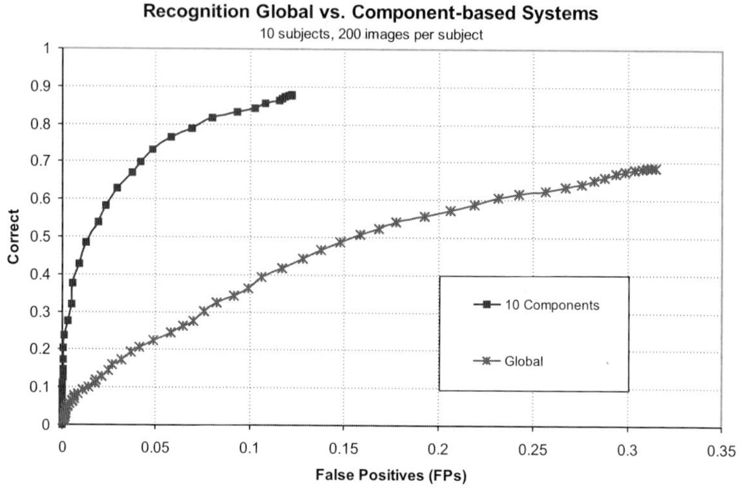

FIGURE 14.9: ROC curves for the component-based and the global face-recognition system. Both systems were trained and tested using the same data.

is largest at the starting point of the ROC curve, i.e., the recognition and false positive (FP) rates are zero. At the endpoint of the ROC curve, the rejection rate is zero, recognition rate and FP rate sum up to 100%. The component-based system achieved a maximal recognition rate of 88%, which is approximately 20% above the recognition rate of the global system. This significant discrepancy in results can be attributed to two main factors: First, the components of a face vary less under rotation than the whole face pattern, which explains why the component-based recognition is more robust against pose changes. Second, performing histogram equalization on the individual components reduces the in-class variations caused by illumination changes.

The error distribution among the ten subjects was highly unbalanced. While nine out of the ten people could be recognized with about 92% accuracy, the recognition rate for the tenth subject was as low as 49%. This might be explained by an inaccurate 3D head model or by the fact that this subject's training and test data were recorded six months apart from each other. Upon visual inspection of the misclassified faces, about 50% of the errors could be attributed to pose, facial expression, illumination, and failures in the component-detection stage. Figure 14.10 shows some of these images. The remaining 50% of the errors could not be explained by visual inspection.

The processing times of the system were evaluated on a different test set of 100 images of size 640×480. Each image included a single face at a resolution between 80×80 to 120×120 pixels. The overall speed was 4 Hz; the hierarchical

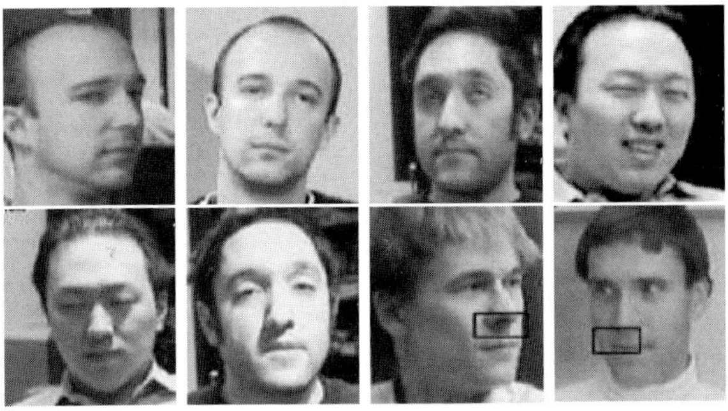

FIGURE 14.10: Examples of misclassified faces in the test set. From top left to bottom right the reasons for misclassification are: Pose, expression, illumination, and failure in detecting the mouth component.

detector took about 81%, the component detection 10%, and the recognition 6.4% of the overall computation time.

14.5 LEARNING COMPONENTS FOR FACE RECOGNITION

In the previous experiment, we used components which were specifically learned for the task of face detection [24]. It is not clear, however, if these components are also a good choice for face recognition. To investigate this issue, we ran the component-learning algorithm proposed in [24] on training and cross-validation (CV) set for face recognition. The training and CV sets were generated from the 3D head models of six of the ten subjects. Approximately 10,900 synthetic faces were rendered at a resolution of 58×58 for the training set. The faces were rotated in depth from 0° to 44° in 2° increments. They were illuminated by ambient light and a single directional light pointing towards the center of the face. The directional light source was positioned between −90° and 90° in azimuth and 0° and 75° in elevation. Its angular position was incremented by 15° in both directions. For the CV set, we rendered 18,200 images with slightly different pose and illumination settings. The rotation in depth ranged from 1° to 45° in 2° steps, the position of the directional light source varied between −112.5° and 97.5° in azimuth and between −22.5° and 67.5° in elevation. Its angular position was incremented by 30° in both directions. In addition, each face was tilted by ±10° and rotated in the image plane by ±5°. For computational reasons, we only used about one third of the CV data which was randomly selected from the 18,200 images.

The component-learning algorithm starts with a small rectangular component located around a preselected point in the face. The set of the initial fourteen components is shown in Figure 14.11. Each component is learned separately.

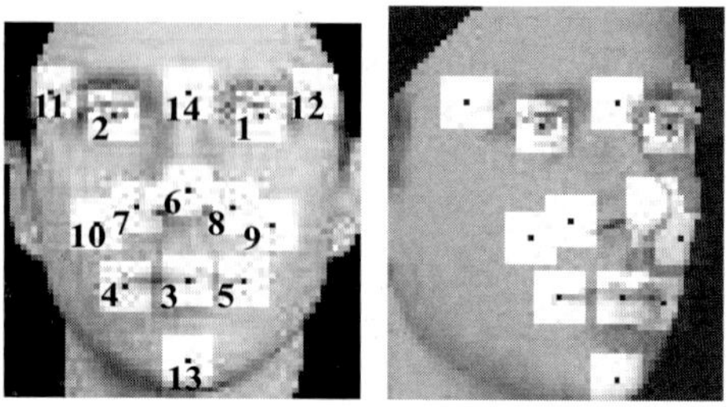

FIGURE 14.11: The initial 14 components for a frontal and rotated face.

We first extracted the initial component, e.g., a 5×5 patch around the center of the left eye, from each face image. The extraction could be done automatically, since the point-to-point correspondence between the synthetic face images was known. We then trained a face recognizer to distinguish between people based on this single component. In our experiments, the face recognizer was a set of SVM classifiers with second-degree polynomial kernels which were trained using the one-versus-all approach. The performance of the face recognizer was measured by computing the recognition rate on the CV set (CV rate).[4] We then enlarged the component by expanding the rectangle by one pixel into one of the four directions: up, down, left, right. As before, we generated training data, trained the SVM classifiers and determined the CV rate. We did this for expansions into all four directions and finally kept the expansion which led to the largest CV rate. We ran 30 iterations of the growing algorithm for each component and selected the component with the maximum CV rate across all iterations as the final choice.

The set of learned components is shown in Figure 14.12; here, the brighter the component the higher its CV rate. The size of the components and their CV rates are given in Table 14.1. The components on the left of the face have a smaller CV rate than their counterparts on the right side. This is not surprising, since the faces in the training and CV set were rotated to the left side only.[5] The components around the eyes and mouth corners achieve the highest recognition rate, followed by the nostril and eyebrow components. The component with the smallest CV rate is located around the tip of the nose, probably because variations in perspective and illumination cause significant changes in the image pattern of the nose. The cheek components are also weak, which can be explained by the lack of structure in this part of the face. It is very likely that the absence of eye movement, mouth

[4]In [24] we used the bound on the expected generalization error computed on the training set to control the growing direction.

[5]In [22] we show that the CV rates of the components decrease with increasing rotation in depth, indicating that frontal views are optimal for our face-recognition system.

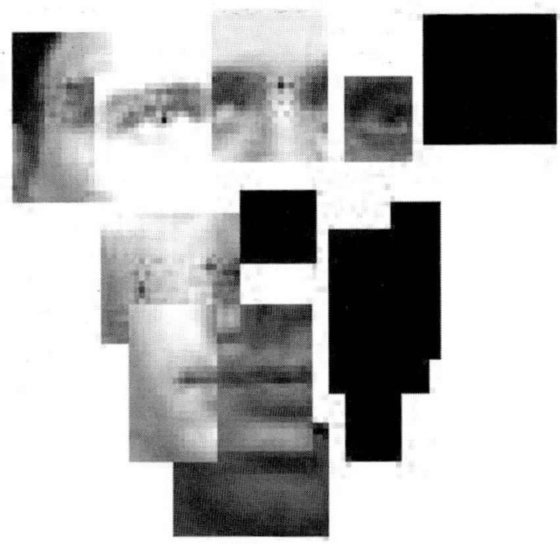

FIGURE 14.12: The final set of 14 learned components for face recognition. Bright components have a high CV rate.

Table 14.1: The dimensions of the 14 components for face recognition and their CV rates. Left and right are relative to the face.

Component	Number	CV rate [%]	Right	Left	Up	Down
Right eye	2	96	12	8	8	12
Right mouth corner	4	88	10	5	13	14
Right nostril	7	85	13	5	9	15
Right eyebrow	11	84	10	7	9	20
Nose bridge	14	83	12	8	13	13
Right cheek	10	82	7	6	10	9
Center of mouth	3	78	16	5	13	12
Left eye	1	75	7	4	7	7
Chin	13	71	19	8	15	4
Left mouth corner	5	55	4	5	19	12
Left cheek	9	52	4	4	15	12
Left eyebrow	12	50	9	14	13	9
Left nostril	8	49	6	11	6	22
Nose tip	6	30	9	4	8	4

movement and facial expressions in the synthetic data has some effect on size and shape of the learned components.

Since component-based face recognition requires the detection and extraction of facial components prior to their classification, an idea worth exploring is to learn a single set of components which is suited for both detection and recognition, e.g., by iteratively growing components which maximize the sum over the CV rates for detection and recognition. One could argue against this idea that the detection and the recognition problems are too different to be combined into one. In face detection, the components should be similar across individuals. In face recognition, on the other hand, the components should carry information to distinguish between individuals. Supporting the idea of learning a single set of components is the notion that both types of component are required to be robust against changes in illumination and pose. Components around the nose, for example, vary strongly under changes in illumination and pose, which makes them weak candidates for both detection and recognition.

14.6 SUMMARY AND OUTLOOK

This chapter explained how 3D morphable models can be used for training a component-based face-recognition system. From only three images per subject, 3D face models were computed and subsequently rendered under varying poses and lighting conditions to build a large number of synthetic images. These synthetic images were then used to train a component-based face recognizer. The face-recognition module was combined with a hierarchical face detector, resulting in a system that could detect and identify faces in video images at about 4 Hz. Results on 2000 real images of ten subjects show that the component-based recognition system clearly outperforms a comparable global face-recognition system. Component-based recognition was at 88% for faces rotated up to approximately half profile in depth. Finally we presented an algorithm for learning components for face recognition given a training and cross-validation set of synthetic face images. Among the fourteen learned components, the components which were located around the eyes and the mouth corners achieved the highest recognition rates on the cross-validation set.

In the following we discuss several possibilities to improve on the current state of our component-based face-recognition system.

3D Morphable Model. Our database for building the morphable model included 199 Caucasian and one Asian face. Even though previous results indicate that the model can successfully reconstruct faces from other ethnic groups [11], the quality of the reconstructions would very likely improve with a more diverse database. Recently, morphable models have been expanded by incorporating the capability to include facial expressions [8]. This new generation of morphable models could generate synthetic training images with various facial expressions and thus increase the expression tolerance of the component-based face-recognition system.

Rendering of the 3D Models. We rendered thousands of training images by varying the rotation in depth in fixed steps across certain intervals. Only a few hundred of these images became support vectors and contributed to the decision function of the classifier. Sampling the viewing sphere in small, equidistant steps can quickly lead to an explosion of the training data, especially if we consider all three degrees of freedom in the space of rotations. A way to keep the data manageable for the SVM training algorithm is to bootstrap the system by starting with a small initial training set and then successively adding misclassified examples to the training data. Another possibility to contain the synthetic training data is to render the 3D head models for application-specific illumination models only.

Tolerance to pose changes. A straightforward approach to increase the tolerance to pose changes is to train a set of view-tuned classifiers. Our current detection/recognition system was tuned to a single pose interval ranging from frontal to half-profile views. To extend the pose range to 90° rotation in depth, we could train a second set of component classifiers on faces rotated from half-profile to profile views.

Localization of the components. Currently, we search for components independently within their respective search regions. Improvements in the localization accuracy can be expected from taking into account the correlations between the components' positions during the detection stage; see, e.g., [7].

Learning facial components. The fourteen components in our first experiment have been specifically chosen for face detection. In the previous section, we learned facial components for identifying people; however, we did not consider the problem of how to localize these components within the face. The next logical step is to learn components which are easy to localize and are well suited for identification.

REFERENCES

[1] J. J. Atick, P. A. Griffin, and A. N. Redlich. Statistical approach to shape from shading: Reconstruction of 3D face surfaces from single 2d images. *Neural Computation* **86**:1321–1340, 1996.

[2] P. Belhumeur, P. Hespanha, and D. Kriegman. Eigenfaces vs fisherfaces: recognition using class specific linear projection. *IEEE Transactions on Pattern Analysis and Machine Intelligence* **19**(7):711–720, 1997.

[3] J.R. Bergen and R. Hingorani. Hierarchical motion-based frame rate conversion. Technical report, David Sarnoff Research Center Princeton NJ 08540, 1990.

[4] D. Beymer and T. Poggio. Face recognition from one model view. In: *Proc. 5th International Conference on Computer Vision*, 1995.

[5] D. Beymer and T. Poggio. Image representations for visual learning. *Science* **272**:1905–1909, 1996.

[6] D.J. Beymer. Face recognition under varying pose. A.I. Memo 1461, Center for Biological and Computational Learning, M.I.T., Cambridge, MA, 1993.

[7] S.M. Bileschi and B. Heisele. Advances in component-based face detection. In: *Proceedings of Pattern Recognition with Support Vector Machines, First International Workshop, SVM 2002*, pages 135–143, Niagara Falls, 2002.

[8] V. Blanz, C. Basso, T. Poggio, and T. Vetter. Reanimating faces in images and video. In: P. Brunet and D. Fellner, editors, *Computer Graphics Forum*, Vol. 22, No. 3 *EUROGRAPHICS 2003*, pages 641–650, Granada, Spain, 2003.

[9] V. Blanz, S. Romdhani, and T. Vetter. Face identification across different poses and illuminations with a 3D morphable model. In: *Proc. of the 5th Int. Conf. on Automatic Face and Gesture Recognition*, pages 202–207, 2002.

[10] V. Blanz and T. Vetter. A morphable model for the synthesis of 3D faces. In: *Computer Graphics Proc. SIGGRAPH'99*, pages 187–194, Los Angeles, 1999.

[11] V. Blanz and T. Vetter. Face recognition based on fitting a 3D morphable model. *IEEE Trans. on Pattern Analysis and Machine Intell.* **25**(9):1063–1074, 2003.

[12] R. Brunelli and T. Poggio. Face recognition: Features versus templates. *IEEE Transactions on Pattern Analysis and Machine Intelligence* **15**(10):1042–1052, 1993.

[13] C.S. Choi, T. Okazaki, H. Harashima, and T. Takebe. A system of analyzing and synthesizing facial images. In: *Proc. IEEE Int. Symposium of Circuit and Systems (ISCAS91)*, pages 2665–2668, 1991.

[14] T.F. Cootes, K. Walker, and C.J. Taylor. View-based active appearance models. In: *Int. Conf. on Autom. Face and Gesture Recognition*, pages 227–232, 2000.

[15] G. Dorko and C. Schmid. Selection of scale invariant neighborhoods for object class recognition. In: *International Conference on Computer Vision*, pages 634–640, 2003.

[16] R.O. Duda, P.E. Hart, and D.G. Stork. *Pattern Classification*. John Wiley & Sons, New York, 2nd edition, 2001.

[17] G.J. Edwards, T.F. Cootes, and C.J. Taylor. Face recogition using active appearance models. In: Burkhardt and Neumann, editors, *Computer Vision – ECCV'98*, Freiburg, 1998. Springer LNCS 1407.

[18] R. Fergus, P. Perona, and A. Zisserman. Object class recognition by unsupervised scale-invariant learning. In: *Proceedings of the IEEE Conference on Computer Vision and Pattern Recognition*, 2003.

[19] M. Fleming and G. Cottrell. Categorization of faces using unsupervised feature extraction. In: *Proc. IEEE IJCNN International Joint Conference on Neural Networks*, pages 65–70, 90.

[20] P.W. Hallinan. A deformable model for the recognition of human faces under arbitrary illumination. PhD thesis, Harvard University, Cambridge, Mass, 1995.

[21] B. Heisele, P. Ho, and T. Poggio. Face recognition with support vector machines: global versus component-based approach. In: *Proc. 8th International Conference on Computer Vision*, volume 2, pages 688–694, Vancouver, 2001.

[22] B. Heisele and T. Koshizen. Components for face recognition. In: *Proc. 6th International Conference on Automatic Face and Gesture Recognition*, pages 153–158, 2004.

[23] B. Heisele, T. Serre, S. Mukherjee, and T. Poggio. Feature reduction and hierarchy of classifiers for fast object detection in video image. In: *Proc. IEEE Conference on Computer Vision and Pattern Recognition* (*CVPR*), volume 2, pages 18–24, Kauai, 2001.

[24] B. Heisele, T. Serre, M. Pontil, T. Vetter, and T. Poggio. Categorization by learning and combining object parts. In: *Neural Information Processing Systems* (*NIPS*), Vancouver, 2001.

[25] M. Jones and T. Poggio. Multidimensional morphable models: A framework for representing and matching object classes. *Int. Journal of Comp. Vision* **29**(2):107–131, 1998.

[26] K. Jonsson, J. Matas, J. Kittler, and Y. Li. Learning support vectors for face verification and recognition. In: *Proc. IEEE International Conference on Automatic Face and Gesture Recognition*, pages 208–213, 2000.

[27] D.J. Kriegman and P.N. Belhumeur. What shadows reveal about object structure. In: Burkhardt and Neumann, editors, *Computer Vision – ECCV'98* Vol. II, Freiburg, Germany, 1998. Springer, Lecture Notes in Computer Science 1407.

[28] A. Lanitis, C.J. Taylor, and T.F. Cootes. Automatic face identification system using flexible appearance models. *Image and Vision Computing* **13**(5):393–401, 1995.

[29] D.G. Lowe. Fitting parametrized three-dimensional models to images. *IEEE Transactions on Pattern Analysis and Machine Intelligence* **13**(5):441–450, 1991.

[30] D.G. Lowe. Distinctive image features from scale-invariant keypoints. *International Journal of Computer Vision* **60**(2):91–110, 2004.

[31] A.M. Martinez. Recognizing imprecisely localized, partially occluded, and expression variant faces from a single sample per class. *IEEE Transactions on Pattern Analysis and Machine Intelligence* **24**(6):748–763, 2002.

[32] A.M. Martinez and A.C. Kak. Pca versus LDA. *IEEE Transactions on Pattern Analysis and Machine Intelligence* **23**(2):228–233, 2001.

[33] B. Moghaddam, W. Wahid, and A. Pentland. Beyond eigenfaces: probabilistic matching for face recognition. In: *Proc. IEEE International Conference on Automatic Face and Gesture Recognition*, pages 30–35, 1998.

[34] A. Mohan, C. Papageorgiou, and T. Poggio. Example-based object detection in images by components. In: *IEEE Transactions on Pattern Analysis and Machine Intelligence*, volume 23, pages 349–361, April 2001.

[35] A.V. Nefian and M.H. Hayes. An embedded HMM-based approach for face detection and recognition. In: *Proc. IEEE International Conference on Acoustics, Speech, and Signal Processing*, volume 6, pages 3553–3556, 1999.

[36] F.I. Parke. A parametric model of human faces. PhD thesis, University of Utah, Salt Lake City, 1974.

[37] A. Pentland, B. Mogghadam, and T. Starner. View-based and modular eigenspaces for face recognition. Technical Report 245, MIT Media Laboratory, Cambridge, 1994.

[38] T. Poggio and S. Edelman. A network that learns to recognize 3D objects. *Nature* **343**:163–266, 1990.

[39] W. H. Press, S. A. Teukolsky, W. T. Vetterling, and B. P. Flannery. *Numerical Recipes in C*. Cambridge University Press, Cambridge, 1992.

[40] S. Romdhani, V. Blanz, and T. Vetter. Face identification by fitting a 3D morphable model using linear shape and texture error functions. In: *Computer Vision — ECCV 2002, LNCS 2353*, pages 3–19, 2002.

[41] H. Schneiderman and T. Kanade. A statistical method for 3D object detection applied to faces and cars. In: *Proc. IEEE Conference on Computer Vision and Pattern Recognition*, pages 746–751, 2000.

[42] L. Sirovich and M. Kirby. Low-dimensional procedure for the characterization of human faces. *Journal of the Optical Society of America* A**4**:519–524, 1987.

[43] M. Turk and A. Pentland. Eigenfaces for recognition. *Journal of Cognitive Neuroscience* **3**:71–86, 1991.

[44] S. Ullman and E. Sali. Object classification using a fragment-based representation. In: *Biologically Motivated Computer Vision* (eds. S.-W. Lee, H. Bulthoff and T. Poggio), pages 73–87 (Springer, New York), 2000.

[45] T. Vetter and T. Poggio. Linear object classes and image synthesis from a single example image. *IEEE Trans. on Pattern Analysis and Machine Intelligence* **19**(7): 733–742, 1997.

[46] L. Wiskott, J.-M. Fellous, N. Krüger, and C. von der Malsburg. Face recognition by elastic bunch graph matching. *IEEE Transactions on Pattern Analysis and Machine Intelligence* **19**(7):775 –779, 1997.

[47] M.-H. Yang. Face recognition using kernel methods. In: *Neural Information Processing Systems* (*NIPS*), pages 215–220 Vancouver, 2002.

[48] W. Zhao, R. Chellappa, J. Phillips, and A. Rosenfeld. "Face Recognition: A Literature Survey," ACM Computing Surveys **35**(4): 399–458, 2003.

CHAPTER 15

MODEL-BASED FACE MODELING AND TRACKING WITH APPLICATION TO VIDEOCONFERENCING

15.1 INTRODUCTION

Animated face models are essential to computer games, filmmaking, online chat, virtual presence, video conferencing, etc. Generating realistic 3D human face models and facial animation has been a persistent challenge in computer vision and graphics. So far, the most popular commercially available tools have utilized laser scanners or structured lights. Not only are these scanners expensive, but also the data are usually quite noisy, requiring hand touchup and manual registration prior to animating the model. Because inexpensive computers and cameras are becoming ubiquitous, there is great interest in generating face models directly from video images. In spite of progress toward this goal, the available techniques are either manually intensive or computationally expensive.

The goal of our work is to develop an easy-to-use and cost-effective system that constructs textured 3D animated face models from videos with minimal user interaction in no more than a few minutes. The user interface is very simple. First, a video sequence is captured while the user turns his/her head from one side to the other side in about 5 seconds. Second, the user browses the video to select a frame with frontal face. That frame and its neighboring one (called two base images) pop up, and the user marks 5 feature points (eye corners, nose top, and mouth corners) on each of them. Optionally, the user marks 3 points under the chin on the frontal view image. After these manual steps, the program produces, in less than a minute, a 3D model having the same face and structure as the user's shows up on the screen and greets the user. The automatic process, invisible to the user, matches points of interest across images, determines the head motion, reconstructs the face in 3D space, and builds a photorealistic texture map from images. It uses many state-of-the-art computer vision and graphics techniques and also several innovative techniques we have recently developed specifically for face modeling. In this chapter, we describe the architecture of our system as well as those new techniques.

The key challenge of face modeling from passive video images is the difficulty of establishing accurate and reliable correspondences across images due to the apparent lack of skin texture. An additional challenge introduced by our choice of system setup is the variation of facial appearance. To acquire multiple views of the head, we ask the user to turn the head instead of moving the camera around the head or using multiple cameras. This greatly reduces the cost; anyone having a video camera (e.g., a webcam) can use our system. On the other hand, the appearance of the face changes in different images due to change in relative lighting condition. This makes image correspondence an even harder problem. Fortunately, as proven by an extensive set of experiments, we observe that, even though it is difficult to extract dense 3D facial geometry from images, it is possible to match a sparse set of corners and use them to compute head motion and the 3D locations of these corner points. Furthermore, since faces are similar across people, the space of all possible faces can be represented by a small number of degrees of freedom. We represent faces by a linear set of deformations from a generic face model. We fit the linear face class to the sparse set of reconstructed corners to generate the complete face geometry. In this chapter, we show that linear classes of face geometries can be used to effectively fit/interpolate a sparse set of 3D reconstructed points even when these points are quite noisy. This novel technique is the key to rapidly generate photorealistic 3D face models with minimal user intervention.

There are several technical innovations to make such a system robust and fast.

1. A technique to generate masking images eliminates most face-irrelevant regions based on generic face structure, face feature positions, and skin colors. The skin colors are computed on the fly for each user.
2. A robust head-motion estimation algorithm takes advantage of the physical properties of the face feature points obtained from manual marking to reduce the number of unknowns.
3. We fit a set of face metrics (3D deformation vectors) to the reconstructed 3D points and markers to generate a complete face geometry. Linear face classes have already been used in [8], where a face is represented by a linear combination of physical face geometries and texture images and is

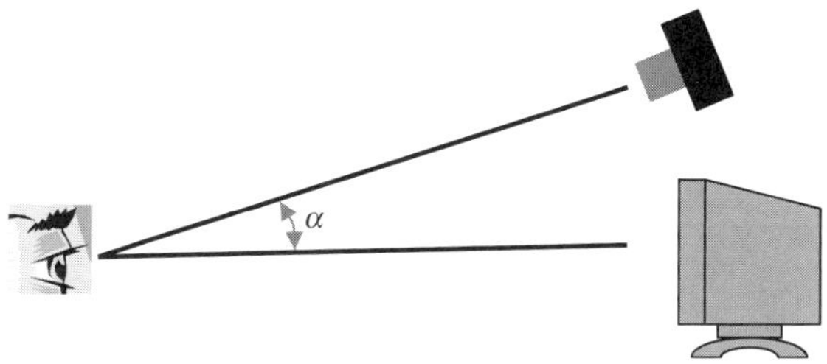

FIGURE 15.1: Camera-screen displacement causes the loss of eye-contact.

reconstructed from images in a morphable model framework. We use linear classes of face metrics to represent faces and to interpolate a sparse set of 3D points for face modeling.

4. Finally, a model-based bundle adjustment refines both the face-geometry and head-pose estimates by taking into account all available information from multiple images. Classical approaches first refine 3D coordinates of isolated matched points, followed by fitting a parametric face model to those points. Our technique directly searches in the face-model space, resulting in a more elegant formulation with fewer unknowns, fewer equations, a smaller search space, and hence a better posed system.

In summary, the approach we follow makes full use of rich domain knowledge whenever possible in order to make an ill-posed problem better behaved.

In a typical desktop videoteleconferencing setup, the camera and the display screen cannot be physically aligned, as depicted in Figure 15.1. A participant looks at the image of the remote party displayed on the monitor but not directly into the camera while the remote party "looks" at her through the camera, thus she does not appear to make eye contact with the remote party. Research [67] has shown that if the divergence angle (α) between the camera and the display is greater than five degrees, the loss of eye-contact is noticeable. If we mount a small camera on the side of a 21-inch monitor, and the normal viewing position is at 20 inches away from the screen, the divergence angle will be 17 degrees, well above the threshold at which the eye contact can be maintained. Under such a setup, the video loses much of its communication value and becomes ineffective compared to the telephone. We will describe a prototype system to overcome this eye-gaze divergence problem by leveraging 3D face models.

3D head-pose information or 3D head tracking in a video sequence is very important for user attention detection, vision-based interface, and head-gesture recognition. This is also true for the eye-gaze correction step to be described in this article. In the past few years, 3D head pose has also been recognized as an essential prerequisite for robust facial-expression/emotion analysis and synthesis, and for face recognition. In multimedia applications, video coding also requires 2D or 3D motion information to reduce the redundant data [19]. When there is no expression change on the face, relative head pose can be solved as a rigid-object tracking problem through traditional 3D vision algorithms for multiple-view analysis [47, 79, 28]. However, in practice, expressional deformation or even occlusion frequently occurs, together with head-pose changes. Furthermore, facial-expression analysis or face recognition also needs to deal with the alignment problem between different head orientations. Therefore, it is necessary to develop effective techniques for head tracking under the condition of expression changes. We have demonstrated that 3D face models help improve the robustness of head tracking. Even better results are obtained with a stereovision system.

This chapter is organized as follows. Related work on face modeling and tracking is reviewed in Section 15.2. Section 15.3 describes the animated face model we use in our system. Since this chapter describes a big vision system, due to space limitation, we only provide an executive summary of our system in Section 15.4, and a guided tour, examining the system step by step, in Section 15.5. References to the known techniques exploited in the system such as robust point matching are given, while details of our new technical contributions enumerated in the last paragraph can be found in [82]. Section 15.6 provides more experimental results on face modeling. Section 15.7 describes our 3D head-pose tracking technique that makes use of the 3D face model built by our system. Section 15.8 shows how we integrate various techniques developed so far to solve the eye-gaze divergence problem in video conferencing. We give the conclusions and perspectives of our system in Section 15.9.

15.2 STATE OF THE ART

In this section, we review some of the previous work related to 3D face modeling, 3D face tracking, and eye-gaze correction for video-conferencing. Because of vast literature in these areas, the survey is far from complete.

15.2.1 3D Face Modeling

Facial modeling and animation has been a computer graphics research topic for over 25 years [16, 36, 37, 38, 39, 40, 50, 54, 55, 56, 58, 69, 71, 73]. The reader is referred to Parke and Waters' book [56] for a complete overview.

Lee et al. [38, 39] developed techniques to clean up and register data generated from laser scanners. The obtained model is then animated by using a physically based approach.

DeCarlo et al. [15] proposed a method to generate face models based on face measurements randomly generated according to anthropometric statistics.

They were able to generate a variety of face geometries using these face measurements as constraints.

A number of researchers have proposed to create face models from two views [1, 31, 12]. They all require two cameras which must be carefully set up so that their directions are orthogonal. Zheng [84] developed a system to construct geometrical object models from image contours, but it requires a turntable setup.

Pighin et al. [57] developed a system to allow a user to manually specify correspondences across multiple images, and use vision techniques to compute 3D reconstructions. A 3D mesh model is then fitted to the reconstructed 3D points. They were able to generate highly realistic face models, but with a manually intensive procedure.

Roy-Chowdhury and Chellappa [61] described a technique of 3D reconstruction from short monocular sequences taking into account the statistical errors in reconstruction algorithms. Stochastic approximation is used as a framework to fuse incomplete information from multiple views. They applied this technique to various applications including face modelling.

Our work is inspired by Blanz and Vetter's work [8]. They demonstrated that linear classes of face geometries and images are very powerful in generating convincing 3D human face models from images. We use a linear class of geometries to constrain our geometrical search space. One main difference is that we do not use an image database. Consequently, the types of skin color we can handle are not limited by the image database. This eliminates the need for a fairly large image database to cover every skin type. Another advantage is that there are fewer unknowns since we need not solve for image database coefficients and illumination parameters. The price we pay is that we cannot generate complete models from a single image.

Kang et al. [33] also use linear spaces of geometrical models to construct 3D face models from multiple images. But their approach requires manually aligning the generic mesh to one of the images, which is in general a tedious task for an average user.

Guenter et al. [26] developed a facial animation capturing system to capture both the 3D geometry and texture image of each frame and reproduce high-quality facial animation. The problem they solved is different from what is addressed here in that they assumed the person's 3D model was available, and the goal was to track subsequent facial deformations.

Fua et al. [22, 23, 24, 21] have done impressive work on face modeling from images, and their approach is the most similar to ours in terms of the vision technologies exploited. There are three major differences between their work and ours: face-model representation, model fitting, and camera motion estimation. They deform a generic face model to fit dense stereo data. Their face model contains a lot of free parameters (each vertex has three unknowns). Although some smoothness constraint is imposed, the fitting still requires many 3D reconstructed points.

With our model, we only have about 60 parameters to estimate (see Section 15.3), and thus only a small set of feature points is necessary. Dense stereo matching is usually computationally more expensive than feature matching. Our system can produce a head model from 40 images in about two minutes on a PC with 366 MHz processor. Regarding camera motion estimation, they use a regularized bundle-adjustment on every image triplet, while we first use a bundle-adjustment on the first two views and determine camera motion for other views using a 3D head model, followed by a model-based bundle adjustment (see [65, 82]).

Zhang et al. [78] employs synchronized video cameras and structured light projectors to record videos of a moving face from multiple viewpoints. A spacetime stereo algorithm is introduced to compute depth maps. The built face models can be animated via key-framing or texture-synthesis techniques.

15.2.2 3D Head Tracking

There is a wide variety of work related to 3D head tracking. Virtually all work on face tracking takes advantage of the constrained scenario: instead of using a generic tracking framework which views the observed face as an arbitrarily object, a model-based approach is favored, which incorporates knowledge about facial deformations, motions, and appearance [14]. Based on the tracking techniques, we classify previous works into the following categories.

Optical flow. Black and Yacoob [7] have developed a regularized optical-flow method in which the head motion is tracked by interpretation of optical flow in terms of a planar two-dimensional patch. Basu et al. [4] generalized this approach by interpreting the optical flow field using a 3D model to avoid the singularities of a 2D model. Better results have been obtained for large angular and translational motions. However, their tracking results were still not very accurate; as reported in their paper, angular errors could be as high as 20 degrees. Recently, DeCarlo and Metaxas [14] used optical flow as a hard constraint on a deformable detailed model. Their approach has produced excellent results. But the heavy processing in each frame makes a real-time implementation difficult. Other flow-based methods include [11, 41].

Features and templates. Azarbeyajani and Pentland [2] presented a recursive estimation method based on tracking of small facial features like the corners of the eyes or mouth using an extended Kalman Filter framework. Horprasert [30] presented a fast method to estimate the head pose from tracking only five salient facial points: four eye corners and the nose top. Other template-based methods include the work of Darrell et al. [13], Saulnier et al. [62], and Tian et al. [70]. The template-based methods usually have the limitation that the same points must be visible over the entire image sequence, thus limiting the range of head motions they can track.

Skin color. Yang et al. [74] presented a technique of tracking human faces using an adaptive stochastic model based on human skin color. This approach is in general very fast. The drawback is that it is usually not very accurate, thus is not sufficient for our applications.

The work by Newman et al. [52] is related to our proposed approach, and falls in the "Features and Templates" category. It also uses a stereo vision system, although the configuration is different (we use a vertical setup for higher disambiguation power in feature matching). Their tracking technique is also different. They first take three snapshots (frontal, 45° to the left, and 45° to the right), and reconstruct up to 32 features selected on the face. Those 3D points, together with the templates extracted from the corresponding snapshots around each feature, are used for face tracking. In our case, we use a much more detailed face model but without a texture map, and features are selected at runtime, making our system more robust to lighting change, occlusion, and varying facial expression.

The work by Lu et al. [49] is also closely related to our work. They develop a head-pose tracking technique exploiting both face models and appearance exemplars. Because of the dynamic nature, it is not possible to represent face appearance by a single texture image. Instead, the complex face-appearance space is sampled by a few reference images (exemplars). By taking advantage of the rich geometric information of a 3D face model and the flexible representation provided by exemplars, that system is able to track head pose robustly under occlusion and/or varying facial expression. In that paper, one camera is used. In this article, we use a stereo setup to achieve even greater accuracy.

15.2.3 Eye-Gaze Correction

Several systems have been proposed to reduce or eliminate the angular deviation using special hardware. They make use of half-silvered mirrors or transparent screens with projectors to allow the camera to be placed on the optical path of the display. A brief review of these hardware-based techniques has been given in [35]. The high cost and the bulky setup prevent them from being used in a ubiquitous way.

On the other track, researchers have attempted to create eye contact using computer-vision and computer-graphics algorithms. Ott et al. [53] proposed to create a virtual center view given two cameras mounted on either side of the display screen. Stereoscopic analysis of the two camera views provides a depth map of the scene. Thus it is possible to "rotate" one of the views to obtain a center virtual view that preserves eye contact. Similarly, Liu et al. [42] used a trinocular stereo setup to establish eye contact. In both cases, they perform dense stereo matching without taking into account the domain knowledge. While they are generic enough to handle a variety of objects besides faces, they are likely to suffer from the vulnerability of brute-force stereo matching. Furthermore, as discussed in the previous section, we suspect that direct dense stereo matching is unlikely to generate satisfactory results, due to the constraint of camera placement imposed by the size of the display monitor—a problem that may be less severe back in the early 1990s when the above two algorithms were proposed.

Cham and Jones at Compaq Cambridge Research Laboratory [10] approached this problem from a machine-learning standpoint. They first register a 2D face model to the input image taken from a single camera, then morph the face model to the desired image. The key is to learn a function that maps the registered face model parameters to the desired morphed model parameters [32]. They achieve this by nonlinear regression from sufficient instances of registered–morphed parameter pairs which are obtained from training data. As far as we know, their research is still in a very early stage, so it is not clear if this approach is capable of handling dramatic facial expression changes. Furthermore, they only deal with the face part of the image; the morphed face image is superimposed on the original image frame, which sometimes leads to errors near the silhouettes due to visibility changes.

The GazeMaster project at Microsoft Research [25] uses a single camera to track the head orientation and eye positions. The Microsoft developers' view synthesis is unique in that they first replace the human eyes in a video frame with synthetic eyes gazing in the desired direction, then texture-map the eye-gaze-corrected video frame to a generic rigid face model rotated to the desired orientation. The synthesized photos they published look more like avatars, probably due to the underlying generic face model. Another drawback is that, as noted in their report, using synthetic eyes sometime inadvertently changes the facial expression as well.

From a much higher level, this GazeMaster work is similar to our proposed approach, in the sense that they both use strong domain knowledge (3D positions of a few predefined feature points) to facilitate tracking and view synthesis. However, our underlying algorithms, from tracking to view synthesis, are very different from theirs. We incorporate a stereo camera pair, which provides the important epipolar constraint that we use throughout the entire process. Furthermore, the configuration of our stereo camera provides much wider coverage of the face, allowing us to generate new distant views without having to worry about occlusions.

A very important component of our work is a real-time full-3D head tracker. It combines both stereovision and a detailed personalized model, and can deal with facial expression and partial occlusion. This technique per se can be used for other applications, and therefore we include here a brief review of previous work in 3D head tracking.

15.3 FACIAL GEOMETRY REPRESENTATION

Before going further, let us describe how a face is represented in our system. A vector is denoted by a boldface lowercase letter such as $\mathbf{p}$ or by an uppercase letter in typewriter style such as `P`. A matrix is usually denoted by a boldface uppercase letter such as **P**.

Faces can be represented as volumes or surfaces, but the most commonly used representation is a polygonal surface because of the real-time efficiency that modern graphic hardware can display [56]. A polygonal surface is typically formed from a triangular mesh. The face-modeling problem is then to determine the 3D coordinates of the vertices by using 3D digitizers, laser scanners, or computer-vision techniques.

If we treat the coordinates of the vertices as free parameters, the number of unknowns is very large, and we need a significant amount of data to model a face in a robust way. However, most faces look similar to each other (i.e., two eyes with a nose and mouth below them). Thus the number of degrees of freedom needed to define a wide rage of faces is limited. Vetter and Poggio [72] represented an arbitrary face image as a linear combination of a few hundred prototypes, and used this representation (called linear object class) for image recognition, coding, and image synthesis. Blanz and Vetter [8] used a linear class of both images and 3D geometries for image matching and face modeling. The advantage of using a linear class of objects is that it eliminates most of the unnatural faces and significantly reduces the search space.

Instead of representing a face as a linear combination of real faces, we represent it as a linear combination of a neutral face and some number of face *metrics* where a metric is vector that linearly deforms a face in a certain way, such as to make the head wider, make the nose bigger, etc. To be more precise, let's denote the face geometry by a vector $\mathcal{S} = (\mathbf{v}_1^{\mathrm{T}}, \ldots, \mathbf{v}_n^{\mathrm{T}})^{\mathrm{T}}$, where $\mathbf{v}_i = (X_i, Y_i, Z_i)^{\mathrm{T}}$ $(i = 1, \ldots, n)$ are the vertices, and a metric by a vector $\mathcal{M} = (\delta\mathbf{v}_1^{\mathrm{T}}, \ldots, \delta\mathbf{v}_n^{\mathrm{T}})^{\mathrm{T}}$, where $\delta\mathbf{v}_i = (\delta X_i, \delta Y_i, \delta Z_i)^{\mathrm{T}}$. Given a neutral face $\mathcal{S}^0 = (\mathbf{v}_1^{0\mathrm{T}}, \ldots, \mathbf{v}_n^{0\mathrm{T}})^{\mathrm{T}}$, and a set of m metrics $\mathcal{M}^j = (\delta\mathbf{v}_1^{j\mathrm{T}}, \ldots, \delta\mathbf{v}_n^{j\mathrm{T}})^{\mathrm{T}}$, the linear space of face geometries spanned by these metrics is

$$\mathcal{S} = \mathcal{S}^0 + \sum_{j=1}^{m} c_j \mathcal{M}^j \quad \text{subject to } c_j \in [l_j, u_j], \tag{1}$$

where the c_j's are the metric coefficients and l_j and u_j are the valid range of c_j. In our implementation, the neutral face and all the metrics are designed by an artist, and it is done once. The neutral face (see Figure 15.2) contains 194 vertices and 360 triangles. There are 65 metrics.

Our facial geometry representation is quite flexible. Since each metric is just a vector describing the desired deformation on each vertex, we can easily add more metrics if we need finer control.

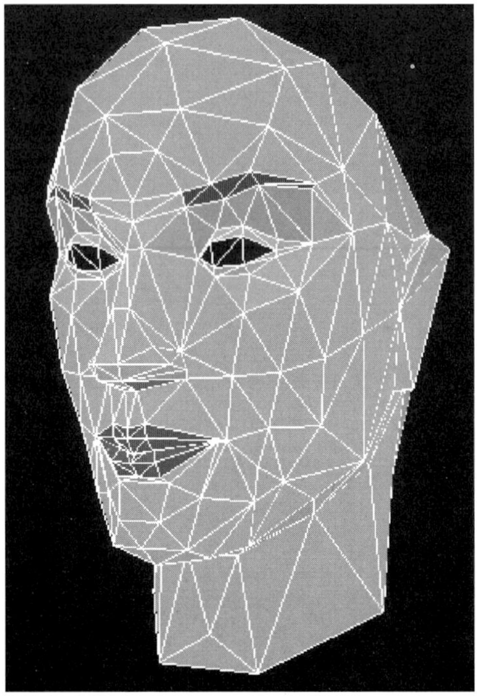

FIGURE 15.2: Neutral face.

15.3.1 Additional Notation

We denote the homogeneous coordinates of a vector $\mathbf{x}$ by $\tilde{\mathbf{x}}$, i.e., the homogeneous coordinates of an image point $\mathbf{p} = (u, v)^T$ are $\tilde{\mathbf{p}} = (u, v, 1)^T$, and those of a 3D point $\mathtt{P} = (x, y, z)^T$ are $\tilde{\mathtt{P}} = (x, y, z, 1)^T$. A camera is described by a pinhole model, and a 3D point $\mathtt{P}$ and its image point $\mathbf{p}$ are related by

$$\lambda\tilde{\mathbf{p}} = \mathbf{APM}\tilde{\mathtt{P}}, \tag{2}$$

where λ is a scale, and $\mathbf{A}$, $\mathbf{P}$, and $\mathbf{M}$ are given by

$$\mathbf{A} = \begin{pmatrix} \alpha & \gamma & u_0 \\ 0 & \beta & v_0 \\ 0 & 0 & 1 \end{pmatrix}, \quad \mathbf{P} = \begin{pmatrix} 1 & 0 & 0 & 0 \\ 0 & 1 & 0 & 0 \\ 0 & 0 & 1 & 0 \end{pmatrix}, \quad \text{and} \quad \mathbf{M} = \begin{pmatrix} \mathbf{R} & \mathbf{t} \\ 0^T & 1 \end{pmatrix}.$$

The elements of matrix $\mathbf{A}$ are the intrinsic parameters of the camera, and matrix $\mathbf{A}$ maps the normalized image coordinates to the pixel image coordinates (see, e.g., [20]). Matrix $\mathbf{P}$ is the perspective projection matrix. Matrix $\mathbf{M}$ is the 3D rigid transformation (rotation $\mathbf{R}$ and translation $\mathbf{t}$) from the object/world coordinate system to the camera coordinate system. For simplicity, we also denote the nonlinear 3D–2D projection function (2) by function ϕ such that

$$\mathbf{p} = \phi(\mathbf{M}, \mathtt{P}). \tag{3}$$

Here, the internal camera parameters are assumed to be known, although it is trivial to add them in our formulation.

When two images are concerned, a prime($'$) is added to denote the quantities related to the second image. When more images are involved, a subscript is used to specify an individual image.

The fundamental geometric constraint between two images is known as the *epipolar constraint* [20, 80]. It states that, in order for a point $\mathbf{p}$ in one image and a point $\mathbf{p}'$ in the other image to be the projections of a single physical point in space, or, in other words, in order for them to be matched, they must satisfy

$$\tilde{\mathbf{p}}'^T \mathbf{A}'^{-T} \mathbf{E} \mathbf{A}^{-1} \tilde{\mathbf{p}} = 0, \tag{4}$$

where $\mathbf{E} = [\mathbf{t}_r]_\times \mathbf{R}_r$ is known as the essential matrix, $(\mathbf{R}_r, \mathbf{t}_r)$ is the relative motion between the two images, and $[\mathbf{t}_r]_\times$ is a skew-symmetric matrix such that $\mathbf{t}_r \times \mathbf{v} = [\mathbf{t}_r]_\times \mathbf{v}$ for any 3D vector $\mathbf{v}$.

15.4 OVERVIEW OF THE 3D FACE-MODELING SYSTEM

Figure 15.3 outlines the components of our system. The equipment includes a computer and a video camera connected to the computer. We assume the camera's intrinsic parameters have been calibrated, a reasonable assumption given the simplicity of calibration procedures (see, e.g., [81]).

The first stage is video capture. The user simply sits in front of the camera and turns his/her head from one side all the way to the other side in about 5 seconds. The user then selects an approximately frontal view. This splits the video into two subsequences referred to as the *left* and *right* sequences, and the selected frontal image and its successive image are called the *base images*.

The second stage is feature-point marking. The user locates five markers in each of the two base images. The five markers correspond to the two inner eye corners, nose tip, and two mouth corners. As an optional step, the user can put three markers below the chin on the frontal view. This additional information usually improves the final face model quality. This manual stage could be replaced by an automatic facial-feature detection algorithm.

The third stage is the recovery of initial face models. The system computes the face mesh geometry and the head pose with respect to the camera frame using the

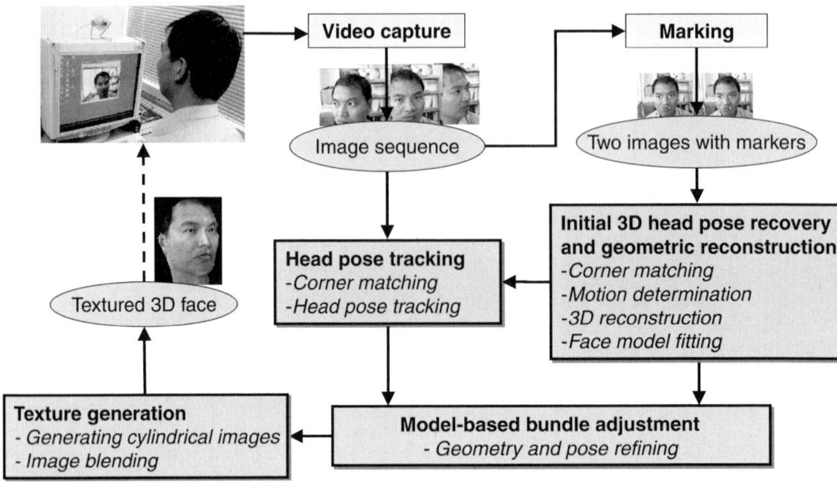

FIGURE 15.3: System overview.

two base images and markers as input. This stage of the system involves corner detection, matching, motion estimation, 3D reconstruction and model fitting.

The fourth stage tracks the head pose in the image sequences. This is based on the same matching technique used in the previous stage, but the initial face model is also used for increasing accuracy and robustness.

The fifth stage refines the head pose and the face geometry using a recently developed technique called *model-based bundle adjustment* [65]. The adjustment is performed by considering all point matches in the image sequences and using the parametric face space as the search space. Note that this stage is optional. Because it is much more time-consuming (about 8 minutes), we usually skip it in live demonstrations.

The final stage blends all the images to generate a facial texture map. This is now possible because the face regions are registered by the head motion estimated in the previous stage. At this point, a textured 3D face model is available for immediate animation or other purposes (see Section 15.6.1).

15.5 A TOUR OF THE SYSTEM ILLUSTRATED WITH A REAL VIDEO SEQUENCE

In this section, we will guide the reader through each step of the system using a real video sequence as an example. The video was captured in a normal room by a static camera while the head was moving in front. There is no control on the head motion, and the motion is unknown, of course. The video sequence can be found at http://research.microsoft.com/˜zhang/Face/duane.avi.

Note that this section intends for the reader to quickly gain knowledge of how our system works with a fair amount of detail. In order not to distract the reader too much by the technical details related to our new contribution, we defer their description to the following sections, but we provide pointers so the reader can easily find the details if necessary.

Although in this section we only demonstrate our system for one head, our system can handle a variety of head shapes, as can be seen later in Figures 15.17, 15.18, and 15.20, and more in [82].

15.5.1 Marking and Masking

The base images are shown in Figure 15.4, together with the five manually picked markers. We have first to determine the motion of the head and match some pixels across the two views before we can fit an animated face model to the images. However, some processing of the images is necessary, because there are at least three major groups of objects undergoing different motions between the two views: background, head, and other parts of the body such as the shoulder. If we do not separate them, there is no way to determine a meaningful head motion. The technique, described in detail in [82], allows us to mask off most irrelevant pixels automatically. Figure 15.5 shows the masked base images to be used for initial face-geometry recovery.

Optionally, the user can also mark three points on the chin in one base image, as shown with three large yellow dots in Figure 15.6. Starting from these dots, our system first tries to trace strong edges, which gives the red, green and blue curves in Figure 15.6. Finally, a spline curve is fitted to the three detected curves with an M-estimator. The small yellow dots in Figure 15.6 are sample points of that curve. Depending on the face shape and lighting condition, the three original curves do

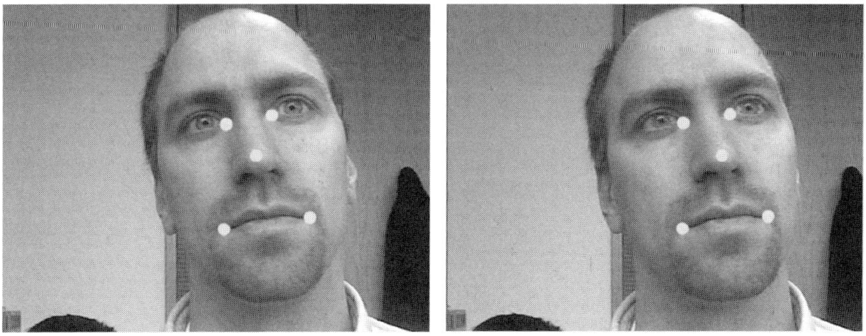

FIGURE 15.4: An example of two base images used for face modeling. Also shown are five manually picked markers indicated by dots.

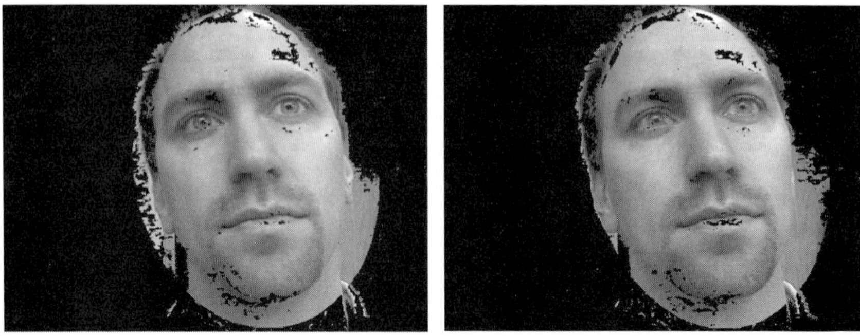

FIGURE 15.5: Masked images to be used in two-view image matching.

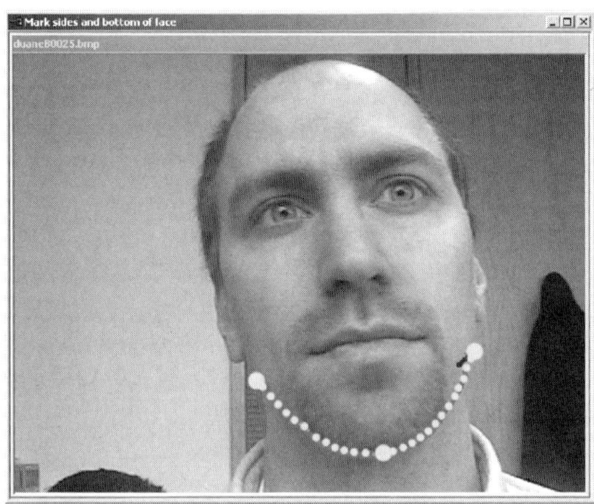

FIGURE 15.6: Marking the lower part of the face with three points (shown in large yellow dots). This is an optional step. See text for explanation. (See also color plate section)

not accurately represent the chin, but the final spline represents the chin reasonably well. The chin curve is used for fitting the face model described in Section 15.5.5.

15.5.2 Matching Between the Two Base Images

One popular technique for image registration uses optical flow [3, 29], which is based on the assumption that the intensity/color is constant. This is not the case in our situation: the color of the same physical point appears to vary between images, because the illumination changes when the head is moving. We therefore resort to a

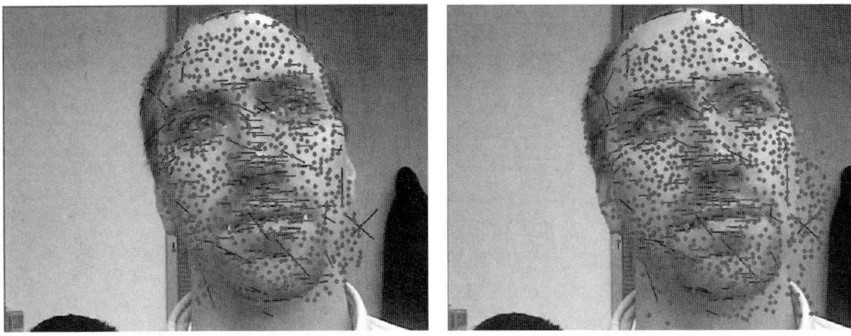

FIGURE 15.7: The set of matches established by correlation for the pair of images shown in Figure 15.4. Red dots are the detected corners. Blue lines are the motion vectors of the matches, with one endpoint (indicated by a red dot) being the matched corner in the current image and the other endpoint being the matched corner in the other image. (See also color plate section)

feature-based approach that is more robust to intensity/color variations. It consists of the following steps: (i) detecting corners in each image; (ii) matching corners between the two images; (iii) detecting false matches based on a robust estimation technique; (iv) determining the head motion; (v) reconstructing matched points in 3D space.

Corner detection. We use the Plessey corner detector, a well-known technique in computer vision [27]. It locates corners corresponding to high curvature points in the intensity surface if we view an image as a 3D surface with the third dimension being the intensity. Only corners whose pixels are white in the mask image are considered. See Figure 15.7 for the detected corners of the images shown in Figure 15.4 (807 and 947 corners detected respectively).

Corner matching. For each corner in the first image, we choose an 11×11 window centered on it, and compare the window with windows of the same size, centered on the corners in the second image. A zero-mean normalized cross-correlation between two windows is computed [20]. If we rearrange the pixels in each window as a vector, the correlation score is equivalent to the cosine of the angle between two intensity vectors. It ranges from -1, for two windows which are not similar at all, to 1, for two windows that are identical. If the largest correlation score exceeds a prefixed threshold (0.866 in our case), then that corner in the second image is considered to be the *match candidate* of the corner in the first image. The match candidate is retained as a *match* if and only if its match candidate in the first image happens to be *the corner being considered*. This symmetric test reduces many potential matching errors.

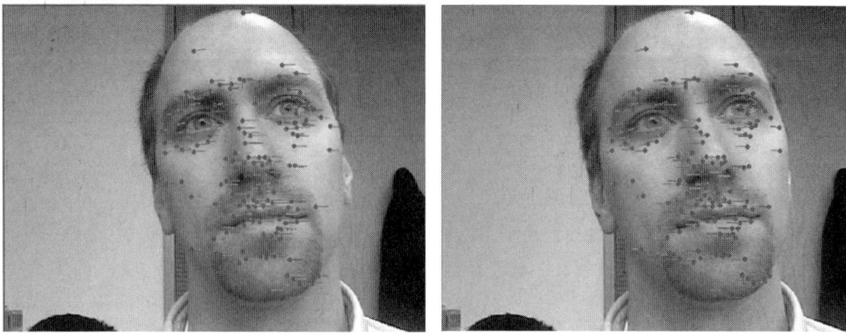

FIGURE 15.8: The final set of matches after automatically discarding false matches for the pair of images shown in Figure 15.4. Green lines are the motion vectors of the matches, with one endpoint (indicated by a red dot) being the matched corner in the current image and the other endpoint being the matched corner in the other image. (See also color plate section)

For the example shown in Figure 15.4, the set of matches established by this correlation technique is shown in Figure 15.7. There are 237 matches in total.

False match detection. The set of matches established so far usually contains false matches because correlation is only a heuristic. The only geometric constraint between two images is the epipolar constraint (4). If two points are correctly matched, they must satisfy this constraint, which is unknown in our case. Inaccurate location of corners because of intensity variation or lack of strong texture features is another source of error. We use the technique described in [80] to detect both false matches and poorly located corners, and simultaneously estimate the epipolar geometry (in terms of the essential matrix $\mathbf{E}$). That technique is based on a robust estimation technique known as the *least median squares* [60], which searches in the parameter space to find the parameters yielding the smallest value for the *median* of squared residuals computed for the entire data set. Consequently, it is able to detect false matches in as many as 49.9% of the whole set of matches.

For the example shown in Figure 15.4, the final set of matches is shown in Figure 15.8. There are 148 remaining matches. Compared with those shown in Figure 15.7, 89 matches have been discarded.

15.5.3 Robust Head-Motion Estimation

We have developed a new algorithm to compute the head motion between two views from the correspondence of five feature points including eye corners, mouth corners and nose top, and zero or more other image-point matches.

If the image locations of these feature points are precise, one could use a five-point algorithm to compute the camera motion. However, this is usually not the case in practice, since the pixel grid plus errors introduced by a human do not result in feature points with high precision. When there are errors, a five-point algorithm is not robust even when refined with a bundle adjustment technique. The key idea of our algorithm is to use the physical properties of the feature points to improve robustness. We use the property of symmetry to reduce the number of unknowns. We put reasonable lower and upper bounds on the nose height and represent the bounds as inequality constraints. As a result, the algorithm becomes significantly more robust. This algorithm is described in detail in [82].

15.5.4 3D Reconstruction

Once the motion is estimated, matched points can be reconstructed in 3D space with respect to the camera frame at the time when the first base image was taken. Let $(\mathbf{p}, \mathbf{p}')$ be a pair of matched points, and $\mathtt{P}$ be their corresponding point in space. 3D point $\mathtt{P}$ is estimated such that $\|\mathbf{p} - \hat{\mathbf{p}}\|^2 + \|\mathbf{p}' - \hat{\mathbf{p}}'\|^2$ is minimized, where $\hat{\mathbf{p}}$ and $\hat{\mathbf{p}}'$ are projections of $\mathtt{P}$ in both images according to (2).

Two views of the 3D reconstructed points for the example shown in Figure 15.4 are shown in Figure 15.9. The wireframes shown on the right are obtained by perform Delaunay triangulation on the matched points. The pictures shown on the left are obtained by using the first base image as the texture map.

15.5.5 Face Model Fitting From Two Views

We now only have a set of unorganized noisy 3D points from matched corners and markers. The face-model fitting process consists of two steps: fitting to the 3D reconstructed points and fine adjustment using image information. The first step consists of estimating both the *pose* of the head and the metric coefficients that minimize the distances from the reconstructed 3D points to the face mesh. The estimated head pose is defined to be the pose with respect to the camera coordinate system when the first base image was taken, and is denoted by $\mathbf{T}_0$. In the second step, we search for silhouettes and other face features in the images and use them, and also the chin curve if available from the marking step (Section 15.5.1), to refine the face geometry. Details of face-model fitting are provided in [82].

Figure 15.10 shows the reconstructed 3D face mesh from the two example images (see Figure 15.4). The mesh is projected back to the two images. Figure 15.11 shows two novel views using the first image as the texture map. The texture corresponding to the right side of the face is still missing.

15.5.6 Determining Head Motions in Video Sequences

Now we have the geometry of the face from only two views that are close to the frontal position. As can be seen in Figure 15.11, for the sides of the face, the texture

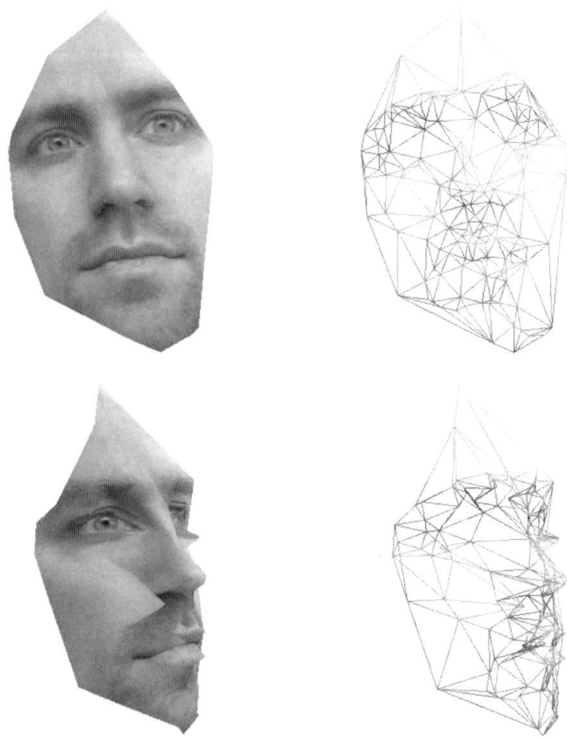

FIGURE 15.9: Reconstructed corner points. This coarse mesh is used later to fit a face model.

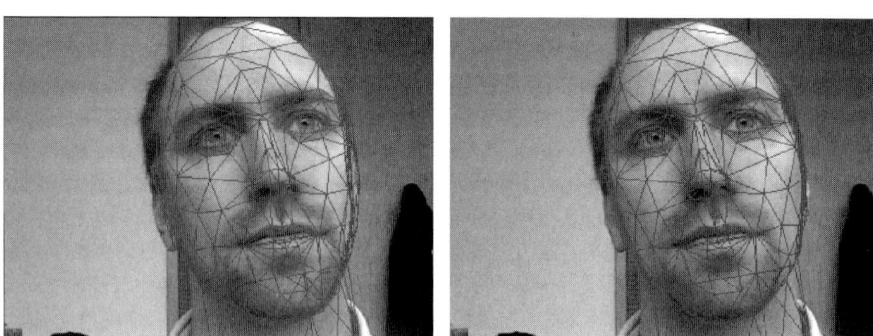

FIGURE 15.10: The constructed 3D face mesh is projected back to the two base images.

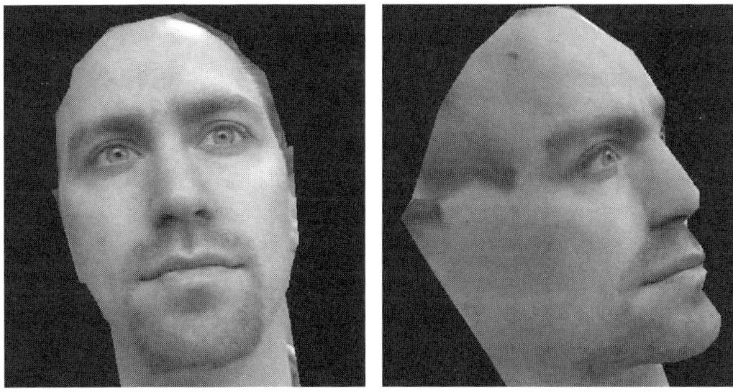

FIGURE 15.11: Two novel views of the reconstructed 3D face mesh with the first base image as texture.

from the two images is therefore quite poor or even not available at all. Since each image only covers a portion of the face, we need to combine all the images in the video sequence to obtain a complete texture map. This is done by first determining the head pose for the images in the video sequence and then blending them to create a complete texture map.

Successive images are first matched using the same technique described in Section 15.5.2. We could incrementally combine the resulting motions to determine the head pose. However, this estimation is quite noisy because it is computed only from 2D points. As we already have the 3D face geometry, a more reliable pose estimation can be obtained by combining both 3D and 2D information, as follows.

Let us denote the first base image by I_0, the images on the video sequences by $I_1, \ldots, I_v$, the relative head motion from I_{i-1} to I_i by $\mathcal{R}_i = \begin{pmatrix} \mathbf{R}_{ri} & \mathbf{t}_{ri} \\ 0^T & 1 \end{pmatrix}$, and the head pose corresponding to image I_i with respect to the camera frame by $\mathbf{T}_i$. The algorithm works incrementally, starting with I_0 and I_1. For each pair of images (I_{i-1}, I_i), we first use the corner-matching algorithm described in Section 15.5.2 to find a set of matched corner pairs $\{(\mathbf{p}_j, \mathbf{p}'_j) | j = 1, \ldots, l\}$. For each $\mathbf{p}_j$ in I_{i-1}, we cast a ray from the camera center through $\mathbf{p}_j$, and compute the intersection $\mathtt{P}_j$ of that ray with the face mesh corresponding to image I_{i-1}. According to (2), $\mathcal{R}_i$ is subject to the following equations

$$\mathbf{A}\mathbf{P}\mathcal{R}_i\tilde{\mathtt{P}}_j = \lambda_j\tilde{\mathbf{p}}'_j \quad \text{for } j = 1, \ldots, l, \tag{5}$$

where $\mathbf{A}$, $\mathbf{P}$, $\mathtt{P}_j$, and $\mathbf{p}'_j$ are known. Each of the above equations gives two constraints on $\mathcal{R}_i$. We compute $\mathcal{R}_i$ with a technique described in [20]. After $\mathcal{R}_i$ is computed, the head pose for image I_i in the camera frame is given by $\mathbf{T}_i = \mathcal{R}_i\mathbf{T}_{i-1}$. The head pose $\mathbf{T}_0$ is known from Section 15.5.5.

In general, it is inefficient to use all the images in the video sequence for texture blending, because head motion between two consecutive frames is usually small. To avoid unnecessary computation, the following process is used to automatically select images from the video sequence. Let us call the amount of rotation of the head between two consecutive frames the *rotation speed*. If s is the current rotation speed and α is the desired angle between each pair of selected images, the next image is selected $\lfloor(\alpha/s)\rfloor$ frames away. In our implementation, the initial guess of the rotation speed is set to 1 degree/frame and the desired separation angle is equal to 5 degrees.

Figures 15.12 and 15.13 show the tracking results for the two example video sequences (The two base images are shown in Figure 15.4). The images from each video sequence are automatically selected using the above algorithm.

15.5.7 Model-Based Bundle Adjustment

We now have an initial face model from two base images, a set of pairwise point matches over the whole video sequence, and an initial estimate of the head poses in the video sequence which is obtained incrementally based on the initial face model. Naturally, we want to refine the face-model and head-pose estimates by taking into account all available information simultaneously. A classical approach is to perform bundle adjustment to determine the head motion and 3D coordinates of isolated points corresponding to matched image points, followed by fitting the parametric face model to the reconstructed isolated points. We have developed a new technique called *model-based bundle adjustment*, which directly searches in the face-model space to minimize the same objective function as that used in the classical bundle adjustment. This results in a more elegant formulation with fewer unknowns, fewer equations, a smaller search space, and hence a better posed system. More details are provided in [82].

Figure 15.14 shows the refined result on the right sequence, which should be compared with that shown in Figure 15.13. The projected face mesh is overlaid on the original images. We can observe clear improvement in the silhouette and chin regions.

15.5.8 Texture Blending

After the head pose of an image is computed, we use an approach similar to Pighin et al.'s [57] to generate a view-independent texture map. We also construct the texture map on a virtual cylinder enclosing the face model. But instead of casting a ray from each pixel to the face mesh and computing the texture blending weights on a pixel by pixel basis, we use a more efficient approach. For each vertex on the face mesh, we compute the blending weight for each image based on the angle between surface normal and the camera direction [57]. If the vertex is invisible, its weight is set to 0.0. The weights are then normalized so that the sum of the

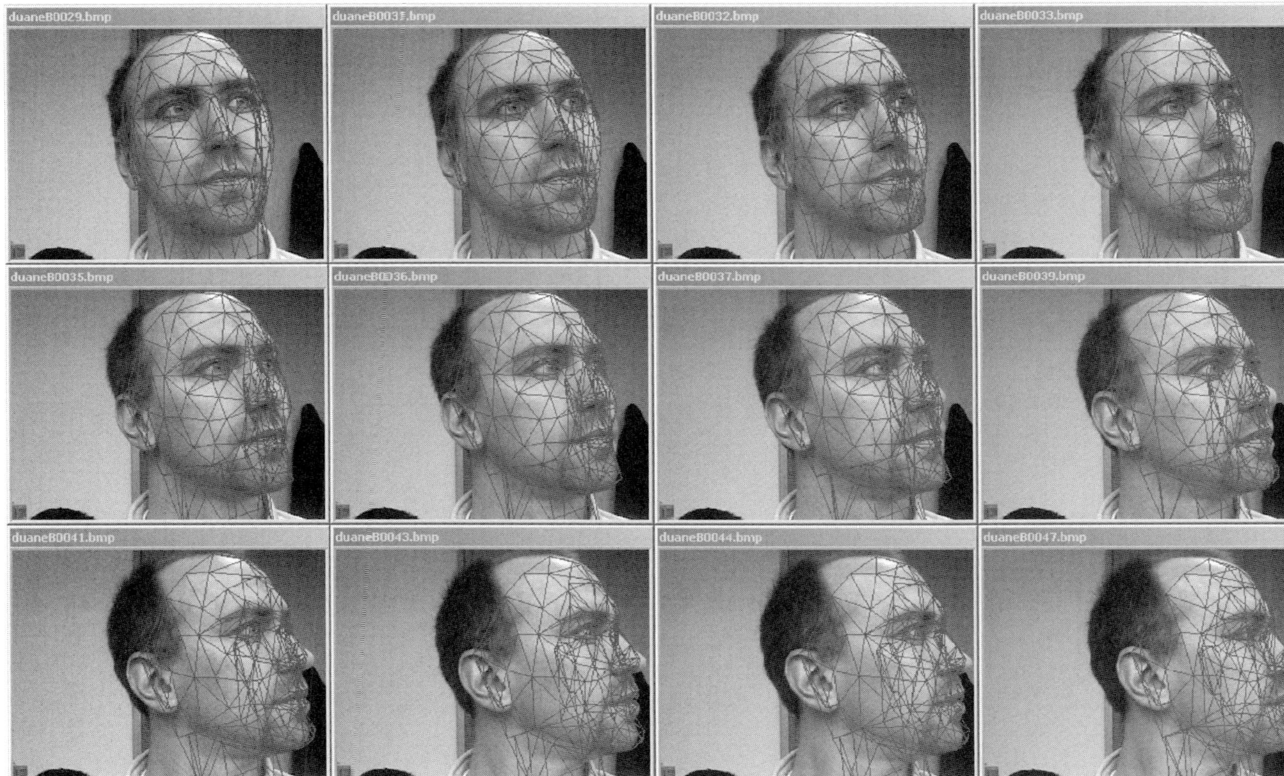

FIGURE 15.12: The face mesh is projected back to the automatically selected images from the video sequence where the head turns to the left.

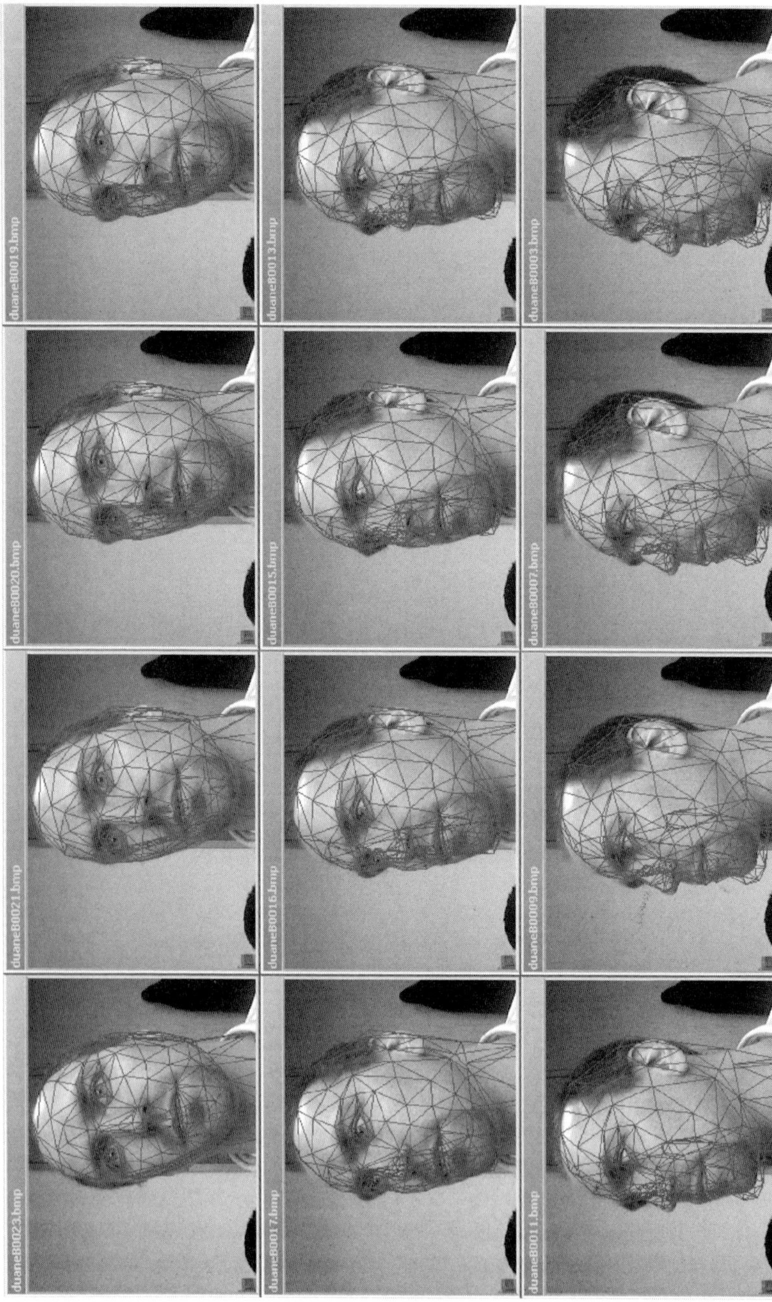

FIGURE 15.13: The face mesh is projected back to the automatically selected images from the video sequence where the head turns to the right.

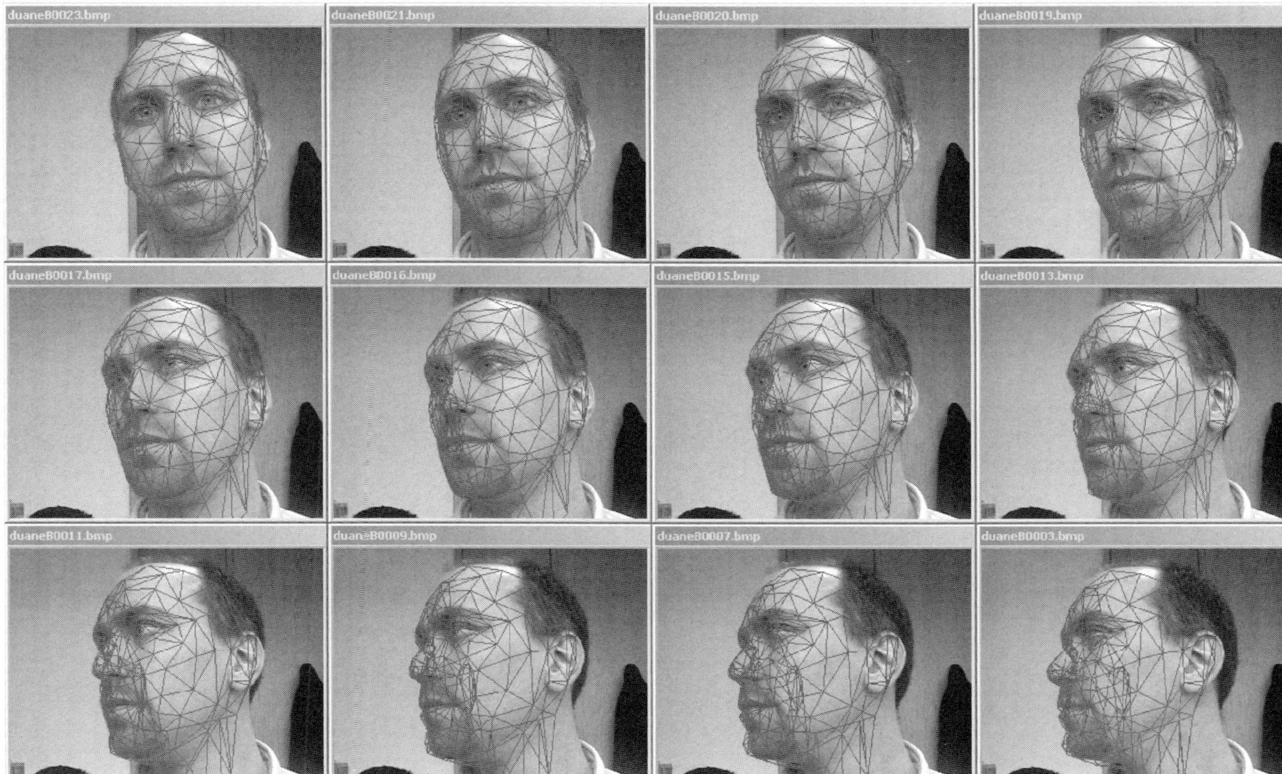

FIGURE 15.14: After model-based bundle adjustment, the refined face mesh is projected back to the automatically selected images from the video sequence where the head turns to the right.

weights over all the images is equal to 1.0. We then set the colors of the vertexes to be their weights, and use the rendered image of the cylindrical mapped mesh as the weight map. For each image, we also generate a cylindrical texture map by rendering the cylindrical mapped mesh with the current image as texture map. Let C_i and W_i $(i = 1, \ldots, k)$ be the cylindrical texture maps and the weight maps. Let C be the final blended texture map. For each pixel (u, v), its color on the final blended texture map is

$$C(u, v) = \sum_{i=1}^{k} W_i(u, v) C_i(u, v). \tag{6}$$

Because the rendering operations can be done using graphics hardware, this approach is very fast.

Figure 15.15 shows the blended texture map from the example video sequences 15.12 and 15.13. Figure 15.16 shows two novels views of the final 3D face model. Compared with those shown in Figure 15.11, we now have a much better texture on the side.

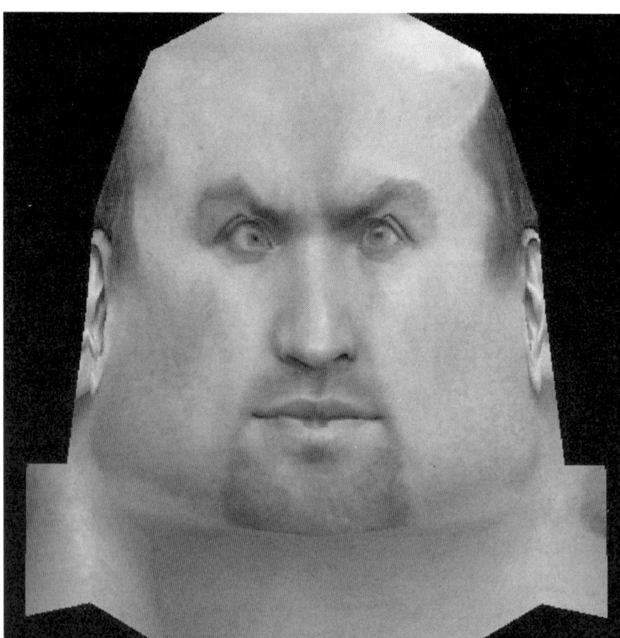

FIGURE 15.15: The blended texture image.

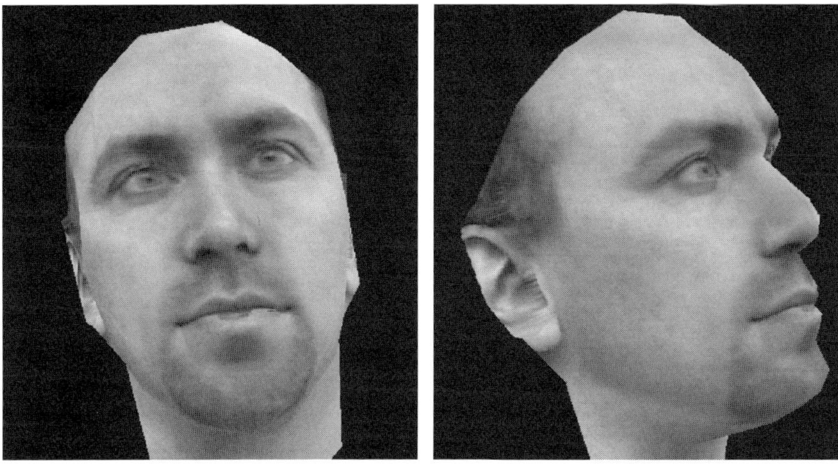

FIGURE 15.16: Two novel views of the final 3D face model.

15.6 MORE FACE-MODELING EXPERIMENTS

We have constructed 3D face models for well over two hundred subjects. We have done live demonstrations at ACM Multimedia 2000, ACM1, CHI2001, ICCV2001 and other events such as the 20th anniversary of the PC, where we set up a booth to construct face models for visitors. At each of these events, the success rate has been 90% or higher. In ACM1, most of the visitors are children or teenagers. Children are usually more difficult to model since they have smooth skins, but our system worked very well. We observe that the main factor for the occasional failure is the head turning too fast.

We should point out that in our live demonstrations, the optional model-based bundle adjustment is not performed because it is quite time-consuming (about 6 to 8 minutes on a 850MHz Pentium III machine). Without that step, our system takes about one minute after data capture and manual marking to generate a textured face model. Most of this time is spent on head tracking in the video sequences. All the results shown in this section were produced in this way.

Figure 15.17 shows side-by-side comparisons of eight reconstructed models with the real images. Figure 15.18 shows the reconstructed face models of our group members immersed in a virtualized environment. In these examples, the video sequences were taken using ordinary video camera in people's offices or in live demonstrations. No special lighting equipment or background was used. Note that the eyes in Figure 15.18 look a little bit ghostly; this is because the eyes have been cropped out and replaced with virtual eye balls. It is still an open problem how to model the eye balls and textures so that they match the face color and shape.

FIGURE 15.17: Side-by-side comparison of the original images with the reconstructed models of various people in various settings.

15.6.1 Applications: Animation and Interactive Games

Having obtained the 3D textured face model, the user can immediately animate the model by incorporating the facial expressions such as smiling, sad, pensive, etc. The model can also perform text to speech.

To accomplish this we have defined a set of vectors, which we call *posemes*. Like the metric vectors described previously, posemes are a collection of artist-designed displacements, corresponding approximately to the widely used action units in the Facial Action Coding System (FACTS) [18]. We can apply these displacements to any face as long as it has the same topology as the neutral face. Posemes are collected in a library of actions, expressions, and visems.

FIGURE 15.18: Face models of our group members in a virtualized environment.

Figure 15.19 shows a few facial expressions which can be generated with our animated face models. Note that the facial expressions shown here are the results of geometric warping, namely, the texture image is warped according to the desired displacement of the vertices. Facial expressions, however, exhibit many detailed image variations due to facial deformation. A simple expression-mapping technique based on ratio images, as described in [43] can generate vivid facial expression.

An important application of face modeling is interactive games. You can import your personalized face model in the games so that you are controlling the "virtualized you" and your friends are seeing the "virtualized you" play.

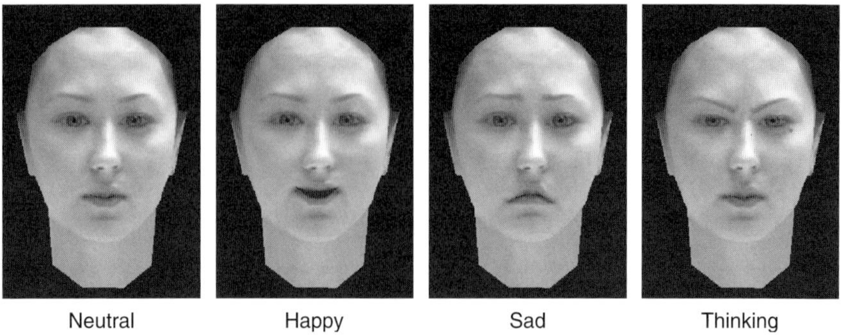

FIGURE 15.19: Examples of facial animation with an animated face model built using our system.

FIGURE 15.20: An online poker game with individualized face models.

This would dramatically enhance the role-playing experience offered in many games. Figure 15.20 shows a snapshot of an online poker game, developed by our colleague Alex Colburn, where the individualized face models are built with our system.

15.7 STEREO 3D HEAD-POSE TRACKING

There have been attempts to address head pose tracking and the eye-contact problem using a single camera (see Section 15.2.3). With a single camera, we found it difficult to maintain both the real-time requirement and the level of accuracy we need with head tracking. Existing model-based monocular head-tracking methods [30, 7, 4, 14, 2] either use a simplistic model, so they could operate in real time but produce less accurate results, or use some sophisticated models and processing to yield highly accurate results but take at least several seconds to compute. A single-camera configuration also has difficulty in dealing with occlusions. The work by Lu et al. [49] which is very similar to ours, also uses a personalized face model, but only with a single camera. As we will see, the results are not as good as with our stereo setup. Considering these problems with a monocular system, we decided to adopt a stereo configuration. The important epipolar constraint in stereo allows us to reject most outliers without using expensive robust estimation techniques, thus keeping our tracking algorithm both robust and simple enough to operate in real time. Furthermore, two cameras usually provide more coverage of the scene.

One might raise the question of why we do not use a dense stereo matching algorithm. We argue that, first, doing a dense stereo matching on a commodity PC in real time is difficult, even with today's latest hardware. Secondly and most importantly, a dense stereo matching is unlikely to generate satisfactory results due to the limitation on camera placement. Aiming at desktop-video teleconferencing applications, we could only put the cameras around the frame of a display monitor. If we put the cameras on the opposite edges of the display, given the normal viewing distance, we have to converge the cameras towards the person sitting in front of the desktop, and such a stereo system will have a long baseline. That makes stereo matching very difficult; even if we were able to get a perfect matching, there would still be a significant portion of the subject which is occluded in one view or the other. Alternatively, if we put the cameras close to each other on the same edge of the monitor frame, the occlusion problem is less severe, but generalization to new distant views is poor because a significant portion of the face is not observed. After considering various aspects, we decided to put one camera on the upper edge of the display and the other on the lower edge, similar to that used in [53], and follow a model-based stereo approach to head pose tracking and eye-gaze correction. The vertical arrangement makes face image matching less ambiguous because faces are mostly horizontally symmetric.

We use the face model described in Section 15.3. We build a personalized face model for each user using the modeling tool described in Section 15.4. Note that, although the face model contains other properties such as textures, we only use the geometric and semantic information in our tracking system.

A camera is modeled as a pinhole, and its intrinsic parameters are captured in a 3×3 matrix. The intrinsic matrices for the stereo pair are denoted by $\mathbf{A}_0$ and $\mathbf{A}_1$, respectively. Without loss of generality, we use the first camera's (camera 0) coordinate system as the world coordinate system. The second camera's (camera 1) coordinate system is related to the first one by a rigid transformation $(\mathbf{R}_{10}, \mathbf{t}_{10})$.

Thus, a point $\mathbf{m}$ in 3D space is projected to the image planes of the stereo cameras by

$$\mathbf{p} = \phi(\mathbf{A}_0\mathbf{m}), \tag{7}$$

$$\mathbf{q} = \phi(\mathbf{A}_1(\mathbf{R}_{10}\mathbf{m} + \mathbf{t}_{10})), \tag{8}$$

where $\mathbf{p}$ and $\mathbf{q}$ are the image coordinates in the first and second camera, and ϕ is a 3D-2D projection function such that $\phi(\begin{bmatrix} u \\ v \\ w \end{bmatrix}) = \begin{bmatrix} u/w \\ v/w \end{bmatrix}$. We use the method in [81] to determine $(\mathbf{A}_0, \mathbf{A}_1, \mathbf{R}_{10}, \mathbf{t}_{10})$.

The face model is described in its local coordinate system. The rigid motion of the head (*head pose*) in the world coordinate system is represented by a 3D rotation matrix $\mathbf{R}$ and a 3D translation vector $\mathbf{t}$. Since a rotation only has three degrees of freedom, the head pose requires 6 parameters. The goal of our tracking system is to determine these parameters.

15.7.1 Stereo Tracking

Our stereo head tracking problem at time instant $t + 1$ can be formally stated as follows:

Given a pair of stereo images $I_{0,t}$ and $I_{1,t}$ at time t, two sets of matched 2D points $S_0 = \{\mathbf{p}=[u, v]^T\}$ and $S_1 = \{\mathbf{q} = [a, b]^T\}$ from that image pair, their corresponding 3D model points $M = \{\mathbf{m} = [x, y, z]^T\}$, and a pair of stereo images $I_{0,t+1}$ and $I_{1,t+1}$ at time t+1,

determine (i) a subset $M' \subseteq M$ whose corresponding $\mathbf{p}$s and $\mathbf{q}$s have matches in $I_{0,t+1}$ and $I_{1,t+1}$, denoted by $S'_0 = \{\mathbf{p}'\}$ and $S'_1 = \{\mathbf{q}'\}$, and (ii) the head pose $(\mathbf{R}, \mathbf{t})$ so that the projections of $\mathbf{m} \in M'$ are $\mathbf{p}'$ and $\mathbf{q}'$.

We show a schematic diagram of the tracking procedure in Figure 15.21.

We first conduct independent feature tracking for each camera from time t to $t + 1$. We use the KLT tracker [66] which works quite well. However, the matched points may be drifted or even wrong. Therefore, we apply the epipolar constraint to remove any stray points. The epipolar constraint states that, if a point $\mathbf{p} = [u, v, 1]^T$ (expressed in homogeneous coordinates) in the first image and a point $\mathbf{q} = [a, b, 1]^T$ in the second image correspond to the same 3D point $\mathbf{m}$ in the physical world, they must satisfy the following equation:

$$\mathbf{q}^T\mathbf{F}\mathbf{p} = 0, \tag{9}$$

where $\mathbf{F}$ is the fundamental matrix[2] that encodes the epipolar geometry between the two images. In fact, $\mathbf{F}\mathbf{p}$ defines the epipolar line in the second image; thus

---

[2]The fundamental matrix is related to the camera parameters as $\mathbf{F} = \mathbf{A}_1{}^{-T}[\mathbf{t}_{10}]_\times \mathbf{R}_{10}\mathbf{A}_0{}^{-1}$.

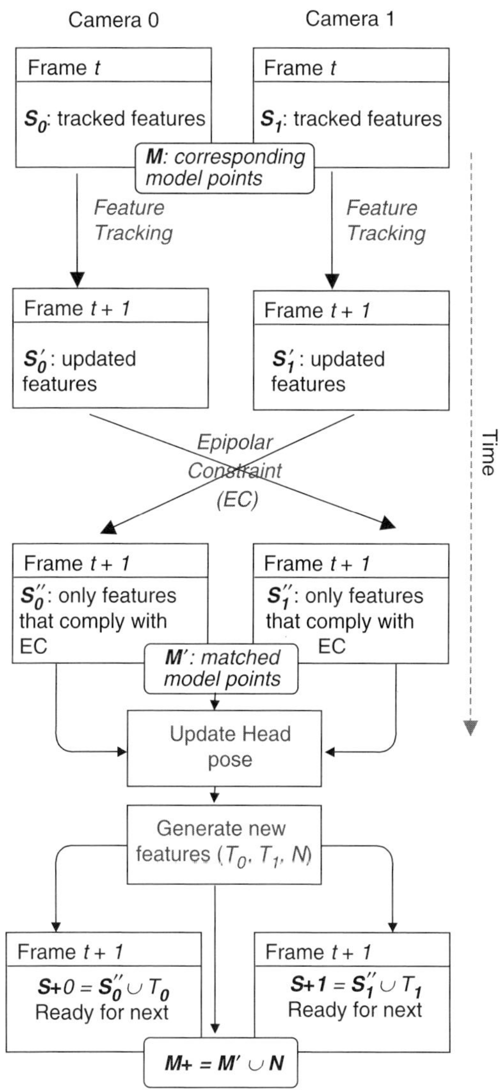

FIGURE 15.21: Model-based stereo 3D head tracking.

Equation 9 means that the epipolar line $\mathbf{Fp}$ must pass through the point $\mathbf{q}$. By transposing (9), we obtain a symmetric equation from the second image to the first image. That is, $\mathbf{F}^{\mathrm{T}}\mathbf{q}$ defines the epipolar line in the first image of $\mathbf{q}$, and (9) says that the corresponding point $\mathbf{p}$ must lie on the epipolar line.

In practice, due to inaccuracy in camera calibration and feature localization, we cannot expect the epipolar constraint to be satisfied exactly. For a triplet $(\mathbf{p}', \mathbf{q}', \mathbf{m})$, if the distance from $\mathbf{q}'$ to the $\mathbf{p}'$'s epipolar line is greater than a certain threshold, this triplet is considered to be an outlier and is discarded. We use a distance threshold of three pixels in our experiments.

After we have removed all the stray points that violate the epipolar constraint, we update the head pose $(\mathbf{R}, \mathbf{t})$ so that the reprojection error of $\mathbf{m}$ to $\mathbf{p}'$ and $\mathbf{q}'$ is minimized. The reprojection error e is defined as

$$e = \sum_i \left(\|\mathbf{p}_i' - \phi(\mathbf{A}_0(\mathbf{R}\mathbf{m}_i + \mathbf{t}))\|^2 + \|\mathbf{q}_i' - \phi(\mathbf{A}_1[\mathbf{R}_{10}(\mathbf{R}\mathbf{m}_i + \mathbf{t}) + \mathbf{t}_{10}])\|^2 \right). \tag{10}$$

We solve $(\mathbf{R}, \mathbf{t})$ using the Levenberg–Marquardt algorithm, and the head pose at time t is used as the initial guess.

15.7.2 Feature Regeneration

After the head pose is determined, we replenish the matched set S_0', S_1', and M' by adding more *good* feature points. We select a *good* feature point based on the following three criteria.

Texture: The feature point in the images must have rich texture information to facilitate the tracking. We first select 2D points in the image using the criteria in [66], then back-project them back onto the face model to get their corresponding model points.

Visibility: The feature point must be visible in both images. We have implemented an intersection routine that returns the first visible triangle given an image point. A feature point is visible if the intersection routine returns the same triangle for its projections in both images.

Rigidity: We must be careful not to add feature points in the nonrigid regions of the face, such as the mouth region. We define a bounding box around the tip of the nose that covers the forehead, eyes, nose, and cheek region. Any points outside this bounding box will not be added to the feature set.

This regeneration scheme improves our tracking system in two ways. First, it replenishes the feature points lost due to occlusions or nonrigid motion, so the tracker always has a sufficient number of features to start with in the next frame. This improves accuracy and stability. Secondly, it alleviates the problem of tracker drift by adding new fresh features at every frame.

15.7.3 Tracker Initialization and Autorecovery

The tracker needs to know the head pose at time 0 to start tracking. In our current implementation, we let the user interactively select seven landmark points in each image, from which the initial head pose can be determined. We show an example of the selected feature points in Figure 15.22, where the epipolar lines in the second image is also drawn. The manual selection does not have to be very accurate. We automatically refine the selection locally to satisfy the epipolar constraint.

The information from this initialization process is also used for tracking recovery when the tracker loses tracking. This may happen when the user moves out of the camera's field of view or rotates her head away from the cameras. When she turns back to the cameras, we prefer to continue tracking with no human intervention. During the tracker recovery process, the initial set of landmark points is used as templates to find the best match in the current image. When a match with a high confidence value is found, the tracker continues the normal tracking.

Furthermore, we also activate the autorecovery process whenever the current head pose is close to the initial head pose. This further alleviates the tracker drift problem, and the accumulative error is reduced after tracker recovery. This scheme could be extended to include multiple templates at different head poses. This is expected to further improve the robustness of our system.

15.7.4 Stereo Tracking Experiments

During our experiments, we captured three sequences with a resolution of 320 × 240 at 30 frames per second. Our current implementation can only run at

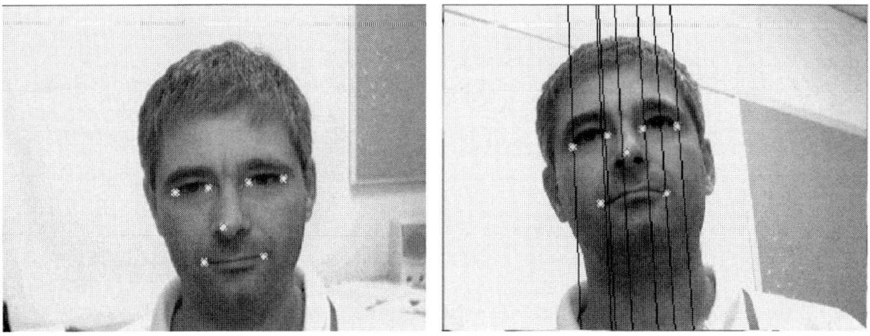

FIGURE 15.22: Manually selected feature points; the epipolar lines are overlayed on the second image.

4 to 5 frames per second. The results shown here are computed with our system in a "step-through" mode. Except the manual initialization performed once at the beginning of the test sequences, the results are computed automatically without any human interaction. We first show some results from the stereo tracking subsystem and compare it against a classic monocular tracker. Our stereo tracker shows significant improvements in both accuracy and robustness. Then we show some gaze-corrected views using the results from stereo tracking and matching, demonstrating the effectiveness of our proposed method.

Figure 15.23 shows the tracking results of the first sequence (A). The 3D face mesh is projected according to the estimated head pose and is overlayed on the input stereo images. This sequence contains large head rotations close to 90º. This type of out-of-plane rotation is usually difficult for head tracking, but we can see that our algorithm determines accurately the head pose, largely due to using the 3D mesh model.

The second sequence (B), shown in Figure 15.24, contains predominantly nonrigid motion (dramatic facial expressions). We also show the original images to better appreciate the nonrigid motion. Because we classify the face into rigid and nonrigid areas and use features from the rigid areas, our tracker is insensitive to nonrigid motion.

Figure 15.25 shows the last sequence (C) in which large occlusions and out-of-plane head motions frequently appear. Our system maintains accurate tracking throughout the 30-second sequence.

15.7.5 Validation

For the purpose of comparison, we have also implemented a model-based monocular tracking technique. Like most prevalent methods, we formulate it as an optimization problem that seeks to minimize the reprojection errors between the

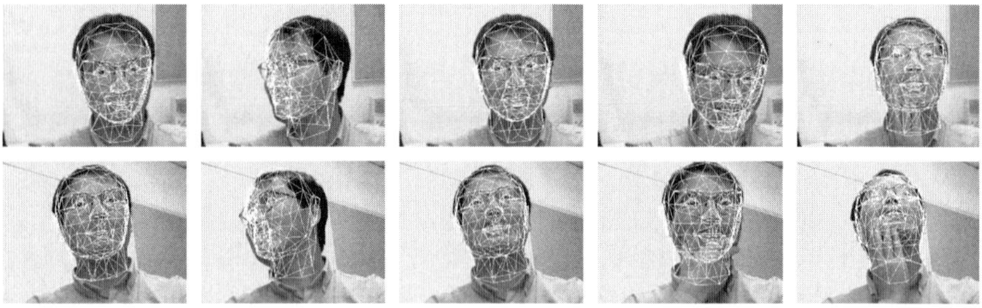

FIGURE 15.23: Stereo tracking result for sequence A (320×240 @ 30 FPS). Images from the first camera are shown in the upper row, while those from the second camera are shown in the lower row. From left to right, the frame numbers are 1, 130, 325, 997, and 1256.

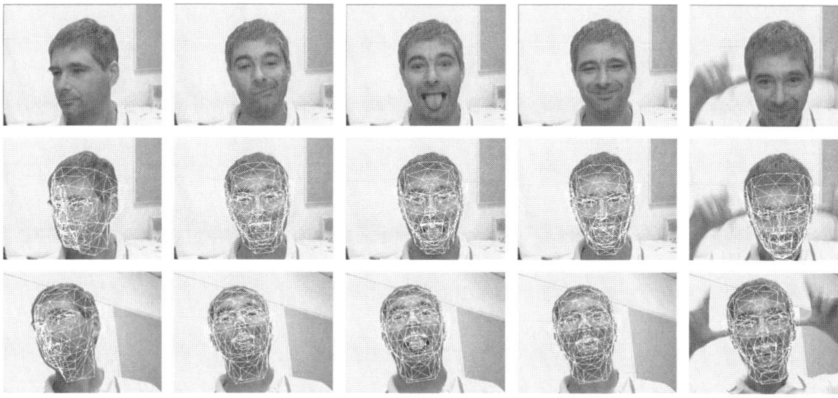

FIGURE 15.24: Stereo tracking result for sequence B (320×240 @ 30 FPS). The first row shows the input images from the upper camera. The second and third rows show the projected face model overlayed on the images from the upper and lower camera, respectively. From left to right, the frame numbers are 56, 524, 568, 624, and 716.

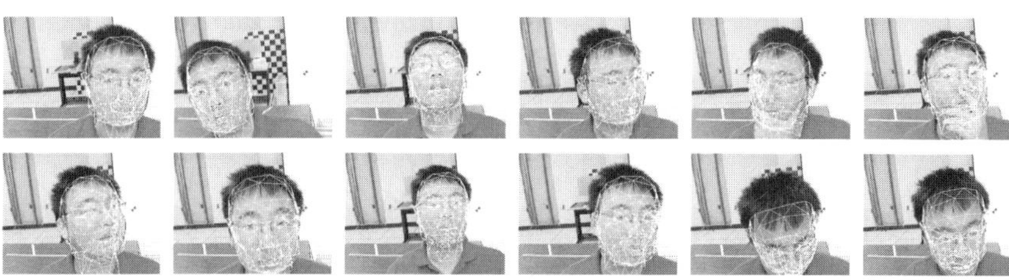

FIGURE 15.25: Stereo tracking result for sequence C (320×240 @ 30 FPS). The frame numbers, from left to right and from top to bottom, are 31, 67, 151, 208, 289, 352, 391, 393, 541, 594, 718, and 737.

projected 3D features points and the actual tracked features. Using the same notion as in (10), the monocular-tracking cost function is defined by

$$e_{\mathrm{m}} = \sum_{i} \|\mathbf{p}_i' - \phi(\mathbf{A}_0(\mathbf{R}\mathbf{m}_i + \mathbf{t}))\|^2. \tag{11}$$

We solve the optimization problem using the Levenberg–Marquardt method.

We run the monocular tracking algorithm over the three sequences. Since we do not know the ground truth of head motions, it is meaningless to compare the absolute values from the two algorithms. Instead, we compare the approximate velocity $\tilde{v} = \|\mathbf{t}_{i+1} - \mathbf{t}_i\|/\delta t$. The head motion is expected to be smooth, and so is the velocity curve. We plot the velocity curves of the three sequences in Figure 15.26. The x axis is the frame number and the y axis is the speed (inches/frame). The velocity curve computed using the monocular algorithm is plotted in red, while that from the stereo in blue. In the red curves, there are several spikes that well exceed the limit of normal head motion (a maximum cap of 3 inches/frame is put in the plots; some of the spikes are actually higher than that). We suspect that they indicate that tracking is lost or the optimization is trapped in a local minimum. On the other hand, the blue curves have significantly fewer spikes or even none. The only spikes in blue curves are in the first sequence (A), which indeed contains abrupt head motions. We also visually compare the results for sequence C between the monocular and stereo tracking method in Figure 15.27. These images are selected corresponding to the spikes in the red curve for sequence C. The top row shows the monocular tracking results and the second row shows the stereo tracking results. For those in the first row, some obviously have lost tracking, while the others have poor accuracy in head-pose estimation.

We should point out that the plots only show the results up to when the monocular tracker reported that the optimization routine failed to converge for 10 consecutive frames. On the other hand, the stereo tracker continued until the end of the sequence. The rich information from the stereo cameras enables the stereo tracker to achieve a much higher level of robustness than the monocular version.

15.8 APPLICATION TO EYE-GAZE CORRECTION

Video teleconferencing, a technology enabling communication with people face-to-face over remote distances, does not seem to be as widespread as predicted. Among many problems faced in videoteleconferencing, such as cost, network bandwidth, and resolution, the lack of eye contact seems to be the most difficult one to overcome [51]. The reason for this is that the camera and the display screen cannot be physically aligned in a typical desktop environment. It results in unnatural and even awkward interaction. Special hardware using half-silver mirrors has been used to address this problem. However this arrangement is bulky and expensive. What's more, as a piece of dedicated equipment, it does not fit well to our familiar computing environment; thus its usability is greatly reduced. In our work, we aim to address the eye-contact problem by synthesizing videos as if they were taken from a camera behind the display screen, thus to establish natural eye contact between videoteleconferencing participants without using any kind of special hardware.

15.8.1 Overview

The approach we take involves three steps: pose tracking, view matching, and view synthesis. We use a pair of calibrated stereo cameras and a personalized face

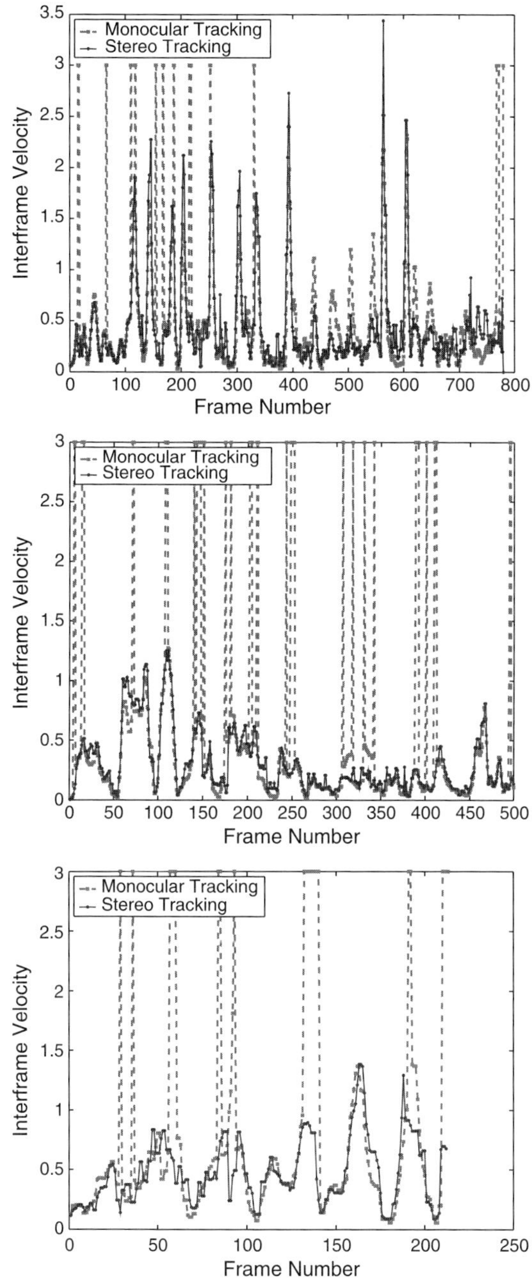

FIGURE 15.26: A comparison between monocular and stereo tracking in terms of the estimated velocity of head motion. Results from sequence A, B, and C are shown from top to bottom. (See also color plate section)

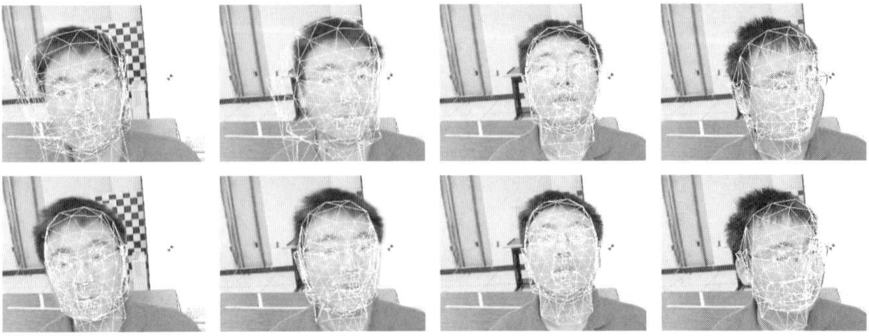

FIGURE 15.27: Visual comparison of the monocular (upper row) versus the stereo (lower row) tracking method. From left to right, the frame numbers are 54, 90, 132, and 207.

model to track the head pose in 3D. One camera is mounted on the upper edge of the display, while the other is on the lower edge of the display. This type of wide-baseline configuration makes stereo matching very hard. The use of strong domain knowledge (a personalized face model) and that of a stereo camera pair greatly increase the robustness and accuracy of the 3D head-pose tracking, which in turn helps stereo matching. The stereo camera pair also allows us to match parts not modelled in the face model, such as the hands and shoulders, thus providing wider coverage of the subject. Finally, the results from head tracking and stereo matching are combined to generate a virtual view. Unlike other methods that only "cut and paste" the face part of the image, our method generates natural-looking and seamless images, as shown in Figure 15.28. We believe our proposed approach advances the state of the art in the following ways.

- By combining a personalized model and a stereo camera pair, we exploit *a priori* domain knowledge with added flexibility from the stereo cameras.
- We present a novel head tracking algorithm that operates in real time and maintains highly accurate full 3D tracking by exploiting both temporal and spatial coherence, even under difficult conditions such as partial occlusions or nonrigid motions.
- We combine a comprehensive set of view-matching criteria to match as many features as possible, including salient point features and object silhouettes. These features are not restricted by the model we use, allowing us to create seamless and convincing virtual views. We believe the quality of our synthesized views is among the best of previously published methods.

Figure 15.29 illustrates the block diagram of our eye-gaze correction system. We use two digital video cameras mounted vertically, one on the top and the other on the bottom of the display screen. They are connected to a PC through 1394 links.

FIGURE 15.28: Eye-gaze correction. The first and the third images were taken from the stereo cameras mounted on the top and bottom sides of a monitor while the person was looking at the screen. The picture in the middle is a synthesized virtual view that preserves eye-contact. Note that the eye gaze in the virtual view appears to be looking forward, as desired.

The cameras are calibrated using the method described in [81]. We choose the vertical setup because it provides wider coverage of the subject and higher disambiguation power in feature matching. Matching ambiguity usually involves symmetric facial features such as eyes and lip contours aligned horizontally. The user's personalized face model is acquired using a rapid face-modeling tool [46]. Both the calibration and model acquisition require little human interaction, and a novice user can complete each task within 5 minutes.

With a calibrated camera pair and a 3D face model, we are able to correct the eye gaze using the algorithm outlined below.

1. Acquire the background model.
2. Initialize the face tracker.
3. For each image pair, perform:
   - background subtraction
   - temporal feature tracking in both images

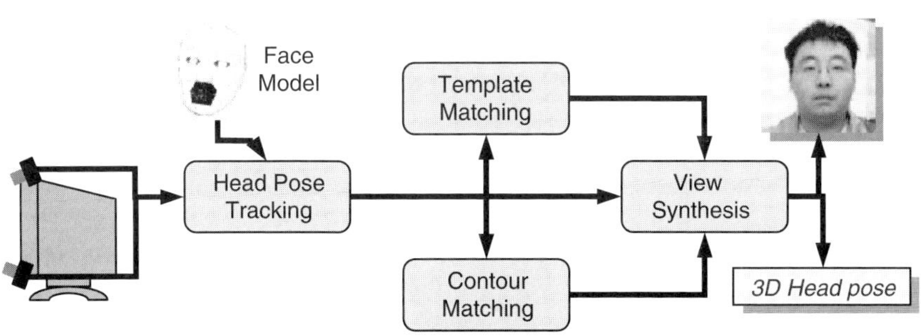

FIGURE 15.29: The components of our eye-gaze correction system.

- updating head pose
- correlation-based stereo feature matching
- stereo silhouette matching
- hardware-assisted view synthesis.

Currently, the only manual part of the system is the face-tracker initialization, which requires the user to interactively select a few markers in the first frame. We are currently working on automatic initialization.

The tracking subsystem includes a feedback loop that supplies fresh salient features at each frame to make the tracking more stable under adverse conditions, such as partial occlusion and facial expression changes. Furthermore, an automatic tracking recovery mechanism is also implemented to make the whole system even more robust over an extended period of time. Based on the tracking information, we are already able to manipulate the head pose by projecting the live images onto the face model. However, we also want to capture the subtleties of facial expressions, as well as foreground objects such as hands and shoulders. So we further conduct correlation-based feature matching and silhouette matching between the stereo images. All the matching information, together with the tracked features, is used to synthesize a *seamless* virtual image that looks as if it were taken from a camera behind the display screen. We have implemented the entire system under the MS Windows environment. Without any effort spent on optimizing the code, our current implementation runs about 4–5 frames per second on a single-CPU 1-GHz PC.

15.8.2 Stereo View Matching

The result from the 3D head-pose tracking gives a set of good matches within the *rigid* part of the face between the stereo pair. To generate convincing and photorealistic virtual views, we need to find more matching points over the entire foreground images, especially along the contour and the nonrigid parts of the face. We incorporate both feature matching and template matching to find as many matches as possible. During this matching process, we use the reliable information obtained from tracking to constrain the search range. In areas where such information is not available, such as hands and shoulders, we relax the search threshold, then apply the disparity-gradient limit to remove false matches.

To facilitate the matching (and later view synthesis in Section 15.8.3), we rectify the images using the technique described in [48], so that the epipolar lines are horizontal.

Disparity and Disparity-Gradient Limit

Before we present the details of our matching algorithm, it is helpful to define disparity, disparity gradient, and the important principle of disparity-gradient limit, which will be exploited throughout the matching process.

Given a pixel (u, v) in the first image and its corresponding pixel (u', v') in the second image, disparity is defined as $d = u' - u$ ($v = v'$ as images have been rectified). Disparity is inversely proportional to the distance of the 3D point to the cameras. A disparity of 0 implies that the 3D point is at infinity.

Consider now two 3D points whose projections are $\mathbf{m}_1 = [u_1, v_1]^T$ and $\mathbf{m}_2 = [u_2, v_2]^T$ in the first image, and $\mathbf{m}'_1 = [u'_1, v'_1]^T$ and $\mathbf{m}'_2 = [u'_2, v'_2]^T$ in the second image. Their disparity gradient is defined to be the ratio of their difference in disparity to their distance in the cyclopean image, i.e.,

$$DG = \left| \frac{d_2 - d_1}{u_2 - u_1 + (d_2 - d_1)/2} \right| \tag{12}$$

Experiments in psychophysics have provided evidence that human perception imposes the constraint that the disparity gradient DG is upper-bounded by a limit K. The limit $K = 1$ was reported in [9]. The theoretical limit for opaque surfaces is 2 to ensure that the surfaces are visible to both eyes [59]. Also reported in [59], less than 10% of world surfaces viewed at more than 26cm with 6.5cm of eye separation will present with disparity gradient larger than 0.5. This justifies use of a disparity-gradient limit well below the theoretical value (of 2) without imposing strong restrictions on the world surfaces that can be fused by the stereo algorithm. In our experiment, we use a disparity-gradient limit of 0.8 ($K = 0.8$).

Feature Matching Using Correlation

For unmatched good features in the first (upper) image, we try to find corresponding points, if any, in the second (lower) image by template matching. We use normalized correlation over a 9×9 window to compute the matching score. The disparity search range is confined by existing matched points from tracking, when available.

Combined with matched points from tracking, we build a sparse disparity map for the first image and use the following procedure to identify potential outliers (false matches) that do not satisfy the disparity gradient limit principle. For a matched pixel $\mathbf{m}$ and a neighboring matched pixel $\mathbf{n}$, we compute their disparity gradient between them using (12). If $DG \leq K$, we register a vote of good match for $\mathbf{m}$; otherwise, we register a vote of bad match for $\mathbf{m}$. After we have counted for every matched pixel in the neighborhood of $\mathbf{m}$, we tally the votes. If the "good" votes are less than the "bad" votes, $\mathbf{m}$ will be removed from the disparity map. This is conducted for every matched pixel in the disparity map; the result is a disparity map that conforms to the principle of disparity gradient limit.

Note that unlike temporal tracking, we consider at this stage all good features including points on the mouth and eye regions. This is because the two cameras are synchronized, and facial deformation does not affect the epipolar constraint in stereo matching.

Contour Matching

Template-matching assumes that corresponding image patches share some similarity and are distinct from their neighbors. This assumption may be wrong at object contours, especially at occluding boundaries. Yet object contours are very important cues for view synthesis because the lack of matching information along object contours will result in excessive smearing or blurring in the synthesized views. So it is necessary to include a module that extracts and matches the contours across views in our system.

The contour of the foreground object can be extracted after background subtraction. It is approximated by polygonal lines using the Douglas–Poker algorithm [17]. The control points (vertices) on the contour, $\mathbf{v}_i$, are further refined to subpixel accuracy using the "snake" technique [34]. Once we have two polygonal contours, denoted by $P = \{\mathbf{v}_i | i = 1..n\}$ in the first image and $P' = \{\mathbf{v}'_i | i = 1..m\}$ in the second image, we use the dynamic programming (DP) technique to find the global optimal matching between them.

Since it is straightforward to formulate contour matching as a DP problem with states, stage, and decisions, we will only discuss in detail the design of the cost functions (the reader is referred to Bellman's book about DP techniques [5]). There are two cost functions, the *matching* cost and the *transition* cost. The matching cost function $C(i,j)$ measures the "goodness" of matching between segment $V_i = \overline{\mathbf{v}_i\mathbf{v}_{i+1}}$ in P and segment $V'_j = \overline{\mathbf{v}'_i\mathbf{v}'_{i+1}}$ in P'. The lower the cost, the better the matching. The transition cost function $W(i,j|i_0,j_0)$ measures the smoothness from segment V_{i_0} to segment V_i, assuming that (V_i, V'_j) and (V_{i_0}, V'_{j_0}) are matched pairs of segments. Usually, V_i and V_{i_0} are continuous segments, i.e., $\|i_0 - i\| \leq 1$. It penalizes matches that are out of order. The scoring scheme of DP, formulated as a forward recursion function, is then given by

$$\begin{aligned} M(i,j) = \min\{&M(i-1, j-1) + C(i, j) + W(i,j|i-1, j-1), \\ &M(i, j-1) + C(i, j) + W(i,j\,|i, j-1), \\ &M(i-1, j) + C(i, j) + W(i, j|i-1, j)\}. \end{aligned}$$

The Matching Cost

The matching cost takes into account the epipolar constraint, the orientation difference, and the disparity gradient limit, described as follows.

The epipolar constraint. We distinguish three configurations, as shown in Figure 15.30 where the red line is the contour in the first (upper) image and the blue line is the contour in the second (lower) image. The dotted lines are the corresponding epipolar lines of the vertices. In Figure 15.30(a), segment $\overline{bc}$ and segment $\overline{qr}$ are being matched, and $C_e = 0$. The epipolar constraint limits $\overline{qr}$ corresponding

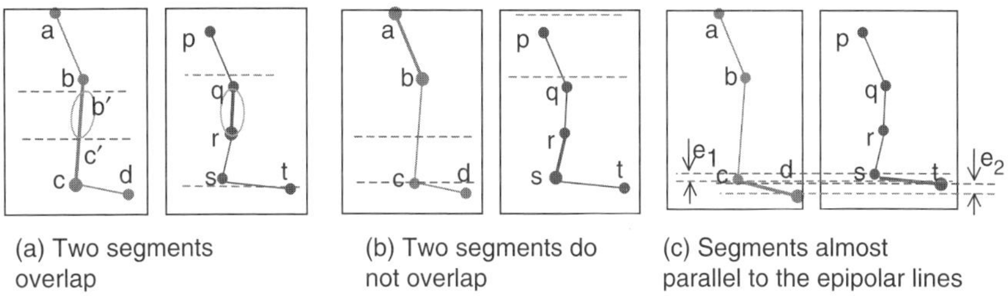

FIGURE 15.30: Applying the epipolar constraint to contour matching.

to segment $\overline{b'c'}$, instead of $\overline{bc}$. In Figure 15.30(b), the epipolar constraint tells that segment $\overline{ab}$ cannot match segment $\overline{rs}$, because there is no overlap. In that case, a sufficiently large cost (T_{highcost}) is assigned to this match. When the orientation of at least one line segment is very close to that of epipolar lines, intersection of the epipolar line with the line segment cannot be computed reliably. In that case, the cost is the average inter-epipolar distance ($d_e = (e_1 + e_2)/2$), as illustrated in the figure. In summary, the epipolar constraint cost for a pair of segment (V_i, V_j') is

$$C_e = \begin{cases} d_e & \text{if } V_i \text{ or } V_j' \text{ is close to horizontal lines;} \\ 0 & \text{if } V_i \text{ or } V_j' \text{ overlaps;} \\ T_{\text{highcost}} & \text{otherwise.} \end{cases} \tag{13}$$

The orientation difference. We expect that matched segments have similar orientation. We define the orientation cost as a power function of the orientation difference between the proposed matching segments (V_i, V_j'). Let a_i and a_j be orientation of V_i and V_j'; then the orientation difference is

$$C_a = \left(\frac{|a_i - a_j|}{T_a} \right)^n, \tag{14}$$

where T_a is the angular-difference threshold, and n is the power factor. We use $T_a = 30°$ and $n = 2$.

The disparity-gradient limit. It is similar to that used in template matching. However, we do not want to consider feature points in matching contour segments, because the contour is on the occluding boundary, where the disparity gradient with respect to the matched feature points is very likely to exceed the limit. On the other hand, it is reasonable to assume that the disparity-gradient limit will be upheld *between* the two endpoints of the segment. We adopt the disparity prediction model in [83]. That is, given a pair of matched points $(\mathbf{m}_i, \mathbf{m}'_i)$, the disparity of a point $\mathbf{m}$ is modeled as

$$d = d_i + D_i n_i, \tag{15}$$

where $d_i = \|\mathbf{m}'_i - \mathbf{m}_i\|$, $D_i = \|\mathbf{m} - \mathbf{m}_i\|$, and $n_i \sim N(0, \sigma^2 I)$ with $\sigma = K/(2-K)$. A pair of matched segments contains two pairs of matched endpoints $(\mathbf{m}_\mathrm{s}, \mathbf{m}'_\mathrm{s})$ and $(\mathbf{m}_\mathrm{e}, \mathbf{m}'_\mathrm{e})$. We use $(\mathbf{m}_\mathrm{s}, \mathbf{m}'_\mathrm{s})$ to predict the disparity of $\mathbf{m}_\mathrm{e}$, and compute the variance of the "real" disparity from the predicted one. Similarly we also compute the variance of the predicted disparity of $\mathbf{m}_s$ using $(\mathbf{m}_\mathrm{e}, \mathbf{m}'_\mathrm{e})$. As suggested in [83], the predicted variance should be less restrictive when the point being considered is away from the matched point, which leads to the following formulas:

$$\sigma_i = [\sigma_{\max} - \sigma_{\min}](1 - \exp(-D_i^2/\tau^2)) + \sigma_{\min}, \tag{16}$$

where the range $[\sigma_{\min}, \sigma_{\max}]$ and τ are parameters. We use $\sigma_{\min} = 0.5, \sigma_{\max} = 1.0$, and $\tau = 30$. Now we can finally write out the disparity gradient cost. Let $d_\mathrm{s} = \|\mathbf{m}'_\mathrm{s} - \mathbf{m}_\mathrm{s}\|$, $d_\mathrm{e} = \|\mathbf{m}'_\mathrm{e} - \mathbf{m}_\mathrm{e}\|$, $D_1 = \|\mathbf{m}_\mathrm{e} - \mathbf{m}_\mathrm{s}\|$, $D_2 = \|\mathbf{m}'_\mathrm{e} - \mathbf{m}'_\mathrm{s}\|$, and $\Delta d = d_\mathrm{e} - d_\mathrm{s}$; σ_e and σ_s are computed by plugging in D_1 and D_2 into (16); the disparity gradient cost is given by

$$C_\mathrm{d} = \Delta d^2/\sigma_\mathrm{e}^2 + \Delta d^2/\sigma_\mathrm{s}^2. \tag{17}$$

Combining all the above three terms, we have the final matching cost as:

$$C = \max(T_{\mathrm{highcost}}, C_\mathrm{e} + w_\mathrm{a} C_\mathrm{a} + w_\mathrm{d} C_\mathrm{d}), \tag{18}$$

where w_a and w_d are weighting constants. The match cost is capped by T_{highcost}. This is necessary to prevent any corrupted segment in the contour from contaminating the entire matching. In our implementation, $T_{\mathrm{highcost}} = 20, w_\mathrm{a} = 1.0$, and $w_\mathrm{d} = 1.0$.

The Transition Cost

In contour matching, when two segments are continuous in one image, we would expect that their matched segments in the other image are continuous too. This is not always possible due to changes in visibility: some part of the contour can only be seen in one image. The transition cost (W) is designed to favor smooth matching from one segment to the next, while taking into account discontinuities due to occlusions. The principle we use is again the gradient disparity limit. For two consecutive segments V_i and V_{i+1} in P, the transition cost function is the same as the one used in matching cost (equation 17) except that the two pairs of matched points involved are now the endpoint of V_i and its corresponding point in P', and the starting point of V_{i+1} and its corresponding point in P'.

15.8.3 View Synthesis

From the previous tracking and matching stages, we have obtained a set of point matches and line matches that can be used to synthesize new views. Note that these matches contain not only the modeled face part, but also other foreground parts such as hands and shoulders. This is yet another advantage of our stereovision-based approach. We could obtain a more complete description of the scene geometry beyond the limit of the face model. Treating these matches as a whole, our view synthesis methods can create seamless virtual imagery.

We implemented and tested two methods for view synthesis. One is based on view morphing [63] and the other uses hardware-assisted multitexture blending. The view-morphing technique allows us to synthesize virtual views along the path connecting the optical centers of the two cameras. A view-morphing factor $c_{\rm m}$ controls the exact view position. It is usually between 0 and 1, whereas a value of 0 corresponds exactly to the first camera view, and a value of 1 corresponds exactly to the second camera view. Any value in-between represents a virtual viewpoint somewhere along the path from the first camera to the second. By changing the view-morphing factor $c_{\rm m}$, we can synthesize correct views with desired eye gaze in real time.

In our hardware-assisted rendering method, we first create a 2D triangular mesh using Delaunay triangulation in the first camera's image space. We then offset each vertex's coordinate by its disparity modulated by the view-morphing factor $c_{\rm m}$, i.e., $[u_i', v_i'] = [u_i + c_{\rm m} d_i, v_i]$. The offset mesh is fed to the hardware render with two sets of texture coordinates, one for each camera image. Note that all the images and the mesh are in the rectified coordinate space. We need to set the viewing matrix to the inverse of the rectification matrix to "un-rectify" the resulting image to its normal view position. This is equivalent to the post-warp in view morphing. Thus the hardware can generate the final synthesized view in a single pass. We also use a more elaborate blending scheme, thanks to the powerful graphics hardware. The weight W_i for the vertex V_i is based on the total area of adjacent triangles and the view-morphing factor, and is defined as

$$W_i = \frac{(1 - c_{\rm m}) \sum S_i^1}{(1 - c_{\rm m}) \sum S_i^1 + c_{\rm m} \sum S_i^2}, \tag{19}$$

where S_i^1 are the areas of the triangles of which V_i is a vertex, and S_i^2 are the areas of the corresponding triangles in the other image. By changing the view-morphing factor $c_{\rm m}$, we can use the graphics hardware to synthesize correct views with desired eye gaze in real time. Because the view synthesis process is conducted by hardware, we can spare the CPU for more challenging tracking and matching tasks.

Comparing these two methods, the hardware-assisted method, aside from its blazing speed, generates crisper results if there is no false match in the mesh. On the other hand, the original view-morphing method is less susceptible to bad matches, because it essentially uses every matched point and line segment to compute the final coloring of a single pixel, while in the hardware-based method only the three closest neighbors are used.

Regarding the background, it is very difficult to obtain a reliable set of matches since the baseline between the two views is very large, as can be observed in the two original images shown in Figure 15.28. In this work, we do not attempt to model the background at all, but we offer two solutions. The first is to treat the background as unstructured, and add image boundary as matches. The result will be ideal if the background has a uniform color; otherwise, it will be fuzzy as shown in the synthesized view shown in Figure 15.28. The second solution is to replace the background by anything appropriate. In that case, view synthesis is only performed for the foreground objects. In our implementation, we overlay the synthesized foreground objects on the image from the first camera. The results shown in the following section were produced in this way.

15.8.4 Eye-Gaze Correction Experiments

We have implemented our proposed approach and tested with several sets of real data. Very promising results have been obtained. We will first present a set of sample images to further illustrate our algorithm, then we will show some more results from different test users. For each user, we built a personalized face model using a face-modeling tool [46]. This process, which takes only a few minutes and requires no additional hardware, only needs to be done once per user. All the parameters in our algorithm are set to be the same for all the tests.

Figure 15.31 shows the intermediate results at various stages of our algorithm. It starts with a pair of stereo images in Figure 15.31(a); Figure 15.31(b) shows the matched feature points, the epipolar lines of feature points in the first image are drawn in the second image. Figure 15.31(c) shows the extracted foreground contours: the red one (typically a few pixels far away from the "true" contour) is the initial contour after background subtraction while the blue one is the refined contour using the "snake" technique. In Figure 15.31(d), we show the rectified images for template matching. All the matched points form a mesh using Delaunay triangulation, as shown in Figure 15.31(e). The last image (Figure 15.31(f)) shows the synthesized virtual view. We can observe that the person appears to look down and up in the two original images but look forward in this synthesized view.

Figure 15.32 shows some synthesized views from sequence A. Note the large disparity changes between the upper and lower camera images, making direct

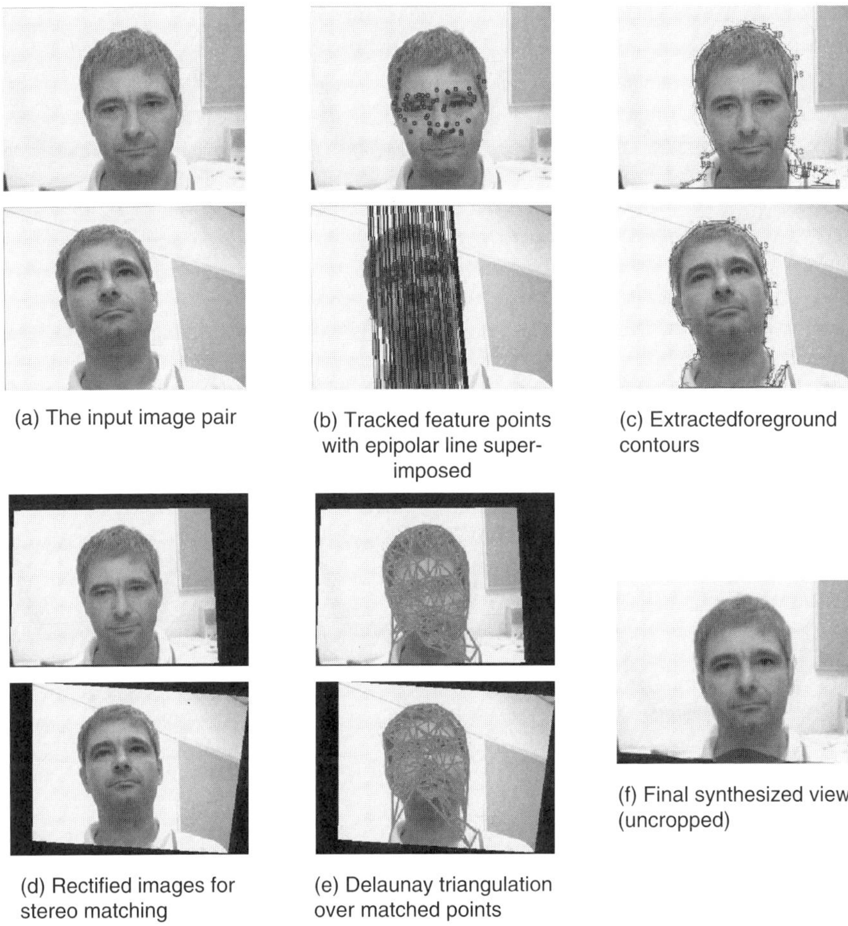

FIGURE 15.31: Intermediate results of our eye gaze correction algorithm. (See also color plate section)

template-based stereo matching very difficult. However, our model-based system is able to accurately track and synthesize photorealistic images under the difficult configuration, even with partial occlusions or oblique viewing angles.

Sequence B is even more challenging, containing not only large head motions, but also dramatic facial expression changes and even hand waving. Results from this sequence, shown in Figure 15.33, demonstrated that our system is both effective and robust under these difficult conditions. Nonrigid facial deformations, as well as the subject's torso and hands, are not in the face model, yet we are still able to generate seamless and convincing views, thanks to our view matching algorithm

FIGURE 15.32: Sample results from sequence A. The top and bottom rows show the images from the top and bottom cameras. The middle row displays the synthesized images from a virtual camera located in the middle of the real cameras. The frame numbers from left to right are 1, 51, 220, 272, and 1010.

that includes a multitude of stereo matching primitives (features, templates, and curves). Template matching finds matching points, as many as possible, in regions that the face model does not cover, while contour matching preserves the important visual cue of silhouettes.

15.9 CONCLUSIONS

We have developed a system to construct textured 3D face models from video sequences with minimal user intervention. With a few simple clicks by the user, our system quickly generates a person's face model which is animated right away. Our experiments show that our system is able to generate face models for people of different races, of different ages, and with different skin colors. Such a system can be potentially used by an ordinary user at home to make their own face models. These face models can be used, for example, as avatars in computer games, online chatting, virtual conferencing, etc.

Besides use of many state-of-the-art computer-vision and computer-graphics techniques, we have developed several innovative techniques including intelligent masking, robust head-pose determination, low-dimensional linear face model fitting, and model-based bundle adjustment. By following the model-based modeling approach, we have been able to develop a robust and efficient modeling system for a very difficult class of objects, namely, faces.

Several researchers in computer vision are working at automatically locating facial features in images [64]. With the advancement of those techniques,

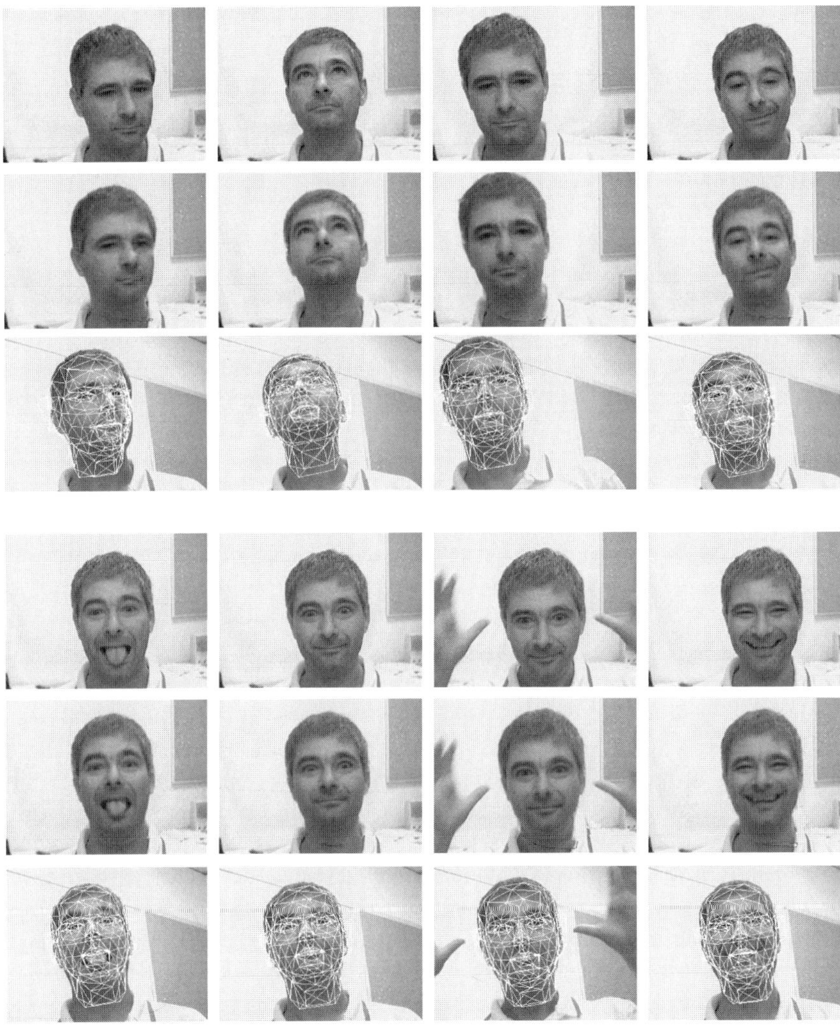

FIGURE 15.33: Sample results from sequence B. The upper and lower rows are the original stereo images, while the middle rows are the synthesized ones. The triangular face model is overlayed on the bottom images. From left to right and from top to bottom, the frame numbers are 159, 200, 400, 577, 617, 720, 743, and 830.

a completely automatic face-modeling system can be expected, even though it is not a burden to click just five points with our current system.

The current face mesh is very sparse. We are investigating techniques to increase the mesh resolution by using higher-resolution face metrics or prototypes. Another possibility is to compute a displacement map for each triangle, using color information.

Additional challenges include automatic generation of eyeballs and eye texture maps, as well as accurate incorporation of hair, teeth, and tongues. For people with hair on the sides or the front of the face, our system will sometimes pick up corner points on the hair and treat them as points on the face. The reconstructed model may be affected by them. Our system treats the points on the hair as normal points on the face.

The animation of the face models is predesigned in our system. In many other applications, it is desirable to understand the facial expression in a real video sequence and use it to drive the facial animation. Some work has been done in that direction [19, 6, 26, 68, 70].

In this chapter, we have also presented a software scheme for maintaining eye contact during videoteleconferencing. We use model-based stereo tracking and stereo analysis to compute a partial 3D description of the scene. Virtual views that preserve eye contact are then synthesized using graphics hardware. In our system, model-based head tracking and stereo analysis work hand-in-hand to provide a new level of accuracy, robustness, and versatility that neither of them alone could provide. Experimental results from real sequences have demonstrated the viability and effectiveness of our proposed approach.

While we believe that our proposed eye-gaze correction scheme represents a large step towards a viable videoteleconferencing system for the mass market, there is still plenty of room for improvements, especially in the stereo view matching stage. We have used several matching techniques and prior domain knowledge to find good matches as many as possible, but we have not exhausted all the possibilities. We believe that the silhouettes in the virtual view could be more clear and consistent across frames if we incorporate temporal information for contour matching. Furthermore, there are still salient curve features, such as hairlines and necklines, that sometimes go unmatched. They are very difficult to match using a correlation-based scheme because of highlights and visibility changes. We are investigating a more advanced curve-matching technique.

ACKNOWLEDGMENT

Springer has granted permission to reuse a substantial part of the contents appeared in [82]. The authors would like to thank D. Adler, M. Cohen, A. Colburn, E. Hanson, C. Jacobs, and Y. Shan for their contributions to the work described in this article. Earlier publications include [45, 44, 46, 65, 76, 75, 82, 77].

REFERENCES

[1] T. Akimoto, Y. Suenaga, and R. S. Wallace. Automatic 3d facial models. *IEEE Computer Graphics and Applications* **13**(5):16–22, September 1993.

[2] A. Azarbayejani, and A. Pentland. Recursive estimation of motion, structure, and focal length. *IEEE Transactions on Pattern Analysis and Machine Intelligence* **17**(6): 562–575, June 1995.

[3] J. Barron, D. Fleet, and S. Beauchemin. Performance of optical flow techniques. *The International Journal of Computer Vision* **12**(1):43–77, 1994.

[4] S. Basu, I. Essa, and A. Pentland. Motion regularization for model-based head tracking. In: *Proceedings of International Conference on Pattern Recognition*, pages 611–616, Vienna, Austria, 1996.

[5] R. Bellman. *Dynamic Programming*. Princeton University Press, Princeton, New Jersey, 1957.

[6] M. Black and Y. Yacoob. Recognizing facial expressions in image sequences using local parametrized models of image motion. *The International Journal of Computer Vision* **25**(1):23–48, 1997.

[7] M. J. Black and Y. Yacoob. Tracking and recognizing rigid and non-rigid facial motions using local parametric model of image motion. In: *Proceedings of International Conference on Computer Vision*, pages 374–381, Cambridge, MA, 1995.

[8] V. Blanz and T. Vetter. A morphable model for the synthesis of 3d faces. In: *Computer Graphics, Annual Conference Series*, pages 187–194. Siggraph, August 1999.

[9] P. Burt and B. Julesz. A gradient limit for binocular fusion. *Science* **208**:615–617, 1980.

[10] T. Cham and M. Jones. Gaze correction for video conferencing. Compaq Cambridge Research Laboratory, http://www.crl.research.digital.com/vision/interfaces/corga.

[11] C. Choi, K. Aizawa, H. Harashima, and T. Takebe. Analysis and synthesis of facial image sequences in model-based image coding. *IEEE Circuits and Systems for Video Technology* **4**(3):257–275, 1994.

[12] B. Dariush, S. B. Kang, and K. Waters. Spatiotemporal analysis of face profiles: detection, segmentation, and registration. In: *Proc. of the 3rd International Conference on Automatic Face and Gesture Recognition*, pages 248–253. IEEE, April 1998.

[13] T. Darrell, B. Moghaddam, and A. Pentland. Active face tracking and pose estimation in an interactive room. In: *In IEEE Computer Vision and Pattern Recognition*, pages 67–72, 1996.

[14] D. DeCarlo and D. Metaxas. Optical flow constraints on deformable models with applications to face tracking. *International Journal of Computer Vision* **38**(2):99–127, July 2001.

[15] D. DeCarlo, D. Metaxas, and M. Stone. An anthropometric face model using variational techniques. In: *Computer Graphics, Annual Conference Series*, pages 67–74. Siggraph, July 1998.

[16] S. DiPaola. Extending the range of facial types. *Journal of Visualization and Computer Animation* **2**(4):129–131, 1991.

[17] D. Douglas and T. Peucker. Algorithms for the reduction of the number of points required to represent a digitized line or its caricature. *Canadian Cartographer* **10**(2):112–122, 1973.

[18] P. Ekman and W. Friesen. *The Facial Action Coding System: a technique for the measurement of Facial Movement*. Consulting Psychologists Press, San Francisco, 1978.

[19] I. Essa and A. Pentland. Coding, analysis, interpretation, and recognition of facial expressions. *IEEE Transactions on Pattern Analysis and Machine Intelligence* **19**(7):757–763, July 1997.

[20] O. Faugeras. *Three-Dimensional Computer Vision: a Geometric Viewpoint*. MIT Press, 1993.

[21] P. Fua. Regularized bundle-adjustment to model heads from image sequences without calibration data. *The International Journal of Computer Vision* **38**(2): 153–171, 2000.

[22] P. Fua and C. Miccio. From regular images to animated heads: a least squares approach. In: *European Conference on Computer Vision*, pages 188–202, 1998.

[23] P. Fua and C. Miccio. Animated heads from ordinary images: A least-squares approach. *Computer Vision and Image Understanding* **75**(3):247–259, 1999.

[24] P. Fua, R. Plaenkers, and D. Thalmann. From synthesis to analysis: Fitting human animation models to image data. In: *Computer Graphics International*, Alberta, Canada, June 1999.

[25] J. Gemmell, C. Zitnick, T. Kang, K. Toyama, and S. Seitz. Gaze-awareness for videoconferencing: a software approach. *IEEE Multimedia* **7**(4):26–35, October 2000.

[26] B. Guenter, C. Grimm, D. Wood, H. Malvar, and F. Pighin. Making faces. In: *Computer Graphics, Annual Conference Series*, pages 55–66. Siggraph, July 1998.

[27] C. Harris and M. Stephens. A combined corner and edge detector. In: *Proc. 4th Alvey Vision Conf.*, pages 189–192, 1988.

[28] R. Hartley and A. Zisserman. *Multiple View Geometry in Computer Vision*. Cambridge University Press, 2000.

[29] B. K. P. Horn and B. G. Schunk. Determining optical flow. *Artificial Intelligence* **17**:185–203, 1981.

[30] T. Horprasert. Computing 3-D head orientation from a monocular image. In: *International Conference of Automatic Face and Gesture Recognition*, pages 242–247, 1996.

[31] H. H. Ip and L. Yin. Constructing a 3d individualized head model from two orthogonal views. *The Visual Computer*, volume **12**, No. 5:254–266, 1996.

[32] M. Jones. Multidimensional morphable models: a framework for representing and matching object classes. *International Journal of Computer Vision* **29**(2):107–131, August 1998.

[33] S. B. Kang and M. Jones. Appearance-based structure from motion using linear classes of 3-d models. Unpublished Manuscript, 1999.

[34] M. Kass, A. Witkin, and D. Terzopoulos. SNAKES: active contour models. *The International Journal of Computer Vision* **1**:321–332, Jan. 1988.

[35] R. Kollarits, C. Woodworth, J. Ribera, and R. Gitlin. An eye-contact camera/display system for videophone applications using a conventional direct-view LCD. *SID Digest*, 1995.

[36] A. Lanitis, C. J. Taylor, and T. F. Cootes. Automatic interpretation and coding of face images using flexible models. *IEEE Transactions on Pattern Analysis and Machine Intelligence* **19**(7):743–756, 1997.

[37] W. Lee and N. Magnenat-Thalmann. Head modeling from photographs and morphing in 3d with image metamorphosis based on triangulation. In: *Proc. CAPTECH'98*, pages 254–267, Geneva, 1998. Springer LNAI and LNCS Press.

[38] Y. C. Lee, D. Terzopoulos, and K. Waters. Constructing physics-based facial models of individuals. In: *Proceedings of Graphics Interface*, pages 1–8, 1993.

[39] Y. C. Lee, D. Terzopoulos, and K. Waters. Realistic modeling for facial animation. In: *Computer Graphics, Annual Conference Series*, pages 55–62. SIGGRAPH, 1995.

[40] J. P. Lewis. Algorithms for solid noise synthesis. In: *Computer Graphics, Annual Conference Series*, pages 263–270. Siggraph, 1989.

[41] H. Li, P. Roivainen, and R. Forchheimer. 3-D motion estimation in model-based facial image coding. *IEEE Pattern Analysis and Machine Intelligence* **15**(6):545–555, June 1993.

[42] J. Liu, I. Beldie, and M. Wopking. A computational approach to establish eye-contact in videocommunication. In: *The International Workshop on Stereoscopic and Three Dimensional Imaging* (*IWS3DI*), pages 229–234, Santorini, Greece, 1995.

[43] Z. Liu, Y. Shan, and Z. Zhang. Expressive expression mapping with ratio images. In: *Computer Graphics, Annual Conference Series*, pages 271–276, Los Angeles, Aug. 2001. ACM SIGGRAPH.

[44] Z. Liu and Z. Zhang. Robust head motion computation by taking advantage of physical properties. In: *Proceedings of the IEEE Workshop on Human Motion* (*HUMO 2000*), pages 73–77, Austin, USA, Dec. 2001.

[45] Z. Liu, Z. Zhang, C. Jacobs, and M. Cohen. Rapid modeling of animated faces from video. In: *Proc. 3rd International Conference on Visual Computing*, pages 58–67, Mexico City, Sept. 2000. Also in the special issue of *The Journal of Visualization and Computer Animation* **12**, 2001. Also available as MSR technical report from http://research.microsoft.com/ zhang/Papers/TR00-11.pdf.

[46] Z. Liu, Z. Zhang, C. Jacobs, and M. Cohen. Rapid modeling of animated faces from video. *The Journal of Visualization and Computer Animation* **12**(4):227–240, 2001.

[47] H. Longuet-Higgins. A computer algorithm for reconstructing a scene from two projections. *Nature* **293**:133–135, 1981.

[48] C. Loop and Z. Zhang. Computing Rectifying Homographies for Stereo Vision. In: *IEEE Conf. Computer Vision and Pattern Recognition*, volume I, pages 125–131, June 1999.

[49] L. Lu, Z. Zhang, H.-Y. Shum, Z. Liu, and H. Chen. Model- and exemplar-based robust head pose tracking under occlusion and varying expression. In: *Proc. IEEE Workshop on Models versus Exemplars in Computer Vision*, pages 58–67, Kauai, Hawaii, Dec. 2001. held in conjunction with CVPR'01.

[50] N. Magneneat-Thalmann, H. Minh, M. Angelis, and D. Thalmann. Design, transformation and animation of human faces. *Visual Computer* **5**:32–39, 1989.

[51] L. Mhlbach, B. Kellner, A. Prussog, and G. Romahn. The importance of eye contact in a videotelephone service. In: *11th International Symposium on Human Factors in Telecommunications*, Cesson Sevigne, France, 1985.

[52] R. Newman, Y. Matsumoto, S. Rougeaux, and A. Zelinsky. Real-time stereo tracking for head pose and gaze estimation. In: *Proceedings of the Fourth IEEE International Conference on Automatic Face and Gesture Recognition* (*FG 2000*), pages 122–128, Grenoble, France, 2000.

[53] M. Ott, J. Lewis, and I. Cox. Teleconferencing eye contact using a virtual camera. In: *INTERCHI' 93*, pages 119–110, 1993.

[54] F. I. Parke. Computer generated animation of faces. In: *ACM National Conference*, November 1972.

[55] F. I. Parke. A parametric model of human faces. PhD thesis, University of Utah, 1974.

[56] F. I. Parke and K. Waters. *Computer Facial Animation*. AKPeters, Wellesley, Massachusetts, 1996.

[57] F. Pighin, J. Hecker, D. Lischinski, R. Szeliski, and D. H. Salesin. Synthesizing realistic facial expressions from photographs. In: *Computer Graphics, Annual Conference Series*, pages 75–84. Siggraph, July 1998.

[58] S. Platt and N. Badler. Animating facial expression. *Computer Graphics* **15**(3): 245–252, 1981.

[59] S. Pollard, J. Porrill, J. Mayhew, and J. Frisby. Disparity Gradient, Lipschitz Continuity, and Computing Binocular Correspondance. In: O. Faugeras and G. Giralt, editors, *Robotics Research: The Third International Symposium*, volume 30, pages 19–26. MIT Press, 1986.

[60] P. Rousseeuw and A. Leroy. *Robust Regression and Outlier Detection*. John Wiley & Sons, New York, 1987.

[61] A. Roy-Chowdhury and R. Chellappa. Stochastic approximation and rate-distortion analysis for robust structure and motion estimation. *The International Journal of Computer Vision* **55**(1):27–53, 2003.

[62] A. Saulnier, M. L. Viaud, and D. Geldreich. Real-time facial analysis and synthesis chain. In: *International Workshop on Automatic Face and Gesture Recognition*, pages 86–91, Zurich, Switzerland, 1995.

[63] S. Seitz and C. Dyer. View Morphing. In: *SIGGRAPH 96 Conference Proceedings*, volume 30 of *Annual Conference Series*, pages 21–30, New Orleans, Louisiana, 1996. ACM SIGGRAPH, Addison Wesley.

[64] T. Shakunaga, K. Ogawa, and S. Oki. Integration of eigentemplate and structure matching for automatic facial feature detection. In: *Proc. of the 3rd International Conference on Automatic Face and Gesture Recognition*, pages 94–99, April 1998.

[65] Y. Shan, Z. Liu, and Z. Zhang. Model-based bundle adjustment with application to face modeling. In: *Proceedings of the 8th International Conference on Computer Vision*, volume II, pages 644–651, Vancouver, Canada, July 2001. IEEE Computer Society Press.

[66] J. Shi and C. Tomasi. Good Features to Track. In *the IEEE Conference on Computer Vision and Pattern Recognition*, pages 593–600, Washington, June 1994.

[67] R. Stokes. Human factors and appearance design considerations of the Mod II PICTUREPHONE station set. *IEEE Trans. on Communication Technology* COM**17**(2), April 1969.

[68] H. Tao and T. Huang. Explanation-based facial motion tracking using a piecewise Bezier volume deformation model. In: *Proceedings of the IEEE Conference on Computer Vision and Pattern Recognition*, volume I, pages 611–617, Colorado, June 1999. IEEE Computer Society.

[69] D. Terzopoulos and K. Waters. Physically based facial modeling, analysis, and animation. In: *Visualization and Computer Animation*, pages 73–80, 1990.

[70] Y.-L. Tian, T. Kanade, and J. Cohn. Recognizing action units for facial expression analysis. *IEEE Transactions on Pattern Analysis and Machine Intelligence* **23**(2): 97–115, Feb. 2001.

[71] J. T. Todd, S. M. Leonard, R. E. Shaw, and J. B. Pittenger. The perception of human growth. *Scientific American* (1242):106–114, 1980.

[72] T. Vetter and T. Poggio. Linear object classes and image synthesis from a single example image. *IEEE Transactions on Pattern Analysis and Machine Intelligence* **19**(7):733–742, 1997.

[73] K. Waters. A muscle model for animating three-dimensional facial expression. *Computer Graphics* **22**(4):17–24, 1987.

[74] J. Yang, R. Stiefelhagen, U. Meier, and A. Waibel. Real-time face and facial feature tracking and applications. In: *Proceedings of AVSP'98*, pages 79–84, Terrigal, Australia, 1998.

[75] R. Yang and Z. Zhang. Eye gaze correction with stereovision for video teleconferencing. In: *Proceedings of the 7th European Conference on Computer Vision*, volume II, pages 479–494, Copenhagen, May 2002. Also available as Technical Report MSR-TR-01-119.

[76] R. Yang and Z. Zhang. Model-based head pose tracking with stereovision. In *Proc. Fifth IEEE International Conference on Automatic Face and Gesture Recognition (FG2002)*, pages 255–260, Washington, DC, May 2002. Also available as Technical Report MSR-TR-01-102.

[77] R. Yang and Z. Zhang. Eye gaze correction with stereovision for videoteleconferencing. *IEEE Transactions on Pattern Analysis and Machine Intelligence* **26**(7):956–960, 2004.

[78] L. Zhang, N. Snavely, B. Curless, and S. Seitz. Spacetime faces: high-resolution capture for modeling and animation. In: *Computer Graphics, Annual Conference Series*, pages 548–558. Siggraph, August 2004.

[79] Z. Zhang. Motion and structure from two perspective views: From essential parameters to Euclidean motion via fundamental matrix. *Journal of the Optical Society of America* **14**(11):2938–2950, 1997.

[80] Z. Zhang. Determining the epipolar geometry and its uncertainty: a review. *The International Journal of Computer Vision* **27**(2):161–195, 1998.

[81] Z. Zhang. A flexible new technique for camera calibration. *IEEE Transactions on Pattern Analysis and Machine Intelligence* **22**(11):1330–1334, 2000.

[82] Z. Zhang, Z. Liu, D. Adler, M. Cohen, E. Hanson, and Y. Shan. Robust and rapid generation of animated faces from video images: a model-based modeling approach. *The International Journal of Computer Vision* **58**(1):93–119, 2004.

[83] Z. Zhang and Y. Shan. A progressive scheme for stereo matching. In: *Springer LNCS 2018: 3D Structure from Images—SMILE 2000*, pages 68–85. Springer-Verlag, 2001.

[84] J. Y. Zheng. Acquiring 3-d models from sequences of contours. *IEEE Transactions on Pattern Analysis and Machine Intelligence* **16**(2):163–178, February 1994.

CHAPTER **16**

A SURVEY OF 3D AND MULTIMODAL 3D+2D FACE RECOGNITION

16.1 INTRODUCTION

Evaluations such as the Face Recognition Vendor Test 2002 [31] make it clear that the current state of the art in face recognition is not yet sufficient for the more demanding biometric applications. However, biometric technologies that currently are believed to offer greater accuracy, such as fingerprint and iris, require much greater explicit cooperation from the user. For example, fingerprinting requires that the person cooperate in making physical contact with the sensor surface. This raises issues of how to keep the surface clean and germ-free in a high-throughput application. Iris imaging currently requires that the person cooperate to carefully position their eye relative to the sensor. This can also cause problems in a high-throughput application. Thus it appears that there is significant potential application-driven demand for improved performance in face-recognition systems.

The vast majority of face-recognition research, and all of the major commercial face-recognition systems, use normal-intensity images of the face. We will refer to these as "2D images." In contrast, a "3D image" of the face is one that represents three-dimensional shape. A distinction can be made between representations that include only the frontal (face) surface of the head and those that include the whole head. We will ignore this distinction here, and refer to the shape of the face surface as 3D. The 3D shape of the face is often sensed in combination with a 2D intensity image. In this case, the 2D image can be thought of as a "texture map" overlaid on the 3D shape. An example of a 2D intensity image and the corresponding 3D shape is shown in Figure 16.1, with the 3D shape rendered both in the form of a range

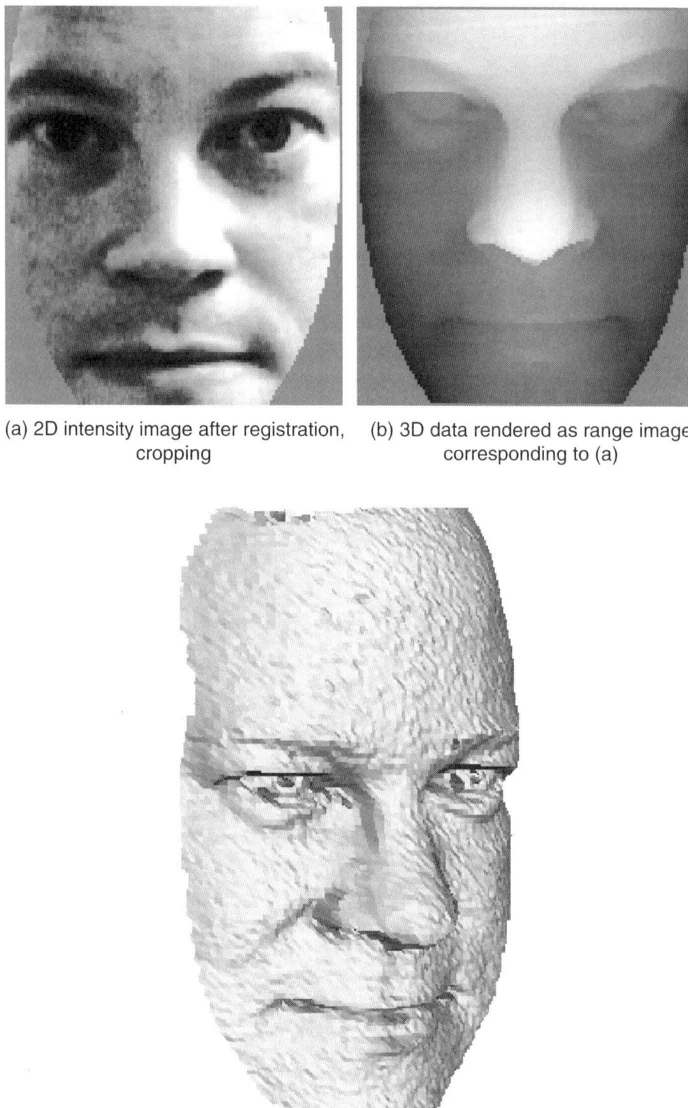

(a) 2D intensity image after registration, cropping

(b) 3D data rendered as range image corresponding to (a)

(c) 3D data in (b), rendered as shaded model from "3/4 view"

FIGURE 16.1: Example of 2D Intensity Image and 3D Shape Data. Parts (a) and (b) depict the data in the forms used by the 2D and 3D PCA-based recognition algorithms, respectively. Part (c) depicts the data in the form used by 3D ICP-based algortihms.

image and in the form of a shaded 3D model. A range image, a shaded model, and a wire-frame mesh are common alternatives for rendering 3D face data.

A recent survey of face-recognition research is given in [43], but it does not include algorithms based on matching 3D shape. Here, we focus specifically on face-recognition algorithms that match the 3D shape of the face to be recognized against the enrolled 3D shape of the face(s) of the known person(s). That is, we are interested in systems that perform person identification or authentication by matching two 3D face descriptions. We use "identification" here to refer to one-to-many matching to find the best match above some threshold, and "authentication" to refer to one-to-one matching used to verify or reject a claimed identity. The term "recognition" is also sometimes used for identification, and the term "verification" is also sometimes used for authentication. A particular research group may present their results in the context of either identification or authentication, but the core 3D representation and matching issues are essentially the same. We do *not* consider here the family of approaches in which a generic, "morphable" 3D face model is used as an intermediate step in matching two 2D images for face recognition [7]. As commonly used, the term *multimodal biometrics* refers to the use of multiple imaging modalities. Strictly speaking, this term may not be applicable when both the 3D and the 2D are acquired using the same sensor. However, we use the term here to refer generally to biometric algorithms that use 3D and 2D images of the face.

We are particularly interested in 3D face recognition because it has been claimed that the use of 3D data has the potential for greater recognition accuracy than the use of 2D face images. For example, one paper states: "Because we are working in 3D, we overcome limitations due to viewpoint and lighting variations" [23]. Another paper describing a different approach to 3D face recognition states: "Range images have the advantage of capturing shape variation irrespective of illumination variabilities" [17]. Similarly, a third paper states: "Depth and curvature features have several advantages over more traditional intensity based features. Specifically, curvature descriptors (1) have the potential for higher accuracy in describing surface based events, (2) are better suited to describe properties of the face in areas such as the cheeks, forehead, and chin, and (3) are viewpoint invariant" [16]. However, on the other hand, there may be useful information in the 2D that is not in the 3D shape, such as skin color, freckles, and other such features. We return to this issue of 3D versus 2D again later in the chapter.

The next section, surveying work on 3D face recognition, is a revised version of material that appeared in an earlier paper presented at the 2004 *International Conference on Pattern Recognition* [9]. Section 3, presenting more detailed results of "eigenface" style 3D and multimodal 3D+2D face recognition, is a revised version of material that appeared in an earlier paper presented at the 2003 *Workshop on Multimodal User Authentication* [15]. We conclude with a section that considers some of the challenges to improved 3D face recognition.

16.2 SURVEY OF 3D AND MULTIMODAL 2D+3D FACE RECOGNITION

Although early work on 3D face recognition was done over a decade ago [11, 16, 20], the number of published papers on 3D and multimodal 2D+3D face recognition is currently still relatively small. Often a research group has published multiple papers as they develop a line of work. In such cases, we discuss only the most recent and/or easily accessible publication from that line. Some important relevant properties of the published algorithms and results in 3D face recognition are summarized in Tables 16.1 and 16.2.

Cartoux et al. [11] approach 3D face recognition by segmenting a range image based on principal curvature and finding a plane of bilateral symmetry through the face. This plane is used to normalize for pose. They consider methods of matching the profile from the plane of symmetry and of matching the face surface, and report 100% recognition for either in a small dataset.

Lee and Milios [20] segment convex regions in the range image based on the sign of the mean and Gaussian curvatures, and create an extended Gaussian image (EGI) for each convex region. A match between a region in a probe image and in a gallery image is done by correlating EGIs. A graph-matching algorithm incorporating relational constraints is used to establish an overall match of probe image to gallery image. Convex regions are believed to change shape less than other regions in response to changes in facial expression. This approach gives some ability to cope with changes in facial expression. However, EGIs are not sensitive to change in object size, taking away one element of potential difference in face shape.

Gordon [16] begins with a curvature-based segmentation of the face. Then a set of features are extracted that describe both curvature and metric size properties of the face. Thus each face becomes a point in a feature space, and matching is done by a nearest-neighbor match in that feature space. Experiments are reported with a test set of three views of each of eight faces, and recognition rates as high as 100% are reported. It is noted that the values of the features used are generally similar for different images of the same face, "except for the cases with large feature-detection error, or variation due to expression" [16].

Nagamine et al. [27] approach 3D face recognition by finding five feature points, using those feature points to standardize face pose, and then matching various curves or profiles through the face data. Experiments are performed for sixteen persons, with ten images per person. The best recognition rates are found using vertical profile curves that pass through the central portion of the face. Computational requirements were apparently regarded as severe at the time this work was performed, as the authors note that "using the whole facial data may not be feasible considering the large computation and hardware capacity needed" [27].

Achermann et al. [4] extend eigenface and hidden-Markov-model approaches used for 2D face recognition to work with range images. They present results for

Table 16.1: Face-recognition algorithms using only 3D data

Author, year, reference	Persons in dataset	Images in dataset	Image size	3D face data	Reported performance	Handles expression variation?
Cartoux 1989 [11]	5	18	?	profile, surface	100%	no
Lee 1990 [20]	6	6	256×150	EGI	none	some
Gordon 1992 [16]	26 train 8 test	26 train 24 test	?	feature vector	100%	no
Nagamine 1992 [27]	16	160	256×240	multiple profiles	100%	no
Achermann 1997 [4]	24	240	75×150	range image	100%	no
Tanaka 1998 [33]	37	37	256×256	EGI	100%	no
Achermann 2000 [3]	24	240	75×150	point set	100%	no
Hesher 2003 [17]	37	222 (6 expr. ea.)	242×347	range image	97%	no
Medioni 2003 [23]	100	700 (7 poses ea.)	?	surface mesh	98%	no
Moreno 2003 [26]	60	420 (3 expr., 2 poses)	avg 2,200 point mesh	feature vector	78%	some
Lee 2003 [21]	35	70	320×320	feature vector	94% at rank 5	no
Xu 2004 [40]	120 (30)	720	?	mesh and feature vector	96% on 30 72% on 120	no
Russ 2004 [32]	200 (30)	468 (60)	640×480	range image	98% V on 200 50% R on 30	no
Lu 2004 [22]	18	113 (varied pose, expression)	240×320	mesh	96% R	no
Chang 2005 [16]	355	3,205 (varied expression)	480×640	mesh	96% same, 77% varying expression	yes

Table 16.2: Multimodal face-recognition algorithms combining use of 3D and 2D data

Author, year, reference	Persons in dataset	Images in dataset	Image size	3D face data	Reported performance	Handles expression variation?
Lao 2000 [19]	10	360	640×480	surface mesh	91%	no
Beumier 2001 [5]	27 gallery 29 probes	240 2D	?	multiple profiles	1.4% EER	no
Wang 2002 [37]	50	300	128×512	feature vector	>90%	yes
Bronstein 2003 [10]	157	?	2250 avg vertices	range image	not reported	yes
Tsalakanidou 2003 [35]	40	80	100×80	range image	99% 3D+color 93% 3D only	no
Chang 2003 [15]	200 (275 for train)	951	640×480	range image	99% 3D+2D 93% 3D only	no
Papatheodorou 2004 [28]	62	806 (11 expr., 2 pose each)	10,000 pt. mesh	point cloud	100% to 66%	no
Tsalakanidou 2004 [34]	50	3,000 (varied time, pose, expr.)	571×752 reduced to 140×200	range image	5% to 8% EER	no

a dataset of 24 persons, with 10 images per person, and report 100% recognition using an adaptation of the 2D face-recognition algorithms.

Tanaka et al. [33] also perform curvature-based segmentation and represent the face using an extended Gaussian image (EGI). Recognition is then performed using a spherical correlation of the EGIs. Experiments are reported with a set of 37 images from a National Research Council of Canada range image dataset, and 100% recognition is reported.

Achermann and Bunke [3] report on a method of 3D face recognition that uses an extension of the Hausdorff distance matching. They report on experiments using 240 range images, 10 images of each of 24 persons, and achieve 100% recognition for some instances of the algorithm.

Hesher et al. [17] explore approaches in the style of principal-component analysis (PCA), using different numbers of eigenvectors and image sizes. The image data set used has 6 different facial expressions for each of 37 persons. The performance figures reported result from using multiple images per person in the gallery. This effectively gives the probe image more chances to make a correct match, and such an approach is known to raise the recognition rate relative to having a single sample per person in the gallery [25].

Medioni and Waupotitsch [23] perform 3D face recognition using iterative-closest-point (ICP) matching of face surfaces. Whereas most of the works covered here used 3D shape acquired through a structured-light sensor, this work uses a stereo-based system. Experiments with seven images each from a set of 100 persons are reported, and an equal error rate of "better than 2%" is reported.

Moreno and co-workers [26] approach 3D face recognition by first performing a segmentation based on Gaussian curvature and then creating a feature vector based on the segmented regions. They report results on a dataset of 420 face meshes representing 60 different persons, with some sampling of different expressions and poses for each person. They report 78% rank-one recognition on the subset of frontal views, and 93% overall rank-five recognition.

Lee and coworkers perform 3D face recognition by locating the nose tip, and then forming a feature vector based on contours along the face at a sequence of depth values [21]. They report 94% correct recognition at rank five, and do not report rank-one recognition.

Russ and coworkers [32] present results of an approach that uses Hausdorff matching on the 2D range images. They use portions of the Notre Dame dataset used in [15] in their experiments. In a verification experiment, 200 persons were enrolled in the gallery, and the same 200 persons plus another 68 imposters were represented in the probe set. A probability of correct verification as high as 98% (of the 200) was achieved at a false alarm rate of 0 (of the 68). In a recognition experiment, 30 persons were enrolled in the gallery and the same 30 persons imaged at a later time were represented in the probe set. A 50% probability of recognition was achieved at a false alarm rate of 0. The recognition experiment uses a subset of the available data "because of the computational cost of the current algorithm" [32].

Xu and coworkers [40] developed a method for 3D face recognition and evaluate it using the database from [5]. The original 3D point cloud is converted to a regular mesh. The nose region is found, and used as an anchor to find other local regions. A feature vector is computed from the data in the local regions of mouth, nose, left eye and right eye. Dimensionality is reduced using principal-components analysis, and matching is based on minimum distance using both global and local shape components. Experimental results are reported for the full 120 persons in the dataset and for a subset of 30 persons, with performance of 72% and 96%, respectively. This illustrates the general point that reported experimental performance can be highly dependent on the dataset size. Most other works have not considered performance variation with dataset size. It should be mentioned that the reported performance was obtained with five images of a person used for enrollment in the gallery. Performance would be expected to be lower with only one image used to enroll a person.

Lao et al. [19] perform 3D face recognition using a sparse depth map constructed from stereo images. Isoluminance contours are used for the stereo matching. Both 2D edges and isoluminance contours are used in finding the irises. In this specific limited sense, this approach is multimodal. However, there is no independent recognition result from 2D face recognition. Using the iris locations, other feature points are found so that pose standardization can be done. Recognition rates of 87% to 96% are reported using a dataset of ten persons, with four images taken at each of nine poses for each person.

Chang et al. [15] compare a PCA-based approach and an ICP-based approach to 3D face recognition, looking in particular at the problem of handling facial expression variation between the enrollment image and the probe image. They experiment with a large dataset, representing over 3000 total 3D scans from over 350 different persons. Individual persons may have multiple scans, taken at different times, with varying facial expressions. They report finding better performance with the ICP-based approach than with the PCA-based approach. They also report finding that a refinement of the ICP-based approach, to focus on only a part of the frontal face, improves performance in the case of varying facial expression.

Beumier and Acheroy [5] approach multimodal recognition by using a weighted sum of 3D and 2D similarity measures. They use a central profile and a lateral profile, each in both 3D and 2D. Therefore, they have a total of four classifiers, and an overall decision is made using a weighted sum of the similarity metrics. Results are reported for experiments using a 27-person gallery set and a 29-person probe set. An equal error rate (EER) as low as 1.4% is reported for multimodal 3D+2D recognition that merges multiple probe images per person. In general, multimodal 3D+2D is found to perform better than either 3D or 2D alone.

Wang et al. [37] use Gabor filter responses in 2D and "point signatures" in 3D to perform multimodal face recognition. The 2D and 3D features together form a feature vector. Classification is done by support-vector machines with a decision directed acyclic graph. Experiments are performed with images from 50 persons, with six images per person, and with pose and expression variations. Recognition rates exceeding 90% are reported.

Bronstein et al. use an isometric-transformation approach to 3D face analysis in an attempt to better cope with variation due to facial expression [10]. One method they propose is effectively multimodal 2D+3D recognition using eigendecomposition of flattened textures and canonical images. They show examples of correct and incorrect recognition by different algorithms, but do not report any overall quantitative performance results for any algorithm.

Tsalakanidou et al. [35] report on multimodal face recognition using 3D and color images. The use of color rather than simply gray-scale intensity appears to be unique among the multimodal work surveyed here. Results of experiments using images of 40 persons from the XM2VTS dataset [24] are reported for color images alone, 3D alone, and 3D + color. The recognition algorithm is PCA style matching, plus a combination of the PCA results for the individual color planes and range image. Recognition rates as high as 99% are achieved for the multimodal algorithm, and multimodal performance is found to be higher than for either 3D or color alone.

Chang et al. [15] report on PCA-based recognition experiments performed using 3D and 2D images from 200 persons. One experiment uses a single set of later images for each person as the probes. Another experiment uses a larger set of 676 probes taken in multiple acquisitions over a longer elapsed time. Results in both experiments are approximately 99% rank-one recognition for multimodal 3D+2D, 94% for 3D alone and 89% for 2D alone. The multimodal result was obtained using a weighted sum of the distances from the individual 3D and 2D face spaces. These results represent the largest experimental study reported to that time, either for 3D face alone or for multimodal 2D+3D, in terms of the number of persons, the number of gallery and probe images, and the time lapse between gallery and probe image acquisition. This work is presented in more detail in the next section.

Papatheodorou and Rueckert [28] perform multimodal 3D+2D face recognition using a generalization of ICP based on point distances in a 4D space (x, y, z, intensity). This approach integrates shape and texture information at an early stage, rather than making a decision using each mode independently and combining decisions. They present results from experiments with 68 persons in the gallery, and probe sets of varying pose and facial expression from the images in the gallery. They report 98% to 100% correct recognition in matching frontal, neutral-expression probes to frontal, neutral-expression gallery images. Recognition drops when the expression and pose of the probe images is not matched to those of the gallery images, for example to 73% to 94% for 45-degree off-angle probes, and to 69% to 89% for smiling expression probes.

Tsalakanidou and a different set of coworkers [34] report on an approach to multimodal face recognition based on an embedded hidden Markov model for each modality. Their experimental data set represents a small number of different persons, but each has 12 images acquired in each of 5 different sessions. The 12 images represent varied pose and facial expression. Interestingly, they report a higher EER for 3D than for 2D in matching frontal neutral-expression probes to frontal neutral-expression gallery images, 19% versus 5%, respectively. They report that "depth data mainly suffers from pose variations and use of eyeglasses" [34]. This work is also unusual in that it is based on using five images to enroll a person in the gallery, and also generates additional synthetic images from those, so that a person is represented by a total of 25 gallery images.

A number of common elements in this survey of past work are representative of problems that still confront this area of research. Research in 3D face recognition is still concerned with problems such as: (1) the computational cost of many of the recognition algorithms, (2) developing the ability to handle variation in facial expression, and (3) the quality of the 3D data and the sensor-dependent noise properties. We can also see several themes or issues represented here that will likely have substantial influence on future work. One is that the combination of results from multiple image modalities, such as 3D+2D, generally improves accuracy over using a single modality. Similarly, the combination of results from several algorithms on the same data sample, or the same algorithm across several data samples, generally improves accuracy over using a single algorithm on a single sample. Another theme represented here is that reported recognition performance has generally been near-perfect but has been based on relatively small datasets, and that reported performance can decrease drastically with increased size of dataset. For example, one study reviewed above reported performance dropping from 96% to 72% when the dataset size grew from 30 persons to 120 persons [40]. Thus it appears problematic to make an direct comparison of performance figures reported based on different image datasets.

16.3 EXAMPLE 3D AND MULTIMODAL 3D+2D FACE RECOGNITION

The experimental results summarized in this section address several points. One point is to test the hypothesis that 3D face data provides better recognition performance than 2D face data, given that essentially the same PCA-based method is used with both. While the results are not conclusive, we find some evidence that recognition using a single 3D face image is more powerful than that using a single 2D face image. However, more work is needed on this issue, especially comparisons that may use different algorithms for 2D and 3D data. Another point is to test the hypothesis that a combination of 2D and 3D face data may provide better performance than either one individually. We find that the multimodal result is statistically significantly better than the result using either alone. However, there is a methodological point to be considered in the fact that the multimodal result is based on using two images to represent a person and either unimodal result is based on using one image to represent a person.

16.3.1 Methods and Materials

Extensive work has been done on face-recognition algorithms based on PCA, popularly known as "eigenfaces" [36]. A standard implementation of the PCA-based algorithm [6] is used in the experiments reported here. The first step in preparing the face image data for use in the PCA based approach is normalization. The main objective of the normalization process is to minimize the uncontrolled variations that occur during image acquisition while at the same time maintaining the variations observed in facial feature differences between individuals. The normalized images are masked to omit the background and leave only the face region, as illustrated in Figure 16.1(a,b). While each person is asked to gaze at the camera during the acquisition, it is inevitable to obtain data with some level of pose variations between acquisition sessions.

The 2D image data is typically treated as having pose variation only around the Z axis: the optical axis. The PCA software [6] uses two landmark points (the eye locations) for geometric normalization to correct for rotation, scale, and position of the face for 2D matching. However, the face is a 3D object, and if 3D data is acquired, there is the opportunity to correct for pose variation around the X, Y, and Z axes.

A transformation matrix is first computed based on the surface normal angle difference in X (roll) and Y (pitch) between manually selected landmark points (two eye tips and center of lower chin) and predefined reference points of a standard face pose and location. Pose variation around the Z axis (yaw) is corrected by measuring the angle difference between the line across the two eye points and a horizontal line. At the end of the pose normalization, the nose tip of every person is transformed to the same point in 3D relative to the sensor. The geometric normalization in 2D gives the same pixel distance between eye locations to all faces. This is necessary because the absolute scale of the face is unknown in 2D. However, this is not the case with a 3D face image, and so the eye locations may naturally be at different pixel locations in depth images of different faces. Thus, the geometric scaling was not imposed to 3D data points as it was in 2D.

When the 3D and 2D images are sensed at the same time by the sensor, and the two images are automatically registered, it is possible to more fully pose-normalize those 2D data than is usually the case. However, we found that missing-data problems with fully pose-corrected 2D images outweighed the gains from the additional pose correction [12]. Thus we use the typical Z-rotation-corrected 2D images.

Problems with the 3D are alleviated to some degree by preprocessing the 3D data to fill in "holes" and remove "spikes." This is done by median filtering followed by linear interpolation using valid data points around a hole. Noise in the 3D data is discussed in more detail later.

16.3.2 Data Collection

Images were acquired at the University of Notre Dame between January and May 2003. Two four-week runs of weekly sessions were conducted, separated by six weeks. The first session is to collect gallery images, and the second session is to collect probe images, with a single-probe study in mind. A gallery image is an image that is enrolled into the system to be identified. A probe image is a test image to be matched against the gallery images. For a study with multiple probes, an image acquired in the first of the eight total acquisition weeks is used as a gallery, and images acquired in later weeks are used as probes. Thus, in the single-probe study, there is a time lapse on at least six and as many as thirteen weeks between the acquisition of gallery image and its probe image, and at least one and as many as thirteen weeks lapse between the gallery and the probe in the multiple probe study. All persons completed a consent form approved by the Internal Review Board at the University of Notre Dame prior to participating in each data acquisition session, making it possible to share this image data with other research groups.[1]

A total of 275 different subjects participated in one or more data acquisition sessions. Among 275 subjects, 200 participated in both a gallery acquisition and a probe acquisition. Thus, there are 200 individuals in the single-probe set, with the same 200 individuals in the gallery, and these are a subset of the 275 individuals in the training set. The training set contains the 200 gallery images plus an additional 75 for subjects whom good data were not acquired in both the gallery and probe sessions. And for the multiple-probe study, 476 new probes are added to the 200 probes, yielding 676 probes in total. The training set of 275 subjects is the same as the set used in the single-probe study.

In each acquisition session, subjects were imaged using a Minolta Vivid 900 range scanner. Subjects stood approximately 1.5 m from the camera, against a plain gray background, with one front-above-center spotlight lighting their face, and were asked to have a normal facial expression ("F_A" in FERET terminology [30]) and to look directly at the camera. Almost all images were taken using the Minolta's "Medium" lens and a small number of images were taken with its "Tele" lens. The height of the Minolta Vivid scanner was adjusted to the approximate height of the subject's face, if needed. The Minolta Vivid 900 uses a projected light stripe to acquire triangulation-based range data. It also captures a color image almost but not quite simultaneously with the range-data capture. As we used the system, the data acquisition results in a 640-by-480 sampling of range data and a registered 640-by-480 color image. The system can also be used in a lower-resolution mode.

16.3.3 Data Fusion

The pixel level provides perhaps the simplest approach to combining the information from multiple-image-based biometrics. The images can simply be concatenated together to form one larger aggregate 2D-plus-3D face image [13]. Score-level fusion combines the match scores that are found in the individual

[1] See http://www.nd.edu/%7Ecvrl/UNDBiometricsDatabase.html for information on the availability of the data.

spaces. Having distance metrics from two or more different spaces, a rule for combination of the distances across the different biometrics for each person in the gallery can be applied. The ranks can then be determined based on the combined distances. In this work, we use score-level fusion.

In score-level fusion, the scores from the different modes generally must be normalized first in order to be comparable. There are several ways of transforming the scores including linear, logarithmic, exponential, and logistic [2]. In the experimental results reported here, the Mahalanobis cosine distance metric was used in each face space [41]. This means that matching scores are automatically distributed on the interval $[-1, 1]$. However, the scores in the different spaces do not necessarily use the same amount of the possible range. Therefore the scores from different modalities are normalized so that the distribution and the range are mapped to the same unit interval.

There are many ways of combining different metrics to achieve the best fused decision, including majority vote, sum rule, multiplication rule, median rule, min rule, average rule, and so on. Depending on the task, a certain combination rule might be better than others. It is known that the sum rule and multiplication rule generally provide plausible results [2, 12, 18, 38].

In our study, a weight is estimated based on the distribution of the top three ranks in each space. The motivation is that a larger distance between first- and second-ranked matches implies greater certainty that the first-ranked match is correct. The level of the certainty can be considered as a weight representing the certainty. The weight can be applied to each metric as the combination rules are applied. The multimodal decision is made as follows. First the 2D probe is matched against the 2D gallery, and the 3D probe against the 3D gallery. This gives a set of N distances in the 2D face space and another set of N distances in the 3D face space, where N is the size of the gallery. A plain sum-of-distances rule would sum the 2D and 3D distances for each gallery subject and select the gallery subject with the smallest sum. We use a confidence-weighted variation of the sum-of-distances rule. For each of 2D and 3D, a "confidence" is computed using the three distances in the top ranks: the confidence weight is the ratio of (a) the second distance minus the first distance to (b) the third distance minus the first distance. If the difference between the first and second match is large compared to the typical distance, then this confidence value will be large. The confidence values are used as weights in distance metric. A simple product-of-distances rule produced similar combination results, and a min-distance rule produced slightly worse combination results.

16.3.4 Experiments

There can be many ways of selecting eigenvectors to accomplish the face-space creation. In this study, at first, one vector is dropped at a time from the eigenvectors of largest eigenvalues, and the rank-one recognition rate is computed using the gallery and probe set again each time, and this continues until a point is reached where the rank-one recognition rate gets worse rather than better. We denote the number of dropped eigenvectors of largest eigenvalues as M. Also, one vector at a time is dropped from the eigenvectors of the smallest eigenvalues, and the rank-one recognition is computed using the gallery and probe set again each time, continuing until a point is reached where the rank-one recognition rate gets worse rather than better. We also denote the number of dropped eigenvectors of smallest eigenvalues as N.

During the eigenvector tuning process, the rank-one recognition rate remains basically constant with from one to 20 eigenvectors dropped from the end of the list. This probably means that more eigenvectors can be dropped from the end to create a lower-dimensional face space. This would make the overall process simpler and faster. The rank-one recognition rate for dropping some of the first eigenvectors tends to improve at the beginning but then starts to decline as M gets larger. After the eigenvectors are tuned, both the 2D and the 3D face spaces are tuned with $M = 3$, and $N = 0$.

The first experiment investigates the performance of individual 2D eigenface and 3D eigenface methods, given (1) the use of the same PCA-based algorithm implementation, (2) the same subject pool represented in training, gallery, and probe sets, and (3) the controlled variation in one parameter—time of image acquisition—between the gallery and probe images. A similar comparison experiment between 2D and 3D acquired using a stereo-based imaging system was performed by Medioni et al. [23].

The cumulative match characteristic (CMC) curves for the first experiment are shown in Figures 16.2 and 16.3. In these results, the rank-one recognition rate is 89.0% for PCA-based face recognition using 2D images, and 94.5% for PCA-based face recognition using 3D data. The data set here is 200 people enrolled in the gallery, and either a single time-lapse probe for each person (Figure 16.2) or multiple time-lapse probes per person (Figure 16.3). Overall, we find that 3D face recognition does perform better than 2D face recognition, but that the difference is near the border of statistical significance.

The next step is to investigate the value of a multimodal biometric using 2D and 3D face images, compared against the individual biometrics. The null hypothesis for this experiment is that there is no significant difference in the performance rate between unibiometrics (2D or 3D alone) and multibiometrics (both 2D and 3D together). Figures 16.2 and 16.3 again show the CMC comparison. The rank-one recognition rate for the multimodal biometric in the single-probe experiment is 98.5%, achieved by combining modalities at the distance metric level. A McNemar's test for significance of the difference in accuracy in the rank-one match between the multimodal biometric and either the 2D face or the 3D face alone shows that multimodal performance is significantly greater, at the 0.05 level.

The multiple-probe dataset consists of 676 probes in total, with subjects having a varying number of time-lapse probes. There are 200 subjects with 1 or more probes,

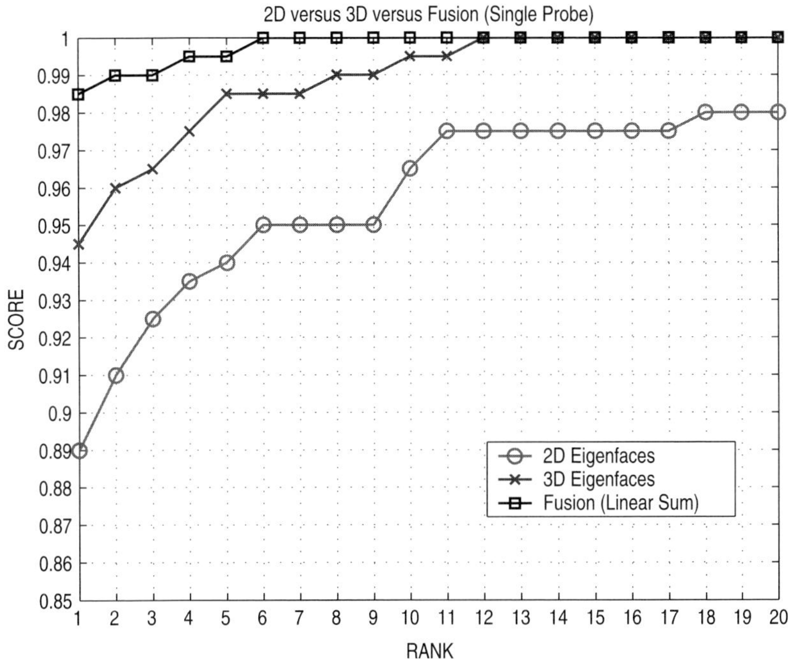

FIGURE 16.2: Performance results in the single-probe study.

166 subjects with 2 or more probes, and so on. The number of probes can be up to 7 per subject. A correct match is measured based on an each individual probe rather than on some function of all probes per subject. The results for the multiple-probe study are shown in Figure 16.3. Again, after combining modalities, we obtain significantly better performance, at 98.8%, than for either 2D or 3D alone. The results of 2D and 3D combination show performance behavior similar to the single-probe study. A McNemar's test for significance of the difference in accuracy in the rank-one match between the multimodal biometric and either the 2D face or the 3D face alone shows that multimodal performance is significantly greater, at the 0.05 level. Thus, significant performance improvement has been accomplished by combining 2D and 3D facial data in both single-probe and multiple-probe studies.

In the fusion methods that we considered, the multiplication rule showed the most consistent performance, regardless of the particular score transformation. The min rule showed lower performance than any other rule in different score transformations. Also, when the distance metrics were weighted based on the confidence level during the decision process, all the rules result in significantly better performance than the individual biometric.

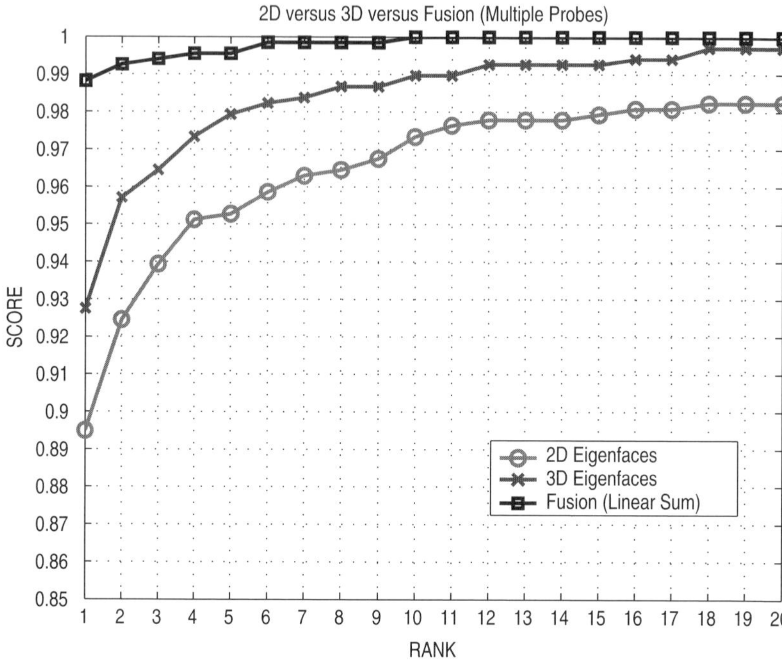

FIGURE 16.3: Performance results in the multiple-probe study.

16.4 CHALLENGES TO IMPROVED 3D FACE RECOGNITION

We can identify three major areas in which advances are required in order for 3D face recognition to become practical for wide application. One area is 3D sensor technology. While the optimism about 3D face data relative to 2D face images may eventually be proved correct, there are still significant limitations in current 3D sensor technology. A second area is improved algorithms. For example, most current algorithms for 3D face recognition do not handle variation in facial expression well. Additionally, current algorithms for multimodal 3D+2D recognition are multimodal only in a weak sense. A third area is experimental methodology. Most published results to date are not based on a large and challenging dataset, do not report statistical significance of observed differences in performance, and make a biased comparison between multimodal results and the baseline results from a single modality.

16.4.1 Improved 3D Sensors

Successful practical application of 3D face recognition would be aided by various improvements in 3D sensor technology. Among these are: (1) reduced frequency and severity of artifacts, (2) increased depth of field, (3) increased spatial and depth resolution, and (4) reduced acquisition time.

It is important to point out that, while 3D shape is *defined* independent of illumination, it is not *sensed* independent of illumination. Illumination conditions *do* affect the quality of sensed 3D data. Even under ideal illumination conditions for a given sensor, it is common for artifacts to occur in face regions such as oily areas that appear specular, the eyes, and regions of facial hair such as eyebrows, mustache, or beard. The most common types of artifact can generally be described subjectively as "holes" or "spikes." A "hole" is essentially an area of missing data, resulting from the sensor being unable to acquire data. A "spike" is an outlier error in the data, resulting from, for example, an interreflection in a projected light pattern or a correspondence error in stereo. An example of "holes" in a 3D face image sensed with the Minolta scanner is shown in Figure 16.4. Artifacts are typically patched up by interpolating new values based on the valid data nearest the artifact.

Another limitation of current 3D sensor technology, especially relative to use with noncooperative subjects, is the depth of field for sensing data. The depth of field for acquiring usable data might range from about 0.3 meter or less for a stereo-based system to about one meter for a structured light system such as the Minolta Vivid 900 [1]. Larger depth of field would lead to more flexible use in application.

There is some evidence suggesting that 3D face-recognition algorithms might benefit from 3D depth resolution accuracy below 1 mm [15]. Many or most 3D sensors do not have this accuracy in depth resolution. Also, in our experience,

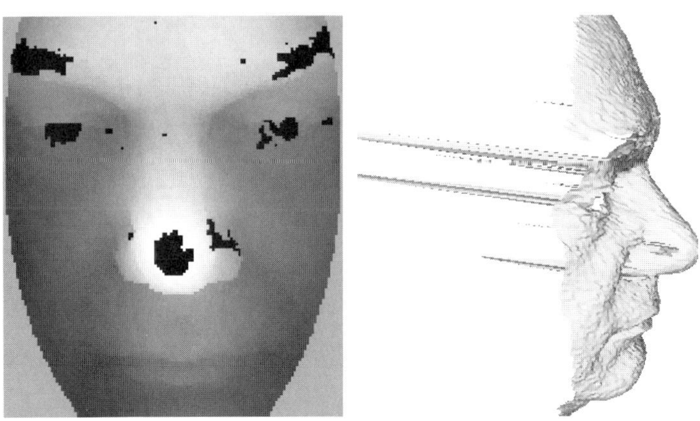

(a) Range image rendering of 3D data with "holes" (in black) apparent

(b) Shaded model rendering of 3D data with "spikes" apparent

FIGURE 16.4: Examples of "hole" and "spike" artifacts in sensed 3D shape.

depth accuracy estimates are not made in any standard way across manufacturers, and so it is not practical to evaluate this other than by experiments in the same image-acquisition setting. Boehnen and Flynn [8] compared five 3D sensors in a face-scanning context. We are not aware of other such sensor evaluations in the literature.

Lastly, the image-acquisition time for the 3D sensor should be short enough that subject motion is not a significant issue. Acquisition time is generally a more significant problem with structured-light systems than with stereo systems. It may be less of an issue for authentication-type applications in which the subjects can be assumed to be cooperative, than it is for recognition-type applications.

Considering all of these factors related to 3D sensor technology, it seems that the optimism sometimes expressed for 3D face recognition relative to 2D face recognition is still somewhat premature. The general pattern of results in the multimodal 3D+2D studies surveyed here suggests that 3D face recognition holds the potential for greater accuracy than 2D. And existing 3D sensors are certainly capable of supporting advanced research in this area. But substantial improvements along the lines mentioned are needed to improve chances for successful use in broad application. Also, those studies that suggest that 3D allows greater accuracy than 2D also suggest that multimodal recognition allows greater accuracy than either modality alone. Thus the appropriate issue may not be 3D versus 2D, but instead be the best method to combine 3D and 2D.

An additional point worth making concerns the variety of types of 3D scanners. A system such as the Minolta Vivid 900/910 takes a single "frontal view" in one image acquisition. In principle, multiple acquisitions can be taken from different viewpoints and then stitched together to get a more complete 3D face model. However, if there is the possibility of facial-expression change between acquisitions, then it may be problematic to create a single model from the multiple views. Other types of scanner may create a more "ear-to-ear" model. For example, the 3Q "Qlonerator" system uses two stereo rigs (see Figure 16.5) with a projected light pattern to create a more complete model of the face with a single acquisition. Examples of data from a "frontal-view" acquisition and from an "ear-to-ear" acquisition are shown in Figure 16.6.

Another point worth making concerns the subjective visual evaluation of 3D shape models. Evaluation of 3D shape should only be done when the color texture is *not* displayed. When a 3D model is viewed with the texture map on, the texture map can hide significant depth-measurement errors in the 3D shape. This is illustrated by the pair of images shown in Figure 16.7. Both images represent the same 3D shape model, but in one case it is rendered with the texture map on and in the other case is rendered as a shaded view from the pure shape. The shape model clearly has major artifacts that are related to the lighting highlights in the image.

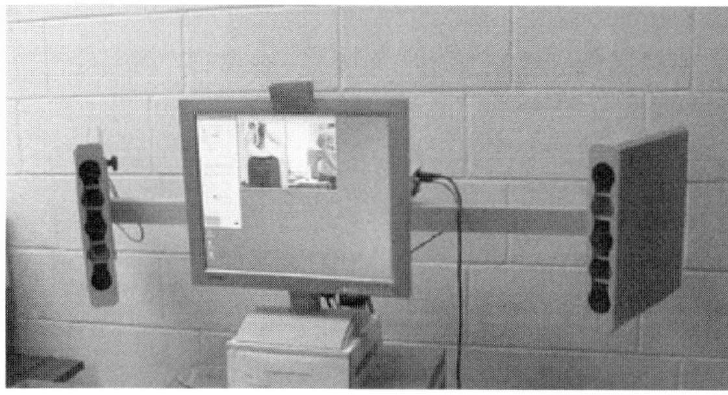

(a) 3Q Qlonerator System - separate stereo systems on right and left

(b) Minolta Vivid 900/910 system - projected light stripe sweeps scene

FIGURE 16.5: Examples of stereo-based and "light-stripe" 3D sensor systems.

16.4.2 Improved Algorithms

One limitation to some existing approaches to 3D face recognition involves sensitivity to size variation. Approaches that use a purely curvature-based representation, such as extended Gaussian images, are not able to distinguish between two faces of similar shape but different size. Approaches that use a PCA-based or ICP-based algorithm can handle size change between faces, but run into problems with change of facial expression between the enrollment image and the image to be recognized.

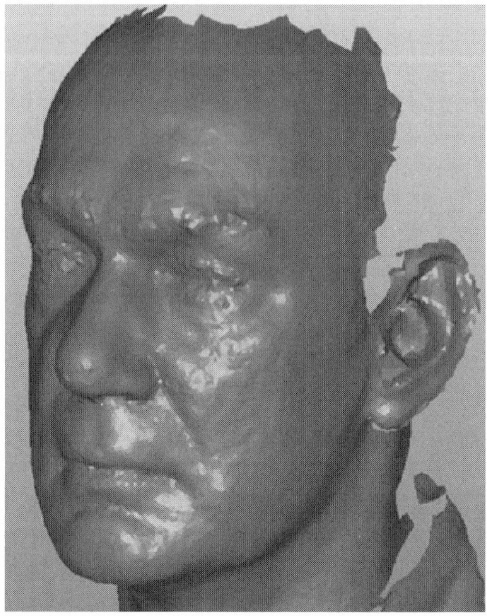

(a) “Ear-to-ear” 3D image taken with the 3Q system

(b) “Front view” 3D image taken with Minolta Vivid 910 system

FIGURE 16.6: Example of “ear-to-ear” and “frontal view” 3D shape acquisition.

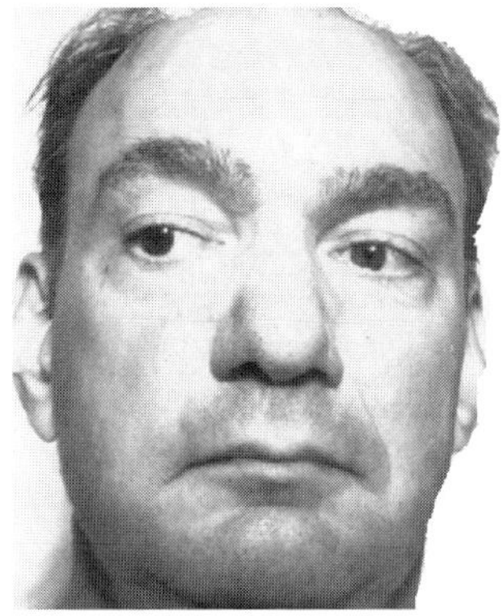

(a) A view of a 3D model rendered with the texture map on

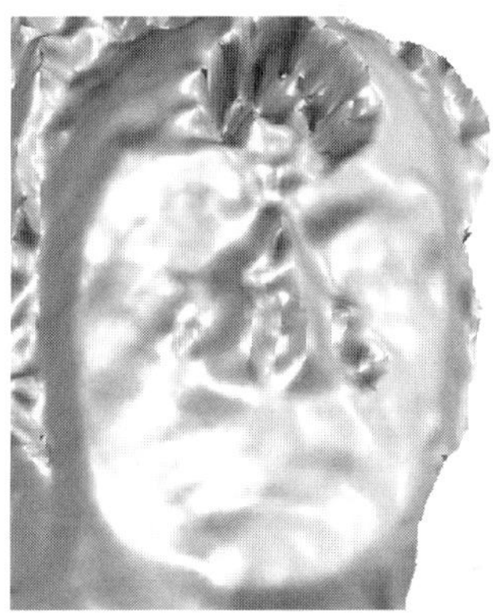

(b) The same 3D model as in (a) but rendered as shaded model without the texture map on

FIGURE 16.7: Example of a 3D shape errors masked by viewing with texture map on.

Approaches that effectively assume that the face is a rigid object will not be able to handle expression change. Handling change in facial expression would seem to require at least some level of part/whole model of the face, and possibly also a model of the range of possible nonrigid motion of the face. The seriousness of the problem of variation in facial expression between the enrollment image and the image to be recognized is illustrated in the results shown in Figure 16.8. This experiment focuses on the effects of expression change. Seventy subjects had their gallery image acquired with "normal expression" one week, a first probe image acquired with "smiling expression" in another week, and a second probe image

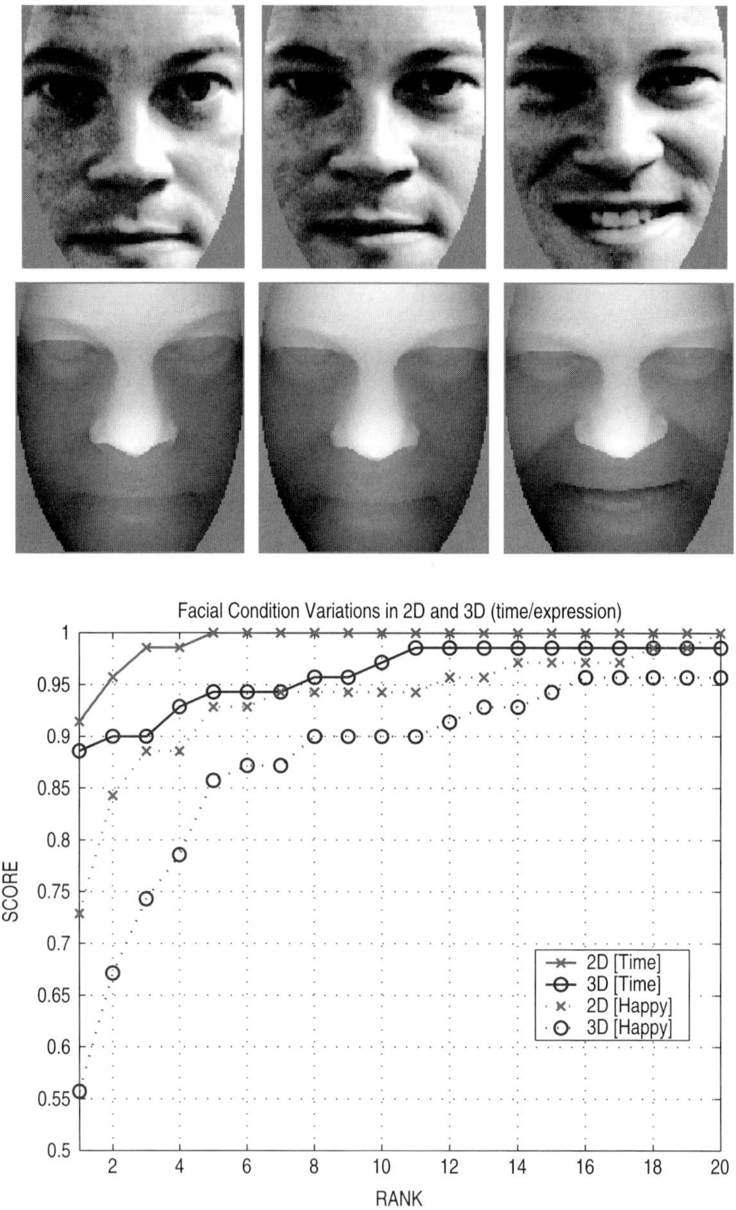

FIGURE 16.8: Effects of expression change on 3D and 2D recognition rates.

acquired with "normal expression" in still another week. Recognition was done with PCA-based 2D and 3D algorithms [15]. The upper CMC curves represent performance with time lapse only between gallery and probe; the lower pair represents time lapse and expression change. With simple time lapse but no expression change between the gallery and probe images, both 3D and 2D result in a rank-one recognition rate around 90%. There is a noticeable drop in performance when expression variation is introduced, to 73% for 2D and 55% for 3D. In this case, where the 3D recognition algorithm effectively assumes the face is a rigid shape, 3D performance is actually more negatively affected by expression change than is 2D performance. The relative degradation between 3D and 2D appears not to be a general effect, but instead is dependent on the particular facial expression. Clearly, variation in facial expression is a major cause of performance degradation that must be addressed in the next generation of algorithms.

In recent work with a larger data set, we have found that an iterative closest point (ICP) approach outperforms the PCA approach in 3D face recognition. This work used a data set representing 449 persons. Using a neutral-expression gallery image for each person, and a single neutral-expression, time-lapse probe image for each person, a PCA-based algorithm such as that described earlier achieved just slightly under 80% rank-one recognition, while an ICP-based approach achieved just over 90%. The performance of either algorithm dropped substantially when the facial expression varied between the gallery and probe images of a person.

In addition to a need for more sophisticated 3D recognition algorithms, there is also a need for more sophisticated multimodal combination. Multimodal combination has so far taken a fairly simple approach. The 3D recognition result and the 2D recognition result are each produced without reference to the other modality, and then the results are combined in some way. It is at least potentially more powerful to exploit possible synergies between the two modalities in the interpretation of the data. For example, knowledge of the 3D shape might help in interpreting shadow regions in the 2D image. Similarly, regions of facial hair might be easy to identify in the 2D image and help to predict regions of the 3D data which are more likely to contain artifacts.

16.4.3 Improved Methodology and Datasets

One barrier to experimental validation and comparison of 3D face recognition is lack of appropriate datasets. Desirable properties of such a dataset include: (1) a large number and demographic variety of people represented, (2) images of a given person taken at repeated intervals of time, (3) images of a given person that represent substantial variation in facial expression, (4) high spatial resolution, for example, depth resolution of 1 mm or better, and (5) low frequency of sensor-specific artifacts in the data. Expanded use of common datasets and baseline algorithms in the research community will facilitate the assessment of the state of the art in this area. It would likely also improve the interpretation of research results if the statistical significance, or lack thereof, was reported for observed performance differences between algorithms and modalities.

Another aspect of improved methodology would be the use, where applicable, of explicit, distinct training, validation and test sets. For example, the "face space" for a PCA algorithm might be created based on a training set of images, with the number of eigenvectors used and the distance metric used then selected based on a validation set, and finally the performance estimated on a test set. The different sets of images would be nonoverlapping with respect to the persons represented in each.

A more subtle methodological point is involved in the comparison of multimodal results to baseline results from a single modality. In the context of this survey, there are several publications that compare the performance of multimodal 3D+2D face recognition to the performance of 2D alone. The multimodal 3D+2D performance is always observed to be greater than the performance of 2D alone. However, this comparison is too simple, and is effectively biased toward the multimodal result. Enrolling a subject in a multimodal system requires two images, a 3D image and a 2D image. The same is true of the information used to recognize a person in a multimodal system. Therefore, a more appropriate comparison would be to a 2D recognition system that uses two images of a person both for enrollment and for recognition. When this sort of controlled comparison is done, the differences observed for multimodal 3D+2D compared to "multisample" 2D are smaller than those for a comparison to plain 2D [14]. Additionally, it is possible that using multiple 2D images may result in higher performance than using a single 3D image.

16.4.4 Summary

As evidenced by the publication dates in Table 16.2, activity in 3D and multimodal 3D+2D face recognition has expanded dramatically in recent years. It is an area with important potential applications. At the same time, there are many challenging research problems still to be addressed. These include the development of more practical and robust sensors, the development of improved recognition algorithms, and the pursuit of more rigorous experimental methodology. The development of improved recognition algorithms will be spurred by more rigorous research methodology, involving larger and more challenging datasets, and more carefully controlled performance evaluations.

In an application where 3D images of the face are acquired, it may be possible to also use 3D ear biometrics. Yan and Bowyer have looked at ear biometrics using 2D and 3D images, and at several different algorithmic approaches to for the 3D images [42]. The combination of 3D ear and 3D face is a form of multibiometric that has not yet, to our knowledge, been explored. Other 3D body features can also be used. Woodard and Flynn [39] describe the use of 3D finger shaper as a biometric, and demonstrate good performance with a correlation-based matcher.

At the time that this is written, a number of research groups are working on 3D face recognition using a large common dataset that incorporate substantial expression variation. Results from some of this work were presented at the Face Recognition Grand Challenge Workshop held in June of 2005 in association with the *Computer Vision and Pattern Recognition* conference. The larger of the two 3D face data sets used in the FRGC program to date contains over 4,000 images from over 400 subjects, with susbtantial facial expression variation represented in the data [29].

ACKNOWLEDGMENTS

This work was supported in part by National Science Foundation EIA 01-30839, Department of Justice grant 2004-DD-BX-1224, and Office of Naval Research (DARPA) grant N000140210410. The authors would like to thank Jonathon Phillips and Gerard Medioni for useful discussions in this area.

REFERENCES

[1] Konica Minolta 3D digitizer. Available at www.minoltausa.com/vivid/, January 2004.
[2] B. Achermann and H. Bunke. Combination of face classifiers for person identification. *International Conference on Pattern Recognition*, pages 416–420, 1996.
[3] B. Achermann and H. Bunke. Classifying range images of human faces with Hausdorff distance. *Fifteenth International Conference on Pattern Recognition*, pages 809–813, September 2000.
[4] B. Achermann, X. Jiang, and H. Bunke. Face recognition using range images. *International Conference on Virtual Systems and MultiMedia*, pages 129–136, 1997.
[5] C. Beumier and M. Acheroy. Face verification from 3D and grey level cues. *Pattern Recognition Letters* **22**:1321–1329, 2001.
[6] R. Beveridge. Evaluation of face recognition algorithms. Available at http://www.cs.colostate.edu/evalfacerec/index.html.
[7] V. Blanz and T. Vetter. Face recognition based on fitting a 3D morphable model. *IEEE Transactions on Pattern Analysis and Machine Intelligence* **25**:1063–1074, September 2003.
[8] C. Boehnen and P. Flynn. Accuracy of 3d scanning technologies in a face scanning context. *Fifth International Conference on 3-D Digital Imaging and Modeling* (*3DIM 2005*), 2005.
[9] K. W. Bowyer, K. Chang, and P. Flynn. A survey of approaches to three-dimensional face recognition. *Seventeenth International Conference on Pattern Recognition*, August 2004.
[10] A. M. Bronstein, M. M. Bronstein, and R. Kimmel. Expression-invariant 3D face recognition. *Audio- and Video-Based Person Authentication (AVBPA 2003)*, LNCS 2688, J. Kittler and M.S. Nixon, eds, Pages 62–70, 2003.

[11] J. Y. Cartoux, J. T. LaPreste, and M. Richetin. Face authentication or recognition by profile extraction from range images. *Proceedings of the Workshop on Interpretation of 3D Scenes*, pages 194–199, November 1989.
[12] K. Chang, K. W. Bowyer, and P. Flynn. Multimodal 2D and 3D biometrics for face recognition. *IEEE International Workshop on Analysis and Modeling of Faces and Gestures*, pages 187–194, October 2003.
[13] K. Chang, K. W. Bowyer, B. Victor, and S. Sarkar. Comparison and combination of ear and face images in appearance-based biometrics. *IEEE Transactions on Pattern Analysis and Machine Intelligence* **25**:1160–1165, 2003.
[14] K. I. Chang, K. W. Bowyer, and P. J. Flynn. An evaluation of multimodal 2d+3d face biometrics. *IEEE Transactions on Pattern Analysis and Machine Intelligence* **27**(4): 619–624, April 2005.
[15] K. I. Chang, K. W. Bowyer, and P. J. Flynn. Face recognition using 2D and 3D facial data. *2003 Multimodal User Authentication Workshop*, pages 25–32, December 2003.
[16] G. Gordon. Face recognition based on depth and curvature features. *Computer Vision and Pattern Recognition (CVPR)*, pages 108–110, June 1992.
[17] C. Hesher, A. Srivastava, and G. Erlebacher. A novel technique for face recognition using range images. *Seventh International Symposium on Signal Processing and Its Applications*, 2003.
[18] J. Kittler, M. Hatef, R. Duin, and J. Matas. On combining classifiers. *IEEE Transactions on Pattern Analysis and Machine Intelligence* **20**(3):226–239, March 1998.
[19] S. Lao, Y. Sumi, M. Kawade, and F. Tomita. 3D template matching for pose invariant face recognition using 3D facial model built with iso-luminance line based stereo vision. *International Conference on Pattern Recognition (ICPR 2000)*, pages II:911–916, 2000.
[20] J. C. Lee and E. Milios. Matching range images of human faces. *International Conference on Computer Vision*, pages 722–726, 1990.
[21] Y. Lee, K. Park, J. Shim, and T. Yi. 3D face recognition using statistical multiple features for the local depth information. *Sixteenth International Conference on Vision Interface*. Available at www.visioninterface.org/vi2003, June 2003.
[22] X. Lu, D. Colbry, and A. Jain. Matching 2.5D scans for face recognition. *International Conference on Biometric Authentication*, pp. 30–36, Hong Kong, July 2004.
[23] G. Medioni and R. Waupotitsch. Face recognition and modeling in 3D. *IEEE International Workshop on Analysis and Modeling of Faces and Gestures (AMFG 2003)*, pages 232–233, October 2003.
[24] K. Messer, J. Matas, J. Kittler, J. Luettin, and G. Maitre. XM2VTSDB: the extended M2VTS database. *Second International Conference on Audio- and Video-based Biometric Person Authentication*, pages 72–77, 1999.
[25] J. Min, K. W. Bowyer, and P. Flynn. Using multiple gallery and probe images per person to improve performance of face recognition. *Notre Dame Computer Science and Engineering Technical Report*, 2003.
[26] A. B. Moreno, Ángel Sánchez, J. F. Vélez, and F. J. Díaz. Face recognition using 3D surface-extracted descriptors. *Irish Machine Vision and Image Processing Conference (IMVIP 2003)*, September 2003.

[27] T. Nagamine, T. Uemura, and I. Masuda. 3D facial image analysis for human identification. *International Conference on Pattern Recognition* (*ICPR 1992*), pages 324–327, 1992.

[28] T. Papatheodorou and D. Reuckert. Evaluation of automatic 4d face recognition using surface and texture registration. *Sixth International Conference on Automated Face and Gesture Recognition*, pages 321–326, May 2004.

[29] P. J. Phillips, P. J. Flynn, T. Scruggs, K. W. Bowyer, J. Chang, K. Homan, J. Marques, J. Min, and W. Worek. Overview of the face recognition grand challenge. *Computer Vision and Pattern Recognition*, June 2005.

[30] J. Phillips, H. Moon, S. Rizvi, and P. Rauss. The FERET evaluation methodology for face-recognition algorithms. *IEEE Transactions on Pattern Analysis and Machine Intelligence* **22**(10):1090–1104, October 2000.

[31] P. J. Phillips, P. Grother, R. J. Michaels, D. M. Blackburn, E. Tabassi, and J. Bone. FRVT 2002: overview and summary. Available at: www.frvt.org, March 2003.

[32] T. D. Russ, K. W. Koch, and C. Q. Little. 3d facial recognition: a quantitative analysis. *Forty-fifth Annual Meeting of the Institute of Nuclear Materials Management* (*INMM*), July 2004.

[33] H. T. Tanaka, M. Ikeda, and H. Chiaki. Curvature-based face surface recognition using spherical correlation principal directions for curved object recognition. *Third International Conference on Automated Face and Gesture Recognition*, pages 372–377, 1998.

[34] F. Tsalakanidou, S. Malassiotis, and M. Strintzis. Integration of 2d and 3d images for enhanced face authentication. *Sixth International Conference on Automated Face and Gesture Recognition*, pages 266–271, May 2004.

[35] F. Tsalakanidou, D. Tzocaras, and M. Strintzis. Use of depth and colour eigenfaces for face recognition. *Pattern Recognition Letters* **24**:1427–1435, 2003.

[36] M. Turk and A. Pentland. Eigenfaces for recognition. *Journal of Cognitive Neuroscience* **3**:71–86, 1991.

[37] Y. Wang, C. Chua, and Y. Ho. Facial feature detection and face recognition from 2D and 3D images. *Pattern Recognition Letters* **23**:1191–1202, 2002.

[38] Y. Wang, T. Tan, and A. K. Jain. Combination of face and iris biometrics for identity verification. *International Conference on Audio- and Video-based Biometric Person Authentication*, June 2003.

[39] D. Woodard and P. Flynn. Personal identification using finger surface features. *Computer Vision and Pattern Recognition*, 2005.

[40] C. Xu, Y. Wang, T. Tan, and L. Quan. Automatic 3d face recognition combining global geometric features with local shape variation information. *Sixth International Conference on Automated Face and Gesture Recognition*, pages 308–313, May 2004.

[41] W. Yambor, B. Draper, and R. Beveridge. Analyzing PCA-based face recognition algorithms: eigenvector selection and distance measures. *Second Workshop on Empirical Evaluation in Computer Vision, Dublin, Ireland*, July 2000.

[42] P. Yan and K. W. Bowyer. Multi-biometrics 2d and 3d ear recognition. *Audio- and Video-based Biometric Person Authentication*, pages 503–512, July 2005.

[43] W. Zhao, R. Chellappa, P. J. Phillips, A. Rosenfeld'. Face recognition: a literature survey. *ACM Computing Surveys* **35**:399–458, December 2003.

CHAPTER 17

BEYOND ONE STILL IMAGE: FACE RECOGNITION FROM MULTIPLE STILL IMAGES OR A VIDEO SEQUENCE

17.1 INTRODUCTION

While face recognition (FR) from a single still image has been extensively studied over a decade, FR based on a group of still images (also referred to as multiple still images) or a video sequence is an emerging topic. This is mainly evidenced by the growing increase in the literature. For instance, a research initiative called Face Recognition Grand Challenge [1] has been organized. One specific challenge directly addresses the use of multiple still images to improve the recognition accuracy significantly [3]. Recently a workshop jointly held with CVPR 2004 was devoted to face processing in video [2]. It is predictable that, with the ubiquity of video sequences, face recognition based on video sequences will become more and more popular.

It is obvious that multiple still images or a video sequence can be regarded as a single still image in a degenerate manner. More specifically, suppose that we have a single-still-image-based FR algorithm $\mathcal{A}$ (or the base algorithm) by some means, we can construct a recognition algorithm based on multiple still images or a video sequence by combining multiple base algorithms denoted by $\mathcal{A}_i$'s. Each A_i takes a different single image $\boldsymbol{y}_i$ as input, coming from the multiple still images or video sequences. The combining rule can be additive, multiplicative, and so on.

Here is a concrete example of such a construction. Suppose that the still-image-based FR uses the nearest-distance classification rule, and the recognition algorithm $\mathcal{A}$ performs the following:

$$\mathcal{A} : \hat{n} = \underset{n=1,2,\dots,N}{\text{argmin}} \; d(\boldsymbol{y}, \mathrm{x}^{[n]}), \tag{1}$$

where N is the number of individuals in the gallery set, d is the distance function, $\mathrm{x}^{[n]}$ represents the nth individual in the gallery set, and $\boldsymbol{y}$ is the probing single still image. Equivalently, the distance function can be replaced by a similarity function s. The recognition algorithm becomes:

$$\mathcal{A} : \hat{n} = \underset{n=1,2,\dots,N}{argmax} \; s(\boldsymbol{y}, \mathrm{x}^{[n]}). \tag{2}$$

In this chapter, we interchange the use of the distance and similarity functions if no confusion arises.

The common choices for d include the following.

- Cosine angle:

$$d(\boldsymbol{y}, \mathrm{x}^{[n]}) = 1 - \cos(\boldsymbol{y}, \mathrm{x}^{[n]}) = 1 - \frac{\boldsymbol{y}^{\mathrm{T}} \mathrm{x}^{[n]}}{||\boldsymbol{y}|| \cdot ||\mathrm{x}^{[n]}||} \tag{3}$$

- Distance in subspace:

$$d(\boldsymbol{y}, \mathrm{x}^{[n]}) = ||\mathtt{P}^{\mathrm{T}}\{\boldsymbol{y} - \mathrm{x}^{[n]}\}||^2 = \{\boldsymbol{y} - \mathrm{x}^{[n]}\}^{\mathrm{T}} \mathtt{P}\mathtt{P}^{\mathrm{T}} \{\boldsymbol{y} - \mathrm{x}^{[n]}\}, \tag{4}$$

  where $\mathtt{P}$ is a subspace projection matrix. Some common subspace methods include principal-component analysis (a.k.a. eigenface [51]), linear-discriminant analysis (a.k.a. Fisherface [7, 19, 57]), independent component analysis [6], local feature analysis [38], intrapersonal subspace [37, 62] etc.
- "Generalized" Mahanalobis distance:

$$k(\boldsymbol{y}, \mathrm{x}^{[n]}) = \{\boldsymbol{y} - \mathrm{x}^{[n]}\}^{\mathrm{T}} \mathtt{W} \{\boldsymbol{y} - \mathrm{x}^{[n]}\}, \tag{5}$$

  where the $\mathtt{W}$ matrix plays a weighting role. If $\mathtt{W} = \mathtt{P}\mathtt{P}^{\mathrm{T}}$, then the 'generalized' Mahanalobis distance reduces to the distance in subspace. If $\mathtt{W} = \Sigma^{-1}$ with Σ being a covariance matrix, then the 'generalized' Mahanalobis distance reduces to the regular Mahanalobis distance.

Using the base algorithm $\mathcal{A}$ defined in (1) as a building block, we can easily construct various recognition algorithms [18] based on a group of still images and a video sequence. By denoting a group of still images and a video sequence by $\{\boldsymbol{y}_t;\ t = 1, 2, ..., T\}$, the recognition algorithm $\mathcal{A}_t$ for $\boldsymbol{y}_t$ is simply

$$\mathcal{A}_t : \hat{n} = \underset{n=1,2,...,N}{argmin}\ d(\boldsymbol{y}_t, \mathrm{x}^{[n]}). \tag{6}$$

Some commonly used combination rules are listed in Table 17.1.

In the above, the nth individual in the gallery set is represented by a single still image $\mathrm{x}^{[n]}$. This can be generalized to use multiple still images or a video sequence $\{\mathrm{x}_s^{[n]};\ s = 1, 2, \ldots, K_s\}$. Similarly, the resulting fused algorithm combinets the base algorithms denoted by the $\mathcal{A}_{ts}$:

$$\mathcal{A}_{ts} : \hat{n} = \underset{n=1,2,...,N}{\text{argmin}}\ d(\boldsymbol{y}_t, \mathrm{x}_s^{[n]}). \tag{7}$$

Even though the fused algorithms might work well in practice, clearly, the overall recognition performance is solely based on the base algorithm, and hence designing the base algorithm $\mathcal{A}$ (or the similarity function k) is of ultimate importance. Therefore, the fused algorithms do not completely utilize additional properties possessed by multiple still images or video sequences.

Three additional properties are available for multiple still images and/or video sequences:

1. [$P1$: Multiple observations]. This property is directly utilized by the assembly algorithms. One main disadvantage of the fused algorithms is the *ad hoc* nature of the combination rule. However, theoretical analysis based on multiple observations can be derived. For example, a set of observations can be summarized using matrix, probability density function, or manifold. Hence, corresponding knowledge can be utilized to match two sets.

Table 17.1: A list of combining rules. The $\mathbb{J}$ function used in majority voting is an indicator function.

Method	Rule
Minimum arithmetic mean	$\hat{n} = \text{argmin}_{n=1,2,...,N} \frac{1}{T} \sum_{t=1}^{T} d(\boldsymbol{y}_t, \mathrm{x}^{[n]})$
Minimum geometric mean	$\hat{n} = \text{argmin}_{n=1,2,...,N} \sqrt[T]{\prod_{t=1}^{T} d(\boldsymbol{y}_t, \mathrm{x}^{[n]})}$
Minimum median	$\hat{n} = \text{argmin}_{\text{n}=1,2,...,N}\{\text{med}_{t=1,2,...,T}\ d(\boldsymbol{y}_t, \mathrm{x}^{[n]})\}$
Minimum minimum	$\hat{n} = \text{argmin}_{n=1,2,...,N}\{\min_{t=1,2,...,T}\ d(\boldsymbol{y}_t, \mathrm{x}^{[n]})\}$
Majority voting	$\hat{n} = \text{argmax}_{n=1,2,...,N} \sum_{t=1}^{T} \mathbb{J}[\mathcal{A}_t(\boldsymbol{y}) = n]$

2. [$P2$: Temporal continuity/Dynamics]. Successive frames in the video sequences are continuous in the temporal dimension. Such continuity, coming from facial expression, geometric continuity related to head and/or camera movement, or photometric continuity related to changes in illumination provides an additional constraint for modeling face appearance. In particular, temporal continuity can be further characterized using kinematics. For example, facial expression and head movement when an individual participates certain activity result in structured changes in face appearance. Modeling of such structured change (or dynamics) further regularizes FR.
3. [$P3$: 3D model]. This means that we are able to reconstruct 3D model from a group of still images and a video sequence. Recognition can then be based on the 3D model. Using the 3D model provides possible invariance to pose and illumination.

Clearly, the first and third properties are shared by multiple still images and video sequences. The second property is solely possessed by video sequences. We will elaborate these properties in Section 17.3.

The properties manifested in multiple still images and video sequences present new challenges and opportunities. On one hand, by judiciously exploiting these features, we can design new recognition algorithms other than those mentioned above. On the other hand, care should be exercised when exploiting these properties. In Section 17.4, we review various face-recognition approaches utilizing these properties in one or more ways. Generally speaking, the newly designed algorithms are better in terms of recognition performance, computational efficiency, etc.

Studying the recognition algorithms from the perspective of additional properties is very beneficial. First of all, we can easily categorize the algorithms available in the literature accordingly. Secondly, we can forecast (as in Section 17.5) new approaches that can be developed to realize the full potentials of multiple still images or video sequences.

There are two recent survey papers [14, 58] on FR in the literature. In [14], face recognition based on a still image was reviewed at length and none of the reviewed approaches were video-based. In [58], video-based recognition is identified as one key topic. Even though it had been reviewed quite extensively, all video-based approaches were not categorized. In this chapter, we attempt to bring out new insights through studying the three additional properties. We proceed to the next section by recapitulating some basics of FR.

17.2 BASICS OF FACE RECOGNITION

We begin this section by introducing three FR tasks: verification, identification, and watch list. We also address the concept of training, gallery, and probe sets and present various recognition settings based on different types of input used in the gallery and probe sets.

17.2.1 Verification, Identification, and Watch List

Face recognition mainly involves the following three tasks [40].

1. Verification. The recognition system determines whether the query face image and the claimed identity match.
2. Identification. The recognition system determines the identity of the query face image by matching it with a database of images with known identities, assuming that the identity is inside the database.
3. Watch list. The recognition system first determines if the identity of the query face image is on the stored watch list and, if yes, then identifies the individual.

Figure 17.1 illustrates the above three tasks and corresponding statistics used for evaluation. Among the three tasks, the watch list task is the most difficult one. The present chapter focuses only on the identification task.

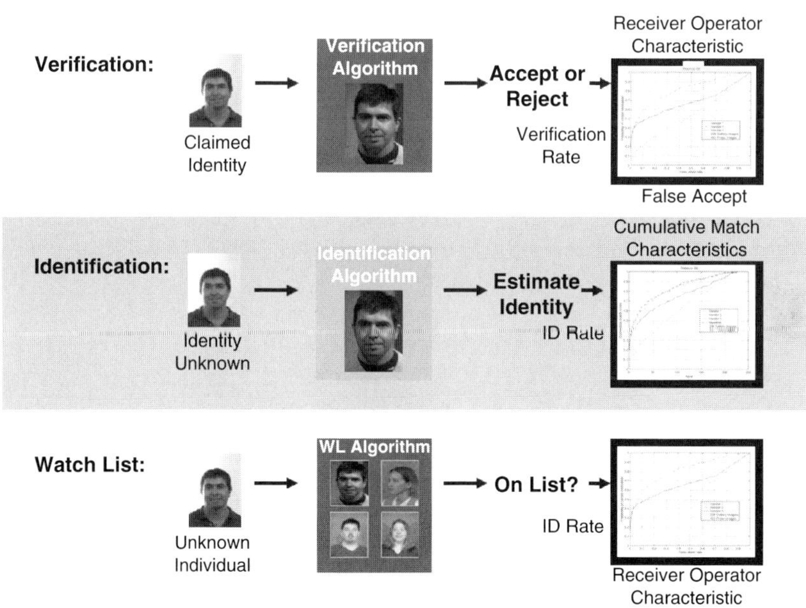

FIGURE 17.1: Three FR tasks: verification, identification, and watch list (courtesy of P.J.Phillips [40]).

17.2.2 Gallery, Probe, and Training Sets

We here follow a FR test protocol FERET [39] widely observed in the FR literature. FERET assumes availability of the following three sets, namely one training set, one gallery set, and one probe set. The gallery and probe sets are used in the testing stage. The gallery set contains images with known identities and the probe set with unknown identities. The algorithm associates descriptive features with the images in the gallery and probe sets and determines the identities of the probe images by comparing their associated features with those features associated with gallery images.

According to the imagery utilized in the gallery and probe sets, we can define the following nine recognition settings as in Table 17.2. For instance, the *mStill-to-Video* setting utilizes multiple still images for each individual in the gallery set and a video sequence for each individual in the probe set. The FERET test investigated the *sStill-to-sStill* recognition setting and the Face Recognition Grand Challenge studies three settings: *sStill-to-sStill*, *mStill-to-sStill*, and *mStill-to-mStill*.

The need for a training set in addition to the gallery and probe sets is mainly motivated by that fact that the *sStill-to-sStill* recognition setting is used in the FERET. We illustrate this using a probabilistic interpretation of the still-image-based FR algorithm exemplified by Eq. (1).

The following two conditions are assumed.

- All classes have the same prior probabilities, i.e.,

$$\pi(n) = \frac{1}{N}, \; n = 1, 2, \ldots, N. \tag{8}$$

- Each class possesses a multivariate density $p^{[n]}(\mathbf{y})$ that shares a common functional form f of the quantity $d(\mathbf{y}, \mathrm{x}^{[n]})$:

$$p^{[n]}(\mathbf{y}) = f(d(\mathbf{y}, \mathrm{x}^{[n]})), \; n = 1, 2, \ldots, N. \tag{9}$$

Table 17.2: Recognition settings based on a single still image, multiple still images, and a video sequence.

Probe\Gallery	A single still image	A group of still images	A video sequence
A single still image	*sStill-to-sStill*	*mStill-to-sStill*	*Video-to-sStill*
A group of still images	*sStill-to-mStill*	*mStill-to-mStill*	*Video-to-mStill*
A video sequence	*sStill-to-Video*	*mStill-to-Video*	*Video-to-Video*

The density f can have various forms. For example, it can be a normal density with a mean $\mathrm{x}^{[n]}$ and an isotropic covariance matrix $\sigma^2 I$:

$$p^{[n]}(\mathbf{y}) \propto \exp\left(\{-\frac{||\mathbf{y}-\mathrm{x}^{[n]}||^2}{2\sigma^2}\}\right), \tag{10}$$

where $f(t) \propto \exp(-t^2/2\sigma^2)$ and $d(\mathbf{y}, \mathrm{x}^{[n]}) = ||\mathbf{y} - \mathrm{x}^{[n]}||$.

With these assumptions, we follow a maximum *a posteriori* (MAP) decision rule to perform classification, i.e.,

$$\begin{aligned}\mathcal{A}: \hat{n} &= \underset{n=1,2,\ldots,N}{\operatorname{argmax}}\ p(n|\mathbf{y}) = \underset{n=1,2,\ldots,N}{\operatorname{argmax}}\ \pi(n)p^{[n]}(\mathbf{y}) \\ &= \underset{n=1,2,\ldots,N}{\operatorname{argmax}}\ p^{[n]}(\mathbf{y}) = \underset{n=1,2,\ldots,N}{\operatorname{argmin}}\ d(\mathbf{y}, \mathrm{x}^{[n]}). \end{aligned} \tag{11}$$

The purpose of the training set is for the recognition algorithm to learn the density function f. For example, in subspace methods, the training set is used to learn the projection matrix $\boldsymbol{P}$. Typically, the training set does not overlap with the gallery and probe sets in terms of identity. This is based on the fact that the same density function f is used for all individuals, and that generalization across the identities in the training and testing stages is possible.

17.3 PROPERTIES

The multiple still images and video sequence are different from a single still image because they possess additional properties not present in a still image. In particular, three properties manifest themselves, which motivated various approaches recently proposed in the literature. Below, we analyze the three properties one by one.

17.3.1 $P1$: Multiple observations

This is the most commonly used feature of multiple still images or a video sequence. If only this property is considered, a video sequence reduces to a group of still images with the temporal dimension stripped out. In other words, every video frame is treated as a still image. Another implicit assumption is that all face images are normalized before subjecting them to subsequent analysis.

As mentioned earlier, the combination rules are rather ad hoc, which leaves room for a systematic exploration of this property. This leads to investigating systematic representations of multiple observations $\{\mathbf{y}_1, \mathbf{y}_2, \ldots, \mathbf{y}_T\}$. Once an appropriate representation is fixed, a recognition algorithm can be designed accordingly.

Various ways of summarizing multiple observations have been proposed. We will review detailed approaches in Section 17.4, including [25, 26, 56, 53, 65, 66]. In terms of the summarizing rules, these approaches can be grouped into four categories, as follows.

One Image or Several Images

Multiple observations $\{\boldsymbol{y}_1, \boldsymbol{y}_2, \ldots, \boldsymbol{y}_T\}$ are summarized into one image $\hat{\boldsymbol{y}}$ or several images $\{\hat{\boldsymbol{y}}_1, \hat{\boldsymbol{y}}_2, ..., \hat{\boldsymbol{y}}_m\}$ (with $m < T$). For instance, one can use the mean or the median of $\{\boldsymbol{y}_1, \boldsymbol{y}_2, \ldots, \boldsymbol{y}_T\}$ as the summary image $\hat{\boldsymbol{y}}$. Clustering techniques can be invoked to produce multiple summary images $\{\hat{\boldsymbol{y}}_1, \hat{\boldsymbol{y}}_2, ..., \hat{\boldsymbol{y}}_m\}$. In terms of recognition, we can simply apply the still-image-based face-recognition algorithm based on $\hat{\boldsymbol{y}}$ or $\{\hat{\boldsymbol{y}}_1, \hat{\boldsymbol{y}}_2, ..., \hat{\boldsymbol{y}}_m\}$. This applies to all nine recognition settings listed in Table 17.2.

Matrix

Multiple observations $\{\boldsymbol{y}_1, \boldsymbol{y}_2, \ldots, \boldsymbol{y}_T\}$ form a matrix[1] $\mathrm{Y} = [\boldsymbol{y}_1, \boldsymbol{y}_2, \ldots, \boldsymbol{y}_T]$. The main advantage of using the matrix representation is that we can rely on the rich literature of matrix analysis. For example, various matrix decompositions can be invoked to represent the original data more efficiently. Metrics measuring similarity between two matrices can be used for recognition.

This applies to the *mStill-to-mStill*, *Video-to-mStill*,*Video-to-mStill*, and *Video-to-Video* recognition settings. Suppose that the nth individual in the gallery set has a matrix $\boldsymbol{X}^{[n]}$, we determine the identity of of a probe matrix Y as

$$\hat{n} = \underset{n=1,2,...,N}{\operatorname{argmin}}\ d(\mathrm{Y}, \mathrm{X}^{[n]}), \tag{12}$$

where d is a matrix distance function.

Probability Density Function (PDF)

In this rule, multiple observations $\{\boldsymbol{y}_1, \boldsymbol{y}_2, \ldots, \boldsymbol{y}_T\}$ are regarded as independent realizations drawn from an underlying distribution. PDF estimation techniques such as parametric, semi-parametric, and non-parametric methods [17] can be utilized to learn the distribution.

In the *mStill-to-mStill*, *Video-to-mStill*, *Video-to-mStill*, and *Video-to-Video* recognition settings, recognition can be performed by comparing distances between PDFs, such as Bhattacharyya and Chernoff distances, Kullback–Leibler divergence, and so on. More specifically, suppose that the nth individual in the

[1] Here we assume that each image $\boldsymbol{y}_i$ is 'vectorized'.

gallery set has a PDF $p^{[n]}(\mathrm{x})$, we determine the identity of a probe PDF $q(\mathbf{y})$ as

$$\hat{n} = \operatorname*{argmin}_{n=1,2,\ldots,N} d(q(\mathbf{y}), p^{[n]}(\mathrm{x})), \tag{13}$$

where d is a probability distance function.

In the *mStill-to-sStill*, *Video-to-sStill*, *sStill-to-mStill*, and *sStill-to-Video* settings, recognition becomes a hypothesis-testing problem. For example, in the *sStill-to-mStill* setting, if we can summarize the multiple still images in query into a pdf, say $q(\mathbf{y})$, then recognition is to test which gallery image $\mathrm{x}^{[n]}$ is mostly likely to be generated by $q(\mathbf{y})$.

$$\hat{n} = \operatorname*{argmax}_{n=1,2,\ldots,N} q(\mathrm{x}^{[n]}). \tag{14}$$

Notice that this is different from the *mStill-to-sStill* setting, where each gallery object has a density $p^{[n]}(\mathbf{y})$, and then, given a probe single still image $\mathbf{y}$, the following recognition check is performed:

$$\hat{n} = \operatorname*{argmax}_{n=1,2,\ldots,N} p^{[n]}(\mathbf{y}). \tag{15}$$

Equation 15 is the same as the probabilistic interpretation of still-image-based recognition, except that the density $p^{[n]}(\mathbf{y})$ for a different n can have a different form. In such a case, we in principle no longer need a training set that is provided to learn the common f function in the *sStill-to-sStill* setting.

Manifold

In this rule, face appearances of multiple observations form a highly nonlinear manifold $\mathcal{P}$. Manifold learning has recently attracted a lot of attention. Examples include [43, 49].

After characterizing the manifold, FR reduces to (i) comparing two manifolds if we are in the *mStill-to-mStill*, *Video-to-mStill*, *Video-to-mStill*, and *Video-to-Video* settings and (ii) comparing distances from one data point to different manifolds if we are in the *mStill-to-sStill*, *Video-to-sStill*, *sStill-to-mStill*, and *sStill-to-Video* settings.

For instance, in the *Video-to-Video* setting, galley videos are summarized into manifolds $\{\mathcal{P}^{[n]}; n = 1, 2, \ldots, N\}$. For the probe video that is summarized into a manifold $\mathcal{Q}$, its identity is determined as

$$\hat{n} = \operatorname*{argmin}_{n=1,2,\ldots,N} d(\mathcal{Q}, \mathcal{P}^{[n]}), \tag{16}$$

where d calibrates the distance between two manifolds.

In the *Video-to-sStill* setting, for the probe still image $\boldsymbol{y}$, its identity is determined as

$$\hat{n} = \underset{n=1,2,\ldots,N}{\operatorname{argmin}} \; d(\boldsymbol{y}, \mathcal{P}^{[n]}), \tag{17}$$

where d calibrates the distance between a data point to a manifold.

17.3.2 $P2$: Temporal Continuity/Dynamics

Property $P1$ strips out the temporal dimension available in the video sequence. In this property $P2$, we bring back the temporal dimension, and hence the property $P2$ only holds for a video sequence.

Successive frames in a video sequence are continuous in the temporal dimension. The continuity arising from dense temporal sampling is two-fold: the face movement is continuous and the change in appearance is continuous.

Temporal continuity provides an additional constraint for modeling face appearance. For example, smoothness of face movement is used in face tracking. As mentioned earlier, it is implicitly assumed that all face images are normalized before utilization of the property $P1$ of multiple observations. For the purpose of normalization, face detection is independently applied on each image. When temporal continuity is available, tracking can be applied instead of detection to perform normalization of each video frame.

Temporal continuity also plays an important role for recognition. Recently psychophysical evidence [27] reveals that moving faces are more recognizable. Computational approaches to incorporating temporal continuity for recognition are reviewed in Section 17.4. In addition to temporal continuity, face movement and face appearance often follow certain kinematics. In other words, changes in movement and appearance are not random. Understanding kinematics is also important for face recognition.

17.3.3 $P3$: 3D Model

This means that we are able to reconstruct a 3D model from a group of still images and a video sequence. This leads to the literature of light-field rendering, which takes multiple still images as input, and structure from motion (SfM), which takes a video sequence as input. Even though SfM has been studied for more than two decades, current SfM algorithms are not reliable enough for accurate 3D model construction. Researchers therefore incorporate or solely use prior 3D face models (that are acquired beforehand) to derive the reconstruction result. In principle, a 3D model provides the possibility of resolving pose and illumination variations.

The 3D model possesses two components: geometric and photometric. The geometric component describes the depth information of the face, and the photometric component depicts the texture map. The SfM algorithm is more focused on recovering the geometric component, whereas the light-field rendering method is more on recovering the photometric component.

Recognition can then be performed directly based on the 3D model. More specifically, for any recognition setting, suppose that, galley individuals are summarized into 3D models $\{\mathcal{M}^{[n]}; n = 1, 2, \ldots, N\}$. For multiple observations of a probe individual that are summarized into a 3D model $\mathcal{N}$, its identity is determined as

$$\hat{n} = \underset{n=1,2,\ldots,N}{\operatorname{argmin}} \; d(\mathcal{N}, \mathcal{M}^{[n]}), \tag{18}$$

where d calibrates the distance between two models.

For one probe still image $\mathbf{y}$, its identity is determined as

$$\hat{n} = \underset{n=1,2,\ldots,N}{\operatorname{argmin}} \; d(\mathbf{y}, \mathcal{M}^{[n]}), \tag{19}$$

where d calibrates the cost of generating a data point from a model.

17.4 REVIEW

In this section, we review the FR approaches utilizing the three properties. Our review mainly emphasizes the technical detail of the reviewed approaches. Other issues, like recognition performance and computational efficiency, are also addressed where relevant.

17.4.1 Approaches Utilizing $P1$: Multiple Observations

Four rules of summarizing multiple observations have been presented. In general, different data representations are utilized to describe multiple observations, and corresponding distance functions based on the presentations are invoked for recognition.

One image or several images

Algorithms designed by representing a set of observations into one image or several images and then applying the combination rules are essentially still-image-based and hence are not reviewed here.

Matrix

Yamaguchi et al. [56] proposed the so-called *mutual subspace method* (MSM). In this method, the matrix representation is used and the similarity function between two matrices is defined as the angle between two subspaces of the matrices (also referred to as principal angle or canonical correlation coefficient). Suppose that the columns of $\mathtt{X}$ and $\mathtt{Y}$ represent two subspaces $\mathtt{U}_\mathtt{X}$ and $\mathtt{U}_\mathtt{Y}$, the principle angle θ between the two subspaces is defined as

$$\cos(\theta) = \max_{\mathtt{u}\in\mathtt{U}_\mathtt{X}} \max_{\mathtt{v}\in\mathtt{U}_\mathtt{Y}} \frac{\mathtt{u}^\mathrm{T}\mathtt{v}}{\sqrt{\mathtt{u}^\mathrm{T}\mathtt{u}}\sqrt{\mathtt{v}^\mathrm{T}\mathtt{v}}}. \tag{20}$$

It can be shown that the principle angle θ is equal to the largest singular value of the matrix $\mathtt{U}_\mathtt{X}^\mathrm{T}\mathtt{U}_\mathtt{Y}$, where $\mathtt{U}_\mathtt{X}$ and $\mathtt{U}_\mathtt{Y}$ are orthogonal matrices encoding the column bases of the $\mathtt{X}$ and $\mathtt{Y}$ matrices respectively.

In general, the leading singular values of the matrices $\mathtt{U}_\mathtt{X}^\mathrm{T}\mathtt{U}_\mathtt{Y}$ define a series of principal angles θ_k:

$$\cos(\theta_k) = \max_{\mathtt{u}\in\mathtt{U}_\mathtt{X}} \max_{\mathtt{v}\in\mathtt{U}_\mathtt{Y}} \frac{\mathtt{u}^\mathrm{T}\mathtt{v}}{\sqrt{\mathtt{u}^\mathrm{T}\mathtt{u}}\sqrt{\mathtt{v}^\mathrm{T}\mathtt{v}}} \tag{21}$$

subject to:

$$\mathtt{u}^\mathrm{T}\mathtt{u}_i = 0, \mathtt{v}^\mathrm{T}\mathtt{v}_i = 0, \ i = 1, 2, \ldots, k-1. \tag{22}$$

Yamaguchi et al. [56] recorded a database of 101 individuals posing variation in facial expression and pose. They discovered that the MSM method is more robust to noisy input image or face normalization error than the still-image-based method that is referred to as conventional subspace method (CSM) in [56]. As shown in Figure 17.2, the similarity function of the MSM method is more stable and consistent than that of the CSM method.

Wolf and Shashua [53] extended computation of the principal angles into a nonlinear feature space $\mathcal{H}$ called the reproducing-kernel Hilbert space (RKHS) [45] induced by a positive definite kernel function. This kernel function represents an inner product between two vectors in the nonlinear feature space that are mapped from the original data space (say $\mathcal{R}^\mathrm{d}$) via a nonlinear mapping function ϕ. For $\mathtt{x}, \boldsymbol{y} \in \mathcal{R}^\mathrm{d}$, the kernel function is given by

$$k(\mathtt{x}, \boldsymbol{y}) = \phi(\mathtt{x})^\mathrm{T}\phi(\boldsymbol{y}). \tag{23}$$

This is called a "kernel trick": once the kernel function k is specified, no explicit form of the ϕ is required. Therefore, as long as we are able to cast all computations into inner product, we can invoke the 'kernel trick' to lift the original data space to an RKHS. Since the mapping function ϕ is nonlinear, the nonlinear characterization of the data structure is captured to some extent. Popular choices for the kernel function k are the polynomial kernel, the radial-basis-function (RBF) kernel, the neural-network kernel, etc. Refer to [45] for their definitions.

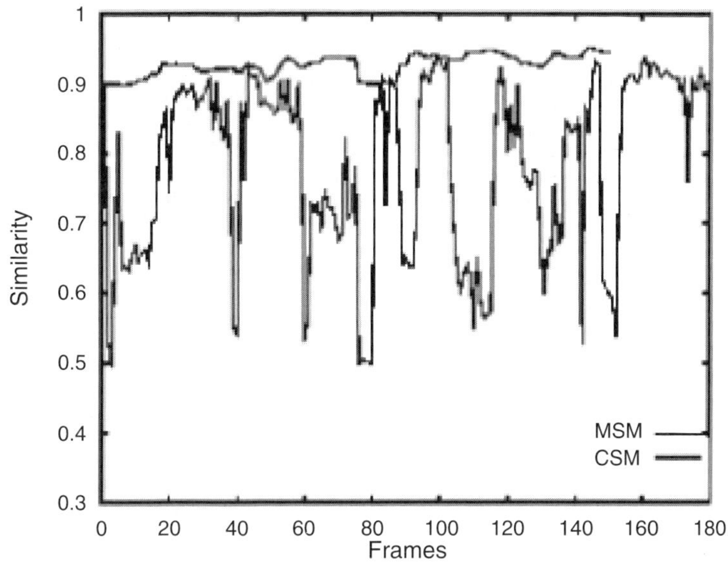

FIGURE 17.2: Comparison between the CSM and MSM methods (reproduced from Yamaguchi et al. [56]).

Kernel-based principal angles between two matrices $\mathtt{X}$ and $\mathtt{Y}$ are then based on their "kernelized" versions $\phi(\mathtt{X})$ and $\phi(\mathtt{Y})$. A "kernelized" matrix $\phi(\mathtt{X})$ of $\mathtt{X} = [\mathtt{x}_1, \mathtt{x}_2, \ldots, \mathtt{x}_n]$ is defined as $\phi(\mathtt{X}) = [\phi(\mathtt{x}_1), \phi(\mathtt{x}_2), \ldots, \phi(\mathtt{x}_n)]$. The key is to evaluate the matrix $\mathtt{U}^{\mathrm{T}}_{\phi(\mathtt{X})}\mathtt{U}_{\phi(\mathtt{Y})}$ defined in RKHS. In [53], Wolf and Shashua showed the computation using the "kernel trick".

Another contribution of Wolf and Shashua [53] is that they further proposed a positive kernel function taking matrices as input. Given such a kernel function, it can be readily plugged into a classification scheme such as support-vector machine (SVM) [45] to take advantage of the SVM's discriminative power. Face recognition using multiple still images coming from a tracked sequence was studied, and using the proposed kernelized principal angles slightly outperforms nonkernelized versions.

Zhou [66] systematically investigated the kernel functions taking matrices as input (also referred to as matrix kernels). More specifically, the following two functions are kernel functions:

$$k_{\bullet}(\mathtt{X}, \mathtt{Y}) = \mathrm{tr}(\mathtt{X}^{\mathrm{T}}\mathtt{Y}), \quad k_{*}(\mathtt{X}, \mathtt{Y}) = \det(\mathtt{X}^{\mathrm{T}}\mathtt{Y}), \tag{24}$$

where tr and det are matrix trace and determinant. They are called matrix trace and determinant kernels. Using them as building blocks, Zhou [66] constructed more kernels based on the column basis matrix, the kernelized matrix, and the column basis matrix of the kernelized matrix.

$$k_{U\bullet}(X, Y) = \mathrm{tr}(U_X^T U_Y), \qquad k_{U*}(X, Y) = \det(U_X^T U_Y), \tag{25}$$

$$k_{\phi\bullet}(X, Y) = \mathrm{tr}(\phi(X)^T \phi(Y)), \qquad k_{\phi*}(X, Y) = \det(\phi(X)^T \phi(Y)), \tag{26}$$

$$k_{U\phi\bullet}(X, Y) = \mathrm{tr}(U_{\phi(X)}^T U_{\phi(Y)}), \qquad k_{U\phi*}(X, Y) = \det(U_{\phi(X)}^T U_{\phi(Y)}). \tag{27}$$

Probability Density Function (PDF)

Shakhnoarovich et al. [46] proposed using a multivariate normal density for summarizing face appearances and the Kullback–Leibler (KL) divergence or relative entropy for recognition. The KL divergence between two normal densities $p_1 = N(\mu_1, \Sigma_1)$ and $p_2 = N(\mu_2, \Sigma_2)$ can be explicitly computed as

$$\mathrm{KL}(p_1 || p_2) = \int_x p_1(x) \log \frac{p_1(x)}{p_2(x)} dx \tag{28}$$

$$= \frac{1}{2} \log \frac{|\Sigma_2|}{|\Sigma_1|} + \frac{1}{2} \mathrm{tr}\left[(\mu_1 - \mu_2)^T \Sigma_2^{-1} (\mu_1 - \mu_2) + \Sigma_1 \Sigma_2^{-1} \right] - \frac{d}{2},$$

where d is the dimensionality of the data. One disadvantage of the KL divergence is that it is asymmetric. To make it symmetric, one can use

$$J_D(p_1, p_2) = \int_x (p_1(x) - p_2(x)) \log \frac{p_1(x)}{p_2(x)} dx = \mathrm{KL}(p_1 || p_2) + KL(p_2 || p_1)$$

$$= \frac{1}{2} \mathrm{tr}\left[(\mu_1 - \mu_2)^T (\Sigma_1^{-1} + \Sigma_2^{-1})(\mu_1 - \mu_2) \right. \tag{29}$$

$$\left. + \Sigma_1 \Sigma_2^{-1} + \Sigma_2 \Sigma_1^{-1}) \right] - d.$$

Shakhnoarovich et al. [46] achieved better performance than the MSM approach by Yamaguchi et al. [56] on a dataset including 29 subjects.

Other than the KL divergence, probabilistic distance measures such as Chernoff distance and Bhattacharyya distance can be used too. The Chernoff distance is defined and computed in the case of normal density as:

$$J_C(p_1, p_2) = -\log \int_{\mathrm{x}} p_1^{\alpha_2}(\mathrm{x}) p_2^{\alpha_1}(\mathrm{x}) d\mathrm{x} \tag{30}$$

$$= \frac{1}{2}\alpha_1\alpha_2(\mu_1 - \mu_2)^{\mathrm{T}}[\alpha_1\Sigma_1 + \alpha_2\Sigma_2]^{-1}(\mu_1 - \mu_2)$$

$$+ \frac{1}{2}\log \frac{|\alpha_1\Sigma_1 + \alpha_2\Sigma_2|}{|\Sigma_1|^{\alpha_1}|\Sigma_2|^{\alpha_2}},$$

where $\alpha_1 > 0$, $\alpha_2 > 0$, and $\alpha_1 + \alpha_2 = 1$. When $\alpha_1 = \alpha_2 = 1/2$, the Chernoff distance reduces to the Bhattacharyya distance.

In [25], Jebara and Kondon proposed probability product kernel function

$$k(p_1, p_2) = \int_{\mathrm{x}} p_1^r(\mathrm{x}) p_2^r(\mathrm{x}) d\mathrm{x}, \; r > 0. \tag{31}$$

When $r = 1/2$, the kernel function k reduces to the so-called Bhattacharyya kernel since it is related to the Bhattacharyya distance. When $r = 1$, the kernel function k reduces to the so-called expected-likelihood kernel. In practice, we can simply use the kernel function k as a similarity function.

However, the Gaussian assumption can be ineffective when modeling a nonlinear face-appearance manifold. To absorb the nonlinearity, mixture models or nonparametric densities are used in practice. For such cases, one has to resort to numerical methods for computing the probabilistic distances. Such computation is not robust since two approximations are invoked: one in estimating the density and the other one in evaluating the numerical integral.

In [65], Zhou and Chellappa modeled the nonlinearity through a different approach: kernel methods. As mentioned earlier, the essence of kernel methods is to combine a linear algorithm with a nonlinear embedding, which maps the data from the original vector space to a nonlinear feature space called the (RKHS). But, no explicit knowledge of the nonlinear mapping function is needed as long as the involved computations can be cast into inner-product evaluations. Since a nonlinear function was used, albeit in an implicit fashion, Zhou and Chellappa [65] obtained a new approach to study these distances and investigate their uses in a different space. To be specific, analytic expressions for probabilistic distances that account for nonlinearity or high-order statistical characteristics of the data are derived. On a dataset involving 15 subjects presenting appearances with pose and illumination variations, the probabilistic distance measures performed better than their nonkernelized counterparts.

Recently, Arandjelović and Cipolla [5] used resistor-average distance (RAD) for video-based recognition:

$$\mathrm{RAD}(p_1, p_2) = [\mathrm{KL}(p_1||p_2)^{-1} + \mathrm{KL}(p_2||p_1)^{-1}]^{-1}. \tag{32}$$

Further, computation of the RAD was conducted on the RKHS to absorb nonlinearity of the face manifold. Some robust techniques such as synthesizing images to account for small localization errors and RANSAC algorithms to reject outliers were introduced to achieve improved performance.

Manifold

Fitzgibbon and Zisserman [20] proposed to compute a joint manifold distance to cluster appearances. A manifold is captured by subspace analysis which is fully specified by a mean and a set of basis vectors. For example, a manifold $\mathcal{P}$ can be represented as

$$\mathcal{P} = \{\mathtt{m_p} + \mathtt{B_p u} | \mathtt{u} \in \mathcal{U}\}, \tag{33}$$

where $\mathtt{m_p}$ is the mean and $\mathtt{B}_p$ encodes the basis vectors. In addition, the authors invoked affine transformation to overcome geometric deformation. The joint manifold distance between $\mathcal{P}$ and $\mathcal{Q}$ is defined as

$$\begin{aligned} d(\mathcal{P}, \mathcal{Q}) = \min_{\mathtt{u,v,a,b}} & \|T(\mathtt{m_p} + \mathtt{B_p u}, \mathtt{a}) - T(\mathtt{m_q} + \mathtt{B_q v}, \mathtt{b})\|^2 \\ & + E(\mathtt{a}) + E(\mathtt{b}) + E(\mathtt{u}) + E(\mathtt{v}), \end{aligned} \tag{34}$$

where $T(\mathtt{x}, \mathtt{a})$ transforms image $\mathtt{x}$ using the affine parameter $\mathtt{a}$ and $E(\mathtt{a})$ is the prior cost incurred by invoking the parameter $\mathtt{a}$.

In experiments, Fitzgibbon and Zisserman [20] performed automatic clustering of faces in feature-length movies. To reduce the lighting effect, the face images are high-pass-filtered before being subjected to clustering step. The authors reported that sequence-to-sequence matching presents a dramatic computational speedup when compared with pairwise image-to-image matching.

Identity surface is a manifold, proposed by Li et al. in [33], that depicts face appearances presented in multiple poses. The pose is parametrized by yaw α and tilt θ. Face image at (α, θ) is first fitted to a 3D point distribution model and an active-appearance model. After the pose fitting, the face appearance is warped to a canonical view to provide a pose-free representation from which a nonlinear discriminatory feature vector is extracted. Suppose that the feature vector is denoted by $\mathtt{f}$; then the function $\mathtt{f}(\alpha, \theta)$ defines the identity surface

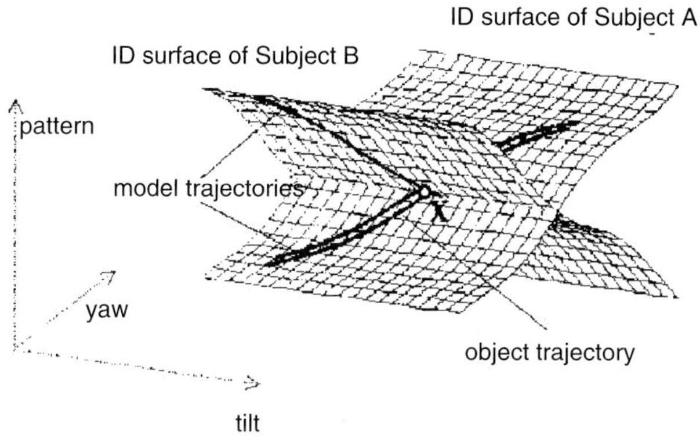

FIGURE 17.3: Identity surface and trajectory. Video-based face recognition reduces to matching two trajectories on identity surfaces (reproduced from Li et al. [33]).

that is pose-parametrized. In practice, since only a discrete set of views are available, the identity surface is approximated by piecewise planes. The manifold distance between two manifolds $\mathcal{P} = \{\mathtt{f}_\mathrm{p}(\alpha, \theta)\}$ and $\mathcal{Q} = \{\mathtt{f}_\mathrm{q}(\alpha, \theta)\}$ is defined as

$$d(\mathcal{Q}, \mathcal{P}) = \int_\alpha \int_\theta w(\alpha, \theta) d(\mathtt{f}_\mathrm{q}(\alpha, \theta), \mathtt{f}_\mathrm{p}(\alpha, \theta)) d\alpha d\theta. \tag{35}$$

where $w(\alpha, \theta)$ is a weight function.

A video sequence corresponds to a trajectory traced out in the identity surface. Suppose that video frames sample the pose space at $\{\alpha_j, \theta_j\}$, the distance $\sum_j w_j d(\mathtt{f}_q(\alpha_j, \theta_j), \mathtt{f}_p(\alpha_j, \theta_j))$ is used for video-based FR. Figure 17.3 illustrates the identity surface and trajectory. In the experiments, 12 subjects were involved and a 100% recognition accuracy was achieved.

17.4.2 Approaches Utilizing $P2$: Temporal Continuity/Dynamics

Simultaneous tracking and recognition is an approach proposed by Zhou et al. [59] that systematically studies how to incorporate temporal continuity in video-based recognition. Zhou et al. modeled two tasks involved, namely tracking and recognition, in one probabilistic framework. A time-series model is used, with the state vector (n_t, θ_t), where n_t is the identity variable and θ_t is the tracking parameter, and the observation $\boldsymbol{y}_t$ (i.e. the video frame). The time-series model is fully specified by the state transition probability $p(n_t, \theta_t | n_{t-1}, \theta_{t-1})$ and the observational likelihood $p(\boldsymbol{y}_t | \theta_t, n_t)$.

The task of recognition is to compute the posterior recognition probability $p(n_t|\mathbf{y}_{0:t})$, where $\mathbf{y}_{0:t} = \{\mathbf{y}_0, \mathbf{y}_2, \ldots, \mathbf{y}_t\}$ a dummy observation $\mathbf{y}_0$ is introduced for convenience in notation. Using time recursion, Markov properties, and statistical independence embedded in the model, one can easily derive

$$
\begin{aligned}
p(n_{0:t}, \theta_{0:t}|\mathbf{y}_{0:t}) &= p(n_{0:t-1}, \theta_{0:t-1}|\mathbf{y}_{0:t-1}) \frac{p(\mathbf{y}_t|n_t, \theta_t)p(n_t|n_{t-1})p(\theta_t|\theta_{t-1})}{p(\mathbf{y}_t|\mathbf{y}_{0:t-1})} \\
&= p(n_0, \theta_0|\mathbf{y}_0) \prod_{s=1}^{t} \frac{p(\mathbf{y}_s|n_s, \theta_s)p(n_s|n_{s-1})p(\theta_s|\theta_{s-1})}{p(\mathbf{y}_s|\mathbf{y}_{0:s-1})} \\
&= p(n_0|\mathbf{y}_0)p(\theta_0|\mathbf{y}_0) \prod_{s=1}^{t} \frac{p(\mathbf{y}_s|n_s, \theta_s)\delta(n_s - n_{s-1})p(\theta_s|\theta_{s-1})}{p(\mathbf{y}_s|\mathbf{y}_{0:s-1})}. \quad (36)
\end{aligned}
$$

Therefore, by marginalizing over $\theta_{0:t}$ and $n_{0:t-1}$, one obtains

$$
p(n_t = l|\mathbf{y}_{0:t}) = p(l|\mathbf{y}_0) \int_{\theta_0} \cdots \int_{\theta_t} p(\theta_0|\mathbf{y}_0) \prod_{s=1}^{t} \frac{p(\mathbf{y}_s|l, \theta_s)p(\theta_s|\theta_{s-1})}{p(\mathbf{y}_s|\mathbf{y}_{0:s-1})} d\theta_t \ldots d\theta_0. \quad (37)
$$

Thus $p(n_t = l|\mathbf{y}_{0:t})$ is determined by the prior distribution $p(n_0 = l|\mathbf{y}_0)$ and the product of the likelihood functions, $\prod_{s=1}^{t} p(\mathbf{y}_s|l, \theta_s)$. If a uniform prior is assumed, then $\prod_{s=1}^{t} p(\mathbf{y}_s|l, \theta_s)$ is the only determining factor that accumulates the evidence.

As this model is nonlinear with nonGaussian noise, the joint posterior distribution of the motion vector and the identity variable is approximated at each time instant using a computationally efficient sequential importance sampling (SIS) algorithm [16N, 25N, 33N] and then propagated to the next time instant. Marginalization over the motion vector yields a robust estimate of the posterior distribution of the identity variable. Empirical results as in Figure 17.4 demonstrate that, due to the propagation of the identity variable over time, a degenerate condition in the posterior probability of the identity variable is achieved to obtain improved recognition.

In the experiments reported in [59], using a database of 30 subjects whose appearances present in the probe videos are very different from those in the gallery set, simultaneous tracking and recognition achieved improvement in both tracking accuracy and recognition performance, compared with pure tracking and still-image-based recognition. The proposed framework in [59] is rather general and accommodates various model choices. The more effective the model choices are, the better is the expected performance in tracking and recognition. In the work of Zhou et al. [60], the enhanced tracking and recognition models are invoked

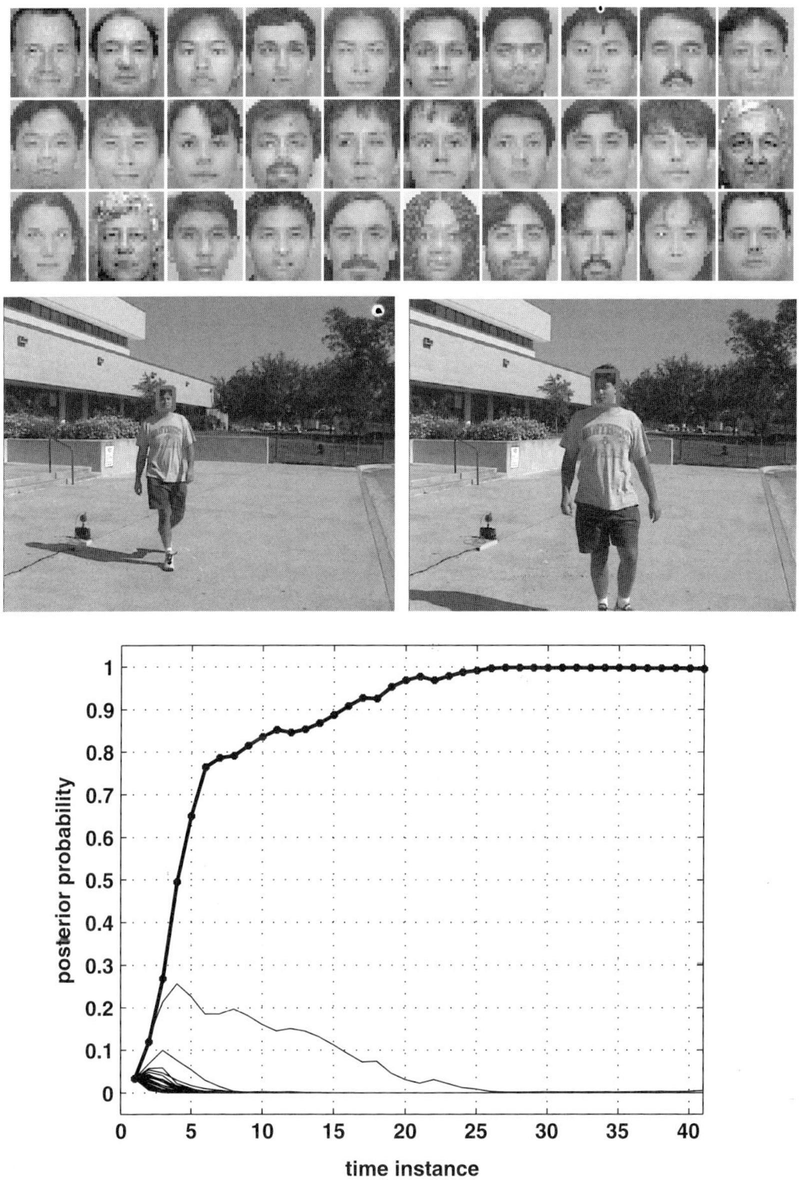

FIGURE 17.4: Top row: the gallery images. Bottom row: the first (left) and the last (middle) frames of the video sequences with tracking results indicated by the box and the posterior probability $p(n_t|\mathbf{y}_{0:t})$.

to handle a challenging dataset with abounding pose and illumination variations which actually fails the earlier approach [59].

In the work of Zhou et al. [59], other than in the case when the gallery consists of one still image per individual, they also extended the approach to handle a video sequence in the gallery set. Representative exemplars are learned from the gallery video sequences to depict individuals. Then simultaneous tracking and recognition was invoked to handle video sequences in the probe set. Li and Chellappa [32] also proposed an approach somewhat similar to [59]. In [32], only tracking was implemented, using SIS, and recognition scores were subsequently derived based on tracking results.

Lee et al. [30] performed video-based face recognition using probabilistic appearance manifolds. The main motivation is to model appearances under pose variation; i.e., a generic appearance manifold consists of several pose manifolds. Since each pose manifold is represented using a linear subspace, the overall appearance manifold is approximated by piecewise linear subspaces. The learning procedure is based on face exemplars extracted from a video sequence. K-means clustering is first applied; then, for each cluster, principal-component analysis is used for a subspace characterization.

In addition, the transition probabilities between pose manifolds are also learned. The temporal continuity is directly captured by the transition probabilities. In general, the transition probabilities between neighboring poses (such as frontal pose to left pose) are higher than those between far-apart poses (such as left pose to right pose). Recognition also reduces to computing a posterior distribution.

In experiments, Lee et al. compared three methods that use temporal information differently: the proposed method with learned transition matrix, the proposed method with uniform transition matrix (meaning that temporal continuity is lost), and majority voting. The proposed method with learned transition matrix achieved a significantly better performance than the other two methods. Recently, Lee and Krigman[29N] extended [30] by learning the appearance manifold from a testing video sequence in an online fashion.

Liu and Chen [34] used the adaptive hidden Markov model (HMM) to depict the dynamics. HMM is a statistical tool to model time series. Usually, the HMM is denoted by $\lambda = (\mathtt{A}, \mathtt{B}, \pi)$, where $\mathtt{A}$ is the state transition probability matrix, $\mathtt{B}$ is the observation PDF, and π is the initial state distribution. Given a probe video sequence $\mathtt{Y}$, its identity is determined as

$$\hat{n} = \underset{1,2,\ldots,N}{\operatorname{argmax}} = p(\mathtt{Y}|\lambda_n), \tag{38}$$

where $p(\mathtt{Y}|\lambda_n)$ is the likelihood of observing the video sequence $\mathtt{Y}$ given the model λ_n. In addition, when certain conditions hold, HMM λ_n is adapted to accommodate the appearance changes in the probe video sequence that results in improved modeling over time. Experimental results on various datasets demonstrated the advantages using of adaptive HMM.

Aggarwal et al. [4] proposed a system identification approach for video-based FR. The face sequence is treated as a first-order autoregressive and moving-averaging (ARMA) random process:

$$\theta_{t+1} = \mathtt{A}\theta_t + \mathtt{v}_t, \;\; \boldsymbol{y}_t = \mathtt{C}\theta_t + \mathtt{w}_t, \tag{39}$$

where $\mathtt{v}_t \sim \mathcal{N}(0, \mathtt{Q})$ and $\mathtt{w}_t \sim \mathcal{N}(0, \mathtt{R})$. System identification is equivalent to estimating the parameters $\mathtt{A}$, $\mathtt{C}$, $\mathtt{Q}$, and $\mathtt{R}$ from the observations $\{\boldsymbol{y}_1, \boldsymbol{y}_2, \ldots, \boldsymbol{y}_T\}$. Once the system is identified or each video sequence is associated with its parameters, video-to-video recognition uses various distance metrics constructed based on the parameters. Promising experimental results (over 90%) were reported when significant pose and expression variations are present in the video sequences.

Facial expression analysis is also related to temporal continuity dynamics, but not directly related to FR. Examples of expression analysis include [9, 50]. A review of face-expression analysis is beyond the scope of this chapter.

17.4.3 Approaches Utilizing $P3$: (3D Model)

There is a large body of literature on SfM. However, the current SfM algorithms do not reliably reconstruct the 3D face model. There are three difficulties in the SfM algorithm. The first lies in the ill-posed nature of the perspective camera model that results in instability of the SfM solution. The second is that the face model is not a truly rigid model, especially when the face presents facial expression and other deformations. The final difficulty is related to the input to the SfM algorithm. This is usually a sparse set of feature points provided by a tracking algorithm that itself has its own flaw. Interpolation from a sparse set of feature points to a dense set is very inaccurate.

To relieve the first difficulty, orthographic and paraperspective models are used to approximate the perspective camera model. Under such approximate models, the ill-posed problem becomes well posed. In Tomasi and Kanade [48], the orthographic model was used, and a matrix factorization principle was discovered. The factorization principle was extended to the paraperspective camera model in Poelman and Kanade [41]. Factorization under uncertainty was considered in [11, 24].

The second difficulty is often resolved by imposing a subspace constraint on the face model. Bregler et al. [12] proposed to regularize the nonrigid face model by using the linear constraint. It was shown that factorization can still be obtained. Brand [11] considered such factorization under uncertainty. Xiao et al. [54] discovered a closed-form solution to nonrigid shape and motion recovery. Figure 17.5 shows the recovered face models obtained by Xiao et al. [54].

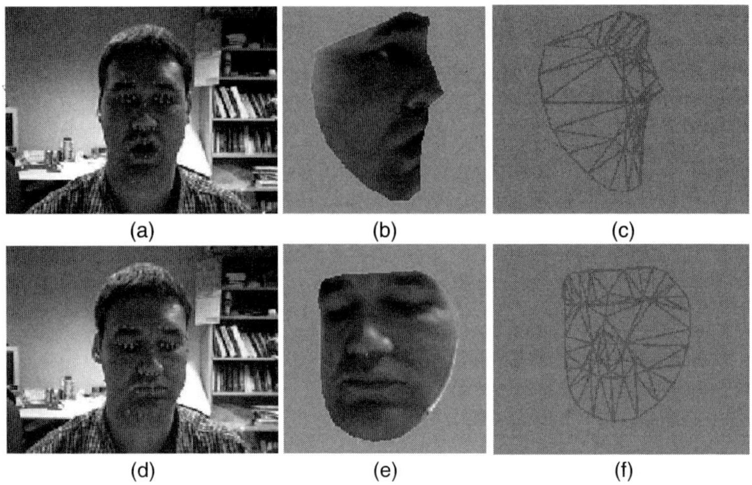

FIGURE 17.5: Nonrigid shape and motion recovery: (a, d) input images, (b, e) reconstructed face shapes seen from novel views; (c, f) the wireframe models demonstrate the recovered facial deformations such as mouth opening and eye closure. (Courtesy of J. Xiao.)

Interpolation from a sparse set to a dense depth map is always a difficult task. To overcome this, a dense face model is used instead of interpolation. However, the dense face model is only a generic model and hence may not be appropriate for a specific individual. Bundle adjustment [21, 47] is a method that adjusts the generic model directly to accommodate the video observation. Roy-Chowdhury and Chellappa [44] took a different approach for combining the 3D face model recovered from the SfM algorithm with the generic prior face model.

The SfM algorithm mainly recovers the geometric component of the face model, i.e., the depth value of every pixel. Its photometric component is naively set to the appearance in one reference video frame. The image-based rendering method, on the other hand, directly recovers the photometric component of the 3D model. Light-field rendering [23, 31] in fact bypasses the stage of recovering the photometric components of the 3D model; rather it recovers the novel views directly. The light-field rendering methods [23, 31] relax the requirement of calibration, by a fine quantization of the pose space and recover a novel view by sampling the captured data that form the so-called light field. The "eigen" light field approach developed by Gross et al. [22] assumes a subspace condition on the light field. In Zhou and Chellappa [63], the light-field subspace and the illumination subspace are combined to arrive at a bilinear analysis. Figure 17.6 shows the rendered images at different views and light conditions. Another line of research relating to

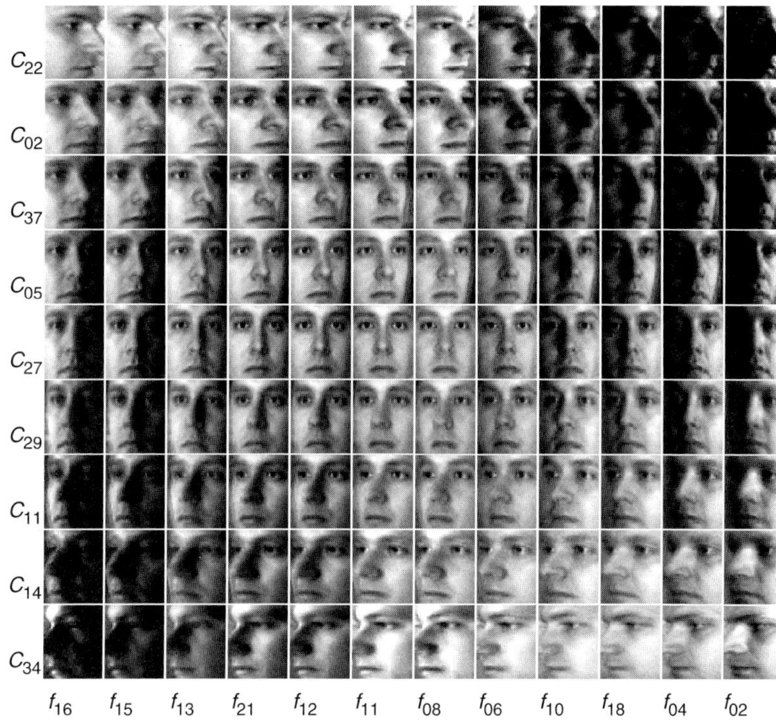

FIGURE 17.6: The reconstruction results using the approach in Zhou and Chellappa [63]. Note that only the images for the row C_{27} are used for reconstructing all images here.

3D model recovery is the visual-hull method [29, 35]. But, this method assumes that the shape of the object is convex, which is not satisfied by the human face, and also requires accurate calibration information. Direct use of visual hull for FR is not found in the literature.

To characterize both the geometric and photometric components of the 3D face model, Blanz and Vetter [10] fitted a 3D morphable model to a single still image. The 3D morphable model uses a linear combination of dense 3D models and texture maps. In principle, the 3D morphable model can be fitted to multiple images. The 3D morphable model can be thought of as an extension of 2D active-appearance model [15] to 3D, but the 3D morphable model uses dense 3D models. Xiao et al. [55] proposed to combine a linear combination of 3D sparse model and a 2D appearance model.

Although there is a lot of interest in recovering the 3D model, directly performing FR using the 3D model is a recent trend [8, 36, 13]. Blanz and Vetter [10] implicitly did so by using the combining coefficients for recognition. Beumier and Acheroy [8] conducted matching based on 2D sections of the facial surface. Mavridis et al. [36] used a 3D color camera to perform face recognition. Bronstein et al. [13] used a 3D face model for compensating the effect of facial expression in face recognition. However, the above approaches use the 3D range data as input. Because in this chapter we are mainly interested in face recognition from multiple still images or video sequence, a thorough review of face recognition based on 3D range data is beyond its scope.

17.5 FUTURE

In the Section 17.4, we reviewed the approaches that utilize the three properties. Although they usually achieved good recognition performance, they have their own assumptions or limitations. For example, the Gaussian distribution used in Shakhnoarovich et al. [46] is easily violated by pose and illumination variations. The hidden Markov model used in Liu and Chen [34] poses a strong constraint on the change of face appearance that is not satisfied by video sequences that contain an arbitrarily moving face.

In this section, we forecast possible new approaches. These new approaches either arise from new representation for more than one still image or extend the capability of the existing approaches.

17.5.1 New Representation

In the matrix representation, multiple observations are encoded using a matrix. In other words, each observation is an image that is "vectorized". The vectorization operator ignores the spatial relationship of the pixels.

To fully characterize the spatial relationship, a tensor can be used in lieu of matrix. Here, a tensor is understood as a 3D array. Tensor representation is used in Vasilescu and Terzopoulos [52] to learn a generic model of the face appearance for all humans, at different views, and under different illuminating conditions, and so on. However, comparing two tensors has not been investigated in the literature.

In principle, the PDF representation is very general. But in the experiment, a certain parametric form is assumed as in Shakhnoarovich et al. [46]. Other PDF forms can be employed. The key is to find an appropriate density that can model face appearance. The same problem happens in the manifold description. With advance in manifold modeling, FR based on manifolds can be improved too.

17.5.2 Using the Training Set

The training set is normally used to provide a generic models of face appearance for all humans. On the other hand, the images in the gallery set are related to modeling the facial appearance of an individual. If there are enough observations, one can build an accurate model of the face for each individual in the gallery, set, and hence knowledge of the training set is not necessary. If the number of images is insufficient, one should combine the knowledge of generic modeling with the individualized modeling to describe the identity signature.

17.5.3 3D Model Comparison

As mentioned in Section 17.4.3, comparison between two 3D models has not been fully investigated yet. In particular, direct comparison of the geometric component of the 3D model is rather difficult, because it is nontrivial to recover the 3D model in the first place, and because the correspondence between two 3D models cannot be easily established.

Current approaches [55] warp the model to frontal view and use the frontal 2D face appearance for recognition. However, these approaches are very sensitive to illumination variation. Generalized photometric stereo [61] can be incorporated into these approaches for a more accurate model.

The most sophisticated 3D model uses statistical description. In other words, both the geometric component g and the texture component f have their distributions, say $p(g)$ and $p(f)$, respectively. Such distributions can be learned from the multiple still images or video sequence. Probabilistic matching can then be applied for FR.

17.5.4 Utilizing More Than One Property

Most of the approaches reviewed in Section 17.4 utilize only one of the three properties. However, these properties are not overlapping, in the sense that more than one property can be unified to achieve further improvements.

Probabilistic identity characterization proposed by Zhou and Chellappa [64] is an instance of integrating the properties $P1$ and $P2$. In [64], FR from multiple still images and FR from video sequences are unified in one framework.

Statistical 3D model is a combination of the properties $P1$ and $P3$, where the PDF part of the property $P1$ is used.

17.6 CONCLUSIONS

We have reviewed the emerging literature on FR using more than one still image, i.e., multiple still images or a video sequence. In a degenerate manner, FR from more than one image can be reduced to recognition based on a single still image. Such a treatment ignores the properties additionally possessed by more than one image. We have also identified three properties that are widely used in the literature, namely *multiple observations*, *temporal continuity/dynamics*, and *3D model*. We then reviewed approaches that utilize the three properties and suggested potential approaches that could realize the full benifits of multiple still images or video sequences.

REFERENCES

[1] Face Recognition Grand Challenge. http://bbs.bee-biometrics.org.

[2] *The First IEEE Workshop on Face Processing in Video.* http://www.visioninterface.net/fpiv04.

[3] F. Fraser. "Exploring the use of face recognition technology for border control applications – Austrian experience," *Biometric Consortium Conference*, 2003.

[4] G. Aggarwal, A. Roy-Chowdhury, and R. Chellappa. "A system identification approach for video-based face recognition," *Proceedings of the International Conference on Pattern Recognition*, Cambridge, UK, August 2004.

[5] O. Arandjelović and R. Cipolla. "Face recognition from face motion manifolds using robust kernel resistor-average distance," *IEEE Workshop on Face Processing in Video*, Washington D.C., USA, 2004.

[5N] O. Arandjelović, G. Shakhnarovich, J. Fisher, R. Cipolla, and T. Darrell. "Face Recognition with Image Sets using Manifold Density Divergence," *Proc. IEEE Conference on Computer Vision and Pattern Recoginition*, vol. 1, pages 581–588, San Diego, USA, June 2005.

[6] M.S. Barlett, H.M. Ladesand, and T.J. Sejnowski. "Independent component representations for face recognition," *Proceedings of SPIE 3299*, pp. 528–539, 1998.

[7] P.N. Belhumeur, J.P. Hespanha, and D.J. Kriegman. "Eigenfaces vs. Fisherfaces: recognition using class specific linear projection," *IEEE Trans. Pattern Analysis and Machine Intelligence* **19**, pp. 711–720, 1997.

[8] C. Beumie and M.P. Acheroy. "Automatic face authentication from 3D surface," *Proc. of the British Machine Vision Conference* pp. 449–458, 1998.

[9] M.J. Black and Y. Yacoob. "Recognizing facial expressions in image sequences using local parametrized models of image motion," *International Journal of Computer Vision* **25**, pp. 23–48, 1997.

[10] V. Blanz and T. Vetter. "Face recognition based on fitting a 3D morphable model," *IEEE Transaction on Pattern Analysis and Machine Intelligence* **25**, pp. 1063–1074, 2003.

[11] M.E. Brand. "Morphable 3D Models from Video," *Proceedings of the IEEE Conference on Computer Vision and Pattern Recognition*, Hawaii, 2001.

[12] C. Bregler, A. Hertzmann, and H. Biermann. "Recovering nonrigid 3D shape from image streams," *IEEE Computer Society Conference on Computer Vision and Pattern Recognition*, 2000.

[13] A.M. Bronstein, M.M. Bronstein, and R. Kimmel. "Expression-invariant 3D face recognition," *Proc. Audio and Video-based Biometric Personal Authentication*, pp. 62–69, 2003.

[14] R. Chellappa, C.L. Wilson, and S. Sirohey. "Human and machine recognition of faces: a survey," *Proceedings of the IEEE* **83**, pp. 705–740, 1995.

[15] T.F. Cootes, G.J. Edwards, and C.J. Taylor. "Active appearance models," *IEEE Trans. on Pattern Analysis and Machine Intelligence* **23**, no. 6, pp. 681–685, 2001.

[16] T.M. Cover and J.A. Thomas, *Elements of Information Theory*. Wiley, 1991.

[16N] A. Doucet, N. de Freitas, and N. Gordon. *Sequential Monte Carlo Methods in Practice*. Springer-Verlag, New-York, 2001.

[17] R.O. Duda, P.E. Hart, and D.G. Stork. *Pattern Classification*. Wiley-Interscience, 2001.

[18] G.J. Edwards, C.J. Taylor, and T.F. Taylor. "Improving idenfication performation by integrating evidence from sequences," *IEEE Intl. Conf. on Computer Vision and Pattern Recognition*, pp. 486–491, Fort Collins, Colorado, USA, 1999.

[19] K. Etemad and R. Chellappa. "Discriminant analysis for recognition of human face images," *Journal of the Optical Society of America A*, pp. 1724–1733, 1997.

[20] A. Fitzgibbon and A. Zisserman. "Joint manifold distance: a new approach to appearance based clustering," *IEEE Computer Society Conference on Computer Vision and Pattern Recognition*, Madison, WI, 2003.

[21] P. Fua. "Regularized bundle adjustment to model heads from image sequences without calibrated data," *International Journal of Computer Vision* **38**, pp. 153–157, 2000.

[22] R. Gross, I. Matthews, and S. Baker. "Eigen light-fields and face recognition across pose," *Proceedings of the International Conference on Automatic Face and Gesture Recognition*, Washington, D.C., 2002.

[23] S.J. Gortler, R. Grzeszczuk, R. Szeliski, and M. Cohen. "The lumigraph," *Proceedings of SIGGRAPH*, pp. 43–54, New Orleans, LA, USA, 1996.

[24] M. Irani and P. Anandan. "Factorization with uncertainty," *European Conference on Computer Vision*, pages 539–553, 2000.

[25] T. Jebara and R. Kondor. "Bhattacharyya and expected likelihood kernels," *Conference on Learning Theory, COLT*, 2003.

[25N] G. Kitagawa. "Monite Carlo filter and smoother for non-gaussian nonlinear state space models," *J. Computational and Graphical Statistics* **5**, pp. 1–25, 1996.

[26] R. Kondor and T. Jebara."A kernel between sets of vectors," *International Conference on Machine Learning, ICML*, 2003.

[27] B. Knight and A. Johnston. "The role of movement in face recognition," *Visual Cognition* **4**, pp. 265–274, 1997.

[28] V. Krueger and S. Zhou. "Exemplar-based face recognition from video," *European Conference on Computer Vision*, Copenhagen, Denmark, 2002.

[29] A. Laurentini. "The visual hull concept for silhouette-based image understanding," *IEEE Trans. Pattern Analysis and Machine Intelligence* **16**, no. 2, pp. 150–162, 1994.

[29N] K. Lee and D. Kriegman. "Online Learning of Probabilistic Appearance Manifolds for Video-based Recognition and Tracking," *IEEE conf. on Computer Vision and Pattern Recognition*, San Diego, USA, June 2005.

[30] K. Lee, M. Yang, and D. Kriegman. "Video-based face recognition using probabilistic appearance manifolds," *IEEE Computer Society Conference on Computer Vision and Pattern Recognition*, Madison, WI, 2003.

[31] M. Levoy and P. Hanrahan. "Light field rendering," *Proceedings of SIGGRAPH*, pp. 31–42, New Orleans, LA, USA, 1996.

[32] B. Li and R. Chellappa. "A generic approach to simultaneous tracking and verification in video," *IEEE Trans. on Image Processing* **11**, no. 5, pp. 530–554, 2002.

[33] Y. Li, S. Gong, and H. Liddell. "Constructing face identity surface for recognition," *Internationl Journal of Computer Vision* **53**, no. 1, pp. 71–92, 2003.

[33N] J.S. Liu and R. Chen. "Sequential monte carlo for dynamic systems," *Journal of the American Statistical Association* **93**, pp. 1031–1041, 1998.

[34] X. Liu and T. Chen. "Video-based face recognition using adaptive hidden Markov models," *Proceedings of the IEEE Computer Society Conference on Computer Vision and Pattern Recognition*, Madison, WI, 2003.

[35] W. Matusik, C. Buehler, R. Raskar, S. Gortler, and L. McMillan. "Image-based visual hulls," *Proceedings of SIGGRAPH*, pp. 369–374, New Orleans, LA, USA, 2000.

[36] M. Mavridis *et al.* "The HISCORE face recognition applicaiton: Affordable desktop face recognition based on a novel 3D camera," *Proc. of the Intl. Conference on Augmented Virtual Environments and 3D Imaging*, 2001.

[37] B. Moghaddam. "Principal manifolds and probabilistic subspaces for visual recognition," *IEEE Trans. Pattern Analysis and Machine Intelligence* **24**, pp. 780–788, 2002.

[38] P. Penev and J. Atick. "Local feature analysis: A general statistical theory for object representation," *Networks: Computations in Neural Systems* **7**, pp. 477–500, 1996.

[39] P.J. Phillips, H. Moon, S. Rizvi, and P.J. Rauss. "The FERET evaluation methodology for face-recognition algorithms," *IEEE Trans. Pattern Analysis and Machine Intelligence* **22**, pp. 1090–1104, 2000.

[40] P.J. Phillips, P. Grother, R.J. Micheals, D.M. Blackburn, E. Tabbssi, and M. Bone. "Face recognition vendor test 2002: evaluation report," *NISTIR 6965*, http://www.frvt.org, 2003.

[41] C. Poelman and T. Kanade. "A paraperpective factorization method for shape and motion recovery," *IEEE Trans. Pattern Analysis and Machine Intelligence* **19**, no. 3, pp. 206–218, 1997.

[42] G. Qian and R. Chellappa. "Structure from motion using sequential Monte Carlo methods," *Proceedings of the International Conference on Computer Vision*, pp. 614–621, Vancouver, BC, 2001.

[43] S.T. Roweis and L.K. Saul. "Nonlinear dimensionality reduction by locally linear embedding," *Science* **290**, pp. 2323–2326, 2000.

[44] A. Roy-Chowdhury and R. Chellappa. "Face reconstruction from video using uncertainty analysis and a generic model," *Computer Vision and Image Understanding* **91**, pp. 188–213, 2003.

[45] B. Schölkopf and A. Smola. *Learning with Kernels Support Vector Machines, Regularization, Optimization, and Beyond*, MIT Press, Cambridge, MA, 2002.

[46] G. Shakhnarovich, J. Fisher, and T. Darrell. "Face recognition from long-term observations," *European Conference on Computer Vision*, Copenhagen, Denmark, 2002.

[47] Y. Shan, Z. Liu, and Z. Zhang. "Model-based bundle adjustment with applicaiton to face modeling," *Proceedings of the International Conference on Computer Vision*, pp. 645–651, Vancouver, BC, 2001.

[48] C. Tomasi and T. Kanade. "Shape and motion from image streams under orthography: a factorization method," *International Journal of Computer Vision* **9**, no. 2, pp. 137–154, 1992.

[49] J.B. Tenenbaum, V. de Silva and J.C. Langford. "A global geometric framework for nonlinear dimensionality reduction," *Science* **290**, pp. 2319–2323, 2000.

[50] Y. Tian, T. Kanade, and J. Cohn. "Recognizing action units of facial expression analysis," *IEEE Trans. Pattern Analysis and Machine Intelligence* **23**, pp. 1–19, 2001.

[51] M. Turk and A. Pentland. "Eigenfaces for recognition," *Journal of Cognitive Neuroscience* **3**, pp. 72–86, 1991.

[52] M.A.O. Vasilescu and D. Terzopoulos. "Multilinear analysis of image ensembles: Tensorfaces," *European Conference on Computer Vision* **2350**, pp. 447–460, Copenhagen, Denmark, May 2002.

[53] L. Wolf and A. Shashua. "Kernel principal angles for classification machines with applications to image sequence interpretation," *IEEE Computer Society Conference on Computer Vision and Pattern Recognition*, Madison, WI, 2003.

[54] J. Xiao, J. Chai, and T. Kanade. "A closed-form solution to non-rigid shape and motion recovery," *European Conference on Computer Vision*, 2004.

[55] J. Xiao, S. Baker, I. Matthews, and T. Kanade. "Real-time combined 2D+3D active appearance models," *IEEE Computer Society Conference on Computer Vision and Pattern Recognition*, Washington, DC, 2004.

[56] O. Yamaguchi, K. Fukui and K. Maeda. "Face recognition using temporal image sequence," *Proceedings of the International Conference on Automatic Face and Gesture Recognition*, Nara, Japan, 1998.

[57] W. Zhao, R. Chellappa, and A. Krishnaswamy. "Discriminant analysis of principal components for face recognition," *Proceedings of the International Conference on Automatic Face and Gesture Recognition*, pp. 361–341, Nara, Japan, 1998.

[58] W. Zhao, R. Chellappa, A. Rosenfeld, and P.J. Phillips. "Face recognition: A literature survey," *ACM Computing Surveys* **12**, 2003.

[59] S. Zhou, V. Krueger, and R. Chellappa. "Probabilistic recognition of human faces from video," *Computer Vision and Image Understanding* **91**, pp. 214–245, 2003.

[60] S. Zhou, R. Chellappa, and B. Moghaddam. "Visual tracking and recognition using appearance-adaptive models in particle filters," *IEEE Trans. Image Processing*, November, 2004.

[61] S. Zhou, R. Chellappa, and D. Jacobs. "Characterization of human faces under illumination variations using rank, integrability, and symmetry constraints," *European Conference on Computer Vision*, Prague, Czech, May 2004.

[62] S. Zhou and R. Chellappa. "Intra-personal kernel subspace for face recognition," *Proceedings of International Conference on Automatic Face and Gesture Recognition*, Seoul, Korea, May 2004.

[63] S. Zhou and R. Chellappa. "Image-based face recognition under illumination and pose variations," *Journal of the Optical Society America, A.* **22**, pp. 217–229, February 2005.

[64] S. Zhou and R. Chellappa. "Probabilistic identity characterization for face recognition," *Proceedings of IEEE Computer Society Conference on Computer Vision and Pattern Recognition*, Washington D.C., USA, June 2004.

[65] S. Zhou and R. Chellappa. "Probabilistic distance measures in reproducing kernel Hilbert space," *SCR Technical Report*, 2004.

[66] S. Zhou. "Trace and determinant kernels between matrices," *SCR Technical Report*, 2004.

CHAPTER **18**

SUBSET MODELING OF FACE LOCALIZATION ERROR, OCCLUSION, AND EXPRESSION

18.1 INTRODUCTION

Object recognition is a core problem in computer vision and other closely related areas of research, such as pattern recognition, cognitive science, and neuroscience [39, 9, 8, 10]. The goal of object recognition in computer vision is two-fold. On the one hand, researchers attempt to build systems that can interact with their environment with the help of a set of cameras. This has many applications as, for example, in human–computer interaction, biometrics, and communications. On the other hand, research is also driven by the interest to better understand how biological systems work. Face recognition is a great window to both of these studies. Faces are so special that some scientists argue that humans have a specialized area in the brain to analyze some of the most commonly seen objects—faces being (possibly) the most common of all [32, 13].

What makes faces so special is that they all look somewhat alike, yet small differences in the location and shape of our facial features or their texture allow people to distinguish among thousands of individuals. A primary goal in the design of computer-vision systems able to classify faces into a large set of classes is to automatically extract those features that vary among people yet are constant for each individual. This is, however, a task much more difficult than one may think. For example, our perception of a person's face changes as he/she speaks and expresses emotions. Facial hair, glasses, or other types of occlusion exacerbate the problem. And, obviously, illumination changes and pose variations also influence the way a face is seen when projected onto our two-dimensional retinas.

Earlier work mainly focused on finding the subset which represents most of the faces of each individual under all possible views and all possible lighting conditions. Structure from motion and photometric stereo are two well-known examples of such modelings [24, 38]. More recently, Belhumeur and Kriegman [1] have shown that the subset of all images generated by any arbitrary point source is a cone in a three dimensional space. Combining their result with photometric stereo, one can construct systems able to synthesize and recognize people's faces seen from distinct viewing angles and under varying illumination from relatively small sample sets [11].

Recently, progress has also been made toward solving the other three problems described above: errors of localization, occlusions, and facial expressions. These problems can be summarized as follows.

Face localization error. Every face-recognition system, whether appearance- or feature-based, requires a localization stage. However, all localization algorithms have an associated error: namely they cannot localize every single face feature with pixel precision. Unfortunately, test feature vectors generated from imprecisely localized faces can be closer to the training vector of an incorrect class. This problem is depicted in Figure 18.1(a). In this figure, we display two classes, one drawn using crosses and the other with pentagons. For each class, there are two learned feature vectors, each corresponding to the same image but accounting for different localization errors. The test image (which belongs to the "cross" class) is shown as a square. Note that while one of the "pentagon" samples is far from the test feature vector, the other corresponds to the closest sample; that is, while one localization leads to a correct classification, the other does not. This point becomes critical when the learning and testing images differ on facial expression

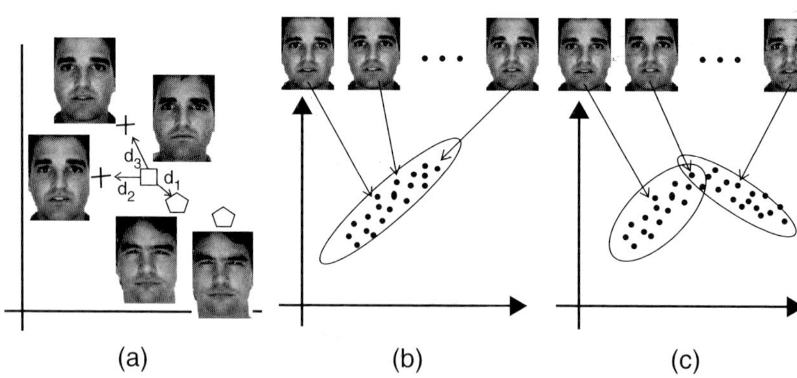

FIGURE 18.1: (a) Different localization results lead to distinct representations in the feature space, which can cause identification failures. (b, c) We model the subspace where different localization lie. From [21] (©2002 IEEE).

or illumination as well as for duplicates. A "duplicate" is an image of a face that is taken at a different time, weeks, months, or even years later.

Occlusions. One of the main drawbacks of any face-recognition system, is its failure to robustly recognize partially occluded faces. Occlusions are unfortunately typical; e.g., facial hair, glasses, clothing, and clutter. Figure 18.2 shows an example, where image (a) is used for learning and image (b) for subsequent recognition.

Expressions. This problem can be formally stated as follows. "How can we robustly identify a person's face for whom the learning and testing face images differ in facial expression?" This problem is illustrated in Figure 18.3. In this figure, three local areas of three face images of the same person, but with different facial expressions, have been highlighted. It is apparent that as the facial expression changes, the local texture of each of these areas becomes quite distinct (even in the forehead). A classification (or identification) algorithm, needs to find where the invariant features are. If this was achieved, we would obtain an expression-invariant face-recognition system. We say that a system is expression-dependent if the image of Figure 18.4(b) is more difficult (or less difficult) to recognize than the image shown in Figure 18.4(c), given that Figure 18.4(a) was used for learning. An expression-invariant algorithm would correspond to one that equally identifies the identity of the subject independently of the facial expression on the training or testing images.

One way to address the problems stated above is to find the subset which represents most of the images under all possible localization errors, occlusions, and expressions for each individual. If this is achieved, the recognition stage will reduce to finding that subset which is closest to the testing vector. In this chapter we show how to achieve this. For illustration we will mainly work within the appearance-based paradigm where each dimension represents the brightness of each image pixel. We will also discuss how our techniques can be used within a feature-based approach where (in general) a set of shape measurements is used instead.

The rest of this chapter is organized as follows. Section 18.2 presents the basic ideas of our modeling and applies it to the problem of localization errors. This formulation is extended in Section 18.3 to incorporate our solutions to the problems of occlusions and expression changes. Experimental results are in Section 18.4. Some of the choices we make in our formulation may seem arbitrary at the time, since other alternatives exist. While this may suffice for some readers, others will want to better understand how such conclusions were reached. In Section 18.5, we show how this was done, either theoretically, experimentally or both. Additional implications of our results and future work are also in this section—where some of the most relevant open problems will be introduced. For example, although the

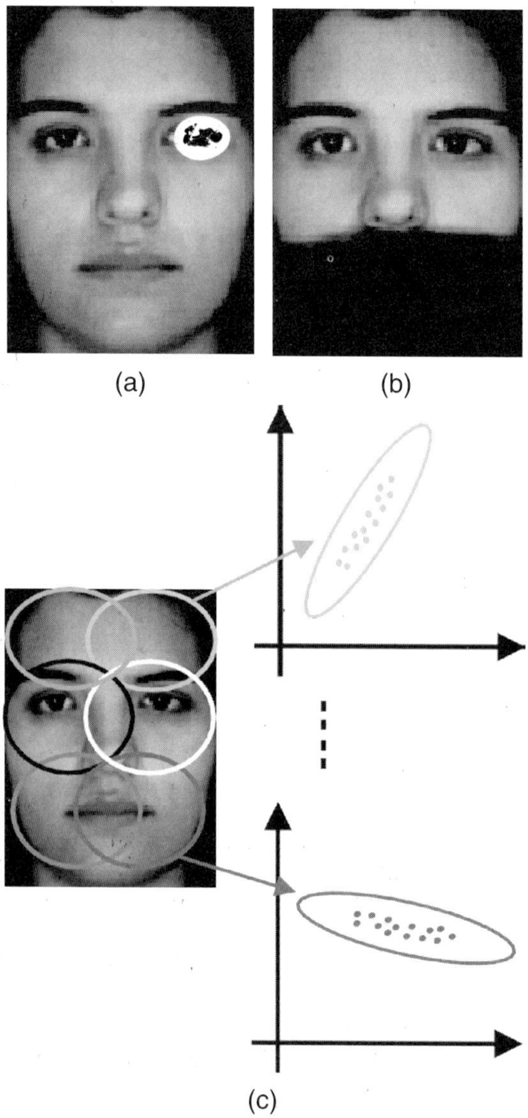

FIGURE 18.2: (a) A learning image. (b) A test image. (c) The local approach. From [21] (©2002 IEEE).

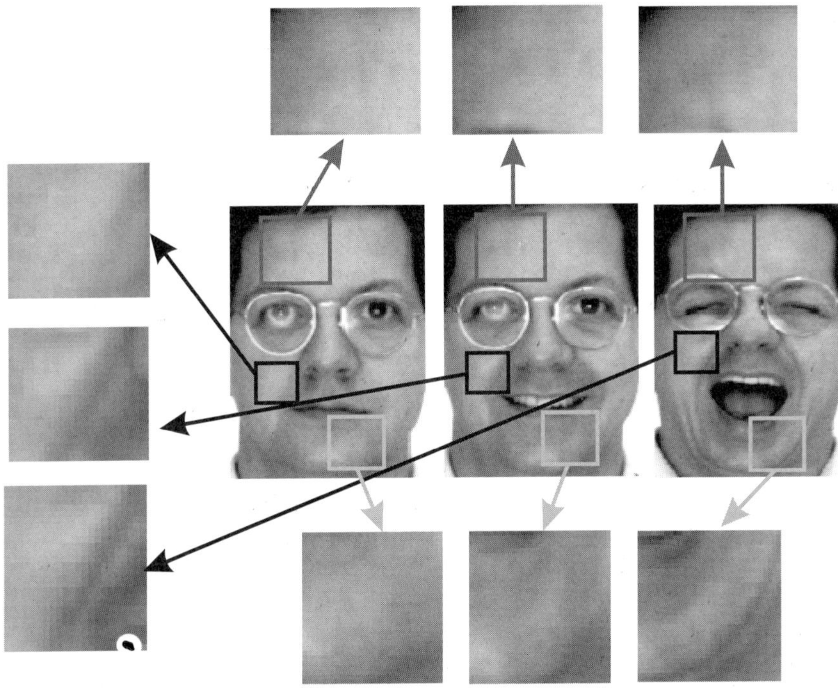

FIGURE 18.3: As the face of a person changes facial expression, the local appearance (texture) of her/his face also changes. In this figure, we compare three local areas of an individual's face under three different expressions (neutral, happy, and screaming).

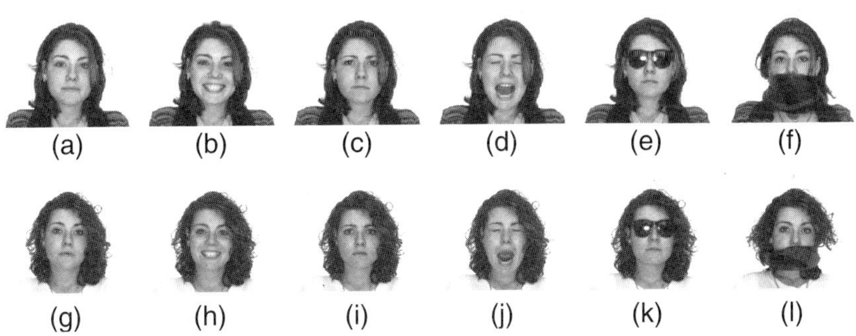

FIGURE 18.4: Images of one subject in the AR face database. The images (a) through (f) were taken during a first session and the images (g) though (l) at a different session 14 days later.

subsets representing each of the above-mentioned variations are assumed to be in a low-dimensional subspace, it is still not known what this dimensionality is. Important challenges lie ahead. We conclude in Section 18.6.

18.2 MODELING THE LOCALIZATION ERROR

18.2.1 Face Localization

Any face-recognition system requires a prelocalization of the face and facial features necessary for recognition. The localization of these facial features is necessary either to construct feature-based representations of the face [5, 39] or to warp all faces to a standard size for appearance-based recognition [3, 12, 21]. However, it would be unrealistic to hope for a system that can localize all the facial features (e.g., eyes, mouth, etc.) with pixel precision. The problem this raises is that the feature representation (i.e., the feature vector) of the correct localized face differs from the feature representation of the actual computed localization, which can ultimately result in an incorrect classification; see Figure 18.1(a).

In order to tackle this problem, some authors have proposed to filter the image before attempting recognition. Filtered images are, in general, less sensitive to the localization problem, but do not fully tackle it. In [21], we proposed to model the subset within the feature space that contains most of the images generated under all possible localizations. An extension of this approach is summarized below.

18.2.2 Finding the subset of localization errors

We propose to model the subset of the localization problem described above by means of a probability density function. Assume at first that the localization errors for both the x and y image axes are known, which we denote as vr_x and vr_y. These two error values tell us that, statistically, the correctly localized face is within the set of images $\hat{\mathbf{x}}_{i,j} = \{\mathbf{x}_{i,j,1}, \ldots, \mathbf{x}_{i,j,r}\}$, where i specifies the class (i.e., $i \in [1, c]$, with c being the total number of classes), j means the jth sample in class i, and r the number of different possible localizations while varying the error value from $-vr_x$ to $+vr_x$ about the x image axis and from $-vr_y$ to $+vr_y$ about the y axis.

We note that r increases exponentially as a function of f (the number of features localized in the face); i.e., $r = O(2^f)$. To be precise,

$$r = \left((2vr_x + 1) \sum_{i=1}^{f_x} \binom{f_x}{i} \right) \left((2vr_y + 1) \sum_{i=1}^{f_y} \binom{f_y}{i} \right)$$

$$= (2vr_x + 1)(2^{f_x} - 1)(2vr_y + 1)(2^{f_y} - 1),$$

where f_x represents the number of features that change their x coordinate value and f_y the ones that change their y coordinate value, and, in our notation, $f = f_x + f_y$. When f is large, this complexity is to be considered. In those cases, given that accounting for the error in only one feature at a time would not imply a significant change of appearance, a subset of all these r images is expected to suffice.

Once the data set $\hat{x}_{i,j}$ has been generated, the subset where the jth sample of class i lies can be readily modeled by means of a Gaussian distribution whose sample mean and sample covariance matrix are

$$\mu_{i,j} = \frac{\sum_{l=1}^{r} \mathbf{x}_{i,j,l}}{r},$$

$$\Sigma_{i,j} = \frac{1}{r-1}(\hat{\mathbf{x}}_{i,j} - \mu_{i,j})(\hat{\mathbf{x}}_{i,j} - \mu_{i,j})^{\mathrm{T}}. \qquad (1)$$

In some cases (especially, when the localization error is large), the subset $\hat{\mathbf{x}}_{i,j}$ cannot be assumed to be Gaussian. In such cases, a mixture of Gaussians is generally a better choice, because it can approximate most convex and nonconvex sets. Better fits come to an extra cost to be considered.

Mixture of Gaussians are usually estimated with the EM (expectation-maximization) algorithm [7], which is an iterative method divided into two steps, the *E step*

$$h_{i,j,l,g}^{[t+1]} = \frac{\left|\Sigma_{i,j,g}^{[t]}\right|^{-1/2} \exp\{-\frac{1}{2}(\mathbf{x}_{i,j,l} - \mu_{i,j,g}^{[t]})^{\mathrm{T}} \Sigma_{i,j,g}^{[t]^{-1}} (\mathbf{x}_{i,j,l} - \mu_{i,j,g}^{[t]})\}}{\sum_{s=1}^{G} \left|\Sigma_{i,j,s}^{[t]}\right|^{-1/2} \exp\{-\frac{1}{2}(\mathbf{x}_{i,j,l} - \mu_{i,j,s}^{[t]})^{\mathrm{T}} \Sigma_{i,j,s}^{[t]^{-1}} (\mathbf{x}_{i,j,l} - \mu_{i,j,s}^{[t]})\}}, \qquad (2)$$

and the *M step*

$$\mu_{i,j,g}^{[t+1]} = \frac{\sum_{l=1}^{r} h_{i,j,l,g}^{[t]} \mathbf{x}_{i,j,l}}{\sum_{l=1}^{r} h_{i,j,l,g}^{[t]}},$$

$$\Sigma_{i,j,g}^{[t+1]} = \frac{\sum_{l=1}^{r} h_{i,j,l,g}^{[t]} (\mathbf{x}_{i,j,l} - \mu_{i,j,g}^{[t+1]})(\mathbf{x}_{i,j,l} - \mu_{i,j,g}^{[t+1]})^{\mathrm{T}}}{\sum_{l=1}^{r} h_{i,j,l,g}^{[t]}}, \qquad (3)$$

where G is the total number of models used in the mixture (i.e., number of Gaussians) and $[t]$ means iteration t. In the formulation above, it is assumed that all models (Gaussians) are equiprobable (i.e., $\pi_{g_1} = \pi_{g_2}, \forall g_1, g_2$; where π_g denotes the probability of the gth model).

Note that the subset modeling presented in this section can be equally applied to those systems using an appearance-based representation and those using a feature-based representation.

18.2.3 Subspace Methods

Although subspace methods are applicable to both feature-based and appearance-based algorithms, they have found most of their success in the latter group. The reason for this is simple. In appearance-based algorithms, the original space corresponds to a dense pixel representation and, hence, its dimensionality is expected to be too large to allow the computation of the subset of the localization error. In such cases, it is convenient to reduce the dimensionality of the original feature space to one where the number of samples per dimension is appropriate. Many subspace techniques have been used to achieve this among which principal-component analysis (PCA) [9], linear-discriminant analysis (LDA) [29, 9] and independent-component analysis (ICA) [16] have arguably been the most popular. Since our n training samples will at most span a space of $n-1$ dimensions, it is reasonable to reduce the original space $\Re^p$ to $\Re^q$ where $q < n$ and generally $n << p$.

Each subspace technique will generate a different projection matrix. In the rest of this chapter we will use Φ_a to refer to the projection matrix obtained with method a; e.g., Φ_{PCA} is the projection matrix obtained using the PCA approach. Φ_{PCA} can be readily computed from the sample covariance matrix, which is given by

$$\Sigma_X = \sum_{i=1}^{c} \sum_{j=1}^{m_i} \sum_{g=1}^{G} \Sigma_{i,j,g}, \tag{4}$$

where m_i is the number of samples in class i. Note that since the rank of these covariance matrices is typically less than the number of dimensions, it is impossible to compute the principal components directly from them. In these cases, we generally calculate the eigenvectors of $\mathbf{X}^{\mathrm{T}}\mathbf{X}$ instead, where $\mathbf{X} = \{\hat{\mathbf{x}}_{1,1}, \ldots, \hat{\mathbf{x}}_{1,m_1}, \ldots, \hat{\mathbf{x}}_{c,m_c}\}$. Thus, if λ_h and $\mathbf{e}_h$ are the hth eigenvalue and eigenvector of $\mathbf{X}^{\mathrm{T}}\mathbf{X}$, then λ_h and $\lambda_h^{-1/2}\mathbf{X}\mathbf{e}_h$ are the eigenvalues and eigenvectors of Equation 4 [28].

While PCA only computes the first and second moments of the data, ICA will use higher moments of the data to find those feature vectors that are most independent from each other [16]. An important advantage of ICA is that it relaxes the Gaussian assumption that PCA imposes upon the underlying distributions of the data. Note that, although we are still assuming that the subset representing the localization error can be represented as a mixture of Gaussians, the underlying causes or latent variables need not be. ICA does not have a general closed-form solution, hence iterative methods are typically required. In our experimental results we have used the Infomax algorithm of [2], but other alternatives exist [16].

While PCA and ICA are unsupervised techniques, LDA is supervised. This can facilitate the task of feature extraction when the number of training images is sufficiently large [20]. LDA selects those basis vectors that maximize the distance between the means of each class and minimize the distance between the samples in each class and its corresponding class mean [29].

After one of these matrices, Φ_a, has been computed, we project each of the Gaussian distributions (describing the subset of the localization error) to the space spanned by Φ_a:

$$\mathcal{G}_{i,j} = \{\widetilde{\Sigma}_{i,j,g}, \widetilde{\mu}_{i,j,g}\}_{g=1}^{G}, \tag{5}$$

where $\widetilde{\Sigma}_{i,j,g} = \Phi_a^{\mathrm{T}} \Sigma_{i,j,g} \Phi_a$, $\widetilde{\mu}_{i,j,g} = \Phi_a^{\mathrm{T}} \mu_{i,j,g}$, and $a = \{PCA, ICA, LDA\}$.

Finally, to identify any previously unseen image, $\mathbf{t}$, we search for the closest model, $\mathcal{G}_i = \{\mathcal{G}_{i,1}, \ldots, \mathcal{G}_{i,m_i}\}$, by means of the Mahalanobis distance in the reduced space. We define the distance from the test vector to each of the mixture of Gaussians as follows

$$\mathrm{Mh}^2(\mathbf{t}, \mathcal{G}_i) = \min_{j,g} (\tilde{\mathbf{t}} - \widetilde{\mu}_{i,j,g})^{\mathrm{T}} \widetilde{\Sigma}_{i,j,g}^{-1} (\tilde{\mathbf{t}} - \widetilde{\mu}_{i,j,g}), \tag{6}$$

where $\tilde{\mathbf{t}} = \Phi_a^{\mathrm{T}} \mathbf{t}$. The closest class is now given by

$$Class_{\mathbf{t}} = \mathrm{argmin}_i \, \mathrm{Mh}^2(\mathbf{t}, \mathcal{G}_i). \tag{7}$$

18.2.4 Estimating the Localization Error

Recall that we still need to determine the error of our localization algorithm. Obviously, this is a difficult task that unfortunately does not have an analytical solution for most face-localization algorithms.

If the correct localization values are known (i.e., the ground truth is known) for a set of s samples, $\mathbf{x}(s) = \{\mathbf{x}_1, \ldots, \mathbf{x}_s\}$, an estimate $E(r; \mathbf{x}(s))$ can be computed, which depends on the number of samples s. It is easy to show that $E(r; \mathbf{x}(s))$ approximates the true value of r (and, thus, those of vr_x and vr_y) as $s \to \infty$. Obviously, we do not have that much data, and therefore, only estimates can be obtained. The reader should keep in mind that, in general, the correct location of every feature has to be manually determined, which is a cost to be considered.

18.3 MODELING OCCLUSIONS AND EXPRESSION CHANGES

When a face is partially occluded or when it changes expression, its global shape and texture vary, making it difficult for feature- and appearance-based methods to work properly. The goal of this section is to define a method robust to such variations. Both problems can be addressed by reformulating our algorithm within a local approach.

We have previously proposed to address the expression problem by assigning low weights to those features of the face that are more affected by facial changes and higher weights to those areas that are more stable [22, 37]. The occlusion problem can be tackled by applying our algorithms in a preselected set of local areas. In such a case, faces are divided into K local regions and then a probabilistic method is used to combine the results of these K local areas [21]. Extensions of these two approaches are described next.

18.3.1 Expression Changes

The problem possed by subspace methods (as defined above) is illustrated in Figure 18.5(a). In this example, we used one training image for each of the four classes to learn a low-dimensional image representation. All training images have a neutral facial expression. A test image with a distinct facial expression is then projected onto this subspace; see Figure 18.5(a). As the figure shows, simple Euclidean distances do not classify the test image into the correct class.

The problem with subspace techniques is that the learned features (dimensions) do not only represent class discriminant information but are also tuned to those specific facial expressions of our training set. Our solution is to first learn which dimensions are *most* affected by this problem when comparing the testing image to each of the training images and, then, build a weighted-distance measure which

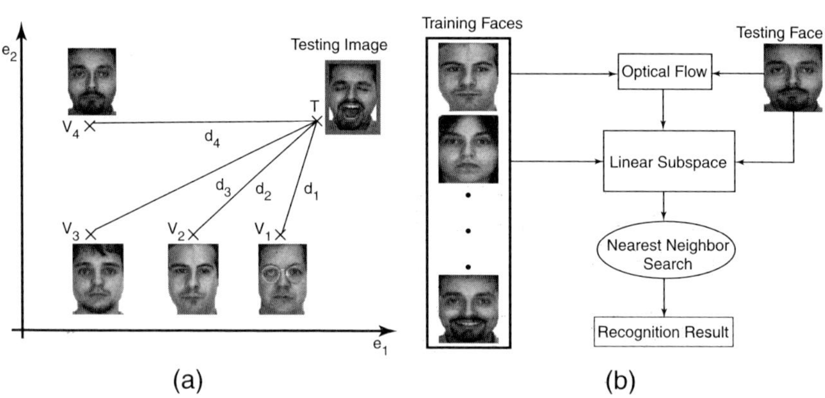

FIGURE 18.5: (a) Expression changes are encoded in the feature spaces or subspaces computed above. To address this problem, we give more importance to those features (dimensions) that are less affected by expression changes. (b) A schematic depiction of our system. The optical-flow algorithm is used to determine correspondences between images.

gives less importance to those dimensions representing facial expression changes and more importance to others. In our Figure 18.5(a), we would need to assign a small weight to the first feature, $\mathbf{e}_1$, and a large weight to the second feature, $\mathbf{e}_2$.

More formally, every test image, $\mathbf{t}$, is classified as belonging to the class of the closest training sample given by the weighted distance

$$\|\widetilde{\mathbf{W}}_i\,(\tilde{\mu}_{i,j} - \tilde{\mathbf{t}})\|^2_{\Sigma}\,, \tag{8}$$

where $\tilde{\mu}_{i,j} = \Phi_a^{\mathrm{T}}\mu_{i,j}$ is the mean feature vector of the jth image of class i projected onto the subspace of our choice, $a = \{\mathrm{PCA}, \mathrm{ICA}, \mathrm{LDA}\}$, $\tilde{\mathbf{t}} = \Phi_a^{\mathrm{T}}\mathbf{t}$, $\widetilde{\mathbf{W}}_i$ is the weighting matrix that defines the importance of each of the basis vectors in the subspace spanned by Φ_a, and $\|\mathbf{a}\|^2_{\Sigma} = \mathbf{a}^{\mathrm{T}}\Sigma\mathbf{a}$ with Σ defining the norm.

Before one can use Equation 8, we need to compute $\widetilde{\mathbf{W}}$. While it may be very difficult to do that in the reduced space spanned by Φ_a, it is easy to calculate this in the original space and then project this result onto the subspace of Φ_a. For example, we can compute the value of the weights in the original space, $\mathbf{W}$, by means of the optical-flow approach as follows (see also Figure 18.2(b)). First we compute the optical flow between the testing image and each of the Gaussian means:

$$\mathbf{F}_{i,j,g} = \text{Optical Flow}\,(\mu_{i,j,g}, \mathbf{t}). \tag{9}$$

Second, we compute $\mathbf{W}$ by assigning large weights to those pixels with small deformations and low weights to those pixels with large deformations,

$$\mathbf{W}_{i,j,g} = \operatorname{diag}\left(\mathbf{F}_{\max} - \|\mathbf{F}_{i,j,g}\|\right), \tag{10}$$

where $\mathbf{F}_{\max} = \max_i \|\mathbf{F}_{i,j,g}\|$.

And, third, we project this result onto our subspace,

$$\widetilde{\mathbf{W}}_{i,j,g} = \Phi_a^{\mathrm{T}}\mathbf{W}_{i,j,g}\Phi_a. \tag{11}$$

A test image is then classified by assigning to it the class label of the closest model. Formally, find the closest model,

$$\widehat{\mathrm{Mh}}^2(\mathbf{t}, \mathcal{G}_i) = \min_{j,g}\,(\tilde{\mathbf{t}} - \tilde{\mu}_{i,j,g})^{\mathrm{T}}\widetilde{\mathbf{W}}^{\mathrm{T}}_{i,j,g}\widetilde{\Sigma}^{-1}_{i,j,g}\widetilde{\mathbf{W}}_{i,j,g}(\tilde{\mathbf{t}} - \tilde{\mu}_{i,j,g})\,, \tag{12}$$

and its corresponding class,

$$\mathit{Class}_{\mathbf{t}} = \operatorname{argmin}_i \widehat{\mathrm{Mh}}^2(\mathbf{t}, \mathcal{G}_i). \tag{13}$$

The reader may have noted that alternatives to Equation 10 exist. Some of these alternatives were defined and tested on a large set of images in [23] of which Equation 10 was found to have the highest performance. These results are summarized in Section 18.5.4.

Finally, we need to select an optical-flow method to be used in Equation 9. In our experimental results, we used Black and Anandan's algorithm [4], because it is robust to motion discontinuities such as the ones produced by facial muscles moving in opposite directions.

18.3.2 Occlusions

One way to deal with partially occluded faces is to use local approaches [5, 15, 25]. In general, local techniques, divide the face into different parts and, then, use a voting space to find the best match. However, a voting technique can easily misclassify a test image because it does not take into account how good each local match is. We will now define a probabilistic approach that is able to address this problem successfully.

Learning Stage

We first divide each face image into K local parts and, then, apply the method defined above. This will require the estimation of the K subsets, and corresponding weighted subspaces, which account for the localization error and facial expression changes. More formally, let us define the set of images of each local area as

$$\mathbf{X}_{i,j,k} = \{\mathbf{x}_{i,j,k,1}, \ldots, \mathbf{x}_{i,j,k,r}\}, \tag{14}$$

where $\mathbf{x}_{i,j,k,l}$ is the kth local area of the jth sample of class i.

We assume that all sample faces have been previously localized and, if necessary, warped to a standard shape [20]. To obtain each of the samples $\mathbf{x}_{i,j,k,m}$, ellipse-shaped segments as defined by $x^2/d_x^2 + y^2/d_y^2 = 1$ are used, Figure 18.2(c). These segments are then represented in vector form. The new mean feature vectors and covariance matrices are given by

$$\mu_{i,j,k} = \frac{1}{r}\sum_{m=1}^{r} \mathbf{x}_{i,j,k,m}, \tag{15}$$

$$\Sigma_{i,k} = \frac{1}{r-1}(\mathbf{X}_{i,j,k} - \mu_{i,j,k})(\mathbf{X}_{i,j,k} - \mu_{i,j,k})^{\mathrm{T}},$$

or by the mixture of Gaussian equations defined in Equations 2 and 3. We note that, in such a case, we would have $\{\Sigma_{i,j,k,g}, \mu_{i,j,k,g}\}_{g=1}^{G}$.

If we require the computation of subspaces, we have to calculate a total of K of them, Φ_{a_k} ($k = \{1, \ldots, K\}$ and $a = \{PCA, LDA, ICA\}$). In such cases, and as described in the previous section, we will then need to project the estimates of our subsets onto each of the above computed subspaces:

$$\mathcal{G}_{i,j,k} = \{\widetilde{\Sigma}_{i,j,k,g}, \widetilde{\mu}_{i,j,k,g}\}_{g=1}^{G}, \tag{16}$$

where $\widetilde{\Sigma}_{i,j,k,g} = \Phi_{a_k}^{\mathrm{T}} \Sigma_{i,j,k,g} \Phi_{a_k}$, and $\widetilde{\mu}_{i,j,k,g} = \Phi_{a_k}^{\mathrm{T}} \mu_{i,j,k,g}$. Note that the set $\mathbf{X}_{i,j,k}$, which could be very large, does not need to be stored in memory. $\mathcal{G}_{i,k} = \{\mathcal{G}_{i,1,k}, \ldots, \mathcal{G}_{i,m_i,k}\}$ is the subset of all images of the kth local area of class i under all possible errors of localization. This learning procedure is summarized in Figure 18.6.

Identification Stage

Let $\mathbf{t}$ be the face image in its vector form—after localization and (if necessary) warping. Next, project each of the local areas of $\mathbf{t}$ (i.e., $\mathbf{t}_k$, $k \in \{1, \ldots, K\}$) to its corresponding subspace

$$\tilde{\mathbf{t}}_k = \Phi_k^{\mathrm{T}} \mathbf{t}_k \,. \tag{17}$$

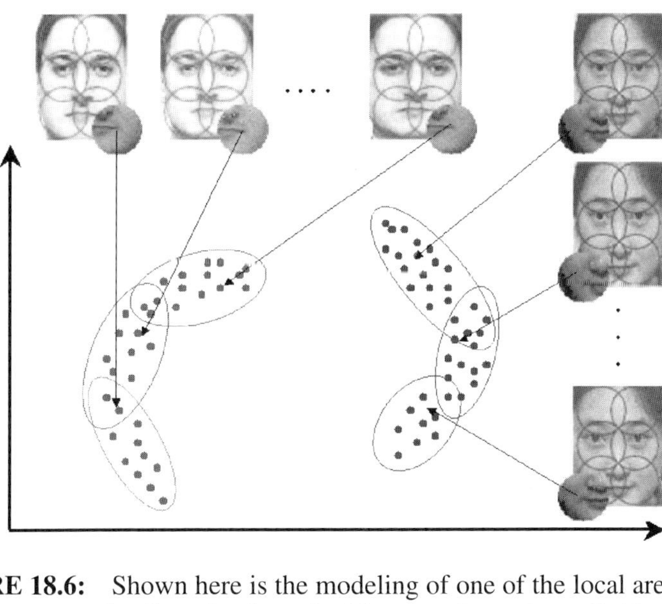

FIGURE 18.6: Shown here is the modeling of one of the local areas of the face for two people of our database. In this example, we have divided the face into six local parts; i.e., $K = 6$. From [36] (©2002 IEEE).

Since the mean feature vector and the covariance matrix of each local subset are already known, the probability of a given local match can be directly associated with a suitably defined distance

$$LocRes_{i,k} = \max_{j,g} (\tilde{\mathbf{t}}_k - \widetilde{\mu}_{i,j,k,g})^{\mathrm{T}} \widetilde{\mathbf{W}}_{i,j,k,g}^{\mathrm{T}} \widetilde{\Sigma}_{i,j,k,g} \widetilde{\mathbf{W}}_{i,j,k,g} (\tilde{\mathbf{t}}_k - \widetilde{\mu}_{i,j,k,g}). \quad (18)$$

We then add all local probabilities:

$$Res_i = \sum_{k=1}^{K} LocRes_{i,k}\,, \quad (19)$$

and search for the maxima,

$$Class = \operatorname{argmax}_i Res_i, \quad (20)$$

where $Class \in [1, c]$.

If a video sequence is available, we keep adding distances (probabilities) for each of the images and only compute Equation (20) at the end of the sequence or when a threshold has been reached [36].

18.4 EXPERIMENTAL RESULTS

18.4.1 Imprecise Localized Faces

Experiment 1

In this test, we want to answer the following question. Assuming that the localization error is precisely known, how well does the above method solve the localization problem? To this end, we manually marked the facial features of all neutral-expression, happy, angry, and screaming face images—see Figure 18.4(a–d)— for 50 randomly selected individuals of the AR-face database. In doing that we took care to be as close to pixel precision as possible. Then we added artificial noise to these data and, hence, the values of vr_x and vr_y will be precisely known. Since the localization error is now known, the recognition rate is not expected to drop as a function of vr_x and vr_y.

To test this, we use the neutral expression image, Figure 18.4(a), of all 50 subjects for learning and all others Figure 18.4(b–d) for testing. Results for vr_x and vr_y, taking the values 2, 4, 6, 8 and 10, are shown in Figure 18.7(a). These results were obtained first for the single-Gaussian (GM) case, $G = 1$, and then for the multiple-Gaussian (MGM) case with $G = 3$. For simplicity both error values, vr_x and vr_y, were assumed to be equal. Results are compared to the classical eigenspace approach that does not account for the localization error. For the

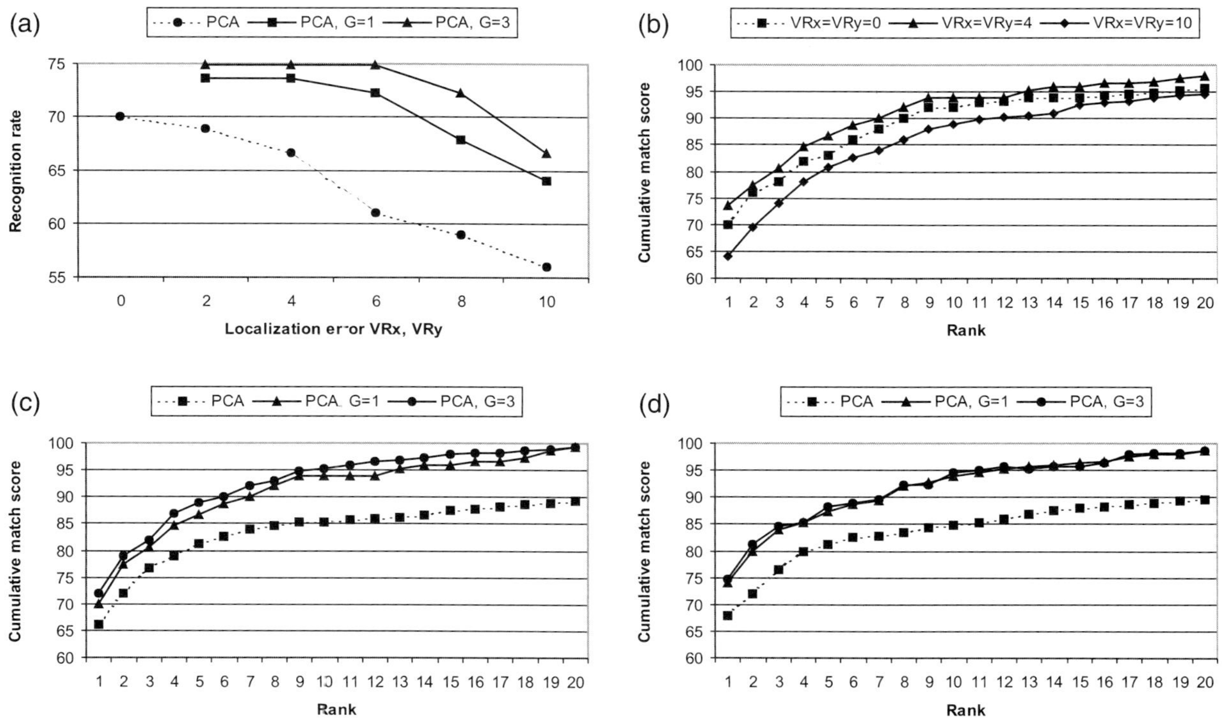

FIGURE 18.7: Summarized here are the results of experiment 1, shown in (a) and (b), and the results of experiment 2, shown in (c) and (d). From [21] (©2002 IEEE).

eigenspace approach, the Euclidean distance is used to find the closest match (class). In the results reported here the original space of pixels was reduced to twenty dimensions by means of the PCA approach.

For vr_x and vr_y values up to 4, neither of the new methods is affected by the localization error. The multiple-Gaussian case obtained slightly better results (as expected) and was not affected by the localization error problem until vr_x and vr_y were set to 8. Obviously, the larger the localization-error, the more probable that two Gaussian models (or mixture of Gaussians) overlap. For our particular case of 20 dimensions and 50 people, this happened at $vr_x = vr_y = 6$ for the single-Gaussian case and at $vr_x = vr_y = 8$ for the multiple-Gaussian case with $G = 3$.

Moreover, note that results of the new proposed methods are superior to the classical PCA approach. This has a simple explanation. In an attempt to overcome the localization-error problem, the Gaussian or mixture of Gaussian distributions give more importance (weight) to those combinations of features that best describe small deformations of the face (small changes of the texture of the face image). It is reasonable to expect to also find these combinations of features more adequate for the recognition of testing face images that express different facial expressions, because the testing images also represent texture changes of the face. It is obvious that these features will by no means be optimal, but they can explain the small increase of recognition rate obtained in our experiment.

It is also interesting to show how the new proposed method behaves as a function of rank and cumulative match score in comparison to the classical PCA approach (which serves as baseline). *Rank* means that the correct solution is within the R nearest neighbors and *cumulative match score* refers to the percentage of successfully recognized images [27]. Figure 18.7(b) shows results for $vr_x = vr_y = 4$ and $vr_x = vr_y = 10$ for case of G=1. We note that, even for localization errors as large as 10, the recognition results do not diverge much from the classical PCA approach (i.e., the $v_x = v_y = 0$ case).

Experiment 2:

We still need to see how well our method performs when the localization error is not known with precision. In this experiment, the localization method described in [21] was used. An approximation for the localization error for this method was also given in [21]; this was $vr_x = 3$ and $vr_y = 4$. The neutral-expression image, Figure 18.4(a), was used for learning while the happy, angry, and screaming faces, Figure 18.4(b–d), were used for recognition. Figure 18.7(c) shows the recognition results for the group of 50 people used to obtain the values of vr_x and vr_y. Figure 18.7(d) shows the results obtained with another (totally different) group of 50 people. Since the values of vr_x and vr_y were obtained from the first group of 50 people, the results of Figure 18.7(c) are expected to be better than

Table 18.1: As the values of vr_x and vr_y are made smaller, the recognition rates obtained progressively approach the baseline value of the PCA algorithm.

	$t = 0$	$t = 1$	$t = 2$	$t = 3$
Recognition rate	70%	68.3%	66.6%	66%

those of Figure 18.7(d). Moreover, it seems to be the case (as for the data shown here) that the more uncertainty we have in the estimation of the localization error, the more independent our results are from G.

It is also expected that the results of the proposed method will approximate the baseline PCA as the uncertainty of vr_x and vr_y increases. To demonstrate this effect, we repeated the first part of this experiment (experiment 2) three more times with $vr_x^{[t+1]} = vr_x^{[t]} - 1$ and $vr_y^{[t+1]} = vr_y^{[t]} - 1$, where $vr_x^{[t]}$ and $vr_y^{[t]}$ represent the localization-error values estimated at time t. We initially set $vr_x^{[0]} = 3$ and $vr_y^{[0]} = 4$ (the values given in [21]). Results are summarized in Table 18.1, with $G = 1$. As expected, the smaller the error values made, the closer the results of our method are to the results of the baseline given by the PCA algorithm.

18.4.2 Expression Variations

For the experiments in this section, we randomly selected a total of 100 people from the AR-face database. Each individual consists of eight images with neutral, happy, angry, and screaming expressions taken during two sessions separated by two weeks time. These images for one of the people in the database were shown in Figure 18.4(a–d) for those images taken during the first session and in Figure 18.4(g–j) for those of the second session.

Experiment 3

In this experiment, only those images taken during the first session were used. The results obtained using the proposed weighted-subspace approaches as well as those of PCA, ICA, and LDA are shown in Figure 18.8(a). In this figure, we also show the results obtained by first morphing all faces to a neutral-expression face image and then building a PCA, ICA, or LDA subspace with the resulting expression-free images. Morphing algorithms have been previously used to address the problem posed by facial deformations with some success [3, 17]. The problem with morphing algorithms is that they only change the shape but not the texture of the face. To be able to change the texture, one would need to be able to first extract the lighting parameters of the environment directly from pictures of faces. If this was achieved, morphing algorithms could render the texture of the

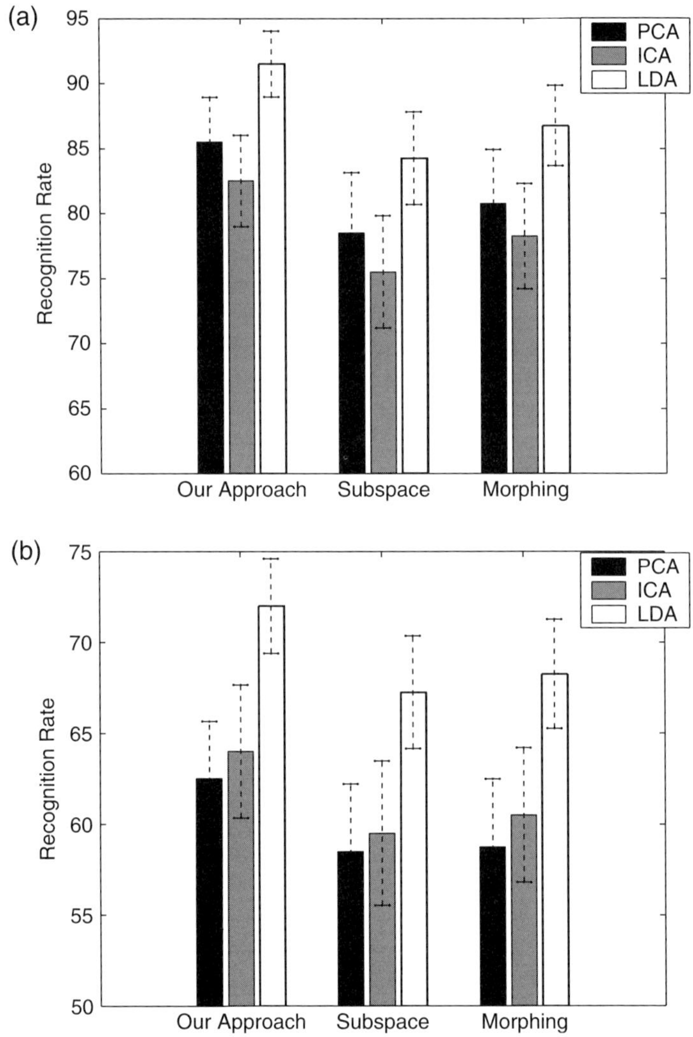

FIGURE 18.8: Recognition rates on the leave-one-expression-out test with: (a) images from the same session (experiment 3) and (b) images from different sessions (experiment 4).

face as it deforms. Yet, to do this with precision, a three-dimensional model of each individual would also need to be available or realizable from a single image. Although much progress has been made toward solving these two problems—extraction of illumination parameters and shape from a single image—much still needs to be done.

Table 18.2: Shown here are the results obtained with Equation 13 in the leave-one-expression-out test of experiments 3 and 4. These results are detailed for each of the expressions that were left out for testing. To compute the results shown here, the LDA subspace method was used.

	First session images				Second session images			
Recognition	Neutral	Happy	Angry	Screaming	Neutral	Happy	Angry	Screaming
rate	96%	97%	90%	83%	74%	77%	75%	62%

The bars in Figure 18.8 show the average recognition rate for each of the methods. All algorithms were tested using the leave-one-expression-out procedure. For example, when the happy face was used for testing, the neutral, angry, and screaming faces were used for training. The standard deviation of the leave-one-expression-out test for each of the methods is represented by the small line at the top of each bar.

Experiment 4

This test is almost identical to the previous one except that, this time, we used the images of the first session for training and those of the second session for testing. That means that, when the happy faces were left out from the training set which includes those images of the first session (i.e., Figures 18.4(a,c,d) were used for training), the happy face images of the second session were used for testing; Figure 18.4(h). These results are summarized in Figure 18.8(b). It is worth mentioning that our weighted-LDA approach worked well for the screaming face shown in Figure 18.4(d) with a recognition rate of about 84%. Other methods could not do better than 70% in this particular case. The results for each of the facial expressions are in Table 18.2.

18.4.3 Occlusions

Experiment 5

The first question we want to address is: how much occlusion can the proposed method handle? To answer this question, the following procedure was followed. The neutral-expression images of 50 randomly selected participants were used for training. For testing, synthetic occlusions were added to the (above used) neutral images and to the smiling, angry, and screaming face images. The occlusion was simulated by setting all the pixels, of a square of size $p \times p$ pixels, to zero. We tested values of p from a low of 5 to a maximum of 50. For each of these values of p, we randomly localize the square in the image 100 times (for each of the four testing images, i.e., neutral, happy, angry, and screaming). Figure 18.9(a) shows the mean and standard deviation of these results for each of the facial-expression groups. Figure 18.9(b) shows the results one would obtain with the classical PCA approach. To increase robustness, one can increment the number of local areas (i.e., K). However, as K increases, the local areas become smaller, and thus the method is more sensitive to the localization problem. At the extreme case, K will equal to the number of pixels in the image; i.e., our method will reduce to a correlation approach. In this case, only those pixels that have not been occluded will be used to compute the Euclidean (norm 2) distance between the testing and training images. The results obtained by means of this approach are shown in Figure 18.9(c). For comparison, we show the results obtained using the classical correlation algorithm in Figure 18.9(d). In general, the most convenient value of K will be dictated by the data or application.

Experiment 6

The experiment reported above was useful to describe how good our algorithm is when identifying partially occluded faces for which the occlusion size and location are not known *a priori*. In addition, it is interesting to know the minimum number of local areas needed to successfully identify a partially occluded face. To study this, the neutral expression images, Figure 18.4(a), of 50 individuals were used for learning, while the smiling, angry, and screaming images were used for testing; see Figure 18.4(b–d). Four different groups of occlusions were considered, denoted as occ_h with $h = \{1, 2, 3, 4\}$. For $h = 1$, only one local area was occluded, for $h = 2$ two local areas were occluded, etc. Occlusions were computed by discarding h of the local areas from the calculation. Given that the face is divided into K local areas, there are many possible ways to occlude h local parts. For example, for $h = 1$, there are K possibilities (in our experiments $k = 6$), for $h = 2$ we have $\sum_{q=1}^{k}(k - q)$ possibilities, and so on. For each value of h, all possibilities were tested and the mean result was computed. Figure 18.9(e) shows the results. occ_0 serves a baseline value (which represents the nonocclusive case). We have not included the results of occ_1 in this figure, because its results were too similar to those of occ_0. We conclude from this that, for our experimental data at least, the suppression of one local part did not affect the recognition rate of the system. The results of occ_2 reflect that our method is quite robust even for those cases where a third of the face is occluded. In this test, $G = 1$.

Experiment 7

The next question we are interested in is: how well does our method handle real occlusions? For this purpose we study two classical occlusions, the sunglasses and the scarf occlusions shown in Figure 18.4(e,f). The neutral expression images, Figure 18.4(a), were used for training, while the occluded images, Figure 18.4(e,f), were used for testing. Here, we used the automatic localization algorithm of [21],

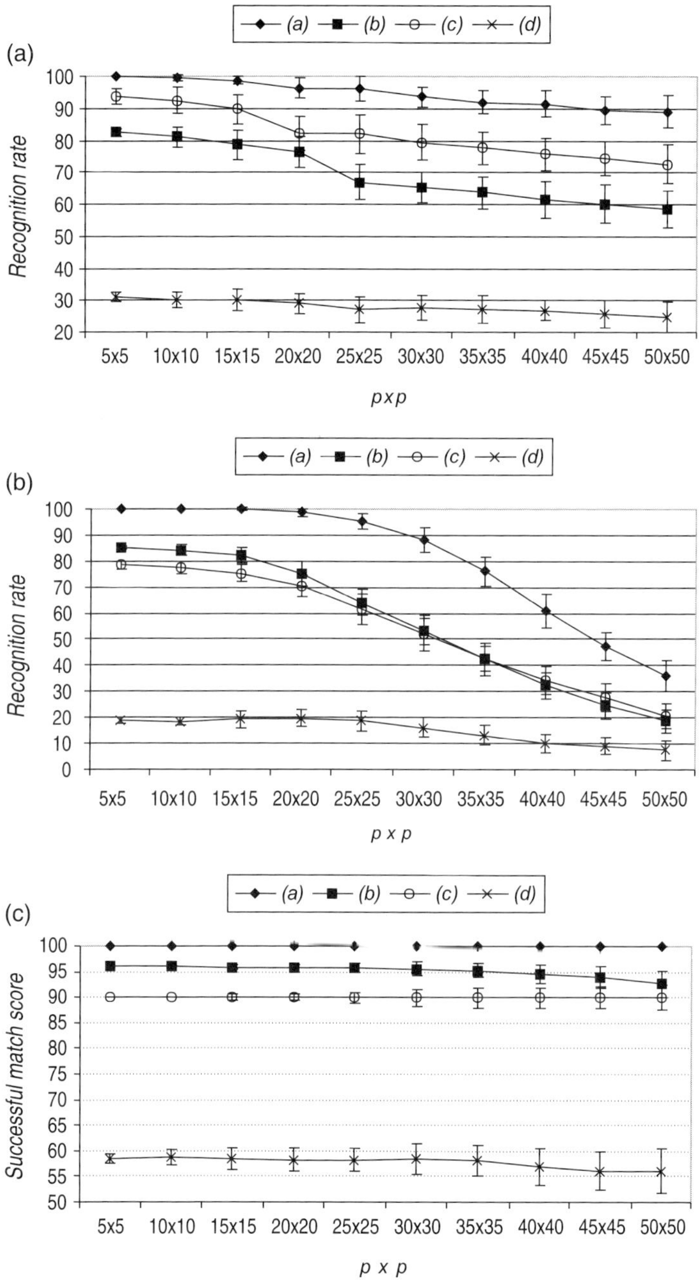

FIGURE 18.9: Experimental results for (a–d) experiment 5, (e) experiment 6, and (f–g) experiment 7. From [21] (©2002 IEEE).

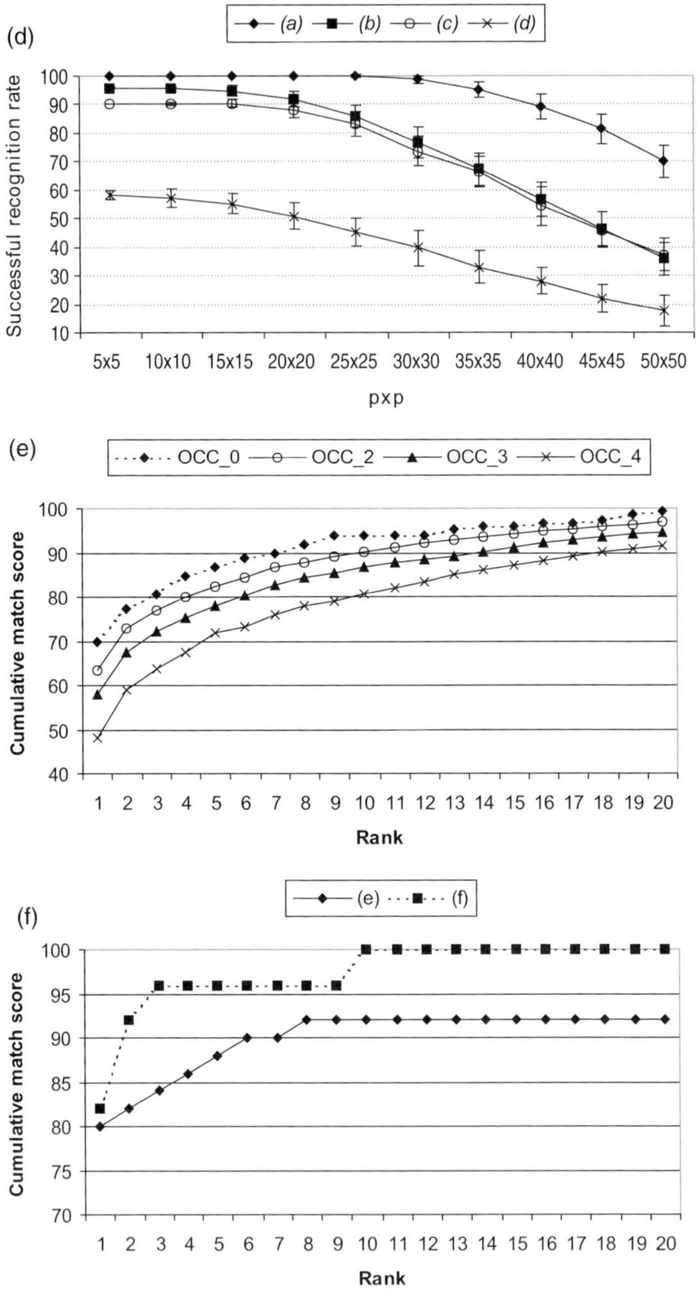

FIGURE 18.9: Continued.

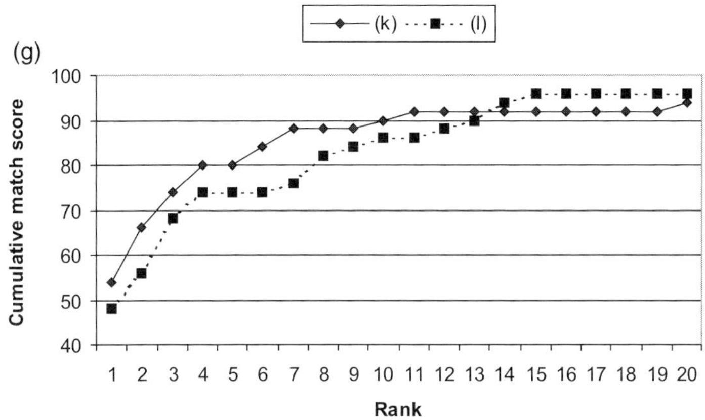

FIGURE 18.9: Continued.

for which $vr_x = 3$ and $vr_y = 4$. Figure 18.9(f) shows the results as a function of rank and cumulative match score. As a final test we repeated this experiment (experiment 7) for the duplicate images, using the neutral expression images of the first session, Figure 18.4(a), for learning and the occluded images of the duplicates for testing; see Figure 18.4(k,l). Figure 18.9(g) shows the results. It seems obvious from Figure 18.9(f) that the occlusion of the eye area affects the identification process much more than the occlusion in the mouth area. This result is to be expected, because it is believed that the eyes (or eye area) carry the most discriminant information of an individual's face. Surprisingly though, this was not the case for the recognition of duplicates, Figure 18.9(g). The occlusion of the eye area led to better recognition results than the occlusion of the mouth area. It seems to be the case that little is still known about how the recognition of duplicate works (or can be made to work). Further studies along these lines are necessary.

18.4.4 Results on all Problems

Experiment 8

Finally, we test how our method performs when attempting to solve all three problems simultaneously; i.e., localization error, occlusion, and expression. To this end, we first model the subset for each of the 100 randomly selected people of the AR-face database using the local weighted approach defined in this chapter. In our first test, we only used the neutral-expression image of the first session for training, Figure 18.4(a). The other images of the first session, Figure 18.4(b–f), were used for testing. These results are summarized in Figure 18.10(a,b), where we used a *rank* versus *cumulative match score* plots. In our second test, we used

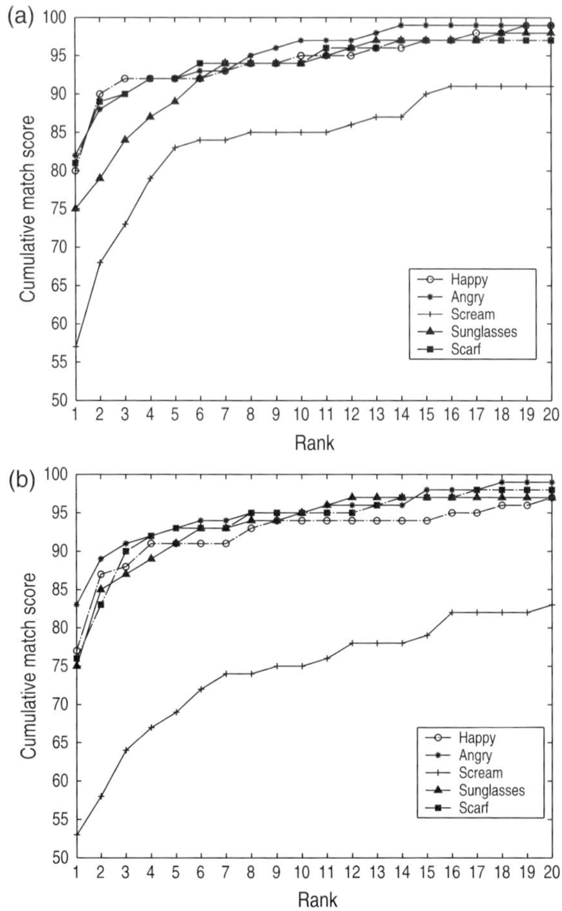

FIGURE 18.10: Summarized here are the results of our method using the images of the AR-database (experiment 8). (a, b) Use the neutral image of the first session for training. The rest of the images of the first session are used for testing. (a) Results obtained with the PCA space; i.e. $a =$ PCA. (b) $a =$ ICA. (c–e) Use all images of the first session for training, and those of the second session for testing. (c) $a =$ PCA. (d) $a =$ ICA. (e) $a =$ LDA. (f) Shows the average results for the plots shown in (c–e).

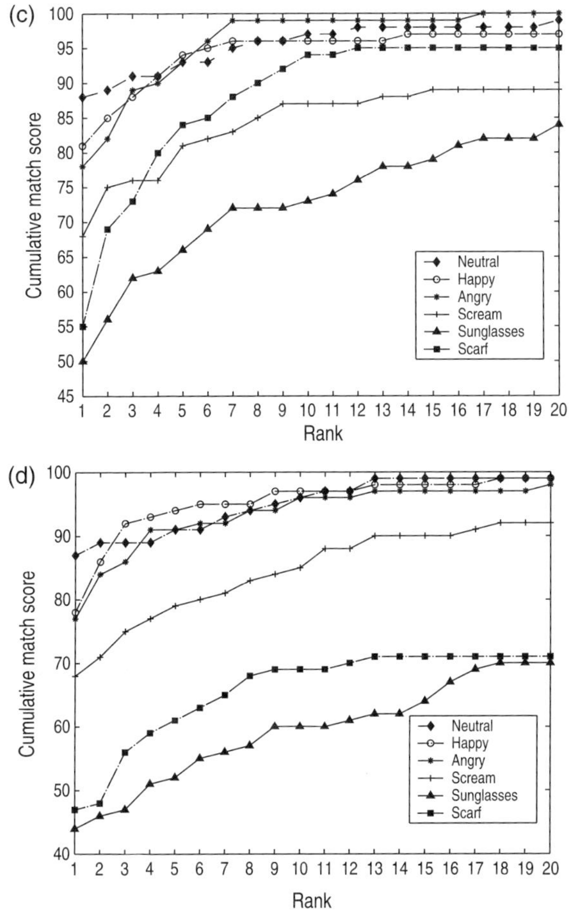

FIGURE 18.10: Continued.

the non-occluded face images of the first session for training, Figure 18.4(a,d), and those of the second session, Figure 18.4(g,l), for testing. These results are in Figure 18.10(c–f). Here $G = 3$. We note that, while in the first test expression changes decrease the recognition accuracy more notably than occlusions, the opposite is true in the second test, where duplicates were used. Once more, we conclude that much still needs to be done to fully understand how the recognition of duplicates works.

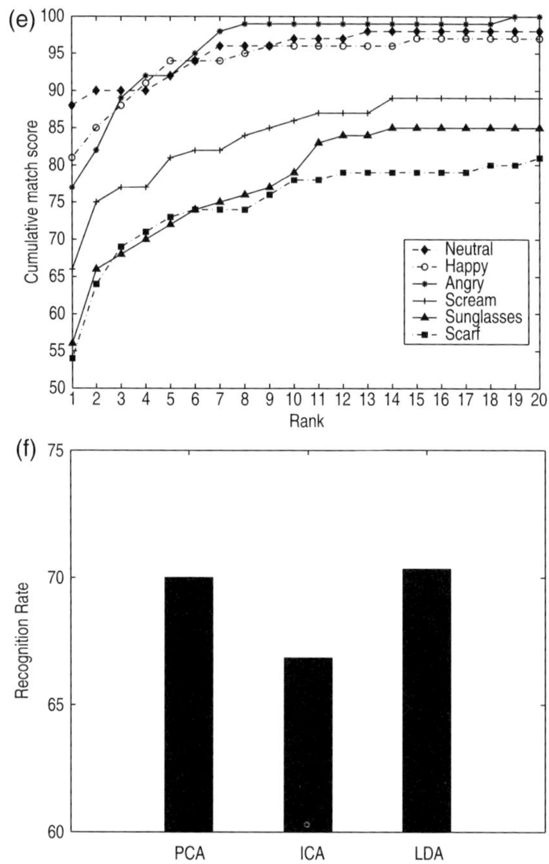

FIGURE 18.10: Continued.

18.4.5 Using Multiple Images for Recognition

As mentioned above, we could also use Equation 19 in those cases where more than one testing image is available. These images may correspond to pictures taken from different angles, with different cameras or from a video sequence. In either case, our algorithm can keep adding probabilities—using Equation 19—until a confident threshold is reached or until all images have been considered.

Experiment 9

To test this, we collected a new database of static images which were used for training, and a set of video sequences that were used for testing. The training set

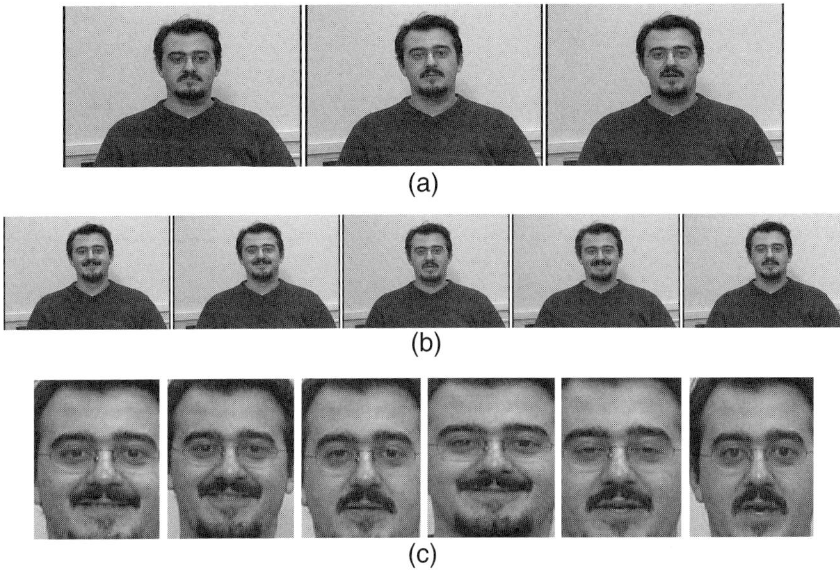

FIGURE 18.11: (a) The three training images for one of the subjects in our database. (b) A few frames of one of the video sequences with random talking. (c) Example of warped faces.

consists of three neutral-face images per person. An example of the training set for one of the subjects is shown in Figure 18.11(a). The testing set includes two sequences of nearly frontal views of people's faces with random speaking; see Figure 18.11(b). Our current database consists of twenty-two people.

Since the localization algorithm used above performed poorly on the much smaller images of our new database, a new localization algorithm was required. For this purpose, we selected the algorithm of Heisel et al. [14]. This algorithm can quite reliably localize the position of the eyes, mouth, and nose of each face. Once these facial features have been localized, using the differences between the x and y coordinates of the two eyes, the original image is rotated until obtaining a frontal-view face where both eyes have the same y value; i.e., $\text{atan}(\|y_1 - y_2\| / \|x_1 - x_2\|)$, where (x_1, y_1) and (x_2, y_2) are the right-eye and left-eye image coordinates. A contour algorithm is used to search for the lateral boundaries of the face (i.e., left and right). The top and bottom of the face are assigned as a function of the above features, and the face is then warped to a final standard face of 120 by 170 pixels. Figure 18.11(c) shows the warping results for the images shown in Figure 18.11(a). The localization error of the algorithm described in [14] was found to be 16 by 16 pixels. In our experiments, $G = 3$.

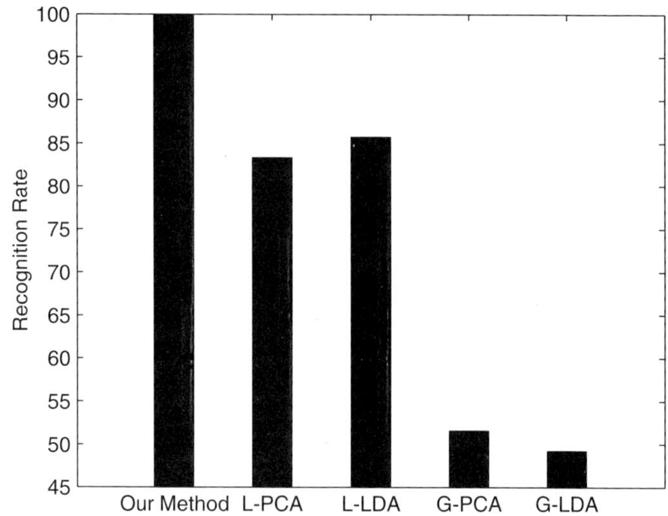

FIGURE 18.12: Shown here are the recognition rates obtained with our probabilistic algorithm and two implementations of PCA and LDA—one local one global.

In Figure 18.12 we show the results of our algorithm on the two video sequences defined above. We have compared our results to those obtained with global PCA and global LDA, where only one frame from the sequence is used for recognition. To obtained these results, we calculate the recognition rate of every frame for each sequence and then compute the average recognition rate. These results are labelled G-PCA and G-LDA in the figure. Additionally, we have also compared our results to the local weighted PCA and LDA approaches used above. While all the methods are expected to be robust to localization errors, partial occlusions and expression changes, the two local weighted algorithms defined earlier only used one single static image for recognition. The average recognition rates obtained with these two single-image methods are labelled L-PCA and L-LDA. These results show the improvement achieved when using our probabilistic approach in a video sequence rather than using a single frame; i.e., multiple versus single frame.

18.5 DISCUSSION AND FUTURE WORK

18.5.1 Why is the Warping Step Necessary?

Most face-recognition systems developed to date use 2D images to find the best match between a set of classes and an unclassified test image of a face. It is important to keep in mind, however, that faces are 3-dimensional objects and, as such, their projection onto the 2D image plane may undergo complex deformations. The frontal face recording scenario is effected by such deformations, and this needs to be considered when developing face-recognition systems.

Frontal face images are generally obtained by placing a camera in front of the subject who is asked to look at the camera while a picture is taken. The resulting 2D image projection obtained using this procedure is called a "frontal" face image. In general, several "frontal" faces are recorded; some to be used for training purposes and others for testing. However, the word "frontal" might be misleading. Since faces are 3D objects, their orientation with respect to a fixed camera is not guaranteed to be the same from image to image—especially for duplicates, because people tend to orient their faces toward distinct locations on different days. For example, when the face is tilted forwards, its appearance (shape and texture) in the 2D image plane changes. This effect is shown on the left part of Figure 18.13. Tilting the face backwards will result in other appearance changes. Also, rotating the face to the right or to the left will impose appearance changes depicted on the right-hand side of Figure 18.13.

These deformations make the identification task difficult, because any combination of the above can be expected. Feature-based approaches might be

FIGURE 18.13: The left part of this figure shows the shape changes that a face undergoes when tilted forwards. The top image corresponds to a totally *frontal* view. The bottom image represents the projection of a slightly tilted face. The bottom face has been re-drawn next to the top one, to facilitate comparison. The right part of this figure shows some of the visible deformations caused by rotating the face to the subject's right. From [21] (©2002 IEEE).

inappropriate if this effect is not properly modeled, because the distance between eyes, eyes and mouth, mouth and chin, etc., are not guaranteed to be the same in different images of the same person. Texture-based approaches will also be affected, because the facial features of a subject can easily change positions in the array of pixels used to identify the face. A pixel-to-pixel analysis is then inappropriate [3, 20, 21]. Additional evidence for this argument comes from studies showing that better recognition rates are obtained when one uses a free-shape representation as opposed to the original shape captured in the 2D array of pixels [34, 6].

To overcome this difficulty, one could use large and representative training sets. By doing this, one expects to collect enough learning data to sample the underlying distribution of the feature space that represents all possible deformations described above. Unfortunately, there are so many possible orientations of the face considered to be a "front-viewed", that the collection of this hypothetical training set will be impossible in most practical situations.

We are therefore left with two main alternatives. The first is to model the subset which includes all images under all possible "frontal" views. This would require an approach similar to that defined in Section 18.2. Alternatively, we can define a model of a face common to all people which cancels out all the effects described above. Although, in general, the former approach is slightly superior, its complexity makes it less attractive.

It has therefore become common practice to warp all faces to a "standard" shape. The localization procedures used earlier incorporated such a mechanism. Note, however, that the warping algorithms do not *deform* the important facial features needed to successfully recognize the identity of the faces in each of the images. The shape of the eyes, mouth, nose, etc., is *not* affected. The warping mechanism will only guarantee that the eyes, nose and chin positions are at the same image coordinate for all individuals, and, that all images (independently of the facial expression displayed on the face) occupy a fixed array of pixels. This allows us to efficiently analyze the texture (appearance) of the face image and to use pixel-to-pixel differences. Caricaturing might also be used to improve the results even further [6].

Although the above discussion is quite convincing with regard to the claim that the warping step is necessary to correctly use appearance-based methods and to successfully identify faces, we still bear the burden of establishing our claim with the help of real data. To prove the validity of such a claim, we have performed a set of experiments [21]. In those, fifty subjects were randomly selected from the AR-face database. Images were then warped to fit a 120×170 array of pixels. A second set of images, which we shall call the *aligned-cropped* set, was also created. For this latter set, each face was aligned only with regard to the horizontal line described by the eyes' centers; which requires rotation of the face image until the y values of both eyes' centers are equal in value (i.e., $\text{atan}(\|y_1 - y_2\| / \|x_1 - x_2\|)$, where (x_1, y_1) are the pixel coordinates of the center of the right eye and (x_2, y_2)

the same for the left eye). These faces were then cropped to fit an array of 120 × 170 pixels. After cropping the horizontal that connects both eyes is at the same y pixel coordinate for all images.

A first (40-dimensional) PCA representation was generated by using the warped neutral-expression face image of each of the 50 selected subjects. A second (40-dimensional) PCA space was generated by using the aligned-cropped neutral-expression images of the same subjects. The neutral-expression image (before warping or aligning-cropping) of one of the participants is shown in Figure 18.4(a). To test recognition accuracy, the happy, angry, and screaming face images, Figure 18.4(b–d), of each participant were used. The warped images were used to test the first learned PCA representation, and the aligned-cropped face images to test the second. The results, as a function of successful recognition rate, are shown in Figure 18.14(a). In this graphical representation of the results, L specifies the

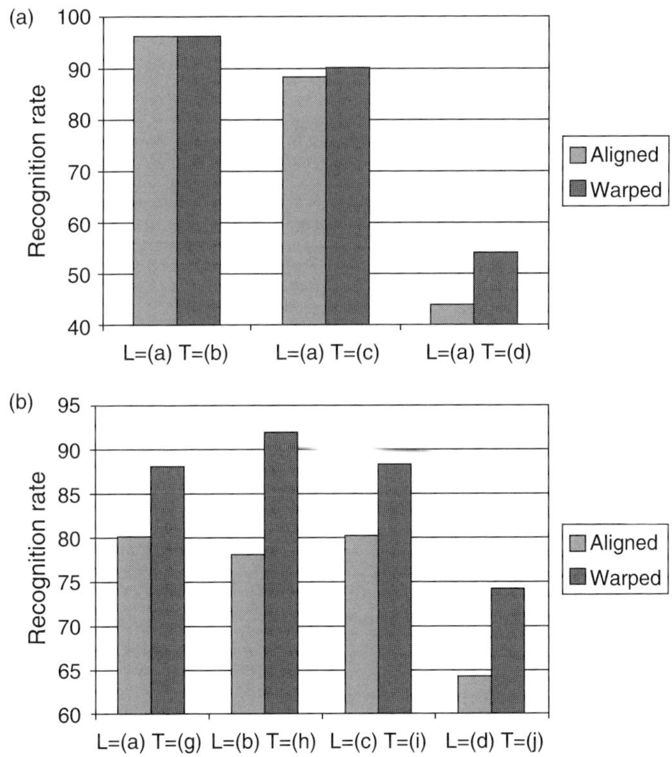

FIGURE 18.14: Shown here are the recognition rates obtained by either using the aligned-cropped or the warped images. From [21] (©2002 IEEE).

image that was used for learning the PCA representation and T the one that was used for testing. For clarity, the results are shown separately for each group of testing images with a common facial expression.

As mentioned above, we should expect this difference to increase for duplicate images—because at different sessions people tend to orient their faces differently. To show this effect, we used the neutral, happy, angry and screaming images taken during the second session, Figure 18.4(g–j), to test the PCA representations of their corresponding duplicate image taken at the first session, Figure 18.4(a–d). The results are shown in Figure 18.14(b). Note that, as expected, the improvement in recognition is now statistically significant.

18.5.2 The Subset of Localization Error

In Section 18.2, we presented a method to estimate the subset of all warped face images under all possible errors of localization. This method was proven to be effective in a set of experiments in Section 18.4.

A problem of our technique is that it needs to first generate all images, under all possible errors of localization, before it can estimate the corresponding subset. Had the dimensionality and form of such subset been known *a priori*, a smaller set of images should have sufficed. Unfortunately, the warped face images are usually generated by means of an affine transformation, and the resulting feature vectors generally lie in a space of large dimensionality. Moreover, this subset may sometimes be nonconvex, making the task of estimating it more challenging.

We believe that, to be able to estimate the subset of localization errors from a smaller set of images, the goal is to define a warping algorithm which generates a set of feature vectors that lie in a convex set of low dimensionality. More theoretical studies are still needed to make the techniques proposed in this chapter more general and readily applicable to other problems in computer vision.

18.5.3 Modeling the Subset of Occlusions

The reader may be wondering why we divided each face into K local areas rather than modeling the occlusion problem in a similar way to that used to model the localization error. This would have worked as follows. First, we would need to synthetically generate all possible occlusions. This could have been readily achieved by drawing occluding squares of increasing size, and placing them at all possible locations of the face. This is similar to what we had done in experiment 5 for testing, but now we would use this idea for training. Second, we would project all the resulting images (or a uniformly distributed set of these) onto the subspace of our choice and then define the subset of all occlusions with Equation 1 or Equations 2 and 3.

This idea works quite well indeed. In Figure 18.15, we show the results obtained using this approach. In this test, the neutral-face image of 100 randomly selected

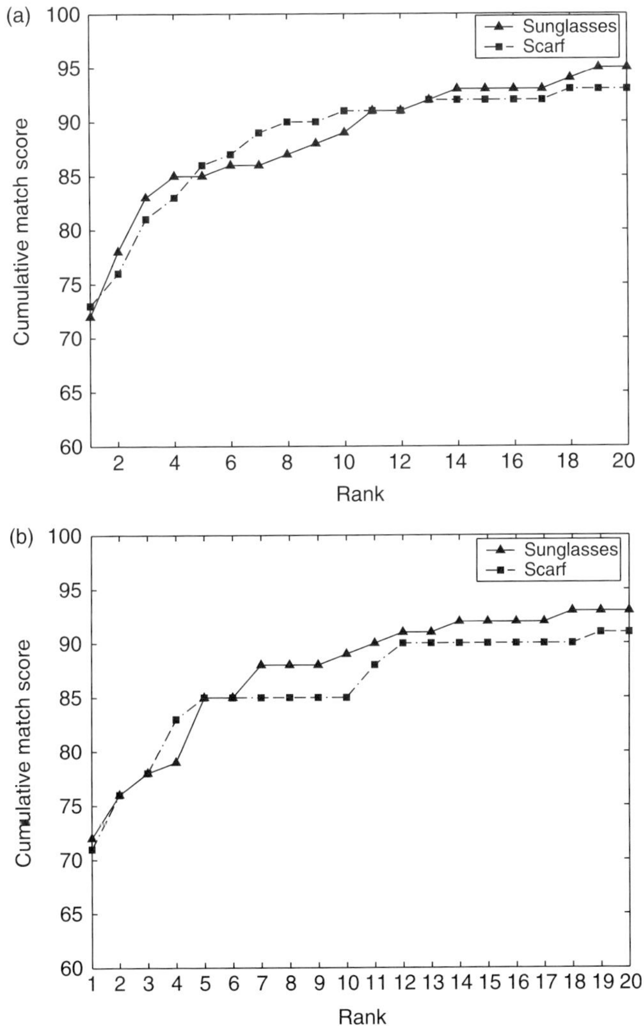

FIGURE 18.15: In this experiment, we learned the subset of occlusions by means of a large number of synthetically generated occlusions. (a) Results obtained with the PCA subspace technique, i.e., a = PCA. (b) a = ICA.

people of the AR-database were used for training. Occluding squares, ranging from 3 by 3 to 17 by 17 pixels were used to generate the training set. The modeling of the subset representing all these occlusions was approximated with Equations 2 and 3 and $G = 3$. As we had done for the localization-error problem, identification is then given by Equation 7. This figure shows results on learning using synthetic occlusions and testing with real occlusions; i.e., sunglasses and scarf; see Figure 18.4(e,f).

Unfortunately, this approach has several important disadvantages when compared to the one proposed earlier. The first (and, arguably, most important), is that we can only learn the subset of those occlusions we synthetically generate. Note that the number of possible occlusions is prohibitive. A second obvious problem is the computational cost associated with generating all the training images. This is similar to the problem stated in the preceding section. And, among others, another problem is that as faces get larger and larger occlusions, the subsets representing each individual will overlap with others. To this, we should add that the results obtained with the approach defined in this section are not superior to those obtained with the local method presented earlier. The approach defined in Section 18.3 is, therefore, a more attractive option.

An open problem is that of finding the dimensionality of the subset of most common occlusions. We note that these subsets may very well be approximated by cones, which are generally easy to estimate [1]. To see that, note that all subsets will have a common point (the origin) representing a full occluded face (i.e., where all pixels of the face are occluded). As more pixels of the face are made visible, the subsets representing each individual will start diverging from each other. More and more feature vectors will become possible as more and more pixels are made visible, which should generate a cone-like set. Although it is obvious to see that the dimensionality of such subsets will be very large, we note that those dimensions (features) that are occluded can be discarded, and therefore the space of all images could, in principle, be approximated with a low number of dimensions. This low-dimensionality argument is the same as used by PCA and other subspace algorithms—techniques that have had success in the past, particularly in addressing the problems defined in this chapter.

18.5.4 Weighting Facial Features for Expression-Variant Recognition

In Equation 10 we gave a linear solution to the problem of assigning weights to each of the dimensions of our feature space, which may have seemed pretty obvious at the time. Nonetheless, alternatives do exist, and some may prove critical for the overall performance of our algorithm. Here, we analyze different alternatives and report comparison results.

Two basic alternatives to that of Equation 10 are as follows. The first one assigns high weights to those pixels that change the least and low weights everywhere else.

This can be expressed mathematically as

$$w_i = \frac{1}{q + \|\mathbf{F}_i\|}, \tag{21}$$

where q is a constant.

The second alternative is to assign low weights only to those pixels that have a very large flow magnitude. Formally,

$$w_i = \frac{1}{\|\mathbf{F}_i\| - (q + MAX_F)} - (q + MAX_F) \,. \tag{22}$$

It is generally interesting to normalize the weights to make the final identification results consistent and comparable to each other. This can be readily achieved as follows: $w_i = w_i/M$, where $M = \sum_{i=1}^{n} w_i$.

In order to test the above described system, we will use two publicly available face datasets, the AR-face database [19] and the JAFFE (Japanese Female Facial Expressions) dataset [18].

In the JAFFE dataset, ten participants posed three examples of six of the basic emotions: happiness, sadness, surprise, anger, disgust, and fear. Three neutral face examples were also photographed. This makes a total of 210 images. Each participant took images of herself while looking through a semi-reflected plastic sheet towards the camera, which should guarantee frontal face views. Hair was tied away from the face to make all areas of the face visible. Light was also controlled. An example of a set of the seven images of a participant is shown in Figure 18.16.

For the AR-test, the different facial expressions used were: happiness, anger, and scream. The fourth image corresponds to a neutral expression. For the experiments reported below, 100 individuals were randomly selected from this database, giving a total of 400 images. These images for one subject were shown in Figure 18.4(a–d).

Since the purpose of our comparison is to determine which weighting function is most appropriate for our formulation, we will use the original space of pixels as our representation. This algorithm was defined in Equation 8.

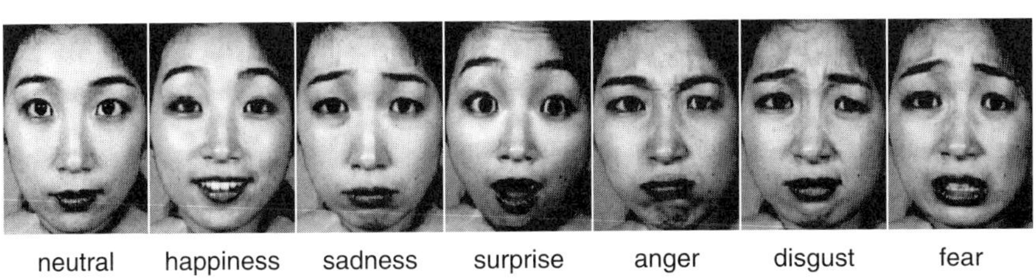

FIGURE 18.16: Examples of images from the JAFFE database.

Basic Emotions with JAFFE

It is interesting to see the extent to which the classification results of the system described above outperform the classical correlation approach when the basic emotions are displayed on the face images by means of facial expressions. For this purpose, the following test was performed. One of the three neutral images of each of the people in the JAFFE database was randomly selected and used as a sample image. All three examples of the six basic expressions in the database (i.e., happiness, sadness, surprise, anger, disgust, fear) were then used for testing. Results on the classification of images by identity using the correlation approach (*corr*) and the weighted distance proposed given in Equation 8 are shown in Figure 18.17(a). Note that our method has three different results (W1, W2, and W3), corresponding to the three possible weighting functions described in Equations 10, 21, and 22 respectively, with $q = 1$. We see that, while Equations 10 and 21 led to superior results than those of the correlation approach, Equation (22) did not and, therefore, is most probably not a good choice. The problem with Equation (22) is that all pixels are used except those with associated large flow, which makes the system little different from the correlation technique.

The AR-face Database Test

The results reported above are useful for analyzing how good or bad our method is in classifying faces when different emotions are displayed on the testing face image. However, the number of people in the JAFFE database is too small to allow conclusive results. The AR-face database allows us to test the expression-variance problem with a much larger group of people. For this experiment, we randomly selected 100 individuals from the AR-face database. The neutral expression images were used as sample images. The other three expressions from the same session (i.e., happiness, anger, scream) were used for testing. Results for the correlation and the weighted-distance approach are shown in Figure 18.17(b). Here, we have added the results one would obtain by using an 80-dimensional PCA representation. As expected [5] the PCA approach performs worse than the correlation method, but requires less storage. In this second test, the difference in performance (between the correlation measure and Equation 8) was more notable than in the first experiment. This is due to the fact that a much larger number of classes were used, leaving room for improvement.

The results shown in Figure 18.17(b) were obtained by using a single image for training (the *neutral expression* image) and the other three images for testing. Alternatively, we can use the smiling or the angry or the screaming faces for training and the other three for testing. This is shown in Figure 18.17(c), where in the x axis we specify the image used for training. The *general mean* represents the expected results when either image is used for training. We see that Equations 10 and 21 are consistently the best.

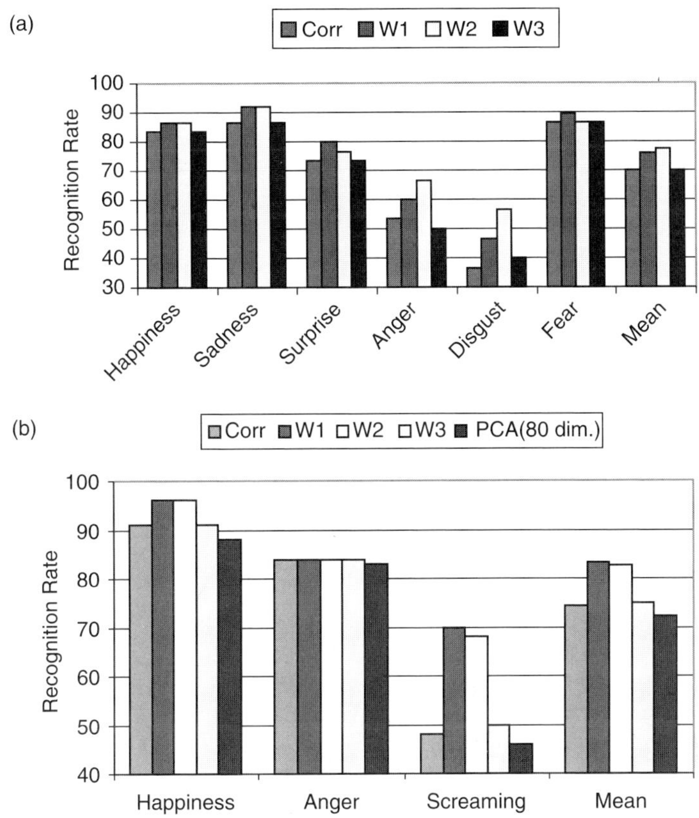

FIGURE 18.17: (a) JAFFE dataset test. (b) AR-face database test using the neutral image for training and the other images from the same session for testing. (c) The x axis specifies the image used for training. the other images from the same session were used for testing. (d) Training was identical as before, but now all the images of the second session were used for testing.

Figure 18.17(d) does the same as 18.17(c) but for the duplicate images. The x axis specifies the image (of the first session) used for training. The y axis indicates the recognition rate achieved with the duplicate images. For testing, the four images of the second session (neutral, happy, angry, and screaming) were used. In this case only Equation 10 produces results superior to simple correlations, and was therefore our choice in Equation 12.

18.5.5 Facial Interfaces

Although many algorithms for the automatic recognition of faces and facial expressions have been proposed to date, few address the question of how adequate these

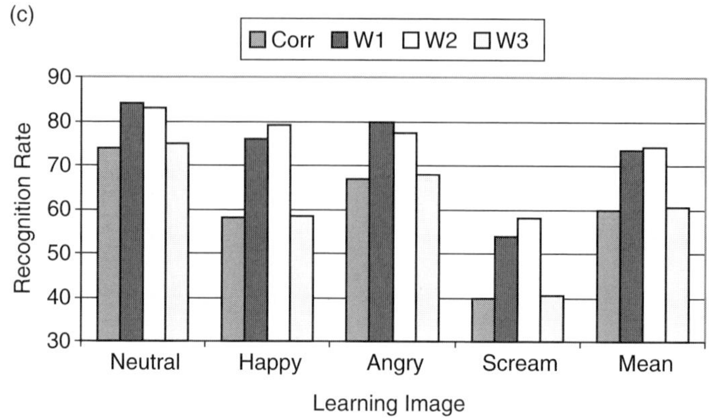

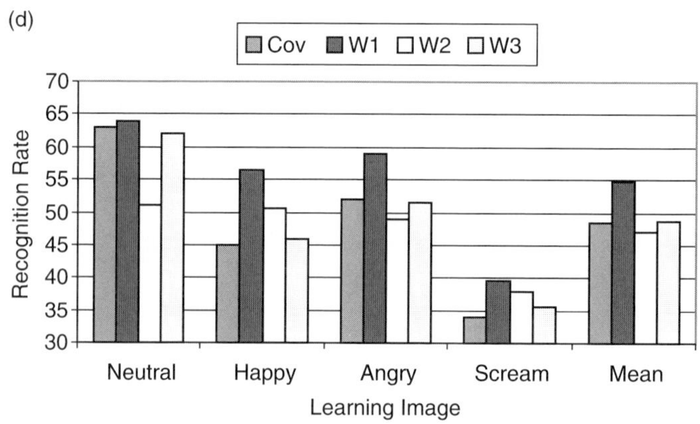

FIGURE 18.17: Continued.

systems are for applications in human–computer interaction (HCI); e.g., perceptual interface. While in face recognition the goal is to design systems that can achieve the highest recognition rate possible, in HCI the goal is to develop systems that "behave" similar to humans. We have shown that the method described above is consistent with these requirements and, therefore, can be used for applications in HCI [22, 23].

The reader may have already noted that the motion field used to weight each of the dimensions of our feature space may also be used to classify images into a predetermined set of facial expressions. We have also shown that this can be achieved well in [22].

18.6 CONCLUSIONS

In this chapter, we have shown that it is possible for an automatic recognition system to compensate for imprecisely localized, partially occluded, and expression-variant faces. First, we have shown that the localization-error problem is indeed important and can sometimes lead to the incorrect classification of faces. We have solved this problem by learning that subset which represents most of the images under all possible errors of localization. We reported results on approximating these subsets by means of a Gaussian or a mixture of Gaussians distribution. The latter led to slightly better results, but needs additional computation cost to be considered. A drawback of this method is that the ground-truth data (i.e., the correct localization of every feature to be localized on each face) for a set of samples is needed in order to estimate the localization error of a given localization algorithm. The problem with this is that the ground-truth data has to be obtained manually, which is labor intensive. Furthermore, while this subset was assumed to lie in a low-dimensional space, the dimensionality of such space is not yet known. More theoretical studies are needed toward solving this problem. We believe that the important question to address is not that of the dimensionality of our subset. Actually, the goal is *to find that warping algorithm* which generates the subset of lowest possible dimensionality.

To solve the occlusion problem, a local approach was defined, where each face is divided into K local parts. A probabilistic approach is then used to find the best match. The probabilities are obtained when one uses the Mahalanobis distance defined by the Gaussian distributions described above (for solving the localization problem). We experimentally demonstrated that the suppression of 1/6 of the face does not decrease accuracy. Even for those cases where 1/3 of the face is occluded, the identification results are very close to those we obtained in the nonoccluded case.

We have also shown that the results of most face-recognition systems depend on the differences between the facial expressions displayed on the training and testing images. To overcome this problem, we have built a weighted-subspace representation that gives more importance to those features that are less affected by the current displayed emotion (of the testing image) and less importance to those that are more affected.

The use of multiple images was also shown to be useful for the recognition of faces under varying conditions. Little research has focused on this issue, but some recent studies suggest that much can be accomplished with such techniques.

The importance of the approach summarized in this chapter is not limited to the problems reported above. Earlier, we argued that the recognition of duplicates is a difficult task that requires further studies. The approach proposed here could be used to study which areas of the face change less over time and why. The local approach could also be used to tackle changes due to local illumination changes.

We could use systems that are robust to illumination changes, and reformulate them within the approach defined above. These and other open problems were discussed in Section 18.5. We hope these will help define new, exciting, and productive projects.

Acknowledgments

The authors would like to thank Mar Jimenez, Avi Kak, and Kim Boyer for useful discussion. The authors are also grateful to Tomaso Poggio and his students for sending us the code of their face-localization algorithm used here, and to Michael Black for making his optical-flow algorithm available. This research was financially supported in part by the National Institutes of Health under grant R01-DC-005241 and by the National Science Foundation under grant 99-05848.

REFERENCES

[1] P.N. Belhumeour and D.J. Kriegman, "What is the set of images of an object under all possible lighting conditions," *Int. J. Comp. Vision*, 1998.

[2] A.J. Bell and T.J. Sejnowski, "An information-maximization approach to blind seperation and blind deconvolution," *Neural Computation* **7**:1129–1159, 1995.

[3] D. Beymer and T. Poggio, "Face recognition from one example view," *Science* **272**(5250), 1996.

[4] M.J. Black and P. Anandan, "The robust estimation of multiple motions: parametric and piecewise-smooth flow fields," *Computer Vision and Image Understanding* **63**:75–104, 1996.

[5] R. Brunelli and T. Poggio, "Face recognition: features versus templates," *IEEE Transactions on Pattern Analysis and Machine Intelligence* **15**(10):1042–1053, 1993.

[6] I. Craw, N. Costen, T. Kato, and S. Akamatsu, "How should we represent faces for automatic recognition?," *IEEE Trans. Pattern Analysis and Machine Intelligence* **21**(8):725–736, 1999.

[7] A.P. Dempster, N.M. Laird, and D.B. Rubin, "Maximum likelihood from incomplete data via the EM algorithm," *Journal Royal Statistical Society* **30**(1):1–38, 1977.

[8] R.S.J. Frackowiak, K.J. Friston, C.D. Frith, R.J. Dolan, and J.C. Mazziotta, *Human Brain Function*, Academic Press, 1997.

[9] K. Fukunaga, *Introduction to Statistical Pattern Recognition* (second edition), Academic Press, 1990.

[10] M.S. Gazzaniga, R.B. Ivry, and G.R. Mangun, Cognitive Neurosceince: the Biology of the Mind, W.W. Norton & Company, 1998.

[11] A.S. Georghiades, P.N. Belhumeur, and D.J. Kriegman, "From few to many: generative models of object recognition," *IEEE Transactions Pattern Analysis and Machine Intelligence* **23**(6):643–660, 2001.

[12] P.L. Hallinan, G.G. Gordon, A.L. Yuille, P. Giblin, and D. Mumford, *Two- and Three-Dimensional Patterns of the Face*, A.K. Peters, 1999.

[13] J.V. Haxby, E.A. Hoffman, and M.I. Gobbini, "The distributed human neural system for face perception," *Trends in Cognitive Science* **4**:223–233, 2000.

[14] B. Heisel, T. Sere, M. Pontil, T. Vetter, and T. Poggio "Categorization by learning and combining object parts," In: *Proc. of Neural Information Processing Systems*, 2001.

[15] C.-Y. Huang, O.I. Camps, and T. Kanungo, "Object recognition using appearance-based parts and relations," In: *Proc. IEEE Computer Vision and Pattern Recognition*, pp. 878–884, 1997.

[16] A. Hyvärinen, J. Karhunen, and E. Oja, *Independent Component Analysis*, John Wiley, 2001.

[17] A. Lanitis, C.J. Taylor and T.F Cootes, "Automatic interpretation and coding of face images using flexible models," *IEEE Trans. Pattern Analysis and Machine Intelligence* **19**(7):743–756, 1997.

[18] M.J. Lyons, J. Budynek and S. Akamatsu, "Automatic classification of single facial images," *IEEE Trans. Pattern Analysis and Machine Intelligence* **21**(12):1357–1362, 1999.

[19] A.M. Martínez and R. Benavente, *The AR face database*, CVC Tech. Rep. #24, June 1998.

[20] A.M. Martínez and A.C. Kak, "PCA versus LDA," *IEEE Transactions on Pattern Analysis and Machine Intelligence* **23**(2):228–233, 2001.

[21] A.M. Martínez, "Recognizing imprecisely localized, partially occluded and expression variant faces from a single sample per class," *IEEE Transactions on Pattern Analysis and Machine Intelligence* **24**(6):748–763, 2002.

[22] A.M. Martínez, "Matching expression variant faces," *Vision Research* **43**(9): 1047–1060, 2003.

[23] A.M. Martínez, "Recognizing expression variant faces from a single sample image per class," In: *Proc. IEEE Computer Vision and Pattern Recognition*, pp. I353–I358, Madison (WI), June 2003.

[24] S. Nayar, K. Ikeuchi, and T. Kanade, "Determining shape and reflectance of hybrid surface by photometric sampling," *IEEE Trans. on Robotics and Automation* **6**(4):418–431, 1990.

[25] K. Ohba and K. Ikeuchi, "Detectability, uniqueness, and reliability of eigen windows for stable verification of partially occluded objects," *IEEE Transactions on Pattern Analysis and Machine Intelligence* **19**(9):1043–1048, 1996.

[26] P.S. Penev and J.J. Atick, "Local feature analysis: a general statistical theory for object representation," *Network: Computation in Neural Systems* **7**(3):477–500, 1996.

[27] P.J. Phillips, H. Moon, P. Rauss, and S.A. Rizvi, "The FERET evaluation methodology for face-recognition algorithms," *Proceedings of the First International Conference on Audio and Video-based Biometric Person Authentification*, Crans-Montana (Switzerland), 1997.

[28] C.R. Rao, *Linear Statistical Inference and Its Applications*, John Wiley, 1973.

[29] C.R. Rao, "Prediction of future observation in growth curve models," *Stat. Science* **2**:434–471, 1987.

[30] L. Sirovich and M. Kirby, "Low-dimensional procedure for the characterization of human faces," *Journal of the Optical Society of America* A**4**:519–524, 1987.

[31] K.-K. Sung and T. Poggio, "Example-based learning for view-based human face detection," *IEEE Transactions on Pattern Analysis and Machine Intelligence* **20**(1):39–51, 1998.

[32] M.J. Tarr and Y.D. Cheng , "Learning to see faces and objects," *Trends in Cognitive Science* **7**(1):23–30, 2003.

[33] M. Turk and A. Pentland, "Eigenfaces for recognition," *Journal Cognitive Neuroscience* **3**(1):71–86, 1991.

[34] T. Vetter and N.F. Troje, "Separation of texture and shape in images of faces for image coding and synthesis," *Journal of the Optical Society of America* A**14**(9): 2152–2161, 1997.

[35] M.-H. Yang, D.J. Kriegman, and N. Ahuja, "Detecting faces in images: a survey," *IEEE Transactions on Pattern Analysis and Machine Intelligence*, **24**:34–58, 2002.

[36] Y. Zhang and A.M. Martínez, "From static to video: face recognition using a probabilistic approach," In: *Proc. IEEE Workshop on Face Processing from Video*, June 2004.

[37] Y. Zhang and A.M. Martínez, "Recognition of expression variant faces Using weighted subspaces," In: *Proc. International Conference on Pattern Recognition*, Cambridge (UK), August 2004.

[38] W. Zhao and R. Chellapa, "SFS based view synthesis for robust face recognition," In: *Proc. IEEE Face and Gesture Recognition*, pp. 285–292, 2000.

[39] W. Zhao, R. Chellappa, J. Phillips, and A. Rosenfeld, "Face recognition in still and video images: a literature surrey," *ACM Computing Surveys* **35**(4):399–458, 2003.

CHAPTER 19

NEAR REAL-TIME ROBUST FACE AND FACIAL-FEATURE DETECTION WITH INFORMATION-BASED MAXIMUM DISCRIMINATION

19.1 INTRODUCTION

Vision-based pattern-recognition algorithms for the continuous analysis of face-related video have to deal with the complexity in appearance and expression of the facial patterns, and whenever possible, the algorithms must use all available knowledge for this analysis. The starting point of all other face analysis algorithms is to detect the presence of faces and accurately track their location in complex environments. This task itself represents an interesting, challenging problem in computer vision. These algorithms are particularly helpful to interactive systems, because they provide the machine with the sense of awareness of the presence of users, thereby triggering alarms, starting applications, initializing systems, etc.

We classify the previous work on face detection into three different categories based on the approach used for dealing with the invariance to scale and rotation. First, a bottom-up approach uses feature-based geometrical constraints [1]. Facial-features are found using spatial filters, then they are combined to formulate hypothetical face candidates which are validated using geometrical constraints. Second, scale- and rotation-invariant face detection has been proposed using multiresolution segmentation. Face candidates are found by merging regions until their shapes approximate ellipses. Next, these candidates are validated using facial-features [2, 3, 4]. Skin-color segmentation has been successfully used to segment faces in complex backgrounds. Last, most face-detection algorithms [8, 9, 10] use multiscale searches with classifiers of fixed size. These classifiers are trained from examples using several learning techniques. Multiscale detection is particularly useful when no color information is available for segmentation and when the faces in the test images are too small to use facial-features.

Most pattern-recognition techniques have been used with multiscale detection schemes for face detection. An early approach uses decision rules based on image intensity to formulate face candidates from a multiscale representation of the input images. Then, these candidates are validated with edge-based facial features [5]. Maximum-likelihood classifiers based on Gaussian models applied on a principal component subspace have also been studied [6] in the context of face detection. Similarly, support-vector machine classifiers have been successfully used [7] for face detection in complex backgrounds.

Neural-network-based systems have also been successfully used for multiscale face detection [8, 9, 10]. More recently, a neural-network-based, rotation-invariant approach to face detection has been proposed [11]. In a multiscale detection setup, subwindows are first tested with a neural network trained to return the best orientation, and then the face detector is applied only at the given orientation. While the use of color has proven to be very helpful for detecting skin [3], the general approach of segmentation and region merging is limited to produce a rough location of the face with almost no detail whatsoever about the features. On the other hand, feature-based face-detection techniques rely on the facial-feature detection which in principle is as difficult as the original problem of face-detection; consequently, these techniques are invariant to rotation and scale only within the range in which the feature detectors are invariant to rotation and scale. Schemes for face-detection that use example-based pattern-recognition techniques and that search at different scales and rotation angles are limited mostly by the computational complexity required to deal with large search spaces. However, the accuracy and performance of example-based recognition techniques allow these systems to locate facial features. Recently Viola and Jones have developed a very fast face-detection system based on simple features using the "integral image". They have used AdaBoost [12] to select a small set of features as weak classifiers to obtain a strong classifier [13]. Robust and fast face detection has been pushed to another level.

The face and facial-feature detection algorithm described in this chapter carries out a multiscale search with a face classifier based on the learning technique described in the next section. Then, nine facial-features are located at positions where faces were found. Finally, candidate validation is further carried out by combining the confidence level of individual facial-feature detection with that of the face detection.

The rest of the chapter is organized as follows: Section 19.2 presents the new learning technique: *information-based maximum discrimination*. Sections 19.3 describes the systems for face and facial-feature detection respectively. Section 19.4 explains in detail the different experiments and data sets used to evaluate the proposed techniques as well as the results of these performance evaluations. Section 19.5 summarizes and provides future directions and concluding remarks.

19.2 INFORMATION-BASED MAXIMUM DISCRIMINATION

An observation is a collection of measurements, or more formally, a d-dimensional vector $\mathbf{x}$ in some space $\mathbf{S}^d$. The unknown nature of the observations, the class Y, is a finite set $\{C_1, C_2, \cdots, C_N\}$ used to label the underlying process that generates these observations. A classifier is the mapping function $g(x) : \mathbf{S}^d \rightarrow \{C_1, C_2, \cdots, C_N\}$, and the question of how to find the best classifier, namely learning, is fundamentally solved by minimizing the probability of classification errors $P[g(X) \neq Y]$.

In practice, the probability of classification errors cannot be computed. Other quantities are used as the optimization criteria for learning. In this section, we present a novel learning technique based on information-theoretic divergence measures, namely, *information-based maximum discrimination*, the most outstanding characteristic of which is its exceptionally low computational requirements in both its training and testing procedures.

19.2.1 The Bayes Classifier

Consider the random pair $(X, Y) \in \mathbf{S}^d \times \{C_1, C_2, \cdots, C_N\}$ with some probability distribution. Classification errors occur when $g(X) \neq Y$. The best classifier g^*, known as the Bayes classifier, is the one that minimizes the probability of classification error.

In a two-class discrimination problem, $Y \in \{0, 1\}$, the Bayes classifier and Bayes error are often expressed in terms of the *a posteriori* probability $P(Y|X)$ as

$$g^*(\mathbf{x}) = \begin{cases} 1 & \text{if } P(Y{=}1|X{=}\mathbf{x}) > P(Y{=}0|X{=}\mathbf{x}), \\ 0 & \text{otherwise,} \end{cases} \tag{1}$$

and

$$\mathbf{L}^* = \mathbf{L}(g^*) = \mathbf{E}\{\min(P(Y|X), 1 - P(Y|X))\}. \tag{2}$$

In practical situations, only approximations of the class-conditional densities $\tilde{f}_0 \approx f_0(\mathbf{x}) = \mathbf{P}(X|Y = 0)$ and $\tilde{f}_1 \approx f_1(\mathbf{x}) = \mathbf{P}(X|Y = 1)$ are available. These, together with the approximations of the *a priori* probabilities of the classes $\tilde{p}_1 \approx p = \mathbf{P}(Y = 1)$ and $\tilde{p}_0 \approx 1 - p = \mathbf{P}(Y = 0)$, are used to construct the following classifier:

$$g(\mathbf{x}) = \begin{cases} 1 & \text{if } \tilde{f}_1(\mathbf{x})/\tilde{f}_0(\mathbf{x}) > \tilde{p}_0/\tilde{p}_1, \\ 0 & \text{otherwise.} \end{cases} \tag{3}$$

For this decision rule, the probability of error is bounded from above [14] by

$$\mathbf{L}(g) - \mathbf{L}^* \leq \int_{\mathbf{S}^d} |(1-p)f_0(\mathbf{x}) - \tilde{p}_0\tilde{f}_0(\mathbf{x})|d\mathbf{x} + \int_{\mathbf{S}^d} |pf_1(\mathbf{x}) - \tilde{p}_1\tilde{f}_1(\mathbf{x})|d\mathbf{x}. \tag{4}$$

Therefore, if the used probability models fit the data well, and if the data is representative of the unknown distribution, then the performance of this approximation is not much different from that of the Bayes classifier. It is also important to note that if the *a priori* class probabilities, $\tilde{p}_1(x)$ and $\tilde{p}_0$, cannot be estimated or are simply assumed equal, then the classifier of Equation 3 turns into a *maximum-likelihood* classifier.

19.2.2 Information-Theory-Based Learning

The probability of error $\mathbf{L}(g)$ is the fundamental quantity to evaluate the discrimination capability of the data distribution; however, many other measures have been suggested and used as the optimization criteria for learning techniques. Nonetheless, the relation between these discrimination measures and the Bayes error has also been studied. For example, in training a support-vector machine [15], an "empirical risk" function is minimized in order to find the best classifier, and a bound of the misclassification error has been found.

In the following section, we briefly describe the information-theoretic discrimination measurements that are of particular interest in developing our learning technique. These have been previously used in different scenarios such as image registration and parameter estimation [16, 17, 18, 19]. The relations between the Bayes error and these divergence measurements have also been established; for further details in this subject, the reader is referred to [14, 20].

Kullback–Leibler divergence

For the remainder of this section, let us consider the random pair (X, Y) with $Y \in \{0, 1\}$, but let $X \in \mathbf{D}^d$ be a d-dimensional vector in some discrete space $\mathbf{D} = \{1, 2, \ldots, M\}$. The Kullback–Leibler divergence between the two classes is then defined as

$$\mathbf{I} = \mathbf{E}\left[\eta(\mathbf{x}) \log \frac{\eta(\mathbf{x})}{1 - \eta(\mathbf{x})}\right], \tag{5}$$

and its symmetric counterpart, the Jeffrey divergence, is defined as

$$\mathbf{J} = \mathbf{E}\left[(2\eta(\mathbf{x}) - 1)\log\frac{\eta(\mathbf{x})}{1-\eta(\mathbf{x})}\right], \tag{6}$$

where $\eta(\mathbf{x}) = \mathbf{P}(Y = 1|\mathbf{x})$ is the *a posteriori* probability of the class $Y = 1$. To understand their meaning, note that both divergences are zero only when $\eta(\mathbf{x}) = 1/2$, and they go to infinity as $\eta(\mathbf{x}) \uparrow 1$ and $\eta(\mathbf{x}) \downarrow 0$.

Once again, in the practical situation where the *a priori* probabilities of the classes cannot be estimated and are assumed equal, one uses only the class-conditional probabilities $\mathbf{P}(X|Y = 0)$ and $\mathbf{P}(X|Y = 1)$, and the Kullback–Leibler divergence and Jeffrey divergence can be computed as

$$\mathbf{H}(\lambda) = \sum_{\mathbf{x}} \{\mathbf{P}(\mathbf{x}|Y = 1) - \lambda\mathbf{P}(\mathbf{x}|Y = 0)\}\log\frac{\mathbf{P}(\mathbf{x}|Y = 1)}{\mathbf{P}(\mathbf{x}|Y = 0)}, \tag{7}$$

where λ is set to zero for the Kullback–Leibler divergence, and set to one for the Jeffrey divergence.

This form of divergence is a non-negative measure, the difference between these two conditional probabilities. It is zero only when the probabilities are identical, in which case the probability of error is $\frac{1}{2}$, and goes to infinity as the probabilities differ from each other, in which case the probability of error goes to zero. Put differently, it measures the underlying discrimination capability of the nature of the observations described by these conditional probability functions. These information-theoretic discrimination measures are particularly attractive as the optimization criteria for our learning technique mainly because of the chain-rule property that the technique exhibits under Markov processes. This property allows us to find a practical solution to the learning problem.

Nonparametric Probability Models

We use nonparametric, or rather discrete probability models to capture the nature of the observation vector $\mathbf{x}$; recall that $\mathbf{x} \in \mathbf{D}^d$ is a d-dimensional vector in some discrete space $\mathbf{D} = \{1, 2, \ldots, M\}$. We have tested several families of models [21, 22]; however, we limit our discussion here to the one with the most outstanding performance, a modified Markov model.

Let us formally define a modified, kth-order Markov process with its probability function

$$\mathbf{P}(x_1,\ldots,x_n|\mathbf{S}) = \mathbf{P}(x_{s_1})\prod_{m=2}^{k}\mathbf{P}(x_{s_m}|x_{s_1},\ldots,x_{s_{m-1}}) \prod_{m=k+1}^{n}\mathbf{P}(x_{s_m}|x_{s_{m-k}},\ldots,x_{s_{m-1}}), \tag{8}$$

where $\mathbf{S} = \{s_i \in [1,n] : i = 1,2,\ldots,n : s_i \neq s_j \; \forall\, i \neq j\}$ is a list of indices used to rearrange the order of the elements of vector $\mathbf{x}$. Note that such a model could be interpreted as a linear transformation, $\mathbf{x}' = \mathbf{Tx}$, followed by a regular Markov model applied in the transformed vector $\mathbf{x}'$, where

$$T_{j,i} = \begin{cases} 1 & \text{if } S_j = i, \\ 0 & \text{otherwise.} \end{cases} \tag{9}$$

Given the two conditional probabilities of these modified Markov processes, $\mathbf{P}(\mathbf{x}|\mathbf{S}, Y = 1)$ and $\mathbf{P}(\mathbf{x}|\mathbf{S}, Y = 0)$ satisfying Equation 8, the divergence can be efficiently computed as

$$\mathbf{H}(\mathbf{S}) = \mathbf{H}(s_1) + \sum_{m=2}^{k}\mathbf{H}(s_m|s_1,\ldots,s_{m-1}) + \sum_{m=k+1}^{n}\mathbf{H}(s_m|s_{m-k},\ldots,s_{m-1}), \tag{10}$$

where

$$\mathbf{H}(i) = \sum_{x_i=1}^{M}\mathbf{P}(x_i|Y=1)\log\frac{\mathbf{P}(x_i|Y=1)}{\mathbf{P}(x_i|Y=0)} \tag{11}$$

and

$$\mathbf{H}(i|j_1,\ldots,j_k) = \sum_{x_i=1}^{M}\sum_{x_{j_1}=1}^{M}\cdots\sum_{x_{j_k}=1}^{M}\left(\mathbf{P}(x_i,x_{j_1},\ldots,x_{j_k}|Y=1)\log\frac{\mathbf{P}(x_i|x_{j_1},\ldots,x_{j_k},Y=1)}{\mathbf{P}(x_i|x_{j_1},\ldots,x_{j_k},Y=0)}\right). \tag{12}$$

Learning Procedure

Based on the previously modified Markov models and the information-theoretic discrimination measurements, we set the goal of the learning procedure: to find the best possible classifier for the given set of training examples.

Let $g(\mathbf{x}, \mathbf{S})$ be a maximum-likelihood classifier such as that in Equation 3, since $\mathbf{x}$ is a d-dimensional vector in some discrete space $\mathbf{D}$:

$$g(\mathbf{x}, \mathbf{S}) = \begin{cases} 1 & \text{if } \mathbf{L}(\mathbf{x}|\mathbf{S}) = \mathbf{P}_1(\mathbf{x}|\mathbf{S})/\mathbf{P}_0(\mathbf{x}|\mathbf{S}) > \mathbf{L}_{\text{th}}, \\ 0 & \text{otherwise,} \end{cases} \tag{13}$$

where $\mathbf{P}_1(\mathbf{x}|\mathbf{S})$ and $\mathbf{P}_0(\mathbf{x}|\mathbf{S})$ are estimations of the first-order modified Markov class conditional probabilities $\mathbf{P}(\mathbf{x}|\mathbf{S}, Y = 1)$ and $\mathbf{P}(\mathbf{x}|\mathbf{S}, Y = 0)$ respectively, and $\mathbf{L}_{\text{th}}$ is an estimation of the ratio of the *a priori* class probabilities $\mathbf{P}(Y = 1)$ and $\mathbf{P}(Y = 0)$.

We use the statistics of the training set, individually and pairwise on the dimensions of the space $\mathbf{D}$, to compute the probabilities and, with them, the divergences $\mathbf{H}(i)$ and $\mathbf{H}(i, j)$ for $i, j = 1, 2, \ldots, d$ using Equations 11 and 12. Then, we set up the optimization procedure

$$\mathbf{S}^* = \arg\max \mathbf{H}(\mathbf{S}), \tag{14}$$

to find the sequence $\mathbf{S}^*$ that maximizes the divergence

$$\mathbf{H}(\mathbf{S}) = \mathbf{H}(s_1) + \sum_{m=2}^{n} \mathbf{H}(s_m|s_{m-1}). \tag{15}$$

If $\mathbf{H}(s_m|s_{m-1})$ is thought to be a distance from vertex s_m to s_{m-1} and $\mathbf{H}(s_1)$ a distance from a fixed starting point (different from any of the n vertices.) to vertex s_1, then the physical meaning of $\mathbf{H}(\mathbf{S})$ is the maximum distance of a path starting from the fixed starting point, traverse each and every vertex exactly once. The optimal solution gives the optimal traversing path $s_1, \cdots, s_n$. Figure 19.1 is an illustration. This is closely related to the NP-complete "travelling salesman problem" (TSP) in graph theory where this optimal path is called the Hamiltonian path [23]. Note that this optimization problem, similarly to the TSP, in practice would not be solved exhaustively. However, a modified version of the Kruskal's algorithm for the minimum-weight spanning tree [23], [24] produces good results.

Once a suboptimal solution $\mathbf{S}'$ is found, the classifier $g(\mathbf{x}, \mathbf{S}')$ is implemented by a look-up table that holds the logarithm of the likelihood ratios $\log \mathbf{L}_{s'_0}$ and $\log \mathbf{L}_{s'_i|s'_{i-1}}$

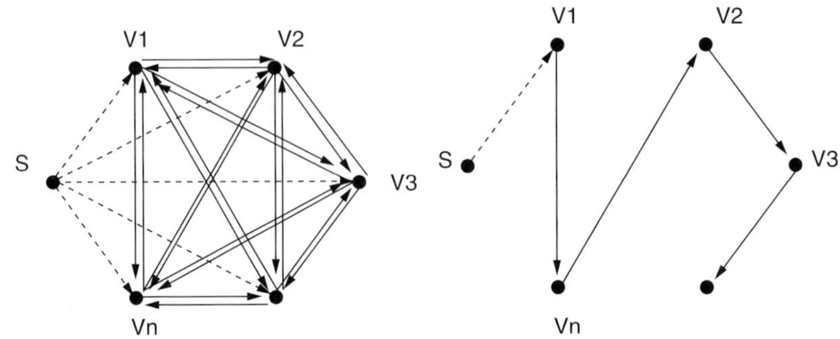

FIGURE 19.1: Finding the optimal path starting from the fixed node S and traversing each and every node exactly once. Dashed-line arrows always start from node S. Solid-line arrows are bidirectional. Each arrow is associated with a weight (i.e., distance). The subgraph on the left with node $V_1, \cdots, V_n$ is fully connected by the solid-line arrows. The one on the right is an illustration of the optimal path.

for $i = 2, \ldots, d$, so that, given an observation vector $\mathbf{x}$, its log-likelihood ratio is computed as

$$\log \mathbf{L}(\mathbf{x}|\mathbf{S}') = \log \mathbf{L}_{s'_0}(x_{s'_0}) + \sum_{m=2}^{n} \log \mathbf{L}_{s'_i|s'_{i-1}}(x_{s_m}|x_{s_{m-1}}). \tag{16}$$

19.2.3 New Insight on the Intuition of Permutation

When modeling a face image as an instance of a random process, rows of the image are concatenated into a long vector. The pixels corresponding to the semantics (e.g., eyes, lips) will be scattered into different parts in the vector. The Markovian property is not easily justified. If some permutation can be found to regroup those scattered pixels (i.e., to put all the pixels corresponding to eyes together, those for lips together), then the Markov assumption is more reasonable. That the permutation of indices leads to a Markovian process is the main intuition and contribution of our previous work [22].

19.2.4 Error Bootstrapping

Consider the problem of object detection as that of a two-class classification. One of the classes, the object class $Y = 1$, corresponds to the object in question, while the other, the background class $Y = 0$, corresponds to the rest of the space. For a given set of observations, the classifier is used to decide which of them corresponds to the desired object, or to put it another way, the classifier detects the object in the observations. In this scenario, the object class can be well represented by the training examples; however, that is not the case with the background class. One tries to use as many and diverse examples as possible to estimate the conditional probability of the background class. Doing so might cause the contribution of the background examples close to the object examples to be unfavorably weighted, resulting in a large probability of false-detection error for this class of observations. One widely used approach to overcome this limitation is called *error bootstrapping*. The classifier is first trained with all the examples in the training set. Then, it is used to further classify the training set, depending on the success of this classification. Then, the samples that were not successfully recognized are used separately in a second stage to reinforce the learning procedure.

Information-based maximum-discrimination learning can also benefit from error bootstrapping. Once the classifier is obtained with all the examples of the training set, and the training set has been evaluated with this classifier, statistics of the correctly classified examples are computed separately from those of the incorrectly classified examples. The new class-conditional probabilities are computed as

$$\begin{aligned} \mathbf{P}(\mathbf{x}|\alpha, Y=1) &= \alpha \mathbf{P}^{\mathrm{i}}(\mathbf{x}|Y=1) + (1-\alpha)\mathbf{P}^{\mathrm{c}}(\mathbf{x}|Y=1) \\ \mathbf{P}(\mathbf{x}|\beta, Y=0) &= \beta \mathbf{P}^{\mathrm{i}}(\mathbf{x}|Y=0) + (1-\beta)\mathbf{P}^{\mathrm{c}}(\mathbf{x}|Y=0), \end{aligned} \tag{17}$$

where $\mathbf{P}^{\mathrm{c}}$ is the probability of the correctly classified examples, $\mathbf{P}^{\mathrm{i}}$ is the probability of the incorrectly classified examples, and α and β are the mixing factor for each of the classes. Using these mixtures and Equation 7, the divergence can be computed as

$$\mathbf{H}(\lambda, \alpha, \beta) = \sum_{\mathbf{x}} \left([\mathbf{P}(\mathbf{x}|\alpha, Y=1) - \lambda \mathbf{P}(\mathbf{x}|\beta, Y=0)] \log \frac{\mathbf{P}(\mathbf{x}|\alpha, Y=1)}{\mathbf{P}(\mathbf{x}|\beta, Y=0)} \right). \tag{18}$$

Then, the new classifier is obtained by solving the maximization in Equation 14.

19.2.5 Fast Classification

The learning technique previously described was developed in the context of discrete observation data, and due to implementation limitations, the number of outcomes M in the discrete space $\mathbf{D} = \{1, 2, \ldots, M\}$ cannot be large. In this section, we discuss these limitations as well as the computational requirements of both learning and testing procedures.

In order to hold the statistics, or the histogram of occurrence, of each pair of dimensions of the observations for each set of training data, M^2d^2 parameters are required. Only one pass through the training data is required to capture the statistics, and the processing requirements for computing the divergence and finding the best sequence $\mathbf{S}$ are negligible. Consequently, the training procedure is incremental; if more data is added to the training set, such as in cases of adaptation, the training procedure does not need to be started from scratch.

Once the classifier has been trained, only $d(M^2 + 1)$ parameters are needed to store its knowledge. And since only the logarithms of the likelihood-ratio functions are needed, and their range of variation is limited, fixed-point parameters can be used. It is also important to mention that only d fixed-point additions are required to perform the classification of an observation; that is, only one operation per dimension of the observation vectors is required to classify these observations.

19.3 IBMD FACE AND FACIAL-FEATURE DETECTION

19.3.1 Multiscale Detection of Faces

Figure 19.2 illustrates the scheme for face detection based on multiscale search with a classifier of fixed size. First, a pyramid of multiple-resolution versions of the input image is constructed. Then, a face classifier is used to test each subwindow in a subset of these images. At each scale, faces are detected depending on the output of the classifier. The detection results at each scale are projected back to the input image with the appropriate size and position. One chooses the scales to be tested depending on the desired range of size variation allowed in the test image.

Each window is preprocessed to deal with illumination variations before it is tested with the classifier. A postprocessing algorithm is also used for face candidate selection. The classification is not carried out independently on each subwindow. Instead, face candidates are selected by analyzing the confidence level of subwindow classifications in neighborhoods so that results from neighboring locations, including those at different scales, are combined to produce a more robust list of face candidates.

In our implementation, we use a face classifier to test subwindows of 16×14 pixels. The preprocessing stage consists of histogram equalization and requantization as follows:

$$[x] = \begin{cases} 0 & x < \frac{3}{4}\overline{x}, \\ 1 & \frac{3}{4}\overline{x} \leq x < \overline{x}, \\ 2 & \overline{x} \leq x < \frac{(255+3\overline{x})}{4} \\ 3 & \text{otherwise}, \end{cases},$$

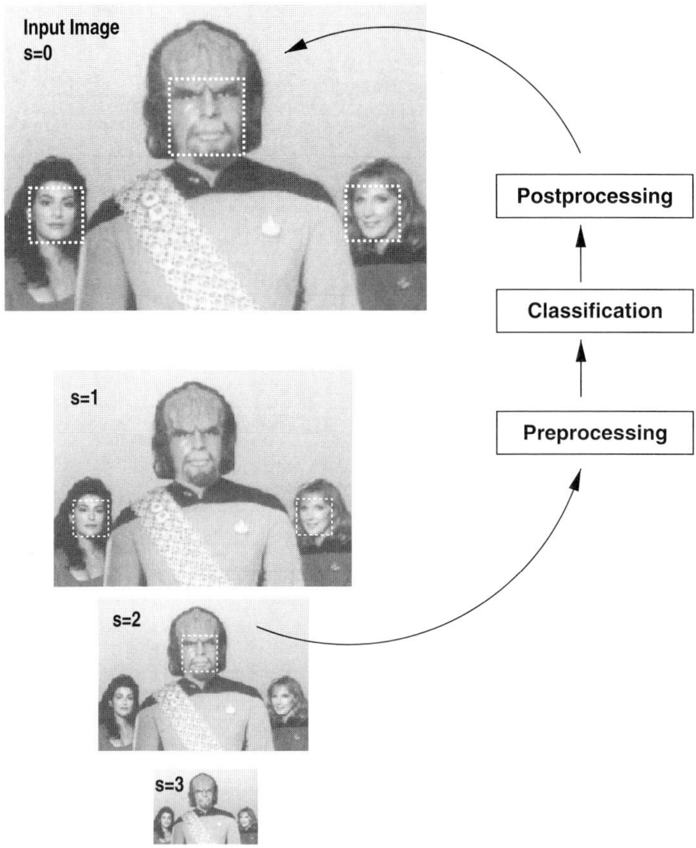

FIGURE 19.2: General scheme for multiscale face detection.

where [ ] represents the quantization operator and $\bar{x}$ is the average of the pixel values within the test window.

Examples of face images preprocessed at four gray levels are shown in Figure 19.3(c). The output of the classifier is the log likelihood of the observations. The greater this value, the more likely the observation corresponds to a face. For each subwindow tested, a log-likelihood map is obtained with the output of the function $\mathbf{L}(\mathbf{x}|\mathbf{S})$ of the classifier in Equation 13. The face-classification decision is not made independently at every subwindow; instead, face candidates are selected by analyzing these log-likelihood maps locally.

The face-detection classifier was trained with a subset of 703 images from the FERET database [25, 26]. Faces were normalized with the location of the outer eye corners. Three rotation angles $\theta = \{8.0, 0.0, -8.0\}$ and three scale factors $s = \{1.000, 0.902, 0.805\}$ were used on each image to produce a total of 6327 examples of faces. Figure 19.3(a–c) shows one example of the images in the FERET database, the scale and rotation normalized images, and the four-gray-level training patterns. On the other hand, 44 images of a variety of scenes and their corresponding scaled versions were used to produce 423,987 examples of background patterns. We used this training data to compute the divergence of each pixel of the $224 = 16 \times 14$ element observation vector. Figure 19.4(a) is a $16^2 \times 14^2$ image that shows the divergence of each pair of pixels as in Equation 12. Similarly, Figure 19.4(b) shows the divergence of each pixel independently; it is computed as in Equation 11. We suboptimally solved the maximization in Equation 14 using a greedy algorithm such as the Kruskal algorithm [23] and obtained a sequence of index pixels with high divergence. Figure 19.4(c, d) shows the divergence of the pixels in the sequence found by our learning algorithm before and after error bootstrapping. Although the sequence itself cannot be visualized

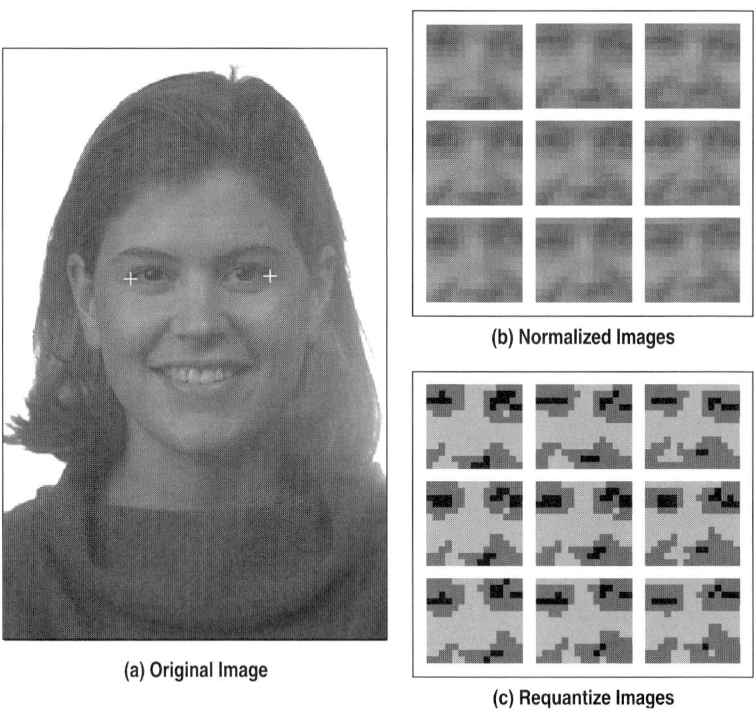

FIGURE 19.3: Examples of the image-preprocessing algorithm.

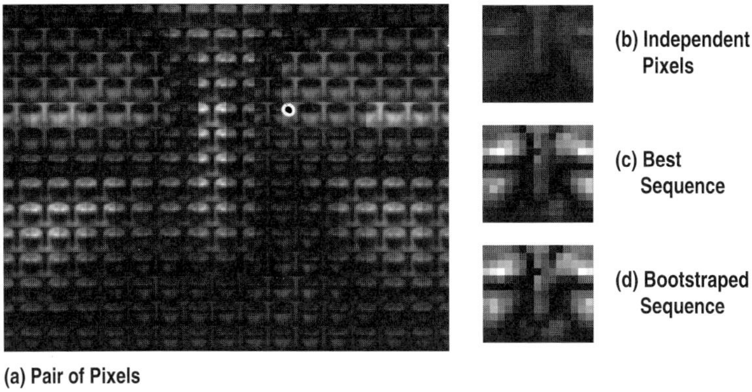

FIGURE 19.4: Divergence of the training data.

from these images, they show the divergence of the facial regions. Note that the eyes, cheeks, and nose are the most discriminative regions of faces.

19.3.2 Facial-Feature Detection

Once the face candidates are found, the facial-features are located using classifiers trained with examples of facial features. Face detection is carried out at a very low resolution with the purpose of speeding up the algorithm. Facial features are located by searching with the feature detectors at the appropriate positions in a higher resolution image, i.e., two or three images up in the multiscale pyramid of images.

In an early approach [22, 21], we used the same preprocessing algorithm and learning scheme described above to train both the face detector and the eye-corner detectors. The facial-feature detection algorithm described here is based on a more complex discrete observation space that combines edges and pixel intensity. This new discrete space is formed by combining the results of three low-level image-processing techniques: (i) binary quantization of the intensity, (ii) three gray-scale level of horizontal edges, and (iii) three gray-scale level of vertical edges. The threshold values used to requantize these low-level feature images are based on a fixed percentage of the pixels in a region in the center of the face. Combining these sets of discrete images, we construct the discrete image $I = I_i + 2I_v + 6I_h$ that is used to locate the facial features. Figure 19.5 shows three sets of examples of these low-level feature images. The left column in Fig. 19.5 indicates the positions of the nine facial-features used to train the facial-feature detectors: the outer corners of each eye, the four corners of the eyebrows, the center of the nostrils, and the corners of the mouth.

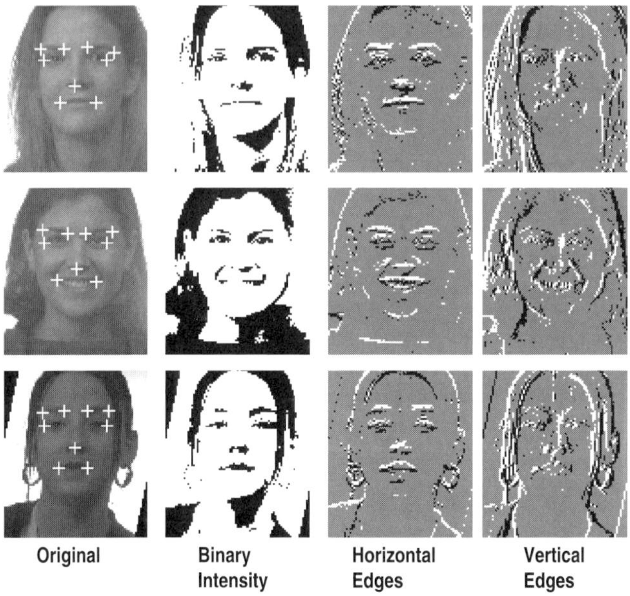

FIGURE 19.5: Low-level features for discrete observations.

We trained the classifiers using 150 images in which the feature positions were located by hand. We used three rotation angles and three scale factors to produce the image examples of facial features. Negative examples were obtained from image subwindows at neighboring locations around the corresponding feature positions. The relative locations of the facial-features in these training images were also used to determine the size and location of the facial-feature search areas of the detection procedure.

Based on the individual performance of the feature detectors, we implemented a hierarchical facial feature detection scheme. First, the nostrils are located and used to center-adjust the search areas of the other facial features. Then, the other facial features are detected. Figure 19.6(a) shows a test image and the location of the facial features as detected by our algorithm. Figure 19.6(b) is the normalized window on which the preprocessing technique is carried out. Figure 19.6(c) shows the low-level feature images. Figure 19.6(d) shows the search areas and the log-likelihood maps of the search of each feature.

19.3.3 Discussion

One of the most important issues in applying information-based maximum-discrimination learning is the technique used for image preprocessing and requantization. This is important because of the mapping between the input-image space

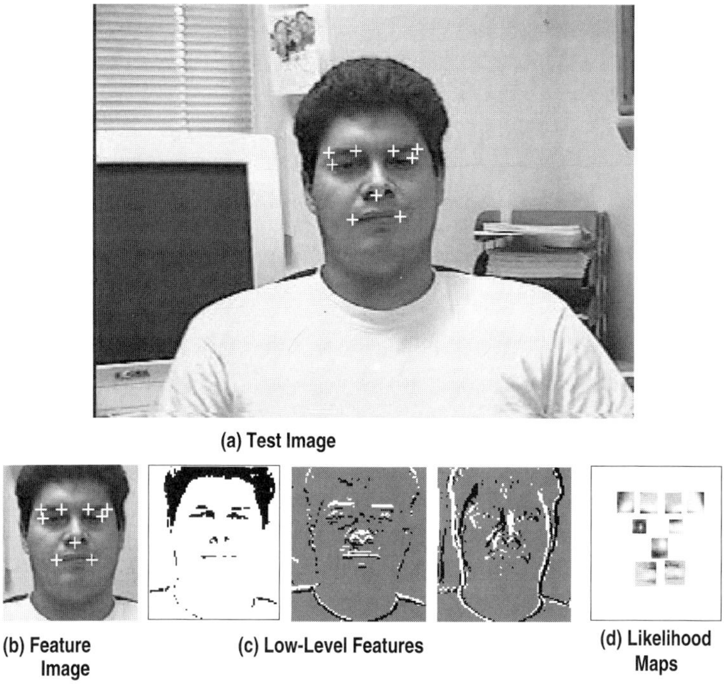

FIGURE 19.6: Example of features detection.

and the discrete-image space. This mapping reduces the number of possible pixel values of the discrete-image space while preserving the information useful for object discrimination so that discrete probability models can be implemented. In this section, we have described two approaches. Four-gray-level intensity was used for face detection while a combination of edges and intensity was used for facial-feature detection.

Another important issue in visual object detection is the choice of size of the image window used for detection. While large windows that include most of the desired object result in high-detection performance, their location accuracy is poor. In our face and facial-detection algorithm, we have presented a hierarchical scheme to improve detection performance and location accuracy. First, faces are detected using low-resolution image subwindows that include most of the face. Then, the facial-features are detected using image subwindows at higher resolution that include only a portion of the facial features being detected.

19.4 EXPERIMENTS AND RESULTS

19.4.1 Image Databases

Examples of images in the following two databases are shown in Fig. 19.7.

FERET Database

The face-recognition technology (FERET) database [25] [26] consists of thousands of facial images. Several pictures per person were taken at different times and views, with different illumination conditions and facial expressions, wearing glasses, make-up, etc.

We use a subset of this database solely with the purpose of extracting information about faces. For this subset of 821 images of frontal view faces, we have labelled the position of the facial-features by hand. We use these locations together with the images to train face and facial feature detectors, and to learn the distribution of the relative positions of the facial features. Examples of these training procedures are given in Figures 19.3 and 19.5.

CMU/MIT Database for Face Detection

The image database has been widely used to test neural-network-based face-detection systems in complex backgrounds [9, 10]. The database consists of three sets of gray-level images, two of which were collected at Carnegie Mellon University (CMU) and the other at the Massachusetts Institute of Technology (MIT). These images are scanned photographs, newspaper pictures, files collected from the World Wide Web, and digitized television shots. The first two sets, from CMU, contain 169 faces in 42 images and 183 faces in 65 images respectively. The set from MIT consists of 23 images containing 155 faces. The images of this database together, with the ground truth location of the faces, are available from http://www.cs.cmu.edu/People/har/faces.html.

19.4.2 Face Detection in Complex Backgrounds

We have tested a simple version of our face-detection algorithm with the CMU/MIT database, and compare our learning technique to that of the neural-network-based face-detection approach reported in [9, 10].

We trained the face-to-background classifier with only 11×11 pixels so that faces as tiny as 8×8 pixels present in this database could be detected. We used face examples from the FERET database and background examples from another collection of general images. Note that these results do not reflect those of the complete system, not only because of the low resolution of the classifier, but also because we turned off the postprocessing of the face candidates and the face validation with the facial-feature detection.

FIGURE 19.7: Images from the FERET and the CMU/MIT database. The 16 images in a 4 × 4 grid on the left half are from FERET. The rest are from CMU/MIT.

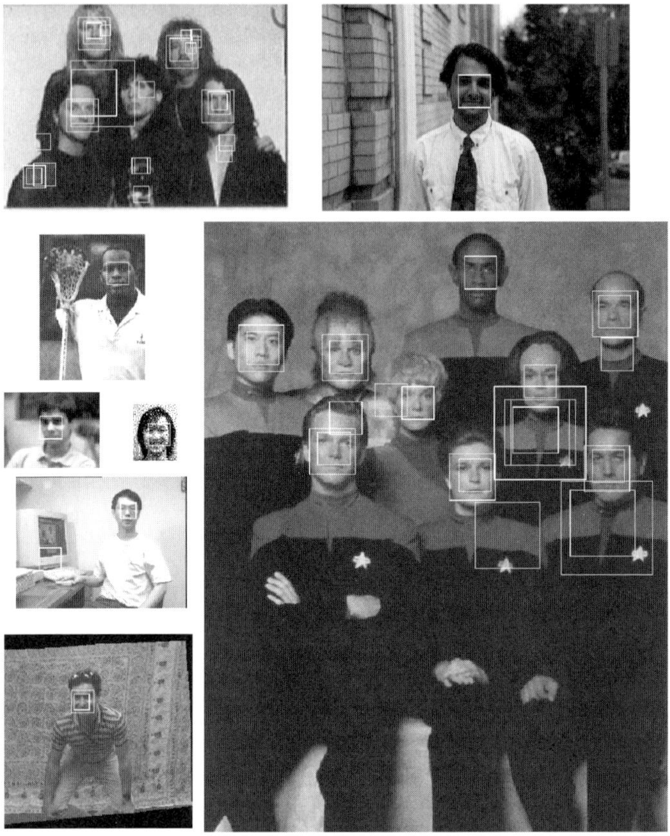

FIGURE 19.8: Examples of the images in the testing set and the face candidates obtained using our algorithm.

In our test, a face candidate obtained from the detection procedure was considered correct if it was in the correct scale and if the error in the position of the face (with respect to the ground truth) was less that 10% of the size of the face. All other candidates are marked as false detections. Figure 19.8 shows several examples of the testing set, and the face candidates obtained using our algorithm. Figure 19.9 shows the receiver operating characteristic (ROC) of the system obtained with the CMU database; the vertical axis shows the correct detection rate, while the horizontal one shows the false detection rate.

Table 19.1 shows the performance of our system together with that of neural-network-based system reported in [27] with two different network architectures. The first has a total of 2905 connections while the second has 4357.

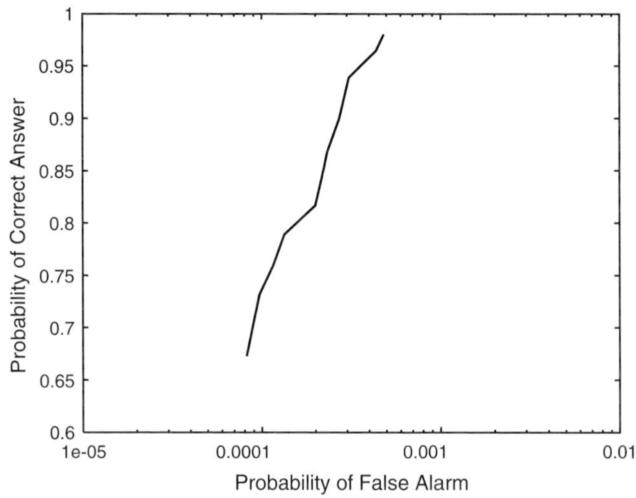

FIGURE 19.9: Face detection ROC with CMU database. Note that the y axis starts from 0.6 instead of 0.0 to show better the characteristics of the curve.

Table 19.1: Comparison between our face-detection system and that reported in [27] based on neural networks.

System description	Detected faces	Detection rate	False faces	False-alarm rate
Ours (121 pixels)	497/507	98.0%	12758	1/2064
Ours (121 pixels)	476/507	93.9%	8122	1/3243
Ours (121 pixels)	456/507	89.9%	7150	1/3684
Ours (121 pixels)	440/507	86.8%	6133	1/4294
NN 1 (2905 conn.)	470/507	92.7%	1768	1/47002
NN 2 (4357 conn.)	466/507	91.9%	1546	1/53751
NN 3 (2905 conn.)	463/507	91.3%	2176	1/38189
NN 4 (4357 conn.)	470/507	92.7%	2508	1/33134

Note that our system produced about three times more false alarms for about the same detection rate. On the other hand, although it is difficult to make a rigorous comparison of the computational requirements of these two face-detection approaches, we roughly estimate our method to be at least two orders of magnitude faster.

Face detection is achieved by testing a large number of subwindows with face classifiers. Our face-detection approach differs from the neural-network-based approach not only in the speed of the subwindow test, but also in the number of subwindows tested. In order to estimate the difference in computational requirements of these two face-detection approaches, we first estimate the ratio between the numbers of operations required by these classifiers to test each subwindow. Then, we estimate the ratio between the total numbers of subwindows that these two classifiers need to test to find faces with the same range of size since the scaling factors in the multiscale search are not equal.

The two network architectures have 4357 and 2905 connections respectively. Neglecting the requirements of the activation functions, and assuming that each connection requires one floating-point multiplication, one floating-point addition, and one "float" (4 bytes) to keep the weight factor, these systems require 8714 and 5810 floating-point operations, and about 17 and 11 kilobytes of memory respectively. On the other hand, assuming that 80% of the 11×11 pixels are used in the likelihood distance, our technique requires about 100 fixed-point additions and 1600 "shorts" (about 3 kilobytes) to hold the pre-computed log likelihood table of Equation 16. Disregarding that floating-point operations require either more hardware or more CPU time than fixed-point operations, and the effect of cache (because the data used by our algorithm is five times smaller), we estimate our classifiers to test each subwindow between 58 and 87 times faster.

Suppose that we use these systems to search for faces between $S_1 = 20$ and $S_2 = 200$ pixels in size, and that the input image is large compared to the subwindow size S_W; that is, $W, H \gg S_W$. In a multiscale-search approach with scale factor α, the total number of windows tested can be approximated by

$$N \approx WH \sum_{k=n_1}^{n_2} (\frac{1}{\alpha})^{2k} \qquad (19)$$

where n_1 and n_2 are computed from $n_k = \ln(S_k/S_W)/\ln \alpha$.

Considering that our system uses the scale factor $\alpha = \sqrt{2}$ and the subwindow size $S_W = 11$, and that the neural-network-based implementation uses $\alpha = 1.2$ and $S_W = 20$, we estimate the ratio between the number of subwindows required to be tested to be 3.26.

Overall, combining the ratio between the number of operations required by these face-detection approaches to test each subwindow with the classifiers (from 58 to 87) and the ratio between the total number of subwindows tested in these approaches (approximately 3.26), we estimate our face-detection system to be between 189 and 283 times faster. A more detailed comparison should also take into account the preprocessing step, which adds additional computation to the neural-network-based system.

19.4.3 Further Comparison Issues

While we have compared the aforementioned detection systems with the same test database, there are a number of issues that prevent this comparison from being truly useful.

1. The training sets used in these two approaches are different in both positive and negative examples. But most importantly, in our case, the training set is far too different from the test set; the face images from the FERET database are noiseless and the faces are in near-perfect frontal-view pose, while the CMU/MIT database includes a wide variety of noisy pictures, cartoons, and faces in different poses. Figure 19.7 shows examples of these databases.
2. The preprocessing algorithms used in these two approaches are far too different. In [10], nonuniform light conditions are compensated by linear fitting. The better the preprocessing algorithm, the better the performance of the learning technique. However, since such algorithms have to be applied to each of the tested subwindows before they are fed to the classifier, complex preprocessing algorithms introduce an extremely large amount of computation, especially in a multiscale search scheme. In our implementation, aimed at near-real-time operation, we used a simple histogram equalization procedure as the preprocessing step and left the classifier with the task of dealing with the variations in illumination.
3. The scale ratios in the multiscale detection schemes in these two techniques, and therefore the scale variations handled by the classifiers, are not the same. With less scale variation, the classifiers are expected to perform better; however, a greater number of scaled images must be used to find faces with similar size, resulting in another increase of computation.
4. The size of the subwindow used to feed the classifiers reflects the amount of information available to make the decision. Larger subwindows are expected to perform better. For this evaluation, however, we used a window of 11×11 pixels, mainly because of the presence of faces as tiny as 8×8 pixels, while the system in [10] was reported to use a window of 20×20 pixels.

19.4.4 Facial-Feature Detection

We have labelled by hand the location of the facial-features in 243 selected frames which show the most distinctive expressions found on the first set of videos of each person in the face video database. We trained the classifiers using one half of the labelled images at three different scales and rotations, and tested them on the other half at randomly selected scales and rotation angles.

In order to evaluate the performance of these classifiers, we measure the inaccuracy of the detector and compute the ROC of the classifiers. The location-error inaccuracy of the detector is measured as the average distance in pixels between the hand-selected location of the features and the peak location of the classifier response in the search areas. Note that this average error does not measure the confidence level of the classification. The performance of the facial-feature detectors, on the other hand, is studied from the ROC of the classifiers. The criterion for errors in the location of the features is 10% of the distance between the eyes.

In addition to computing the ROC of the classifiers, we extracted two measures to ease the comparison between different classifiers: (i) the maximum detection rate, or the top-1 detection performance, and (ii) the average detection rate for a given range of false-detection rates. Note that the former is commonly used to measure the performance of an object detector regardless of the confidence level of the detection. The later measure, obtained from the area under the ROC plot on the region where the system is actually operated, is more powerful in selecting the classifier that is best for the application in question; in our case, the near-real-time face and facial-feature tracker operates at a 1% false detection rate.

Three sets of experiments were conducted to evaluate the information-based maximum-discrimination learning technique. The first set is used to compare classifiers trained with different values of the bootstrapping mixing factor β. This mixing factor reflects the weight given to the error bootstrapping of the negative examples. Table 19.2 shows the results obtained with each of the feature detectors in terms of the inaccuracy, top-1 detection performance, and the average detection rate for a 1% false-alarm rate respectively. The second set of experiments is used to compare classifiers trained with different values of the bootstrapping mixing factor α. The mixing factor α reflects the weight given to the error bootstrapping of the positive examples of the features. Table 19.3 shows the results obtained with each of the feature detectors in terms of the inaccuracy, top-1 detection performance, and the average detection rate for 1% false alarm respectively. The third set of experiments is used to compare classifiers trained with different values of the weight λ in the computation of the divergence given in Equation 18. Table 19.4 shows the results obtained with each of the feature detectors in terms of the inaccuracy, top-1 detection performance and the average detection rate for 1% false alarm respectively.

From the performance comparisons of these classifiers, several observations can be made. First, the nostrils are by far the easiest facial features to detect, followed by the outer corners of the eyes. We take advantage of this by detecting the facial features in a hierarchical scheme in which the center of the nostrils is first found with a larger search area, and then its position is used to help detect the rest of the facial features. This hierarchical scheme improves the overall performance and speeds up the overall facial-feature detection by reducing the search areas of most of the features. From the comparison of the performance under different values of the mixing factor α and β of the error bootstrapping learning step, it is clear and consistent among all facial features that the bootstrapping

Table 19.2: Feature location inaccuracy, top-1 detection performance, and the average detection rate for 1% false alarm w.r.t the error bootstrapping of background examples.

β	Reye	Leye	RRbrw	RLbrw	LRbrw	LLbrw	Nose	Rmth	Lmth
0.00	0.226	0.381	0.904	1.884	2.302	2.318	1.945	1.126	0.649
0.05	0.275	0.400	0.944	1.803	2.388	2.373	1.965	1.153	0.665
0.10	0.218	0.365	0.851	1.813	2.285	2.435	2.000	1.220	0.629
0.15	0.215	0.391	0.834	1.671	2.313	2.466	2.061	1.251	0.661
0.20	0.188	0.356	0.857	1.721	2.370	2.449	2.086	1.207	0.622
0.25	0.182	0.354	0.922	1.763	2.293	2.514	2.016	1.143	0.588
0.30	0.212	0.388	0.918	1.838	2.283	2.455	2.007	1.193	0.630
0.35	0.229	0.372	0.928	1.770	2.328	2.483	2.026	1.162	0.616
β	Reye	Leye	RRbrw	RLbrw	LRbrw	LLbrw	Nose	Rmth	Lmth
0.00	99.45	99.18	96.39	95.06	97.12	95.88	100.00	97.21	95.34
0.05	99.09	99.13	96.57	94.33	95.98	95.24	100.00	96.71	95.24
0.10	99.50	99.41	96.39	95.43	96.02	96.25	100.00	97.26	95.20
0.15	99.27	99.77	96.20	96.07	97.26	95.84	100.00	96.66	95.84
0.20	99.50	99.59	96.30	95.52	97.26	95.88	100.00	96.48	95.56
0.25	99.22	99.63	96.25	95.84	97.21	95.88	100.00	97.71	95.52
0.30	99.22	99.18	96.94	96.25	96.94	95.56	100.00	97.71	95.88
0.35	99.18	99.59	96.39	96.34	97.03	95.20	100.00	97.76	94.92
β	Reye	Leye	RRbrw	RLbrw	LRbrw	LLbrw	Nose	Rmth	Lmth
0.00	0.960	0.906	0.869	0.814	0.818	0.870	0.997	0.939	0.825
0.05	0.943	0.901	0.855	0.814	0.842	0.849	0.996	0.924	0.830
0.10	0.958	0.923	0.889	0.816	0.851	0.852	0.997	0.948	0.842
0.15	0.952	0.918	0.890	0.841	0.899	0.884	0.998	0.945	0.847
0.20	0.954	0.923	0.886	0.858	0.885	0.864	0.998	0.939	0.848
0.25	0.962	0.941	0.856	0.860	0.899	0.869	0.998	0.952	0.857
0.30	0.950	0.916	0.857	0.876	0.908	0.844	0.998	0.953	0.869
0.35	0.957	0.924	0.863	0.908	0.912	0.834	0.998	0.961	0.859

step of the positive examples, i.e., $\alpha > 0$, does not improve the performance, while bootstrapping negative examples, $\beta > 0$, does show significant improvement. We believe that this is because the negative examples are more sparse in the space of images, and the mixture of the two probabilities models does help in fitting the data. Since the value of β that maximizes this improvement is not consistent among all the classifiers, an iterative training algorithm with some validation routine is required to take full advantage of this form of error bootstrapping. Similarly, from the comparison of the results of the use of different weighting values λ in the computation of the divergence, we found consistent improvement

Table 19.3: Feature-location inaccuracy, top-1 detection performance, and the average detection rate for 1% false alarm w.r.t the error bootstrapping of feature examples.

α	Reye	Leye	RRbrw	RLbrw	LRbrw	LLbrw	Nose	Rmth	Lmth
0.00	0.226	0.381	0.904	1.884	2.302	2.318	1.945	1.126	0.649
0.05	0.257	0.393	1.034	1.925	2.323	2.231	1.892	1.246	0.755
0.10	0.243	0.459	1.075	1.893	2.336	2.256	1.972	1.322	0.850
0.15	0.242	0.465	1.071	1.903	2.206	2.236	2.029	1.363	0.738
0.20	0.247	0.439	1.158	1.981	2.232	2.222	1.918	1.327	0.787
0.25	0.282	0.425	1.084	1.962	2.289	2.322	1.981	1.233	0.681
0.30	0.363	0.481	1.091	1.880	2.256	2.326	1.953	1.310	0.753
0.35	0.452	0.521	1.115	1.903	2.219	2.207	1.993	1.335	0.755
α	Reye	Leye	RRbrw	RLbrw	LRbrw	LLbrw	Nose	Rmth	Lmth
0.00	99.45	99.18	96.39	95.06	97.12	95.88	100.00	97.21	95.34
0.05	99.04	98.54	94.24	94.38	96.20	95.61	100.00	96.07	95.38
0.10	98.63	97.62	93.42	94.15	97.26	94.65	100.00	93.83	95.75
0.15	98.54	98.22	94.88	93.42	95.88	95.70	100.00	95.43	94.88
0.20	98.35	97.85	95.38	94.74	96.75	94.83	99.95	95.66	95.61
0.25	98.63	98.45	93.46	95.06	97.07	94.51	99.95	94.42	95.61
0.30	98.86	97.90	91.50	94.60	94.65	94.19	99.95	96.02	95.84
0.35	98.17	97.99	90.86	93.64	95.02	94.88	99.95	94.70	96.34
α	Reye	Leye	RRbrw	RLbrw	LRbrw	LLbrw	Nose	Rmth	Lmth
0.00	0.960	0.906	0.869	0.814	0.818	0.870	0.997	0.939	0.825
0.05	0.942	0.884	0.855	0.786	0.816	0.858	0.995	0.908	0.812
0.10	0.924	0.881	0.856	0.770	0.818	0.867	0.997	0.878	0.842
0.15	0.943	0.874	0.886	0.781	0.778	0.901	0.998	0.920	0.838
0.20	0.943	0.877	0.883	0.802	0.768	0.888	0.997	0.922	0.850
0.25	0.958	0.885	0.854	0.815	0.868	0.886	0.997	0.915	0.862
0.30	0.953	0.892	0.824	0.829	0.793	0.888	0.998	0.944	0.865
0.35	0.956	0.891	0.846	0.810	0.802	0.887	0.998	0.916	0.873

with $\lambda \in [.10, .30]$. Note that this bias towards the distribution of the positive examples in the computation of the divergence from Equation 18 consistently tells us that the distribution of the negative examples does not fit the training data well, and that other techniques such as error bootstrapping can be used to improve it.

19.4.5 Latest Development in a Person-Recognition System

The original system in [22] has been successfully ported from the initial SGI Onyx 10-CPU parallel platform to a Pentium III 600 MHz processor so that the detection

Table 19.4: Feature-location inaccuracy, top-1 detection performance, and the average detection rate for 1% false alarm w.r.t the divergence weight.

λ	Reye	Leye	RRbrw	RLbrw	LRbrw	LLbrw	Nose	Rmth	Lmth
0.05	0.194	0.392	0.893	1.901	2.358	2.375	1.951	1.166	0.614
0.10	0.194	0.385	0.943	1.846	2.343	2.342	1.933	1.124	0.615
0.15	0.203	0.378	1.000	1.858	2.339	2.408	1.975	1.172	0.584
0.20	0.160	0.369	0.944	1.810	2.357	2.431	1.850	1.134	0.595
0.25	0.199	0.363	0.889	1.817	2.331	2.435	1.932	1.190	0.620
0.30	0.192	0.399	0.943	1.857	2.354	2.420	1.918	1.099	0.615
0.35	0.167	0.340	0.920	1.783	2.340	2.465	1.951	1.227	0.660
0.40	0.182	0.400	0.993	1.852	2.420	2.388	1.909	1.098	0.642
λ	Reye	Leye	RRbrw	RLbrw	LRbrw	LLbrw	Nose	Rmth	Lmth
0.05	99.36	99.09	97.49	94.51	96.48	96.48	100.00	97.94	96.43
0.10	99.27	99.13	97.21	95.88	97.62	96.57	99.95	97.26	96.80
0.15	99.36	99.09	97.62	96.11	97.49	96.07	99.95	97.44	95.98
0.20	99.18	99.22	96.71	96.48	97.99	95.56	100.00	97.21	95.88
0.25	99.59	98.99	96.20	95.93	97.67	96.16	99.95	97.81	96.11
0.30	99.22	99.09	96.07	96.11	97.76	96.48	100.00	97.17	96.39
0.35	99.18	98.58	96.20	96.57	97.21	95.84	100.00	97.44	96.75
0.40	99.27	99.09	96.30	96.75	97.53	95.56	100.00	97.26	96.57
λ	Reye	Leye	RRbrw	RLbrw	LRbrw	LLbrw	Nose	Rmth	Lmth
0.05	0.957	0.907	0.876	0.808	0.812	0.865	0.998	0.946	0.832
0.10	0.957	0.913	0.911	0.831	0.822	0.887	0.997	0.942	0.853
0.15	0.964	0.917	0.904	0.839	0.847	0.926	0.998	0.938	0.830
0.20	0.974	0.912	0.881	0.840	0.835	0.888	0.998	0.934	0.834
0.25	0.965	0.917	0.902	0.839	0.869	0.916	0.998	0.939	0.838
0.30	0.956	0.914	0.888	0.837	0.844	0.913	0.997	0.936	0.848
0.35	0.968	0.918	0.850	0.850	0.856	0.902	0.998	0.939	0.835
0.40	0.961	0.917	0.839	0.843	0.833	0.908	0.998	0.942	0.847

speed can be evaluated and compared with other systems. For size 320×240 real-time video, the system in [22] achieves a detection rate of $5 \sim 6$ frames per second (fps). That system has been known for its near-real-time performance. Our recent effort speeds it up to obtain an 11–12 fps detection without sacrificing any detection robustness. In comparison, Viola and Jones's face detection based on Adaboost achieves a detection rate of 15 fps for 384×288 video on a Pentium III 700 MHz processor [13].

The face and facial-feature detection system is used as the kernel of a recently developed multimodal person-identification system, although only face detection and outer eye corner detection are used for face-recognition tasks, since an accurate outer-eye-corner detection can already serve the purpose of getting the normalized faces for recognition. The omission of the detection of other facial-features also makes the system faster. The computer interface of the new system and its description are shown in Figure 19.10.

We obtain the results of face detection in near-real-time video by turning off the functionality of face recognition and speaker ID in Figure 19.10. Forty-two

FIGURE 19.10: Computer interface of the multimodal person identification system. The upper-left corner displaces the near-real-time video captured by a digital camcorder mounted on a tripod behind the computer screen. The upper-center displays text or digits for the user to read in order to do speaker identification. At the upper-right corner there are three buttons titled "start testing", "add user", "delete user" which indicate three functionalities. Two bar charts are in the lower-left corner displaying the face-recognition and speaker-ID likelihood respectively for each user. In the lower-center, icon images of users that are currently in the database are shown in black and white and the recognized person has his/her image enlarged and shown in color form. The lower-right displays all the names of the users currently in the database. In the status bar at the bottom of the window, the processing speed in fps is also shown.

subjects have been tested for face detection in near-real-time video. Several video sequences of each subject have been taken on different days under different lighting conditions. Different sizes of faces have also been tested. Each sequence is between 1 minute and 5 minutes long. Face detection results consistently show high detection accuracy ($\sim 95\%$) and very low false-alarm rate ($< 0.1\%$). Given the near-real-time face detection, a pattern of face tracking by the green square (face region) and the red crosses (outer eye corners) can be observed.

19.5 CONCLUSIONS AND FUTURE WORK

In this chapter, we have described a computer-vision and pattern-recognition algorithm for face and facial-feature detection with applications to human–computer intelligent interface. We have reported solutions to face and facial-feature detection. In this chapter, we have also dealt with the issues regarding the implementation and integration of these algorithms into a fully automatic, near real-time face and facial-feature detection system.

Our face detection has problems detecting faces of people wearing very reflective glasses. What's more important, like other systems ours can only handle small degree of in-plane rotation. A relatively large rotation of the user's head degrades the detection results. We will research on these issues in the future.

REFERENCES

[1] K. Yow and R. Cipolla, "Feature-based human face-detection," *Image and Vision Computing* **15**, no. 9, pp. 713–735, 1997.

[2] R. Qian, A unified scheme for image segmentation and object recognition, Ph.D. thesis, University of Illinois at Urbana-Champaign, Feb. 1996.

[3] M.-H. Yang and N. Ahuja, "Detecting human faces in color images," in: *International Conference on Image Processing*, 1998.

[4] J. Sobottka and I. Pittas, "Segmentation and tracking of faces in color images," in: *International Conference on Automatic Face and Gesture Recognition*, 1996.

[5] G. Yang and T. S. Huang, "Human face-detection in a complex background," *Pattern Recognition* **27**, pp. 53–63, 1994.

[6] B. Moghaddam and A. Pentland, "Maximum likelihood detection of faces and hands," in: *International Conference on Automatic Face and Gesture Recognition*, 1996.

[7] E. Osuna, R. Freund, and F. Girosi, "Training support vector machines: an application to face detection," in: *Proc. IEEE Computer Society of Computer Vision and Pattern Recognition*, 1997.

[8] F. Soulie, E. Viennet, and B. Lamy, "Multi-modular neural network architectures: pattern-recognition applications in optical character recognition and human face recognition," *Intl. Journal of pattern-recognition and Artificial Intelligence* **7**, no. 4, 1993.

[9] K. Sung and T. Poggio, "Example-based learning for view-based human face-detection," *IEEE Trans. on Pattern Analysis and Machine Intelligence* **20**, no. 1, pp. 39–51, 1998.
[10] H. Rowley, S. Baluja, and T. Kanade, "Human face-detection in visual scenes," *Advances in Neural Information Processing Systems*, vol. 8, 1998.
[11] H. Rowley, S. Baluja, and T. Kanade, "Rotation invariant neural network-based face detection," in: *Proc. Computer Vision and Pattern Recognition*, 1998.
[12] Y. Freund and R. E. Schapire, "A decision-theoretic generalization of on-line learning and an application to boosting," *Journal of Computer and System Sciences* **55**, no. 1, pp. 119–139, 1997.
[13] P. Viola and M. Jones, "Robust real-time object detection," *Second International Workshop on Statistical and Computational Theories of Vision—Modeling, Learning, Computing and Sampling*, July 2001, Vancouver, Canada.
[14] L. Devroye, L. Györfi, and G. Lugosi, *A Probabilistic Theory of Pattern Recognition*, Springer, New York, 1996.
[15] V. Vapnik, *The Nature of Statistical Learning Theory*, Springer-Verlag, New York, 1995.
[16] G. Potamianos and H. Graf, "Discriminative training of HMM stream exponents for audio-visual speech recognition," in: *Proc. Intl. Conf. Acoustic Signal and Speech Processing*, 1998.
[17] Y. Chow, "Maximum mutual information estimation of HMM parameters for continuous speech recognition using the N-best algorithm," in: *Proc. Intl. Conf. Acoustic Signal and Speech Processing*, 1990, pp. 701–704.
[18] R. Gray, *Entropy and Information Theory*, Springer-Verlag, New York, 1990.
[19] J. Kapur and H. Kesavan, *The Generalized Maximum Entropy Principle*, Sandford Educational Press, Canada, 1987.
[20] G. Toussaint, "On the divergence between two distributions and the probability of misclassification of several decision rules," in: *Proc. 2nd Int. Joint conf. on Pattern Recognition*, 1974, pp. 27–34.
[21] A. J. Colmenarez and T. S. Huang, "Pattern detection with information-based maximum discrimination and error bootstrapping," in: *Proc. Intl. Conf. on Pattern Recognition*, 1998.
[22] Antonio J. Colmenarez and Thomas S. Huang, "Face detection with information-based maximum discrimination," in: *Proc. Computer Vision and Pattern Recognition*, 1997.
[23] T. H. Cormen, C. E. Leiserson, and R. L. Rivest, *Introduction to algorithms*, McGraw-Hill, 1990.
[24] J. B. Kruskal, "On the shortest spanning subtree of a graph and the traveling salesman problem," *Proc. American Mathematical Society* **7**, pp. 48–50, 1956.
[25] P. J. Phillips, H. Moon, S. Rizvi, and P. Rauss, "The Feret evaluation," in: *Proc. NATO-ASI on Face Recognition: From Theory to Applications*, 1997, pp. 244–261.
[26] P. J. Phillips, H. Moon, S. Rizvi, and P. Rauss, "The Feret evaluation methodology for face-recognition algorithms," in: *Proc. Computer Vision and Pattern Recognition*, 1997, pp. 137–143.
[27] H. Rowley, S. Baluja, and T. Kanade, "Neural network-based face-detection," in: *Proc. Computer Vision and Pattern Recognition*, 1996.

CHAPTER 20

CURRENT LANDSCAPE OF THERMAL INFRARED FACE RECOGNITION

20.1 INTRODUCTION

From its inception now more than thirty years ago to its present state, face recognition by computer has steadily transitioned from very controlled scenarios to increasingly more realistic ones. This progression has seen the introduction of nuisance factors such as pose, illumination, occlusion, and facial expression as integral components of the standard face-recognition problem. A number of methods have been developed to cope with subsets or combinations of these complicating factors, in order to bring automated face recognition to the point of handling unconstrained viewing conditions. As part of that effort, the use of thermal infrared imagery, by itself or in combination with other modalities, has been proposed as an alternative means of handling the problem of variable illumination conditions.

Variation in illumination conditions between enrollment and testing is one of the major problems for visible-spectrum-based face recognition [3, 55]. Since the radiance sensed by a visible camera at a given image location is proportional to the product of object albedo and incident light, changes in illumination can have dramatic effects on object appearance. In terms of faces, this makes modeling the distribution of appearances of a single person under multiple lighting conditions very difficult. Cast shadows, specularities and other non-Lambertian phenomena make the problem even harder. Multiple techniques have been developed to handle this issue [8, 63, 42, 55, 22], all of which improve recognition performance by explicitly taking into account the effect of illumination on facial appearance. An alternative route taken by other researchers was to explore the potential of thermal infrared imagery for face recognition. The primary advantage of this imaging modality, further discussed in Section 20.2, is that changes in ambient illumination have little or no influence on facial appearance. Thus, instead of incorporating the large variability in appearance caused by lighting variation into a model, a new imaging modality is chosen so that such variability is simply not present.

We know from personal everyday experience that face recognition in the visible spectrum is possible. That is, the appearance of human faces in the visible spectrum is diverse and distinctive enough to allow for recognition of individuals. This fact is not *a priori* true for the thermal appearance of human faces. For this reason, the early research in thermal face recognition [61, 15, 37] was aimed at validating the imaging modality as useful for biometrics applications.

While thermal imagery provides us with the advantages of illumination invariance and no-light operation, it is not without downsides. Of particular importance is the fact that thermal emissions from the face are dependent on ambient temperature and wind conditions, as well as on metabolic activity of the subject. These issues are discussed in more detail in Section 20.2. Additionally, the fact that the lenses of most glasses are opaque in the thermal infrared means that a large portion of the population have partial occlusions in the infrared images of their face. This is an important issue that must be addressed by any deployable thermal face-recognition system. Fortunately, most of the situations that hamper recognition performance with thermal imagery are not a problem for visible imagery, and vice-versa. For this reason, systems using a combination of both modalities have proved to be superior to those using either modality separately.

In Section 20.2, we review the nature of thermal infrared imagery of the human face. This provides motivation for the use of such imagery in biometric applications, as well as indicate some of the strengths and weaknesses of the modality. The rest of the chapter is structured to reflect the nature of the recognition task, and the historical development of the field. We progress from same-session recognition experiments in Section 20.3, where we mention the earliest and simplest experimental setups used to validate the use of thermal imagery for biometrics, to more complex and realistic scenarios in Sections 20.4 and 20.5, where we explore the effect of time passage and unconstrained outdoor illumination. Section 20.6 focuses on the problem of eye localization in thermal infrared imagery, as it pertains to recognition using thermal imagery alone (not in combination with visible imagery). Most of the research highlighted in this chapter is due to work at Equinox Corporation, http://www.equinoxsensors.com, although we mention the efforts of other groups.

20.2 PHENOMENOLOGY

The potential for illumination-invariant face recognition using thermal IR imagery has been recognized in the past [38, 61]. This invariance can be qualitatively observed in Figure 20.1 for a coregistered long-wave infrared (LWIR) and visible video camera sequence of a face under three different illumination conditions. For this sequence a single 60 Watt light bulb mounted in a desk lamp illuminates a face in an otherwise completely dark room and is moved into different positions. The left column of visible video imagery shows dramatic changes in the appearance of the face, which is a well-known problem for face-recognition algorithms [54, 3]. The right column shows LWIR imagery, which unlike its visible counterpart, appears to be remarkably invariant across different illuminations.

All objects above absolute zero temperature emit electromagnetic radiation. In the early 1900s Planck was the first to characterize the spectral distribution of this radiation for a *black body*, which is an object that completely absorbs electromagnetic radiation at all wavelengths [43]. Very few objects are nearly

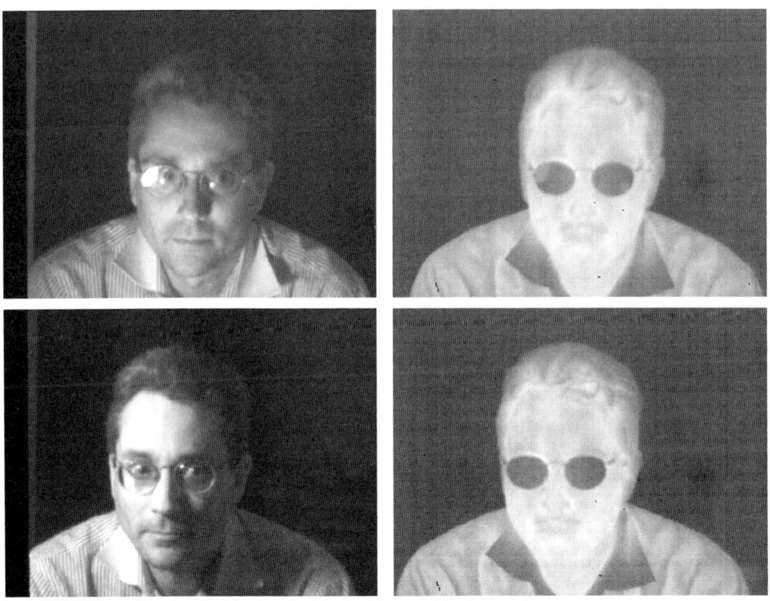

FIGURE 20.1: Two pairs of simultaneously acquired visible/LWIR images taken with different illuminations. Note the large variation in the visible images compared to that of the LWIR images.

perfect energy absorbers, particularly at all wavelengths. The proportional amount of energy emission with respect to a perfect absorber is called the *emissivity*, $\epsilon(T, \lambda, \psi)$, with values in the range $[0, 1.0]$. In addition to temperature T and, wavelength λ, emissivity can also be a function of emission angle ψ. Kirchoff's law states that the emissivity at a point on an object is equal to the *absorption* $\alpha(T, \lambda, \psi)$, namely:

$$\epsilon(T, \lambda, \psi) = \alpha(T, \lambda, \psi) .$$

This is a fundamental law that effectively asserts the conservation of energy. Blackbody objects are therefore the most efficient radiators and, for a given temperature T, emit the most energy possible at any given wavelength. Under most practical conditions, 2D imaging-array thermal-IR sensors measure simultaneously over broadband wavelength spectra, as opposed to making measurements at narrow, nearly monochromatic wavelengths (e.g., an IR spectrophotometer which measures only one point in a scene). With a staring-array sensor it is possible to measure *average emissivity* over a broadband spectrum (e.g., 3–5 microns, 8–14 microns).

For objects that are opaque at certain wavelengths, emissivity is the arithmetic complement of reflectance with respect to summing to total energy. The more reflective the object, the lower its emissivity and vice versa. Using terminology adapted from the computer-vision literature, emissivity is a *thermal albedo* which is complementary to the more familiar reflectance albedo. For instance, a Lambertian reflector can appear white or gray depending on its efficiency for reflecting light energy. The more efficient it is in reflecting energy of a given wavelength (more reflectance albedo) the less efficient it is in thermally emitting energy at that same wavelength respective to its temperature (less thermal albedo).

While the nature of face imagery in the visible domain is well studied, particularly with respect to illumination dependence [3], its thermal counterpart has received less attention. In [14], the authors find some variability in thermal emission patterns during time-lapse experiments, and properly blame it for decreased recognition performance. Figure 20.2 shows comparable variability within our data. The first image in each pair shows enrollment images while the second image in the pair shows test images from the same subject acquired days later. We can plainly see how emission patterns are different around the nose, mouth, and eyes. Weather conditions during our data collection were quite variable, with some days being substantially colder and windier than others. In addition, some subjects were imaged indoors immediately after coming from outside, while others had as much as twenty minutes of waiting time indoors before being imaged. These conditions contribute to a fair amount of variability in the thermal appearance of the face, as can be seen in the example images. When exposed to cold or wind, capillary vessels at the surface of the skin contract, reducing the effective blood flow and

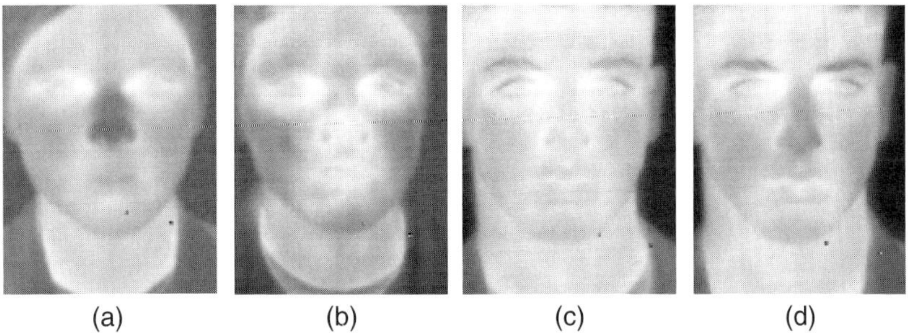

FIGURE 20.2: Variation in facial thermal emission from two subjects in different sessions. The first image in each pair (a, c) is the enrollment image, and the second image in each pair (b, d) is the test image.

thereby the surface temperature of the face. When a subject transitions from a cold outdoor environment to a warm indoor one, a reverse process occurs, whereby capillaries dilate, suddenly flushing the skin with warm blood in the body's effort to regain normal temperature.

Additional fluctuations in thermal appearance are unrelated to ambient conditions, but are rather related to the subject's metabolism. Figure 20.3 shows images of the same test subject under different conditions: at rest, after jogging and after exposure to severe cold. It is clear that physical exertion changes the pattern of thermal emission from the subject's face, as does cold exposure. In addition to the slow appearance of changes due to metabolic state, high-temporal-frequency thermal variation is associated with breathing. The nose or mouth will appear cooler as the subject is inhaling and warmer as he or she exhales, since exhaled air is at

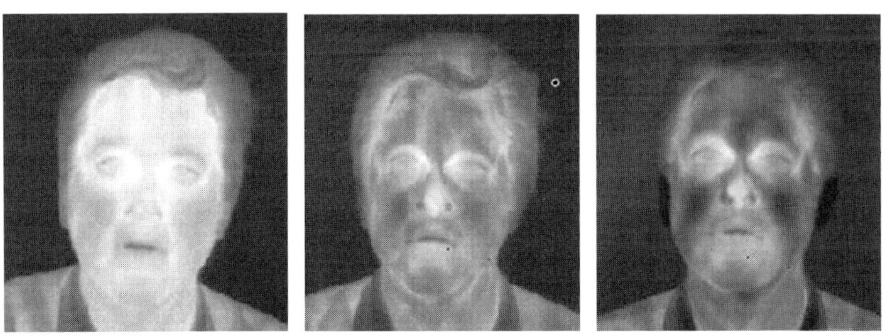

FIGURE 20.3: Images of the same subject under different conditions: at rest, after jogging, and after exposure to severe cold.

core body temperature, which is several degrees warmer than skin temperature. These problems are complementary to the illumination problem in visible face recognition, and must be addressed by any successful system.

Much like recognition from visible imagery is affected by illumination, recognition with thermal imagery is affected by a number of exogenous and endogenous factors. And while the appearance of some features may change, their underlying shape remains the same and continues to hold useful information for recognition. Thus, much like in the case of visible imagery, different algorithms are more or less sensitive to image variations. As is well known in the visible face-recognition realm, proper compensation for illumination prior to recognition has a favorable effect on recognition performance. Clearly, the better algorithms for thermal face recognition will perform equivalent compensation on the infrared imagery prior to comparing probe and gallery samples.

20.3 SAME-SESSION RECOGNITION

The first studies in thermal infrared face recognition were aimed at determining whether the imaging modality held any promise for human identification. In this context, simple experiments were designed, where enrollment and testing images of multiple subjects were acquired during a short period of time. This type of scenario is often referred to as same-session recognition, since it tests the ability to recognize as such images of the same subject acquired during the same data-collection session. This is clearly the easiest data-collection scenario from an organizational viewpoint, since it only requires that subjects be imaged on one day, thus greatly reducing the scheduling overhead.

In [61], the authors use a pyroelectric sensor to collect indoor and outdoor imagery during a single session. They show that recognition performance is roughly comparable between both modalities, and that using a simple fusion strategy to combine both modalities greatly increases performance. The work in [15] uses a low-sensitivity MWIR sensor and the eigenfaces algorithm, and concludes that recognition performance is equivalent to that attainable in the visible spectrum. In [46], a database of approximately 90 subjects, collected during a single session and containing controlled variations in illumination is used to show that recognition in the LWIR spectrum outperforms visible-based recognition with two different algorithms. More recently, [20] shows that recognition rates achieved with thermal imagery on a pose-and-expression-variant same-session database are higher than those achieved with comparable visible imagery. As part of a time-lapse study, the authors of [14] find that same-session recognition in the LWIR spectrum outperforms visible recognition, when both of them use a PCA-based algorithm.

The most comprehensive study to date on same-session thermal infrared recognition is [51]. The database used comprises ninety ethnically diverse subjects of both genders imaged during a single day. All images were collected with a custom sensor capable of simultaneously imaging visible and LWIR images through a common aperture. This provides the authors a unique opportunity to compare performance on imagery which differs only in modality, but is alike in all other respects, such as pose, illumination and expression. In order to gauge the effect of illumination variation on recognition performance, the authors collected imagery under three controlled lighting conditions, and for a variety of facial expressions. A sample of this imagery can be seen in Figure 20.4, where we see matched pairs in both modalities. We should note (see [51] for more details), that all eye coordinates were manually located on the visible images and transfered to the LWIR ones via their coregistered nature.

The study in [51] compares recognition performance for visible and thermal imagery across multiple data-representation algorithms, including PCA, LDA, ICA, and LFA. Each of these representations was coupled with a number of distance measures including L^1, L^2, and angle. Figures 20.5 and 20.6 show cumulative match curves for PCA- and LDA-based recognition for visible and thermal imagery under a variety of distance measures.[1] These performance curves were generated

FIGURE 20.4: Sample visible and LWIR imagery from the study in [51].

[1] We reproduce only the PCA and LDA results, since one of them is the standard by which all other algorithms are measured, and the other showed best performance in the study.

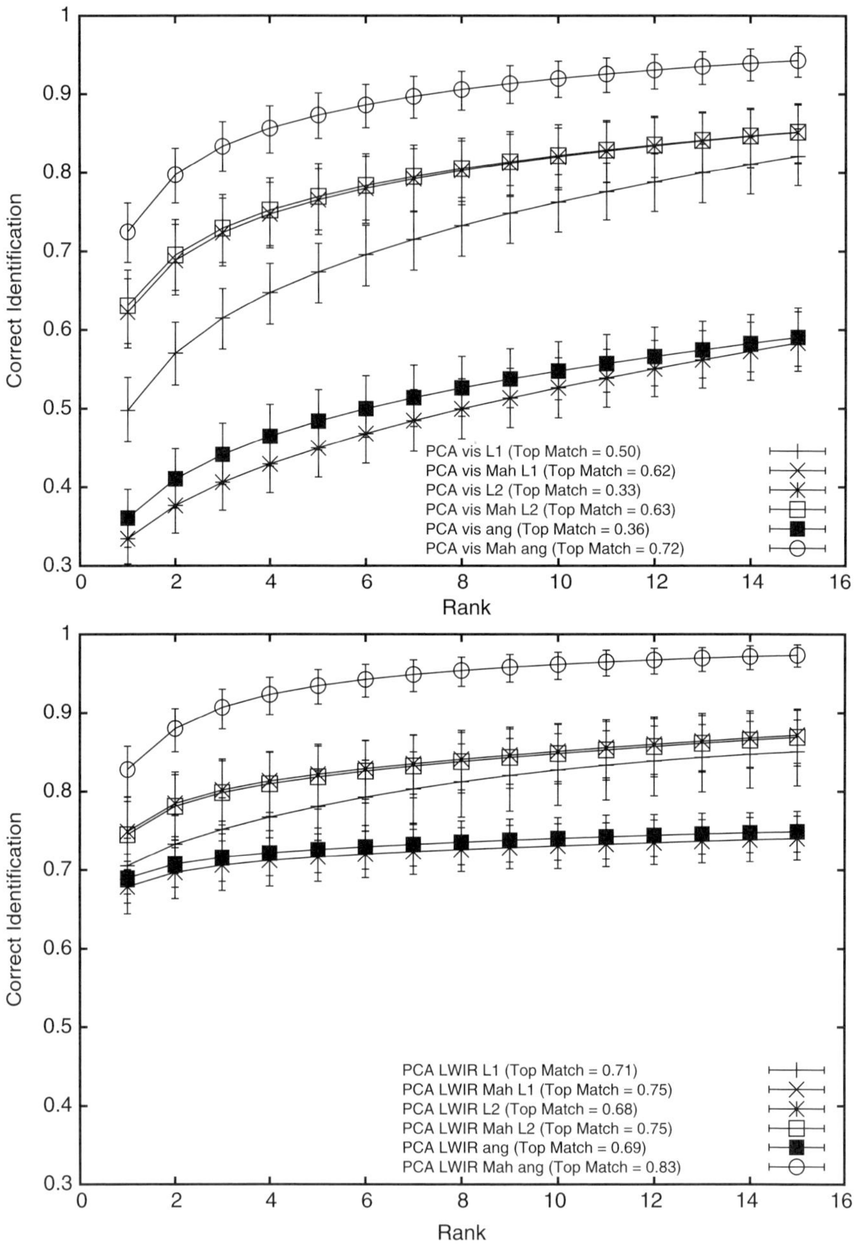

FIGURE 20.5: Cumulative recognition rates for PCA-based identification algorithms on visible (top) and LWIR (bottom) imagery.

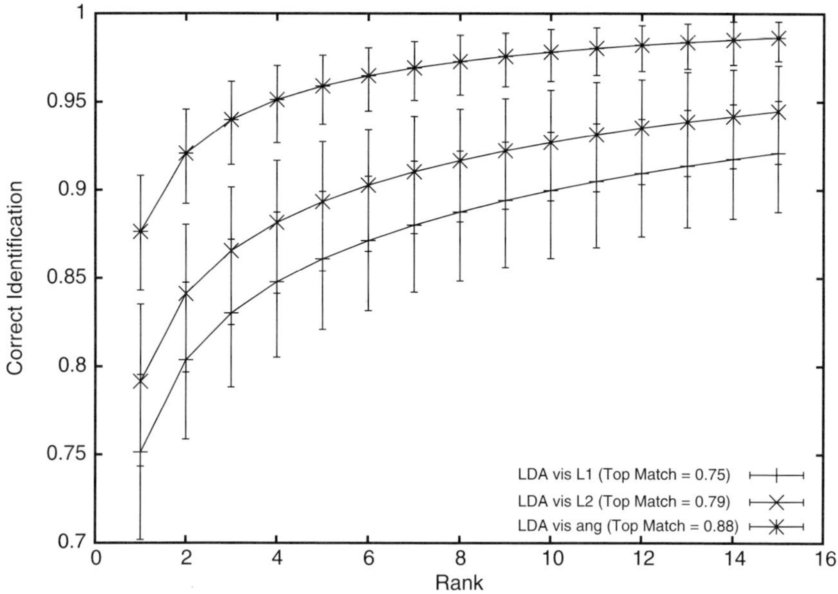

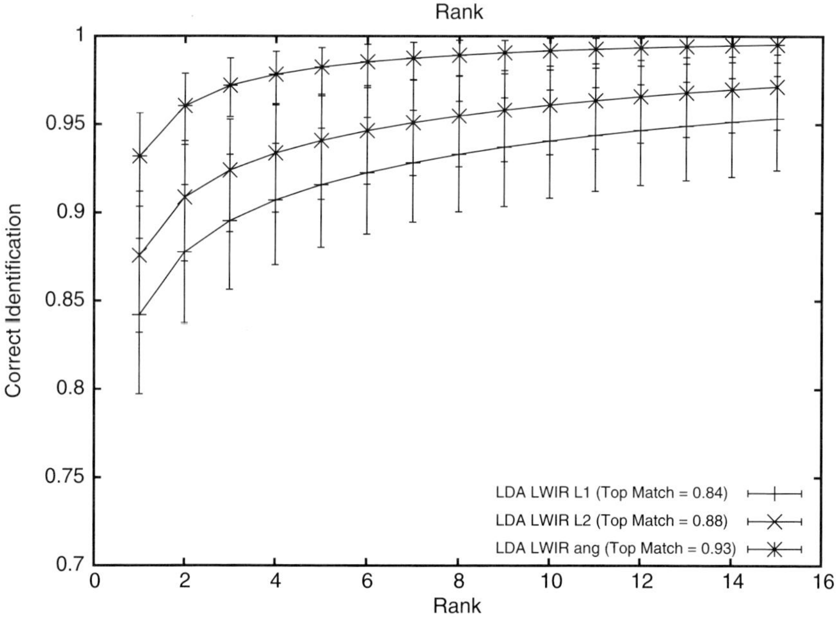

FIGURE 20.6: Cumulative recognition rates for LDA-based identification algorithms on visible (top) and LWIR (bottom) imagery.

using a Monte-Carlo sampling procedure based on the work in [25, 26] and [30]. Thousands of randomly drawn gallery/probe set pairs were chosen from a database of over 20,000 images per modality in order to produce estimates of mean recognition performance as well as variance. The curves in Figures 20.5 and 20.6 show 95% confidence intervals derived from the Monte-Carlo analysis. We feel this is an important feature of any complete statistical analysis of any biometric algorithm, since variations that are normally attributed to better algorithm (or modality) performance are often the result of mere random variation, and this cannot be detected without variance estimates.

The general conclusion to be drawn from these performance graphs is that in a same-session recognition scenario, thermal imagery is superior to visible imagery, at least if the illumination is not carefully controlled. In fact, even if the illumination is kept under strict control, thermal recognition performance is higher than its visible counterpart (this is not shown in the graphs), although the difference is not so pronounced. This clearly indicates that the pattern of thermal emission from a person's face is distinct, and bears strict correlation with their identity. A second definitive conclusion of the study is that using a combination of visible and thermal imagery greatly outperforms the use of visible (or LWIR) imagery alone. This is supported by further studies by other researchers [14, 13]. Table 20.1 shows absolute and relative performance differences between visible, LWIR, and fused results, for the study in [51]. Fusion results were obtained by simply adding individual modality scores (distances), and using the composite number as a new distance.

The major caveat to this, and all other same-session studies, is that this scenario is not a realistic reflection of the use of biometrics in the real world. A deployed biometric identification system is normally tasked with recognizing or verifying the identity of an individual at the current time based on a pattern acquired at an

Table 20.1: Lower bounds for average identification performance difference between top-performing norms (angular) with p values for paired Wilcoxon test (p-values for paired t test are smaller), along with reduction of error with respect to visible performance.

	LWIR versus Visible			Fusion versus Visible		
	Performance difference		Error reduction	Performance difference		Error reduction
PCA	10.0%	$(p < 10^{-8})$	37.4%	14.8%	$(p < 10^{-8})$	54.8%
LFA	9.7%	$(p < 10^{-6})$	29.0%	15.6%	$(p < 10^{-10})$	46.2%
ICA	9.8%	$(p < 10^{-8})$	36.9%	14.6%	$(p < 10^{-7})$	54.1%
LDA	5.3%	$(p < 10^{-7})$	44.9%	9.0%	$(p < 10^{-6})$	74.6%

earlier, possibly quite distant time. Therefore, high performance in a same-session experiment is not immediately indicative of good performance in a real-world application.

20.4 TIME-LAPSE RECOGNITION

As we point out in the previous section, it is not possible to extrapolate results from a same-session experiment to a time-lapse situation. It is well known that performance of visible face-recognition algorithms degrades as time elapsed between training and testing images increases [36, 35]. This is especially true if other parameters such as venue and lighting also change. This is, of course, not a surprising fact to anyone who has noticed that old friends look different after a long period of absence. Facial appearance changes with weight fluctuations, facial hair, makeup, aging, exposure to sunlight and many other factors. Unfortunately, the task of a face-recognition system is to ignore as many of those exogenous factors as possible, and still home in on the underlying identity variable, which is unchanged. In order to do that, a stable signature must be extracted from the changing facial appearance. To the extent that an imaging modality is able to provide a more stable and discriminating signature over time, it can be considered better for facial recognition. The only way to evaluate this stability is through the time-consuming process of collecting data over an extended period at regular intervals.

A few researchers have recently undertaken studies to measure the relative stability of visible and thermal (LWIR) imagery for time-lapse face recognition. The authors of [14, 13] collected a database specifically designed to measure the effect of time. Visible and LWIR images of 240 distinct subjects were acquired under controlled conditions, over a period of ten weeks. During each weekly session, each subject was imaged under two different illumination conditions (FERET and mugshot), and with two different expressions ("neutral" and "other"). Visible images were acquired in color and with a 1200×1600 pixel resolution. Thermal images were acquired at 320×240 resolution and at a depth of 12 bits per pixel. Eye coordinates used for geometric normalization of all images, both visible and thermal, were manually located independently in each modality.

In their studies [14, 13], the authors find that a PCA-based recognition algorithm using LWIR imagery outperformed the same algorithm on visible data in a same-session scenario, yet underperformed it when the test imagery was acquired a week or more after the enrollment imagery. Furthermore, the loss in performance suffered when using LWIR imagery was more severe than the corresponding loss when using visible imagery. They attribute these results to the noticeable variability in thermal appearance of subjects faces imaged at different times. An example of this variation can be seen within our own data in Figure 20.2, where we see enrollment and testing images of two subjects acquired on different sessions. As explained in Section 20.2, this is due to a combination of endogenous and exogenous factors that cannot normally be controlled. The immediate conclusion of [14, 13] is that thermal infrared imagery is less suitable than visible imagery for face-recognition applications, due to its instability over time. However, the study also notes that when used in concert with visible imagery as part of a fused system, overall performance is superior to even state-of-the-art commercial (visible only) systems.

While the time-lapse studies in [14, 13] are well designed and certainly fulfill the requirements for testing the effect of time passage on recognition performance, they are limited in scope by the choice of algorithms. When evaluating an imaging modality for face recognition, we really evaluate the joint performance of the modality and recognition algorithm, and therefore are only able to obtain a lower bound in performance (there could always be a better algorithm out there). The assumption is that, by using the same underlying algorithm on both modalities, we are zeroing in on the intrinsic contribution of the data itself. This is a dangerous assumption, especially for widely disparate modalities with low correlation. There is no guarantee that the geometry of the data is similar for both modalities, and thus the fact that a certain classifier is able to model the distribution of data in one modality is no indication that it should be able to do so in the other. Using the same data as the authors of [14, 13], we conducted a new study [49] aimed at correcting this problem.

In order to evaluate recognition performance with time-lapse data, we performed the following experiments. The first-week frontal illumination images of each subject with neutral expression were used as the gallery. Thus the gallery contains a single image of each subject. For all weeks, the probe set contains neutral expression images of each subject, with mugshot lighting. The number of subjects in each week ranges from 44 to 68, while the number of overlapping subjects with respect to the first week ranges from 31 to 56. We computed top-rank recognition rates for each of the weekly probe sets with both modalities and algorithms. In addition to the PCA algorithm used in [14, 13], we also tested the proprietary Equinox algorithm. This algorithm works directly on visible–thermal image pairs, and is capable of using either or both modalities as its input. Experimental results are shown in Figures 20.7 and 20.8. Note that the first data point in each graph corresponds to same-session recognition performance.

Focusing for a moment on the performance curves, we notice that there is no clear trend for either visible or thermal modalities, encompassing weeks two through ten. That is, we do not see a clearly decreasing performance trend for either modality. This appears to indicate that, whatever time-lapse effects are responsible for performance degradation in comparison with same-session results, the results are roughly constant over the ten-week trial period. Other studies, have shown that, over a period of years, face-recognition performance degrades linearly with time [35]. Our observation here is simply that the slope of the degradation line is small enough as to be nearly flat over a ten-week period (except for

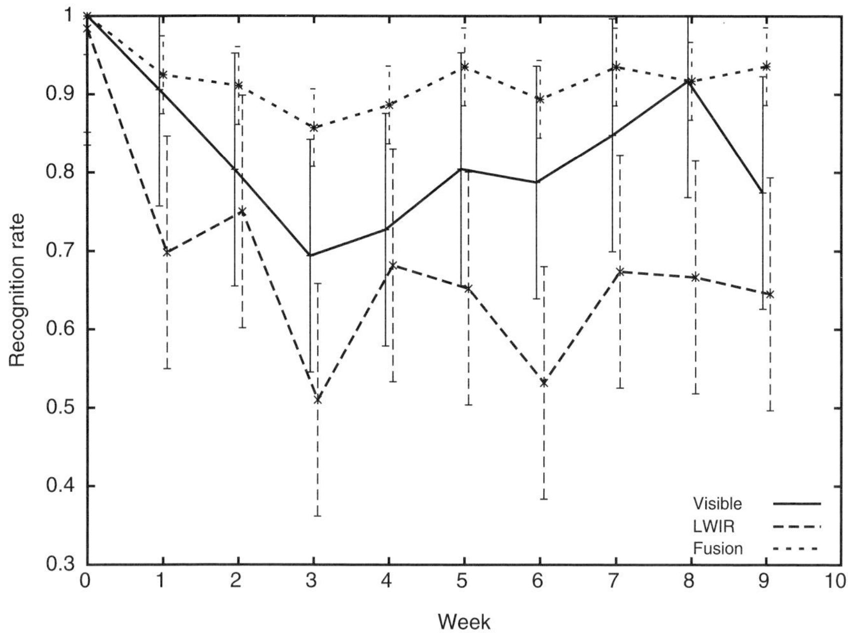

FIGURE 20.7: Top-rank recognition results for visible, LWIR and fusion as a function of weeks elapsed between enrollment and testing, using PCA. Note that the x coordinate of each curve is slightly offset in order to better present the error bars.

the same-session result, of course). Following that observation, we assume that weekly recognition performances for both algorithms and modalities are drawn independently and distributed according to a (locally) constant distribution, which we may assume to be Gaussian. Using this assumption, we estimate the standard deviation of that distribution, and plot error bars at two standard deviations.

Figure 20.7 shows the week-by-week recognition rates using PCA-based recognition. We see that, consistently with the results in [14, 13], thermal performance is lower than visible performance. In fact, for at least six out of nine time-lapse weeks, that difference is statistically significant. Table 20.2 shows mean recognition rates over weeks two through nine for each algorithm and modality. As shown in the last column, we see that mean visible performance is higher than the mean thermal performance by 2.17 standard deviations. This clearly indicates that thermal face recognition with PCA under a time-lapse scenario is significantly less reliable than its visible counterpart.

Turning to Figure 20.8, we see the results of running the same experiments with the Equinox algorithm. Firstly, we note that overall recognition performance

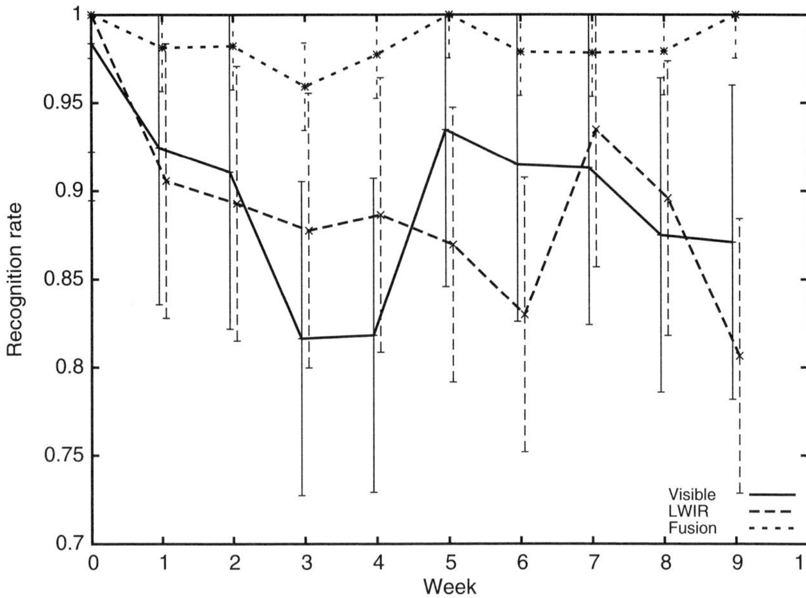

FIGURE 20.8: Top-rank recognition results for visible, LWIR, and fusion as a function of weeks elapsed between enrollment and testing, using the Equinox algorithm. Note that the x coordinate of each curve is slightly offset in order to better present the error bars.

Table 20.2: Mean top-match recognition performance for time-lapse experiments with both algorithms.

	Vis	LWIR	Fusion	Δ/σ Vis versus LWIR
PCA	80.67	64.55	91.04	2.17
Equinox	88.65	87.77	98.17	0.21

is markedly improved in both modalities. More importantly, we see that weekly performance curves for both modalities cross each other many times, while remaining within each other's error bars. This indicates that the performance difference between modalities using this algorithm is not statistically significant. In fact, looking at Table 20.2, we see that the difference between mean performances for the modalities is only 0.21 standard deviations, hardly a significant result. We should also note that the mean visible time-lapse performance with this algorithm is 88.65%, compared to approximately 86.5% for the FaceIt algorithm, as reported in [14]. This shows that the Equinox algorithm is competitive with the commercial state of the art on this data set, and therefore provides a fair means of evaluating thermal recognition performance, as using a poor visible algorithm for comparison would make thermal recognition appear better.

Figures 20.7 and 20.8, as well as Table 20.2 also show the result of fusing both imaging modalities for recognition. Following [51] and [14] we simply add the scores from each modality to create a combined score. Recognition is performed by a nearest-neighbor classifier with respect to the combined score. As many previous studies have shown [47, 51, 14], fusion greatly increases performance.

When this study is taken in context with [14], it shows that care must be exercised when evaluating imaging modalities based solely on the outcome of classification experiments. In those cases, one should keep in mind that all we are obtaining is an upper bound on the Bayes error for each modality, and that there is no indication how tight that bound may be. In fact, the tightness of the bounds may not be equal for each modality. Specifically referring to face recognition with visible and thermal imagery, the study shows that there is no significant difference in recognition performance when a state-of-the-art algorithm is used for both modalities. Thus thermal face recognition is a viable alternative for realistic deployments. We should also note that the imagery used in this study has carefully controlled lighting, and even though two illumination conditions were used, they are so similar that they are hard to differentiate by a human observer (as noted by the authors of [14]). This makes the equivalence statement all the more powerful, since the comparison is being made against visible imagery acquired in the most favorable conditions for visible face recognition.

20.5 OUTDOOR RECOGNITION

Face recognition in outdoor conditions is known to be a very difficult task. It has been highlighted in the Face Recognition Vendor Test 2002 [35] as one of the major challenges to be addressed by researchers in years to come. This is primarily due to the dramatic illumination effects caused by unconstrained outdoor lighting. Thermal imaging has a unique advantage in this context, given the very high emissivity of human skin, and the resulting nearly complete illumination invariance of face imagery in the LWIR spectrum. While a number of algorithmic techniques have been proposed for tempering the effect of illumination on visible face recognition, all reported results have been on databases acquired indoors under controlled illumination. No large-scale studies prior to our own and [35] have been reported on outdoor recognition performance.

The majority of the imagery used in this study was collected during eight separate day-long sessions spanning a two week period. A total of 385 subjects participated in the collection. Four of the sessions were held indoors in a room with no windows and carefully controlled illumination. Subjects were imaged against a plain background some seven feet from the cameras, and illuminated by a combination of overhead fluorescent lighting and two photographic lights with umbrella-type diffusers positioned symmetrically on both sides of the cameras and about six feet up from the floor. Due to the intensity of the photographic lights, the contribution of the fluorescent overhead lighting was small. The remaining four sessions were held outdoors at two different locations. During the four outdoor sessions, the weather included sun, partial clouds, and moderate rain. All illumination was natural; no lights or reflectors were added. Subjects were always shaded by the side of a building, but were imaged against an unconstrained natural background which included moving vehicles, trees and pedestrians. Even during periods of rain, subjects were imaged outside and uncovered, in an earnest attempt to simulate true operational conditions. For each individual, the earliest available video sequence in each modality is used for gallery images and all subsequent sequences in future sessions are used for probe images.

For all sessions, subjects were cooperative, standing about seven feet from the cameras, and looking directly at them when so requested. An example pair of visible images is shown in Figure 20.9. On half of the sessions (both indoors and outdoors), subjects were asked to speak while being imaged, in order to introduce some variation in facial expression into the data. For each subject and session, a four second video clip was collected at ten frames per second in two simultaneous imaging modalities. We used a sensor capable of acquiring coregistered visible and longwave thermal infrared (LWIR) video. The visible camera, a Pulnix 6710, has a spatial resolution of 640 $\times$ 480 pixels, and 8 bits of spectral resolution. The thermal sensor is an Indigo Merlin uncooled microbolometer, and has 12 bits of depth, sensing between 8μ and 14μ at a resolution of 320 $\times$ 240 pixels.

FIGURE 20.9: Example visible images of a subject from indoor and outdoor sessions.

Faces were automatically detected in all acquired indoor and outdoor frames, using a system based on the algorithm described in [45]. No operator intervention was required for this step. Recall that, since visible and thermal images are coregistered, eye locations in one modality give us those in the other. Thermal images were appropriately calibrated, and visible images were preprocessed through a simple yet effective illumination compensation method to boost performance. For details see [49]. We should emphasize the fact that all data used for the experiments below was processed in a completely automatic fashion, once again in an attempt to simulate true operational conditions.

Using a methodology akin to that described in Section 20.3 we performed randomized Monte-Carlo trials in order to estimate mean performance and corresponding variances. Enrollment images were taken from indoor sessions and test images from indoor or outdoor ones (separately). This reflects the most realistic scenario where a subject is enrolled under controlled conditions in an office environment (for example while obtaining an identification card), and is subsequently identified outdoors, for instance, while seeking access to a secure building.

A summary of top-match recognition performance is shown in Tables 20.3 and 20.4. A quick glance yields some preliminary observations. Under controlled indoor conditions, two of the visible algorithms are probably showing saturated performance on the data, which indicates that the test is too easy according to [34]. This may also be the case for the best thermal algorithm. Across the board, for both indoor and outdoor conditions, fusion of both modalities improves performance

Table 20.3: Top-match recognition results for indoor probes.

	Vis	LWIR	Fusion
PCA	81.54	58.89	87.87
LDA	94.98	73.92	97.36
Equinox	97.05	93.93	98.40

Table 20.4: Top-match recognition results for outdoor probes.

	Vis	LWIR	Fusion
PCA	22.18	44.29	52.56
LDA	54.91	65.30	82.53
Equinox	67.06	83.02	89.02

over either one separately. Comparing indoor versus outdoor performance shows that the latter is considerably lower with visible imagery, and significantly so even with thermal imagery. Fusion of both modalities improves the situation, but performance outdoors is statistically significantly lower than indoors, even for fusion. This difference, however, is much more pronounced for the lower-performing algorithms, which is simply a reflection of the fact that the better algorithms have superior performance with more difficult data, without sacrificing performance on the easy cases.

Looking at the results from the outdoor experiments in Figure 20.10, we see clear indication of the difficulty of outdoor face recognition with visible imagery. All algorithms have a difficult time in this test, and even the best performer achieves only about 84% recognition at rank 10. Thermal performance is also lower for all methods than with indoor imagery, but not so much as in the visible case. However, in this case, the performance difference between the modalities is very significant for all three algorithms. It is clear from this experiment, as from those in [35], that face recognition outdoors with visible imagery is far less accurate than when performed under fairly controlled indoor conditions. For outdoor use, thermal imaging provides us with a considerable performance boost. Fusion of both imaging modalities improves performance under all tests and algorithms. This supports previous results reported in [51, 14]. It is interesting to note that while even for the best performing algorithm there is a statistically significant difference between fusion performance outdoors and indoors, that significance is smaller the better the algorithm. This is a reflection of the fact that all methods perform well with easy data, but only the better methods perform well in difficult conditions.

The value of thermal imagery for outdoor face recognition is undoubtable. When used in combination with visible imagery, even though the latter is a poor performer, combined accuracy is high enough to make a system useful for a number of realistic applications that could not be accomplished with visible imagery alone. The ability to handle large unconstrained variations in lighting is a key requisite of any realistically deployable system, and the addition of thermal imaging puts us several steps closer to that goal.

20.6 RECOGNITION IN THE DARK WITH THERMAL INFRARED

One of the main advantages of thermal imagery is that it does not require the presence of any ambient illumination. For face recognition, this means that a system based on thermal imagery could potentially operate in complete darkness. While this is true in principle, certain obstacles must first be overcome. The first step in any automated face-recognition task is the detection and localization of the subject's face. There is a large literature on the subject of face detection in the

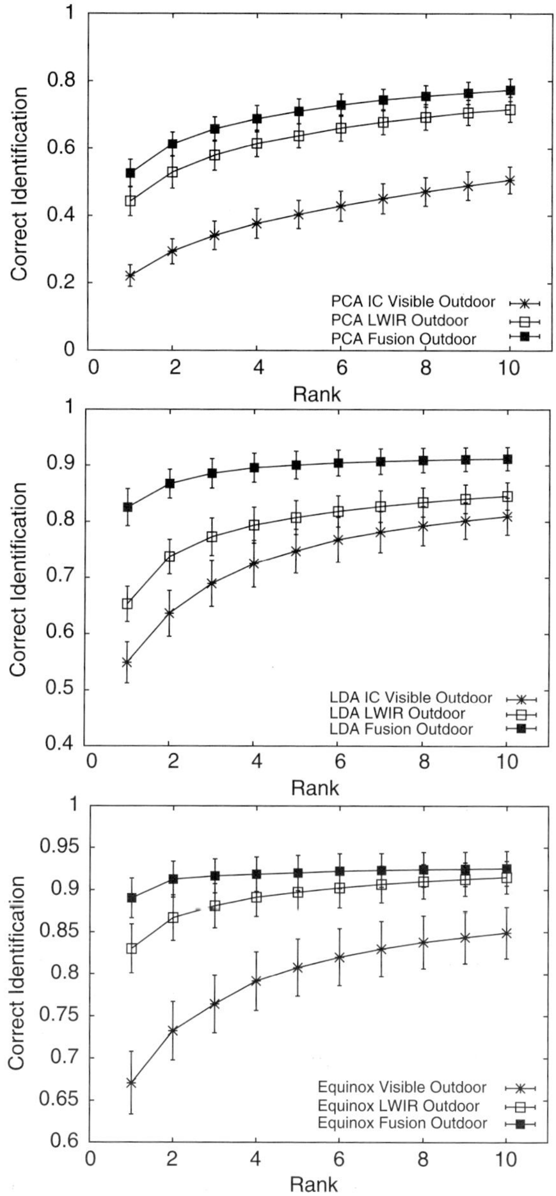

FIGURE 20.10: Recognition results by algorithm for indoor enrollment and outdoor testing. Note that the vertical scales are different in each graph. Top: PCA with illumination compensation. Center: LDA with illumination compensation. Bottom: Equinox algorithm.

visible spectrum (see for example [58] and the references therein). Face detection and tracking using a combination of visible and thermal imagery is addressed in [19], where the author develops a system capable of detecting and tracking faces using either one or both modalities together. A system based on this work was used to obtain the results in Section 20.5.

The second major step after face localization, is geometric normalization. This normally entails the detection of two or more points on the face, and the subsequent affine mapping of the acquired image onto a canonical geometry. By doing this, variations in size and in-plane rotation are normalized. Normalization for out-of-plane rotation (rotation in depth) is more complex and will not be addressed in this section (see for example [55]). In the visible spectrum, geometric normalization is often achieved by locating the centers of both eyes, and affinely mapping them to standard locations. Most reported experimental results in face recognition take advantage of manual eye locations, where the authors of the study (or more often their students) painstakingly locate eye coordinates manually for all training, gallery, and testing imagery. While this method is useful for isolating the recognition and detection parts of the problem, it is not feasible for an automated system.

Automated location of eyes (and pupils) in visible imagery is a well-studied problem [17, 64], both passively and with active methods. Active methods exploit the reflection of a narrow beam of near infrared light off the cornea, and could potentially be used in darkness. However, as they are active introducing a (strobing) illuminator into the environment, they can be detected. In fact, a standard CCD camera is sensitive in the near infrared, and will easily "see" the active illumination. If one wants a completely passive, undetectable system, facial landmark localization must rely exclusively on the thermal image itself.

It is easy to see that thermal images of human faces have fewer readily localizable landmarks, even by a human operator. The eyes themselves are completely uniform, with no distinction whatsoever between pupil, iris, and sclera. The authors of [14] performed experiments with manual localization of eye centers in LWIR imagery, and they report that, due to the lack of detail in such imagery around the eyes, it was difficult to obtain precise measurements. Furthermore, they go on to perform a perturbation analysis, whereby manually located eye coordinates are randomly perturbed prior to recognition by a fixed algorithm. They note that recognition performance with thermal imagery decays more rapidly as a function of incorrect eye localization than does recognition with visible images. The authors of [20] use a combination of filtering and thresholding to detect the center point of the eyebrows as landmarks. They do not provide experimental results as to the accuracy of the procedure, but they claim that a recognition system using those landmarks performed well.

We undertook a study [48] to determine the feasibility of a fully automated face-recognition system operating exclusively in the thermal domain, that is without the aid of a coregistered visible sensor. As we pointed out above, face detection has already been tackled in this context, we focused our efforts on detection of eyes and comparative performance analysis versus the equivalent process with visible imagery.

In order to detect eyes in thermal images, we rely on the face location detected using the face detection and tracking algorithm in [19, 50]. We then look for the eye locations in the upper half of the face area using a slightly modified version of the object detector provided in the Intel Open Computer Vision Library [1]. Before detection, we apply an automatic-gain-control algorithm to the search area. Although LWIR images have 12-bit spectral resolution, the temperature of different areas of a human face has a range of only a few degrees and thus is generally well represented by at most 6 bits. We improve the contrast in the eye region by mapping the pixels in the interest area to an 8-bit interval, from 0 to 255.

The detection algorithm is based on the scheme for rapid object detection using a boosted cascade of simple feature classifiers introduced in [57] and extended in [28]. The OpenCV version of the algorithm [1] extends the Haar-like features by an efficient set of 45°-rotated features and uses small decision trees instead of stumps as weak classifiers. Since we know that there is one and only one eye on the left and right halves of the face, we force the algorithm to return the best guess regarding its location. Figure 20.11 shows an example of face and eyes detected in a thermal infrared image.

The drawback of the algorithm, and of eye detection in thermal infrared in general, is that it fails to detect the eye center locations for subjects wearing glasses. Glasses are opaque in the thermal infrared spectrum and therefore show up black in thermal images, blocking the view of the eyes. (See Figure 20.12.) In these images, the glasses can be easily segmented and the eye center location can be inferred

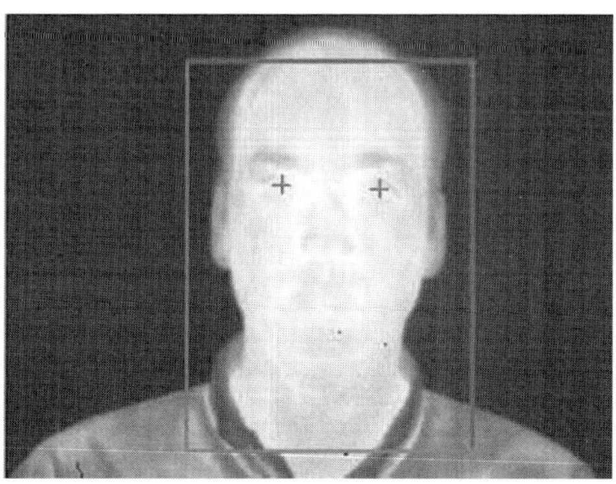

FIGURE 20.11: Automatic detection of the face and eyes in a thermal infrared image.

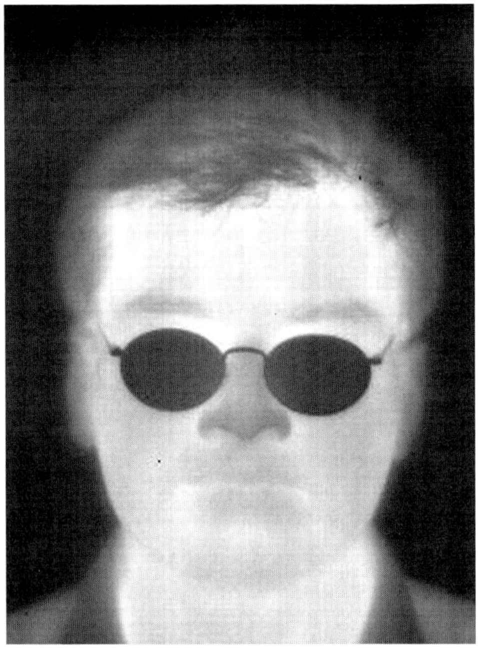

FIGURE 20.12: Thermal infrared image of a person wearing glasses.

from the shape of the lens. Unfortunately, the errors incurred by such inference are rather large. For the experiments outlined below, only images of subjects without glasses were used. Proper normalization of thermal images of subjects wearing glasses is an area of active research. A recent paper [23] introduces a method for detecting and segmenting glasses on infrared facial imagery. Additionally, they report recognition results after applying a method for filling in the area obscured by eyeglasses with an average eye-region image. Using the FaceIt commercial face-recognition system, they observe marked improvement after removal of eyeglasses. Our own experiments on the subject (which predate [23]) show that, when using an algorithm capable of handling occlusion, removal of eyeglasses does not provide a statistically significant performance boost.

In order to evaluate the viability of thermal-only face recognition, we must compare its performance against a similar visible-only system. It is clear that the visible-only system would not function in the dark, but it does provide us with a suitable performance baseline. We do not use the OpenCV object-detection algorithm for eye detection in visible images. While this method does work reasonably well, we can obtain better localization results with the algorithm outlined below. This is simply because we can take advantage of clear structure within the eye region and model it explicitly, rather than depend on a generic object detector. We simply search for the center of the pupil of the open eye. The initial search area relies again on the position of the face as returned by a face detector [19]. Within this region, we look for a dark circle surrounded by lighter background using an operator similar to the Hough transform widely used for detection in the iris recognition community [17]:

$$\max_{(r,x_0,y_0)} \left| G_\sigma(r) * \frac{\partial}{\partial r} \oint_{r,x_0,y_0} \frac{I(x,y)}{2\pi r} ds \right|. \tag{1}$$

This operator searches over the image domain (x, y) for the maximum in the smoothed partial derivative with respect to increasing radius r, of the normalized contour integral of $I(x, y)$ along a cicrcular arc ds of radius r and center coordinates (x_0, y_0). The symbol $*$ denotes convolution and $G_\sigma(r)$ is a smoothing function such as a Gaussian of scale σ.

We performed localization and recognition experiments using a large database of over 3700 images of 207 subjects not wearing glasses. Images were collected during several sessions, both indoors and outdoors. All thermal imagery was collected with an uncooled LWIR sensor at 320×240 resolution, and coregistered visible imagery was acquired for all frames. Recognition performance was evaluated both using a PCA-based algorithm, and the proprietary Equinox algorithm. For more details see [48].

Table 20.5 shows the mean absolute error and the standard deviation of the error in the x and y coordinates for the left and right eye, for detection in the visible domain, while Table 20.6 shows the equivalent quantities for the LWIR domain. While the number of outliers is much larger in the visible than in LWIR, the means and standard deviations of the visible errors stay below 1 pixel.[2] The means of the

Table 20.5: Means and standard deviations of visible eye detection errors.

	Mean	Standard deviation
Left x	0.56735	1.1017
Left y	0.55006	0.83537
Right x	0.59527	1.1364
Right y	0.56735	0.83601

[2]Obviously, removing the outliers reduces the standard deviation, so to some extent the lower variance is due to the large number of outliers.

Table 20.6: Means and standard deviations of IR eye detection errors.

	Mean	Standard deviation
Left x	1.9477	2.0254
Left y	1.5738	1.6789
Right x	2.8054	2.0702
Right y	1.5338	1.6821

Table 20.7: Performance of PCA algorithm with eyes detected in the visible domain.

Gallery/Probe	Visible	LWIR
Indoor/indoor	89.47	72.11
Outdoor/outdoor	38.41	72.85
Indoor/outdoor	28.66	53.50

absolute LWIR errors go up to 2.8 pixels, a 4.7 times increase over visible, and the standard deviations go up to 1.75, a 1.77 times increase over visible. We have to keep in mind though that, at the resolution of our images, the average size of an eye is 20 pixels wide by 15 pixels high. So although the error increase from visible to LWIR is large, LWIR values still stay within 15% of the eye size, quite a reasonable bound. We will see below how this error increase affects recognition performance.

Top match recognition performances are shown in Tables 20.7– 20.10. Recognition performance with LWIR eye locations is followed in parentheses by the percentage of the corresponding performance with visible eye locations that this represents. PCA performs very poorly on difficult data sets (outdoor probes) and the performance decreases even more when the eyes are detected in LWIR. The decrease in performance is about the same in both modalities (performance with LWIR eye locations is about 70% of the performance with visible eye locations). This is in contrast with the observation in [14], but is probably due to the difficulty of the data set as well as a lower error in the eye center location. The Equinox algorithm performs much better than PCA in general. The decrease in performance when using LWIR eyes is not as steep as in the case of PCA (about 90% of the visible eyes performance in both modalities).

Table 20.8: Performance of PCA algorithm with eyes detected in the LWIR domain. In parentheses. percentage of corresponding performance with eyes detected in the visible domain.

Gallery/Probe	Visible	LWIR
Indoor/indoor	68.95 (77)	48.42 (67)
Outdoor/outdoor	33.53 (87)	65.89 (90)
Indoor/outdoor	14.01 (49)	33.76 (63)

Table 20.9: Performance of the Equinox algorithm with eyes detected in the visible domain.

Gallery/Probe	Visible	LWIR
Indoor/indoor	99.47	95.79
Outdoor/outdoor	88.74	96.03
Indoor/outdoor	87.90	90.45

Table 20.10: Performance of Equinox algorithm with eyes detected in the LWIR domain. In parentheses percentage of corresponding performance with eyes detected in the visible domain.

Gallery/Probe	Visible	LWIR
Indoor/indoor	95.80 (96)	87.90 (92)
Outdoor/outdoor	82.78 (93)	92.72 (97)
Indoor/outdoor	73.25 (83)	78.34 (87)

Thermal infrared face-recognition algorithms, so far, have relied on eye center locations that were detected either manually or automatically in a coregistered visible image. In an attempt to solve the problem of real-time nocturnal face recognition, we performed eye detection on thermal infrared images of faces and used the detected eye center locations to geometrically normalize the images prior to applying face-recognition algorithms. Results show that, while the localization error is greater than when using visible imagery, the absolute error is still within 15% of the size of the eye.

We analyzed the impact of eye locations detected in the visible and thermal infrared domains on two face-recognition algorithms. It is clear that recognition performance drops, for both algorithms, when eyes are detected in the LWIR. For the (better performing) Equinox algorithm, the performance loss is on the order of 10%. Notably, our experiments show that the decay in performance due to poor eye localization is comparable across modalities. Based on our results, we believe that, using the right algorithm, thermal infrared face recognition is a viable biometric tool not only when visible light is available, but also in the dark. These results can probably be improved by locating more features, and using a least-squares, or RANSAC approach for geometrically normalizing the face image. We have experimented with nose detection using a similar approach to the one outlined here, and while results are significantly better than chance, they are considerably more error-prone than those for eye detection.

20.7 CONCLUSION

Face recognition with thermal infrared imagery has become a subject of research interest in the last few years. It has emerged as both an alternative and a complementary modality to visible imagery. The primary strength of thermal imagery for face recognition is its near invariance to illumination effects. This sidesteps a major difficulty normally encountered when dealing with visible face recognition, and opens up new possibilities for deployment in adverse environments. The main drawback of thermal imaging for widespread use has always been the cost of the sensors. While this is still a major issue, prices have steadily decreased, and it is now possible to purchase thermal sensors for fraction of the price they commanded a decade ago. There will always be a cost differential with visible cameras, as these are sold in greatly larger volumes, but as this gap diminishes, thermal cameras will see their way into many more applications.

Over the last few years, work by a number of research groups has considerably expanded our knowledge of thermal face recognition. We have gone from early studies merely able to show the viability of thermal imaging for biometric applications to comprehensive experimental designs aimed at testing realistic technology deployments. No longer is research restricted to the laboratory, but rather it has leaped out into outdoor scenarios and live demonstrations. Research on time-lapse recognition has been instrumental in determining under what conditions thermal systems can and should be used. The notion of fusing visible and thermal imagery has been constant throughout this line of research. Under all circumstances considered by all groups working in the problem, fusion has surfaced as a certain way to increase recognition performance and increase the range of viable operating conditions. Interestingly, even very simple fusion schemes yield a large payout, suggesting that, as our intuition would indicate, facial appearance in the two spectral regions is quite uncorrelated. Most recently, we demonstrated that, under certain conditions (mainly lack of eyeglasses), a thermal-only recognition system is possible, as all the pieces, from face detection to normalization to recognition are now available and sufficiently accurate.

Many challenges remain for future work. While current state-of-the-art thermal recognition algorithms are able to cope with a fair degree of endogenous variation (metabolic and otherwise), one can never have too much invariance to nuisance factors. More robust algorithms are needed to maintain consistently high performance levels through wide variations in thermal appearance caused by extreme metabolic states and weather conditions. To some extent, this may be achieved by thorough thermal radiative modeling of the human head. In addition, a more complete database of face images under such extreme states is necessary, the collection of which will be quite involved. It is hard, however, to overemphasize the importance of data properly representative of those one would expect to see under operational conditions, and therefore this should be a high priority for future work.

As work by ourselves and other researchers indicates, the variable presence of eyeglasses is a surmountable problem in terms of recognition. Through a variety of techniques we are able to minimize the effect of the large occlusion presented by eyeglasses, and obtain good performance even in mixed glasses/no-glasses situations. This is probably due to the fact that, unlike in visible imagery, the area around the eyes does not appear to carry so much discriminant information in the thermal infrared. The main problem with eyeglasses is that they hamper our ability to find facial landmarks necessary for geometric normalization. Since registration is a large portion of the problem for face recognition, eyeglasses do present a serious challenge. As we pointed out above, it is possible to infer the location of the eyes from a segmented contour of the eyeglasses, but the resulting accuracy is rather poor, barely exceeding that of simply picking average eye locations based on the position of the head. Other facial landmarks are either hard to localize or often occluded. Thus normalization of faces with eyeglasses remains a challenging problem. Fortunately, detection of eyeglasses is very easy and reliable in the thermal infrared. Therefore, in an access control scenario, one can envision a system that requests the user to remove his glasses if necessary. Unfortunately this is not possible in a more unconstrained surveillance scenario.

Most outstanding problems in visible face recognition, such as dealing with pose variation, facial expressions and occlusions are also valid for the thermal domain. We can expect that some of the solutions developed in the visible realm will carry over readily to thermal imagery, while others will require specialized techniques. Most importantly, what the last few years of effort have shown us is that developing or modifying algorithms for thermal face recognition is well worth the trouble. Work in this field has been quite fruitful, and we can expect it to remain so.

ACKNOWLEDGMENT

This research was supported by the DARPA Human Identification at a Distance (HID) program, contract # DARPA/AFOSR F49620-01-C-0008.

REFERENCES

[1] Open computer vision library. http://sourceforge.net/projects/opencvlibrary/.
[2] A. J. Bell and T. J. Sejnowski. An information-maximization approach to blind separation and blind deconvolution. *Neural Computation* **7**(6):1129–1159, 1995.
[3] Yael Adini, Yael Moses, and Shimon Ullman. Face recognition: the problem of compensating for changes in illumination direction. *IEEE Transactions on Pattern Analysis and Machine Intelligence* **19**(7):721–732, July 1997.
[4] B. Achermann and H. Bunke. Combination of classifiers on the decision level for face recognition. Technical Report IAM-96-002, Institute of Computer Science and Applied Mathematics, University of Bern, Bern, Switzerland, January 1996.
[5] R. Chellappa, B. S. Manjunath, and C. von der Malsburg. A feature based approach to face recognition. In: *Computer Vision and Pattern Recognition 92*, pages 373–378, 1992.
[6] M. Bartlett. *Face Image Analysis by Unsupervised Learning*, volume 612 of *Kluwer International Series on Engineering and Computer Science*. Kluwer, Boston, 2001.
[7] P. Belhumeur, J. Hespanha, and D. Kriegman. Eigenfaces vs. Fisherfaces: recognition using class specific linear projection. *IEEE Transactions PAMI* **19**(7):711–720, July 1997.
[8] P. N. Belhumeur and D. J. Kriegman. What is the set of images of an object under all possible illumination conditions. *International Journal of Computer Vision* **28**(3):245–260, July 1998.
[9] J. R. Beveridge and K. She. Fall 2001 update to CSU PCA versus PCA+LDA comparison. Technical report, Colorado State University, Fort Collins, CO, December 2001. Available at http://www.cs.colostate.edu/evalfacerec/papers.html.
[10] J. Bigun, B. Duc, S. Fischer, A. Makarov, and F. Smeraldi. Multi-modal person authentication. In: *Proceedings of Face Recognition: from Theory to Applications*, Stirling, 1997. NATO Advanced Study Institute.
[11] R. Brunelli and T. Poggio. Hyperbf networks for real object recognition. In: *International Joint Conference on Artificial Intelligence 91*, pages 1278–1284, 1991.
[12] C. Liu and H. Wechsler. Comparative assesment of independent component analysis (ICA) for face recognition. In: *Proceedings of the Second Int. Conf. on Audio- and Video-based Biometric Person Authentication*, Washington, DC, March 1999.
[13] X. Chen, P. Flynn, and K. Bowyer. PCA-based face recognition in infrared imagery: Baseline and comparative studies. In: *International Workshop on Analysis and Modeling of Faces and Gestures*, Nice, France, October 2003.
[14] X. Chen, P. Flynn, and K. Bowyer. Visible-light and infrared face recognition. In: *Proceedings of the Workshop on Multimodal User Authentication*, Santa Barbara, CA, December 2003.
[15] R. Cutler. Face recognition using infrared images and eigenfaces. Technical report, University of Maryland, April 1996. Available at http://www.cs.umd.edu/~rgc/pub/ireigenface.pdf.

[16] J. Schuler D. Scribner, P. Warren. Extending color vision methods to bands beyond the visible. In: *Proceedings of IEEE Workshop on Computer Vision Beyond the Visible Spectrum: Methods and Applications*, Fort Collins, CO, June 1999.

[17] J. Daugman. How iris recognition works. *IEEE Transactions on Circuits and Systems for Video Technology* **14**(1),January 2004.

[18] K. Etemad and R. Chellappa. Discriminant analysis for recognition of human face images. *Journal of the Optical Society of America* **14**:1724–1733, August 1997.

[19] C.K. Eveland. Utilizing visible and thermal infrared video for the fast detection and tracking of faces. PhD thesis, University of Rochester, Rochester, NY, December 2002.

[20] G. Friedrich and Y. Yeshurun. Seeing people in the dark: face recognition in infrared images. In: *Biologically Motivated Computer Vision 2002*, page 348 ff., 2002.

[21] G. Gaussorgues. *Infrared Thermography*. Microwave Technology Series 5. Chapman and Hall, London, 1994.

[22] Ralph Gross, Iain Matthews, and Simon Baker. Fisher light-fields for face recognition across pose and illumination. In: *Proceedings of the German Symposium on Pattern Recognition (DAGM)*, September 2002.

[23] J. Heo, S. G. Kong, B. R. Abidi, and M. A. Abidi. Fusion of visual and thermal signatures with eyeglass removal for robust face recognition. In: *Proceedings of CVPR Workshop on Object Tracking and Classification Beyond the Visible Spectrum (OTCBVS) 2004*, Washington, DC, June 2004.

[24] A. Hyvärinen. Fast and robust fixed-point algorithms for independent component analysis. *IEEE Transactions on Neural Networks* **10**(3):626–634, 1999.

[25] J. R. Beveridge, K. She, B. A. Draper, and G. H. Givens. Parametric and nonparametric methods for the statistical evaluation of human ID algorithms. In: *Proceedings of the Third Workshop on Empirical Evaluation Methods in Computer Vision*, Kauai, HI, December 2001.

[26] J. R. Beveridge K. Baek, B. A.Draper, and K. She. PCA vs. ICA: a comparison on the FERET data set. In: *Proceedings of the International Conference on Computer Vision, Pattern Recognition and Image Processing*, Durham, NC, March 2002.

[27] T. Kanade. Computer recognition of human faces. In: *Birkhauser*, 1977.

[28] R. Lienhart and J. Maudt. An extended set of Haar-like features for rapid object detection. In: *Proceedings Int. Conf. on Image Processing 2002*, volume 1, pages 900–903, 2002.

[29] A. M. Martinez and A. C. Kak. PCA versus LDA. *IEEE Transaction on Pattern Analysis and Machine Intelligence* **23**(2):228–233, February 2001.

[30] R. J. Micheals and T. Boult. Efficient evaluation of classification and recognition systems. In: *Proceedings of Computer Vision and Pattern Recognition*, Kauai, HI, December 2001.

[31] P. Comon. Independent component analysis: a new concept? *Signal Processing* **36**(3):287–314, 1994.

[32] P. J. Bickel and K. A. Doksum. *Mathematical Statistics*. Prentice Hall, Englewood Cliffs, NJ, 1977.

[33] P. Penev and J. Attick. Local Feature Analysis: A general statistical theory for object representation. *Network: Computation in Neural Systems* **7**(3):477–500, 1996.

[34] J. Phillips and E. Newton. Meta-analysis of face recognition algorithms. In: *Proceedings of IEEE International Conference on Automatic Face and Gesture Recognition*, Washington D.C., May 2002.

[35] P. J. Phillips, D. Blackburn, M. Bone, P. Grother, R. Micheals, and E. Tabassi. Face recognition vendor test 2002 (FRVT 2002). Available at http://www.frvt.org/FRVT2002/default.htm, 2002.

[36] P. Jonathon Phillips, Hyeonjoon Moon, Syed A. Rizvi, and Patrick J. Rauss. The FERET evaluation methodology for face-recognition algorithms. Technical Report NISTIR 6264, National Institiute of Standards and Technology, 7 January 1999.

[37] F. J. Prokoski. Method for identifying individuals from analysis of elemental shapes derived from biosensor data. US Patent 5,163,094, November 1992.

[38] F. J. Prokoski. History, current status, and future of infrared identification. In: *Proceedings of the IEEE Workshop on Computer Vision Beyond the Visible Spectrum: Methods and Applications*, Hilton Head, 2000.

[39] R. M. McCabe. Best practice recommendations for the capture of mugshots. Available at http://www.itl.nist.gov/iaui/894.03/face/bpr_mug3.html, 23 September 1997.

[40] S. Edelman D. Reisfeld, and Y. Yeshurun. Learning to recognize faces from examples. In: *European Conference on Computer Vision 92*, pages 787–791, 1992.

[41] S. A. Rizvi, P. J. Phillips, and H. Moon. The FERET verification testing protocol for face recognition algorithms. Technical Report 6281, National Institute of Standards and Technology, October 1998.

[42] A. Shashua and T. R. Raviv. The quotient image: class based re-rendering and recognition with varying illuminations. *IEEE Transaction on Pattern Analysis and Machine Intelligence* **23**(2):129–139, 2001.

[43] R. Siegel and J. Howell. *Thermal Radiation Heat Transfer*. Taylor and Francis, Washington, DC, third edition, 1992.

[44] T. Sim and T. Kanade. Illuminating the face. Technical report, The Robotics Institute, Carnegie Mellon University, September 2001. Available at http://www.ri.cmu.edu/pubs/pub-3869.html.

[45] D. Socolinsky, J. Neuheisel, C. Priebe, and J. DeVinney. A boosted cccd classifier for fast face detection. In: *35th Symposium on the Interface*, Salt Lake City, UT, March 2003.

[46] D. Socolinsky, L. Wolff, J. Neuheisel, and C. Eveland. Illumination invariant face recognition using thermal infrared imagery. In: *Proceedings IEEE Conference on Computer Vision and Pattern Recognition*, Kauai, HI, December 2001.

[47] D. A. Socolinsky and A. Selinger. A comparative analysis of face recognition performance with visible and thermal infrared imagery. In: *Proceedings of the International Conference on Pattern Recognition*, Quebec, Canada, August 2002.

[48] D. A. Socolinsky and A. Selinger. Face recognition in the dark. In: *Proceedings of CVPR Workshop on Object Tracking and Classification Beyond the Visible Spectrum (OTCBVS) 2004*, Washington, DC, June 2004.

[49] D. A. Socolinsky and A. Selinger. Thermal face recognition over time. In *Proceedings of the International Conference on Pattern Recognition 2004*, Cambridge, UK, August 2004.

[50] D.A. Socolinsky, J.D. Neuheisel, C.E. Priebe, D. Marchette, and J.G. DeVinney. A boosted cccd classifier for fast face detection. *Computing Science and Statistics*

35, Interface 2003: Security and Infrastructure Protection 35th SYMPOSIUM ON THE INTERFACE.

[51] D.A. Socolinsky and A. Selinger. Face recognition with visible and thermal infrared imagery. *Computer Vision and Image Understanding*, July - August 2003.

[52] D. L. Swets and J. J. Weng. Using discriminant eigenfeatures for image retrieval. *IEEE Transactions on Pattern Analysis and Machine Intelligence*, **18**(8):831–836, August 1996.

[53] T. K. Ho, J. J. Hull, and S. N. Srihari. Decision combination in multiple classifier systems. *IEEE Transactions on Pattern Analysis and Machine Intelligence*, **16**(1):66–75, January 1994.

[54] M. Turk and A. Pentland. Eigenfaces for recognition. *J. Cognitive Neuroscience* **3**:71–86, 1991.

[55] S. Romdhani V. Blanz and T. Vetter. Face identification across different poses and illuminations with a 3D morphable model. In: *Automatic Face and Gesture Recognition 02*, pages 192–197, 2002.

[56] S. Romdhani V. Blanz and T. Vetter. Face identification by fitting a 3d morphable model using linear shape and texture error functions. In: *European Conference on Computer Vision 02*, page IV:3 ff., 2002.

[57] P. Viola and M. Jones. Rapid object detection using a boosted cascade of simple features. In: *Computer Vision and Pattern Recognition 01*, pages I:511–518, 2001.

[58] P. Viola and M. Jones. Robust real-time face detection. In: *International Conference on Computer Vision 01*, page II: 747, 2001.

[59] W. S. Yambor, B. A. Draper, and J. R. Beveridge. Analyzing PCA-based face recognition algorithms: eigenvector selection and distance measures. In: *Proceedings of the 2nd Workshop on Empirical Evaluation in Computer Vision*, Dublin, Ireland, 1 July 2000.

[60] W. Y. Zhao, R. Chellappa, and A. Krishnaswamy. Discriminant analysis of principal components for face recognition. In: *Automatic Face and Gesture Recognition 98*, pages 336–341, 1998.

[61] Joseph Wilder, P. Jonathon Phillips, Cunhong Jiang, and Stephen Wiener. Comparison of visible and infra-red imagery for face recognition. In: *Proceedings of 2nd International Conference on Automatic Face & Gesture Recognition*, pages 182–187, Killington, VT, 1996.

[62] L. Wolff, D. Socolinsky, and C. Eveland. Quantitative measurement of illumination invariance for face recognition using thermal infrared imagery. In: *Proceedings Computer Vision Beyond the Visual Spectrum*, Kauai, December 2001.

[63] W. Zhao and R. Chellappa. Robust face recognition using symmetric shape-from-shading. Technical report, Center for Automation Research, University of Maryland, College Park, 1999. Available at http://citeseer.nj.nec.com/zhao99robust.html.

[64] Zhiwei Zhu, Kikuo Fujimura, and Qiang Ji. Real-time eye detection and tracking under various light conditions. In: *Proceedings of the Symposium on Eye Tracking Research and Applications*, pages 139–144. ACM Press, 2002.

CHAPTER 21

MULTIMODAL BIOMETRICS: AUGMENTING FACE WITH OTHER CUES

21.1 INTRODUCTION

Face is an attractive biometric characteristic because it is easily collectible, and has wide social acceptance. Since the face image can be obtained from a distance and without any cooperation from the user, face-recognition technology is well suited for surveillance applications. However, the appearance-based facial features commonly used in most of the current face-recognition systems, are found to have limited discrimination capability and further they change over a period of time. A small proportion of the population can have nearly identical appearances due to genetic factors (e.g., father and son, identical twins, etc.) making the face-recognition task even more challenging.

The current state-of-the-art face-recognition systems have good recognition performance when the user presents a frontal face with neutral expression under consistent lighting conditions. The performance falls off drastically with variations in pose, illumination, background, aging, and expression. The Face Recognition Vendor Test (FRVT) 2002 [1] reported that the most accurate face-recognition system had a false nonmatch rate (FNMR) of 10% at a false match rate (FMR) of 1% when tested in an indoor environment. Under outdoor conditions, the FNMR was as high as 50% at the same value of FMR. In contrast to this situation, the results of the Fingerprint Vendor Technology Evaluation (FpVTE) 2003 [2] and the Fingerprint Verification Competition (FVC) 2004 [3] indicated that the state-of-the-art fingerprint verification systems had a FNMR of 0.1% and 2.54%, respectively, corresponding to an FMR of 1%. Iris recognition systems have been found to be more accurate than face and fingerprint systems ([4] p. 121). Fingerprint and iris patterns are also more stable over time when compared to the facial appearance. Due to these reasons, current face-recognition systems are not as competitive, in terms of recognition accuracy, compared to the biometric systems that are based on fingerprint or iris. Though fingerprint and iris identification systems are more accurate than face-recognition systems, they have their own limitations. Both fingerprint and iris are not truly universal [5]. The National Institute of Standards and Technology (NIST) has reported that it is not possible to obtain a good-quality fingerprint from approximately two percent of the population (people with hand-related disabilities, manual workers with many cuts and bruises on their fingertips, and people with very oily or dry fingers) [6]. Hence, such people cannot be enrolled in a fingerprint-verification system. Similarly, persons having long eye-lashes and those suffering from eye abnormalities or diseases like glaucoma, cataract, aniridia, and nystagmus cannot provide good-quality iris images for automatic recognition [7]. This leads to failure-to-enroll (FTE) and/or failure-to-capture (FTC) errors in a biometric system.

Apart from face, fingerprint, and iris, many other physiological and behavioral characteristics are being used for biometric recognition [8]. A good biometric trait should satisfy all the seven requirements that are listed below:

- Universality: every individual in the target population should possess the trait;
- Distinctiveness: ability of the trait to sufficiently differentiate between any two persons;
- Persistence: the trait should be sufficiently invariant (with respect to the matching criterion) over a period of time;
- Collectability: the trait should be easily obtainable (or measurable);
- Performance: high recognition accuracy and speed should be achievable with limited resources in a variety of operational and environmental conditions;
- Acceptability: the biometric identifier should have a wide public acceptance and the device used for measurement should be harmless;
- Circumvention: spoofing of the characteristic using fraudulent methods should be difficult.

A comparison of some of the common characteristics used for biometric recognition based on these seven requirements is shown in Table 21.1. Since no biometric trait satisfies all the requirements in all operational environments, it is fairly reasonable to claim that there is no optimal biometric characteristic. As a result, systems based on a single biometric trait may not achieve the stringent performance guarantees required by many real-world applications.

In order to overcome the problems faced by biometric systems based on a single biometric identifier, more than one biometric trait of a person can be used for recognition. Systems that use two or more biometric characteristics for person identification are called multimodal biometric systems. Apart from reducing the

Table 21.1: Comparison of various biometric characteristics based on the table presented in [8]. High, medium, and low values are denoted by H, M, and L, respectively. (©IEEE 2005.)

Biometric identifier	Universality	Distinctiveness	Persistence	Collectability	Performance	Acceptability	Circumvention
DNA	H	H	H	L	H	L	L
Ear	M	M	H	M	M	H	M
Face	H	L	M	H	L	H	H
Fingerprint	M	H	H	M	H	M	M
Gait	M	L	L	H	L	H	M
Hand geometry	M	M	M	H	M	M	M
Iris	H	H	H	M	H	L	L
Palmprint	M	H	H	M	H	M	M
Signature	L	L	L	H	L	H	H
Voice	M	L	L	M	L	H	H

FTE/FTC rates of the biometric system, the use of multiple biometric traits also has other advantages. Combining the evidence obtained from different modalities using an effective fusion scheme can significantly improve the overall accuracy of the biometric system. A multimodal biometric system can provide more resistance against spoofing because it is difficult to simultaneously spoof multiple biometric traits. Multimodal systems can also provide the capability to search a large database in an efficient and fast manner. This can be achieved by using a relatively simple but less accurate modality to prune the database before using the more complex and accurate modality on the pruned database to perform the final identification task. However, multimodal biometric systems also have some disadvantages. They are more expensive and require more resources for computation and storage than unimodal biometric systems. Multimodal systems generally require more time for enrollment and verification causing some inconvenience to the user. Finally, the system accuracy can actually degrade compared to the unimodal system if a proper technique is not followed for combining the evidence provided by different modalities. However, the advantages of multimodal systems far outweigh the limitations, and hence such systems are being increasingly deployed in critical applications.

Face recognition is usually an important component of a multimodal biometric system. Face is a convenient biometric trait, because the facial image of a person can be easily acquired without any physical contact with the biometric system. Further, humans recognize each other using the facial characteristics, and face images have been traditionally collected in a variety of applications (e.g., driver's licenses, passports, identification cards, etc.) for identification purposes. Hence, face is a socially accepted biometric characteristic with relatively few privacy concerns. Several large face databases collected under a variety of operating conditions are already available for testing a face-recognition system. Smaller size and very low power requirements of the cameras used for acquiring the face images increase their applicability and deployment in resource-constrained environments like mobile phones. Due to these reasons, a number of multimodal biometric systems have been proposed that include face as one of the modalities. Before analyzing the multimodal systems that use face along with other biometric modalities for person identification, the issues involved in designing a multimodal biometric system must be considered. Specifically, issues like what information from each modality should be combined and how the evidence is integrated must be carefully addressed in order to maximize the advantages of a multimodal biometric system in terms of accuracy, speed, and population coverage.

21.2 DESIGN OF A MULTIMODAL BIOMETRIC SYSTEM

Multimodal biometric systems differ from one another in terms of their architecture, the number and choice of biometric modalities, the level at which the evidence is accumulated, and the methods used for the integration or fusion of information. Generally, these design decisions depend on the application scenario, and they have a profound influence on the performance of the multimodal biometric system. This section presents a brief discussion on the various options that are available when designing a multimodal biometric system.

21.2.1 Architecture

Multimodal biometric system architecture can be either serial or parallel (see Figure 21.1). In the serial or cascade architecture, the processing of the modalities takes place sequentially, and the outcome of one modality affects the processing of the subsequent modalities. In the parallel design, different modalities operate independently and their results are combined using an appropriate fusion scheme. Both of these architectures have their own advantages and limitations. The cascading scheme can improve the user convenience as well as allow fast and efficient searches in large-scale identification tasks. For example, when a cascaded multimodal biometric system has sufficient confidence on the identity of the user after processing the first modality, the user may not be required to provide the other modalities. The system can also allow the user to decide which modality he/she would present first. Finally, if the system is faced with the task of

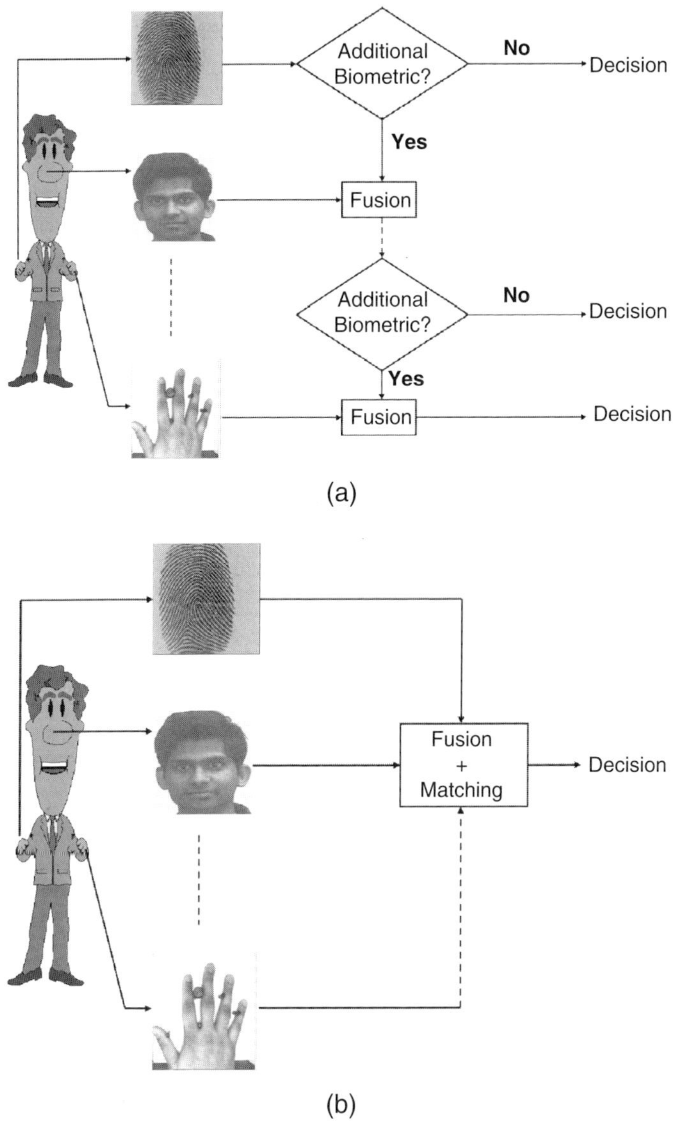

FIGURE 21.1: Architecture of multimodal biometric systems; (a) serial and (b) parallel.

identifying the user from a large database, it can utilize the outcome of each modality to successively prune the database, thereby making the search faster and more efficient. Thus, a cascaded system can be more convenient to the user and generally requires less recognition time when compared to its parallel counterpart. However, it requires robust algorithms to handle the different sequence of events. A multimodal system designed to operate in the parallel mode generally has a higher accuracy because it utilizes more evidence about the user for recognition. The choice of the system architecture depends on the application requirements. User-friendly and less critical applications like bank ATMs can use a cascaded multimodal biometric system. On the other hand, parallel multimodal systems are more suited for applications where security is of paramount importance (e.g., access to military installations).

21.2.2 Choice of Biometric Modalities

The number of biometric modalities that are available for combination is rather large. Figure 21.2 shows most of the modalities that are currently available for fusion [8]. The choice of biometric traits to be used in a multimodal biometric system is entirely application-specific. However, certain combinations of biometric traits are quite popular due to various reasons. Fingerprint and face-recognition technologies are often deployed together because these technologies complement each other well. Face overcomes the nonuniversality problem of fingerprints, while

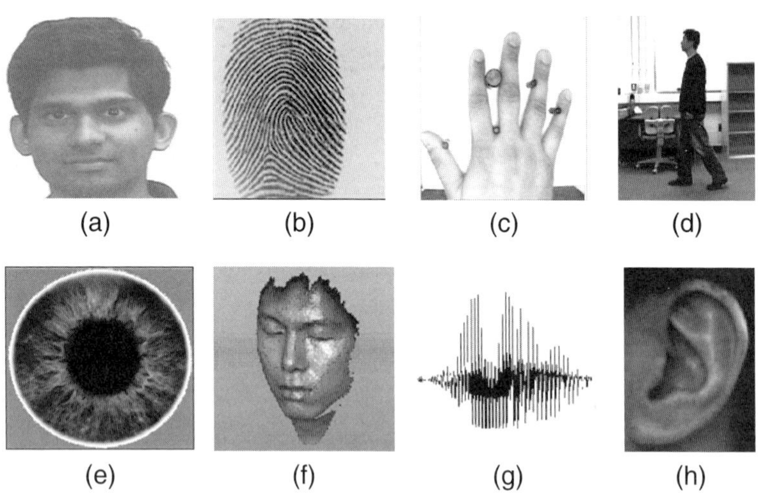

FIGURE 21.2: Biometric modalities; (a) face; (b) fingerprint; (c) hand-geometry; (d) gait; (e) iris; (f) 3D face; (g) voice; (h) ear structure.

fingerprint recognition has much higher accuracy than face recognition. Integrated face and speaker recognition is popular because both face and voice biometric traits are user-friendly. It is relatively easy for a person to face the camera and utter a few words. Since both face and gait biometric traits can be obtained from noncooperative users, this combination is ideal for applications that require recognition at a distance. Apart from these combinations, several other combinations of biometric modalities have also been studied and deployed.

21.2.3 Level of Fusion

Information can be integrated at various levels in a multimodal biometric system. Fusion can be done either prior to matching or after applying the matchers on the input data. At the sensor and feature levels, information is integrated before any matching algorithm is applied. When fusion is done at the matching-score or rank or abstract level, the outputs of the matchers acting on the individual modalities are combined. When the raw data from the sensors of the different modalities are directly combined, it is known as sensor level fusion. Sensor-level fusion is extremely rare in multimodal biometric systems, because the data obtained from the various sensors are not usually compatible. Fusion at the feature level, the confidence or matching-score level, the rank level, and the abstract or decision level are quite common in multimodal biometric systems.

Feature-level fusion refers to combining the feature vectors of the different modalities into a single feature vector. This is generally achieved by a simple concatenation of the feature vectors of the various modalities. Integration of information at the feature level is believed to be more effective than at the matching-score or abstract levels, because the features contain richer information about the input biometric data than the matching scores or class ranks/labels obtained after matching the features. However, proper care must be taken during feature-level fusion to examine the relationship between the feature spaces that are combined to remove the highly correlated features. Further, concatenating feature vectors increases the dimensionality of the new feature space. At the same time, the cost and privacy issues involved in biometric data collection limit the availability of data for training the multimodal biometric system. This can lead to the "curse of dimensionality" [9]. Finally, for proprietary reasons, most commercial biometric vendors do not provide access to their feature vectors. This makes it very difficult to apply feature-level fusion in multimodal biometric systems that are built using commercial off-the-shelf (COTS) unimodal biometric systems.

Abstract-level fusion refers to a combination technique where the only information available for fusion is the decision made by each modality (e.g., "accept" or "reject" in a verification scenario). Methods like majority voting [10], behavior knowledge space [11], weighted voting based on the Dempster-Shafer theory of evidence [12], AND rule and OR rule [13], etc. can be used to combine the individual decisions at the abstract level and arrive at the final decision. Fusion is done at the rank level if each modality outputs a set of possible matches along with the corresponding ranks. Ho et al. [14] proposed the use of highest rank, Borda count, or logistic regression to combine the ranks assigned by the different modalities.

The most common level of fusion in a multimodal system is the matching-score level. In matching score fusion, each biometric modality returns a matching score indicating the similarity of the user's input biometric data (feature vector) to his/her template stored in the database. Integration at the matching-score level offers the best trade-off in terms of information content and ease of fusion. Fusion at the matching-score level generally requires a normalization technique to transform the scores of the individual modalities into a common domain and an appropriate fusion strategy to combine the transformed scores.

21.2.4 Fusion Strategies

Kittler et al. [15] proposed a generic classifier combination framework in which the probabilities of a pattern being in a specific class are fused using rules such as sum rule, product rule, max rule, and min rule. In this framework, a pattern X is classified into one of m possible classes (there are only two classes in a verification system, namely, "genuine user" and "impostor") using the outputs of R classifiers or matchers. Given the pattern X, let $\vec{x}_i$ denote the feature vector obtained from the ith classifier and let $P(\omega_j|\vec{x}_i)$ be the posterior probability of class ω_j given the feature vector $\vec{x}_i$. If the outputs of the individual classifiers are $P(\omega_j|\vec{x}_i)$, then selecting the class $c^* \in \{1, 2, \cdots, m\}$ associated with the input pattern X can be achieved by applying one of the following rules.

Product rule. Assign the pattern X to the class c^*, where

$$c^* = \text{argmax}_j \prod_{i=1}^{R} P(\omega_j|\vec{x}_i).$$

Here, the feature vectors $\vec{x}_1, \vec{x}_2, ..., \vec{x}_R$ are assumed to be statistically independent. This assumption is generally true for different biometric traits of an individual (e.g., face, iris, fingerprint).

Sum rule. Assign the pattern X to the class c^*, where

$$c^* = \text{argmax}_j \sum_{i=1}^{R} P(\omega_j|\vec{x}_i).$$

Max rule. Assign the pattern X to the class c^*, where

$$c^* = \text{argmax}_j \max_i \; P(\omega_j|\vec{x}_i).$$

Min rule. Assign the pattern X to the class c^*, where

$$c^* = \text{argmax}_j \min_i \; P(\omega_j|\vec{x}_i).$$

Note that these fusion rules can be applied only if the posterior probabilities $P(\omega_j|\vec{x}_i)$ are known. In general, it is difficult to accurately estimate these probabilities from the matching scores [16]. Therefore, it would be better to combine the matching scores directly using an appropriate method without converting them into probabilities. Modifications can be made to the sum rule, max rule, and min rule so that they can be applied to the matching scores. Let s_k^i denote the matching score output by the ith modality ($i = 1, \cdots, R$) for user k ($k = 1, \cdots, N$). Then the combined matching score f_k for user k can be obtained by applying one of the following rules.

Sum Score. The combined score for user k is the sum (or average) of the scores of the individual modalities:

$$f_k = \sum_{i=1}^{R} s_k^i.$$

Max Score. The combined score is the maximum of the scores of the individual modalities:

$$f_k = \max\{s_k^1, s_k^2, \cdots, s_k^R\}.$$

Min Score. The combined score is the minimum of the scores of the individual modalities:

$$f_k = \min\{s_k^1, s_k^2, \cdots, s_k^R\}.$$

The basic sum-score rule can be further modified by assigning different weights to the scores of various matchers, and this is generally referred to as the weighted-sum rule.

21.2.5 Normalization Techniques

The sum-score, max-score, and min-score rules are meaningful only if the matching scores of various modalities are compatible. Hence, the matching scores output by the different matchers must be normalized before they can be combined. Score normalization is necessary because (i) the matching scores can have different characteristics (similarity versus dissimilarity), (ii) the scores can have different numerical ranges, (iii) the statistical distribution of the scores may be different. To make the matching scores compatible, different normalization techniques have been proposed. We present three most commonly used normalization schemes [17].

Min-max. Assuming that the set of matching scores corresponding to the ith matcher is bounded by the values MIN^i and MAX^i, the normalized score can be computed as

$$ns_k^i = \frac{s_k^i - MIN^i}{MAX^i - MIN^i},$$

where $MIN^i = \min\{s_1^i, s_2^i, \cdots, s_N^i\}$ and $MAX^i = \max\{s_1^i, s_2^i, \cdots, s_N^i\}$. This normalization maps all the matching scores to the range [0, 1].

Z-score. Assuming that the arithmetic mean μ^i and standard deviation σ^i of the matching scores output by the ith matcher are known (or can be estimated from the training data), the scores can be mapped to a distribution with mean 0 and standard deviation 1 as follows:

$$ns_k^i = \frac{s_k^i - \mu^i}{\sigma^i}.$$

Tanh. This method was originally proposed by Hampel et al. [18] as a robust normalization technique. The normalized score is computed as

$$ns_k^i = \frac{1}{2}\left\{\tanh\left(0.01\left(\frac{s_k^i - \mu_{\mathrm{H}}^i}{\sigma_{\mathrm{H}}^i}\right)\right) + 1\right\},$$

where μ_{H}^i and σ_{H}^i are the Hampel estimates of the mean and standard deviation of the matching scores output by the ith matcher, respectively.

21.2.6 System Evaluation

The performance metrics of a biometric system such as accuracy, throughput, and scalability can be estimated with a high degree of confidence only when the system is tested on a large representative database. For example, face [1] and fingerprint [2] recognition systems have been evaluated on large databases (containing samples from more than 25,000 individuals) obtained from a diverse population under a variety of environmental conditions. In contrast, current multimodal systems have been tested only on small databases containing fewer than 1000 individuals. Further, multimodal biometric databases can be either true or virtual. In a true multimodal database (e.g., XM2VTS database [19]), different biometric cues are collected from the same individual. Virtual multimodal databases contain records which are created by consistently pairing a user from one unimodal database with a user from another database. The creation of virtual users is based on the assumption that different biometric traits of the same person are independent. This assumption of independence of the various modalities has not been explicitly investigated till date. However, Indovina et al. [20] attempted to validate the use of virtual subjects. They randomly created 1000 sets of virtual users and showed that the variation in performance among these sets was not statistically significant. Recently, NIST has released a true multimodal database [21] containing the face and fingerprint matching scores of a relatively large number of individuals.

21.3 EXAMPLES OF MULTIMODAL BIOMETRIC SYSTEMS

We now present a number of multimodal biometric systems proposed in the literature that use face as one of the modalities. Figure 21.3 shows some of the traits that have been used to augment facial information in multimodal biometric systems. Table 21.2 presents a comparison of these multimodal systems in terms of the design parameters and recognition performance.

Hong and Jain [22] proposed an identification system that integrates face and fingerprint modalities. After noting that face recognition is relatively fast but not very reliable, and fingerprint recognition is reliable but slow (hence not very feasible for database retrieval), the authors cite that these two modalities can be combined to design a system that can achieve both high performance and acceptable response time. Their face-recognition module was based on the eigenfaces method [36] and an elastic-string minutiae-matching algorithm [37] was used for fingerprint recognition. The multimodal system operates by finding the top n identities using the face-recognition system alone and then verifying these identities using the fingerprint-verification subsystem. Hence, this system had a serial architecture. Impostor distributions for fingerprint and face subsystems were estimated and used for selecting at most one of the n possible identities as the genuine identity, hence the system may not always correctly retrieve an identity from the database. In their experiments, Hong and Jain used a database of 1500 images from 150 individuals with 10 fingerprints each. The face database contained a total of 1132 images of 86 individuals, resampled to size 92×112. A total of 64 individuals in fingerprint database were used as the training set, and the remaining 86 users were used as the test set. Virtual subjects were created by assigning an individual

Table 21.2: Comparison of multimodal biometric systems. The third column shows the lowest error rates of the multimodal (unimodal) biometric system. FRR denotes the false reject rate, FAR denotes the false acceptance rate, TER denotes the total error rate (FAR + FRR), and ROA denotes the rank-one accuracy.

Authors	Traits used to augment face	Accuracy	Architecture	Fusion level	Normalization technique	Fusion strategy
Hong and Jain [22]	Fingerprint	FRR: 4.4%(6.9%) at 0.1% FAR	Serial	Matching score/rank	N/A	Bayes
Jain, Hong, and Kulkarni [23]	Fingerprint, voice	FRR: 3%(15%) at 0.1% FAR	Parallel	Matching score	N/A	Neyman–Pearson
Jain and Ross [24]	Fingerprint, hand geometry	FRR: 4%(18%), at 0.1% FAR	Parallel	Matching score	N/A	Weighted sum score
Ross and Jain [25]	Fingerprint, hand geometry	FRR: 1%(18%) at 0.1% FAR	Parallel	Matching score	N/A	Sum score, decision trees, linear discriminant function
Snelick et al. [17]	Fingerprint	FRR: 5%(18%) at 0.1% FAR	Parallel	Matching score	Min-max, z-score, MAD, tanh	Sum score, min score, max score, sum rule, product rule
Snelick et al. [26]	Fingerprint	FRR: 1%(3.3%) at 0.1% FAR	Parallel	Matching score	Min-max, z-score, tanh, adaptive	Sum score, min score, max score, weighted sum score
Brunelli and Falavigna [27]	Voice	FRR: 1.5% at 0.5% FAR (FRR: 8% at 4% FAR)	Parallel	Matching score/ rank	tanh	Geometric weighted average / HyperBF

Bigun et al. [28]	Voice	FRR: 0.5% at < 0.1% FAR (FRR: 3.5% at < 0.1% FAR)	Parallel	Matching score	N/A	Model based on Bayesian theory
Verlinde and Chollet [29]	Voice	TER: 0.1% (TER: 3.7%)	Parallel	Matching score	N/A	k-NN, decision tree, logistic regression
Chatzis et al. [30]	Voice	FRR: 0.68% at 0.39% FAR (FRR: 0.0% at 6.70% FAR)	Parallel	Matching score	N/A	Fuzzy k-means, fuzzy vector quantization, median radial basis function
Ben-Yacoub et al. [31]	Voice	TER: 0.6% (TER: 1.48%)	Parallel	Matching score	N/A	SVM, multilayer perceptron, C4.5 decision tree, Fisher's linear discriminant, Bayesian
Frischholz and Diechmann [32]	Voice, lip movement	N/A	Parallel	Matching score/ abstract	N/A	Majority voting, weighted-sum score
Wang et al. [33]	Iris	TER: 0.27% (TER: 0.3%)	Parallel	Matching score	N/A	Sum score, weighted sum score, Fisher's linear discriminant, neural network
Shakhnarovich et al. [34]	Gait	ROA: 91% (ROA: 87%)	Parallel	Matching score (probabilities)	N/A	sum rule
Kale et al. [35]	Gait	ROA: 97% (ROA: 93%) for cascade mcde	Cascade/parallel	Matching score (probabilities)	N/A	sum rule, product rule

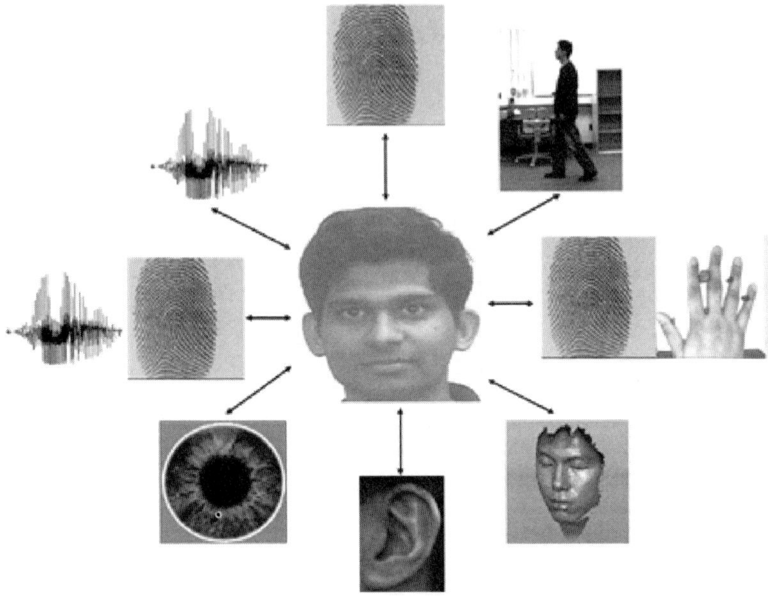

FIGURE 21.3: Other biometric modalities that have been used to augment face.

from the fingerprint database to an individual from the face database consistently. The face-recognition system retrieved the top five ($n = 5$) matches among the 86 individuals and the fingerprint system provided the final decision. False-rejection rates (FRR) of unimodal face and fingerprint systems and the multimodal system were 42.2%, 6.9%, and 4.4%, respectively, at the false-acceptance rate (FAR) of 0.1%. Figure 21.4 shows the associated receiver operating characteristic (ROC) curves. These results indicate that a multimodal system can significantly improve the performance of a face-recognition system.

Jain et al. [23] combined face, fingerprint, and speech modalities at the matching-score level. This specific set of traits was chosen because these traits are frequently used by law enforcement agencies. The parallel fusion scheme submits the matching scores corresponding to these three modalities as inputs to the Neyman–Pearson decision rule to arrive at the verification result. The face-recognition subsystem was based on the eigenfaces approach. The fingerprint verification was based on minutiae-based elastic-string matching [37]. Linear-prediction coefficients (LPC) were extracted from the speech signal and modeled using a hidden Markov model (HMM). The speaker verification was text-dependent (four digits, 1,2,7, and 9 were used). The training database consisted

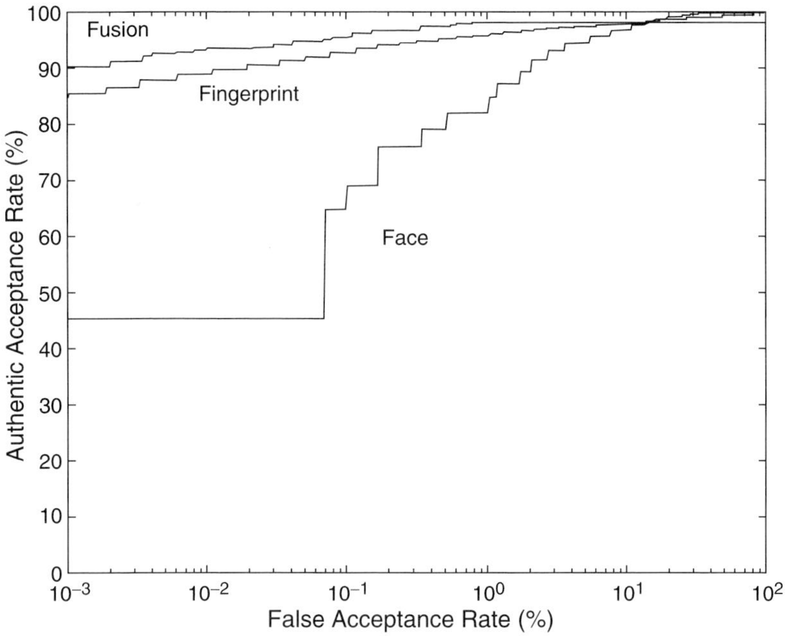

FIGURE 21.4: ROC curves for unimodal and multimodal systems in [22] (©IEEE 2005).

of 50 individuals, each one providing 10 fingerprint images, 9 face images, and 12 speech samples. The test database consisted of 15 fingerprint images, 15 face images, and 15 speech samples collected from 25 individuals. The fused system attained nearly 98% genuine-acceptance rate (GAR) at a FAR of 0.1%. This translates to nearly 12% improvement in GAR over the best individual modality (fingerprint) at 0.1% FAR. The associated ROC curves are shown in Figure 21.5.

Jain and Ross [24] proposed algorithms for estimating user-specific decision thresholds and weights associated with individual matchers for a face–fingerprint–hand geometry based a parallel fusion system. The face module used the eigenface approach. The algorithm proposed in [37] was used for fingerprint verification. The hand-geometry subsystem [38] used 14 features comprised of lengths and widths of fingers and palm widths at several locations of the hand. The user-specific thresholds for each modality were computed with the help of cumulative impostor scores. The weights for individual modalities were found by an exhaustive-search algorithm: all three weights were varied over the range [0,1] with increments of 0.1, and the best combination resulting in the smallest total error rate (sum of false accept and false reject rates) was selected for each user. The database used

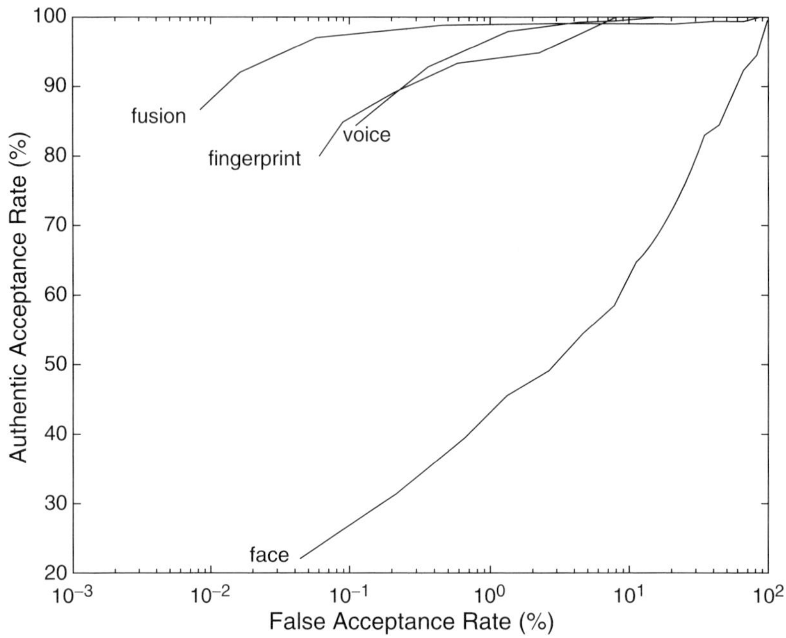

FIGURE 21.5: ROC curves for unimodal and multimodal systems in [23].

in these experiments had 50 users, 40 of these users provided 5 samples of each biometric, and 10 users provided around 30 samples. One third of the samples were used in the training phase, while the remaining were used in the testing phase. At a FAR of 0.1%, user-specific thresholds resulted in nearly 2% GAR improvement; at the same operating point, user-specific weights improved the GAR by nearly 4%. Figure 21.6 shows these performance improvements. Ross and Jain [25] further investigated the effect of different fusion strategies on the multimodal system proposed in [24]. They employed three methods of fusion: namely, sum rule, decision trees, and linear discriminant function. The simple sum fusion outperformed the other two methods, resulting in nearly 17% GAR improvement at a FAR of 0.1%. Figure 21.7 shows the associated ROC curves.

Snelick et al. [17] fused matching scores of commercial face and fingerprint verification systems. They considered the min-max, z-score, MAD (median absolute deviation), and tanh techniques for normalizing the matching scores. In the fusion stage, they investigated the sum-score, min-score, max-score, sum-of-probabilities, and the product-of-probabilities rules. The database consisted of 1005 individuals, each one providing 2 face and 2 fingerprint images. The face images were selected from the FERET database [39], but the authors did not

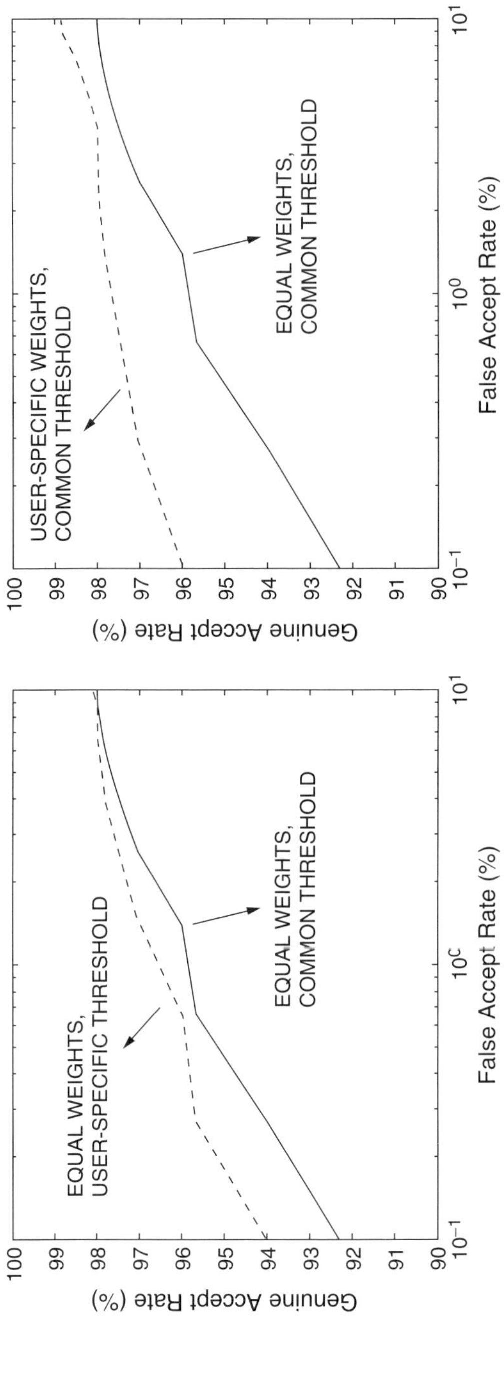

FIGURE 21.6: ROC curves for user-specific thresholds and user-specific weights in a multimodal system [24] (©IEEE 2005).

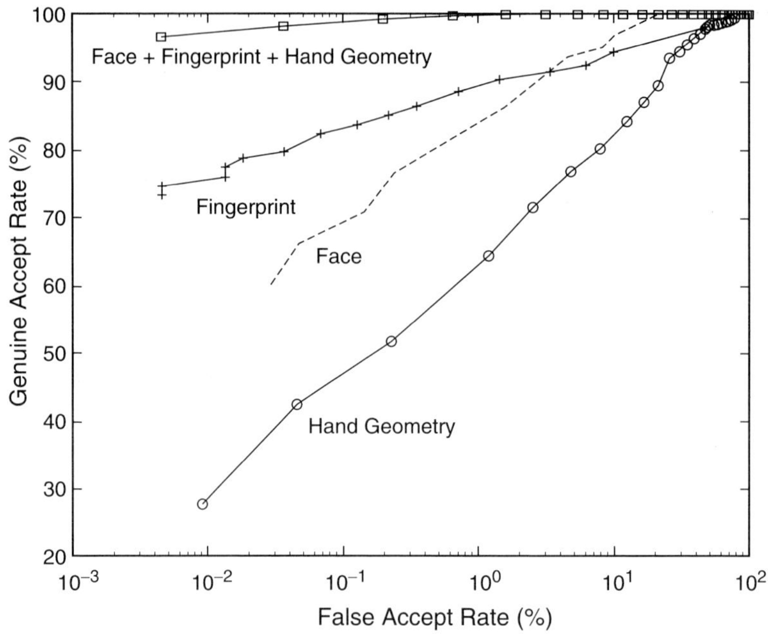

FIGURE 21.7: ROC curves for unimodal and multimodal systems in [25].

provide any information about the characteristics of the fingerprint images. Their results showed that, while every normalization method resulted in performance improvement, the min-max normalization outperformed the other methods. Further, the sum score fusion gave the best performance among all the fusion methods considered in this study. At 0.1% FAR value, the min-max normalization followed by the sum-score fusion rule resulted in a GAR improvement of nearly 13% compared to the best performing individual modality (fingerprint) at the same operating point. Also, the authors reported a considerable decrease in the number of falsely rejected individuals (248 for face, 183 for fingerprint and 28 for multimodal system), indicating that multimodal systems have the potential to increase user convenience by reducing false rejects, as well as reducing false acceptances.

Snelick et al. [26] again used commercial face and fingerprint systems from four vendors in a parallel matching-score fusion framework. They experimented with several normalization and fusion techniques to study their effect on the performance for a database of 972 users. A number of normalization techniques, including min-max, z-score, and tanh schemes were considered along with a novel adaptive normalization technique. The adaptive normalization scheme transforms the min-max normalized scores with the aim of increasing the separation between the genuine and impostor score distributions. The fusion techniques considered in this study included sum of scores, max score, min score, matcher-weighted sum rule, and weighted sum of scores using user-specific weights. The authors used the relative accuracy of individual matchers, indicated by their equal-error rates (EER), to determine the matcher weights. Their user-weighting scheme made use of the wolf–lamb concept [40] originally proposed in speaker-recognition community. The set of weights for each user was found by considering the chance of false accepts for the respective (user, matcher) pairs. The results indicated that min-max and adaptive normalization techniques outperformed the other normalization methods; sum score, max score, and matcher-weighted sum score outperformed other fusion methods. The multimodal system had nearly 2.3% GAR improvement at a FAR of 0.1%. Comparing these improvements with their earlier work [17], the authors note that highly accurate commercial face and fingerprint authentication systems (compared to their academic counterparts) decrease the room for improvement achievable through multimodal biometrics.

Brunelli and Falavigna [27] presented a person-identification system combining acoustic and visual (facial) features. A rejection option was also provided in the system using two different methods. A HyperBF network was used as the rank/measurement level integration strategy. The speaker-recognition subsystem was based on vector quantization of the acoustic-parameter space and included an adaptation phase of the codebooks to the test environment. Two classifiers were used for static and dynamic acoustic features. Face identification was achieved by analyzing three facial components: eyes, nose, and mouth. The basic template-matching technique was applied for face matching. Since the matching scores obtained from the different classifiers were nonhomogeneous, the scores were normalized based on the corresponding distributions. The normalized scores were combined in two different ways: a weighted geometric average and a HyperBF network. The acoustic and visual-cue-based identification achieved 88% and 91% correct recognition rates individually, while the fusion achieved 98% accuracy.

Bigun et al. [28] introduced a new model based on Bayesian theory for combining the matching scores of face and voice-recognition systems. Experiments on the M2VTS database [41] showed that their model resulted in higher accuracy than the sum score rule. For a false-acceptance rate of less than 0.1%, the Bayesian model accepted 99.5% of the genuine users. This was substantially better than the accuracy of the unimodal face and speaker recognition systems that were reported to be 94.4% and 96.5%, respectively.

Verlinde and Chollet [29] formulated the multimodal verification as a classification problem. The inputs were the matching scores obtained from the individual modalities, and the output was a label belonging to the set {reject, accept}. The k-nearest-neighbor classifier using vector quantization, the decision-tree-based classifier, and the classifier based on a logistic regression model were applied to this classification problem. The modalities were based on profile face image,

Table 21.3: Characteristics of the five modalities used in [30].

Algorithm	Features	GAR (%)	FAR (%)
Morphological dynamic link architecture (MDLA)	gray-level and shape	91.9	10.4
Profile shape matching (PSM)	shape	84.5	4.6
Gray-level matching (GLM)	gray-level	73.7	1.3
Gabor dynamic link architecture (GDLA)	Gabor features	92.6	3.7
Hidden markov models (MSP)	speech	100	6.7

frontal face image, and speech. The experiments were carried out on the multimodal M2VTS database [41], and the total error rate (sum of the false accept and false reject rates) of the multimodal system was found to be 0.1% when the classifier based on a logistic regression model was employed. The total error rates of the individual modalities were 8.9% for profile face, 8.7% for frontal face, and 3.7% for speaker verification. Hence, the multimodal system was more accurate than the individual modalities by an order of magnitude.

Chatzis et al. [30] used classical algorithms based on k-means clustering, fuzzy clustering, and median radial basis functions (MRBFs) for fusion at the matching score level. Five methods for person authentication that were based on grey-level and shape information of face image and voice features were explored. Among the five modalities, four used the face image as the biometric, and the remaining one utilized the voice biometric. Table 21.3 shows the algorithms used in this work along with the features used and the rates of genuine and false acceptance. Each algorithm provided a matching score and a quality metric that measures the reliability of the matching score. Results from the five algorithms were concatenated to form a 10-dimensional vector. Clustering algorithms were applied on this 10-dimensional feature vector to form two clusters, namely, genuine and impostor. The M2VTS database was used to evaluate the fusion algorithms. Clustering of the results obtained from MDLA, GDLA, PSM, and MSP algorithms by the k-means method had the best genuine accept rate of 99.32% at a FAR of 0.39%.

Ben-Yacoub et al. [31] considered the fusion of different modalities as a binary classification problem, i.e., accepting or rejecting the identity claim. A number of classification schemes were evaluated for combining the multiple modalities, including support-vector machine (SVM) with polynomial kernels, SVM with Gaussian kernels, C4.5 decision trees, multilayer perceptron, Fisher linear discriminant, and Bayesian classifier. The experiments were conducted on the XM2VTS database [19] consisting of 295 subjects. The database included four recordings of each person obtained at one-month intervals. During each session, two recordings were made: a speech shot and a head rotation shot. The speech shot was composed of the frontal face recording of each subject during the dialogue. The two modalities utilized in the experiments were face image and speech. The face recognition was performed by using elastic graph matching (EGM) [42]. Two different approaches were used for speaker verification. A sphericity measure [43] was used for text-independent speaker verification. Hidden Markov models (HMM) were used for text-dependent speaker verification. The total error rate of 0.6% achieved by the Bayesian classifier was significantly lower than the total error rate of 1.48% achieved by the HMM based speaker recognition system, which was the best individual modality in terms of total error rate.

Frischholz and Diechmann [32] developed a commercial multimodal identification system utilizing three different modalities: face, voice, and lip movement. Unlike other multimodal biometric systems, this system not only included the static features such as face images, but also a dynamic feature, namely, the lip movement. The face was located in an image using an edge-based Hausdorff distance metric. The lip movement was calculated by the optical-flow approach. The synergetic computer [44] was used as the learning classifier for the "optical" biometrics, namely, face and lip movement. Vector quantization was applied for acoustic biometric-based recognition. The input sample was rejected when the difference between the highest and the second highest matching scores was smaller than a given threshold. The sum rule and majority voting served as the two fusion strategies according to the security level of the application. The proposed system was tested on a database of 150 subjects for three months, and the false-acceptance rate was reported to be less than 1%. However, the corresponding genuine-acceptance rate was not reported.

Wang et al. [33] studied the usefulness of combining face and iris biometric traits in identity verification. Iris-recognition systems generally have a relatively high failure-to-enroll rate [7], so using face as an additional biometric trait can reduce the FTE rate of the multimodal system. Further, some of the commercial iris-acquisition equipment can also capture the face image of the user. Therefore, no additional sensor is required for obtaining the face image along with the iris image. The authors used the eigenface approach for face recognition and developed a wavelet-based approach that identifies local variations in the iris images [45]. Fusion was carried out at the matching-score level using strategies like the sum rule, the weighted-sum rule, Fisher's discriminant-analysis classifier, and neural-network classifier using radial basis functions (RBFNN). Both matcher weighting and user-specific weighting of the modalities were attempted for fusion using the weighted-sum rule. Fusion using learning-based methods like weighted-sum rule, discriminant analysis, and RBFNN were found to perform better in terms of their ability to separate the genuine and impostor classes. Since the iris-recognition module was highly accurate (total error rate of 0.3%), the error rate was not reduced significantly after fusion.

Biometric characteristics like fingerprint, voice, and to some extent iris cannot be obtained when the user is far away from the sensor and/or noncooperative. Hence, the multimodal biometric systems that combine face with these modalities cannot be used in surveillance applications. Since the gait of a person can be obtained without user interaction, a combination of gait and face biometric traits can be used for surveillance. However, the gait of a person is best observed from a profile view, and face-recognition systems work well only when the frontal view is used. These two views are called the canonical views, and this canonical-view requirement is a major obstacle in building a multimodal system for integrated face and gait recognition from arbitrary views. Three approaches have been proposed to solve this problem. One approach is to build robust 3D models for both face and gait which alleviates the need for canonical views. Other possible solutions are synthesizing the canonical views from arbitrary views and employing view-invariant algorithms for gait and face recognition.

Shakhnarovich et al. [34] synthesized canonical views of face and gait from arbitrary views. They proposed a view-normalization approach that computes an image based visual hull from a set of monocular views. The visual hull was used to render the frontal face image and a side view of the person. Silhouette-extent analysis proposed by Lee [46] was used for gait recognition and the face recognition was based on eigenfaces (trained using frontal face images). The face and gait scores were transformed into probability measures by applying a simple transformation (distance scores were converted into similarities and normalized such that they sum to unity) and combined using the sum rule. Based on experiments on a very small database of 12 users walking in arbitrary directions, they reported a rank-one recognition rate of 80% for the face-recognition system, 87% for the gait-recognition system, and 91% for the integrated face-and-gait-recognition system. Shakhnarovich and Darrell [47] further extended the study conducted in [34] to analyze the effect of different fusion rules. In this work, the genuine and impostor score distributions were modeled as exponential distributions, and these models were used for converting the scores into *a posteriori* probabilities. Sum, max, min, and product rules were applied to combine these probabilities. An empirical study of these rules on a database of 26 users indicated that the product rule performed better than the other fusion rules. They reported a significant improvement (the total error rate was about 15% less) in the leave-one-out test performance of the integrated face-and-gait system over the individual modalities.

A view-invariant gait-recognition algorithm [48] and a probabilistic algorithm for face recognition [49] were employed by Kale et al. [35] to build an integrated recognition system that captures a video sequence of the person using a single camera. They explored both cascade and parallel architectures. In the cascaded system, the gait-recognition algorithm was used as a filter to prune the database and pass a smaller set of candidates to the face-recognition algorithm. In the parallel architecture, the matching scores of the two algorithms were combined using the sum and product rules. Experiments were conducted on the NIST database consisting of outdoor face and gait data of 30 subjects. No recognition errors were observed when the multimodal biometric system operated in the parallel mode. In the cascade mode, the rank-one accuracy was 97% and the number of face comparisons was reduced to 20% of the subjects in the database.

In addition to the above multimodal biometric systems that utilize face along with other biometric traits, researchers have also proposed a few other multimodal systems. Chang et al. [50] proposed a combination of face and ear for person identification. The benefits of utilizing soft biometric characteristics like gender, ethnicity, and height along with more reliable biometric traits like face and fingerprint were demonstrated by Jain et al. [51],[52]. Although the soft biometric characteristics do not have sufficient discriminative ability and stability over a period of time to uniquely identify an individual, they can provide some useful information that can assist in recognizing a person. Hence, they can be used to improve the recognition accuracy of a face-recognition system. The primary and soft biometric information were integrated using a Bayesian framework and experiments on a database of 263 users showed the additional soft biometric information can improve the rank-one accuracy of a face-recognition system by about 5%.

A recent trend in multimodal biometrics is the combination of 2D and 3D facial information. Beumier and Acheroy [53], Wang et al. [54], and Chang et al. [55] have proposed systems that employ fusion of 2D and 3D facial data. Lu and Jain [56] proposed an integration scheme to combine the surface matching and appearance-based matching for multiview face recognition. All these studies show that the multimodal 2D–3D face recognition can achieve a significantly higher accuracy compared to the current face-recognition systems operating on either 2D or 3D information alone.

21.4 CONCLUSIONS AND FUTURE RESEARCH DIRECTIONS

Even though face is one of the most commonly used biometric for authentication, multimodal systems, where multiple sources of evidence originating from different modalities (e.g., face, fingerprint, voice, iris) are combined, can offer significant performance improvements. Further, multimodal fusion may add other functionalities (e.g., fast database retrieval) to a biometric system that are not always feasible with any single biometric. In this chapter, we have summarized architectures for such combinations and presented multimodal systems reported in the literature. These systems vary in the choice of modalities, fusion levels, normalization, and fusion techniques. The characteristics of face-recognition systems, such as high matching speed, database pruning capability, low sensor cost, wide applicability, and wide public acceptability, make them an important building block for future multimodal systems.

Most of the current multimodal biometric systems operate either in the serial mode or in the parallel mode. The serial mode is computationally efficient whereas the parallel mode is more accurate. By carefully designing a hierarchical (tree-like) architecture it is possible to combine the advantages of both cascade and parallel architectures. This hierarchical architecture can be made dynamic so that it is robust and can handle problems like missing biometric samples that are possible in biometric systems. Hence, the structural design of a multimodal biometric system is an important research topic to be addressed in future. Currently, multimodal systems have been tested only on relatively small databases that are not truly multimodal. Large-scale evaluation on true multimodal databases is required to carry out a cost–benefit analysis of multimodal biometric systems. Recent release of a large multimodal (face and fingerprint) database by NIST is likely to result in a more careful and thorough evaluation of multimodal systems. Finally, the security aspects of a biometric system including resistance against spoofing (liveness detection), template protection, and biometric cryptography also require considerable attention.

REFERENCES

[1] P. J. Philips, P. Grother, R. J. Micheals, D. M. Blackburn, E. Tabassi, and J. M. Bone. FRVT2002: Evaluation Report. Available at http://www.frvt.org/DLs/FRVT_2002_Evaluation_Report.pdf, December 2002.

[2] C. Wilson, A. R. Hicklin, H. Korves, B. Ulery, M. Zoepfl, M. Bone, P. Grother, R. J. Micheals, S. Otto, and C. Watson. Fingerprint vendor technology evaluation 2003: summary of results and analysis report. NIST Internal Report 7123; available at http://fpvte.nist.gov/report/ir_7123_summary.pdf, June 2004.

[3] D. Maio, D. Maltoni, R. Cappelli, J. L. Wayman, and A. K. Jain. FVC2004: Third fingerprint verification competition. In: *Proceedings of the International Conference on Biometric Authentication*, pages 1–7, Hong Kong, China, July 2004.

[4] R. M. Bolle, J. H. Connell, S. Pankanti, and N. K. Ratha. *Guide to Biometrics*. Springer, 2003.

[5] A. K. Jain, L. Hong, S. Pankanti, and R. Bolle. An identity authentication system using fingerprints. *Proceedings of the IEEE* **85**(9):1365–1388, 1997.

[6] NIST Report to the United States Congress. Summary of NIST standards for biometric accuracy, tamper resistance, and interoperability. Available at ftp://sequoyah.nist.gov/pub/nist_internal_reports/NISTAPP_Nov02.pdf, November 2002.

[7] BBC News. Long lashes thwart ID scan trial. Available at urlhttp://news.bbc.co.uk//hi/uk_news/politics/3693375.stm, May 2004.

[8] A. K. Jain, A. Ross, and S. Prabhakar. An introduction to biometric recognition. *IEEE Transactions on Circuits and Systems for Video Technology, Special Issue on Image- and Video-Based Biometrics* **14**(1):4–20, January 2004.

[9] R. O. Duda, P. E. Hart, and D. G. Stork. *Pattern Classification.* John Wiley & Sons, 2001.

[10] L. Lam and C. Y. Suen. A theoretical analysis of the application of majority voting to pattern recognition. In: *Proceedings of the* 12*th International Conference on Pattern Recognition*, pages 418–420, Jerusalem, 1994.

[11] L. Lam and C. Y. Suen. Optimal combination of pattern classifiers. *Pattern Recognition Letters* **16**:945–954, 1995.

[12] L. Xu, A. Krzyzak, and C. Y. Suen. Methods for combining multiple classifiers and their applications to handwriting recognition. *IEEE Transactions on Systems, Man, and Cybernetics* **22**(3):418–435, 1992.

[13] J. Daugman. Combining multiple biometrics. Available at http://www.cl.cam.ac.uk/users/jgd1000/combine/combine.html.

[14] T. K. Ho, J. J. Hull, and S. N. Srihari. Decision combination in multiple classifier systems. *IEEE Transactions on Pattern Analysis and Machine Intelligence* **16**(1): 66–75, January 1994.

[15] J. Kittler, M. Hatef, R. P. Duin, and J. G. Matas. On combining classifiers. *IEEE Transactions on Pattern Analysis and Machine Intelligence* **20**(3):226–239, March 1998.

[16] A. K. Jain, K. Nandakumar, and A. Ross. Score normalization in multimodal biometric systems. *Pattern Recognition (To appear)*, 2005.

[17] R. Snelick, M. Indovina, J. Yen, and A. Mink. Multimodal Biometrics: Issues in design and testing. In: *Proceedings of Fifth International Conference on Multimodal Interfaces*, pages 68–72, Vancouver, Canada, November 2003.

[18] F. R. Hampel, P. J. Rousseeuw, E. M. Ronchetti, and W. A. Stahel. *Robust Statistics: The Approach Based on Influence Functions.* John Wiley & Sons, 1986.

[19] K. Messer, J. Matas, J. Kittler, J. Luettin, and G. Maitre. XM2VTSDB: The Extended M2VTS Database. In: *Proceedings of the Second International Conference on AVBPA*, pages 72–77, March 1999.

[20] M. Indovina, U. Uludag, R. Snelick, A. Mink, and A. K. Jain. Multimodal biometric authentication methods: a COTS approach. In: *Proceedings of the Workshop on Multimodal User Authentication*, pages 99–106, Santa Barbara, USA, December 2003.

[21] National Institute of Standards, The Image Group of the Information Access Division and Technology. Biometric scores set: Release 1. Available at http://www.itl.nist.gov/iad/894.03/biometricscores, September 2004.

[22] L. Hong and A. K. Jain. Integrating faces and fingerprints for personal identification. *IEEE Transactions on Pattern Analysis and Machine Intelligence* **20**(12):1295–1307, December 1998.

[23] A. K. Jain, L. Hong, and Y. Kulkarni. A multimodal biometric system using fingerprint, face and speech. In: *Proceedings of the Second International Conference on AVBPA*, pages 182–187, Washington D.C., USA, March 1999.

[24] A. K. Jain and A. Ross. Learning user-specific parameters in a multibiometric system. In: *Proceedings of the IEEE International Conference on Image Processing*, pages 57–60, New York, USA, September 2002.

[25] A. Ross and A. K. Jain. Information fusion in biometrics. *Pattern Recognition Letters* (Special issue on multimodal biometrics) **24**(13):2115–2125, September 2003.

[26] R. Snelick, U. Uludag, A. Mink, M. Indovina, and A. K. Jain. Large scale evaluation of multimodal biometric authentication using state-of-the-art systems. *IEEE Transactions on Pattern Analysis and Machine Intelligence* **27**(3):450–455, March 2005.

[27] R. Brunelli and D. Falavigna. Person Identification using multiple cues. *IEEE Transactions on Pattern Analysis and Machine Intelligence* **12**(10):955–966, October 1995.

[28] E. S. Bigun, J. Bigun, B. Duc, and S. Fischer. Expert conciliation for multimodal person authentication systems using Bayesian statistics. In: *Proceedings of First International Conference on AVBPA*, pages 291–300, Crans-Montana, Switzerland, March 1997.

[29] P. Verlinde and G. Cholet. Comparing decision fusion paradigms using k-NN based classifiers, decision trees and logistic regression in a multi-modal identity verification application. In: *Proceedings of the Second International Conference on AVBPA*, pages 188–193, Washington D.C., USA, March 1999.

[30] V. Chatzis, A. G. Bors, and I. Pitas. Multimodal decision-level fusion for person authentication. *IEEE Transactions on Systems, Man, and Cybernetics, Part A: Systems and Humans* **29**(6):674–681, November 1999.

[31] S. Ben-Yacoub, Y. Abdeljaoued, and E. Mayoraz. Fusion of face and speech data for person identity verification. *IEEE Transactions on Neural Networks* **10**(5):1065–1075, 1999.

[32] R. Frischholz and U. Dieckmann. BioID: A multimodal biometric identification system. *IEEE Transactions on Computers* **33**(2):64–68, February 2000.

[33] Y. Wang, T. Tan, and A. K. Jain. Combining face and iris biometrics for identity verification. In: *Proceedings of Fourth International Conference on AVBPA*, pages 805–813, Guildford, UK, June 2003.

[34] G. Shakhnarovich, L. Lee, and T.J. Darrell. Integrated face and gait recognition from multiple views. In: *Proccedings of the IEEE Computer Society Conference on Computer Vision and Pattern Recognition*, pages 439–446, Hawaii, USA, December 2001.

[35] A. Kale, A. K. RoyChowdhury, and R. Chellappa. Fusion of gait and face for human identification. In: *Proceedings of the IEEE International Conference on Speech, Acoustics, and Signal Processing*, Philadelphia, USA, March 2005.

[36] M. Turk and A. Pentland. Eigenfaces for recognition. *Journal of Cognitive Neuroscience* **3**(1):71–86, 1991.

[37] A. K. Jain, L. Hong, and R. Bolle. On-line fingerprint verification. *IEEE Transactions on Pattern Analysis and Machine Intelligence* **19**(4):302–314, 1997.

[38] A. K. Jain, A. Ross, and S. Pankanti. A prototype hand geometry-based verification system. In: *Proceedings of the Second International Conference on AVBPA*, pages 166–171, Washington D.C., USA, March 1999.

[39] P. J. Phillips, H. Moon, P. J. Rauss, and S. Rizvi. The FERET evaluation methodology for face-recognition algorithms. *IEEE Transactions on Pattern Analysis and Machine Intelligence*, **22**(10):1090–1104, October 2000.

[40] G. Doddington, W. Liggett, A. Martin, M. Przybocki, and D. Reynolds. Sheeps, goats, lambs and wolves: A statistical analysis of speaker performance in the NIST 1998 Speaker Recognition Evaluation. In: *Proceedings of International Conference*

on Spoken Language Processing, pages 1351–1354, Sydney, Australia, November 1998.

[41] S. Pigeon and L. Vandendrope. M2VTS Multimodal Face Database Release 1.00. Available at http://www.tele.ucl.ac.be/PROJECTS/M2VTS/m2fdb.html, 1996.

[42] M. Lades, J. Vorbruggen, J. Buhmann, J. Lange, C. V. D. Malburg, and R. Wurtz. Distortion invariant object recognition in the dynamic link architecture. *IEEE Transactions on Computers* **42**(3):300–311, March 1993.

[43] F. Bimbot, I. Magrin-Chagnolleau, and L. Mathan. Second-order statistical measure for text-independent speaker identification. *Speech Communication* **17**(1/2):177–192, 1995.

[44] R. W. Frischholz, F. G. Boebel, and K. P. Spinnler. Face recognition with the synergetic computer. In: *Proceedings of the International Conference on Applied Synergetics and Synergetic Engineering*, pages 107–110, Erlangen, Germany, 1994.

[45] L. Ma, T. Tan, Y. Wang, and D. Zhang. Personal identification based on iris texture analysis. *IEEE Transactions on Pattern Analysis and Machine Intelligence* **25**(12):1519–1533, December 2003.

[46] L. Lee. Gait dynamics for recognition and classification. In: *MIT Technical Report AIM-2001-019*, September 2001.

[47] G. Shakhnarovich and T.J. Darrell. On probabilistic combination of face and gait cues for identification. In: *Proceedings of the Fifth IEEE International Conference on Automatic Face and Gesture Recognition*, pages 169–174, Washington D.C., USA, May 2002.

[48] A. Kale, A. K. R. Chowdhury, and R. Chellappa. Towards a view-invariant gait recognition algorithm. In: *Proceedings of IEEE International Conference on Advanced Video and Signal based Surveillance*, pages 143–150, Miami, USA, 2003.

[49] S. Zhou, V. Krueger, and R. Chellappa. Probabilistic recognition of human faces from video. *Computer Vision and Image Understanding* **91**:214–245, July-August 2003.

[50] K. Chang, K. W. Bowyer, S. Sarkar, and B. Victor. Comparison and combination of ear and face images in appearance-based biometrics. *IEEE Transactions on Pattern Analysis and Machine Intelligence* **25**(9):1160–1165, September 2003.

[51] A. K. Jain, S. C. Dass, and K. Nandakumar. Soft biometric traits for personal recognition systems. In: *Proceedings of the International Conference on Biometric Authentication, LNCS 3072*, pages 731–738, Hong Kong, China, July 2004.

[52] A. K. Jain, K. Nandakumar, X. Lu, and U. Park. Integrating faces, fingerprints and soft biometric traits for user recognition. In: *Proceedings of Biometric Authentication Workshop, LNCS 3087*, pages 259–269, Prague, Czech Republic, May 2004.

[53] C. Beumier and M. Acheroy. Automatic face verification from 3D and grey level clues. In: *Eleventh Portuguese Conference on Pattern Recognition*, Porto, Portugal, May 2000.

[54] Y. Wang, C. Chua, and Y. Ho. facial feature detection and face recognition from 2D and 3D images. *Pattern Recognition Letters* **23**(10):1191–1202, August 2002.

[55] K. I. Chang, K. W. Bowyer, and P. J. Flynn. Multimodal 2D and 3D biometrics for face recognition. In: *Proceedings of the IEEE International Workshop on Analysis and Modeling of Faces and Gestures*, pages 187–194, Nice, France, October 2003.

[56] X. Lu and A. Jain. Integrating range and texture information for 3D face recognition. In: *Proceedings of IEEE Computer Society Workshop on Application of Computer Vision*, Breckenridge, USA, January 2005.

INDEX

B

F

G

J

K

L

Q

R